Proceedings

IEEE Computer Society
Annual Symposium on

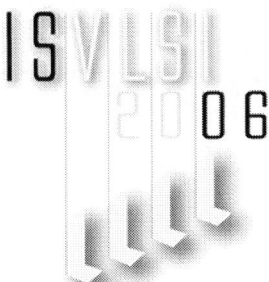

Emerging VLSI Technologies and Architectures

Proceedings

IEEE Computer Society
Annual Symposium on

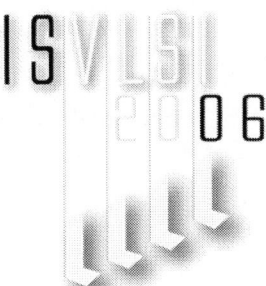

Emerging VLSI Technologies and Architectures

2-3 March 2006 • **Karlsruhe, Germany**

Edited by
Jürgen Becker, Andreas Herkersdorf, Amar Mukherjee, and Asim Smailagic

Sponsored by
IEEE Computer Society Technical Committee on VLSI

Los Alamitos, California

Washington • Tokyo

Copyright © 2006 by The Institute of Electrical and Electronics Engineers, Inc.
All rights reserved.

Copyright and Reprint Permissions: Abstracting is permitted with credit to the source. Libraries may photocopy beyond the limits of US copyright law, for private use of patrons, those articles in this volume that carry a code at the bottom of the first page, provided that the per-copy fee indicated in the code is paid through the Copyright Clearance Center, 222 Rosewood Drive, Danvers, MA 01923.

Other copying, reprint, or republication requests should be addressed to: IEEE Copyrights Manager, IEEE Service Center, 445 Hoes Lane, P.O. Box 133, Piscataway, NJ 08855-1331.

The papers in this book comprise the proceedings of the meeting mentioned on the cover and title page. They reflect the authors' opinions and, in the interests of timely dissemination, are published as presented and without change. Their inclusion in this publication does not necessarily constitute endorsement by the editors, the IEEE Computer Society, or the Institute of Electrical and Electronics Engineers, Inc.

IEEE Computer Society Order Number P2533
ISBN-13: 978-0-7695-2533-4
ISBN-10: 0-7695-2533-4
Library of Congress Number 2006920140

Additional copies may be ordered from:

IEEE Computer Society	IEEE Service Center	IEEE Computer Society
Customer Service Center	445 Hoes Lane	Asia/Pacific Office
10662 Los Vaqueros Circle	P.O. Box 1331	Watanabe Bldg., 1-4-2
P.O. Box 3014	Piscataway, NJ 08855-1331	Minami-Aoyama
Los Alamitos, CA 90720-1314	Tel: + 1 732 981 0060	Minato-ku, Tokyo 107-0062
Tel: + 1 800 272 6657	Fax: + 1 732 981 9667	JAPAN
Fax: + 1 714 821 4641	http://shop.ieee.org/store/	Tel: + 81 3 3408 3118
http://computer.org/cspress	customer-service@ieee.org	Fax: + 81 3 3408 3553
csbooks@computer.org		tokyo.ofc@computer.org

Individual paper REPRINTS may be ordered at: reprints@computer.org

Editorial production by Stephanie Kawada

Cover art production by Joe Daigle/Studio Productions

Printed in the United States of America by Applied Digital Imaging

IEEE Computer Society
Conference Publishing Services
http://www.computer.org/proceedings/

Proceedings

Table of Contents

Message from the General and Program Chairs xiii

Symposium Committees xiv

Keynotes

Advanced Channel Decoding Algorithms and Their Implementation
for Future Communication Systems 3
    Norbert Wehn

Multiprocessor Systems-on-Chips 4
    Wayne Wolf

Regular Papers

Intellectual Property and Design

Floorplanning Based on Particle Swarm Optimization 7
    Tsung-Ying Sun, Sheng-Ta Hsieh, Hsiang-Min Wang, and Cheng-Wei Lin

Enhanced Dual Strategy Based VLSI Architecture for Computing Pseudo Inverse
of Channel Matrix in a MIMO Wireless System 12
    Zahid Khan, Tughrul Arslan, John S. Thompson, and Ahmet T. Erdogan

Adaptive Porting of Analog IPs with Reusable Conservative Properties 18
    Takashi Nojima, Shigetoshi Nakatake, Toru Fujimura, Koji Okazaki,
    Yoji Kajitani, and Nobuto Ono

v

VLSI Design Exchange with Intellectual Property Protection in FPGA Environment
Using Both Secret and Public-Key Cryptography 24
  Wael Adi, Rolf Ernst, Bassel Soudan, and Abdulrahman Hanoun

Physical Design

Metal Fix and Power Network Repair for SOC 33
  Qing K. Zhu and Paige Kolze

Multi-SP: A Representation with United Rectangles for Analog Placement
and Routing 38
  Ning Fu, Shigetoshi Nakatake, and Mitsutoshi Mineshima

Formulating the Empirical Strategies in Module Generation of Analog
MOS Layout 44
  Tan Yan, Takashi Nojima, and Shigetoshi Nakatake

An Integer Linear Programming Based Approach to Simultaneous Memory
Space Partitioning and Data Allocation for Chip Multiprocessors 50
  O. Ozturk, G. Chen, M. Kandemir, and M. Karakoy

High Performance Circuits

High Speed Low Swing Dynamic Circuits with Multiple Supply and
Threshold Voltages 59
  Zhiyu Liu and Volkan Kursun

High Performance Service-Time-Stamp Computation for WFQ IP Packet
Scheduling 65
  C. McKillen, S. Sezer, and Xin Yang

Synthesis of Pipelined SRSL Circuits 71
  Rashad Oreifej, Abdelhalim Alsharqawi, and Abdel Ejnioui

An Efficient Hardware Implementation of a Self-Adaptable Equalizer
for WCDMA Downlink UMTS Standard 77
  Romualdo Begale Prudêncio, Leandro Soares Indrusiak, and Manfred Glesner

Reconfigurable Systems Integration

Autonomous Realization of Boeing/JPL Sensor Electronics Based on Reconfigurable
System-on-Chip Technology 85
  Evangelos F. Stefatos, Tughrul Arslan, Didier Keymeulen, and Ian Ferguson

Defect-Aware Design Paradigm for Reconfigurable Architectures 91
  Rahul Jain, Anindita Mukherjee, and Kolin Paul

New 2-Dimensional Partial Dynamic Reconfiguration Techniques
for Real-Time Adaptive Microelectronic Circuits 97
    Michael Hübner, Christian Schuck, Matthias Kühnle, and Jürgen Becker

A "Soft++" eFPGA Physical Design Approach with Case Studies
in 180 nm and 90 nm 103
    Victor Aken'Ova and Resve Saleh

QUKU: A Two-Level Reconfigurable Architecture...109
    Sunil Shukla, Neil W. Bergmann, and Jürgen Becker

Mixed-Signal Design and Analysis

Space-Saving Layout for Passive Components 117
    Päivi H. Karjalainen and Pekka Heino

A Novel Low Power Multilevel Current Mode Interconnect System.......................................122
    Supreet Joshi and Dinesh Sharma

The Design of Analog Front-End Circuitry for 1X HD-DVD PRML Read Channel................128
    Sheng-Jang Lin, I-Shun Chen, Bo-Wei Chen, and Feng-Hsiang Lo

Adaptive Signal Processing in Mixed-Signal VLSI with Anti-Hebbian Learning 133
    Miguel Figueroa, Esteban Matamala, Gonzalo Carvajal, and Seth Bridges

Test and Verification

Verification of Scheduling in High-Level Synthesis...141
    C. Karfa, C. Mandal, D. Sarkar, S. R. Pentakota, and Chris Reade

An Efficient Wrapper Scan Chain Configuration Method for
Network-on-Chip Testing..147
    Ming Li, Wen-Ben Jone, and Qing-An Zeng

An Efficient Data-Independent Technique for Compressing Test Vectors
in Systems-on-a-Chip 153
    Xiaoyu Ruan and Rajendra Katti

Methods for Run-Time Failure Recognition and Recovery in Dynamic
and Partial Reconfigurable Systems Based on Xilinx Virtex-II Pro FPGAs 159
    Katarina Paulsson, Michael Hübner, Markus Jung, and Jürgen Becker

Low Power System Design

Design and Analysis of a Low Power VLIW DSP Core 167
    Chan-Hao Chang and Diana Marculescu

High-Performance Noise-Robust Asynchronous Circuits 173
    Pankaj Golani and Peter A. Beerel

A Low Power Lookup Technique for Multi-hashing Network Applications 179
    Ilhan Kaya and Taskin Kocak

A Low Power Pipelined Maximum Likelihood Detector for 4x4 QPSK MIMO
Wireless Communication Systems 185
    J. H. Han, A. T. Erdogan, and T. Arslan

System-on-Chip

Optimal Periodical Memory Allocation for Logic-in-Memory Image Processors 193
    Masanori Hariyama, Michitaka Kameyama, and Yasuhiro Kobayashi

Globally Asynchronous Locally Synchronous Wrapper Circuit Based on Clock Gating...................199
    Esmail Amini, Mehrdad Najibi, and Hossein Pedram

Connection-Oriented Multicasting in Wormhole-Switched Networks on Chip 205
    Zhonghai Lu, Bei Yin, and Axel Jantsch

A Virtual Channel Network-on-Chip for GT and BE Traffic...211
    Nikolay Kavaldjiev, Gerard J. M. Smit, Pierre G. Jansen, and Pascal T. Wolkotte

Delay-Insensitive On-Chip Communication Link Using Low-Swing Simultaneous
Bidirectional Signaling 217
    Ethiopia Nigussie, Juha Plosila, and Jouni Isoaho

Nano Electronics

Nanowire Addressing in the Face of Uncertainty..225
    Eric Rachlin and John E. Savage

Si Nanocrystal MOSFET with Silicon Nitride Tunnel Insulator for High-Rate
Random Number Generation..231
    Ryuji Ohba, Daisuke Matsushita, Koichi Muraoka, Shinichi Yasuda,
    Tetsufumi Tanamoto, Ken Uchida, and Shinobu Fujita

Finite State Machine Implementation with Single-Electron Tunneling Technology............................237
    Jialin Mi and Chunhong Chen

PLAs in Quantum-dot Cellular Automata 242
    Xiaobo Sharon Hu, Michael Crocker, Michael Niemier, Minjun Yan,
    and Gary Bernstein

Reconfigurable System Design and Technologies

Dynamic Hardware Multiplexing: Improving Adaptability with a Run Time
Reconfiguration Manager 251
    P. Benoit, L. Torres, G. Sassatelli, M. Robert, G. Cambon, and J. Becker

Regular Routing Architecture for a LUT-Based MPGA 257
    *Francisco-Javier Veredas, Michael Scheppler, Bumei Zhai, and
    Hans-Joerg Pfleiderer*

A New Multilevel Hierarchical MFPGA and Its Suitable Configuration Tools.......................263
    Zied Marrakchi, Hayder Mrabet, and Habib Mehrez

New Non-volatile FPGA Concept Using Magnetic Tunneling Junction....................................269
    Nicolas Bruchon, Lionel Torres, Gilles Sassatelli, and Gaston Cambon

Complexity and System Organization

Profile Directed Instruction Cache Tuning for Embedded Systems 277
    Kugan Vivekanandarajah, Thambipillai Srikanthan, and Christopher T. Clarke

Complexity and Low Power Issues for On-Chip Interconnections
in MPSoC System Level Design 283
    Yuriy Sheynin, Elena Suvorova, and Felix Shutenko

Fast Configuration of an Energy-Efficient Branch Predictor 289
    P. Hallschmid and R. Saleh

Exploiting Software Pipelining for Network-on-Chip Architectures 295
    Feihui Li, Mahmut Kandemir, and Ibrahim Kolcu

System Level and Circuit Analysis

An Efficient Algorithm for the Analysis of Cyclic Circuits 303
    Osama Neiroukh, Stephen A. Edwards, and Xiaoyu Song

Improving System Level Design Space Exploration by Incorporating
SAT-Solvers into Multi-objective Evolutionary Algorithms..309
    Thomas Schlichter, Martin Lukasiewycz, Christian Haubelt, and Jürgen Teich

System Level Design

Optimisation of the SHA-2 Family of Hash Functions on FPGAs...317
    *Robert P. McEvoy, Francis M. Crowe, Colin C. Murphy, and
    William P. Marnane*

A Novel Approach to Performance-Oriented Datapath Allocation and
Floorplanning 323
    Vijay Sundaresan and Ranga Vemuri

CHESS: A Comprehensive Tool for CDFG Extraction and Synthesis
of Low Power Designs from VHDL...329
    Nagarajan Ranganathan, Ravi Namballa, and Narender Hanchate

System Exploration of SystemC Designs 335
    Christian Genz and Rolf Drechsler

Power Aware VLSI Design

Reliability-Aware SOC Voltage Islands Partition and Floorplan.........................343
    Shengqi Yang, Wayne Wolf, N. Vijaykrishnan, and Yuan Xie

Ultra-Low Energy Computing with Noise: Energy-Performance-Probability
Trade-Offs 349
    Pinar Korkmaz, Bilge E. S. Akgul, and Krishna V. Palem

Delay and Energy Efficient Data Transmission for On-Chip Buses.....................355
    Madhu Mutyam, Melvin Eze, N. Vijaykrishnan, and Yuan Xie

Power-Oriented Delay Budgeting for Combinational Circuits..............................361
    Jialin Mi and Chunhong Chen

VLSI Circuits and Optimization

Routing-Tree Construction with Concurrent Performance, Power and
Congestion Optimization 367
    Cengiz Alkan and Tom Chen

Clock Gated Static Pulsed Flip-Flop (CGSPFF) in Sub 100 nm Technology 373
    A. S. Seyedi, S. H. Rasouli, A. Amirabadi, and A. Afzali-Kusha

Performance and Power Analysis of Globally Asynchronous Locally
Synchronous Multi-processor Systems...378
    Zhiyi Yu and Bevan M. Baas

Implementing Register Files for High-Performance Microprocessors
in a Die-Stacked (3D) Technology 384
    Kiran Puttaswamy and Gabriel H. Loh

VLSI Circuits and Technologies

Leakage-Aware SPM Management 393
    Guangyu Chen, Feihui Li, Ozcan Ozturk, Guilin Chen, Mahmut Kandemir,
    and Ibrahim Kolcu

Dependability Analysis of Nano-scale FinFET Circuits 399
    Feng Wang, Yuan Xie, Kerry Bernstein, and Yan Luo

A Low-Power 2-Dimensional Bypassing Multiplier Using 0.35 μm
CMOS Technology 405
    Chua-Chin Wang and Gang-Neng Sung

Poster Papers

Multi-level Buffer Block Planning and Buffer Insertion for Large
Design Circuits 411
    Ali Jahanian and Morteza Saheb Zamani

Effect of Glitches on the Efficiency of Components' Region-Constrained
Placement as a Fast Approach to Reduce FPGA's Dynamic Power Consumption 416
    Seyed E. Esmaeili, Nabil I. Khachab, and Moustafa Y. Ghannam

Towards a Faster Simulation of SystemC Designs 418
    Ali Habibi, Haja Moinudeen, Amer Samarah, and Sofiène Tahar

An Optimized BIST Architecture for FPGA Look-Up Table Testing 420
    Mahnaz Sadoughi Yarandi, Armin Alaghi, and Zainalabedin Navabi

Variation Aware Placement for FPGAs 422
    Suresh Srinivasan and Vijaykrishnan Narayanan

A Regular Layout Approach for ASICs 424
    Cláudio Menezes, Cristina Meinhardt, Ricardo Reis, and Reginaldo Tavares

Evaluating the Impact of Data Encoding Techniques on the Power Consumption
in Networks-on-Chip 426
    José C. S. Palma, Leandro Soares Indrusiak, Fernando G. Moraes,
    Alberto Garcia Ortiz, Manfred Glesner, and Ricardo A. L. Reis

Dual-Mode High-Speed Low-Energy Binary Addition 428
    Johannes Grad and James E. Stine

A Flexible Architecture for Block Turbo Decoders Using BCH or
Reed-Solomon Components Codes 430
    Erwan Piriou, Christophe Jego, Patrick Adde, and Michel Jezequel

xi

Transparent Management of Reconfigurable Hardware in Embedded
Operating Systems 432
    Krzysztof Kościuszkiewicz, Fearghal Morgan, and Krzysztof Kępa

An Open-Source Tool for Simulation of Partially Reconfigurable Systems
Using SystemC 434
    Alisson V. de Brito, Elmar U. K. Melcher, and Wilson Rosas

Partial and Dynamic Reconfiguration of FPGAs: A Top Down Design
Methodology for an Automatic Implementation 436
    Florent Berthelot and Fabienne Nouvel

Self-Timed Thermally-Aware Circuits 438
    David Fang, Filipp Akopyan, and Rajit Manohar

A New Protocol Stack Model for Network on Chip 440
    *Masood Dehyadgari, Mohsen Nickray, Ali Afzali-kusha, and
    Zainalabedin Navabi*

A Robust Synchronizer 442
    Jun Zhou, David Kinniment, Gordon Russell, and Alex Yakovlev

Low Power Layered Space-Time Channel Detector Architecture
for Mimo Systems 444
    T. Takahashi, A. T. Erdogan, T. Arslan, and J. H. Han

Sensor-Driven Power Management: Enhancing Performance and Reliability
of Autonomously Powered Systems 446
    Josef Haid and Dietmar Scheiblhofer

Reducing Memory Requirements through Task Recomputation in Embedded
Multi-CPU Systems 448
    H. Koc, S. Tosun, O. Ozturk, and M. Kandemir

Compiler-Directed Management of Leakage Power in Software-Managed Memories 450
    G. Chen, F. Li, M. Kandemir, O. Ozturk, and I. Demirkiran

A Parallel Architecture for Hardware Face Detection 452
    T. Theocharides, N. Vijaykrishnan, and M. J. Irwin

A VLSI GFP Frame Delineation Circuit 454
    Ciaran Toal, Sakir Sezer, and Xin Yang

Effects of Parameter Variations and Crosstalk Noise on H-Tree Clock
Distribution Networks 456
    Itisha Chanodia and Dimitrios Velenis

Author Index 459

Message from the General and Program Chairs

This book contains the papers presented at the IEEE Computer Society Annual Symposium on VLSI, held on 2-3 March 2006 in Karlsruhe, hosted by the Universität Karlsruhe (TH), Germany.

The Symposium covers a wide variety of topics from system-on-chip issues to novel technologies and innovative nano devices. Field Programmable Gate Arrays were used to evaluate new design methods but also to integrate and exploit adaptive mechanisms for run-time reconfigurable systems. Design methodologies enable the optimization of analog and digital chips for electronic systems of the future. Leading scientists and researchers from academia and industry are able to discuss their ideas for technology filling the gap to new solutions for future systems. All this is possible by the submission of papers from scientists all over the world. Researchers from 27 different countries submitted their contributions to the Program Committee of ISVLSI 2006. A critical review process for the 151 submitted papers enabled us once again to offer a high-quality conference and proceedings. A total number of 64 selected regular papers and 28 poster papers stretch a field of competence and knowledge for all important issues of VLSI.

We would like to thank all the authors for submitting the first and final versions of their papers. We also gratefully acknowledge the tremendous review work done by the Program Committee members and many additional reviewers who contributed their time and expertise toward the compilation of this volume. We would also like to thank the members of the Organizing Committee for their competent guidance and work in the last month. Special thanks go to the invited speakers for their contributions to the technical program.

Many thanks to the Steering Committee for their outstanding support. Especially, we acknowledge the assistance of Michael Hübner, Oliver Sander, and Michael Ullmann from Universität Karlsruhe (TH), and Walter Stechele and Christopher Claus from Technical University Munich for their valuable work in managing many technical, local, and financial issues regarding the ISVLSI organization. We are indebted to Richard van de Stadt, the author of CyberChair. This extraordinary free software facilitated the submission and review process.

We wish you all an interesting VLSI Symposium with stimulating discussion, valuable information, and fruitful cooperation for your future work.

Jürgen Becker and Amar Mukherjee
General Co-Chairs

Andreas Herkersdorf and Asim Smailagic
Program Co-Chairs

Karlsruhe, December 2005

Symposium Committees

General Chair

Jürgen Becker, *Universität Karlsruhe (TH), Germany*

General Co-Chair

Amar Mukherjee, *University of Central Florida, USA*

Program Committee Chair

Andreas Herkersdorf, *TU München, Germany*

Program Committee Co-Chair

Asim Smailagic, *Carnegie Mellon University, USA*

Steering Committee

A. Mukherjee, *University of Central Florida, USA*
D. W. Bouldin, *University of Tennessee, USA*
V. Narayanan, *Penn State University, USA*
N. Ranganathan, *University of South Florida, USA*
A. Smailagic, *Carnegie Mellon University, USA*

Treasurer

Michael Ullmann, *Universität Karlsruhe (TH), Germany*

Publicity Chairs

Don Bouldin, *University of Tennessee, USA*
Reiner Hartenstein, *TU Kaiserslautern, Germany*

Local Arrangements Chairs

Michael Hübner, *Universität Karlsruhe (TH), Germany*
Oliver Sander, *Universität Karlsruhe (TH), Germany*

Program Committee

Tughrul Arslan, *University of Edinburgh, UK*
Magdy Bayoumi, *University of Louisiana, USA*
Jürgen Becker, *Universität Karlsruhe (TH), Germany*
Neil Bergmann, *University of Queensland, Australia*
Don Bouldin, *University of Tennessee, USA*
Jay Brockman, *University of Notre Dame, USA*
Thomas Buechner, *IBM, Germany*
Peter Y. K. Cheung, *Imperial College, London, UK*
L. Richard Carley, *Carnegie Mellon University, USA*
Andreas Döring, *IBM Rüschlikon, Switzerland*
Rolf Ernst, *University of Braunschweig, Germany*
Christian Gamrat, *CEA-Saclay, France*
Manfred Glesner, *TU Darmstadt, Germany*
Rajesh Gupta, *University of California, San Diego, USA*
Joerg Henkel, *Universität Karlsruhe (TH), Germany*
Andreas Herkersdorf, *Techn. Universität München, Germany*
Mike Hutton, *Altera, USA*
Ricardo Jacobi, *Universidade de Brasilia, Brazil*
Kevin Kornegay, *Cornell University, USA*
Ram Krishnamurthy, *Intel Corporation, USA*
Rudy Lauwereins, *IMEC, Leuven, Belgium*
Philip Leong, *Imperial College, London, UK*
Patrick Lysaght, *Xilinx, USA*
Nihar Mahapatra, *Michigan State University, USA*
Jef Van Meerbergen, *Philips, Netherlands*
Toshiaki Miyazaki, *University of Aizu, Japan*
Klaus D. Müller-Glaser, *Universität Karlsruhe (TH), Germany*
Vijaykrishnan Narayanan, *Penn State University, USA*
Kunle Olukotun, *Stanford, USA*
Hidetoshi Onodera, *Kyoto University, Japan*
Marios Papaefthymiou, *University of Michigan, USA*
Pierre Paulin, *SGS-THOMSON Microelectronics, Grenoble, France*
Nagarajan Ranganathan, *University of South Florida, USA*
Ricardo Reis, *Universidade Federal do Rio Grande do Sul, Brazil*
Sergei Sawitzki, *Philips Research, Netherlands*
Matthias Schöbinger, *Infineon, Germany*
Sakir Sezer, *Queens University Belfast, UK*
Kaijian Shi, *Synopsis Inc., USA*
Mircea Stan, *University of Virginia, USA*
Mitch Thornton, *SMU Dallas, USA*
Vivek Tiwari, *Intel, USA*
Lionel Torres, *LIRMM Montpellier, France*
Stamatis Vassiliadis, *Delft University of Technology, Netherlands*
Miroslav Velev, *Reservoir Labs, USA*
Norbert Wehn, *TU Kaiserslautern, Germany*
Wayne Wolf, *Princeton University, USA*

Keynotes

Advanced Channel Decoding Algorithms and Their Implementation for Future Communication Systems

Norbert Wehn
Microelectronic System Design Research Group
University of Kaiserslautern
wehn@eit.uni-kl.de
www.eit.uni-kl.de/wehn

Today's information society demands access to huge amount of data anywhere and data any time. Hence wireless communications is a key technology. In such systems bandwidth and transmission power are critical resources. Thus advanced communications systems have to rely on sophisticated error correction schemes (FEC). Turbo- and LDPC codes belong to the most efficient error correction schemes known by now. However the iterative decoding nature of these codes imply big implementation challenges w.r.t. throughput, latency and low energy. Moreover future communication systems require flexibility. Thus we have to trade-off flexibility versus implementation costs. In this talk we discuss the design space for Turbo- and LDPC decoders and compare different implementation alternatives. Moreover we show the need of a design methodology which have to consider code-design and architecture development in a unified way. We discuss this challenge on the base of 3GPP Turbo-decoders and DVB-S2 LDPC decoders.

Multiprocessor Systems-on-Chips

Wayne Wolf
Dept of Electrical Engineering
Princeton University, NJ

Moore's Law has reached the point at which we can build single-chips with multiple processors and significant amounts of memory. Multiprocessor systems-on-chips (MPSoCs) have opened up new application areas, such as low-power and real-time embedded systems. This talk will review the architectures of multiprocessor systems-on-chips and the design methodologies used to create them. MPSoCs make use of advanced processors, memory systems, and on-chip networks, often delivered as intellectual property modules. The design methodologies required to design these complex systems build on earlier VLSI techniques but must address many new problems as well.

Regular Papers

Intellectual Property and Design

Floorplanning Based on Particle Swarm Optimization

Tsung-Ying Sun, *Member, IEEE*, Sheng-Ta Hsieh, *Student Member, IEEE*, Hsiang-Min Wang
and Cheng-Wei Lin

Department of Electrical Engineering
National Dong Hwa University
Hualien, Taiwan, R.O.C.

E-mail: sunty@mail.ndhu.edu.tw

Abstract

This paper presents a floorplanning method based on particle swarm optimization (PSO). We adopted the B*-tree floorplan structure to generate an initial stage with overlap free for placement and utilized PSO to find out the potential optimal placement solution. Unlike other related research, our method can avoid the solution from falling into the local minimal and has ability of more efficiency and robustness for explored solution space. Experiments employing MCNC and GSRC benchmarks show that the performance of our method for placement by the ability of exploring better solutions. The proposed approach exhibited rapidly convergence and led to more optimal solutions than other related approach.

1. Introduction

Floorplanning has been an important stage in VLSI design as a means to manage circuit complexity and deep-submicron effects. Floorplanning in VLSI design is to arrange the modules on a chip under the constraint that no two modules are overlap while controlling the area, wire length, and other performance indices to be optimal. Today, automatic floorplanning is encouraged by the growing adoption of embedded memories and IP blocks in *System on Chips* (SoCs) deigns.

The physical placement of circuits in VLSI chips or SoCs has been given sustained attention in the recent years. Early research on the placement problem applied force to reduce the overlap betweens cells [1]. [2-4] shows the generation of overlap free placements, and [5] compare various floorplan representations which cooperate with simulated annealing (SA). Adopting a floorplan representations could easily apply to different applications and various requirements through modify objective functions. The drawback of adopting SA is that the system must be close to equilibrium throughout the process, which demands a careful adjustment of the annealing schedule parameters. [6-7] shows the generated layouts with cell overlaps. While allowing overlaps during the process of placement was shown to obtain a better floorplanning solution, this process could not guarantee the entire elimination of overlaps.

[8] introduced an integer linear programming (ILP) formula for finding the optimal module orientations in macrocell placement. The proposed method handles multi-terminal nets based on the Manhattan metric of the minimum bounding box of the pin positions which is more accurate than some earlier approaches which can only adapt to two-terminal nets based on the Euclidean metric for wire length estimation.

As opposed to these previously mentioned methods, we adopt a non-slicing structure of representation B*-tree with particle swarm optimization (PSO) algorithm in this paper. PSO is a swarm intelligence method that roughly models the social behavior of swarms. The consequence of modeling this social behavior is that the search process allows particles to stochastically return toward previously successful regions in the search space. It has proved to be efficient on many problems in science and engineering. Our method can reduce much of computational time, obtains rapid convergence and better solutions. Furthermore, our floorplanner can widely explore the solution space and prevent the solution from falling into the local minimal

We divided this paper into five sections. Section II describes the original PSO methodology. Section III presents B*-tree representation and our proposed methods for floorplanning. Section IV exhibits the experiment results. Finally, the conclusion is in section V.

2. Particle swarm optimization

The PSO is a population based optimization technique that was proposed by Kennedy and Eberhart [9] in 1995, which the population is referred to as a *swarm*. The particles express the ability of fast convergence to local and/or global optimal position(s) over a small number of generations.

A swarm in PSO consists of a number of particles. Each particle represents a potential solution of the optimization task. All of the particles iteratively discover the probable solution. Each particle generates a position according to the new velocity and the previous positions of the cell, and is compared with the best position which is generated by previous particles in the cost function and keeps the best one; i.e., each particle accelerates the directions not only the local best solution but also the global best position. If a particle discovers a new probable solution, other particles will move closer to it so as to explore the region more completely in the process [10].

Let s denote the swarm numbers. In general, there are three attributes, current position x_i, current velocity v_i and local best position y_i, for particles in the search space to present their features. Each particle in the swarm is iteratively updated according to the aforementioned attributes. Assuming that the function f is to be minimized so that the dimension consists of n particles and the new velocity of every particle is updated by (1).

$$v_{i,j}(t+1) = wv_{i,j}(t) + c_1 r_{1,i,j}(t)[y_{i,j}(t) - x_{i,j}(t)] \\ + c_2 r_{2,i,j}(t)[\hat{y}_j(t) - x_{i,j}(t)] \quad (1)$$

where $v_{i,j}$ is the velocity of the ith particle of the jth swarm for all $j \in 1 \ldots s$, w is the inertia weight of velocity, c_1 and c_2 denote the *acceleration coefficients*, r_1 and r_2 are elements from two uniform random sequences in the range $(0, 1)$, and t is the number of generations. The new position of the ith particle is calculated as follows:

$$x_i(t+1) = x_i(t) + v_i(t+1) \quad (2)$$

The local best position of each particle could be updated by (3), and the global best position $\hat{y}$ was found from all particles by (4).

$$y_i(t+1) = \begin{cases} y_i(t), & \text{if } f(x_i(t+1)) \geq f(y_i(t)) \\ x_i(t+1), & \text{if } f(x_i(t+1)) < f(y_i(t)) \end{cases} \quad (3)$$

$$\hat{y}(t+1) = \arg\min_{y_i} f(y_i(t+1)), \quad 1 \leq i \leq n \quad (4)$$

3. PSO for floorplanning

3.1. The B*-tree representation

In this paper, we adopted the B*-tree representation to model a floorplan [4]. A B*-tree is an ordered binary tree for modeling non-slicing floorplans. Figure 1 shows a packing of floorplan and its corresponding B*-tree, where the tree nodes n_i are directly mapping to the placement blocks b_i. The root of B*-tree which denotes as n_0 is corresponding to the block b_0 on the bottom-left corner of the placement. The construction of a B*-tree is starting from the root, and then the first recursively create the left subtree, finally is the right subtree. Let R_i be the set of blocks located on the right-hand side and adjacent to b_i. The left child of the node n_i corresponds to the lowest, unvisited block in R_i. The right child of n_i represents the lowest block located above and with its x-coordinate equal to that of b_i and its y-coordinate less than that of the top boundary of the module on the left-hand side and adjacent to b_i.

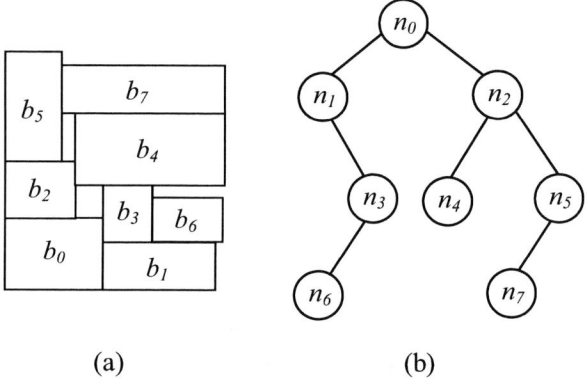

(a)   (b)

Figure 1. A packing represented by B*-tree

Each nodes combination of B*-tree corresponds to a floorplan. Therefore, the solution space consists of all B*-trees with the given nodes (blocks). In order to find a next better solution, the B*-tree is disturbed by following operations to get another nodes combination:

1. **Node movement**
   This operation will delete a node and insert it into other place of the tree.

2. **Nodes (blocks) swap**
   This operation will swap two blocks of the tree.

3. **Block rotation**
   This operation just rotate blocks without change the tree structure.

3.2. Handling Floorplanning Using Particle Swarm Optimization

At the beginning move, each particle move will random pick up one operation that mentioned above, after that, the particles movement will inherit the pervious experience, i.e. past best solution and global best solution to guide them to select a suitable operation for finding a better solution. In the searching space, each particle move will lead the solution toward global best solution. Figure 2 illustrates the particle movement behavior for getting a better solution.

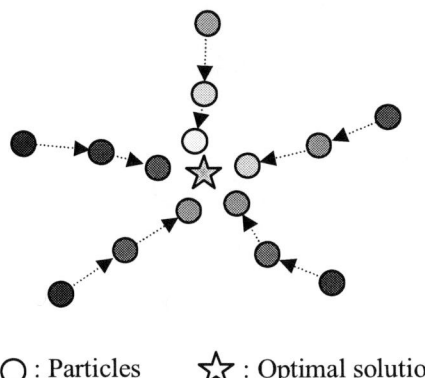

$\bigcirc$: Particles $\quad\bigstar$: Optimal solution

Figure 2. Particle movement behavior

Adopting more than one particle will explore more potential better solutions and each particle would provide its experience for other particles to prevent any particle's solution from trapping into local minimal and jump out to find better solution. The amount of particles would not influence the computation time directly. Due to more amounts of particles will find more solution in current generation and spend more time on computation. Therefore, it would spend less generation for getting an acceptable solution. On the contrary, in case of adopting small quantity particles will reverse the result. After a number of generations, each module will get closer, i.e. the chip size is gets smaller. The requirement of minimized chip area is estimated by the objective function f. The amount of particles in the swarm is defined as $1 \le i \le m$. The x_i represents the particle's current position in solution space and the initial states of $v_{i,j}$, y_i and $\hat{y}$ were set as 0. As soon as particles moved following (1) and (2), the new local best position and the global best position would be updated by (3) and (4) respectively. The particle would keep moving to find a better solution until it reached the goal or met the termination condition [11][12]. The pseudo code of our method is presented in Figure 3.

Create and initiate a B*-tree and an N-dimensional PSO: P
Repeat:
    Execute PSO to update P by (1) and (2)
    Perturb the B*-tree
    for each particle $i \in [1...m]$
        if $f(x_i) < f(y_i)$ or $f(x_i)$ is acceptable
        then $y_i = x_i$
        if $f(y_i) < f(\hat{y})$
        then $\hat{y} = y_i$
    endfor
Until Termination condition reached

Figure 3. Pseudo code

TABLE 1 EXPERIMENTAL RESULTS

Circuit	# of modules	SA with B*-tree [4]		Our method	
		Area (mm²)	Time (sec)	Area (mm²)	Time (sec)
apte	9	47.30	58	46.92	23
xerox	10	20.47	69	19.55	6
hp	11	9.57	213	9.22	87
ami33	33	1.36	1821	1.28	614
ami49	49	43.34	5762	41.01	3710
n_30	30	0.247	178	0.238	101
n_50	50	0.243	769	0.233	154

4. Experiments result

The experiments in this study employed GSRC and MCNC benchmarks [13] for the proposed floorplanner and compare with [4]. All the cells were set as hard IP modules. The simulation programs were written in MATLAB [14], and the results were obtained on a Pentium 4 1.7 GHz with 512MB RAM. The PSO experiments with w, $c1$ and $c2$ initializations were 1, 0.1 and 0.1, respectively. The particle number is set as five. We ran the both floorplanner 10 times and calculated their average outcomes of chip area and run time.

The experiment results of both floorplanner are shown in Table 1. Compare with [4], our method can find a better placement solution in even less computation time. Under the same tree structure, that is to say, our method has more efficiency and solution searching ability for floorplan. Although the SA in [4]

0-7695-2533-4/06 $20.00 © 2006 IEEE

adopted three the same operations that mentioned above, but it would randomly pick up the operation (somewhat like a kind of trial and error strategy) but not following the previous experience while trying to find another better solution. This will result in the floorplanner waste too much time on trapping the solution into local minimal and harder to get a better solution. Our method can overcome these drawbacks. Thus, the acceptable solution can find out in shorter computational time. The convergence curves of both methods are shown in Figure 4. Relative to both methods, our method possesses more robustness to prevent the solution from falling into local minimal. It would be beneficial to find a better solution in shorter time. Figures 5-7 show three benchmark results of hard modules packing for n50, ami33 and xerox.

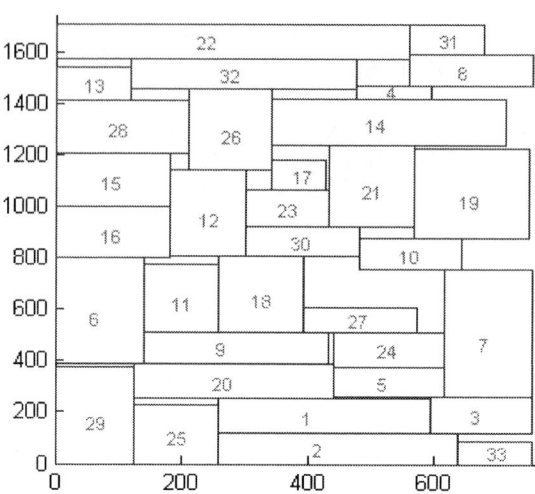

Figure 6. Placement result in GSRC ami33

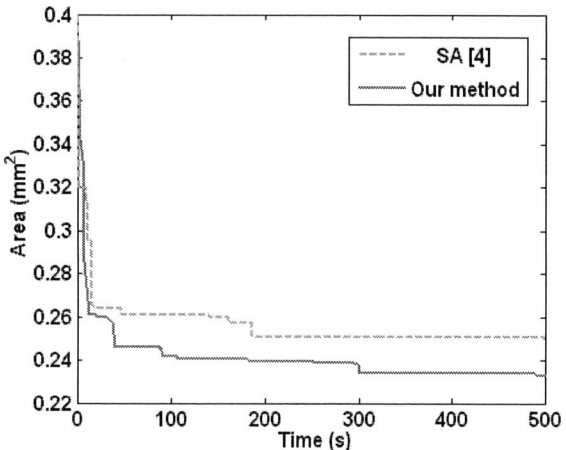

Figure 4. Convergence curves

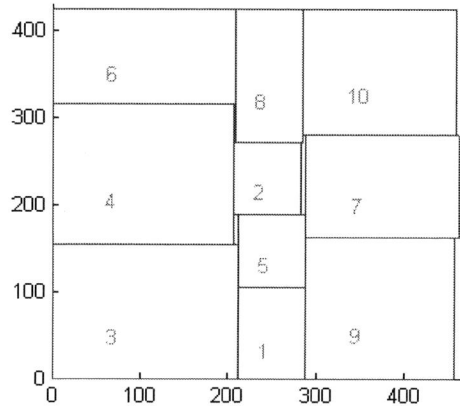

Figure 7. Placement result in GSRC xerox

5. Conclusion

In this paper, we proposed a floorplanner based on the PSO with B*-tree structure for placing blocks. PSO exhibits the ability of searching the solution space more efficiency than SA, Furthermore, PSO can save more computation time for finding an acceptable solution. The experiment results proved that the proposed PSO method can lead to a more optimal and reasonable solutions on the hard IP modules placement problem.

6. Future works

Our future works will focus on finding ways to apply to different representations for enhancing the efficiency of the floorplanning, and dealing with soft IP modules placement problem.

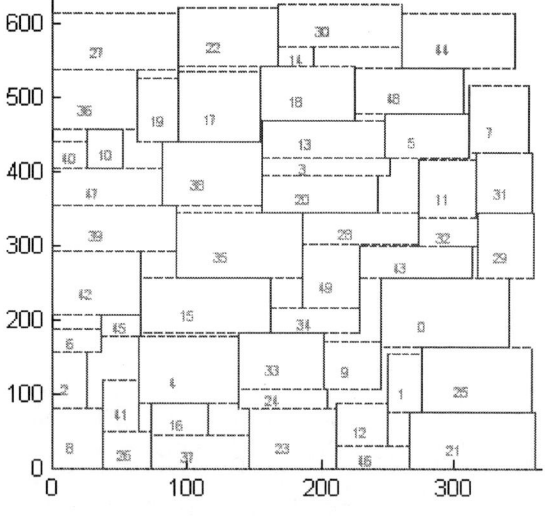

Figure 5. Placement result in MCNC n50

0-7695-2533-4/06 $20.00 © 2006 IEEE   10

7. References

[1] N. Quinn, and M. Breuer, "A forced directed component placement procedure for printed circuit boards," IEEE Trans. on Circuits and Systems, vol. 26, pp. 377-388, June 1979.

[2] H. Murata, K. Fujiyoshi, S. Nakatake, and Y. Kajitani, "VLSI module placement based on rectangle-packing by the sequence-pair," IEEE Trans. on Computer Aided Design, vol. 15, pp. 1518-1524, Dec. 1996.

[3] P.N. Guo, T. Takahashi, C. K. Cheng, and T. Yoshimura, "Floorplanning using a tree representation," IEEE Trans. on Computer-Aided Design, vol. 20, pp. 281-289, Feb. 2001.

[4] Y. C. Chang, Y. W. Chang, G. M. Wu, and S. W. Wu, "B*-Trees: A new representation for nonslicing floorplans," Design Automation Conference, pp. 458-463. 2000.

[5] B. Yao et al., "Floorplan Representations: Complexity and Connections," ACM Trans. on Design Automation. of Electronic Systems 8(1), pp. 55-80, 2003.

[6] G. Sigl, K. Doll, and F. M. Johannes, "Analytical placement: A linear or a quadratic objective function," Design Automation Conference, pp. 427-432, 1991.

[7] F. Mo, A. Tabbara, and R. K. Brayton, "A force-directed macro-cell place," Computer-Aided Design Conference, pp. 177-180, 2000.

[8] J. C. Jeong and C. M. Kyung, "Finding optimal module orientations in macrocell placement," Electronics Letters, vol. 27, pp. 804-805, May 1991.

[9] R. C. Eberhart and J. Kennedy, "A new optimizer using particle swarm theory," in Proc. 6th Int. Symp. Micro Machine and Human Science, Nagoya, Japan, pp. 39-43, 1995.

[10] V. G. Gudise and G. K. Venayagamoorthy, "Comparison of Particle Swarm Optimization and Backpropagation as Training Algorithms for Neural Networks." IEEE Swarm Intelligence Symposium, pp. 110-117, Apr. 2003.

[11] S. T. Hsieh, C. W. Lin and T. Y. Sun, "Particle Swarm Optimization for Macrocell Overlap Removal and Placement," in Proc. of IEEE Swarm Intelligence Symposium (SIS'05), pp. 177-180, June 2005

[12] T. Y. Sun, S. T. Hsieh and C. W. Lin "Particle Swarm Optimization Incorporated with Disturbance for Improving the Efficiency of Macrocell Overlap Removal and Placement," in Proc. of The 2005 International Conference on Artificial Intelligence (ICAI'05), pp. 122-125, June 2005

[13] http://www.cse.ucsc.edu/research/surf/GSRC/progress.html

[14] http://www.mathworks.com/

0-7695-2533-4/06 $20.00 © 2006 IEEE

Enhanced Dual Strategy based VLSI Architecture for Computing Pseudo Inverse of Channel Matrix in a MIMO Wireless System

[1]Zahid Khan, [1,2]Tughrul Arslan, [1]John S. Thompson, [1,2]Ahmet T. Erdogan

[1]System Level Integration Group
School of Engineering and Electronics
The University of Edinburgh,
Edinburgh, Scotland, UK

[2]Institute of System Level Integration
The ALBA Campus,
Livingston, Edinburgh,
Scotland, UK

Abstract

Multiple Input Multiple Output (MIMO) wireless technology involves highly complex signal processing which is directly related to increased power and area consumption in VLSI architecture. This paper proposes an enhanced dual strategy based VLSI architecture developed for computing the pseudo inverse of augmented channel matrix used in MIMO systems. The architecture concurrently addresses algorithmic optimization of number of multipliers while at the same time allowing for intelligent selective clock gating to disable the clock to those portions of the architecture that remain inactive during period of computation. Results indicate overall 36% power and 31% area reduction compared to previous architecture without degrading the BER performance.

1. Introduction

Multiple Input - Multiple Output (MIMO) wireless communication is a new technology that promises to remove the limits of wireless networks by providing spectral efficiency near Shannon's bound [1]. MIMO multiplies range, reliability and data speed of existing wireless systems [2]. Because of its advantages, MIMO is entering into almost every wireless network such as CDMA2000, and WCDMA as an example [2]. However, such benefits of MIMO come at the expense of highly complex signal processing which directly contributes to high power consumption [3].

Communication is becoming more wireless and portable. The weight and size are the bottlenecks of portable electronic systems. Power supplies are a major, if not dominant factor contributing to the weight and size of portable electronic systems and these are directly affected by the power dissipation due to electronic circuits.

In CMOS, sources of power consumption include short circuits, leakage currents and switching. The switching or dynamic power equation is described as

$$P = kC_L V^2 f \quad [4]$$

where k represents the switching activity factor, C_L the total physical capacitance, V the supply voltage and f the frequency of operation. Algorithmic power optimization includes reduction of both physical capacitance and switching activity factor. Physical capacitance can be reduced by reducing the area of hardware through efficient implementation [5]. Switching activity reduction either comes from area reduction that reduces the number of nodes or from reducing the switching frequency of nodes. One of the algorithmic optimizations is reducing redundancy [6] from a design. By reducing the redundant operations or hardware, unnecessary switching of the clock as well as other signals can be avoided, thereby saving power.

VBLAST (Vertical Bell Labs Layered Space Time) is a MIMO detection algorithm [7] that provides a good trade-off between BER (Bit Error Rate) performance and computational complexity compared to its counter parts. Zero Forcing (ZF) and Minimum Mean Square Error (MMSE) detectors [8] are computationally less expensive than VBLAST; however, they provide inferior BER performance compared to VBLAST. The optimal solution is maximum likelihood (ML) [8] detection which provides best BER performance. However, it is highly expensive regarding computational complexity. This increases exponentially with the number of antennas used and is prohibitively high for antennas more than 4. Therefore, the ML algorithm cannot be implemented on mobile platforms due to its high overhead of area and power [3].

VBLAST can provide BER performance close to ML at a computational complexity much less than ML [8]. In VBLAST itself, the bottlenecks are repeated pseudo inverse calculation required to compute optimal ordering and nulling vectors. This repeated computation leads to numerical instability in hardware implementation and complexity which can be reduced using alternative algorithms.

The pseudo inverse can be computed using the singular value decomposition method (SVD) [9]. However, pseudo inverse computation through SVD is expensive both in silicon and power consumption. For equal transmit and receive antennas *(M=N)*, the

0-7695-2533-4/06 $20.00 © 2006 IEEE

complexity of pseudo inverse through SVD in MMSE-VBLAST is $(27/4)M^4$ [10]. The complexity grows as the fourth power of M which is quite huge. The square root algorithm [10] not only computes pseudo inverse but also avoids repeated pseudo inverse computation and reduces the computational complexity of VBLAST to $O(M^3)$ without degrading performance [10].

A VLSI architecture for computing the pseudo inverse module through the square root algorithm has been devised in [11] in which a 2-CORDIC based supercell has been used. The architecture presented in [11] is an improved version of the architecture presented in [12] which is a straight forward implementation without regard to area and power optimizations. In this paper, two strategies have been suggested to the architecture proposed in [11]. The first is the use of two multipliers in the MAC (Multiply and Accumulate) unit instead of the conventional four multipliers used to carry out multiplication and accumulation of complex numbers. The second strategy is based on disabling the clock to some of the portions of the architecture that remain inactive for some time during period of computation.

2. MIMO System Model and Square Root Algorithm for VBLAST

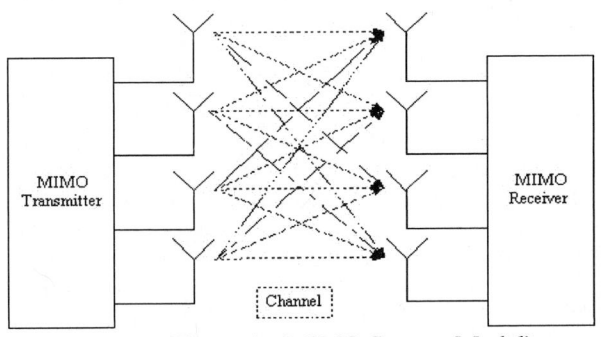

Figure 1: (MIMO System Model)

In MIMO communication systems, more than one antenna is used at the transmitter to transmit symbols and more than one antenna is used at the receiver to receive them. In the diagram of Figure 1, spatial multiplexing is used and M transmit antennas transmit M symbols simultaneously while each symbol is received by the N receive antennas. Each symbol transmitted is received by all the receiving antennas thus making multiple channel paths. These paths, if combined, make a matrix of channel elements. Each symbol makes N channel paths and is received by N receive antennas. Since there are M symbols transmitted simultaneously, the channel becomes a NxM matrix.

If $s = [s_1, s_2, s_3, s_4 ....., s_M]^T$ denotes the symbol vector transmitted, H denotes the NxM channel matrix between the receive and transmit antenna array, and v denotes the AWGN independent and identically distributed noise vector, then the corresponding receive vector r at the input of the MIMO receiver is given by

$$r = Hs + v \qquad (1)$$

To recover the transmitted symbol vector s, it is necessary to invert the channel matrix H. The inversion can be done depending upon the detection method used. For Zero Forcing, the channel matrix is simply inverted and the detector output is given by (2)

$$z = s + H\dagger v \qquad (2)$$

where $\dagger$ is the pseudo inverse. For MMSE, the channel matrix is augmented by the noise variance (α) and the detector output is as in (3)

$$s = (\alpha I + H^* H)^{-1} H^* r \qquad (3)$$

where * represents complex conjugate transpose.

In VBLAST, successive nulling and cancellation is used to detect the transmitted symbols. The channel matrix is first inverted and then reordered to detect that symbol first which has the highest post detection Signal to Noise ratio (SNR). This corresponds to the row of the inverted channel matrix having minimum Euclidean distance. The symbol after detection is subtracted from the received symbol vector. The corresponding column of the H matrix is zeroed down and the process is repeated with the deflated channel matrix until all the symbols are detected. In this research, MMSE is used for channel inversion. The pseudo inverse of a generic matrix H is given by

$$H^+ = (H^* H)^{-1} H^* = R^{-1} Q^* \qquad (4)$$

The pseudo inverse can be computed using either singular value decomposition (SVD) or QR decomposition. The square root algorithm [10] is developed for MMSE-VBLAST and computes the QR decomposition of the augmented channel matrix.

$$\begin{bmatrix} H^{NxM} \\ \sqrt{\alpha} I^{MxM} \end{bmatrix} = QR = \begin{bmatrix} Q_a^{NxM} \\ x \end{bmatrix} R^{MxM} \qquad (5)$$

here x denotes the entries that are not relevant. The algorithm first decomposes the channel matrix into QR $a_r + ja_i$ and then computes $P^{1/2} = R^{-1}$. Once Qa and $P^{1/2}$ are computed, the repeated pseudo inverse can be avoided. Both Qa and $P^{1/2}$ are computed together in a

0-7695-2533-4/06 $20.00 © 2006 IEEE   13

series of unitary transformations. The algorithm is described below:

Compute Qa and $P^{1/2}$ using equation (6).

1. $$B \in_i = X \quad (6)$$

$$B = \begin{bmatrix} 1 & H_i P^{1/2}_{i\,|\,i-1} \\ 0^{Mx1} & P^{1/2}_{|\,i-1} \\ -e_i^{Nx1} & B_{i-1} \end{bmatrix} \qquad X = \begin{bmatrix} x & 0^{1xM} \\ x & P^{1/2}_{|\,i} \\ x & B_i \end{bmatrix}$$

Here i represents iterations and $i=1,\ldots\ldots,N$
B is the prearray matrix and has dimension of $(1+M+N) \times (1+M)$ and $P^{1/2}_{|0} = 1/\sqrt{\alpha}I, B_0 = 0_{NxM}$
$e_i^{Nx\,1}$ is the i-th unit vector of dimension N and Θ_i is any unitary transformation (Jacobi rotation) that block lower triangularizes the pre-array denoted by M. After N iterations,

$$P^{1/2}_{|0} = P^{1/2}_{|N} \text{ and } Q_a = B_N \quad (7)$$

Equations (6) and (7) are used in pseudo inverse computation. For the rest of the algorithm, the reader is referred to [10].

3. Sequence of operations in Pseudo Inverse computation

A MatLab program has been developed to model computation of pseudo inverse using the square root algorithm. The program first generates independent and identically distributed (iid) complex channel elements and assumes a particular value for the Signal to Noise Ratio (SNR). After generating channel elements, the rest of the code models pseudo inverse the way it will be implemented in hardware.

Computation of pseudo inverse is done in N iterations. Each iteration consists of three steps. The first step starts with forming/updating the prearray matrix B which for a $M=N=4$ antenna system is a matrix of dimension $9x5$. The weighted channel elements in the first row described by $H_i P^{1/2}_{i\,|\,i-1}$ in which H_i is a $1x4$ vector and $P^{1/2}_{|i-1}$ is a $4x4$ matrix is obtained using complex multiplication and addition. In the second step, each complex weighted channel element in the first row is made real by rotating its imaginary part to zero [13] using Jacobi's rotation. Jacobi's rotation is carried out using CORDICs (Coordinate Rotation Digital Computers). Jacobi's rotation is described by equations (9) and (10) [9].

$$\begin{bmatrix} \cos(\theta) & \sin(\theta) \\ -\sin(\theta) & \cos(\theta) \end{bmatrix}\begin{bmatrix} a_r \\ a_i \end{bmatrix} = \begin{bmatrix} r \\ 0 \end{bmatrix} \quad (9)$$

$$\theta = -\tan^{-1}(a_i/a_r) \quad (10)$$

Given a complex number, the angle is first calculated using equation (10) and then the complex number is rotated using equation (9) to make its imaginary part 0. Both rotation and angle calculation are carried out using CORDICs.

The first row of the prearray matrix is given by (11)

$$[1 \quad h_{11} \quad h_{12} \quad h_{13} \quad h_{14}] \quad (11)$$

where h_{11}, h_{12}, h_{13} and h_{14} are the weighted channel elements. These are the first elements of their respective columns which consist of 9 elements per column in the case of $M=N=4$ antenna system. The first element of each column (described in 11) inside the prearray is made leader with others as followers. The leaders as in (11) are applied to the CORDIC for angle calculation. The leaders together with their corresponding followers are applied to the CORDIC for rotation. The effect of this rotation on the leader is to make its imaginary part equal to zero while its effect on the followers is to change their phase by the phase angle of the leader. At the end of this step, all leaders become real and the followers of each leader have their phase angle changed.

The third step involves zeroing the leaders. However, before proceeding to this step, it is necessary to explain the parallelism inside Jacobi's rotation in which the process of zeroing elements can be done in parallel. Each Jacobi's rotation involves either two rows or two columns. For the prearray B two Jacobi's rotations can be done in parallel and leaders can be zeroed in 3 instead of 4 iterations as shown in Figure 2. Only leaders are shown and all symbols represent real numbers. The vertical axis represents iteration number.

1	h_{11}	h_{12}	h_{13}	h_{14}	1	h_{11}	h_{12}	h_{13}	h_{14}
1	α	0	h_{13}	h_{14}	1	0	β_{12}	0	α_{14}
2	β	.0	h_{13}	h_{14}	1	0	0	0	λ_{14}
3	χ	0	0	h_{14}	α_{11}	0	0	0	0
4	λ	0	0	0					
	(a)						(b)		

Figure 2: (Sequential (a) and Parallel (b) Jacobi's rotation)

In the sequential case described in Figure 2(a), $(1, h_{11})$ is rotated which zeros h_{11} and changes 1 to α where α is any real number. Then (α, h_{12}) is rotated to $(\beta,0)$. After this, (β,h_{13}) is rotated to $(\chi,0)$ and lastly (χ,h_{14}) is rotated to $(\lambda,0)$. In the parallel case described in 2(b), (h_{11}, h_{12}) and (h_{13}, h_{14}) are rotated to $(0, \beta_{12})$ and $(0, \alpha_{14})$ simultaneously. After this, there are only three columns left, rotation is done sequentially and $(\beta_{12}, \alpha_{14})$ is rotated to $(0, \lambda_{14})$. Lastly, $(1, \lambda_{14})$ is rotated to $(\alpha, 0)$. This implies that one iteration can be removed if rotation is carried out in parallel.

The leaders are then zeroed out using the parallel Jacobi rotation and this completes one iteration. After $(N=4)$ such iterations, $P^{1/2}$ and Qa give the pseudo inverse of the channel matrix. Initially the three pipelined CORDIC based pseudo inverse module proposed in [12] and shown in Figure 3, was implemented in MatLab. The module consists of a Θ-CORDIC to calculate the angle as according to equation *(10)* and two Φ-CORDICs to perform rotation as according to equation *(9)*.

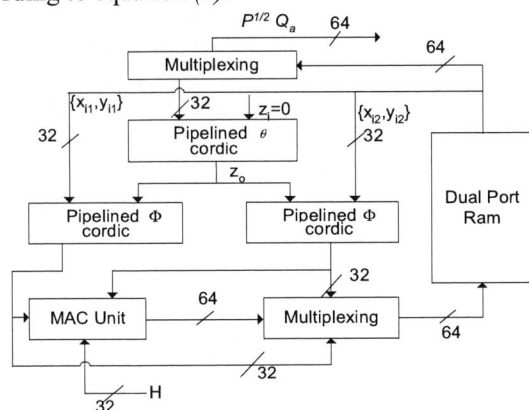

Figure 3: (Previous pseudo inverse module)[12]

In this implementation the number of iterations as well as the first step in each iteration is the same as described above. In step 2, all the four leaders are applied to the Θ-CORDIC in pipeline. When the angle comes out of the Θ-CORDIC, the leaders and their corresponding followers are applied to the two Φ-CORDICs for making leaders real and changing the phase angle of the followers. In step 3 and in the process of zeroing the leaders, the Θ-CORDIC is fed first with (h_{11}, h_{12}) and then (h_{13}, h_{14}). Thus two angles need be calculated. When the angle for (h_{11}, h_{12}) rotation comes out of the Θ-CORDIC, it is applied to the two Φ-CORDICs together with columns 2 and 3 of the prearray. The two columns are applied in parallel. After all elements of columns 2 and 3 are applied to the Φ-CORDICs, the next angle is applied for rotating h_{13} to zero, thus 4th and 5th columns are applied in parallel for rotation.

The output is stored in the prearray. With this process, two of the four leaders are zeroed out. This corresponds to iteration 1 in Figure 2. In iteration 2, the leaders β_{12} and α_{14} are applied for zeroing β_{12}. First these are applied to the Θ-CORDIC and after calculating the angle, the angle together with columns 3 and 5 of the prearray are applied to the Φ-CORDICs. The output from the Φ-CORDICs is stored in the

prearray with β_{12} rotated to zero and α_{14} modified to λ_{14}. In the last iteration $(1, \lambda_{14})$ are applied to the Θ-CORDIC, the angle is then applied to the Φ-CORDICs for rotation and this rotates the last leader to zero. This completes one iteration. After four such iterations, $P^{1/2}$ and Qa give the pseudo inverse of the augmented channel matrix.

4. Proposed Pseudo Inverse module

The proposed architecture 11] for the pseudo inverse (Figure 4) consists of two independent and generic pipelined CORDICs and a dual port ram to support the two CORDICs. Each CORDIC is developed to have 13 micro-cells. The number system used is 16Q8 with 8 bits for precision, 7 bits for dynamic range and 1 bit for sign. The process of computing the pseudo inverse is the same as described in section 3; however, the difference is that the two CORDICs are used both in vectoring as well as in rotation mode. The control unit of each micro-cell is designed to configure the cell for either vector or rotation mode. The inputs to the control unit of each micro-cell are the sign bits of x, y and z inputs as well as CORDIC mode signal. The CORDIC mode signal determines the vectoring or rotation mode for the micro-cell. This signal is propagated together with x, y and z data from the first micro-cell to the last one. Initially the two CORDICs are in the vectoring mode and calculate the angles. The angles are then applied as input to both CORDICs and both CORDICs perform rotation together in the way described in section 3. The rotated vector needs scale correction in which each coordinate of the rotated vector is multiplied with the scale correction constant which is 0.6057 [13] A simple shift and add circuit instead of multiplier is used to perform this correction. The scale correction in 16Q8 format is 10011011 which is equal to $2^7+2^4+2^3+2^1+2^0$ and can be implemented.

5. Dual Strategy in Proposed Architecture

5.1 Algorithmic Optimization of number of multipliers

The function of the MAC unit inside the pseudo inverse module is to perform the complex matrix multiplication given by $H_i P_{|i-1}^{1/2}$ where H_i *(1x4)* is the ith row of the channel matrix, $P_{|i-1}^{1/2}$ (is the *4x4*) inverse triangular matrix and $H_i P_{|i-1}^{1/2}$ is 1x4 vector. The MAC unit shown in Figure 5b computes one value of $H_i P_{|i-1}^{1/2}$ in 4 clock cycles using 4 multipliers and

0-7695-2533-4/06 $20.00 © 2006 IEEE

the circuitry as shown in Figure 5b by performing the multiplication as shown in Figure 5a. Each complex multiplication h_ip_i takes one clock cycle.

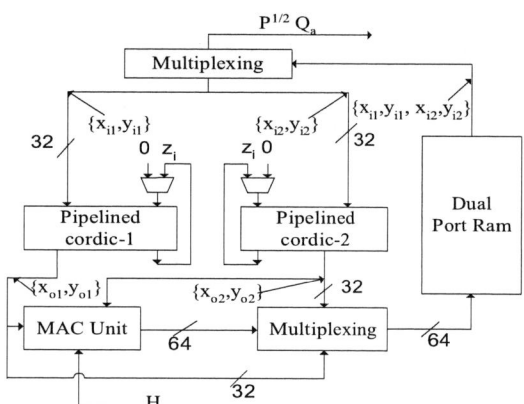

Figure 4:(Proposed pseudo inverse module)[11]

The multiplication h_ip_i can be performed in two clock cycles using the circuit shown in Figure 6. Let $h_i=r_1+ij_1$ and $p_i=r_2+ij_2$ then $Real(h_ip_i)=r_1r_2-j_1j_2$, $Imag(h_ip_i)=r_1j_2+r_2j_1$. The Ram in Figure 6 stores the four values of pi as provided by the pipelined cordic unit. As soon as the first complex element is stored in the Ram, it is read in the next clock cycle for multiplication with the complex channel elements. In the first cycle $Real(h_ip_i)=r_1r_2-j_1j_2$ is computed and stored in the register, reg1. In the next cycle, $Imag(h_ip_i)=r_1j_2+r_2j_1$ is computed. Both $Real(h_ip_i)$ and $Imag(h_ip_i)$ are applied to the Accumulator in the second clock cycle. In this way, it takes 8 clock cycles to compute h_w. In this design latency is traded for both area and power reduction. Trading latency against power and area in computing $H_iP_{|i-1}^{1/2}$ is chosen as it did not produce any increase in latency in computing the pseudo inverse of the augmented channel matrix.

5.2 Selective Clock Gating

The pseudo inverse module takes 460 clock cycles to compute the pseudo inverse of the channel matrix. Not all modules are busy in computation during this period. Those parts which are not active can be stopped from switching either by using clock gating or holding the inputs at their previous states. Clock gating is used here to disable the pipelined Cordic 2 for the period of time during which it remains idle. The clock to the flipflop inside the pipelined Cordic has been gated using a control signal. The control signal is asserted for a specific period of time during which the Cordic remains idle. This clock gating which is done at RTL level can also be extended to other modules like MAC and the second Cordic unit.

$$\begin{bmatrix} h_1 & h_2 & h_3 & h_4 \end{bmatrix} \begin{bmatrix} p_1 \\ p_2 \\ p_3 \\ p_4 \end{bmatrix} = h_w$$

(a)

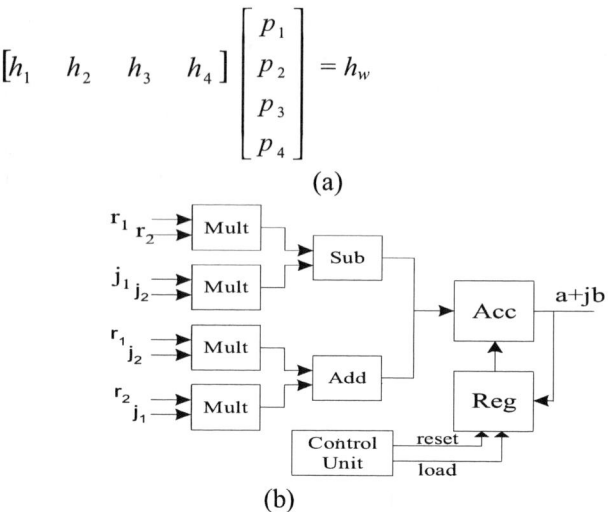

(b)

Figure 5: (a: a complex weighted channel element, b: Conventional MAC Module)

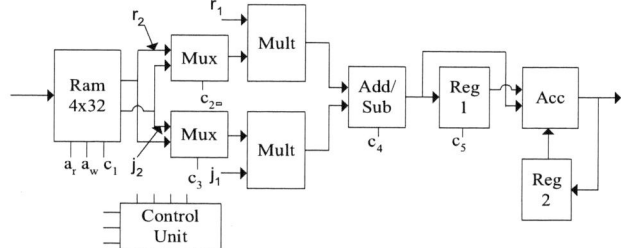

Figure 6: (Modified MAC Unit)

6. Simulation and Synthesis results

The pseudo inverse architectures (previous, proposed and modified proposed based on the dual strategies) have been synthesized using Synopsys Design Compiler and mapped to 0.18um CMOS technology. The area comparison is given in Table 1 which shows area figures for the previous (Figure 3), proposed (Figure 4) and the modified proposed architectures. The table shows an overall saving of about 31% over the previous architecture. The architecture has been simulated at 100MHz for power comparison which is recorded in Table 2. From the architecture, it is clear that the CORDIC and the multipliers (in the MAC unit) are the major power consuming units. Modifying the MAC module and using clock gating has reduced both area and power consumption further. Inside the MAC unit, the multipliers are the major power and area consuming blocks. Reducing the total number of multipliers from 4 to 2 reduced the area of the MAC module by 20%. Ideally the reduction should be about 40%, however, the overhead of Ram, Control Unit, programmable Add/Sub and a 16-bit register reduced the area reduction to 20%. The power reduction is 18.4%

0-7695-2533-4/06 $20.00 © 2006 IEEE   16

compared to the power consumption of the conventional MAC module.

The clock gating is just a simple ANDing of the clock signal with a control signal. Clock gating has produced negligible overhead in terms of area. By clock gating, the power consumption of Cordic-2 has been reduced from 29.409mw to 20.808mw which is a reduction of about 29%. With the two modifications, the overall power and area has been reduced by 36% and 31% respectively compared to the architecture proposed in [12].

7. Conclusion

The authors have presented an area and power efficient VLSI architecture for computing pseudo inverse through square root algorithm. The architecture reduces area and power consumption by reducing the redundant hardware as much as possible and achieves 31% area and 36% power efficiency. The architecture exploits parallelism inherent in Jacobi's rotation. The architecture is simulated within MMSE-VBLAST system and imposes no degradation in performance as well as latency.

8. References

[1] G. J. Foschini, "Layered Space-Time architecture for wireless communication in fading environments when using multiple antennas," Bell Labs Tech. J., vol 2, Autumn 1996

[2] G. Lawton, "Is MIMO the future of wireless communications?", computer, vol: 37, issue 7, July 2004 Pages 20-22

[3] D. Garrett, L. Davis, St. Brink, B. Hochwald, G. Knagge, "Silicon complexity for maximum likelihood MIMO detection using spherical decoding", Solid-State Circuits, IEEE Journal, vol. 39, Issue 9, Sept. 2004, Pp(s):1544-52

[4] M. Chandrakasam, M. Potkonjak, R.Mehra, J. Rabaey, and R. Broderson, "Optimizing power using transformations", IEEE Transactions on Computer-Aided Design of Integrated Circuits and Systems, vol. 14, no.1 pp. 12-31, January 1995

[5] M.Pedram,"Power Minimisation in IC Design: Principles and Applications", ACM Transactions on Design Automation of Electronic Systems, vol. 1, no. 1, pp. 3-56, January 1996.

[6] X. Wu, M. Pedram, L. Wang, "Multi-code state assignment for low power design", Circuits, Devices and Systems, IEE Proceedings vol. 147, Issue 5, Oct. 2000 Page(s):271 - 275

[7] P.W. Wolniansky, G.J. Foschini, G.D. Golden, and R.A. Valenzuela, "V-BLAST: an architecture for realizing very high data rates over the rich-scattering wireless channel", Proc. ISSSE'98, Sept. 1998

[8] A. Adjoudani, E.C. Beck, A.P. Burg, G.M. Djuknic, T.G. Gvoth, D. Haessig, S. Manji, M.A. Milbrodt, M. Rupp, D. Samardzija," Prototype experience for MIMO

BLAST over third-generation wireless system", Selected Areas in Communications, IEEE Journal on Vol. 21, Issue 3, April 2003 Page(s):440 - 451

[9] Gene. H. Golub, Charless F. Van Loan, "Matrix Computation"

[10] B. Hassibi, "An efficient square-root algorithm for BLAST", Acoustics, Speech, and Signal Processing, 2000. ICASSP '00. Proceedings. 2000 IEEE International Conference on, Volume 2, 5-9 June 2000 Page(s):II737 - II740 vol.2

[11] Zahid Khan, Tughrul Arslan, John S. Thompson, Ahmet T. Erdogan, "Area and Power efficient VLSI Architecture for Computing Pseudo Inverse of Channel Matrix in a MIMO Wireless System", accepted for publication in IEEE-VLSI conference to be held in India on Jan 3, 2006

[12] Z.Guo, P.Nilsson, "A VLSI implementation of MIMO detection for future wireless communications", Personal, Indoor and Mobile Radio Communications, 2003. PIMRC 2003. 14th IEEE Proceedings on, vol. 3, 7-10 Sept. 2003 Pages:29-49

[13] C.M. Rader, "VLSI systolic arrays for adaptive nulling", (radar) Signal Processing Magazine, IEEE , vol. 13 , Issue: 4 , July 1996, Pages:29 – 49

	Previous Pseudo Inverse Module	Area in μm^2	Proposed Pseudo Inverse Module	Area in μm^2	Modified Pseudo Inverse Module	Area in μm^2
1	Θ-CORDIC	250120	CORDIC-1	253659	CORDIC-1	253659
2	Φ-CORDIC	258920	CORDIC-2	253659	CORDIC-2	253659
3	Φ-CORDIC	258920				
4	MAC Unit	147611	MAC Unit	147611	MAC Unit	93651
5	Control Unit and duram	49210	Control Unit and duram	66015	Control Unit and duram	66015
	Total	964781	Total	720944	Total	666984
	% Saving	%		24%		31%

Table 1: (Area figures for comparison)

	Previous Pseudo Inverse Module	P:ower in mw	Proposed Pseudo Inverse Module	Power in mw	Modified Pseudo Inverse Module	Power in mw
1	Θ-CORDIC	31.278	CORDIC-1	33.608	CORDIC-1	33.608
2	Φ-CORDIC	33.548	CORDIC-2	29.409	CORDIC-2	20.808
3	Φ-CORDIC	33.463	---------		---------	
4	MAC Unit	28.484	MAC Unit	28.449	MAC Unit	20.449
5	Control Unit and duram	12.225	Control Unit and duram	14.329	Control Unit and duram	14.329
	Total	138.998	Total	105.795	Total	89.194
	% Saving			25%		36%

Table 2: (Power figures for comparison)

0-7695-2533-4/06 $20.00 © 2006 IEEE

Adaptive Porting of Analog IPs with Reusable Conservative Properties

Takashi Nojima[†,††] Shigetoshi Nakatake[†] Toru Fujimura[†] Koji Okazaki[†] Yoji Kajitani[†] Nobuto Ono[††]

† Dept. of Information and Media Sciences   †† Research and Development Division
The University of Kitakyushu   Jedat Innovation Inc.
1–1, Hibikino, Wakamatsu-ku,   2–5, Hibikino, Wakamatsu-ku,
Kitakyushu, Fukuoka, 808–0135, Japan   Kitakyushu, Fukuoka, 808–0135, Japan
Email: {nojima.takashi, ono.nobuto}@jedat.co.jp, {nakatake, kajitani}@env.kitakyu-u.ac.jp

Abstract

Analog layout automation is one of the most challenging subjects that has to cope with trade-offs among analog specific requirements such as noise, linearity, gain, supply-voltage, speed, power consumption, etc. This paper proposes a novel porting methodology that guides the reuse of analog IPs, followed by an automation system. The methodology introduces a concept of conservative properties that are necessary and sufficient for the configuration of the high quality layout. The properties are extracted from schematics and the past layouts, and then are represented in terms of module configurations and topological constraints imposed on devices. In experiments, our porting system is applied to several industrial analog circuits. In the design of an A/D converter, we ported the layout on 0.20μm/3.3V technology to that on 0.18μm/1.8V technology. The result not only met the required performance, but also achieved the comparable quality with the manual layout. The design time was reduced drastically.

1 Introduction

Recent trend of SoCs has been switched to the mixed signal ones. In order to progress manufacturability, the layout designs for both digital and analog are required not only to achieve high quality layout with high integration and low power but also to shorten the design-time-to-market. To bring about a revolution in the designs which will be inevitable in future, a solution to manage the difficulty has been believed in reuse of the past designs, *intellectual properties*(IPs), and efforts have been devoted [6, 7].

In digital design, reuse of IPs has already been established. It allows industrial suppliers to distribute circuit blocks available on different technologies. It is reported that IPs can reduce the design time drastically in several instances [6].

In analog design, however, designs reusing IPs have not matured yet because they have not been successful to cope with trade-offs among analog specific requirements depending on technologies, such as noise, linearity, gain, supply-voltage, speed, power consumption, etc. There are no consensus to the choice of data to be reused. When the technology changes, some differences will follow on the structure of devices, design rules, supply voltage, etc., even if the specifications and the schematic are unchanged.

From the view point of expert designers, they find out *conservative properties* inside the past layout and its schematic. Such conservative properties usually include module configurations and placement constraints. The module configurations imply analog layout techniques such as diffusion merging, common-centroid and well island, while placement constraints are for symmetry placement and device matching. In other words, these properties are available on different technologies, and then a layout with the same properties shall achieve high quality. Thus, for an effective methodology to reuse analog IPs, we have to answer with a way to handle following two issues; 1) Extraction of the conservative properties. 2) Expression of the conservative properties. They are so-called knowledge transfers, each corresponds to *reuse* and *migration*, so that we call the combination *porting*.

Automatic tools based on the above concepts are required to use geometry-free representations to represent a layout while the conservative properties are available on different technologies and different instances. The traditional analog CAD tools [1, 4, 5] have limited their available constraints, however, it is difficult to express conservative properties.

Besides, there exist several works related to the automatic migration tools [11, 14, 15]. IPRAIL [11] and its development [15] extract topological relations between devices and wires from the past layout, and impose them on the targeting design as constraints. The relations also include symmetry relations. All the topological relations are represented by vertical and horizontal constraint graphs for compaction. However, such compaction-based tools are failed in the case when the devices do not scale down. When migration uses device scaling to drop supply-voltage, the channel width of a transistor scales up by K_v^2 if the scaling factor is K_v [10]. In this case, some constraints are redundant for compaction, so that the layout is apt to have much dead space. The designer is needed to scrap and build the layout. That is the reason why their constraints and strategies are not geometry-free.

On the other hand, since mid-1990's, several topology-based placements have been proposed such as Sequence-Pair [2], BSG [3], etc. They use geometry-free representations to represent a placement. Also, recent works introduce efficient techniques to cope with symmetry constraints or cluster

0-7695-2533-4/06 $20.00 © 2006 IEEE

constraints based on the Sequence-Pair [8, 12, 13]. Such constraints are included in the conservative properties.

In this paper, we propose a novel porting methodology. Our conservative properties are expressed in terms of topological constraints and module configurations that are extracted from schematics and the past layouts. The topological constraints consist of cluster and symmetry constraints, all of which are translated into formulations on the Sequence-Pair. The module configurations include diffusion merging and well island. Once these conservative properties are extracted, we can reuse them again and again as long as the schematic is same.

In experiments, to show our conservative properties are necessary and sufficient for porting, we prepared several industrial analog circuits with the manual layouts. First, we placed devices under the topological constraints and the module configurations on the same technology, and compared with the manual placements. Our resultant placements were comparable to the manual layouts. The results show that these conservative properties are sufficient not redundant.

Next, we applied our porting methodology to an A/D converter from the design on $0.20\mu m/3.3V$ technology to that on $0.18\mu m/1.8V$ technology. Then, we showed our methodology is effective for reusing analog IPs. In this porting, the channel width of a transistor scales up four times and the cell width was reduced half. Nevertheless, our result is comparable to the manual layout with respect to the cell area and routing area. Furthermore, we applied post-layout simulation to the both layouts. Observing the waveforms, we confirm our layout met the required performance. The results convince us the conservative properties are also necessary. In addition, our porting system reduced the design time drastically compared with the manual design.

The rest of this paper is organized as follows. Section 2 introduces previous works and our methodology. Section 3 describes topological representations and Sequence-Pair. Section 4 describes how to deal with module configurations and constraints in our system. Section 5 shows the experimental results. Section 6 concludes contributions and future works.

2 Previous Works and Our Methodology

The previous works related to migration [11, 14, 15] introduce compaction-based strategies. They extract topological relations between devices and wires from the past layout, and impose them on the targeting layout as constraints. The relations also include symmetry relations. All the topological relations are represented by vertical and horizontal constraint graphs for compaction.

In migration of analog IPs, the devices do not always scale down. When supply-voltage is dropped, the channel width of a MOS transistor scales up by K_v^2, where K_v is the scaling factor [10]. In this case, however, they output layouts with much dead space.

We tested a migration of an A/D converter from the layout on $0.20\mu m/3.3V$ technology to that on $0.18\mu m/1.8V$ technology, which are shown in Fig.1 A and B, respectively. In this migration, the supply voltage was reduced half thus the channel width of a MOS transistor scales up four times. The devices were placed keeping topological relations in the last layout. As shown in the figure B, there are much dead space. The dead space is caused because the constraints are redundant and are restricted to their compaction strategy.

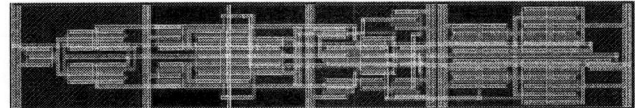

A : Manual layout on $0.20\mu m/3.3V$ technology

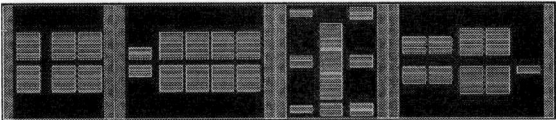

B : Migrated to the layout on $0.18\mu m/1.8V$ technology

Fig. 1. Migration of an A/D converter

We propose a novel porting methodology based on Sequence-Pair[2]. Our conservative properties are expressed in terms of topological constraints and module configurations that are extracted from schematics and the past layouts. The topological constraints consist of cluster and symmetry constraints, all of which are translated into formulations on the Sequence-Pair. We show that the constraints are necessary and sufficient not redundant, and do not restrict the placement strategy.

3 Topological Representations and Sequence-Pair

The traditional analog CAD tools have been used *absolute representations* to solve device-level placement problems [1, 4, 5]. The absolute representations make optimization engines, for instance, Simulated Annealing, to explore a huge search space. The penalty cost of the optimization is associated with the total infeasible overlaps, and this penalty must be driven to zero in the optimization process. The main disadvantages of using the absolute representations need a lot running time and sometimes apt to lead low-quality placements where the designers are needed to scrap and build.

An alternative approach is to use *topological representations* decoding the position of devices without overlapping. They show topological relations, the *ABLR-relations*, between every pair of devices in the form that "one is above, below, left-of or right-of the other". The optimization engines need a little running time to reach feasible and high-quality placements by changing the topological relations between devices. The topological representations include two types of structures, *slicing structure* and *non-slicing structure*. Since the slicing structure does not have all topological representations, some solutions are still low-quality.

In mid-1990's, two kinds of representation for device-level placement, Sequence-Pair [2] and BSG [3], were invented. They proposed remarkable encoding systems for the non-slicing structure which has all topological representa-

0-7695-2533-4/06 $20.00 © 2006 IEEE

tions. Furthermore, some methodologies based on Sequence-Pair how to represent analog constraints such as symmetry, cluster and schema-driven constraints have been proposed [8, 12, 13]. The algorithm based on Simulated Annealing outputs a highly compacted placement under the constraints. In this paper, we apply the Sequence-Pair to represent a placement and constraints.

3.1 ABLR-Relations from Sequence-Pair

Given n modules, the Sequence-Pair(SP) is an ordered pair of permutations Γ_+ and Γ_- of module names. The k-th module in Γ_* ($*$ is $+$ or $-$) is denoted as $\Gamma_*(k)$. While the position of module x in Γ_* is denoted as $\Gamma_*^{-1}(x)$. An SP shows ABLR-relations as follows;

ABLR-relations from an SP : For a pair of modules $\{a, b\}$,

- If $\Gamma_+^{-1}(a) < \Gamma_+^{-1}(b)$ and $\Gamma_-^{-1}(a) < \Gamma_-^{-1}(b)$, a is left-of b.
- If $\Gamma_+^{-1}(a) > \Gamma_+^{-1}(b)$ and $\Gamma_-^{-1}(a) < \Gamma_-^{-1}(b)$, a is below b.

Fig.2 shows a placement of seven modules derived from $(\Gamma_+; \Gamma_-) = (ABCDEFG; BADCGFE)$.

Note that the ABLR-relations derived from the SP are all satisfied. Once ABLR-relations are given, it is easy to obtain a placement by using vertical and horizontal constraint graphs. It is proved that any SP corresponds to a placement, and any placement with the minimum area has the corresponding SP. Hence, searching the placements for optimization is equivalent to searching SP's. It is also proved that any SP has its corresponding placement that can be obtained very quickly in $O(n \log \log n)$ time [9].

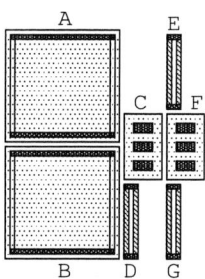

Fig. 2. A placement derived from ($ABCDEFG$; $BADCGFE$)

4 Module Configurations and Topological Constraints

Given a set of devices such as transistors, capacitors and resistors, and a net-list interconnecting terminals on these devices, a placement engine needs to select an optimal position of each device where the layout area is minimal and all the nets can be routed in the routing phase. The following constraints are imposed on our device-level placements.

4.1 Module Configurations

The module configurations imply analog layout techniques such as diffusion merging, common-centroid and well island. The system merges two or more devices into one module. The devices are corresponding to the multiplier devices and the devices referred to the matching constraints. The common-well devices are also merged here. Other de-

vices are dealt with a module one by one.

A matching group is a set of devices for which an accurate ratio of device characteristics is required. The devices in the matching constraints correspond to the devices of a current-mirror or differential pair. The matching groups have the following specifications for the device-level placement [1, 5]:

- All devices which belong to the same matching group must equal orientations.
- The placement engine has to determine the positions and the distance between the matched devices to satisfy the constraints. The matching degree are selected by the view of its influence on the circuit performances.

To satisfy such specifications, our system merges such matching devices into one module. The devices are placed adjacently with specified distance and same orientations. An example of module configuration for a current-mirror is shown in Fig.3. Three transistors M1, M2, and M3 are included by a current-mirror, and placed adjacently and same orientations after the module modification.

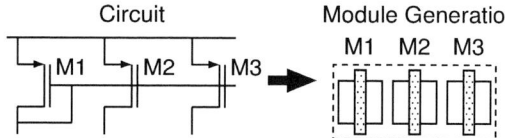

Fig. 3. Module configuration for a current-mirror

4.2 Symmetry Constraints

In high performance analog circuits, it is often required that groups of devices are placed symmetrically with respect to one or more symmetry axes [1, 5]. Symmetric placement allows for symmetric routing and results in matched parasitics. Symmetry constraints can be formulated in terms of couples, *self-symmetric devices* and *symmetry groups*. Two devices which are placed symmetrically with respect to an axis form a couple. A self-symmetric device is a device which is placed on a symmetry axis. A symmetry group is a collection of couples and self-symmetric devices which share the same symmetry axis. A symmetry group shown in Fig.4, consists of the couples (M2, M3) and (M4, M5) and the self-symmetric device M1.

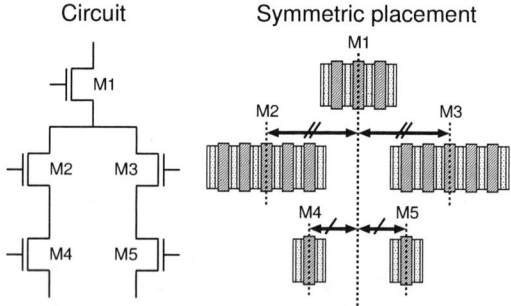

Fig. 4. Differential circuit and its symmetric placement

More than one symmetry group can be specified for a circuit. The symmetry groups have the following specifications for the device-level placement:

0-7695-2533-4/06 $20.00 © 2006 IEEE   20

Table 1. Site constraints between $x \in A$ and $y \in B$

Topological relation between clusters A and B	Site constraints between $x \in A$ and $y \in B$, and their Sequence-Pair(SP)	
	straight boundary	jagged boundary
A is left-of B	x must be left-of y	x must not be right-of y
	$SP : \{\Gamma_+^{-1}(x) < \Gamma_+^{-1}(y)\} \wedge \{\Gamma_-^{-1}(x) < \Gamma_-^{-1}(y)\}$	$SP : \{\Gamma_+^{-1}(x) < \Gamma_+^{-1}(y)\} \vee \{\Gamma_-^{-1}(x) < \Gamma_-^{-1}(y)\}$
A is right-of B	x must be right-of y	x must not be left-of y
	$SP : \{\Gamma_+^{-1}(y) < \Gamma_+^{-1}(x)\} \wedge \{\Gamma_-^{-1}(y) < \Gamma_-^{-1}(x)\}$	$SP : \{\Gamma_+^{-1}(y) < \Gamma_+^{-1}(x)\} \vee \{\Gamma_-^{-1}(y) < \Gamma_-^{-1}(x)\}$
A is above B	x must be above y	x must not be below y
	$SP : \{\Gamma_+^{-1}(x) < \Gamma_+^{-1}(y)\} \wedge \{\Gamma_-^{-1}(y) < \Gamma_-^{-1}(x)\}$	$SP : \{\Gamma_+^{-1}(x) < \Gamma_+^{-1}(y)\} \vee \{\Gamma_-^{-1}(y) < \Gamma_-^{-1}(x)\}$
A is below B	x must be below y	x must not be above y
	$SP : \{\Gamma_+^{-1}(y) < \Gamma_+^{-1}(x)\} \wedge \{\Gamma_-^{-1}(x) < \Gamma_-^{-1}(y)\}$	$SP : \{\Gamma_+^{-1}(y) < \Gamma_+^{-1}(x)\} \vee \{\Gamma_-^{-1}(x) < \Gamma_-^{-1}(y)\}$

- Two devices which are specified as a couple must be placed symmetrically with respect to an axis and must have identical variants and mirrored orientations.
- A device which is specifies as self-symmetric must be placed on a symmetry axis.
- Couples and self-symmetric devices that belong to the same symmetry group must share the same symmetry axis.

There have been several researches about symmetric placement. F.Balasa and K.Lampaert proposed a symmetry placement based on Sequence-Pair [8].

However, its primitive application will not be effective for some cases such that a constraint is imposed on few tens modules to be on one symmetry axis. In these cases, most of the solutions derived from Sequence-Pair might be infeasible. Then they would have to consider appropriate adjustments among different constraints to achieve a high quality placement. Furthermore, since their focus is restricted locally to one pair of symmetry halves, there would exist a difficulty in adjusting the bisector of two halves of modules to one axis. We have overcome these difficulties, at least empirically on industrial data and applied them to our placement.

4.3 Cluster and Site Constraints

An analog circuit consists of several elemental functions, I/Os, amplifiers, etc. Modules belonging to the same function are grouped as a cluster. In an example shown in Fig.5, the circuit is partitioned into four clusters. The modules in a cluster are required to be placed closely to each other. We call the constraints cluster constraints.

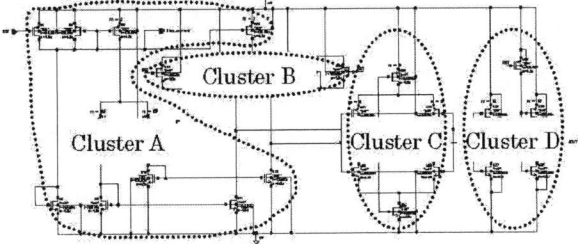

Fig. 5. Clusters on schematic

In floorplanning, the given area is partitioned into slots of the number of the clusters consulting with the topological relations between clusters. Each cluster is assigned to the corresponding slot. The topological relations between clusters are abstracted here.

In the placement phase, the relation between clusters is translated into the constraints between modules. We call the constraints site constraints. Assume that a cluster A is to be placed below cluster B. If the boundary between them is straight, the site constraint is that "a module in A must be placed below any module in B". On the other hand, if the boundary is jagged, we apply the concept introduced in [12], then the site constraint that "any module in A must not be placed above any module in B". This ABLR-relation can be led by De Morgan's principle as follows;

The relation "A is above B" is

$$\{\Gamma_+^{-1}(A) < \Gamma_+^{-1}(B)\} \wedge \{\Gamma_-^{-1}(B) < \Gamma_-^{-1}(A)\}$$

then the relation "A is not above B" is

$$\overline{\{\Gamma_+^{-1}(A) < \Gamma_+^{-1}(B)\} \wedge \{\Gamma_-^{-1}(B) < \Gamma_-^{-1}(A)\}}$$
$$= \{\Gamma_+^{-1}(B) < \Gamma_+^{-1}(A)\} \vee \{\Gamma_-^{-1}(A) < \Gamma_-^{-1}(B)\}.$$

Other site constraints on Sequence-Pair are obtained in the same way. All the site constraints are summarized in Table 1.

4.4 Schema-Driven Constraints

A schematic of analog circuit is often drawn considering the layout so that the placement crafted by experts resemble to the schematic. [13] introduced a new type of constraints that generates a layout along this concept. In this paper, we apply the concept to the schema-driven constraints.

First, we extract eight topological relations between modules according to their positions on the schematic. They are eight types of topological relations, left-of, right-of, above, below, left-above, right-below, right-above and left-above. Fig.6 shows an example to extract the topological relations between modules on the schematic.

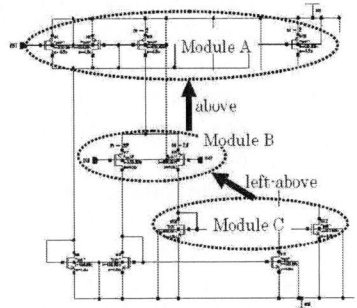

Fig. 6. Topological relations on schematic

Next, we converge these observed relations into the schema-driven constraints. If the relation between A and B is "A is above B", this relation is converged into the schema-driven constraint "A must not be placed below B". This can be represented on Sequence-Pair by following the way introduced in Section 4.3. If the relation between B and C is "B is left-above C", the relation is converged into the schema-driven constraint "B must not be placed right-below C". The relation is interpreted as "B is not right-of C and B is not below C". This relation on Sequence-Pair can be led by De Morgan's principle as follows;

$$\overline{\{\Gamma_+^{-1}(C) < \Gamma_+^{-1}(B)\} \wedge \{\Gamma_-^{-1}(C) < \Gamma_-^{-1}(B)\}} \wedge$$
$$\overline{\{\Gamma_+^{-1}(C) < \Gamma_+^{-1}(B)\} \wedge \{\Gamma_-^{-1}(B) < \Gamma_-^{-1}(C)\}}$$
$$= \quad \Gamma_+^{-1}(B) < \Gamma_+^{-1}(C).$$

Other schema-driven constraints on Sequence-Pair are obtained in the same way. All constraints are listed in Table 2.

Table 2. Schema-driven constraints between x and y

x is "..." y on schematic	schema-driven constraint on Sequence-Pair
left-of	$\{\Gamma_+^{-1}(x) < \Gamma_+^{-1}(y)\} \vee \{\Gamma_-^{-1}(x) < \Gamma_-^{-1}(y)\}$
below	$\{\Gamma_+^{-1}(y) < \Gamma_+^{-1}(x)\} \vee \{\Gamma_-^{-1}(x) < \Gamma_-^{-1}(y)\}$
right-of	$\{\Gamma_+^{-1}(y) < \Gamma_+^{-1}(x)\} \vee \{\Gamma_-^{-1}(y) < \Gamma_-^{-1}(x)\}$
above	$\{\Gamma_+^{-1}(x) < \Gamma_+^{-1}(y)\} \vee \{\Gamma_-^{-1}(y) < \Gamma_-^{-1}(x)\}$
left-below	$\Gamma_-^{-1}(x) < \Gamma_-^{-1}(y)$
right-below	$\Gamma_+^{-1}(y) < \Gamma_+^{-1}(x)$
right-above	$\Gamma_-^{-1}(y) < \Gamma_-^{-1}(x)$
left-above	$\Gamma_+^{-1}(x) < \Gamma_+^{-1}(y)$

4.5 Geometrical Constraints

The blocks are specified by a system level floorplan. To minimize system level performance degradation, a target aspect ratio, fixed height/width of the outline, fixed position of rails for power/ground and intermediate voltage, fixed position of terminals, etc. may be specified by a floorplanning tool or by a designer. These additional geometrical constraints also have to be taken into account during placement. Every device must be placed not to violate the constraints.

5 Experiments

First, we tested our placement algorithm for six instances of analog circuits to show our conservative properties are sufficient for porting. We prepare a manual design by experts for each instance to compare. We imposed constraints on devices and placed the devices on the same technology as the manual placements. The number of devices, nets, modules, and clusters, and the results by our algorithm and those by manual are shown in Table3. Our placements achieve a comparable quality to manual placements. The results show that our conservative properties are sufficient not redundant.

Next, we applied our porting methodology to an A/D converter from the design on $0.20 \mu m/3.3V$ technology to that on $0.18 \mu m/1.8V$ technology. The A/D converter is corresponding to the data B as shown in Table3. Its schematic is shown in Fig.7. The required specification is shown in Table 4. The

Table 3. Data information and placement results
area ratio = our area / manual area

	#devices	#nets	#constraints		area ratio	time [sec]
			#mod.	#clus.		
A	6	8	0	4	0.997	0
B	26	18	24	4	0.977	60
C	29	18	3	5	0.974	11
D	44	26	5	5	1.000	43
E	51	32	4	3	0.966	123
F	186	93	9	9	1.082	3
avg.					0.999	

table shows that the supply voltage was reduced half thus the channel width of a MOS transistor scales up four times. The cell width was changed from $21.0 \mu m$ to $13.0 \mu m$.

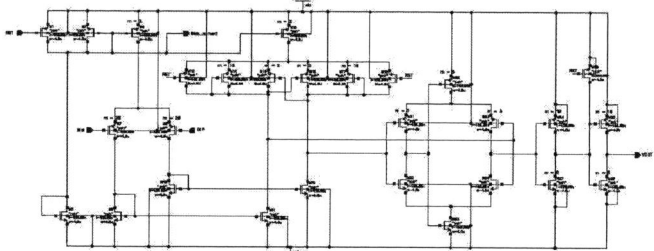

Fig. 7. Schematic of A/D converter

In this porting, the matching, symmetry and cluster constraints were reused. Fig.8A shows the layout by our system. On the other hand, we prepared the manual layout on $0.18 \mu m/1.8V$ technology by an expert to compare with our result. The layout is shown in Fig.8B. The comparisons with respect to the layout size of cell, the layer usage of wiring, and the design time are shown in Table 5. Both layouts meet the required cell width. As for the cell area and wiring area, our layout is comparable to the manual layout.

Furthermore, we extracted RC parameters from the layouts by a commercial LPE tool and applied the post-layout simulation to the both layouts. The waveforms and that of the schematic simulation at the input and output are shown in Fig.9. It is observed that there exists a delay of $1 ns$ in our result. This delay is attributed to the usage of the wiring layer where the 1st and 2nd metal layers are dominant used. In the manual layout, wiring on the 3rd metal layer is aggressively used to meet the timing, while most of the wires are routed on the 1st and 2nd metal layers in our result. This is observed in Table 5. The resistive of the 3rd metal layer is lower than that of the 1st and 2nd metal layers. If we had been able to use the 3rd metal layer more, we might have decreased the delay. However, in our system layout, we followed a strategy that a better wiring leads less usage of the 3rd metal layer. As the result, a small delay occurred. But the performance of the ADC(7bit resolution and 30MHz) is satisfied. This is why we regard the delay is not significant degradation at all. The analysis of the waveforms confirmed us that our result also met the required performances as well as the manual design did. The results show that our conservative properties are necessary.

0-7695-2533-4/06 $20.00 © 2006 IEEE

Table 4. Specifications of A/D converter

technology	0.20 μm	0.18 μm
$V_{DD}/2$	3.3 V	1.8 V
frequency	30 MHz	30 MHz
resolution	7 bit	7 bit
channel width of a transistor	5.0 μm	22.5 μm
cell width	21.0μm	13.0μm

Table 5. Layout results manual vs. ours

	area [μm^2]	layer usage of wiring [μm^2]			time [hr]
		Metal1	Metal2	Metal3	
manual	1471	665	48	152	11
ours	1437	630	56	19	2.5
improving	2.4%	5.3%	-16.7%	87.5%	77.2%

6 Conclusions

We introduced a porting methodology of analog IPs. The main contribution is how to reuse conservative properties. To embed an IP onto the chip adaptively, we imposed constraints on the layout using Sequence-Pair. In experiments, we applied our porting system to several industrial analog circuits and compared with the manual placements. In the design of an A/D converter, we ported the layout on 0.20μm/3.3V technology to that on 0.18μm/1.8V technology. The result not only met the required performance, but also achieved the comparable quality with the manual layout. The design time was reduced drastically. We could show our conservative properties are necessary and sufficient for porting, and our methodology is effective for analog circuit design.

As future works, we will enhance our porting system to deal with conservative properties for routing. Furthermore, we will develop an ECO and fast back-annotation in order to cooperate the layout design with the simulation.

Acknowledgments

We appreciate Mr. Masaki Mastui, Toppan Technical Design Center Co. for his valuable discussion to develop our algorithms

References

[1] J. M. Cohn, D. J. Garrod, R. A. Rutenbar, and L. R. Carley, "Analog Device-Level Layout Automation", Kluwer Academic Publishers, 1994

[2] H. Murata, K. Fujiyoshi, S. Nakatake, and Y. Kajitani, "Rectangle-Packing-Based Module Placement", ICCAD 1995, pp.472–479.

[3] S. Nakatake, K. Fujiyoshi, H. Murata, and Y. Kajitani, "Module Placement on BSG-Structure and IC Layout Applications", ICCAD 1996, pp.484–pp.491, 1996.

[4] H. Chang, E. Charbon, U. Choudhury, A. Demir, E. Felt, E. Liu, E. Malavasi, A. Sangiovanni-Vincentelli, and I. Vassiliou, "A Top-Down Constraint-Driven Design Methodology for Analog Integrated Circuits" Kluwer Academic Publishers, 1997

[5] K.Lampaert, G.Gielen, and W.Sansen "Analog Layout Generation for Performance and Manufacturability", Kluwer Academic Publishers, 1999

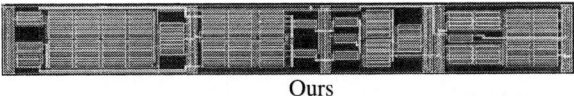

Ours

Manual

Fig. 8. Layouts of A/D converter on 0.18μm technology

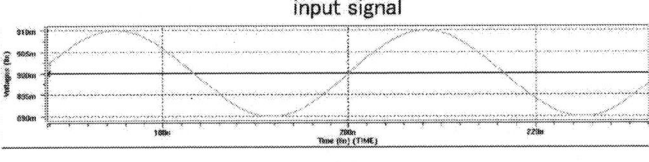

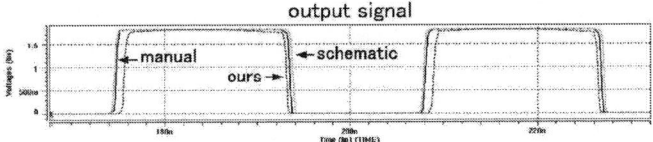

Fig. 9. Waveforms of input and output. schematic : pre-layout simulation, ours and manual : post-layout simulation.

[6] M. Keating and P. Bricaud, "Reuse Methodology Manual for System-On-A-Chip Designs" Kluwer Academic Publishers, 1999

[7] R. Seepold and A. Kunzmann, "Reuse Techniques for VLSI Design", Kluwer Academic Publishers, 1999

[8] F. Balasa and K. Lampaert, "Module Placement for Analog Layout Using the Sequence-Pair Representation", DAC 1999, pp.274–279.

[9] X. Tang, and D.F. Wong, "FAST-SP : A Fast Algorithm for Block Placement based on Sequence-Pair", ASPDAC 2001, pp.521–526.

[10] C. Galoup-Montoro, M. C. Schneider, and R. M. Coitinho, "Resizing Rules for MOS Analog-Design Reuse", IEEE Design and Test of Computers 2002, Vol.19(2) pp.50–58.

[11] N. Jangkrajarng, S. Bhattacharya, R. Hartono, and C-J. R. Shi, "IPRAIL – Intellectual Property Reuse-Based Analog IC Layout Automation", Integration The VLSI Journal vol.36 2003, pp.237–262

[12] T. Nojima, X. Zhu, Y. Takashima, S. Nakatake, and Y. Kajitani, "Multi-Level Placement with Circuit Schema Based Clustering in Analog IC Layouts", ASPDAC 2004, pp.406–411.

[13] T. Nojima, Y. Takashima, S. Nakatake, and Y. Kajitani, "A Device-Level Placement with Multi-Directional Convex Clustering", GLSVLSI 2004, pp.196–201.

[14] A. Savio, L.Colalongo, Zs. M. Kovacs-Vajna and M. Quarantelli "Scaling Rules and Parameter Tuning Procedure for Analog Design Reuse in Technology Matching", ISCAS 2004, pp.V-117–V-120.

[15] S. Bhattacharya, N. Jangkrajarng, R. Hartono, and C-J. R. Shi, "Correct-by-Construction Layout-Centric Retargeting of Large Analog Designs", DAC 2004, pp.139–144.

VLSI Design Exchange with Intellectual Property Protection in FPGA Environment Using both Secret and Public-Key Cryptography

Wael Adi
Technical University of Braunschweig, Braunschweig, Germany
wadi@ieee.org

Rolf Ernst
Technical University of Braunschweig, Braunschweig, Germany
r.ernst@tu-bs.de

Bassel Soudan
University of Sharjah, Sharjah, United Arab Emirates
bsoudan@sharjah.ac.ae

Abdulrahman Hanoun
Technical University Hamburg, Harburg, Germany
a.hanoun@tu-harburg.de

Abstract

With the advent of multi-million gate chips, Field Programmable Gate Arrays (FPGAs) have achieved high usability for design verification, exchange, test and even production. Adding to this is the possibility of reusing readily available licensed IP to shorten the design cycle. A major concern for IP owners is the possible over-deployment of the IP into more devices than originally licensed. In this paper, we propose a system based on both public-key and secret-key cryptography embedded in a secured design exchange protocol for protecting the rights of the IP owner. The system consists of hardware-supported design encryption and secured device authentication protocols. Design encryption based on secured device identification ensures that the IP can only be deployed into explicitly identified and agreed upon devices. The system uses a combination of secret and public-key cryptographic functions devised for an uncomplicated trustable design exchange scenario. The public-key functions use modular squaring (Rabin Lock) on the FPGA chip instead of exponentiation to reduce the hardware complexity.

1. Introduction

FPGAs have become the tool of choice for fab-less design houses. Especially those who deal with designs that don't require cutting-edge speed or complicated implementations requiring customized layout. These design houses typically depend on licensing ready-made cores from IP providers to be downloaded into their FPGAs.

The IP core providers have an interest in protecting their rights to the licensed IP against three possible forms of attack. Interception during IP transfer to the customer (or device), duplication by cloners after the IP has been deployed into the market, and possible over-deployment by the customer or the customer's out-sourced device programmer.

Recently, major FPGA device manufacturers have implemented different ciphering mechanisms to ensure that the details of the design cannot be meaningfully intercepted en-route to the device or accessed once on the device. This helps protect against interception and cloning. However, it does not address over-deployment.

With the current methodologies, there is no way to limit the number of deployments once the IP arrives at the programming site. To counter the loss of control over the scale of deployment of licensed IP cores, IP providers have resorted to a business model based on huge up-front licensing fees coupled with overt and covert customer audits [1]. While the upfront fees might be agreeable to customers with huge installation bases, they are prohibitively costly for customers with limited profit margins. While the concept of licensing ready-made IP cores for FPGAs is particularly suited for small-scale operations, the current pricing strategy ways heavily against such customers. Since the IP core providers cannot limit the number of deployment instances, they cannot differentiate in pricing between the two classes of customers [2].

So far, there has not been wide spread violation of IP core licenses as the market for IP cores remains small and concentrated amongst a reasonably small number of players. The size of the market has remained relatively small due in large to the high expense of IP licenses. However, this high expense makes an eventual violation a very lucrative endeavor. Eventually, one of the players (or their surrogates) will leak/over-use an IP core and the whole cycle will unravel. Unless a mechanism for preventing the IPR violation is devised.

This paper discusses a secure mechanism for protecting against IP over-deployment. Our proposed solution allows the IP to be licensed for specific uniquely identifiable devices. It is impossible to deploy the IP into any additional device without the explicit involvement of the IP owner (or provable illegal collaboration of the device manufacturer).

Even more, the device gets an additional identity which belongs to the IP owner rather than manufacturer and device owner

The concept introduced in this paper differs from our recently introduced concept in [3] in that no Smart Card is necessary and the number of data exchanges on the open network is reduced to two way instead of four-way exchanges. In difference to the earlier design, the system is also conceptually unbreakable even if the manufacturer and IP customer collaborate.

The rest of this paper is organized as follows: section 2 discusses current FPGA-based IPR protection mechanisms, section 3 describes the proposed IPR protection technology and protocols, section 4 discusses security threats and possible attacks and section 5 presents a summary and conclusion.

2. Current FPGA-based IPR Protection Mechanisms

An IP is most vulnerable when it is outside the control of the IP owner. It is vulnerable when released to the customer for programming and while it resides on the device itself after the programming. Mechanisms have been proposed for preventing IPR violation and detecting the violation when it occurs.

2.1. Methods for Detecting IPR Violation

Several techniques based on water marking have been proposed to detect IPR violations such as those described in reference [4] and [5]. However, the sheer number of designs continuously appearing on the market, makes attempting to check for the inclusion of a supplier's IP in all of them a prohibitively expensive proposition. While these methods may be successful in proving that an IPR violation has occurred, they cannot prevent the violation from occurring in the first place.

2.2. Methods for Preventing IPR Violation

The most basic method for preventing IPR violations is license agreements. License agreements are excellent tools for extracting an IP owner's rights in court once a violation is detected and proven. However, their strength is mainly built on the assumption of goodwill between the IP owner and IP customer. The concern is the possible development of a black (or gray) market in unlicensed IP cores similar to the market that exists for software knockoffs [6].

Most FPGA manufacturers have incorporated design stream decryption cores into their devices to guard against IP bitstream interception and cloning [3]. Having an encrypted bitstream makes intercepting the design bitstream useless and cloning impossible without knowing the exact decryption key. However, in current solutions the IP owner must pre-program the devices with the decryption key before sending them to the out-sourced or end user's programming facility. If the IP owner is going to bring the devices in-house for key pre-programming, they might as well download the IP into the devices themselves. Most IP owners don't want (or are not equipped) to be directly involved in the programming of the devices, even for key insertion. They are simply design houses that would prefer to leave dealing with the devices to the end user or the contract programming facility.

In addition, Actel's use of Flash ROM elements instead of SRAM for the configuration memory has eliminated the need for an on-board memory to allow autonomous re-configuration after power interruption. Therefore, there is no bit stream to intercept or clone in the deployed product [7].

A pay-per-use methodology based on secret-key cryptography was previously proposed [8]. The methodology requires an e-commerce server at a *Trusted-External-Party* (TEP) that maintains a device database built through extracting device details (including secret device IDs) from the manufacturer. The server also maintains an IP core database containing information for each IP core available from each IP core provider. Upon purchasing an IP core, the IP customer uses trusted software that communicates with the server at the TEP to decrypt and then re-encrypt the licensed IP for programming into the device.

This methodology suffers from several shortcomings. First, the whole methodology is centered around the TEP server which now presents a single point of complete failure for the system. The TEP server is being entrusted with a lot of sensitive information about the chips, the cores, the device manufacturers, the IP providers, and the customers. The TEP is also expected to handle payments of customers to IP providers. The TEP is expected to maintain data on every chip manufactured by every chip manufacturer and every core provided by every provider. The data about the chips needs to be kept permanently to allow for later chip re-configuration. There is a strong possibility of data explosion at the TEP. The IP provider ships an encrypted design bit

0-7695-2533-4/06 $20.00 © 2006 IEEE

stream to the customer. This bit stream is decrypted at the customer's site and is then re-encrypted with a token received from the TEP. This means that the raw decrypted design bit stream will exist for some instance of time on the customer's premises exposing it to possible compromise. Lastly, the proposed scheme requires too many parties to be involved in the deployment of a licensed IP.

In our recent paper [3], a pure secret-key IPR protection system was introduced, which included a Smart Card function to be produced by the FPGA manufacturer to allow device verification. The system exhibits public key features, however needs complex traffic and transactions between the IP owner and IP customer to generate a shared secret that allows secured design exchange. Figure 1 shows the main protocol exchange transactions for that design. The main drawbacks of this system are the complex interaction and that the customer and manufacturer can easily collaborate and reveal the design.

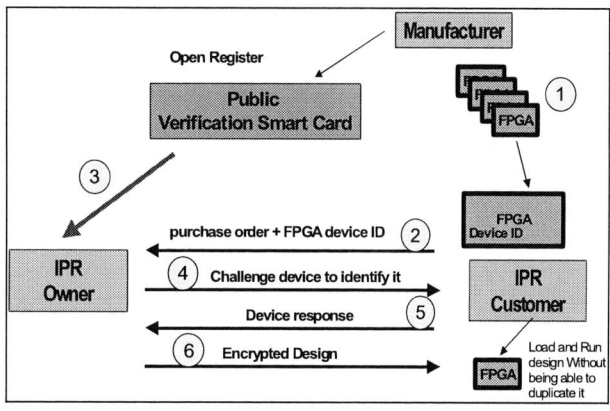

Figure 1. Main protocol transactions of a secret-key FPGA secured design transfer

3. Proposed Public-Key IPR protection Mechanism

To simplify the process of IP deployment, the number of parties involved should be reduced to the absolute minimum. Compared to our earlier work, the aim of this proposal is to eliminate the involvement of the device manufacturer and minimize his influence on the system's operational security. The basic expected secured design exchange scenario is shown in Figure 2. A main advantage of the proposed system is that it reduces the amount of interaction required between the IP owner, IP customer and device programmer. It also eliminates the need for the smart card required for the earlier design.

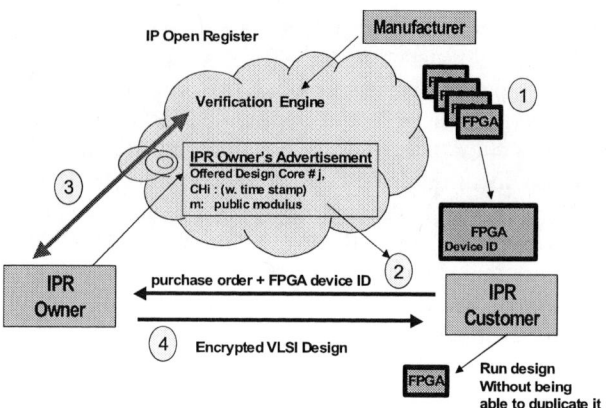

Figure 2. Main protocol transactions of the proposed combined secret and public-key FPGA secured design transfer

3.1. Requirements for Proposed Mechanism

The proposed IPR protection mechanism is designed to satisfy the following requirements:

- The IP should be securely distributable over an open channel like the Internet.
- The system security should be based on known unbroken ciphering technology.
- The system response time and hardware complexity must be kept as low as possible.
- The IPR owner should not need to pre-program the devices on-site with a decryption key.
- The IPR owner must be able to limit the number of system deployments as part of a business agreement.
- The design bit stream must not be extractable from the device after programming to prevent illicit replication.
- The system should not require the sharing of secrets between IPR owner and customer or FPGA programmer such as an encryption/decryption key.
- Collaboration between FPGA manufacturer and IP customer should not lead to breaking the system.

3.2. Hardware required

The proposed system implementation includes essentially one hardware component:

Device Identity Module **DIM:** The manufacturer should guarantee the essential physical uniqueness of each device. It should reside in the FPGA in a tamper-proof area where no attack would be easily possible in any operating mode. Figure 3 represents a simplified functional block diagram for the proposed DIM.

0-7695-2533-4/06 $20.00 © 2006 IEEE   26

3.3. Manufacturer Device Initiation:

The manufacturer needs to initiate and personalize the FPGA's at the time the device is manufactured. For each device, the manufacturer establishes a unique public *device identity* (DI), which can be branded on the device itself and/or stored in a readable area in the device. Then the manufacturer implements in each device a DIM like the one shown in Figure 3.

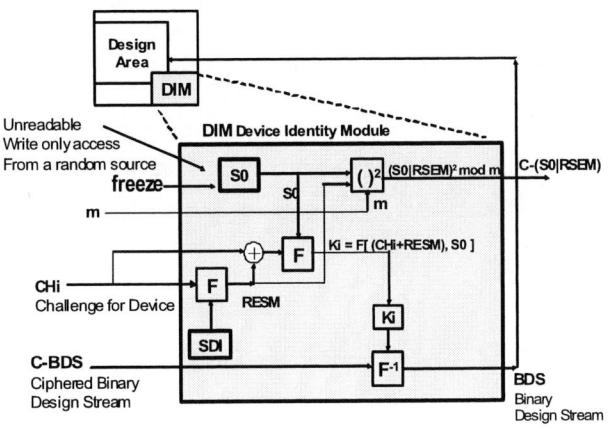

SDI : Secret Device Identity (Manufacturer)
S0= Any initial random contents

$C\text{-}(S0|RESM) = (S0|RESM)^2 \bmod m$
$Ki = F(CH + RSEM, S0)$

Figure 3. Architecture of a Possible FPGA Device Identity Module (DIM)

The DIM includes the following elements:

- A non-volatile unreadable *memory* location or a pre-configurable fuse array or laser trimmed non-volatile storage to hold the *secret device identity* (SDI). The SDI register must not be modifiable without knowledge of its current contents. SDI is mapped from DI such that no key collision is possible:

$$SDI = F(DI, SMK) \qquad (1)$$

Where SMK is the manufacturer's Secret Master Key for that particular FPGA type. The manufacturer should embed the SDI value before delivering any device and be responsible for the uniqueness of all SDI's and DI's. The manufacturer is also responsible to keep SMK secret.

- A second write-once-register S0 whose contents should be fully random and not necessarily known to the FPGA manufacturer or the FPGA user.

- A modular squaring block $(\ )^2$ modulo m, where m is the product of two large primes (p and q) say

each 500 - 1000 bits depending on the security level required using our technology in [9].

- An hardware decipher block F^{-1}.

The function F (and its inverse F^{-1}) should be a strong cipher with a size of 128-bits such as AES, or any secure hash function [10] – [12]. The size of all registers, secret keys and other vectors should be 128 bits the same as the cipher block size. Whenever the mapping Y = F (X, K) is used, we mean that X corresponds to the clear text, K to the key and Y to the cipher text of a secret key block cipher in the sequence given above.

The device manufacturer creates SDI using a secret master key SMK and a strong known block cipher F' not necessarily the same F used in the DIM.

The IPR owner publishes on his Internet page his design offer together with a random challenge CHi, which he should change frequently with a time stamp defining its lifetime. The IPR owner also selects two secret primes p and q and computes m = p * q and publishes m on his home page.

On its web-page, the manufacturer makes public a function that generates RESM (manufacturer's response) based on any combination of DI (device identity) and a challenge (CHi) as shown in Figure 4. This implements a public challenge-response "Verification Engine (VE)" that can be used to verify devices owned by the IP customer as shown in Figure 2. It should be guaranteed that the response time to deliver RESM is much larger than the lifetime of CHi otherwise the IPR owner should use a manufacturer smart card to be sure that the customer did not generate RESM from the manufacturer public verification engine.

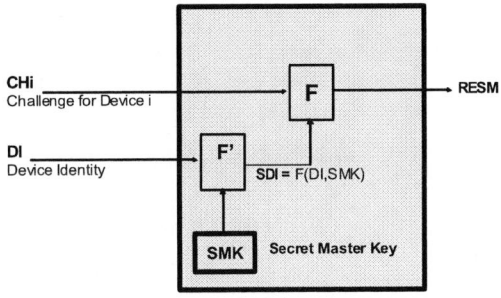

Figure 4. The "Verification Engine" published by manufacturer to verify any device identity

3.4. Secured Design Transfer Protocol:

The secured design transfer can be expressed as follows (refer to the 4-step protocol in Figure 5):

1. The IP customer obtains the challenge CHi and m from the IPR owner's web page. CHi and m are fed into the devices on which the design will eventually be downloaded. This operation will freeze the value of S0 in each device and cause the device to produce the ciphered concatenation C-(S0|RSEM) as shown in expression (2). In addition, the operation causes an internal device-unique secret key Ki to be generated inside each device according to expression (3).

$$C\text{-}(S0|RESM) = (S0|RESM)^2 \bmod m \qquad (2)$$

The customer collects the device identity DI and C-(S0|RESM) for each device and sends them to the IPR owner along with the purchase order.

$$K_i = F\,[\,(CHi + RESM)\,,\,S0\,] \qquad (3)$$

2. The IPR owner uses the verification engine (VE) published on the device manufacturer's web page to generate a response RESM' based on the IPR owner's CHi and the DI received from the IP customer.

The IPR owner uses p and q to compute the square roots of C-(S0|RESM). The number of square roots over the ring Z_m is 4. The correct one is the one that corresponds to the concatenation (S0|RESM). That is the one with RESM located in the LSB side. The IPR owner can extract S0 from the concatenation and generate a matching Ki for each device following the operation in expression (3). The IPR owner is the only party in the entire exchange that knows the exact values of S0 and therefore Ki.

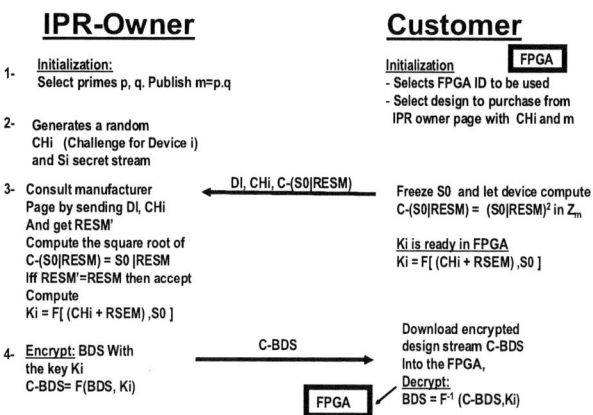

Figure 5. Secured design transfer protocol

3. The IPR owner uses the device-unique Ki to encrypt the Binary Design Stream (BDS) for the particular purchased VLSI design and get a unique

C-BDS for each device as shown in the following expression.

$$C\text{-}BDS = F\,(BDS,\,Ki) \qquad (4)$$

The resulting C-BDS is sent to the customer for the device with the matching DI.

4. The customer feeds C-BDS to the device with the matching DI. C-BDS will be automatically decrypted as the right Ki has already been generated in the device in step 1 at the beginning of the process.

$$BDS_i = F^{-1}\,(C\text{-}BDS,\,K_i) \qquad (5)$$

The above protocol can run in parallel for all required device identities.

4. Analysis

4.1. Advantages of the system

- Given that the C-BDS is encrypted using a particular device's SDI, the IP owner can be sure that the resulting C-BDS cannot be used to program any other device. Short of cloning the actual devices (including the unique SDI and ransom S0), the customer (or any other party) cannot over-deploy a licensed IP.

- While the interaction between the IPR owner and customer may seem a bit complex, it has the advantage that **"only"** the IPR owner and customer are involved in the exchange. The complexity is required as two identical versions of the secret encryption key need to be created in isolation and without the IP customer ever having direct access to the key.

- From the customer's point of view, the proposed mechanism allows the customer to pay only for the number of instances the IP is downloaded into a device. This should be a major improvement for low volume customers over current pricing strategies.

4.2. Security threats and possible attacks

The system is breakable in only two possible scenarios:

1. The system depends on a unique DI/DIM pair and a randomly generated S0 for each device. Even if the device manufacturer cheats by generating devices with duplicate DI/DIM pairs the system would continue to work. The only way the system

can be broken is if the manufacturer generates devices with duplicate DI/DIM pairs **and** makes the S0 generator act in a deterministic way. The only party in the whole exchange that knows the exact value of S0 for a specific device is the IPR owner. Therefore, the IPR owner can easily determine and prove such collusion.

2. The device manufacturer may build a backdoor into the device where the customer can access the decrypted BDS directly. This is a highly unlikely scenario as it would completely destroy the manufacturer's reputation and would be easily provable.

Both cases can be traced if the design layout is published and the devices are sealed adequately. The system security appears to be as good as the security of the used block cipher and the Rabin Lock which is based on factoring problem known today to be still unbreakable. The fact that S0 is unpredictable to all parties gives the system a special strong security level. Combining secret and public-key systems in a joint security function appears also to be one additional argument in favor of its security.

More attack scenarios have to be studied. The authors invite other researchers to show attacks that can break the system.

5. Summary and Conclusion

The paper proposed a mechanism for the authenticated transfer of design information that prevents IPR violations in an FPGA design environment. The novel security technology uses combined public and secret-key cryptographic mechanisms. The resulting system appears to offer a high level of security while still being reasonably easy to handle. The proposed mechanism allows design distribution over public Internet media without loss of security. The mechanisms employed are based on trustable cryptographic primitives well known in secret-key and public-key cryptography, but utilizes low complexity functions to simplify the implementation and save chip area. In particular, the selected public-key technology employs squaring in a ring results in the simplest public key system known to date.

The system appears to be unbreakable even if the device manufacturer collaborates with the end customer. The design transfer does not involve the manufacturer on-line and does not need the IPR owner to have any contact with the FPGA devices. All transactions can run over any open communication network without prior secret sharing. The FPGA manufacturer should however be trusted to manufacture according to the specified hardware security architecture and have to be trusted that he did not build any backdoors in the FPGA architecture.

References

[1] Technical discussions with Tim Holden, Director of EDA Relations, ARM Limited, November 2005.

[2] T. Kean, "Cryptographically Enforced Pay-Per-Use Licensing of FPGA Design Intellectual Property", Proceedings International Workshop on IP Based Design 2002.

[3] B., Soudan, W. Adi, and A. Hanoun, "Novel Secret-Key IPR Protection in FPGA Environment," Proceedings System-on-Chip Conference, September 2005.

[4] Kahng , A. B., Kirovski , D., Mantik, S., Potkonjak, M., and Wong, J. L., "Copy Detection for Intellectual Property Protection of VLSI Design." Proc. IEEE/ACM Intl. Conference on Computer-Aided Design, November 1999, pp. 600-604.

[5] Newbould, R. D., Carothers, J. D., Rodriguez, J. J., and HolmanW. T., "A Hierarchy Of Physical Design Watermarking Schemes For Intellectual Property Protection Of IC Designs," Proceedings of the International Symposium on Circuits and Systems, 2002, Vol. IV, pp. 862 – 865

[6] Technical discussions with Rich Goldman, Vice President, Strategic Alliances, Synopsys. November 2005.

[7] Actel, "ProASIC3/E Security," Application Note available at http://www.actel.com, cited on 14/4/2005.

[8] T. Kean, "Cryptographically Enforced Pay-Per-Use Licensing of FPGA Design Intellectual Property", Proceedings International Workshop on IP Based Design 2002.

[9] W. Adi , Fuzzy Modular Arithmetic for Cryptographic Schemes with Applications to Mobile Security EUROCOM 2000

[10] Technical Specification 3G Security, Security Architecture 3G TS 33.102 V. 3.2.0 from 10.1999

[11] Specification of the MILENAGE Algorithm Set. 3GPP TS 35.206 V5.0.0 ETSI, http://www.3gpp.org

[12] AES, Advanced Encryption Standard, Federal Information Processing Standards Publication, FIPS 197, 2001. Or http://csrc.nist.gov/publications/.

Physical Design

Metal Fix and Power Network Repair for SOC

Qing K. Zhu
Matrix Semiconductor Inc.
7848 Pineville Circle
Castro Valley, CA, USA
qkzhu@yahoo.com

Paige Kolze
Xilinx Inc.
2100 Logic Drive
San Jose, CA, USA

Abstract

This paper shows the design flowchart to do metal fix in Chameleon CS2112. CS2112 is an integrated SOC system in one chip with a fabric circuit of 84 processors, 48 local memories, ARC processor, memory controller, PCI controller, DMA configuration subsystems and programmed I/Os. The first silicon failed due to the hold time problem at program I/Os and large IR drop at the center of the chip. The advantage of metal fixes will reduce the mask cost for bug fixing found in silicon. There are a couple of problems to be addressed in the post-silicon metal fixing. Make sure the simulation result in the old design to match the silicon measurement and the simulation setup provides the accuracy for metal fix to predict the performance in new silicon.

1. Introduction

The Chameleon CS2112 is an integrated system on a chip that incorporates a proprietary 32-bit arithmetic reconfigurable processing fabric of 84 processors, 24 multipliers, 48 local memories and 12 control PLAs [1]. In addition, it provides for very high bandwidth programmable I/O, and a 32-bit embedded processor subsystem for control processing. The chip is targeted at very high-speed digital signal processing (DSP) applications. The CS2112 incorporates a PCI 2.1 compliant 32-bit PCI Bus. The PCI bus supports master PCI DMA transfers and target single transaction host/debug accesses. The CS2112 provides a 64-bit memory bus that supports SDRAM, SSRAM, and Flash memory types, and 160 pins of programmable I/O for implementing flexible, high-bandwidth physical layer interfaces. The reconfigurable processing fabric is divided into slices, the basic unit of reconfiguration. The CS2112 includes four slices, each of which can be independently reconfigured. Each slice consists of three tiles. Figure 1 shows the CS2112 architecture.

The parallel multiple DPUs in fabric architecture uses a lot of power. We observed 5.3A supply current in the CS2112 silicon tests. It will be better to use flip chip for this high current. But we decide to use wire-bonding package for cost reason. Therefore, power pins are on the periphery of the chip with long route distances to the middle of the chip. The first tape-out of the chip failed due to the significant IR drop (~0.8V between Vss and Vdd nets) across the chip, because of the wire bonding technology used in this chip. The old design has an IR drop (Vdd – Vss) about 30% of the nominal voltage. We can observe significantly low voltage at the center of the chip. We describe a metal fix methodology to fix this problem. We added a dedicated M5 for Vdd and Vss straps to reduce the IR drop. A bug has been captured due to the hold time violation at the input flip flop in programmable I/Os for this SOC. The hold time requirement is the minimum length of the time that the data input signal must remain stable after the active edge of the clock to ensure correct function. Hold time violations are usually caused by the short delays in data paths compared to clock delays. To fix the hold problem, a thoughtful analysis in clock and data delays to flip-flops are needed in varied process/voltage/temperature conditions [6].

2. Metal fix

Metal fix refers to the general term for repairing the silicon-found bugs based on metal layer changes. No device layers will be changed so we can use the same device masks and untouched metal masks in next spin of silicon manufacturing. This is due to the continually increasing cost of masks in deep sub-micron manufacturing. It is estimated to be $1 million for a whole set of masks in 90nm process node. The metal fix method is used to repair clock distribution and hold time bugs in our design.

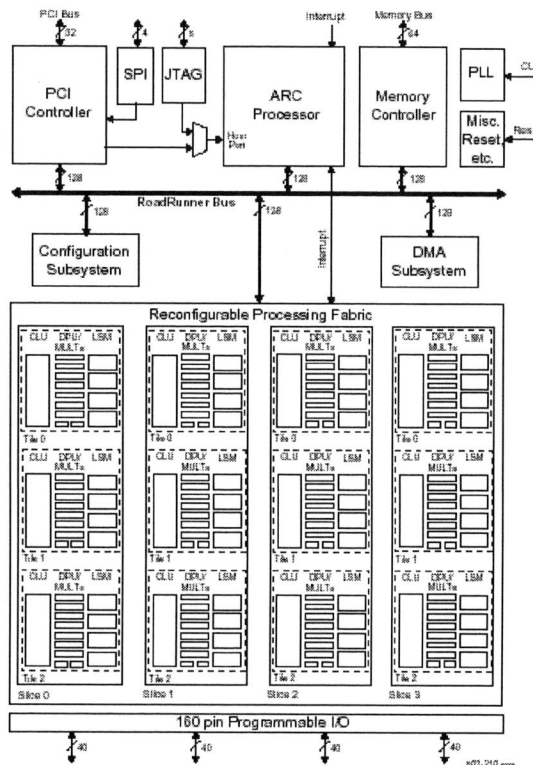

Figure 1. CS2112 architecture [1].

Figure 2 shows a part of the programmable I/O circuit used in the SOC chip as described in Figure 1. DFFHQX1 is the flip-flop with the hold time violation in the first silicon. We could re-connect BUFX2 in the data path and clock path at the same time to fix the hold time violation. The re-connect is done in one metal layer and therefore we can change one mask layer in the new silicon. The advantage of metal fixes will reduce the mask cost for bug fixing in new silicon. There are a couple of problems to be addressed in the post-silicon metal fixing. Make sure the simulation result in the old design to match the silicon measurement. Here are the general steps to quantify the metal fix for improving the performance. The method can be slightly varied to do the power network repair to be discussed in Section III.

1. Simulate the old circuit to match the silicon measurement result in various design corners (process/temperature/voltage). The simulation environment is tuned to match the measurement.
2. Repair the circuit in metal layers. Use as little as possible of metal layers changed to fix circuit bugs.
3. Estimate interconnect RC model due to metal changes in Step 2.
4. Simulate the new circuit to predict the performance based on the same simulation environment in Step 1.
5. If the performance result is not satisfied in Step 4, we need to loop back to Step 2 for optimization.

The interconnect model is needed to estimate the wire resistance and capacitance, when we change the metal routing. Since layout based RC extraction is a time long process, RC estimation based on equations will be a practical solution for routing optimization. Shown in (1), we sum up the unit-length metal capacitance C in each layer i in multiple components [3,4]. C_{dfi} and C_{ufi} are the fringe capacitances to upper and lower metals. T_{di} is the thickness of the dielectric to lower metal and T_{ui} is the thickness of the dielectric to upper metal. C_c is the coupling capacitance between neighboring metal lines. C_c is the function of the spacing (S_i) and metal width (W_i). C_p is the plate or area capacitance to the upper and lower metals. The resistance modeling includes the metals and vias [4]. All the parameters (α, β, γ, etc.) in (1) are based on the process technology model from TSMC. We have the device model plus the estimated RC tree model for each net. So, we can run the circuit simulation in design corners shown in Step 4. After couple of iterations in Step 5 to finalize the metal fix, we could run the layout extraction for the accurate RC data and post-layout circuit simulation [4].

$$C_{dfi}=\alpha(1-exp(\beta(T_{di}+\gamma)))$$
$$C_{ufi}=\alpha(1-exp(\beta(T_{ui}+\gamma)))$$
$$C_c=\eta(exp(\sigma-\delta S_i)+\phi/exp(\varphi*log(S_i)))exp(\kappa log(W_i)) \ (1)$$
$$C_p=\theta W_i$$
$$C=2(C_{dfi}+C_{ufi}+C_{ufi}+C_p)$$

Step 1 is crucial to make sure the simulation environment matches the post-silicon measurement result. We need to adjust couple of times in the circuit and interconnect models for simulation. Table 1 and Table 2 shows the mismatch of simulation result to the tester hold time measurement for the circuit shown in Figure 2(a). The design corners in supply voltages and process corners are experimented as shown in Table 1 and Table 2. With the satisfied simulation environment, we can simulate the metal fixed circuit to quantify the

0-7695-2533-4/06 $20.00 © 2006 IEEE

hold time numbers in the new silicon. Table 3 shows the simulation result in various design corners after the metal fix. The minimum hold time is defined as the minimum length of time that the data input at DFFHQX1 (flip-flop in Figure 2) must remain stable after the active edge of the clock input to DFFHQX1. The worst-case minimum hold time is defined as the maximum for the minimum hold time needed for DFFHQX1 at various design corners. The worst-case minimum hold time is reduced from 1.90ns shown in Table 1 to 0.39ns shown in Table 3 after metal fixing. We have to simulate as many design corners as possible to estimate the worst-case hold time value in the circuit. Although more time is spent in the analysis, we can guarantee the design robustness in the new silicon.

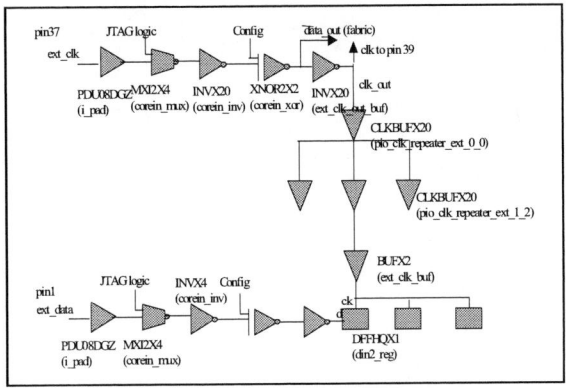

(a)

(b)
Figure 2. Hold time violation and metal fixing.

3. Power network repair

We show the metal repair technique for power network using one dedicated M5 layer to reduce the IR drop in design. The dedicated metal layers for power

and ground meshes have been reported in an IBM microprocessor [2]. Figure 3 shows the design flowchart to do the dedicated M5 repair on the power grid. The first step is to quantify the post-layout power grid simulation accuracy. The simulation should achieve the same range of the accuracy as silicon measurement, so we have the confidence to guide the metal repair. Otherwise, we have no criteria to see if our IR drop goal has been reached. The simulation tool will estimate the currents of devices. The device switching currents have significant over-estimation due to the static algorithm nature [7,8].

Table 1. Hold time simulation and mismatch to measurement (slow process corner).

Core and IO Voltages	Clk_in at DFFHQXZ1 (rise/fall delay, rise/fall time)	Data_in at DFFHQX1 (rise/fall delay, rise/fall time)	Minimum hold time in simulation /tester's measurement, mismatch (%)
2.25/3.0V	2.44/2.23ns, 0.13/0.08ns	0.78/0.72ns, 0.08/0.05ns	1.6/1.66ns, 4%
2.50/3.3V	2.08/2.02ns, 0.12/0.08ns	0.75/0.64ns, 0.07/0.04ns	1.7/1.96ns, 13%
2.75/3.6V	1.98/1.87ns, 0.11/0.07ns	0.74/0.58ns, 0.07/0.04ns	1.9/1.91ns, 1%

Table 2. Hold time simulation and mismatch to measurement (fast process corner).

Core and IO Voltages	Clk_in at DFFHQXZ1 (rise/fall delay, rise/fall time)	Data_in at DFFHQX1 (rise/fall delay, rise/fall time)	Minimum Hod time in simulation /tester's measurement, mismatch (%)
2.25/3.0V	1.55/1.42ns, 0.09/0.06ns	0.60/0.44ns, 0.05/0.04ns	1.2/1.43ns, 16%
2.50/3.3V	1.50/1.32ns, 0.08/0.06ns	0.61/0.39ns, 0.05/0.03ns	1.5/1.84ns, 18%
2.75/3.6V	1.47/1.23ns, 0.08/0.05ns	0.62/0.34ns, 0.05/0.03ns	1.7/1.861ns, 3%

Table 3. Hold time Simulation after metal fix.

Core and IO Voltages	Slow process corner, 0°C	Fast process corner, 0°C	Slow process corner, 85°C	Fast process corner, 85°C
2.25/3.0V	0.29ns	0.29ns	0.39ns	0.29ns
2.50/3.3V	0.29ns	0.29ns	0.39ns	0.29ns
2.75/3.6V	0.29ns	0.29ns	0.39ns	0.29ns

To compensate the inaccuracy of the current estimation, we need to add a current scaling factor to match the current from the silicon measurement. Table 4 shows the current scaling factors in the simulation for the Vdd and Vss nets in order to match the measured current consumption (4A) from the same chip. Notice

0-7695-2533-4/06 $20.00 © 2006 IEEE

that some logic circuits may be changed by metal fix at the same time when we repair the power grid.

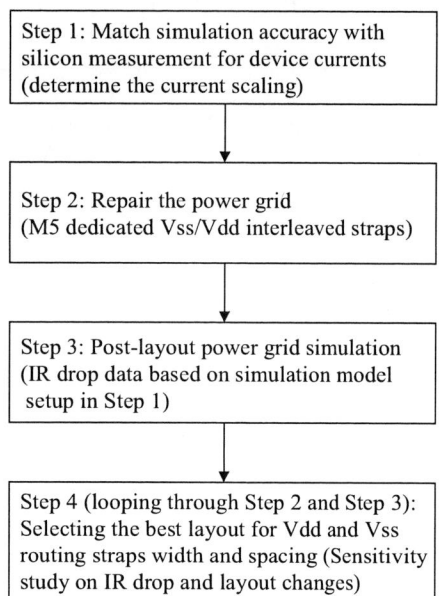

Figure 3. Design flowchart for power grid metal fix.

Table 4. Current scaling factors in simulation.

	Post-layout simulated current	Silicon measured current	Current scaling factor
Vdd	4.22e+03A	4A	0.00095
Vss	5.25e+03A	4A	0.00076

Table 5 shows the voltage drops across the chip for Vdd and Vss networks by using the separation 40μm and 75μm between two adjacent Vdd lines and adjacent Vss lines, as well as the original design without the M5 power straps. The worst case IR drop calculation is the sum of the voltage drops in Vdd and Vss nets. About 67% IR drop reduction is observed by adding power straps on the M5 layer for this chip, and 0.25V IR drops is within the required supply voltage ranges (nominal voltage: 2.5V) for the correct device timing, which is about 10% of the nominal voltage.

We developed a CAD flow to analyze the power grid based on VoltageStorm[TM] tool from Cadence Design Systems Inc.. The IR drop analysis flow is summarized in Figure 4. The simulation time can be long over 24 hours in a large chip. Due to the sensitivity study involving couple of changes on the power grid design selection in M5 layer, we would like to use the layout partition and incremental simulation to speed up the analysis. We partition the chip into multiple straps and then extract the device current model and resistance

model in each strap separately. If we change a portion of the power grid and the circuit, we only need to re-extract this portion in particular straps for modeling. Vdd and Vss may have different simulation results due to the model difference between two nets. But the measurement will be the same for Vdd and Vss. The worst-case voltage drop is estimated as the sum of the worst-case Vdd and Vss voltage drops in the chip as reported in Table 5.

Table 5. Power grid metal repair results.

	Vdd IR drop	Vss IR drop	Vdd + Vss drop
Original chip : No M5 power straps	0.356V	0.434V	0.79V
M5 power straps : 75um Vdd/Vss pairs separation	0.177V	0.146V	0.26V
M5 power straps : 40um Vdd/Vss pairs separation	0.13V	0.124V	0.25V

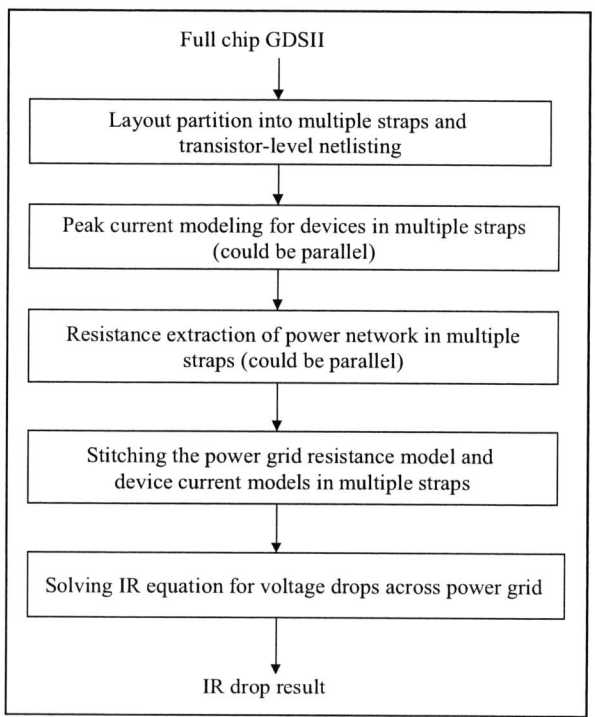

Figure 4. Static IR analysis flow [5,7].

The IR simulation slows down significantly with many floating metals and floating nodes in the power network layout as we use the metal repair on M5. The disk space and memory are required very large in a large-chip power network. The turnaround time in the SOC chip is about 24 hours and the automation of the running flow as shown in Figure 4, helps us to submit the job overnight and do the power grid improvement

0-7695-2533-4/06 $20.00 © 2006 IEEE

during the next day. To reduce extraction and analysis time, we can separate power networks for I/O ring and core power networks using different Vdd/Vss labels. Multiple machines are used to do the resistance extraction in multiple straps in parallel. The IR equation solving shown in Figure 4 takes the most of the computational time in the flow. We run this command in multiple 64-bit machines to run the full-chip level. The peak current estimation and full chip level resistance modeling is needed once and for metal fixing we do the incremental changes to the model only needed for affected layout straps. A step, shown in Figure 4, stitches multiple straps together for the power grid model.

4. Conclusion

One annoying issue in every VLSI product is silicon bugs not captured in the design. Due to the cost of mask making, we want to fix bugs in the least changed layers. Metal fix is a common technique to fix circuit bugs by re-wiring the gates. We discuss a metal fix method in a complex SOC chip. The first step is to verify the simulation environment with the measurement data from the failure silicon. In our case, bugs are eliminated in new silicon based on the proposed flow for silicon failures due to IR drop problem and hold time violation in I/Os. The power network design has a global impact on circuit performance. A pre-layout power network planning method will be useful [10].

References

[1] "Datasheet of CS2112 Reconfigurable Communications Processor", Chameleon Systems Inc., Aug. 2001.

[2] P. E. Gronowski, W. J. Bowhill, R. P. Preston, M. K. Gowan, and R. L. Allmon, "High-Performance Microprocessor Design", IEEE Journal of Solid-State Circuits, Vol. 33, No. 5, 1998, p. 676.

[3] N. D. Arora, K. V. Raol, R. Schumann, and L. M. Richardson, "Modeling and Extraction of Interconnect Capacitances for Multilayer VLSI Circuits," IEEE Transactions on Computer-Aided Design of Integrated Circuits and Systems, Vol. 15, No. 1, 1996, p. 56.

[4] Q. K. Zhu, "Interconnect RC and Layout Extraction for VLSI", Trafford Publishing, 2002.

[5] Q. K. Zhu, "Power Distribution Network Design for VLSI", John & Wiley, 2004.

[6] Q. K. Zhu, "High-Speed Clock Network Design", Kluwer Academic Publishers, 2003.

[7] "VoltageStorm Transistor-Level PGS User Guide", Cadence Design Systems, Inc., 2002.

[8] Dharchoudhury, R. Panda, D. Blaauw, R. Vaidyanathan, B. Tutuianu, and D. Bearden, "Design and Analysis of Power Distribution Networks in PowerPC Microprocessors", Proc. of 35th Design Automation Conference, 1998, p. 738.

[9] A. Chandrakasan, W. J. Bowhill, and F. Fox, edited, Design of High-Performance Microprocessor Circuits, IEEE Press, 2000.

[10] Q. K. Zhu and D. Ayers, "Pre-Layout Modelling and Planning for On-Chip Power Grid Design", Proc. of the 12th International Conference on Mixed Design of Integrated Circuits and Systems, 2005, p. 163.

0-7695-2533-4/06 $20.00 © 2006 IEEE

Multi-SP: A Representation with United Rectangles for Analog Placement and Routing

Ning Fu[†,††] Shigetoshi Nakatake[†] Mitsutoshi Mineshima[††]

† Dept. of Information and Media Sciences
The University of Kitakyushu
1–1, Hibikino, Wakamatsu-ku,
Kitakyushu, Fukuoka, 808–0135, Japan

†† Research and Development Division
Jedat Innovation Inc.
2–5, Hibikino, Wakamatsu-ku,
Kitakyushu, Fukuoka, 808–0135, Japan

Abstract

In analog layout, the placement and routing are closely connected with each other in the optimization of the area, the parasitics, the routability and so on. In this paper, we introduce a common data-structure to the placement and multi-layer routing, where devices and wires are represented by united rectangles. It is called Multi-Layer Sequence-Pair (Multi-SP). We also provide a bi-directional translation between a layout and a Multi-SP under align-constraint. Since the Multi-SP is geometry free, it enables us to manage diversified methodologies such as device sizing and technology migration. Also, it gives a way to take the parasitics between devices and wires of optimizing placement and routing simultaneously. We implemented a compaction tool based on Multi-SP and showed its potential experimentally.

1 Introduction

A typical SoC (System-on-a-Chip) becomes a mixed-signal type, which consists of digital part and analog part. While design automation of the digital part has made progress well, that of the analog part is still a bottleneck in the turn around time of the mixed-signal SoC design. The difficulty of analog layout problem is due to the dependency of performance on physical design.

The traditional design flow of analog layout includes placement, routing and compaction. Ordinarily, automation tools are developed separately, and executed sequentially. A placer determines the position of devices, a router finishes wire connections and a compactor removes excess spaces and erase design rule errors. The representations of element configurations are different in placement, routing and compaction.

However, the decomposition of placement and routing lacks global view of the interaction of placement and routing. For example, a router may not obtain a feasible result considering parasitics due to non-consideration of wire parasitics in prior placer. Also, ignoring parasitics between devices and wires causes performance degradation. A traditional compactor can not guarantee a good performance neither since it does not include optimization portion.

A solution of simultaneous placement and routing is desired. Accordingly, a united representation of devices and wires is requested.

The channeled-BSG [9] has been proposed as a common data-structure to placement and routing. It guarantees that a global routing on the grid-based data-structure can be translated a detailed routing on 2-layer without any violation of the design rule. However, it often causes much waste area since it gives the upper-bound of the routing space.

[4] provided a data-structure to express placement and routing based on SP. SP was developed for rectangle packing originally. It is a popular representation in analog placers for its clear topological system and its ability of solving analog constraints such as alignment[10], symmetry[11] and so on. Those features also are available in optimizing placement and routing simultaneously considering analog performance. In [4], each wire path is dissected into a set of rectangles. The topological relation of all rectangles of wire segments and device outlines are represented by an SP. [4] also showed simultaneous optimization of placement and routing on a single layer. It is said that this representation can be enhanced to multi-layer technology easily. But the layout model is not able to deal with the case that a wire passes through a cell between its terminals. Also, analog constraints like align constraint was not considered.

This paper proposes a data structure, named Multi-Layer Sequence-Pair (Multi-SP), based on the concept introduced in [4]. It is a three dimensional data structure corresponding to multi-layer technology. We generate an SP for each routing layer and an SP for all non-routing layer. Moreover, we add constraints to keep relative position of objects on neighboring layers.

We provide the way to extract a Multi-SP from a multi-layer layout and the way to generate a multi-layer layout from a Multi-SP. Furthermore, we develop the Multi-SP to handle complex analog design rules and analog constraints such as align-constraints.

In experiments, we apply the Multi-SP to an analog layout compactor as its first application. It uses a constraint graph compaction approach[1]. In each layer, constraints between two objects are obtained from the corresponding SP, since the topological relationships between pairs of two objects are represented by its SP. Thus, we need not generate or store any constraint graph. Also, it is easy to remove, actually ignore, redundant constraints during the compaction, while it takes $O(n^2 \log n)$ in a per-procedure of compaction in a traditional way[13]. It also handles wire-length minimization easily.

Since Multi-SP is geometry free and each SP holds a two dimensional topological information of each layer. It stores a horizontal constraint graph and a vertical constraint graph for a two-dimensional compaction simultaneously. Any reconstruction of constraint graphs is not necessary, even though devices become bigger or wires become wider after sizing.

The rest of this paper is organized as follows. A layout model used in Multi-SP is introduced in section 2. Section 3 gives a definition of Multi-SP. Section 4 and section 5 are devoted to a bi-directional translation between a layout and a Multi-SP. Experimental results on compaction problem are shown in section 6, and section 7 is for conclusion and future works.

2 Rectangle-based Layout Model

In this section, we introduce the layout model used in Multi-SP. In this model, all outlines of cells and segments of wires are represented by rectangles. Furthermore, those rectangles are assigned to some virtual layers for multi-layer routing.

2.1 Cell Model

We use "cell" to denote devices, via, blocks and so on which have reference cells in a layout. A cell is represented by multiple rectangles as shown in Figure 1. We use *multiple outline circuitry model* [4] to represent cells. Furthermore, we make an extension, in section 2.3, to handle a layout with multi-layer routing.

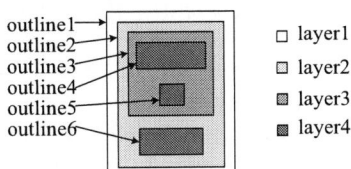

Fig. 1. a cell with multiple outlines

Here, we give a brief description of *multiple outline circuitry model* for the sake of consistency. A set of cells is denoted by $C = \{c_i | i = 1, 2, \ldots, n\}$, and a set of the outlines of c_i is denoted by $O(c_i) = \{o_{c_i,j} | j = 1, 2, \ldots\}$. A set of layers is denoted by $L = \{l_k | k = 1, 2, \ldots, m\}$. The width and height of an outline $o_{c_i,j}$ are denoted by $w(o_{c_i,j})$ and $h(o_{c_i,j})$, respectively. The coordinate of the reference point of cell c_i are denoted by $x(c_i)$ and $y(c_i)$, respectively, while $x(o_{c_i,j})$ and $y(o_{c_i,j})$ stand for relative coordinates from the left-bottom point of $o_{c_i,j}$ to the reference point of c_i. Accordingly, the left and right border of $o_{c_i,j}$ on the plane has the coordinate of $x(c) + x(o_{c_i,j})$ and $x(c) + x(o_{c_i,j}) + w(o_{c_i,j})$, the bottom and top border, $y(c) + y(o_{c_i,j})$ and $y(c) + y(o_{c_i,j}) + h(o_{c_i,j})$.

2.2 Wire Model

We adopt *rectangle-based wire model*, which is introduced in [4], to handle wire simultaneously with cells. It dissects wires into rectangles (called wire-rects) along the path, where the corner part is assigned to the later rectangle as shown in Figure 2.

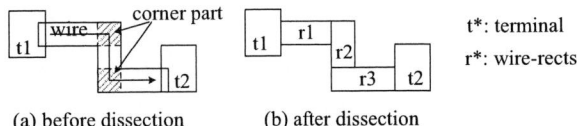

(a) before dissection   (b) after dissection

Fig. 2. Rectangle-based wire model

Each rectangles has just one outline. The border of a wire-rect is variable along the wire path, for the length of the segment is stretchable.

2.3 Multiple Topological Layer Model

For simplicity, cells and wire-rects are referred to be objects. An SP is used to represent 'topological relations' of objects to other objects such as 'above', 'below', 'left-of' and 'right-of'. However, taking multi-layer routing into account, we can not define such a relation between objects in the case that a wire passes through a cell, or wires cross each other on different layers, as shown in Figure 3(a).

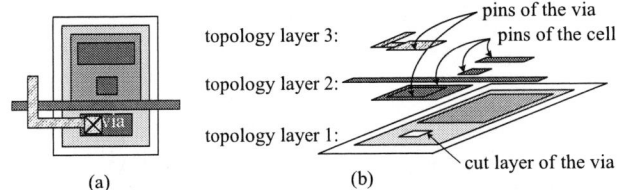

(a)   (b)

Fig. 3. An example of multiple topological layer model

It is a critical limitation in the development of a common data-structure to placement and multi-layer routing. To release the layout model from this limitation, we introduce *topological layer*. A topological layer is a collection of one or more real layers. An object just belongs to one topological layer. Any pair of objects on a same topological layer can not overlap each other, while overlapping is allowed among different topological layers. Topological relations between objects are generated on each topological layer.

Based on this definition, we introduce a *multiple topological layer model*. Since a crossing of wires is not allowed on same layer, we make a topological layer for every routing layer. All other layers compose a topological layer, called base layer.

Because a cell may belong to more than one topological layer, we decompose a cell into multiple objects. Every outline on a routing layer is regarded as an object, called decomposed cell. The rest of outlines compose an object, belonging to a base layer, which is called original cell. The relations between an original cell and its decomposed cells are stored to keep their relative positions. Thus, objects in this model are cells, parts of cells and wire-rects.

In the example in Figure 3(b), there are two routing lay-

0-7695-2533-4/06 $20.00 © 2006 IEEE   39

ers, therefore there are three topological layers. A transistor is decomposed to four cells, one original cell and three decomposed cells. A via for adjacent layers is decomposed to three cells, one original cell and two decomposed cells.

3 Definition of Multi-SP

3.1 Sequence-Pair

Sequence-Pair, abbreviate to SP, is a topological representation for placement. It is a pair of sequences of blocks, (α, β), specifying a topological relation, that is above, below, left-of and right-of, between every pair of objects as follows.

$\alpha = (\ldots a \ldots b \ldots), \beta = (\ldots a \ldots b \ldots) \Rightarrow a$ is left of b
$\alpha = (\ldots b \ldots a \ldots), \beta = (\ldots a \ldots b \ldots) \Rightarrow a$ is below b

An equivalent description of SP is as follows. α and β are defined as the sequences of blocks satisfying the property that all blocks can be removed one after another along $\alpha(\beta)$ in the above-left (below-left) direction without touching the remaining blocks. The abbreviation SP is also used to denote such a pair of sequence(α, β). The inverse function $\alpha^{-1}(a)$ denotes the index of a in α sequence, and $\beta^{-1}(a)$ the index of a in β sequence.

Note that, an SP only can represent either horizontal or vertical relationship between two objects. Accordingly, there may be more than one SP corresponding to a layout, and an SP can not represent two crossing objects.

3.2 Multi-SP

Based on the *multiple topological layer model* introduced in section 2. this paper introduces an SP-based representation of simultaneous placement and multi-layer routing, which is called Multi-SP.

In Multi-SP, we extract an SP on each topological layer. The relationship between SP's in adjacent layers is derived from relationships between original cells and their decomposed cells. Those relationships are represented by align-constraints, which we will discuss in section 5.2.

4 From Layout to Multi-SP

In this section, we introduce an extraction algorithm to get a corresponding Multi-SP from a layout. We extract an SP on a topological layer independently of the other topological layers. The extraction follows the definition of SP (See section 3.1). α sequence is extracted from the left-top to the right-bottom, while β sequence from the left-bottom to the right-top.

The algorithm is shown in Algorithm 1. The extraction keeps the feasibility of separation constraint, align-constraint and wire-connectivity, which we will introduce in the following.

4.1 Separation constraint

In analog layout, the design rule imposes a separation constraint on a pair of objects. For extracting SP satisfying separation constraints, we introduce *rule borders* of each object against other objects. Rule borders of an object against different objects might be different. A rule border of an

---

Algorithm 1 Extraction of α sequence under separation constraint and align-constraint

```
1:  for i = 1 to N do
2:      α(i) = −1;
3:  end for
4:  n = 0;
5:  SortHAlignByXcoord(INCREASE);
6:  SortVAlignByYcoord(DECREASE);
7:  while n ≤ N do
8:      prev_n = n;
9:      for i = 1 to N do
10:         if α(i) < 0 && IsFirstInAlign(obj_i) then
11:             for j = 1 to N do
12:                 if α(j) == −1 then
13:                     (l_i, b_i, r_i, t_i) = GetRuleBorder(obj_i, obj_j);
14:                     (l_j, b_j, r_j, t_j) = GetRuleBorder(obj_j, obj_i);
15:                     if b_i ≤ t_j and l_j ≤ r_i then
16:                         goto LABEL a;
17:                     end if
18:                 end if
19:             end for
20:             α(i) = n;
21:             RemoveFromAlign(obj_i);
22:             n + +;
23:             break;
24:         end if
25:         LABEL a: continue;
26:     end for
27:     if n == prev_n then
28:         failed and exit.
29:     end if
30: end while
```

---

object(obj_i) against another object(obj_j) is determined one side after another according to the following way: (1) Set the border of each object to be the minimal rectangle including all its outlines. (2) Determine the left side by putting obj_j to the left of obj_i as close as possible without separation violation with respect to each layer, and expanding their borders until they share a line in the middle. (3) Determine the other three sides similarly.

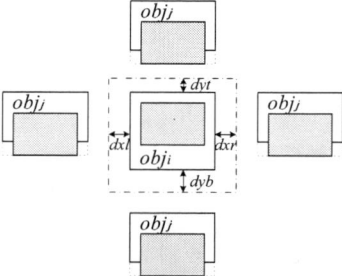

Fig. 4. Rule border of obj_i **against** obj_j

We give an example of determining rule borders in Figure 4. Four sides of the rule border of cell obj_i against cell obj_j are determined.

We define a *leader region* of an object obj_i in α extraction. It is an intersection region of two regions; one is the left region of its right rule border (ri) and the other is the upper region of its bottom rule border (b_i). From line 11 to line 19 in Algorithm 1, it checks if there is any other un-extracted object overlaps the *leader region* of the current object obj_i. If not, obj_i is extracted and given an index in α sequence. After all objects have been extracted, the algorithm finishes.

The algorithm of extracting β sequence is similar. Only the word "upper" in the definition of *leader region* should be replace by "lower", "bottom" by "top" and "b_i" by "t_i". Accordingly, the condition in line 15 should be modified as $b_j \le t_i$ and $l_j \le r_i$.

4.2 Analog-Constraint

Align-constraint is often used in analog layout. It is classified into H(orizontal)-align and V(ertical)-align.

A layout might correspond to two or more SP's. But not all of them are feasible to align-constraints. For the example shown in Figure 5, there are two corresponding SP's of the layout in (a). But SP2 is not feasible to H-align, and SP1 is not feasible to V-align. It is obvious that a vertical relationship is not allowable for H-align. Neither is a horizontal relationship for V-align.

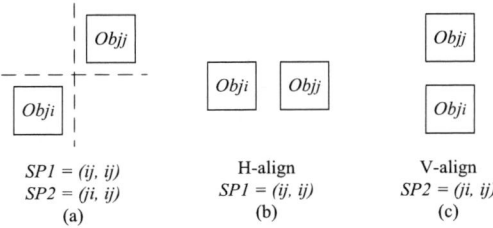

SP1 = (ij, ij)   H-align   V-align
SP2 = (ji, ij)   SP1 = (ij, ij) SP2 = (ji, ij)
(a)   (b)   (c)

Fig. 5. Feasible SP's for different align-constraints. (a) a layout with two possible SP's, (b) a layout of H-align and its feasible SP, (c) a layout of V-align and its feasible SP.

Lemma 1 *With respect to any pair of objects(obj_i, obj_j), an SP is feasible to the H-align if the SP satisfies that $(\alpha^{-1}(i) - \alpha^{-1}(j))(\beta^{-1}(i) - \beta^{-1}(j)) > 0$. Analogously, an SP is feasible to an V-align, if it satisfies $(\alpha^{-1}(i) - \alpha^{-1}(j))(\beta^{-1}(i) - \beta^{-1}(j)) < 0$.*

To obtain a feasible SP to align-constraints, we introduce *align-constraint-sorting* as follows. As for α sequence, objects in an H-align are sorted in increasing order of the x-coordinates, while those in V-align are sorted in decreasing order of y-coordinates. As for β sequence, objects in an H-align and V-align are sorted in increasing order of x-coordinates and y-coordinates, respectively. An object can not be extracted before any previous object in the same align-constraint.

Lemma 2 *An SP extracted in the order of objects in an align-constraint sorted by align-constraint-sorting is feasible for this align-constraint.*

In the Algorithm 1, the line 5, 6, 10 and 21 are for align-constraints.

Theorem 1 *Let a layout be L, separation constraints be C_s and align-constraints be C_a. For a given L, Algorithm 1 outputs an SP such that it keeps the topology between any pair of objects in L and it is feasible for C_s and C_a, if such an SP exists.*

We can prove Theorem 1 by employing Lemma 1 and Lemma 2, but the proof is omitted here for the space.

4.3 Adaptation to Wire-Connectivity

We introduced wire model in section 2.2, in which wires are dissected into rectangular objects with directional variable borders. In the translation from a layout to an SP, the length of the border is the original length of its rectangle. In the inverse translation, to get a compact layout, the length of the border is regarded as zero during the calculation of objects' positions, and finally extended according to its connection to keep the wire-connectivity. The extension may fail from an unfeasible SP, like the SP1 in Figure 6. SP2 is feasible for wire-connectivity in this figure.

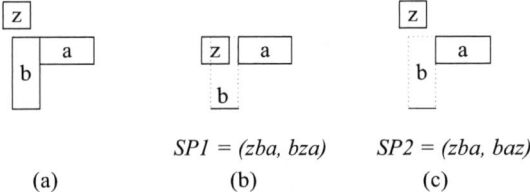

SP1 = (zba, bza) *SP2 = (zba, baz)*

(a)   (b)   (c)

Fig. 6. Feasibility of SP for wire-connectivity

Wire-connectivity consists of J(og)-, T(erminal)- and S(teiner point)-connectivity, which are the connectivity of two consecutive wire segments at the corner of a wire path, a wire path and a terminal, and two wire paths at the Steiner point, respectively [4]. An SP extracted by Algorithm 1 satisfies T- and S-connectivity. We provide a revision algorithm to get an SP satisfying J-connectivity.

There are eight patterns of adjacent rectangles at a jog.

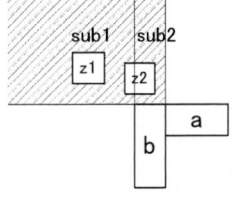

Fig. 7. Interference region for the connection of a and b

We discuss the left-down pattern here, see Figure 7. It is introduced in [4] that an SP satisfying one of the following conditions is feasible to J-connectivity: (1) $\beta^{-1}(z) < \beta^{-1}(b)$, (2) $\beta^{-1}(z) > \beta^{-1}(a)$ and (3) $\alpha^{-1}(z) > \alpha^{-1}(a)$, where a and b are objects of adjacent wire segments and z is any other object. If z is out of the shaded region in the Figure 7, the corresponding SP is satisfied condition (1) or (3) definitely. However, for an object included in the shaded region, the SP might be infeasible for J-connectivity. For this case, We propose a revision to SP to satisfy the condition (2).

An SP can be regarded as an assignment of objects on an

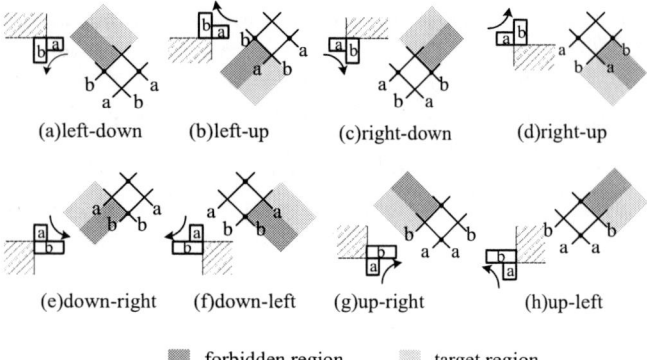

(a)left-down   (b)left-up   (c)right-down   (d)right-up

(e)down-right   (f)down-left   (g)up-right   (h)up-left

▨ forbidden region   ▨ target region

Fig. 8. SP revision in eight patterns

oblique line grid [6]. Figure 8(a) shows the feasible and infeasible region of z on the oblique line grid for J-connectivity of a and b, which are deep shaded and light shaded, respectively. We move all objects in the infeasible region to the feasible region so that only the order between a and those objects in β sequence are changed. This move does not destroy the relation between those objects to any other object. Thus, an SP $(z_1...z_i ba, b z_1...z_i a)$ is revised to $(z_1...z_i ba, \underline{b a z_1...z_i})$, which is feasible for J-connectivity. The revisions in other patterns are applied in similar ways. Figure 8 (b)-(h) are illustrations for those revisions.

5 From Multi-SP to Layout

The mapping of a Multi-SP to a layout is executed by a graph-based compaction. It is to find the longest path in a horizontal and a vertical constraint graph which are denoted by $G_h = (V_h, E_h)$ and $G_v = (V_v, E_v)$, respectively. The key is to generate constraint graphs excluding transitive edges and being capable of incorporating layout constraints. We describe the generation of the horizontal constraint graph in the following. The vertical constraint graph is handled similarly.

A vertex in V_h is corresponding to the x-coordinate of the reference point of an object. The reference point of a decomposed cell or a wire-rect is set to the left-bottom point of its outline. An edge denoted by $e(v_i, v_j)$ is the minimum distance constraint from v_i to v_j, which is derived from a separation constraint, an align-constraint or a bound-constraint. Actually, we do not generate any real graph, but use the relations represented by Multi-SP.

5.1 Separation Constraint

A separation constraint is used for keeping design rules. Since constraints are transitive, not all the constraints between pairs of vertices are necessary. Shadow-propagation approach is a conventional way to remove redundant constraints. Pan et al [13] also provide a constraint reduction approach to get an optimal reduction of the original constraint graph. But it is time consuming (with time complexity of $O(n^2 \log n)$).

For using Multi-SP not a real constraint graph, we do not need the constraint graph reduction phase before the compaction, but ignore the redundant constraints during the longest path calculation on each topological layer. Topological relationships between pairs of objects can be derived from the corresponding SP on each topological layer, and these relationships are transitive. For example, if v_i is to the right of b and b is to the right of a, then v_i is to the right of a. The idea to ignore $e(a, v_i)$(known as transitive edges), where v_i is any vertex to the right of b, as show in Figure 9(a). The case in G_v is also shown in Figure 9(b).

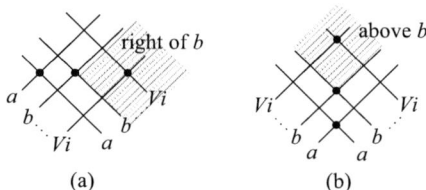

(a)   (b)

Fig. 9. Region of objects having redundant separation constraints with a

5.2 Align-Constraint

Algin-constraint is used for the purpose of keeping several cells in one row or column, or fixing relative positions between cells, which often occurs in analog layouts for the device matching. It is also used to keep the connection between terminals and the connected wire-rects, and to fix relative positions between original cells and their terminals.

We show an example contains three kinds of align-constraints in Figure 10. To satisfy align-constraints, constraints must be generated in pairs. Then the weights are calculated according to the layout distance, shown as the value of arcs in this figure.

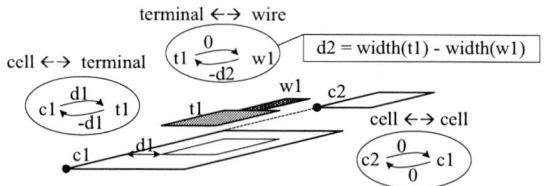

Fig. 10. Align-constraints

5.3 Bound-Constraint

The connection of wires at a Steiner point is kept, if a wire segment end at a Steiner point, say seg-a, is bounded in the range of the other wire segment, say seg-b. Thus, we add a bound-constraint between seg-a and either object connected to the end of seg-b. An example is shown in Figure 11. The object d is bounded between the left border of a and the right border of c.

6 Experimental results

We have implemented a compaction tool based on Multi-SP in C language. The time complexity of the worst case of extracting a Multi-SP from a layout is $O(n^3)$, where n is the number of objects. Since there are many cycles generated by align-constraints in each constraint graph, the worst case

0-7695-2533-4/06 $20.00 © 2006 IEEE   42

Table 1. Comparison of two system

input					result by a commercial tool		result by our system	
data	#cell	#net	#align-const	DRC error	area ratio	DRC error	area ratio	DRC error
S5	58	58	5	97	78.62%	36	77.16%	1
S5'	58	58	5	0	77.53%	3	74.53%	0
T	105	105	9	0	92.2%	0	95.51%	0

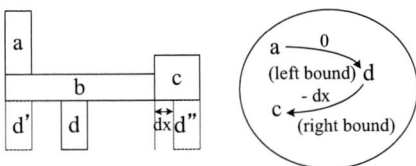

Fig. 11. Bound-constraints

of getting a compact layout from a Multi-SP takes $O(n^3)$. However, from the experimental results, shown in Table 2, we get a curve of the time and the number of objects as $time = 3.0 \times 10^{-10}n^3 - 3.0 \times 10^{-7}n^2 + 1.5 \times 10^{-3}n + 3.5$. The coefficients of n^3 and n^2 are small enough to be ignored. We also employ some techniques to detect positive cycles during the longest path calculating.

Table 2. Computation time

No.	#object	#instance	#nets	time(sec)
1.	838	69	60	1
2.	1446	83	61	3
3.	2966	151	105	13
4.	5362	310	123	42
5.	7454	438	160	115
6.	9060	527	192	217
7.	11501	661	245	427

We also compared our system with a commercial compactor. The results are shown in Table 1. It is shown that our system can get more compact layouts with less violations of design rules. Besides, the results of our system satisfy all the align-constraints while the commercial tool ignore them. (In some of results, our system output a larger area than that of the commercial compactor for this reason.)

7 Conclusion

We proposed Multi-SP, which is a common data-structure for placement and multi-layer routing, where devices and wires are represented by united rectangles. We showed the way of a bi-directional translation that includes extraction of a Multi-SP from a layout under align-constraints and generation of a layout from the Multi-SP under those constraints. This translation is of significance in the point of how to manage the diversified methodologies such as device sizing and technology migration.

In experiments, more compact layouts with less violations of design rules are generated by our tool based on Multi-SP compared with a commercial compactor. We also showed the ability of de-compaction based on Multi-SP. Furthermore, our (de-)compaction is applicable to an analog block with

hundred of cells in practical time.

Future works are to consider symmetry constraints in the bi-directional translation and to develop simultaneous placement and routing taking the parasitics.

References

[1] D. G. Boyer "Symbolic Layout Compaction Review" 25th. DAC pp.383-389 1988

[2] R. Okuda, T. Sato, H. Onodera, and K. Tamaru "An Efficient Algorithm for Layout Compaction Problem with Symmtery Constraints" ICCAD89 pp.148-151 1989

[3] E. Felt, E. Malavasi, E. Charbon, R. Totaro, and A. Sangiovanni-Vincentelli "Performance-driven compaction for analog integrated circuits" Proc. IEEE Custom Integrated Circuits Conference, pp. 1731-1735, May 1993.

[4] Y. Kubo, S. Nakatake, Y. Kajitani, and M. Kawakita "Explicit Expression and Simultaneous Optimization of Placement and Routing for Analog IC Layouts" Proc. 7th ASPDAC pp.467-472 2002

[5] S.Nakatake, H.Murata, K.Fujiyoshi, and Y.Kajitani "Module Packing Based on the BSG-Structure and IC Layout Applications," IEEE Trans. on Computer Aided Design of IC's and Systems, Vol.17, No 6, pp.519-530, 1998.

[6] H.Murata, K.Fujiyoshi, S.Nakatake and Y.Kajitani, "VLSI Module Placement Based on Rectangle-Packing by the Sequence-Pair," IEEE Trans. on Computer-Aided Design of IC's and Systems, Vol.15 No.12 pp.1518-1524, 1996.

[7] P.Guo, C.Cheng and T.Yoshimura, "An O-Tree Representation of Non-Slicing Floorplan and Its Applications," Proc. 36th Design Automation Conference, pp.268-273, 1999.

[8] Yun-Chih Chang, Yao-Wen Chang, Guang-Ming Wu, and Shu-Wei Wu, "B*-Trees: A New Representation for Non-Slicing Floorplans" Proc. 37th Design Automation Conference, pp.458-463, 2000.

[9] S.Nakatake, K.Sakanushi, Y.Kajitani and M.Kawakita, "The Channeled-BSG: A Universal Floorplan for Simultaneous Place/Route with IC Applications," Proc. ACM/IEEE Intl. Conf. of CAD, pp.418-425, 1998.

[10] X. Tang, and D. F. Wong "Floorplanning with Alignment and Performance Constraints" Proc. ACM/IEEE Design Automation Conf., pp.848-853, 2002.

[11] F. Balasa, and K. Lampaert "Symmetry within the sequence-pair representation in the context of placement for analog design" IEEE Tran. on CAD of ICs and Systems, vol.19, no.7, pp721-731, 2000

[12] J. K. Ousterhout, "Corner stitching: a data-structuring technique for VLSI layout tools," IEEE Trans. on CAD, Vol. 3, no.1, pp. 87-100, 1984.

[13] P. Pan, S. Dong and C.L. Liu, "Optimal Graph Constraint Reduction for Symbolic Layout Compactioin," DAC93, pp. 401-406, 1993

0-7695-2533-4/06 $20.00 © 2006 IEEE

Formulating the Empirical Strategies in Module Generation of Analog MOS Layout

Tan Yan[†]   Takashi Nojima[‡ §]   Shigetoshi Nakatake[§]

† R&D Center of Semiconductor Design
§ Dept. of Information and Media Sciences
The University of Kitakyushu
1-1, Hibikino, Wakamatsu-ku
Kitakyushu, Fukuoka, 808-0135, Japan
{yantan, nojima, nakatake}@env.kitakyu-u.ac.jp

‡ R&D Division
Jedat Innovation Inc.
IT Advancement Center
2-5, Hibikino, Wakamatsu-ku
Kitakyushu, Fukuoka, 808-0135, Japan
nojima.takashi@jedat.co.jp

Abstract

In module generation for analog cell layout, it is necessary to incorporate the designers' empirical techniques to achieve small area as well as high performance. This paper presents the formulation of two types of module generation problems, faithful to expert's empirical knowledge, to ease such incorporation. One is series module generation problem, where we introduce equivalent circuit modifications to maximize the diffusion merging and minimize the number of diffusion contacts. The other is the common-centroid module generation problem which is formulated as a mathematical optimization problem taking coincidence, symmetry, dispersion and compactness into consideration. A greedy algorithm is also presented to solve this problem.

1. Introduction

In a typical design flow of analog cell layout, transistors requiring high accuracy are usually realized into modules before the cell-level layout design. The layout inside the module greatly affects the circuit performance. Therefore, generating such modules, which is also called *module generation*, is a critical step in analog layout design [8].

For many years, module generation has been done manually by expert designers, using human intelligence and experience to achieve small area as well as good performance. However, in recent analog mixed signal LSIs, the number of transistors within a module is becoming larger and larger. To cope with the low supply voltage, modern transistors usually have large width-to-length ratios(W/L) and are therefore realized by smaller transistors connected in parallel. Such an increase of transistors makes the manual design

a more time consuming work than ever. Design automation of module generation is therefore becoming more and more important for analog layout design.

However, it is hard to incorporate the experts experience into automated tools because human experience are usually quite subjective, providing few guidelines for EDA tools. This is why, in this paper, we start with formulating the experts' experiences as optimization problems, and then provides algorithms to solve the problems. This paper presents two types of module generation problems. One is the *series module generation* problem, where transistors are connected in series and each transistor is divided into several parallel connected sub-transistors, due to its large W/L. These sub-transistors are usually implemented in a stacking structure. There exist many studies to maximize the diffusion merging of the stacking assuming that the topology of the circuit is fixed [2, 6, 11, 14]. However, experts sometimes reorder the transistors in the series circuit to achieve maximal diffusion merging [1]. In addition, experts minimize the number of diffusion contacts [3] by modifying the circuit topology. Consequently, they could achieve a smaller area and better performance. In this work, we formulate the 1-dimensional transistor placement problem taking these techniques into consideration and propose an algorithm to output the optimum.

The other problem discussed is the *common-centroid module generation* problem. In analog circuit designs, the relative accuracy of the transistors' characteristics significantly influences the circuit performance. Accordingly, in module generation, we must preserve the relative accuracy inside the module against the process variation or thermal effect. Several models are provided to estimate the mismatching [4, 9, 12]. However, they can hardly provide any guideline for automating the module generation. Experts

0-7695-2533-4/06 $20.00 © 2006 IEEE

design modules according to empirical rules such as *coincidence, symmetry, dispersion* and *compactness* [7, 10, 11, 13] instead of those estimation models. In this work, we translate these empirical rules into optimization problems. A greedy algorithm for this problem is also presented.

The rest of this paper is organized as follows. Section 2 describes the series module generation problem. Section 3 describes the common-centroid module generation problem and Section 5 concludes this paper.

2. Module Generation for Series MOS circuit

In this sections, we describe how experts modifie the circuit by some equivalent transformation of the circuit to achieve more diffusion merging and less number of diffusion contacts. These modifications are described in an algorithmic way so that they are easy to implement.

2.1. Previous Works

To cope with low supply voltage, transistors in modern analog circuits usually have large W/Ls and are therefore realized by several smaller *sub-transistors* that are parallel connected for better performance as well as more flexible layout options. All the sub-transistors of a series MOS circuit or sub-circuit are usually implemented in the stacking structure. The stacking structure is to place the sub-transistors in a row and the diffusions of the adjacent sub-transistors can be merged if they are connected in the circuit. Many researches have been done on optimizing the order of the sub-transistors in the stacking [2, 14]. On the topic of maximizing diffusion merging, all the works are based on the *diffusion graph* [14]. The diffusion graph is derived from the circuit. Each node of the graph corresponds to a net and the edge connecting two nodes corresponds to the sub-transistor jointing two nets. A stacking structure with all the adjacent diffusions merged could be constructed by finding an Euler trail in this graph and ordering the sub-transistors according to the order of the visited edges in the Euler trail. Fig. 1 gives an illustration of utilizing the diffusion graph.

2.2. Our Approach

Noticing that in some cases the diffusion graph is not an Euler graph(see Fig. 2), complete diffusion merging cannot be achieved. Therefore the resultant layout must contain one gap between two diffusions which is inevitable according to the circuit topology. Moreover, since multiple sub-transistors must be connected in parallel, wires must be drawn from all the diffusions, letting the number of contacts be the same as the number of the diffusions. In order to generate layout with smaller size and better performance, experts usually change the topology of the circuit with out

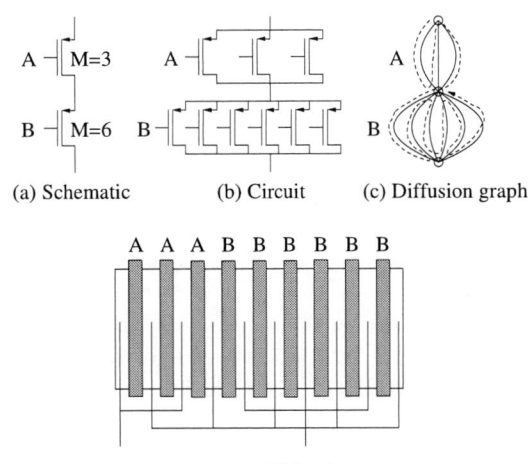

(a) Schematic   (b) Circuit   (c) Diffusion graph

(d) Layout

Figure 1. Maximizing the diffusion merging by diffusion graph.

affecting the circuit performance [1, 3]. Such changes are formulated as *equivalent circuit modifications (ECM)* in this paper.

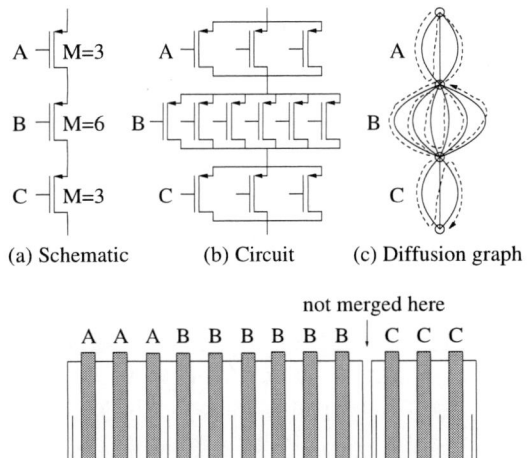

(a) Schematic   (b) Circuit   (c) Diffusion graph

(d) Layout

Figure 2. Some circuit provides no chance of complete diffusion merging.

An interesting observation is that a graph in the form of Fig. 4 is always an Euler graph because only the degrees of the two black vertex are odd and that of the rest are even. Therefore its corresponding circuit can always be laid out with complete diffusion merging. To get such a diffusion graph, we introduce our first ECM: *Series Reordering*. We pick out all the transistors with odd number of sub-transistors and connect them in series to form a chain of transistors with only odd number of sub-transistors. Such

0-7695-2533-4/06 $20.00 © 2006 IEEE   45

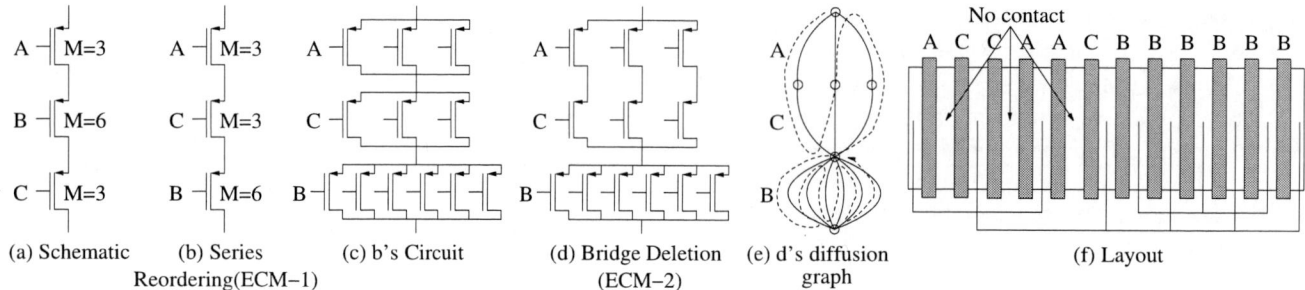

(a) Schematic (b) Series Reordering(ECM−1) (c) b's Circuit (d) Bridge Deletion (ECM−2) (e) d's diffusion graph (f) Layout

Figure 3. After two ECMs, the corresponding layout of the equivalent circuit achieves complete diffusion merging and has less necessary contacts than the original one.

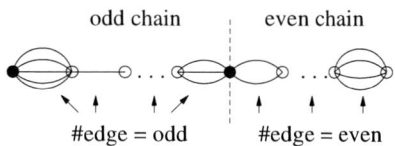

Figure 4. A structure guaranteed to have Euler trails.

a chain is called *odd chain*. The remaining transistors with even number of sub-transistors are also connected in series to form an *even chain*. We then connect the head of the even chain to the tail of the odd chain. The resultant circuit should have the same form as Fig. 4 and is therefore an Euler graph. Noticing that the ordering of the transistors in a series circuit does not affect the performance, the resultant circuit is equivalent to the original one.

Another observation is that when a connection has only two terminals(see nets marked by the ellipses in Fig. 5), it can be realized by diffusion merging instead of metal wires and then the contact to the diffusion can be saved. However, the original series parallel structure of the circuit implies no 2-terminal nets. In this case, experts sometimes introduce another ECM: *Bridge Deletion* which turns the original parallel sub-circuits in series into series sub-circuits in parallel [3](see Fig. 5). This transformation leads to a

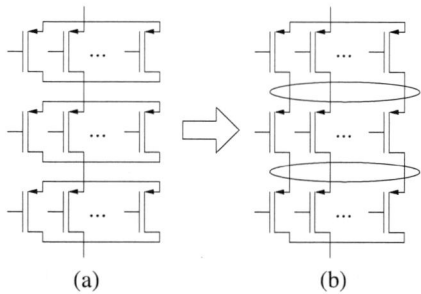

Figure 5. Removing the bridging edge reduces the contact number.

reduction in the transistor connections and the number of contacts is reduced consequently. To do so, we first sort the odd and even chains according to the sub-transistor number of the transistors so that transistors with the same number of sub-transistors are adjacent in both chains. Then consecutive transistors with the same number of sub-transistors are grouped with each group forming a bridge circuit. Finally, the bridging edges are removed. By this ECM, the number of the connections and the number of corresponding contacts are minimized. It is easy to examine that by our bridge deletion, maximal number of series sub-circuits can be obtained and therefore the number of contacts is minimized. It is obvious that the resultant diffusion graph of this transformation is still an Euler graph. Besides, noticing that ideally the bridging edge should have no current in a balanced bridge circuit, removing it does not affect the performance of the circuit. Therefore, the resultant circuit is equivalent to the original one.

The design flow of our proposed approach is shown in Fig. 3. It can be seen that the complete diffusion merging is achieved and the number of contacts is minimized.

3. Common-centroid Module Generation

Mismatching caused by process variation may greatly tamper the performance of analog circuits because some transistors, for example, current mirror, require high level of matching. Many design techniques are proposed to reduce mismatching, among which common-centroid layout is the most widely used strategy. Usually, a common-centroid layout should satisfy the following rules [7]:

- **Coincidence:** The centroids of the matched transistors are expected to coincide as much as possible.
- **Symmetry:** The array should be symmetric around both the x- and the y-axes.
- **Dispersion:** The array should exhibit the highest possible degree of dispersion. The sub-transistors of one transistor should be distributed throughout the array as uniformly as possible.

0-7695-2533-4/06 $20.00 © 2006 IEEE   46

- **Compactness:** The array should be as compact as possible. Ideally, it should form a square.

In our layout scheme, the module is a sub-transistor array with the size given by the high-level placer. Therefore, the compactness rule is handled by the high-level placer instead of the module generator. The axial symmetry rule somehow conflicts with the coincidence rule for 2-D array, especially when the number of sub-transistors cannot be divided by 4. An obvious counter example is that the popular cross-coupled pair pattern[1] is symmetric around neither x- nor y- axes. In practical designs, the coincidence rule is much more prior to the axial symmetry rule [1]. Therefore, the axial symmetry rule is replaced with our centrosymmetric rule that will be explained later. Since these rules are proposed as guidelines for human designers, their description are very subjective and vague which makes them difficult to be applied to design automation. To build a common-centroid module generator, we must first formulate the rules in a way that computer can understand. Here, we present our formulation which treats the *coincidence* and *dispersion* rules as optimization problems.

3.1. Our Formulation

Before we present our formulation, we first give some definitions. To describe the locations of the sub-transistors, we first build a coordinate system with the origin placed at the center of the array and the x-axis spans across the width and the y-axis across the height. The location of a sub-transistor is represented by the coordinate of its center(see Fig. 6) because the gradient effect inside a sub-transistor is negligible [4]. A z-axis is used to represent the gradient direction which is assumed to form an angle of θ degrees with the x-axis.

For a sub-transistor A_i of a transistor A, it is easy to calculate its z coordinate from its location (x_i, y_i) through simple coordinate transformation:

$$z_i = x_i \cos \theta + y_i \sin \theta$$

The z-coordinates of transistor A is then defined as the set of z-coordinates of all its sub-transistors:

$$Z_A = \{z_i | sub\text{-}transistor_i \in A\}$$

The z-centroid of A is defined as the average of Z_A:

$$z_A^c = (\sum_{z_i \in Z_A} z_i)/\|Z_A\|$$

By ordering the elements in Z_A so that they follow an increasing order, we could get an ordered list:

$$Z'_A = \{z'_i | z'_i \in Z_A, z'_i \leq z'_{i+1}\}$$

---

[1]The cross-coupled pair pattern for two transistors A and B with two sub-transistors each should be $\begin{smallmatrix} A B \\ B A \end{smallmatrix}$.

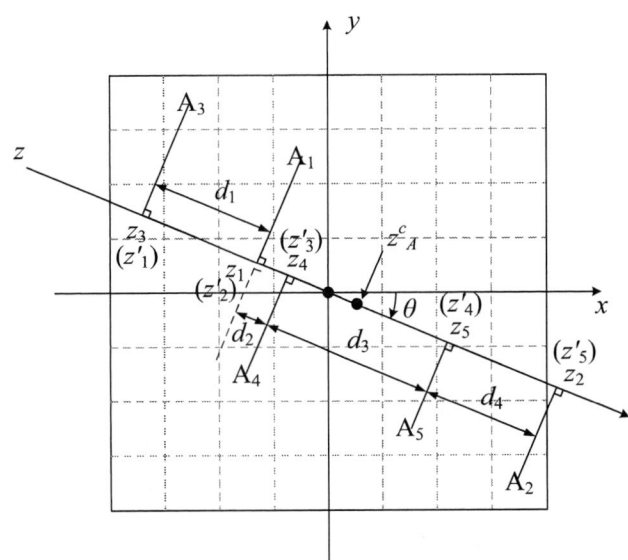

Figure 6. Our formulation.

The *difference* of Z_A, which is denoted as D_{Z_A}, is defined on this list:

$$D_{Z_A} = \{d_i | d_i = z'_{i+1} - z'_i, i = 1, \ldots, n-1\}$$

An example of these definitions could be found in Fig. 6.

In [7], the *coincidence* rule is proposed to cancel the linear component of the gradient by making the centroids of all the transistors coincide at the center of the layout array. It is trivial to examine that the linear component of the gradient could be completely canceled if and only if the z centroids of all the transistors are zero. Therefore, we could formulate the *coincidence* rule as minimizing the maximal z centroid of all the transistors(T_k):

$$\text{Minimize} \quad \max_k(|z_{T_k}^c|) \qquad (1)$$

To alleviate the influence of nonlinear components of the gradient, the *dispersion* rule in [7] claims that the sub-transistors of one transistor should be distributed throughout the array as uniformly as possible along the gradient direction. The uniformity of the distribution could then be measured by the deviation(σ) of D_{Z_A} in our metric and the *dispersion* rule could be formulated as minimizing the deviation of D_{Z_A}. To guarantee the dispersion of all the transistors, the one with maximal deviation should be minimized:

$$\text{Minimize} \quad \max_k(\sigma(D_{Z_{T_k}})) \qquad (2)$$

Since the gradient direction is not obtainable in module generation [5], we must consider the worst case for all possible directions, which means we should minimize the maximal value of Eq. (1) and Eq. (2) with respect to all θ. Therefore, the formulations should be modified to:

$$\text{Minimize} \quad \max_{\theta} \max_{k} (|z^c_{T_k}|) \tag{3}$$

$$\text{Minimize} \quad \max_{\theta} \max_{k} (\sigma(D_{Z_{T_k}})) \tag{4}$$

3.2. Centrosymmetric Assignment

We have formulated the common-centroid problem as optimization problems. However, it is difficult to handle the numerical optimization problems directly. Therefore, we propose a greedy algorithm to optimize Eq. (1). It is obvious that when the sub-transistors are centrosymmetric across the layout array, the z centroid of this transistor is zero. In practical designs, a transistor usually has even number of sub-transistors, giving us the chance to implement this strategy. However, there are still cases that a transistor has odd number of sub-transistors. In that case, we remove one sub-transistor from each transistor with odd number of sub-transistors. The removed sub-transistors are then collected and placed at the center of the array. Since there might be more than one removed sub-transistors, sub-transistors from transistor with smaller number of sub-transistors have higher priority and are placed closer to the center of the array. Similar greedy strategy could be found in [13]. Such a strategy is called "centrosymmetric assignment".

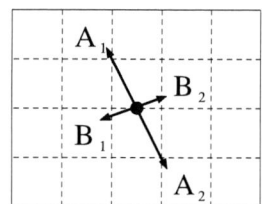

Figure 7. The centrosymmetric assignment for common-centroid layout.

3.3. Transistors with Different Priorities

In practical design, some transistors may require higher level of matching than others. Therefore, to fully utilize the approximate linearity of the local gradient, those transistors are expected to be placed as compact as possible. Fig. 8 gives such an example. In this module, transistor A and B each has 2 sub-transistors and transistor C has 16 sub-transistors. A and B form a current mirror and therefore require higher level of matching than C. In this case, designers often sacrifice the global centroid coincidence for better compactness of A and B as shown in Fig. 8(a) [1]. The layout in Fig. 8(b) is not preferable in this situation even though

it is perfectly common-centroid. To handle such cases, we

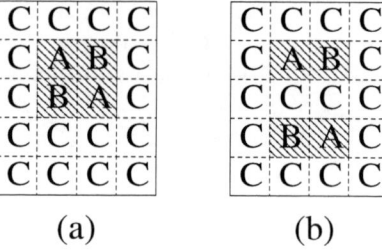

Figure 8. An example when some transistors have higher priority than others. (a) is more preferable than (b).

divide the whole module area into two parts: the central area for high priority transistors and the peripheral area for low priority transistors. The shape of the central area is close to a circle so that it is compact. We first assign the high priority transistors to the central area using centrosymmetric assignment and then assign those of low priority to the peripheral area also by centrosymmetric assignment.

Please note that the center of the central area might not coincide with the module center(Fig. 8(a) gives an example). After the central area is assigned, there might be some slots left without corresponding centrosymmetric slots(the shaded ones in Fig. 9). We call such slots *lonely slots*. These lonely slots will cause the transistors' centroids to deviate from the module center. To minimize such deviation, we have to carefully assign the lonely slots before we assign the peripheral area. First, transistors with more sub-transistors should be assigned to slots with larger distance to the module center(*the distance rule*). In addition, if there are two axial symmetric lonely slots, they should be assigned to sub-transistors from the same transistor so that the centroid lies at the x- or y- axes(*the symmetry rule*). Therefore, we first sort the lonely slots according to their distances to the module center and then assign sub-transistors to the lonely slots following the above two rules. The maximum z-centroid of the transistors is expected to be optimized in this step.

4. Disperse the Sub-Transistors

Dispersion rule is handled after the coincidence is optimized. By swapping the locations of two sub-transistors of the same priority, say A_1 and B_1, we could change the deviations of D_{Z_A} and D_{Z_B}. However, since we don't expect the optimized coincidence to be tampered by this change, the two centrosymmetric sub-transistors are also swapped. See Fig. 7, if we exchange A_1 with B_1 and A_2 with B_2, we

0-7695-2533-4/06 $20.00 © 2006 IEEE   48

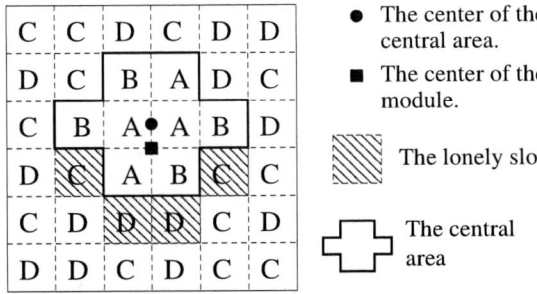

- ● The center of the central area.
- ■ The center of the module.
- ▨ The lonely slot
- ✛ The central area

Figure 9. A common-centroid layout with consideration of transistor priority.

could change the deviations of D_{Z_A} and D_{Z_B}. If the change is negative, which means the deviations are reduced, we accept this change and try to swap another sub-transistors. Otherwise, we swap the sub-transistors back to their original position and try to swap another sub-transistors. When no swapping could improve the deviation, the whole process is terminated. The maximum deviation is expected to be optimized by this algorithm.

4.1. An Example

Fig. 9 gives an example to explain our strategy. Suppose we want to assign transistors A, B, C and D into a module with 6×6 slots. High priority transistors A and B each has 4 sub-transistors and low priority transistors C and D each has 14 sub-transistors. We try to assign the sub-transistors following the rules introduced above so that the mismatch is minimized. First, we assign sub-transistors of A and B to the central area so that they are placed compactly and the mismatch between them is minimized. The assignment follows the centrosymmetric way so that the centroids of A and B coincide at the center of this bounded area. After that, we assign the sub-transistors to the lonely slots according to the distance and symmetry rules. Finally, we assign the remaining sub-transistors to the peripheral area also using the centrosymmetric assignment. After this, we could disperse the sub-transistors using the greedy algorithm.

5. Conclusions & Future Work

In this work, we provide the formulations of two modules generation problems. The formulations are faithful to the experts' empirical knowledge. We also propose algorithms to solve them.

In the future work, we will incorporate these algorithms into an automatic placement tool, where configurations of the modules are changed dynamically to optimize the whole

cell area during the placement. We will also verify the performance of our results by post layout simulation.

References

[1] Personal communication with NEC Microsystems.

[2] B. Basaran and R. A. Rutenbar. An $O(n)$ algorithm for transistor stacking with performance constraints. In *Proc. DAC96*, pages 221–226. ACM/IEEE, July 1996.

[3] D. Clien. *CMOS IC Layout Concepts, Methodologies and Tools*. Newnes, 1999.

[4] X. Dai, C. He, H. Xing, D. Chen, and R. Geiger. An Nth order central symmetrical layout pattern for nonlinear gradients cancellation. In *Proc. ISCAS05*, pages 4835–4838. IEEE, May 2005.

[5] E. Felt, A. Narayan, and A. Sangiovanni-Vincentelli. Measurement and modeling of mos transistor current mismatch in analog ic's. In *Proc. ICCAD'94*, pages 272–277. ACM/IEEE, November 1994.

[6] P. Gopalakrishnan and R. A. Rutenbar. Direct transistor-level layout for digital blocks. In *Proc. ICCAD01*, pages 577–584. ACM/IEEE, November 2001.

[7] A. Hastings. *The Art of Analog Layout*. Prentice Hall, 2000.

[8] J. Kampe, C. Wisser, and G. Scarbata. Module generators for a regular analog layout. In *Proc. ICCD'96*, pages 280–285. IEEE, October 1996.

[9] M.-F. Lan, A. Tammineedi, and R. Geiger. Current mirror layout strategies for enhancing matching performance. *Analog Integrated Circuits and Signal Processing*, 28(1):9–26, July 2001.

[10] D. Long, X. Hong, and S. Dong. Optimal two-dimension common centroid layout generation for mos transistors unit-circuit. In *Proc. ISCAS05*, pages 2999–3002. IEEE, May 2005.

[11] R. Naiknaware and T. S. Fiez. Automated hierarchical cmos analog circuit stack generation with intramodule connectivity and matching considerations. *IEEE Journal of Solid-State Circuits*, 34(3):304–317, March 1999.

[12] M. Pelgrom, A. Duimnaijer, and A. Welbers. Matching properties of MOS transistors. *IEEE Journal of Solid-State Circuits*, 24(5):1433–1440, October 1989.

[13] D. E. Sayed and M. Dessouky. Automatic generation of common-centroid capacitor arrays with arbitrary capacitor ratio. In *Proc. DATE02*, pages 576–579. ACM/IEEE, March 2002.

[14] T. Uehara and W. M. VanCleemput. Optimal layout of CMOS functional arrays. *IEEE Trans. on Computers*, C-30(5):305–312, May 1981.

0-7695-2533-4/06 $20.00 © 2006 IEEE

An Integer Linear Programming Based Approach to Simultaneous Memory Space Partitioning and Data Allocation for Chip Multiprocessors*

O. Ozturk, G. Chen, M. Kandemir
Computer Science and Engineering Department
Pennsylvania State University
University Park, PA 16802, USA
{ozturk, gchen, kandemir}@cse.psu.edu

M. Karakoy
Department of Computing
Imperial College
London, SW 2AZ, UK
m.karakoy@ic.ac.uk

Abstract

The trends in advanced integrated circuit technologies require us to look for new ways to utilize large numbers of gates and reduce the effects of high interconnect delays. One promising research direction is chip multiprocessors that integrate multiple processors on the same die. Among the components of a chip multiprocessor, its memory subsystem is maybe the most critical one, since it shapes both power and performance characteristics of the resulting design. Motivated by this observation, this paper addresses the problem of decomposing (partitioning) on-chip memory space across parallel processors and allocating data across memory components in an integrated manner. In the most general case, the resulting memory architecture is a hybrid one, where some memory components are accessed privately, whereas the others are shared by two or more processors. The proposed approach for achieving this has two complementary components: an optimizing compiler and an ILP (integer linear programming) solver. The role of the compiler in this approach is to analyze the application code and detect the interprocessor data sharing patterns, given the loop parallelization information. The job of the ILP solver, on the other hand, is to determine the sizes of the on-chip memory components, how these memory components are shared across multiple processors in the system, and what data each component holds. In other words, we address the problem of integrated memory space partitioning and data allocation for chip multiprocessors.

1 Introduction

A chip multiprocessor integrates multiple processors on the same die. Recent research [18, 11, 13, 14, 19] discusses several advantages of these architectures over complex single-processor based designs. These advantages include capability of exploiting both high level (loop/thread level) and low level (instruction level) parallelism, better performance and power consumption profiles, and easier verification. In particular, it has been reported [17] that,

on applications with high-level parallelism (e.g., embedded multimedia codes that make frequent use of large arrays), a chip multiprocessor can perform better than a wide superscalar architecture.

A critical component of a chip multiprocessor is its memory subsystem. This is because both power and performance behavior of a chip multiprocessor is largely shaped by its on-chip memory [6, 16]. While it is possible to employ conventional memory designs such as pure private memory or pure shared memory, such designs are very general and rigid, and may not generate the best behavior for a given embedded application. Our belief is that, for embedded systems that repeatedly execute the same application, it makes sense to design a customized, software-managed on-chip memory architecture. Such a memory architecture should be a hybrid one that contains both private and shared components. Figure 1 depicts the high-level views of pure private, pure shared, and sample hybrid memory architectures. Note that, in the hybrid architecture case, while some processors have private memories, others do not have one. Similarly, the different processor groups can share memory in different fashions. For example, a memory component can be shared by two processors, whereas another component can be shared by three processors.

Designing such a customized hybrid memory architecture is not trivial because of at least three main reasons. First, since the memory architecture to be designed changes from one application to another, a hand-waived approach is not suitable, as it can be extremely time consuming and error-prone to go through the same complex process each time we want to design a memory system for a new application. Therefore, we need an automated strategy that comes up with the most suitable design for a given application. Second, the design of such a memory needs to be guided by a tool that can extract the data sharing exhibited by the application at hand. After all, in order to decide how the different memory components need to be shared by parallel processors, one needs to capture the data sharing patterns across the processors. Third, data allocation in a hybrid memory system is not a trivial problem, and should be carried out along with data partitioning if we want obtain the best results.

---

*This work is supported in part by NSF Career Award #0093082 and a fund from GSRC.

0-7695-2533-4/06 $20.00 © 2006 IEEE

In this paper, we propose a strategy for designing application-specific on-chip hybrid memories for chip multiprocessors that employs both an optimizing compiler and an ILP (integer linear programming) solver (see Figure 2). The role of the compiler in this approach is to analyze the application code and detect the data sharing patterns across processors, given the loop parallelization information. The job of the ILP solver, on the other hand, is to determine the sizes of the memory components, how these memory components are shared across multiple processors in the system, and what data each memory component is to hold. Note that, the ILP based solution can be used not only for designing hybrid memories, but also as an upper bound against which future heuristic solutions can be compared. Our focus is on array-based application codes that occur frequently in embedded image/video processing. It needs also be noted that, our approach can be targeted at different objectives such as maximizing performance, reducing power/energy consumption, and minimizing the memory space occupied.

The rest of this paper is organized as follows. Section 2 presents the compiler analysis necessary to identify and characterize the data sharings across parallel processors. Section 3 gives the details of our ILP formulation. Section 4 gives an example. Section 5 concludes the paper with a summary.

2 Compiler Analysis for Identifying Shared and Privately-Accessed Data

As mentioned earlier, our ILP solver takes as input the data accessed by each processor and the data shared by processor groups. While there are several ways of obtaining this data (e.g., through simulation or static analysis of the application code), in this work we employ a compiler-based approach. More specifically, the compiler analyzes the application source code and extracts the interprocessor data sharing information. To achieve this, the proposed compiler support employs a polyhedral tool called the Omega Library [10]. Basically, the Omega Library provides several functions that operate on Presburger formulas. Presburger formulas are a class of logical formulas which can be built from affine constraints over integer variables, the logical connectives ($\vee$, $\wedge$, and $\neg$), and the existential and universal quantifiers ($\exists$ and $\forall$). The Omega Library manipulates integer tuple relations and sets, which are described using Presburger formulas. Specifically, the conditions describing a set or tuple can be described by a Presburger formula. Relations and sets can be combined using functions (operators) such as composition, intersection, union, and difference.

In our work, we express the set of elements accessed by processors and the set of elements shared among processors using Presburger formulas. As an example, consider the nested loop depicted in Figure 3. The iteration space of this loop nest (i.e., the set of iteration points that will be executed by the nest) can be expressed using the following Presburger formula:

$$\mathcal{J} = \{(j_1, j_2) \mid L_1 \leq j_1 \leq U_1 \wedge L_2 \leq j_2 \leq U_2\}.$$

However, when the loop nest is parallelized[1], each processor typically executes a subset of the iteration points in the nest. In the loop nest shown above, assuming that the outer loop (j_1) is parallelized across P processors, the pth processor ($0 \leq p < P$) is assigned the iterations captured by the following Presburger set:

$$\mathcal{J}(p) = \{(j_1, j_2) \mid (L_1 + p(U_1 - L_1 + 1)/P \leq j_1 \\ < L_1 + (p+1)(U_1 - L_1 + 1)/P) \wedge (L_2 \leq j_2 \leq U_2)\}.$$

Note that,

$$\mathcal{J} = \bigcup_{p \in \{P\}} \mathcal{J}(p),$$

where $\bigcup$ denotes the set union operator and $\{P\}$ represents the set of processors in the system. Note also that, we assumed, for simplicity, $(U_1 - L_1 + 1)$ is evenly divided by P.

The set of array elements accessed by processor p (based on this parallelization) can be calculated as the union of the set of elements accessed by each reference within the loop nest. In mathematical terms, for our example nest in Figure 3, we have:

$$\mathcal{D}(p, X) = \mathcal{D}(p, X[j_1, j_1 + j_2]) \cup \mathcal{D}(p, X[j_2 - 1, j_1 + 3]) \\ \cup \mathcal{D}(p, X[j_2 + 2, j_2 + j_1]),$$

where

$$\mathcal{D}(p, X[j_1, j_1 + j_2]) = \{(d_1, d_2) \mid \exists(j_1, j_2) \text{ such that} \\ (d_1 = j_1 \wedge d_2 = j_1 + j_2) \wedge (j_1, j_2) \in \mathcal{J}(p)\}$$
$$\mathcal{D}(p, X[j_2 - 1, j_1 + 3]) = \{(d_1, d_2) \mid \exists(j_1, j_2) \text{ such that} \\ (d_1 = j_2 - 1 \wedge d_2 = j_1 + 3) \wedge (j_1, j_2) \in \mathcal{J}(p)\}$$
$$\mathcal{D}(p, X[j_2 + 2, j_2 + j_1]) = \{(d_1, d_2) \mid \exists(j_1, j_2) \text{ such that} \\ (d_1 = j_2 + 2 \wedge d_2 = j_2 + j_1) \wedge (j_1, j_2) \in \mathcal{J}(p)\}.$$

We can also express the array elements shared among a set of processors using the set intersection operator. For example, let $\{P'\}$ be a subset of $\{P\}$, i.e., $\{P'\} \subseteq \{P\}$. For our example, the set of data elements shared by all the processors in $\{P'\}$ can be expressed as:

$$\mathcal{S}(\{P'\}, X) = \bigcap_{p \in \{P'\}} \mathcal{D}(p, X).$$

As will be discussed later in detail, the set of elements shared among processors helps us determine the sizes of the on-chip memory components shared among processors. To determine the size of the on-chip private memories, on the other hand, we need the set of array elements accessed exclusively by processor p. This can be expressed as follows:

$$E(p, X) = \mathcal{D}(p, X) \setminus \bigcup_{[p \in \{P'\}] \wedge [\{p\} \neq \{P'\}]} \mathcal{S}(\{P'\}, X),$$

where $\setminus$ denotes the set subtraction (set difference) operator. In informal terms, what this last expression says that an

---

[1]In this paper, we do not assume a specific loop parallelization strategy. A loop nest can be parallelized either through user-specified annotations or via automatic compiler analysis.

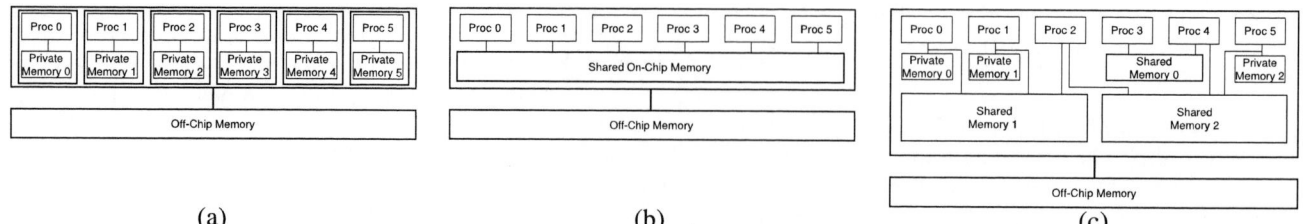

(a)   (b)   (c)

Figure 1. (a) Pure private memory. (b) Pure shared memory. (c) Hybrid memory. Although not shown for clarity, in (a) and (c) there is an on-chip interconnect to allow a processor access a data item which resides in a memory that is not directly accessible by that processor.

array element belongs to the set $E(p, X)$ if and only if this item is not accessed by any processor other than p.

An important point to note here is that, while the analysis above is given in terms of a single nest accessing a single array, it is straightforward to extend it to multiple nest and multiple array cases. For example, if a loop nest accesses q different arrays, namely, $X_0, X_1, ..., X_{q-1}$, the total set of elements accessed by processor p can be computed as: $\mathcal{D}(p) = \mathcal{D}(p, X_0) \cup \mathcal{D}(p, X_1) \cup ... \cup \mathcal{D}(p, X_{q-1})$. The other sets can be computed in a similar fashion. A similar extension can be written for the multiple nest case as well.

Capturing only the array elements shared or privately accessed by the processors is sufficient to perform memory partitioning, but not sufficient for data allocation across the memory components. This is because it may be necessary in many cases to store a data item in a memory location that belongs to a component that is not shared by a processor that accesses that element. In such a case, we consume energy in both memory access and extra interconnect access if the processor in question wants to access that item. To decide which data needs to be stored remotely and which data locally, we need a mechanism to *rank* the different data items based on their importances (criticalities). For example, from the viewpoint of processor p, not all the data items in $\mathcal{S}(\{p, p'\}, X)$, the set of shared elements between processors p and p', have the same importance; some of them can be more important than the others. In this work, we use the *number of accesses* as the metric using which we can rank the different data items. While a polyhedral analysis, similar to the one conducted above, can be used for capturing the number of accesses to each individual data item, the associated overheads can be too much to tolerate. Instead, we calculate the number of accesses at a set (of data items) granularity. As an example, for the loop nest discussed above, we compute the number of accesses to the elements in sets $E(p, X)$ and $\mathcal{S}(\{P'\}, X)$, for all $p \in \{P\}$ and for all $\{P'\} \subseteq \{P\}$. Note that, we can use polyhedral arithmetic to achieve this. For example, the loop iterations that access the elements in $\mathcal{S}(\{P'\}, X)$ can be captured as:

$$
\begin{aligned}
\mathcal{J}(\{P'\}, X) = \{&(j_1, j_2) \mid \exists (d_1, d_2) \text{ such that} \\
&(j_1, j_2) \in \mathcal{J}(p) \wedge ((d_1 = j_1 \wedge d_2 = j_1 + j_2) \vee \\
&(d_1 = j_2 - 1 \wedge d_2 = j_1 + 3) \vee (d_1 = j_2 + 2 \wedge \\
&d_2 = j_2 + j_1)) \wedge \\
&(\forall p \in \{P'\} : (d_1, d_2) \in \mathcal{D}(p, X))\}.
\end{aligned}
$$

One can similarly compute a set $\mathcal{J}(p, X)$, the set of loop iterations that access the elements in $E(p, X)$. An impor-

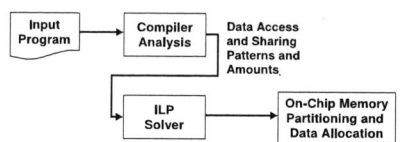

Figure 2. High-level view of our approach.

$$
\begin{aligned}
&for(j_1 = L_1; j_1 \leq U_1; j_1 + +) \\
&\quad for(j_2 = L_2; j_2 \leq U_2; j_2 + +) \\
&\quad\quad \cdots = X[j_1, j_1 + j_2] + X[j_2 - 1, j_1 + 3] + X[j_2 + 2, j_2 + j_1];
\end{aligned}
$$

Figure 3. An example loop nest.

tant issue that needs to be clarified at this point is how one can enumerate and count the elements in the Presburger sets defined above, i.e., in sets such as $E(p, X)$, $\mathcal{S}(\{P'\}, X)$, $\mathcal{J}(p, X)$, and $\mathcal{J}(\{P'\}, X)$. This is important since our ILP solver needs the number of elements in these sets, not the sets themselves. We address this problem using the "codegen" utility provided by the Omega Library. Codegen generates code to traverse the points in a given Presburger set in lexicographical order. After generating this code, what we do is to insert a counter variable in the code that keeps track of the number of points in the code, and execute the resulting code at compile time. The final value of the counter variable at the end of this execution gives us the number we want to determine (i.e., the number of elements in the set). A similar approach has been employed for an entirely different problem in [7]. It is to be noted that, the elements in the sets discussed above can be counted using other existing methods as well such as [5] and [1], among the others. In the rest of this paper, we use $|E(p)|$ and $|\mathcal{S}(\{P'\})|$ to denote, respectively, the set of privately accessed elements by processor p and the set elements shared by processors in $\{P'\}$. Similarly, $|\mathcal{J}(p, X)|$ and $|\mathcal{J}(\{P'\}, X)|$ give the number of elements in sets $\mathcal{J}(p, X)$ and $\mathcal{J}(\{P'\}, X)$, respectively.

3 ILP Formulation

In this paper, 0-1 ILP is used to determine the sizes of the memory components and how they are shared across multiple processors in the system. Table 1 gives the constant terms used in our ILP formulation. Note that, the sizes given in this table and used in the following discussion are

0-7695-2533-4/06 $20.00 © 2006 IEEE

Constant	Definition
P	Number of processors
L	Total memory size in terms of slices
D	Maximum possible number of data sets ($D = 2^P - 1$)
E_{Comm}	Communication energy for a non-local data access
$E_{port,s}$	Energy consumed by accessing a memory component with $port$ number of ports and of size s
$F_{p,d}$	Frequency of accesses to data set d by processor p
S_d	Size of data set d
$Local_{m,p}$	Indicates whether memory component m is directly accessible by p
$ports(m)$	Number of ports that memory component m has

Table 1. The constant terms used in our ILP formulation. These are either architecture specific or program specific.

in terms of units (*slices*), which represents the minimum amount of space that can be allocated to a memory component (i.e., the unit of allocation).

Our objective is to partition the available on-chip memory space into memory components that can be shared among different processors or privately accessed by a processor and perform data allocation across these memory components to minimize the overall energy consumption. We determine the existence and sizes of memory components based on the data access frequency of the processors and the sizes of the data sets using 0-1 variables. For each possible memory component size, we define 0-1 variables. The sizes of memory components are restricted by the total data size. There are possibly $2^P - 1$ memory components on the chip multiprocessor, where P is the number of processors. For example, if there are two processors (p_0 and p_1), possible memory components can be a private memory for p_0, a private memory for p_1, and a shared memory that can be accessed by p_0 and p_1. Therefore, the available on-chip memory space can potentially be partitioned into these three ($2^2 - 1$) memory components in an energy minimizing fashion. Note that, it is possible, in the final design (memory partitioning), to have one of these memory components alone, or only two of them, or all of them together. That is, the final on-chip memory space partitioning determined by our approach can contain any subset of these three components. Similarly, there are possibly $D = 2^P - 1$ data sets. The size of a data set, denoted by S_d in Table 1, is determined by the number of elements in the data set. For example, if we have three processors (denoted p_0, p_1, and p_2), we have: $S_1 = |E(p_0)|$, $S_2 = |E(p_1)|$, $S_3 = |E(p_2)|$, $S_4 = |\mathcal{S}(p_0, p_1)|$, $S_5 = |\mathcal{S}(p_0, p_2)|$, $S_6 = |\mathcal{S}(p_1, p_2)|$, and $S_7 = |\mathcal{S}(p_0, p_1, p_2)|$. Similarly, the access frequency of a processor p to a data set d, denoted by $F_{p,d}$ in Table 1, is obtained from $|\mathcal{J}(p, X)|$ and $|\mathcal{J}(\{P'\}, X)|$ defined earlier.

We can use 0-1 variables to specify the size (*size*) of a memory component. Specifically, we have:

- $size_{m,s}$: indicates whether memory component m is of size s.

We use a variable for each one of the possible sizes. If this 0-1 variable is 1, this indicates that the corresponding memory component size is s. If this size is 0, then we conclude that this memory component does not exist.

We use another 0-1 variable for assigning each data set to memory component(s):

- map_{d,d_s,m,m_s} : indicates whether data set d of size d_s is located in memory component m of size m_s.

Note that, we allow a data set to be divided among the different memory components. If we were to restrict a data set to be located only in one memory component, removing the second parameter in the subscript above (d_s), would be sufficient.

If a data set (d_s) does not reside in one of the memory components that is directly accessible by the processor, then accessing that data set would incur an extra on-chip communication energy due to accessing the interconnect. To capture the communication energy, we use $comm_{p,d,d_s,m,m_s}$:

- $comm_{p,d,d_s,m,m_s}$: indicates whether accessing the data set d of size d_s located in memory component m of size m_s by processor p would require communication cost.

Each processor's (p) energy consumption due to accesses to data sets can be identified using a variable A_p which is defined as follows:

- A_p : the energy consumed by processor p due to data accesses.

Also, the energy consumption due to interconnect accesses is captured by the $C_{p,m}$ variable.

- $C_{p,m}$: the energy consumed by processor p due to communication to access data in memory component m.

It should be noted that, access energy A_p and communication energy $C_{p,m}$ are not 0-1 variables. They are simply used to calculate the total energy consumption. After defining the variables in our ILP formulation, now we explain the necessary constraints.

The total memory space (L) should be equal to the sum of the sizes of the individual memory components. This can be captured as:

$$L = \sum_{i=1}^{2^P - 1} \sum_{j=0}^{L} size_{i,j} \times j \qquad (1)$$

In this expression, index variable i iterates over the $2^P - 1$ memory components. On the other hand, index variable j iterates over the possible sizes from 0 up to L.

A memory component can have one and only one size. We capture this constraint as follows:

$$\sum_{j=0}^{L} size_{m,j} = 1, \quad \forall m. \qquad (2)$$

A data set (d) can be divided among memory components:

$$\sum_{j=0}^{S_d} \sum_{k=1}^{2^P - 1} \sum_{l=0}^{L} map_{d,j,k,l} \times j = S_d, \quad j \leq l, \quad \forall d. \qquad (3)$$

In the above formulation, index variable j iterates over the possible data set sizes from 0 up to S_d (the data set size). On

0-7695-2533-4/06 $20.00 © 2006 IEEE

the other hand, k iterates over the memory components, and similarly, l is used to identify the size of the corresponding memory component. The sum of these data set portions should be equal to the total data set size (S_d).

A data set (d) can at most have one size allocated within a memory component (m). Although this constraint does not affect the result of the ILP, it prevents having two separate space assignments for the same data set within a memory component. For example, instead of two memory spaces with sizes s_1 and s_2, with this constraint the ILP solution will result in a single memory space of $s_1 + s_2$ within the same memory component.

$$\sum_{j=0}^{S_d} map_{d,j,m,l} \leq 1, \quad j \leq l, \quad \forall d, m, m_s. \tag{4}$$

If a memory component is used by a data set, this memory component has to exist, which can be captured by:

$$size_{m,m_s} \geq map_{d,d_s,m,m_s}, \quad d_s \leq m_s, \quad \forall d, d_s, m, m_s. \tag{5}$$

The total data size stored in a memory component must be less than or equal to the size of the memory component itself:

$$\sum_{i=1}^{D} \sum_{j=0}^{S_d} map_{i,j,m,m_s} \times j \leq m_s, \quad \forall m, m_s. \tag{6}$$

In this formulation, i iterates over the data sets, whereas, j iterates over the possible data set sizes.

A communication cost will be incurred, if the data is mapped to a memory component (m) and the memory component is not local to the processor (p) accessing it (i.e., it is not one of the components to which p has a direct access). As explained earlier, we use $comm_{p,d,d_s,m,m_s}$ to denote a 0-1 variable that captures the existence of communication. We have:

$$comm_{p,d,d_s,m,m_s} \geq map_{d,d_s,m,m_s} - Local_{m,p},$$
$$d_s \leq m_s, \forall p, d, d_s, m, m_s.$$

In the above expression, $Local_{m,p}$ is a parameter given to the ILP solver based on the memory component in question. For example, if there are two processors (p_0 and p_1), the possible memory components can be m_0 (a private memory for p_0), m_1 (a private memory for p_1), and m_2 (a shared memory between p_0 and p_1). For m_0, this parameter will be given as 0 for p_1. Similarly, for m_1, it will be set to 0 for p_0. On the other hand, for m_2, it will be set to 1 for both processors.

Having specified the necessary constraints in our ILP formulation, we next give our objective function. In our execution model, there are two components of the total memory energy consumption:

• *access:* the energy consumed when a memory component is accessed.

• *communication:* the extra interconnect energy consumed when a remote memory component is accessed.

Each processor's memory access cost, A_p can be formulated as follows:

$$A_p = \sum_{i=1}^{D} \sum_{j=0}^{S_d} \sum_{k=1}^{2^P-1} \sum_{l=0}^{L} map_{i,j,k,l} \times F_{p,i} \times j \times E_{ports(k),l} \quad j \leq l, \quad \forall p. \tag{7}$$

Data Set	Size	Proc. 0	Proc. 1	Proc. 2	Proc. 3	Location
D_0	2K	10%	-	-	-	M_0
D_1	1K	-	5%	-	-	M_1
$D_{0,1}$	4K	5%	15%	-	-	$M_{0,1}$
$D_{2,3}$	3K	-	-	5%	10%	$M_{2,3}$
$D_{0,1,2,3}$	5K	5%	10%	10%	25%	$M_{1,2,3}$

(a) $\Rightarrow$ (b)

Table 2. (a) An example data set with access frequencies. (b) The resulting locations (placement) for the data sets.

Each processor's communication cost due to accessing the interconnect can be formulated using $C_{p,d}$:

$$C_{p,d} = \sum_{j=0}^{S_d} \sum_{k=1}^{2^P-1} \sum_{l=0}^{L} comm_{p,d,j,k,l} \times F_{p,d} \times j \times E_{Comm}, \quad j \leq l, \quad \forall p, d. \tag{8}$$

In the last two expressions, indices i, j, k, and l iterate over the data sets, data set sizes, memory components, and memory component sizes, respectively. $F_{p,i}$ denotes the frequency of the accesses to data set i by processor p. $E_{ports(k),l}$ is the unit access energy consumption for a memory component of a size l with $ports(k)$ number of ports. $ports(k)$, the number of ports required for the memory component, is obtained based on the number of processors accessing it (k). The unit communication energy for a remote access is E_{Comm}. Using these two cost expressions (A_p and $C_{p,d}$), we can express the total energy consumption due to memory accesses (E) as follows:

$$E = \sum_{i=1}^{P} A_i + \sum_{i=1}^{P} \sum_{j=1}^{D} C_{i,j}. \tag{9}$$

Based on this formulation, our 0-1 ILP problem can formally be defined as one of "minimizing E under constraints (1) through (8)."

Let us explain the operation of our software-managed hybrid memory architecture. There are three scenarios for the outcome of a memory access (request) in our hybrid on-chip memory architecture, depending on where the requested item is located:

• *Local Hit:* When the processor finds the data in one of the memory components it has direct access to.

• *Remote Hit:* In this case, the lookup amongst its assigned component(s) fails, but the data is found in another (on-chip) component that is not directly connected to the processor that issued the memory request.

• *On-Chip Miss:* In this case, the data is not in any of the on-chip components, and requires an off-chip access. The access cost in this case will involve the cost of the off-chip access.

4 Example

An example data access frequency and data set size information for a case with 4 processors is given in Table 2(a). We assume, for the sake of explanation, that the data sets can exactly fit into the available on-chip memory space (15K). The processors that share a data set are indicated in

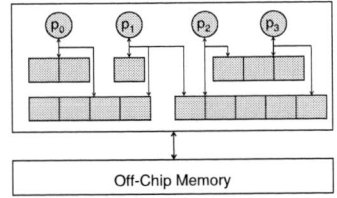

Figure 4. The hybrid memory.

the subscript of that data set. For example, D_0 is a data set privately accessed by processor p_0. Similarly, $D_{2,3}$ denotes a data set shared by processors p_2 and p_3. The second column of Table 2 gives the size of each data set. The next four columns indicate the access frequencies (of processors). In this example, we assume that if a memory component is private to a processor, it has a single port. If the memory component is shared by two or three processors, it has two ports. Finally, if it is shared by more than three processors, it has three ports.

In Figure 4, the resulting hybrid on-chip memory partitioning returned by our approach is shown. Processors p_0 and p_1 have access to both private and shared memory components. In comparison, the other processors have access to shared components only. The location (memory component) of each data set is shown in Table 2(b). As it can be seen, except for the data set $D_{0,1,2,3}$, all of the data sets are in the memory components in accordance with their data set sharing information. In other words, every processor is directly connected to the memory from which it needs data. The only exception is that the memory component that holds data set $D_{0,1,2,3}$ is accessed by only three processors instead of all four. The main reason for this is that, allowing one more processor access to that component would increase the overall energy consumption in this example (we do not give here explicitly the energy values used in the example). Overall, except for accesses by processor p_0 to the memory component that hold data set $D_{0,1,2,3}$, all the memory accesses are local. This example shows how our approach comes up with a hybrid on-chip memory architecture.

5 Concluding Remarks

Chip multiprocessors are suitable for executing data-intensive embedded applications with source-level parallelism. Ensuring that most of data accesses are satisfied from on-chip memories is a critical problem for chip multiprocessors, as cost of an off-chip access is very high. Particularly, multiple cores that need to access the off-chip memory system may contend with each other for the same buses/pins to get there. While it is possible to structure on-chip memory space as shared memory or private memory, each of these has its own drawbacks. An important observation here is that, to reach minimum energy consumption, both memory space partitioning and data allocation need to be optimized in a coordinated manner. In an attempt to achieve lower power consumption than the conventional

on-chip memory architectures, this paper proposes an ILP-based strategy that comes up with an application-specific hybrid memory architecture that has both shared and private components. Our strategy also determines the optimal data allocation (placement) for the resulting hybrid memory.

References

[1] R. Bagnara. The Parma Polyhedra Library. *Seminar given in Departement de Mathematiques et Informatique in St Denis de La Reunion,* France, Indian Ocean, May 2002.

[2] L. A. Barroso, K. Gharachorloo, R. McNamara, A. Nowatzyk, S. Qadeer, B. Sano, S. Smith, R. Stets, and B. Verghese. Piranha: a scalable architecture based on single chip multiprocessing. In *Proc. ISCA,* 2000.

[3] CACTI 3.0. http://research.compaq.com/wrl/people/jouppi/CACTI.html

[4] Berkeley Predictive Technology Model, U.C. Berkeley, http://www-devices.eecs.berkeley.edu/ptm/

[5] P. Clauss. Counting solutions to linear and nonlinear constraints through Ehrhart polynomials: applications to analyze and transform scientific programs. In *Proc. ICS,* May 1996.

[6] F. Gharsalli, S. Meftali, F. Rousseau, and A. A. Jerraya. Automatic Generation of Embedded Memory Wrapper for Multiprocessor SoC. In *Proc. DAC,* New Orleans, Louisiana, 1999.

[7] S. Ghosh, M. Martonosi, and S. Malik. Precise miss analysis for program transformations with caches of arbitrary associativity. In *Proc. ASPLOS,* San Jose, CA, October 1998.

[8] U. Ko, P. Balsara, and A. Nanda. Power and Performance Optimization for On-Chip Multi Level Cache Hierarchies in Microprocessors. *IEEE Transactions on VLSI Systems,* pp. 299–308, June 1998.

[9] M. Kandemir, O. Ozturk, and M. Karakoy. Dynamic on-chip memory management for chip multiprocessors, by In *Proc. CASES,* Washington D.C., September 2004.

[10] W. Kelly and W. Pugh. Finding legal reordering transformations using mappings. In *Proc. LCPC Workshop,* pp. 107–124, 1994.

[11] V. Krishnan and J. Torrellas. A Chip Multiprocessor Architecture with Speculative Multi-threading. *IEEE Transactions on Computers, Special Issue on Multi-threaded Architecture,* September 1999.

[12] C. Liu, A. Sivasubramaniam, and M. Kandemir. Organizing the Last Line of Defense Before Hitting the Memory Wall for CMPs. In *Proc. the International Symposium on High-Performance Computer Architecture,* Madrid, Spain, February 2004.

[13] MAJC-5200. http://www.sun.com/microelectronics/MAJC/5200wp.html

[14] MP98: A Mobile Processor. http://www.labs.nec.co.jp/MP98/top-e.htm.

[15] W. Pugh and D. Wonnacott. Constraint-based array dependence analysis. *ACM TOPLAS,* 20(3):635-678, May 1998.

[16] S. Meftali, F. Gharsalli, F. Rousseau, and A. A. Jerraya. An optimal memory allocation for application-specific multiprocessor system-on-chip. In *Proc. ISSS,* Montreal, Canada, 2001.

[17] K. Olukotun, B. A. Nayfeh, L. Hammond, K. Wilson, and K. Chang. The Case for a Single Chip Multiprocessor. In *Proc. ASPLOS,* 1996.

[18] S. Richardson. MPOC: a chip multiprocessor for embedded systems. In *HP Laboratories Technical Report HPL-2002-186,* Palo Alto, CA, July 2002.

[19] S. F. Smith. Performance of a GALS single-chip multiprocessor. In *Proc. PDPTA,* 2004.

[20] G. E. Suh, L. Rudolph, and S. Devadas. Dynamic partitioning of shared cache memory. *Journal of Supercomputing,* 2002.

0-7695-2533-4/06 $20.00 © 2006 IEEE

High Performance Circuits

High Speed Low Swing Dynamic Circuits with Multiple Supply and Threshold Voltages

Zhiyu Liu and Volkan Kursun

Department of Electrical and Computer Engineering
University of Wisconsin – Madison
Madison, Wisconsin 53706 - 1691

Abstract — **A new low voltage swing circuit technique based on a dual threshold voltage CMOS technology is presented in this paper for simultaneously reducing active and standby mode power consumption and enhancing evaluation speed and noise immunity in domino logic circuits. The proposed circuit technique modifies both the upper and lower boundaries of the voltage swing at the dynamic node. Meanwhile, full voltage swing signals are maintained at inputs and outputs for robust and high speed operation. Power supply, ground, and threshold voltages are simultaneously optimized to minimize the power-delay product (PDP). The proposed technique reduces the PDP by up to 51.9% as compared to standard full-swing circuits in a 45nm CMOS technology. The active mode power consumption is reduced by up to 40.4% due to lower switching power required to charge/discharge the dynamic node. Furthermore, the evaluation speed and noise immunity are enhanced by up to 19.4% and 39.1%, respectively, as compared to standard full-swing circuits. The proposed low swing technique also reduces the idle mode leakage power consumption by up to 84.2% in the high fan-in domino gates.**

Index Terms — **Domino Logic, Low Voltage Swing, Dual Threshold Voltage, Gate Oxide Leakage, Subthreshold Leakage.**

I. INTRODUCTION

Domino logic circuits have been extensively applied in modern high performance microprocessors because of the superior speed and area characteristics of dynamic circuits as compared to static CMOS circuits [1]-[3]. However, domino gates typically consume higher dynamic switching and leakage power and display weaker noise immunity as compared to static CMOS gates. The low power and error free operation of domino logic circuits is a major challenge in the current CMOS technologies [4]-[5].

Several low swing techniques for reducing power consumption in domino logic circuits are presented in [6]-[8]. These circuit techniques reduce the dynamic switching power consumption by lowering the voltage swing at the output node of a domino logic gate. However, the evaluation speed is significantly degraded due to the lower gate overdrive of the pull-down network transistors in the fan-out gates driven by these low-swing circuits.

Another circuit technique is presented in [9] to reduce the voltage swing at the dynamic node of a domino gate, as shown in Fig. 1. The dynamic node voltage swing is from $V_{gnd} + |V_{tp}|$ to $V_{DD} - V_{tn}$ (V_{tn} and V_{tp} are the threshold voltages of N1 and P1 in Fig. 1, respectively). This technique can reduce the

energy required to charge/discharge the dynamic node of a domino gate. However, PMOS and NMOS transistors within the output inverter are simultaneously turned on, producing a significant short-circuit current during both the active and idle modes of operation with this technique. Short-circuit current produced by output inverter diminishes the active mode power savings and increases the idle mode power consumption. Furthermore, propagation delay is increased due to significantly degraded gate overdrive of both PMOS and NMOS transistors in the output inverter.

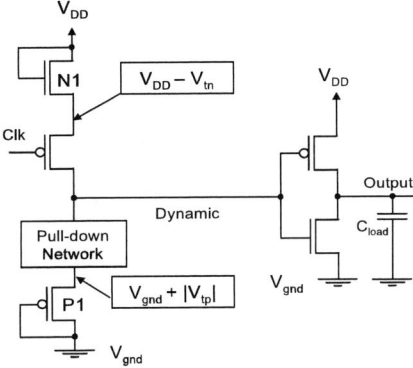

Fig. 1. Dynamic node low voltage swing domino logic circuit [9].

A novel dynamic node low voltage swing domino logic circuit technique based on dual power supply and ground voltages and a dual threshold voltage (dual-V_t) CMOS technology is proposed in this paper to reduce active mode power consumption. The circuit technique modifies both the upper and lower boundaries of the voltage swing at the dynamic node by utilizing dual power supply and ground voltages. Switching power consumed to charge/discharge the dynamic node is, therefore, quadratically reduced with the voltage swing. Meanwhile, full voltage swing signals are maintained at the inputs and outputs for high speed and robust operation. In order to suppress the short-circuit current produced by the output inverter, high threshold voltage (high-V_t) transistors are employed in the output inverter.

A quantitative study is presented in this paper to find the optimum dynamic node voltage swing range and optimum high-V_t for reducing PDP and enhancing noise immunity as compared to standard full-swing circuits. It is shown that employing a dual-V_t CMOS technology significantly enhances the effectiveness of the bidirectional dynamic node low swing domino circuit technique. The results indicate that an asymmetric voltage swing range of 0.1V-to-0.6V along with an optimized high-V_t of 0.28V achieve the lowest PDP in

---

* This research was supported in part by a grant from the Wisconsin Alumni Research Foundation (WARF).

0-7695-2533-4/06 $20.00 © 2006 IEEE

various types of domino logic circuits in a 45nm CMOS technology. The circuit technique reduces the PDP by up to 51.9% and increases the noise immunity by up to 39.1% as compared to standard full-swing circuits. The active mode power consumption is reduced by up to 40.4% and the evaluation speed is enhanced by up to 19.4%.

Effectiveness of the circuit technique for reducing standby mode power consumption is also evaluated. Both subthreshold and gate dielectric leakage currents produced by the pull-down network transistors in an idle domino gate are suppressed with the proposed technique. The low swing dual-V_t circuit technique reduces the idle mode leakage power consumption by up to 84.2% as compared to standard full-swing circuits.

The paper is organized as follows. The dynamic node low voltage swing dual-V_t domino logic circuit technique is described in Section II. Active mode power, evaluation delay, and PDP characteristics of various low swing domino circuits are evaluated in section III. Noise immunity characteristics are discussed in section IV. Leakage power consumption is presented in section V. Some conclusions are offered in Section VI.

II. DYNAMIC NODE LOW VOLTAGE SWING DUAL THRESHOLD VOLTAGE DOMINO LOGIC CIRCUITS

The proposed dynamic node low swing dual-V_t domino circuit technique is presented in this section. Operation of the proposed circuit technique is explained in Section A. Quadratic reduction of dynamic node switching power with the proposed technique is described in Section B. Leakage current characteristics of the low swing circuits are discussed in Section C.

A. Dynamic Node Low Voltage Swing Domino Logic

A low swing circuit based on dual power supply and ground voltages is shown in Fig. 2. The output inverter is connected to a higher power supply and a lower ground voltage (V_{DD} and V_{gnd}) as in standard domino logic circuits. Alternatively, the pull-down network, the keeper, and the precharge transistor are connected to a lower power supply and a higher ground voltage ($V_{DDL} < V_{DD}$ and $V_{gndH} > V_{gnd}$), as illustrated in Fig. 2.

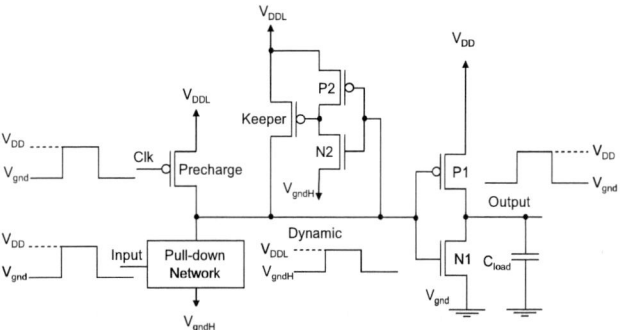

Fig. 2. The dynamic node low voltage swing domino circuit technique with dual power supplies and ground voltages. $V_{DDL} < V_{DD}$. $V_{gndH} > V_{gnd}$.

Advantage of employing a dual-V_t CMOS technology for reducing the short-circuit current overhead at the output inverter is also explored in this paper. Schematics of two

flavors of the proposed dual-V_t low swing circuits are shown in Fig. 3. In Fig. 3a, both the NMOS and PMOS transistors in the output inverter have a high-V_t. Alternatively, in Fig. 3b, only the NMOS transistor in the output inverter has a high-V_t.

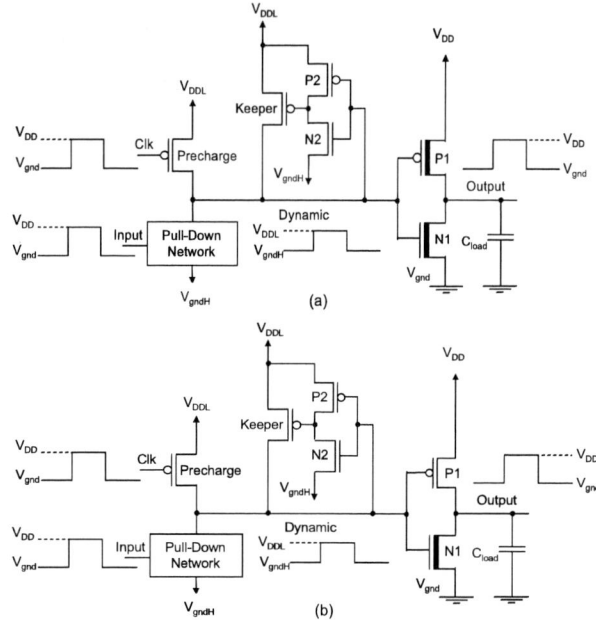

Fig. 3. The dynamic node low voltage swing dual-V_t domino circuit technique with dual power supplies and ground voltages. (a) Both the NMOS and PMOS transistors in the output inverter have high-V_t. (b) Only the NMOS transistor in the output inverter has high-V_t. $V_{DDL} < V_{DD}$. $V_{gndH} > V_{gnd}$. High-V_t transistors are represented by a thick line in the channel region.

The operation of the proposed circuit technique behaves in the following manner. In the precharge phase, clock is low. Dynamic node is charged to V_{DDL} by the precharge transistor. Output node is discharged to V_{gnd} by N1. P1 is weakly inverted. Weak inversion current produced by P1 depends on the upper boundary of voltage swing (V_{DDL}) at the dynamic node and the threshold voltage of P1 (V_{tp1}). In this paper, both V_{DDL} and V_{tp1} (high-V_t) are considered as optimization parameters to minimize the power consumption.

Evaluation phase begins when the clock transitions high. Dynamic node is discharged to V_{gndH} through the pull-down network provided that the inputs are high. Output node is charged to V_{DD} by P1. N1 is weakly inverted. Weak inversion current conducted by N1 depends on the lower boundary of voltage swing (V_{gndH}) at the dynamic node and the threshold voltage of N1 (V_{tn1}). Similar to V_{DDL} and V_{tp1}, V_{gndH} and V_{tn1} (high-V_t) are also optimized to minimize the power consumption.

The voltage swing at the dynamic node is from V_{gndH} to V_{DDL} while the voltage at the output, clock, and input nodes is maintained full swing between V_{gnd} and V_{DD} for high speed operation with the proposed circuits. With the circuit technique shown in Fig. 3a, short-circuit current produced by the output inverter is suppressed during both the precharge and evaluation phases of operation since the NMOS and PMOS transistors in the output inverter have high-V_t. However, evaluation speed is also degraded due to the weaker pull-up strength of the high-V_t PMOS transistor. With the circuit technique shown in Fig. 3b, only the NMOS transistor in the output inverter has high-V_t. Short-circuit current is, therefore, reduced only in the evaluation phase if the inputs are high. Evaluation speed of the circuit in Fig. 3b is higher as compared to the circuit technique shown in Fig. 3a.

Active mode power consumption of the proposed low swing dual-V_t circuit technique is reduced since the power required to charge/discharge the dynamic node is lower as compared to standard full-swing domino logic circuits. Noise immunity is enhanced since the gate-to-source voltage (V_{gs}) of the pull-down network transistors is negative (a higher noise signal is needed to activate the transistors in the pull-down network). Evaluation speed can also be enhanced with an asymmetric low voltage swing range which reduces the charge stored at the dynamic node, thereby making the dynamic node easier to be discharged.

B. Quadratic Reduction of Dynamic Node Switching Power

When dual power supply and ground voltages are utilized, the dynamic node switching power of a low swing domino circuit (Figs. 2 and 3) is

$$P = \alpha C_L \left(V_{DDL} - V_{gndH} \right)^2 f = \alpha C_L V_{swing}^2 f , \quad (1)$$

where P, α, and f are the switching power consumed for charging/discharging the dynamic node, switching activity factor, and clock frequency, respectively. C_L is the capacitive load at the dynamic node. V_{swing} is the dynamic node voltage swing ($V_{swing} = V_{DDL} - V_{gndH}$). The switching power consumed for charging/discharging the dynamic node has a quadratic dependence on the dynamic node voltage swing with the proposed technique.

Alternatively, the circuit technique presented in [9] operates with the standard power supply and ground voltages as shown in Fig. 1. The topologies of the pull-up and pull-down networks are modified to lower the dynamic node voltage swing with the technique presented in [9]. The switching power consumed for charging/discharging the dynamic node of the low swing circuit in Fig. 1 is

$$P = \alpha C_L V_{DD} V_{swing} f . \quad (2)$$

The dynamic node switching power has a linear dependence on the dynamic node voltage swing with the technique in [9]. The power savings offered with the technique presented in this paper is, therefore, significantly higher as compared to the technique presented in [9], even if V_{DDL} and V_{gndH} are selected to provide the same dynamic node voltage swing as compared to the technique presented in [9].

The technique in Fig. 1 places a PMOS transistor in series with the pull-down network to modify the lower boundary of dynamic node voltage swing. Similarly, an NMOS transistor is placed in series with the precharge transistor to modify the upper boundary of dynamic node voltage swing. In order to maintain similar precharge and evaluation delay as compared to the circuit in Fig. 2, the size of transistors in the pull-up and pull-down networks must be increased with the technique in Fig. 1. The low swing technique based on dual power supply and ground voltages, therefore, also reduces the parasitic capacitance at the dynamic and internal nodes as compared to the technique presented in [9].

The third important advantage of the proposed technique is the ability to optimize the upper and lower boundaries of the voltage swing at the dynamic node. V_{DDL} and V_{gndH} can be optimized to minimize the total power consumption by trading the dynamic node switching power reduction with the output inverter weak inversion current overhead intrinsic to the dynamic node low voltage swing technique. Alternatively, the voltage swing of dynamic node is fixed with the technique in Fig. 1. Trading the switching power reduction at the low swing dynamic node with the short-circuit current overhead at the output inverter for minimizing the total power consumption is, therefore, not possible with the technique presented in [9].

C. Leakage Power Reduction

The proposed circuit technique can also lower power consumption during the idle mode. It is assumed that clock is gated high in an idle domino logic circuit. When clock is gated high, precharge transistor is turned off, ensuring that no short-circuit current conduction path exists between V_{DDL} and V_{gndH} (provided that the inputs are high in footless domino logic circuits). Dynamic node is cyclically charged to V_{DDL} every clock period. Therefore, provided that the inputs are low after the clock is gated, the dynamic node is maintained at V_{DDL} in the idle mode. Alternatively, provided that the inputs are high after the clock is gated, the dynamic node is discharged to V_{gndH} and the output transitions to V_{DD}.

In the idle mode, if the inputs are maintained low (dynamic node voltage is V_{DDL}), V_{gs} is negative and drain-to-source voltage (V_{ds}) is smaller in the pull-down network transistors. Since the pull-down transistors are strongly cut-off, subthreshold leakage current is reduced as compared to the standard full-swing circuits. Furthermore, gate oxide leakage currents produced by N1, N2, and the pull-down network transistors are decreased due to smaller voltage difference across the gate dielectric layer as compared to the full-swing domino circuits. However, P1 is weakly inverted. Subthreshold leakage current produced by P1 depends on the upper boundary of voltage swing (V_{DDL}) at the dynamic node and the threshold voltage of P1 (V_{tp1}).

Alternatively, if the inputs are maintained high (dynamic node voltage is V_{gndH}), gate oxide leakage current produced by the pull-down transistors is reduced due to smaller voltage difference across the gate dielectric layer. Furthermore, subthreshold leakage current produced by the precharge transistor is also reduced since V_{gs} is positive and V_{ds} is lower as compared to a standard full-swing domino gate. However, N1 operates in the weak inversion region. Subthreshold leakage current conducted by N1 depends on the lower boundary of voltage swing (V_{gndH}) at the dynamic node and the threshold voltage of N1 (V_{tn1}).

III. ACTIVE MODE DELAY AND POWER

Dynamic power consumption, evaluation delay, PDP, noise immunity, and leakage power consumption change with the voltage swing, threshold voltages, fan-in, and circuit type. A quantitative study is presented in this paper to identify the optimum dynamic node voltage swing and threshold voltages with lower PDP and higher noise immunity as compared to standard full-swing circuits.

BSIM4 device models are used for an accurate estimation of the performance of various circuits in a 45nm CMOS technology [11] ($V_{tn} = |V_{tp}| = 0.22$V, $V_{tn-high} = |V_{tp-high}| = 0.28$ or 0.35V (high-V_t is an optimization parameter), $V_{DD} = 0.8$V, clock frequency = 2GHz with a 50% duty cycle, and temperature = 110 °C). Following circuits are simulated: 2-input and 4-input domino AND gates (AND2 and AND4), 2-input, 4-input, and 8-input domino OR gates (OR2, OR4, and OR8, respectively), and 16-bit domino multiplexer (MUX16). All of the circuits (except MUX16) are composed of three stages. Each gate drives a fan-out of four. The circuits are designed with the following four techniques: single threshold voltage standard full-swing domino (single-V_t, standard full swing), single-V_t low swing domino circuit technique shown in Fig. 2 (single-V_t, low swing), the circuit technique shown in Fig. 3a (dual-V_t-N&P), and the circuit technique shown in Fig. 3b (dual-V_t-N). Each test circuit is sized same with the four circuit techniques. To minimize power consumption and PDP, a complete set of possible dynamic node voltage swings are tested with the minimum voltage swing difference of 50 mV.

Reducing the dynamic node voltage swing beyond a certain value can significantly degrade the output node voltage swing. Voltage swing of output signals must be maintained close to the full V_{DD} - V_{gnd} rail for high speed and robust operation of fan-out gates. In this paper, the minimum acceptable output voltage swing is assumed to be 700mV. The minimum acceptable dynamic node voltage swing for a voltage swing higher than 700 mV at the output node with different low swing circuit techniques is listed in Table I. The minimum acceptable dynamic node voltage swing does not necessarily provide the lowest active mode power consumption due to the increased short-circuit current in the output inverter as the dynamic node voltage swing is reduced. Dynamic node voltage swings with the lowest active mode power consumption and the lowest PDP are also listed in Table I.

TABLE I

MINIMUM ACCEPTABLE, MINIMUM POWER, AND MINIMUM PDP DYNAMIC NODE VOLTAGE SWINGS WITH THE LOW SWING CIRCUIT TECHNIQUES

Circuit Type	Minimum Acceptable Voltage Swing (V)	Voltage Swing (V) for Lowest Power	Voltage Swing (V) for Lowest PDP
Low Swing Single-V_t	0.25 – 0.55	0.15 – 0.65	0.1 – 0.55
Dual-V_t-N&P, High-V_t = 0.35 V	0.2 – 0.55	0.2 – 0.55	0.2 – 0.55
Dual-V_t-N&P, High-V_t = 0.28 V	0.25 – 0.55	0.25 – 0.6	0.1 – 0.6
Dual-V_t-N, High-V_t = 0.35 V	0.35 – 0.65	0.35 – 0.7	0.15 – 0.65
Dual-V_t-N, High-V_t = 0.28 V	0.3 – 0.6	0.25 – 0.65	0.1 – 0.6

The active mode power consumption of circuits is compared in Fig. 4. The dynamic node voltage swing which achieves the lowest active mode power consumption is typically higher than the minimum acceptable voltage swing due to the short-circuit current overhead of the output inverter. For single-V_t low swing circuits, the minimum acceptable

dynamic node voltage swing is 0.25V-to-0.55V. However, the lowest active mode power consumption and PDP of single-V_t low swing circuits are achieved with the 0.15V-to-0.65V and 0.1V-to-0.55V voltage swings, respectively, as listed in Table I. This result indicates that the technique presented in [9] with a fixed voltage swing of $V_{swing} \approx 0.2$V-to-0.7V provides neither the lowest power consumption nor the lowest PDP.

Power required to charge/discharge the dynamic node dominates the total power consumption in the high fan-in gates. Power reduction provided by the low swing circuit technique is, therefore, more significant in the high fan-in circuits. High-V_t transistors in the output inverter reduce the short-circuit current overhead of dual-V_t-N&P and dual-V_t-N circuits. Short-circuit current of dual-V_t-N&P circuits is reduced during both the evaluation and precharge phases. Alternatively, short-circuit current of dual-V_t-N circuits is reduced only in the evaluation phase. The dual-V_t-N&P technique, therefore, consumes lower active mode power as compared to the other techniques with similar voltage swing range. As shown in Fig. 4, the lowest active mode power consumption among the low swing circuit techniques is achieved by the dual-V_t-N&P technique at the 0.2V-to-0.55V low swing range with a high-V_t of 0.35V. The active mode power consumption is reduced by 46.3% (AND2) to 67.2% (MUX16) as compared to standard full-swing circuits.

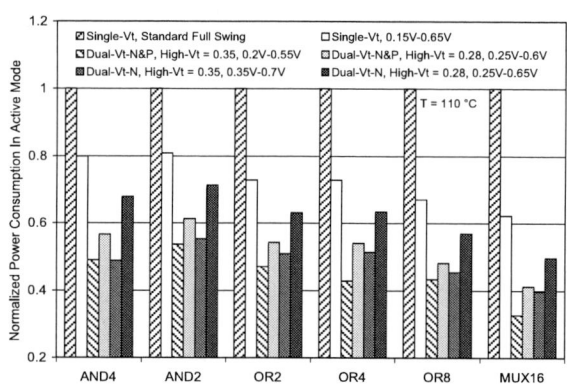

Fig. 4. Comparison of the lowest active mode power consumption achieved by single-V_t and dual-V_t low swing domino logic circuits. For each circuit, power consumption of a low swing configuration is normalized to the power consumption of standard full-swing configuration.

The lowest PDPs achieved by each low swing domino circuit and the PDPs of the voltage swings which achieve the lowest power consumption (Table I) are compared in Fig. 5. Among the low swing circuits, the lowest PDP is achieved with the dual-V_t-N technique for the asymmetric 0.1V-to-0.6V voltage swing with a high-V_t of 0.28V (except MUX16). PDP is reduced by 23.3% (AND2) to 51.9% (MUX16) as compared to standard full swing circuits. Dual-V_t-N&P circuits, which achieve the lowest power consumption, typically do not produce the lowest PDP due to the significantly degraded evaluation speed with the dual-V_t-N&P technique.

Power consumption and evaluation delay characteristics of low swing domino circuits which achieve the lowest PDP with each technique (Table I) are shown in Figs. 6 and 7, respectively. For the dual-V_t-N&P technique (high-V_t =

0-7695-2533-4/06 $20.00 © 2006 IEEE

0.35V), the lowest PDP and the lowest power consumption are achieved by the same low swing range of 0.2V-to-0.55V. The dual-V_t-N technique with an asymmetric 0.1V-to-0.6V voltage swing range and a high-V_t of 0.28V reduces the active mode power consumption by 19.1% (AND2) to 40.4% (MUX16) as compared to standard full-swing circuits, as illustrated in Fig. 6.

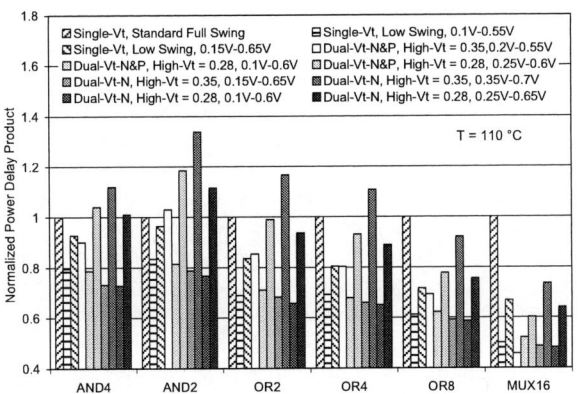

Fig. 5. Comparison of the lowest PDP achieved by single-V_t and dual-V_t low swing domino logic circuits. For each circuit, PDP of a low swing configuration is normalized to the PDP of standard full-swing configuration.

The pull-up strength of output inverter is degraded since gate overdrive of P1 is reduced as the lower boundary of dynamic node voltage swing (V_{gndH}) is increased in the low-swing circuits. Employing a higher threshold voltage transistor further degrades the pull-up strength of output inverter in dual-V_t-N&P circuits. As shown in Fig. 7, evaluation speed of the dual-V_t-N&P circuits are typically slower as compared to the other low swing circuits with a similar voltage swing range.

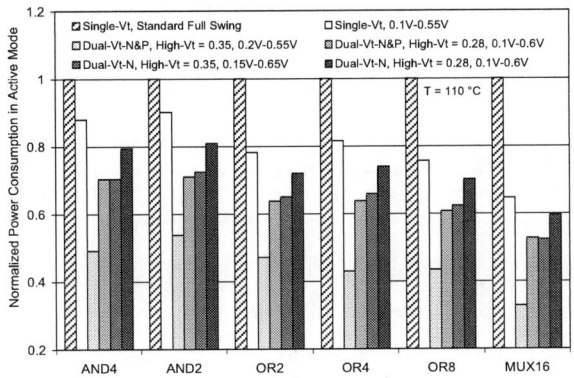

Fig. 6. Comparison of power consumption of single-V_t and dual-V_t low swing domino logic circuits which achieve the lowest PDP with each technique. For each circuit, power consumption of a low swing configuration is normalized to the power consumption of standard full-swing configuration.

Alternatively, reducing the upper boundary of dynamic node voltage swing (V_{DDL}) tends to enhance the evaluation speed due to the smaller amount of charge stored at the dynamic node. The shortest evaluation delay is achieved by the single-V_t technique with a 0.1V-to-0.55V dynamic node voltage swing. The evaluation delay is reduced by 7.2% (AND2) to 22.7% (MUX16) as compared to standard full swing circuits. The dual-V_t-N configuration offering the

minimum PDP characteristics (0.1V-to-0.6V with high-V_t = 0.28V) also reduces the evaluation delay by 5.2% (AND2) to 19.4% (MUX16) because of the relatively modest degradation in output inverter pull-up strength and the smaller amount of charge stored at the dynamic node as compared to standard full swing circuits.

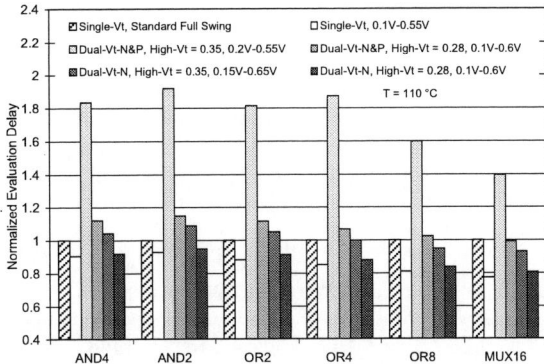

Fig. 7. Comparison of evaluation delay of single-V_t and dual-V_t low swing domino logic circuits which achieve the lowest PDP with each technique. For each circuit, evaluation delay of a low swing configuration is normalized to the delay of standard full-swing configuration.

IV. NOISE IMMUNITY

Noise immunity of domino logic circuits is evaluated in this section. The noise immunity criterion used in this paper is similar to the criterion described in [12]. The noise margin is the voltage amplitude of noise signal applied to the inputs that produces a signal with the same amplitude at the output of a domino logic circuit. Since the unidirectional low swing configuration (0V-to-V_{DDL}, $V_{DDL} < V_{DD}$) degrades the noise immunity, the unidirectional voltage swing range is not considered in this paper. Since the minimum PDP is achieved for high-V_t = 0.28V, the high-V_t is assumed to be 0.28V in the rest of the paper. Variation of the noise immunity of various circuits with the dynamic node voltage swing is illustrated in Fig. 8.

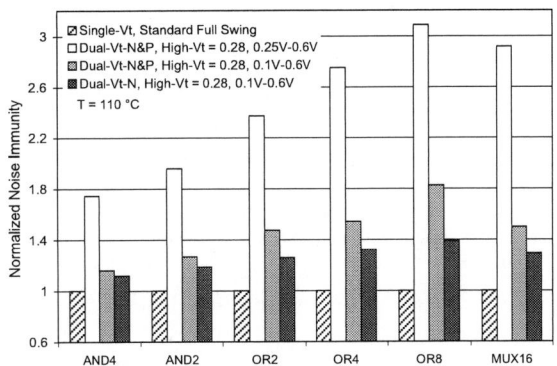

Fig. 8. Comparison of noise immunity of single-V_t and dual-V_t low swing domino logic circuits. For each circuit, noise margin of a low swing configuration is normalized to the noise margin of standard full-swing configuration.

A higher noise signal is needed to turn on the pull-down network transistors due to negative gate-to-source voltage when the inputs are low in the low swing circuits. Noise immunity of the proposed dynamic node low voltage swing

0-7695-2533-4/06 $20.00 © 2006 IEEE

dual-V_t-N&P circuit technique is further enhanced because of the high-V_t pull-up transistor in the output inverter. For the same voltage swing configuration, the dual-V_t-N&P technique displays a higher noise immunity as compared to the dual-V_t-N technique because the high-V_t PMOS transistor in the output inverter of a dual-V_t-N&P gate is more difficult to be turned on as the dynamic node is discharged by a noise signal. When the voltage swing is 0.25V-to-0.6V with the dual-V_t-N&P technique, the noise immunity is enhanced by 74.6% (AND4) to 208.7% (OR8) as compared to standard full swing circuits. The dual-V_t-N configuration offering the minimum PDP characteristics (0.1V-to-0.6V, high-V_t = 0.28V) also enhances the noise immunity by 11.7% (AND4) to 39.1% (OR8) as compared to standard full swing circuits.

V. LEAKAGE POWER CONSUMPTION

In this section, leakage power characteristics of dual-V_t low swing circuits with the lowest PDP (dual-V_t-N, 0.1V-to-0.6V, high-V_t = 0.28V) and the lowest active mode power consumption (dual-V_t-N&P, 0.2V-to-0.55V, high-V_t = 0.35V) are evaluated. Circuits are assumed to be operating at a worst case high temperature of 110 °C before the beginning of idle mode. Furthermore, it is assumed that the idle mode is short. Total leakage power consumption of low swing domino circuits at 110 °C (assuming the die temperature does not significantly change during the short idle period) is shown in Fig. 9. For the asymmetric 0.1V-to-0.6V and 0.2V-to-0.55V dynamic node voltage swing configurations, the inputs are maintained high to reduce the subthreshold leakage current overhead of the output inverter in the idle mode.

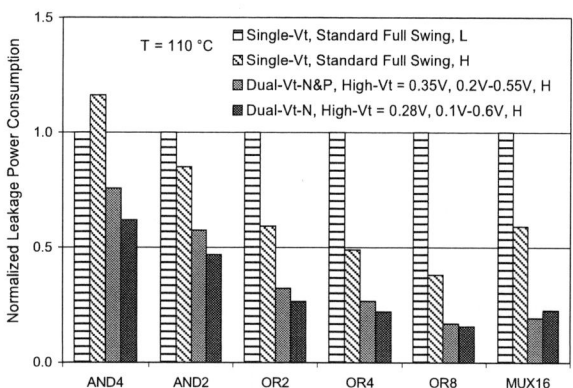

Fig. 9. Comparison of total leakage power consumption of single-V_t and dual-V_t low swing domino logic circuits at 110 °C. For each circuit, leakage power of a low swing configuration is normalized to the leakage power of standard full-swing configuration with a low (L) input vector. H: high input vector.

The single-V_t low swing circuit technique reduces the leakage current produced by the pull-down network and precharge transistor while increasing the subthreshold leakage current produced by the output inverter. The proposed dual-V_t-N&P and dual-V_t-N techniques reduce the subthreshold leakage current conducted by the output inverter while providing similar leakage reduction in the pull-down and pull-up transistors as compared to the single-V_t low swing circuit technique. Effectiveness of the proposed low swing dual-V_t circuit technique for leakage reduction depends on the relative contribution of pull-down network, precharge transistor, and

output inverter to the total leakage power consumption. As shown in Fig. 9, the dual-V_t-N&P circuits with a high input vector reduce the total leakage power consumption by 24.3% (AND4) to 82.9% (OR8) as compared to standard single-V_t full-swing circuits with a low input vector. Similarly, the dual-V_t-N technique reduces the total leakage power by 38.1% (AND4) to 84.2% (OR8).

VI. CONCLUSIONS

A novel dynamic node low voltage swing domino circuit technique based on dual power supply, dual ground, and dual threshold voltages is presented in this paper. The proposed circuit technique quadratically reduces the dynamic node switching power consumption by lowering the voltage swing at the dynamic node. Meanwhile, full voltage swing signals are maintained at input, clock, and output nodes for high speed operation. Dual threshold voltage transistors are employed for suppressing the short-circuit current overhead intrinsic to the dynamic node low voltage swing technique. Power supply, ground, and threshold voltages are simultaneously optimized to minimize PDP and enhance noise immunity.

It is shown that an asymmetric low voltage swing of 0.1V-to-0.6V along with an optimized high-V_t of 0.28V achieve the lowest PDP in various types of domino logic circuits in a 45nm CMOS technology. The proposed circuit technique reduces PDP and evaluation delay by up to 51.9% and 19.4%, respectively, while enhancing the noise immunity by up to 39.1% as compared to standard full-swing circuits. Furthermore, the circuit technique reduces idle mode leakage power by up to 84.2% as compared to standard full-swing circuits.

REFERENCES

[1] S. Rusu and G. Singer, "The First IA-64 Microprocessor," *IEEE Journal of Solid-State Circuits*, Vol. 35, No. 11, pp. 1539 - 1544, November 2000.
[2] P. E. Gronowski *et al.*, "High-Performance Microprocessor Design," *IEEE Journal of Solid-State Circuits*, Vol. 33, No. 5, pp. 676 - 686, May 1998.
[3] K. J. Nowka and T. Galambos, "Circuit Design Techniques for a Gigahertz Integer Microprocessor," *Proceedings of the IEEE International Conference on Computer Design: VLSI in Computers and Processors,* pp. 11-16, October 1998.
[4] S. Borkar, "Low Power Design Challenges for the Decade," *Proceedings of the IEEE/ACM International Design Automation Conference,* pp. 293-296, June 2001.
[5] P. Srivastava, A. Pua, and L. Welch, "Issues in the Design of Domino Logic Circuits," *Proceedings of the ACM/SIGDA Great Lakes Symposium on VLSI,* pp. 108-112, February 1998.
[6] A. Rjoub, O. Koufopavlou, and S. Nikolaidis, "Low Power/Low Swing Domino CMOS Logic," *Proceedings of the IEEE International Symposium on Circuits and Systems,* Vol. 2, pp. 13-16, May 1998.
[7] V. Kursun and E. G. Friedman, "Low Swing Dual Threshold Voltage Domino Logic," *Proceedings of the ACM/SIGDA Great Lakes Symposium on VLSI,* pp. 47-52, April 2002.
[8] V. Kursun and E. G. Friedman, "Dual Threshold Voltage and Low Swing Domino Logic Circuits," United States Patent, No. 6, 900, 666, May 31, 2005.
[9] R. Mader and I. Kourtev, "Reduced Dynamic Swing Domino Logic," *Proceedings of the ACM/SIGDA Great Lakes Symposium on VLSI,* pp. 33-35, April 2003.
[10] Z. Liu and V. Kursun, "Low Swing Domino Logic Circuits with Multiple Supply and Ground Voltages," United States Patent Pending.
[11] Berkeley Predictive Technology Model (BPTM), http://www.device.eecs .berkeley.edu/~ptm/download.html.
[12] V. Kursun and E. G. Friedman, "Node Voltage Dependent Subthreshold Leakage Current Characteristics of Dynamic Circuits," *Proceedings of the IEEE/ACM International Symposium of Quality Electronic Design,* pp. 104-109, March 2004.

High performance service-time-stamp computation for WFQ IP packet scheduling

C. McKillen, S. Sezer and Xin Yang

Institute of Electronics, Communications and Information Technology
colm.mckillen@ee.qub.ac.uk

Abstract

In this paper the design and implementation of a unique service-time-stamp computation circuit, called the finishing tag, for WFQ based packet scheduling is presented. The implementation is based on UMC 130nm standard cell technology, and placed and routed using Cadence SoC encounter. The design targets the development of programmable IP packet scheduling circuits for next generation network processing platforms for line-rates beyond 200Gbps.

1. Introduction

Internet Service Providers (ISPs) are striving to deploy new solutions that adequately address the demands being placed on them by their customers and investors. Demand is rapidly increasing for integrated data, voice, and video services with on-demand provisioning of bandwidth and services. More innovative services and network intelligence, namely security, storage and adaptability, will be key to the development and roll out of the next generation network.

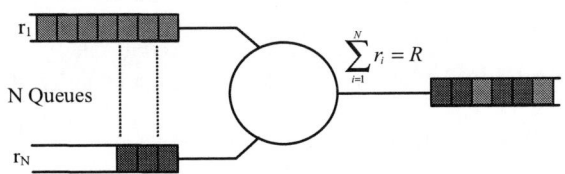

Figure 1. A typical packet scheduler

Quality of Service (QoS) refers to the capability of a network to provide better service to selected network traffic over various technologies. The solutions used to provide QoS will constitute the elemental building blocks making up the future ISP networks. Fundamentally QoS is provided by either raising the priority of a flow, or limiting the priority of another flow. At a router a flows priority is altered by queuing, and then servicing these queues in different ways.

The queues at a router are serviced using a scheduler, as shown in Figure 1. This scheduler is a lossless, work-conserving, and non-pre-emptive link with rate R that services N First-Come-First-Served (FCFS) queues. Each queue represents a different class of traffic, with its own individual rate or weight parameter specified. A scalable, high speed and efficient scheduler architecture is required to meet the demands of the future Internet network.

2. Packet Fair Queuing (PFQ) Algorithms

FCFS is the simplest scheduling algorithm, transmitting the packets in the order that they arrive to the scheduler. However, it provides only one priority level and therefore cannot give performance guarantees to the connections that it serves. To improve upon FCFS scheduling, Round Robin [1] schedulers were developed. These have a number of distinct FCFS queues, each of which has an associated priority level. Each service level has a time interval during which it receives service, after which the scheduler proceeds to the next lower priority level. Weighted Round Robin allows each session to receive a number of services equal to its allocated weight. The weight value assigned to a session is proportional to its reserved bandwidth. The Round Robin family of schedulers are easily implemented but only suitable when fairness requirements are loose and the packet sizes are small.

Generalised Processor Sharing (GPS) is the ideal scheduling discipline [2]. A server with N connections has each connection characterized by a positive real number, ϕ_1, ϕ_2,... to ϕ_N, representing its weight. The connections receive service in proportion to this weight whenever data is present in their queue. GPS guarantees a minimum session throughput, a bounded delay and worst-case network queuing delay. However, GPS cannot transmit packets as full entities. It assumes the scheduler can serve multiple sessions simultaneously and that the traffic is infinitely divisible, which is not the case.

0-7695-2533-4/06 $20.00 © 2006 IEEE

PFQ algorithms were developed to approximate the GPS discipline. They use a priority queue mechanism based on the notion of a virtual time, *V(t)*, function. They differ in there choice of *V(t)* and also in there packet selection policies. Different *V(t)* functions result in trade-offs between accuracy and complexity in the various PFQ schemes. Weighted Fair Queuing (WFQ) [3] simulates GPS on the side to determine the service order. It serves the packets independently of the departure events in the GPS simulation using a smallest finishing tag first packet selection policy. WFQ is within one packet transmission time of approximating GPS regardless of the packet arrival pattern. WFQ requires that a per connection scheduler state is stored. This is complex and expensive when serving a lot of connections. The computational complexity associated with the evaluation of *V(t)* has also been seen as prohibitive. Calculations must be performed in real time for the simulation to remain accurate.

Self clocked schedulers use a *V(t)* function dependent on the progress of work in the actual packet based queuing system. In Self-Clocked Fair Queuing (SCFQ) [4] and Start-time Fair Queuing (SFQ) [5] the *V(t)* at any moment *t* is simply estimated from the finish or start tag of the packet receiving service at time *t* respectively. This self contained approach is much less complex but has considerably reduced fairness.

Frame-based Fair Queuing (FFQ) and Starting Potential-based Fair Queuing (SPFQ) are two Rate Proportional Service (RPS) scheduling algorithms [6]. The service received by each session *i* is represented by a function $P_i(t)$. When a session is backlogged its potential increases exactly by the normalised service that it receives. The system potential function, $P(t)$, keeps track of the progress of the total work done by the scheduler. Among the backlogged sessions, sessions with the minimum potential at time *t* are serviced. RPS algorithms differ in the way they update the potentials of idle sessions. Furthermore recalibration of $P(t)$ is the key to providing bounded fairness. FFQ uses a framing approach to recalibrate $P(t)$ periodically where fairness depends on the frame size. SPFQ recalibrates $P(t)$ at the end of transmission of every packet, requiring more state information to be maintained, resulting in a more complex implementation.

The service provided by a WF^2Q [7] scheduler is almost identical to that of GPS. WF^2Q uses a different packet selection policy compared to WFQ, the Smallest Eligible Finishing-tag First policy. A packet is eligible at time *t* if its start tag is no greater than *V(t)*. The WF^2Q scheduling algorithm provides connections with tight delay bounds. WF^2Q is also defined with respect to the GPS system resulting in it having to simulate GPS on the side.

The DiffServ Internet service model has been under extensive investigation, resulting in two different types of service. Absolute DiffServ can guarantee hard QoS which is necessary for some applications, while relative DiffServ is more controllable and less complex. Absolute DiffServ aims to achieve performance measures similar to IntServ, but without keeping per-flow state within core routers. Virtual Leased Line service and Assured Service [8] are two implementations, but they are extremely complex. Rate-based scheduling algorithms and simpler static-priority schedulers [9] have also been used to provide absolute DiffServ. These can derive delay bounds without specific traffic flow information, instead they use the utilisation of links along the path of a flow.

In relative DiffServ, the network assures that higher classes will offer better QoS than lower classes. The Proportional Delay Differentiation (PDD) [10] model controls the ratios of the average queuing delays between classes based on specified delay parameters. The aforementioned paper also includes three scheduling schemes for implementing the PDD model: Proportional Average Delay (PAD), Waiting Time Priority (WTP) and Hybrid Proportional Delay (HPD). PAD equalises the normalised average delays among all classes, but it does not always give tight delay bounds. WTP assigns priorities that increase proportionally with the packet's waiting time. The class with the maximum normalised head waiting time is serviced. WTP outperforms PAD by providing higher classes with lower delays over short timescales, but accurate delay differentiation is only achieved when system utilisation is almost 100%. Combining PAD and WTP scheduling, a HPD scheduler is produced. HPD provides predictable service over short timescales but has large errors in moderate load conditions.

Adaptive WTP (AWTP) [11] utilises WTP scheduling along with an adaptive control algorithm. Real time measurement and adaptation techniques track the arrival rate of each flow and adjust the control parameters to maintain the target waiting-time ratios. AWTP maintains more accurate delay differentiation than WTP but there are still large deviations over short timescales and in light and moderate load conditions.

3. The Architecture

3.1 The WFQ Algorithm

The calculations required to implement a WFQ scheduler will now be examined in more detail. In WFQ, and other packet fair queuing algorithms, the start and finish tags for each and every packet are calculated using equations (1) and (2) respectively:

$$S_i^k = \max\left\{F_i^{k-1}, V\left(a_i^k\right)\right\} \tag{1}$$

$$F_i^k = S_i^k + \frac{L_i^k}{\phi_i} \qquad (2)$$

A start tag is calculated for each packet on its arrival. S_i^k is the starting tag given to the kth packet in session i, F_i^{k-1} is the finishing tag of the previous packet in session i and $V(a_i^k)$ is the virtual time in the GPS scheduler when the kth packet from session i arrives. After a start tag has been assigned to a packet, its finish tag can then be calculated. F_i^k is the finish tag given to the kth packet in session i, L_i^k is the length of the kth packet in session i, and ϕ_i is the session i weight.

To keep track of the GPS simulation in WFQ a method based on a single virtual time measure is utilised. An event is defined as the arrival or departure of a packet from the simulation where t_j represents the time at which the jth event occurs (simultaneous events are ordered arbitrarily). The set of sessions that are busy in an interval (t_{j-1}, t_j) is fixed and denoted as the set B_j. The virtual time, $V(t)$, is defined to be zero for all times when the server is idle. For an interval τ with a constant set of backlogged flows within any busy period, $V(t)$ evolves as follows:

$$V(t_{j-1} + \tau) = V(t_{j-1}) + \frac{\tau}{\sum_{i \in B_j} \phi_i} \qquad (3)$$

$$\tau \le t_j - t_{j-1}, j = 2, 3, \ldots$$

Next(t) is the next point in real time at which the set of backlogged flows may change as a result of a packet departure, thus having an effect on the slope of $V(t)$. *Next(t)* is calculated using the equation shown below assuming that there are no further arrivals of packets in the interval *(t, Next(t))*, where F_{MIN} is the minimum value of finish tag yet to depart the GPS simulation:

$$Next(t) = t + (F_{MIN} - V(t)) \sum_{i \in B_j} \phi_i \qquad (4)$$

Using these four equations the algorithm for the full implementation of a WFQ scheduler can be broken down into two distinct parts:

i) Packet arrival
   a. If system idle before packet arrival set $V(t)$ and all finishing tags in sessions to zero. Otherwise $V(t)$ updated with Equation (3).
   b. Calculate start and finish tags for packet using Equations (1) and (2) respectively.

   c. If session not backlogged before then added to set B_j. Sum of backlogged sessions is modified.
   d. Calculate real time of the next packet departure from the GPS system, i.e. *Next(t)* from Equation (4).
ii) Packet departure (At $t = Next(t)$)
   a. Update the value of $V(t)$ using Equation (3).
   b. De-queue the packet with the smallest finish tag, F_{min}.
   c. If a session is no longer backlogged when the packet is de-queued then it is removed from set B_j.
   d. Calculate *Next(t)* using Equation 4.

An examination of our innovative scheduler architecture will now be made, in particular looking at the implementation of a WFQ scheduler architecture.

3.2 Our high level scheduler

The model that we have developed for a QoS scheduler is composed of three distinct structures as is shown in Figure 2: (1) the Weighted Fair Queuing (WFQ) scheduler, (2) finishing tag sorting circuit and (3) the shared buffer. When a packet arrives at the scheduler its data payload is sent to the shared buffer and is assigned an associated finishing tag by the WFQ scheduler based on its priority of service. The smallest tag values are serviced first. This tag is then passed on to the tag sorter. At the scheduler output the lowest tag currently in the scheduler is received from the tag sorter and the associated payload retrieved from the shared buffer. The correct packet can then be transmitted.

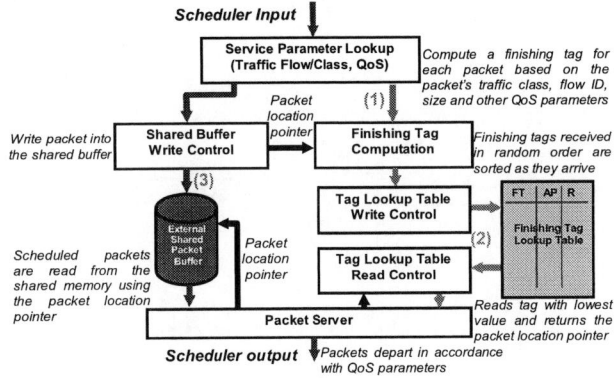

Figure 2. The high level scheduler architecture

3.3 The Decomposition

Figure 3 shows the top architecture of our model to implement the WFQ algorithm. It can be seen that there are three separate dual port memories utilised in the implementation. Each of these memories is composed

0-7695-2533-4/06 $20.00 © 2006 IEEE   67

of 2 dual port 8 bit by 8k RAM's resulting in a 16 bit by 8k RAM. Each of these RAM's is addressed by the 13 bit Flow ID associated with each packet by the classifier and the output of these RAM's is a 16 bit session parameter integral to the WFQ calculation. One RAM stores each individual sessions weight, another stores the previous finishing tags, and the third one is used to keep a per session count of how many packets are yet to leave the server.

With this implementation it is possible to service eight thousand different service classes each with there own individual QoS agreements.

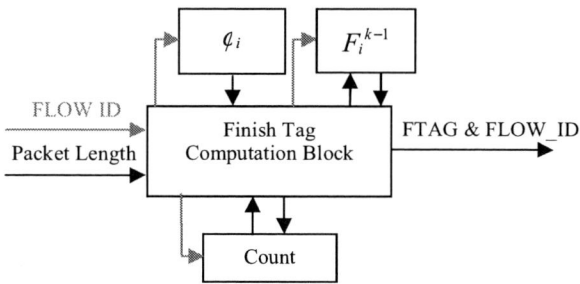

Figure 3. Block level description

Figure 4 shows the finish tag computation block in greater detail, indicating the different responses of the system to packet arrivals and departures. This results in a tag computation data path capable of computing one finishing tag value every clock cycle. To achieve this, the architecture must be parallel and pipelined. The flow specific session parameters that can be seen being utilised by three of the computation blocks in Figure 4.

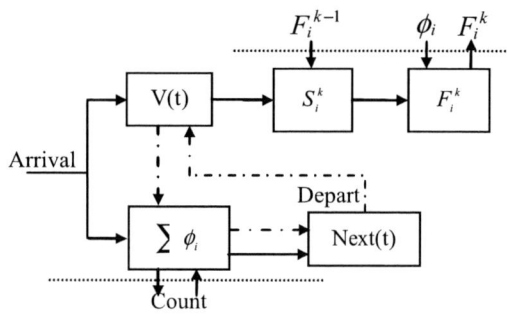

Figure 4. Finish tag computation block

3.3.1 Virtual Time

Figure 5 depicts a pipelined virtual time function. The first computation of $V(t+\tau)$ requires 19 clock cycles, whilst subsequent values can be calculated on each clock cycle. The block is primarily composed of a 16 bit pipelined binary divider that requires 17 clock cycles to compute a division, while 16 additional divisions can be calculated at the same time.

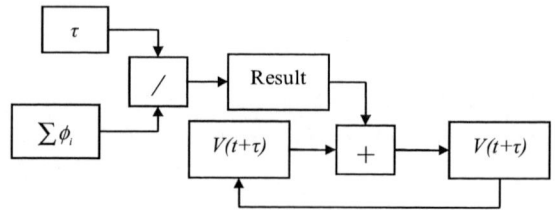

Figure 5. The virtual time computation block

Between events a counter keeps track of time. When an event occurs the counter value is forwarded to the divider and the counter is reset to zero. The value in τ is divided by the sum of the weights. Before the next event occurs the sum of the weights must be updated if a new session has become active or idle as a result of the previous event. The result from the pipelined divider is added to the previous $V(t)$ value to produce the updated virtual time value. The updated value immediately replaces the previous value of virtual time and is then forwarded to the start and finish tag computation block.

3.3.2 Start and Finish tags

Figure 6 shows the circuit architecture of the start and finish tag computation block. $V(t+\tau)$ and the previous finishing tag for a packet in that session are compared, resulting in the start tag being set to the greater of the two values. The correct previous finishing tag value is set 19 clock cycles after an arrival in order that the correct comparison can take place.

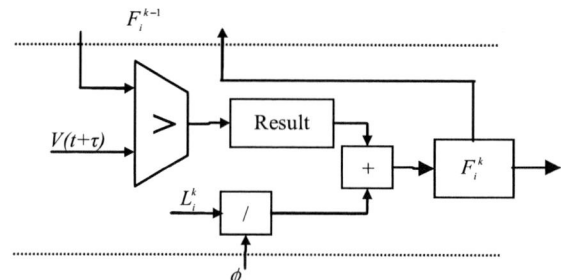

Figure 6. Start & finish tag computation block

The start tag is added to the result of the division of the packets length by the session's weight, giving a finish tag in 20 clock cycles. The length of a packet is obtained from the arriving packets header and is pipelined along with the session weight in order that the result is produced at the correct time.

The previous finishing tag memory block must be immediately updated with the new finish tag value after a computation. Due to the pipelined tag computation procedure the next packet on the next clock cycle may belong to the same flow. Finally, the finish tag value is combined with its Flow ID and the address the packet was assigned in the shared memory. This 64 bit value is passed to the tag sorter which orders the packets by the finish tags.

3.3.3 Backlogged Session Block

The architecture of the backlogged session block is shown in Figure 7. This count value represents per session the number of packets assigned a finish tag by the scheduler, but are yet to depart the GPS simulation. When a packet arrives if its session count equals zero, then its session weight is added to the sum of the weights. If the count is greater than zero then the session was already active, so the sum of the weights remains unchanged. In both cases the session count value is increased by one. On a packet departure that particular sessions count is decreased by one. If the count value then equals zero, that session has become idle and its weight is subtracted from the sum of the weights. If count is greater than zero after the subtraction then the session still has packets yet to be serviced and the sum of the weights is not modified.

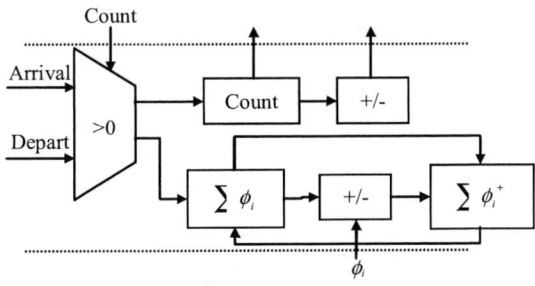

Figure 7. The sum of weights computation block

3.3.4 Next(t) Computation

Figure 8 shows the *Next(t)* computation block. F_{MIN} is obtained from the tag sorter along with its Flow ID. The updated virtual time is subtracted from F_{MIN} and the result is then multiplied with the updated sum of the weights. A 16 bit pipelined multiplier is able to carry out this task in 17 clock cycles. To keep the design technology independent and retarget able to a silicon implementation, the pipelined divider was developed by hand instead of using an embedded hard macro multiplier.

A counter, *t*, is set to zero when all sessions are idle. Once packets arrive this counter keeps track of the time that has elapsed since the scheduler became active. When an event occurs the value held in *t* at that instant

is pipelined and then added to the multiplier result at the correct instant to produce the new value of *Next(t)*. When *t* is found to be greater than or equal to *Next(t)* then the departure signal is set to one and signifies to the system that a packet has departed the GPS simulation. With departure set to one, the virtual time and sum of the weights is updated using the Flow ID of F_{MIN}. *Next(t)* is calculated for the newest F_{MIN}.

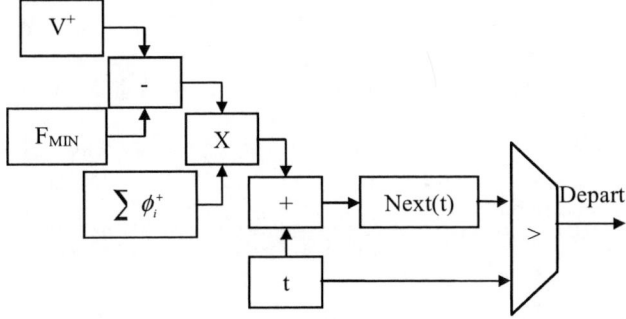

Figure 8. The Next(t) computation block

4. The Implementation

The ASIC layout for the WFQ scheduler is shown below in Figure 8. The investigated finishing tag computation has been implemented using UMC 130nm standard cell technology. The design was captured in VHDL and synthesised using Synopsis Physical Compiler. The layout is placed and routed using Cadence SoC Encounter. The post layout synthesis results are given in Table 1 and the chip layout is shown in Figure 8.

Table 1. Post layout synthesis results

Cadence Encounter – UMC – 130nm	
Clock Frequency	200MHz
Total Area	9.8mm^2
Total Throughput	200Gbps
Total Power	110mW
Internal Power	85.3mW
Switching Power	24.2mW
Leakage Power	0.5mW

The embedded dual port memory has been compiled using a UMC memory compiler from IMEC [12]. Six 8x8Kbit dual-port embedded memory blocks are being used, supporting up to 8,000 independent sessions. This is equivalent to 8,000 independent virtual queues. The chip area required by the implementation is 9.8mm^2 with a core area equal to 5.3mm^2. The total power usage of the circuit equals 110mW. The clock period of our implementation is 5.8ns giving a frequency of 200 MHz. The pipelined nature of our implementation means that 200 million finishing tags can be assigned by this scheduling architecture each and every second. Assuming a mean IP packet length of 1000 bits (125 Bytes), the presented implementations is able process a route output link (or shared backplane bus) of up to 200Gbps.

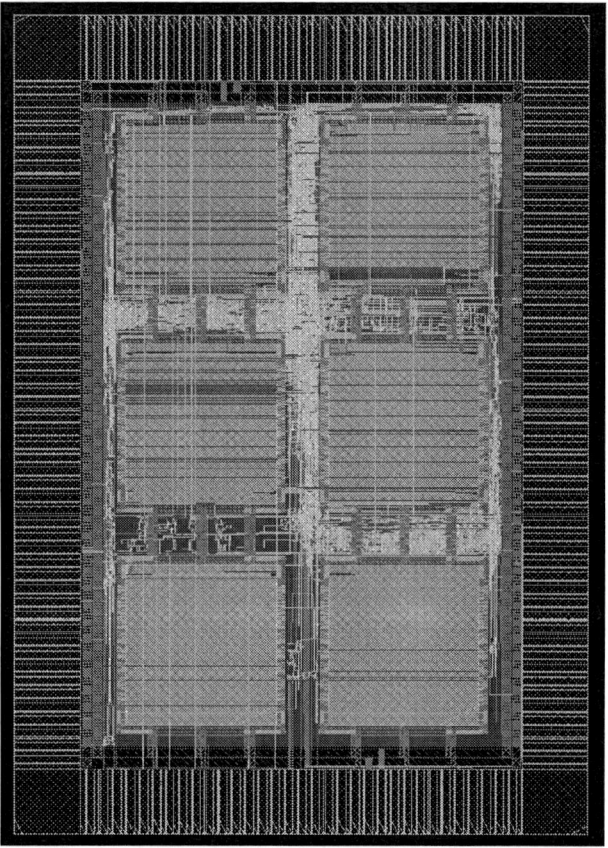

Figure 9. The overall system layout

5. Conclusion and Future Work

This paper presents the architecture, design and implementation of a tag computation circuit for high performance packet scheduling. By utilising a highly pipelined and customised parallel processing architecture an extremely fast tag computation is achieved. The implementation has been proven to have a relatively small hardware cost for a standard cell based design. The embedded memory used in this implementation was dimensioned to support up to 8000 sessions, without the use of external memory, with each session supporting individual QoS agreements. However, the architecture is already dimensioned to support up to 64,000 different sessions without

modification if sufficient embedded or external memory for session parameter storage can be accommodated.

The presented architecture offers enormous potential for next generation terabit router design with link bandwidth beyond 200Gbps and router bandwidth beyond several terabits. With the introduction and presumable rapid expansion of HDTV and other high bandwidth multimedia services, this solution provides an answer to the needs of internet traffic management, control and scheduling in the next ten years.

References

[1] J. Nagle, "On Packet Switches with Infinite Storage," IEEE Transactions on Communications, Volume 35, pp 435-438, 1987.

[2] A. K. Parekh and R. G. Gallager, "A Generalized Processor Sharing Approach to Flow Control in Integrated Services Networks: The Single-Node Case," ACM/IEEE Trans. Networking, vol. 1, pp. 344–357, June 1993.

[3] Demers, S. Keshav, and S. Shenker, "Analysis and Simulation of a Fair Queuing Algorithm," J. Internetworking Res. Experience, pp. 3–26, Oct. 1990; also in Proc. ACM SIGCOMM'89, pp. 3–12.

[4] S. J. Golestani, "A Self-Clocking Fair Queuing Scheme for Broadband Applications", INFOCOM'94 Proceedings, pp 636-646, 1994.

[5] P. Goyal, H. Vin, and H. Chen, "Start-time fair queuing: A scheduling algorithm for integrated services packet switching networks," Proc. ACM SIGCOMM'96, Sept. 1996, pp. 157–169.

[6] Dimitrios Stiliadis and Anujan Varma, "Efficient Fair-Queuing Algorithms for Packet-Switched Networks," IEEE/ACM Transactions on Computers, vol. 6, no. 2, pp 175-185, 1998.

[7] J. C. R. Bennett and H. Zhang, "WF^{2}Q: Worst-case Fair Weighted Fair Queuing," In Proc. of IEEE INFOCOM'96, pages 120-128, San Francisco, CA, March 1996.

[8] S. Sahu, P. Nain, D. Towsley, C. Diot, and V. Firoiu, "On achievable service differentiation with token bucket marking for TCP," in Proc. ACM SIGMETRICS, June 2000.

[9] S. Wang, D. Xuan, R. Bettati and W. Zhao, "Providing Absolute Differentiated Services for Real-Time Applications in Static-Priority Scheduling Networks", IEEE/ACM Transactions on networking, Vol 12, No 2, April 2004.

[10] C. Dovrolis, D. Stiliadis, and P. Ramanathan. Proportional differentiated services: Delay differentiation and packet scheduling. In Proceedings of SIGCOMM, pages 109–120, 1999.

[11] M. K. Leung, J. C. Lui, and D. K. Yau. Adaptive proportional delay differentiated services: Characterization and performance evaluation. IEEE/ACM Transactions on Networking, 9(6):801–817, December 2001.

[12] http://www.imec.be/

0-7695-2533-4/06 $20.00 © 2006 IEEE 70

Synthesis of Pipelined SRSL Circuits

Rashad Oreifej, Abdelhalim Alsharqawi
University of Central Florida
Dept. of Electrical and Computer Engineering
Orlando, Florida 32816-2450
rashad@mail.ucf.edu, aalsharq@mail.ucf.edu

Abdel Ejnioui
University of South Florida
Dept. of Information Technology
Lakeland, Florida 33803-7096
aejnioui@lakeland.usf.edu

Abstract

In this paper, we propose a new design methodology for clockless circuits based on the present methodology of clocked circuits. This methodology takes advantage of the maturity of current CAD tools to synthesize new clockless pipelines without disrupting their design flow. Currently, there is no established design methodology to support the design and verification of clockless circuits. As a case in study, the proposed design methodology targets the synthesis of new pipelines based on a recently introduced clockless design technique called self-resetting stage logic (SRSL). The synthesis of SRSL pipelines starts from a synthesized gate netlist to satisfy a specified data rate by minimizing overall pipeline area. Since this synthesis problem is formulated as a large integer programming problem, an efficient two-phase heuristic algorithm is proposed to solve this problem. Experimental results show that SRSL pipelines can reach throughputs in the GHz range and are highly suitable for coarse-grain datapaths.

1. Introduction

The increasing gravity of the clocking problem in current chips is motivating designers to seek clockless circuits as alternatives to the prevalent clocked circuit paradigm. Although a number of clockless circuits have been proposed in the past, there does not seem to be a unified design methodology to support the design and verification of these circuits. Because the dominant design methodology relies on current CAD tools for clocked circuits, it would make sense to leverage the widespread usage of these tools by adopting as much as possible of their design methodology to implement clockless circuits. However, implemented mostly as fine-grain pipelines, the majority of these clockless circuits are not suitable to synthesize large datapaths [1]. Faced with this difficulty, it would make sense to (i) either select specific coarse-grain pipelining techniques among previously proposed clockless techniques, or (ii)

propose novel coarse-grain clockless pipelining techniques that seem supportable by existing CAD tools. A few attempts have been already undertaken in pursuing the former alternative [1]. However, if the latter alternative is pursued, the best place to transform a clocked design into a clockless one is at the gate level where minimum disruption of the design flow supported by existing CAD tools is achievable. By doing so, the synthesis step of the clocked gate netlist from the initial RTL model in the design flow is completely preserved. The obtained clockless gate netlist can be mapped using technology mappers packaged in existing CAD tools and standard cell libraries as shown in Figure 1.

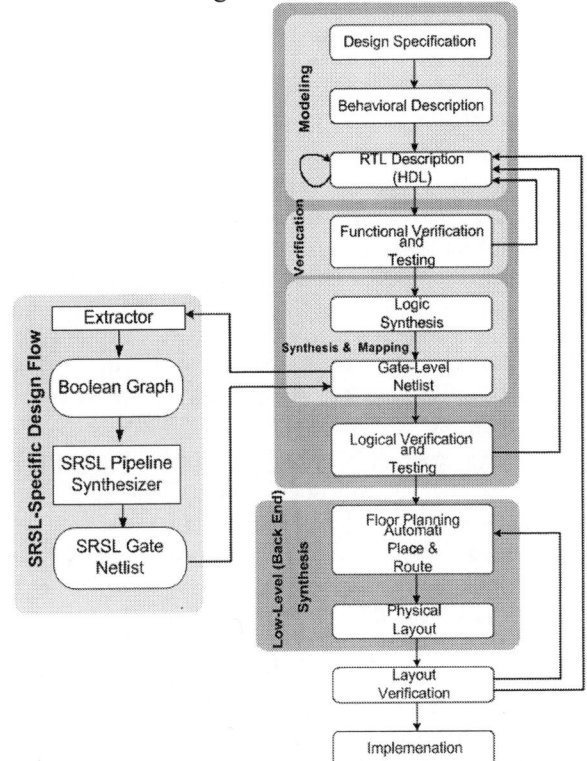

Figure 1. SRSL design methodology.

In addition, the same gate netlist can be simulated using any existing simulators. Furthermore, the proposed clockless design technique is of sufficient

0-7695-2533-4/06 $20.00 © 2006 IEEE

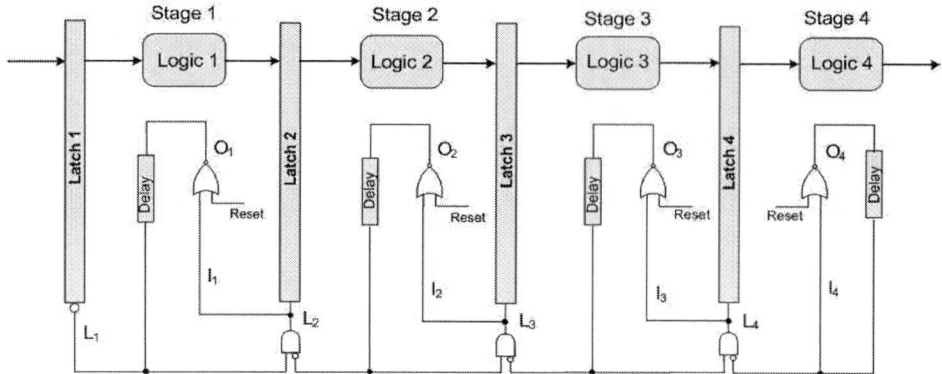

Figure 2. A four-stage SRSL pipeline.

granularity as to not impose a high area overhead. Based on the rationale of the second alternative, this paper reviews a previously introduced clockless design technique, called *self-resetting stage logic* (SRSL), highly adaptable for existing CAD tools. This design technique is used to develop a coarse-grain pipelining techniques, which can be used to transform a clocked gate netlist into a highly pipelined clockless gate netlist based on data rate and area cost specifications [2].

2. SRSL Pipelines

In SRSL, a stage consists of a *reset* network and a *combinational* network. In Figure 2, the reset network of each pipeline stage consists of a NOR gate whose output O feeds one of its inputs. The other input is tied to a reset line. As long as the reset input is asserted, O remains 0. When the reset is de-asserted, O oscillates from 0 to 1 and vice versa. The oscillation frequency is controlled by the delay block embedded in the loop between the NOR output and its input. When O is 0, the reset network is in the *reset* phase. Later, when O switches to 1, the reset network is in the *evaluate* phase. As such, a reset network can oscillate between these two phases in an autonomous fashion. The period of the reset network consists of the two phases: reset and evaluate. Based on this oscillation, a reset network can be embedded in a pipeline stage forcing the stage to oscillate between two phases. This oscillation can be used to synchronize data transfer between neighboring stages in a pipeline. Driven by this oscillation, data flows from one stage to another through a latch in the linear pipeline. To insure proper data flow across stages, data is transferred from the current stage to the next one if the current stage is in the evaluate phase while the next stage is in the reset phase at which time the latch separating both stages is enabled. This enable signal is the output of the AND gate that triggers the latch. During one cycle, if the first stage is in the reset phase, every odd-numbered stage in the pipeline will also be in the reset phase. During the same cycle, every even-numbered stage in the pipeline will be in the evaluate phase. In the next cycle, the odd-numbered stages transition to the evaluate phase while the even-numbered stages transition to the reset phase.

3. Synthesis of SRSL Pipelines

The synthesis of SRSL pipelines consists of transforming a gate netlist into an SRSL pipeline characterized by a data rate and an area cost. This area consists primarily of (i) latches located between pipeline stages, and (ii) delay elements needed for the reset network of each stage. These components can be characterized by area and delay attributes. To transform a gate netlist into an SRSL pipeline, the following problem is considered:

P1: *Given a gate netlist and a data rate, transform the gate netlist into an SRSL pipeline by incurring the smallest area cost.*

3.1. Preliminaries

In order to transform a gate netlist into an SRSL pipeline, a gate netlist can be abstracted into a Boolean graph as shown in Figure 3 [3].

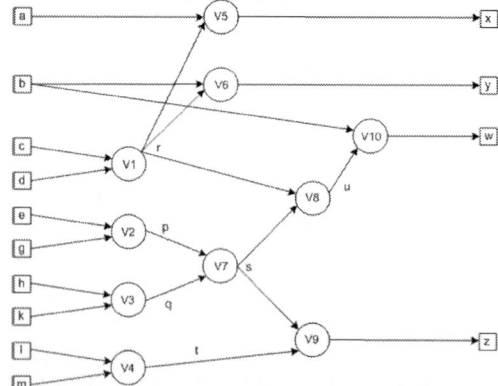

Figure 3. Example of a Boolean graph.

Definition 1: A *Boolean graph* $G(V, E)$ is a directed graph where: (i) the vertices in V represent gates, primary inputs, or primary outputs; (ii) the edges in E represents the decomposition of the multi-terminal nets of the gate netlist into two-terminal nets.

Let $S = \{s_1, s_2, ..., s_{|S|}\}$ be the set of pipeline stages where the size of this set, $|S|$, is some positive integer. Let $V = \{v_i ; i = 1, 2, ..., |V|\}$ and $E = \{(v_i, v_j) ; i, j = 1, 2, ..., |E|\}$. Since each vertex in V has a delay, $D = \{d_i ; i = 1, 2, ..., |V|\}$.

Definition 2: The delay of a path p in a graph G, denoted by d_p, is the sum of the delays of the vertices in p, i.e., $d_p = \sum\limits_{i:v_i \in p} d_i$.

Definition 3: Let Π be the set of all paths in a Boolean graph $G(V, E)$. A *critical path* in G is a path π whose delay is the largest path delay in Π, i.e., $d_\pi = \max \{d_p : p \in \Pi\}$.

3.2. Formulation of P1

In P1, a date rate f is given and the objective is to minimize the area cost incurred by partitioning the Boolean graph into stage partitions. The period P of a single stage can be obtained from f as $P = \dfrac{1}{f}$. Surely, there is a critical path π in the Boolean graph G whose delay is d_π. An upper bound on the number of stages in the pipeline, called *maximum pipeline depth*, can be obtained from P and d_π. If $|S|$ is the cardinality of S, the maximum pipeline depth is $|S| = \left\lceil \dfrac{d_\pi}{P} \right\rceil = \lceil d_\pi f \rceil$.

Definition 4: A binary variable $x_{i,s}$ is associated with each vertex v_i in V of $G(V, E)$ where (i) $x_{i,s} = 1$ iff the gate i, represented by v_i, is assigned to stage s; (ii) $x_{i,s} = 0$ otherwise.

In order to realize a correct partitioning, it is imperative that each vertex in the Boolean graph be assigned to a single stage:

$$\sum_{s=1}^{|S|} x_{i,s} = 1, \quad i = 1, 2, ..., |V| \quad (1)$$

It also imperative to observe that the successor of a vertex should be assigned to (i) the same stage as its predecessor, or (ii) a stage located after the stage of its predecessor:

$$\sum_{s=1}^{|S|} s x_{j,s} - \sum_{s=1}^{|S|} s x_{i,s} \geq 0, \ \forall (v_i, v_j) \in E \quad (2)$$

Since P can be obtained from the given data rate, it is important that the delay through each stage does not exceed P:

$$\sum_{i:v_i \in \pi} d_i x_{i,s} \leq P, \ s = 1, 2, ..., |S| \quad (3)$$

The partitioning of the gate netlist into stages requires the insertion of (i) latches to separate neighboring stages, and (ii) delay elements to realize the reset network of each pipeline stage. In general, the number of latches inserted between two adjacent vertices, $(v_i, v_j) \in E$, depend on the stages, s_k and $s_l \in S$, to which both vertices are assigned respectively.

Definition 5: If two adjacent vertices, $(v_i, v_j) \in E$, are assigned to stages s_k and $s_l \in S$ respectively, the *pipeline distance* between v_i and v_j, denoted by $\delta_{i,j}$, is $\delta_{i,j} = l - k$.

Depending on the bit width of the combinational network in a given stage, latches of different bit widths can be used to separate a stage from its neighbor. It would make sense to quantify the area of the inter-stage latches by multiplying the area of a single-bit latch by the number of output bit lines crossing from stage to stage. These lines correspond to edges in the Boolean graph. Assume that a_l is the area of a single-bit latch. If n bit lines are crossing from a stage to another, n latches are needed adding up to an area of na_l. Using the definition of pipeline distance, the number of 1-bit latches between two adjacent vertices can be determined as:

$$\delta_{i,j} = \sum_{s=1}^{|S|} s x_{j,s} - \sum_{s=1}^{|S|} s x_{i,s}, \ (v_i, v_j) \in E \quad (4)$$

The latch area needed to support the stages between v_i and v_j is $\delta_{i,j} a_l$. By considering all the edges in the Boolean graph, the total latch area needed in an entire pipeline can be determined as follows:

$$\sum_{\forall (v_i, v_j) \in E} a_l \left(\sum_{s=1}^{|S|} s x_{j,s} - \sum_{s=1}^{|S|} s x_{i,s} \right) \quad (5)$$

Beside the insertion of latches, the insertion of delay elements, implemented as buffers, is also needed to realize the reset network of a stage. If the area and delay of an elementary buffer is a_{buf} and d_{buf} respectively, the area of the matching delay in a stage s, denoted by a_s, can be computed as follows:

$$a_s = \left\lceil \frac{\sum\limits_{i:v_i \in \pi} d_i x_{i,s}}{d_{buf}} \right\rceil a_{buf}, \ s = 1, 2, ..., |S| \quad (6)$$

By considering all the stages in the pipeline, the total area of matching delays can be determined as:

$$\sum_{s=1}^{|S|}\left(\sum_{i:v_i\in\pi}a_ix_{i,s}\right)\quad(7)$$

By summing the total area needed for latches shown in (5), and matching delays shown in (6), the minimization of the area cost can be expressed as the following objective function:

$$\min a_l\sum_{\forall(v_i,v_j)\in E}\left(\sum_{s=1}^{|S|}sx_{j,s}-\sum_{s=1}^{|S|}sx_{i,s}\right)+\sum_{s=1}^{|S|}\left(\sum_{i:v_i\in\pi}a_ix_{i,s}\right)\quad(8)$$

Note that in this formulation, the variables are:

$$x_{i,s}\in\{0,1\},\ i=1,2,...,|V|,\ s=1,2,...,|S|\quad(9)$$

In summary, P1 can be formulated as the *integer programming* (IP) problem in which the objective function (8) is subject to the constraints (1), (2), (3), and (9).

4. Proposed Solution for P1

Although it is possible to solve P1 using standard combinatorial approaches suitable for general IPs, using such methods may not be efficient due to the size of P1's IP formulation in some cases. For example, the IP formulation of C6822 circuit, which is an ISCAS benchmark circuit consisting of 6,656 gates and a 245-gate long critical path, can generate 6,656 constraints (1), 9,082 constraints (2), and 245 constraints (3). In total, the matrix of the IP has 15,983 rows and 245 columns. As a result, a two-phase efficient heuristic solution is proposed to obtain its solution instead. The first phase is a stage-assignment algorithm which assigns each gate to a single stage by partitioning the Boolean graph of the gate netlist into subgraphs that meet specific timing constraints. On the other hand, the second phase is a vertex-shuffling algorithm which minimizes the area occupied by inter-stage latches through the shuffling of nearby vertices from the Boolean graph between adjacent stages without violating timing constraints.

4.1. Stage Assignment Phase

The following procedure is used in the stage assignment phase:

```
Input:  G(V, E), D = {d_i ; i = 1, 2, …, |V|}
        A = {a_i ; i = 1, 2, …, |V|}, f
Output: Partitioned graph G'(V', E')

1. Let P = 1 / f;
2. While there are unassigned vertices in V
3.     Select a vertex v in V whose
           predecessors are all assigned to the
           current or preceding partition;
4.     Get the critical path of the vertices
           in the current partition including v;
5.     If the path delay is less than or
```

```
                equal to P
6.         Assign v to the current partition;
7.     Else
8.         Add another partition;
9.         Assign v the newly added partition;
10.    Endif
11. Endwhile
12. The final obtained partitioned graph is
        G'(V', E');
```

In line 1, the stage delay is obtained. The algorithm starts with partition 1 which does not contain any vertices at this point. Line 2 shows a loop which looks for vertices in V which have not been assigned to any partition. Line 3 shows that the first step in assigning a vertex from V to the vertex set of the current partition is to select a vertex whose predecessors have been already assigned to the vertex set of the current partition. Next, the critical path of the Boolean graph including vertex v is obtained in line 4. In line 5 through 10, the algorithm checks if the critical path of the Boolean graph obtained in line 4 is less than or equal to the period of the partition. If the check result is true the selected vertex is added to the vertex set of the current partition. Otherwise, a new graph partition is created to which the selected vertex is subsequently added. The algorithm repeats the line 3 through 11 until there no unassigned vertices in V. At the end, each vertex in V is assigned to a distinct vertex set V_i which belongs to a subgraph $G_i (V_i, E_i)$.

4.2. Vertex Shuffling Phase

The pseudocode of the vertex shuffling phase is as follows:

```
Input:  G'(V', E') SRSL pipelined graph that
        meets p
        D = {d_i ; i = 1, 2, …, |V|}
        A = {a_i ; i = 1, 2, …, |V|}

Output: Partitioned graph G''(V'', E'') with
        minimum number of latches between each
        pair of partitions.

1. For every pair of adjacent partitions in
     G'(V', E')
2.     While the minimum move cost function
           in the current pass is less than the
           minimum move cost function in the
           previous pass
3.         While there are unmarked vertices
               in the left and right cut vertex
               sets
4.             For every unmarked vertex in
                   this cut vertex set
5.                 Compute its gain function;
6.             Endfor
7.             Get the vertex with the next
                   highest gain function and whose
                   delay does not violate the
                   period constraint in its
                   opposite partition;
8.                 Compute the move cost
```

```
                        function of this vertex;
9.              Mark this vertex and insert
                        it into a queue;
10.         Endwhile
11.         For every cut vertex in the queue
                starting from the first vertex to
                the vertex with the minimum move
                cost function
12.             If this vertex is a left cut
                    vertex
13.                 Move it to the right cut
                        vertex set;
14.                 Perform the set of its
                        induced moves;
15.             Else
16.                 Move it to the left cut
                        vertex set;
17.                 Perform the set of its
                        induced moves;
18.             Endif
19.         Endfor
20.         For every cut vertex in the queue
                starting from the vertex following
                the minimum move cost function
                vertex to the last vertex
21.             Unmark this vertex;
22.         Endfor
23.     Endwhile
24. Endfor
25. For each edge in E'
26.     Compute the pipeline distance δ;
27.     Add δ vertices to V'';
28.     Add δ edges to E'';
29. Endfor
30. For each partition in V';
31.     Get the critical path in the current
            partition;
32.     Duplicate the path with buffers and
            insert it into the current partition;
33. Endfor
34. The final obtained partitioned graph is
        G''(V'', E'');
```

Line 1 shows that this procedure executes for every pair of adjacent partitions in $G'(V', E')$. A minimum cost function from a given cut vertex, that is selected to be moved from one partition to another, is computed in every pass of the procedure, whereby a pass consists of the pseudocode shown in lines 2 through 23. This cost function is measured by the number of cut vertices located in the left graph partition. Note that each stage is represented by a graph partition. The objective is to minimize the number of cuts across each pair of partitions. As long as this cost function is less than the cost function computed in the previous pass as shown in line 2, another pass is executed. In line 3, all the unmarked vertices in the left and right cut vertex sets will be processed. This processing starts first by computing the gain function for each vertex in these two sets as shown in lines 4 through 6. The gain function represents the difference between the numbers of inter-partition cut edges and intra-partition internal edges. Next, the move cost function of the vertex with the highest gain function is computed as shown in lines

7 and 8, after which the vertex is marked and inserted in a queue as shown in line 9. This procedure is repeated for every unmarked vertex with the next highest gain function until there are no more unmarked vertices in the left and cut vertex sets as shown in line 3 through 10. As shown in lines 11 through 19, every vertex in the queue, starting from the vertex in the first entry of the queue until the vertex with the minimum move cost function in the queue, is moved to the opposite partition followed by the completion of the set of its induced moves. The remaining vertices in the queue are unmarked as shown in lines 20 through 22 to be processed in another pass starting from line 2. After latches are minimized for each pair of partitions in $G'(V', E')$, the next step consists of adding vertices between the partitions to represent latches between pipeline stages as shown in line 25 through 29. This step is followed by a second step in which the portion of the critical path contained in a partition is duplicated and added to that partition as shown in line 30 through 33. This duplicated path represents the matching delay of the reset network which will be attached to the combinational network of the stage represented by the partition. At the end, the final graph $G''(V'', E'')$ is obtained.

5. Experimental Results

The two-phase solution has been implemented and applied on a set of six circuits shown in Table1.

Table 1. Experimental circuits.

Circuit	Gates	Critical Path Delay (ps)
C6288	6656	25355
C7552	3569	4957
C5135	2332	6026
16_Bit_Multiplier	1456	12658
32_Bit_Adder	160	18850
16_Bit_Adder	80	9380

In order to study how SRSL pipelining affects the throughput of a circuit, the two-phase procedure is applied to the six circuits for different pipeline depths measured in number of stages as shown in Figure 4. For each circuit, the pipeline depth is increased until the circuit ceases to operate correctly. In Figure 4, one stage represents the circuit in its non-pipelined version. This figure shows that the throughput of a circuit can increase significantly depending on the pipeline depth. Indeed, for a shallow circuit, such as C7552, the throughput goes from 201 Mega-operations/sec in its non-pipelined version to 1327.79 Mega-operations/sec in its 10-stage SRSL pipeline. This increase is equivalent to a 6.6 times improvement in throughput. This improvement is even more pronounced in deep circuits.

0-7695-2533-4/06 $20.00 © 2006 IEEE

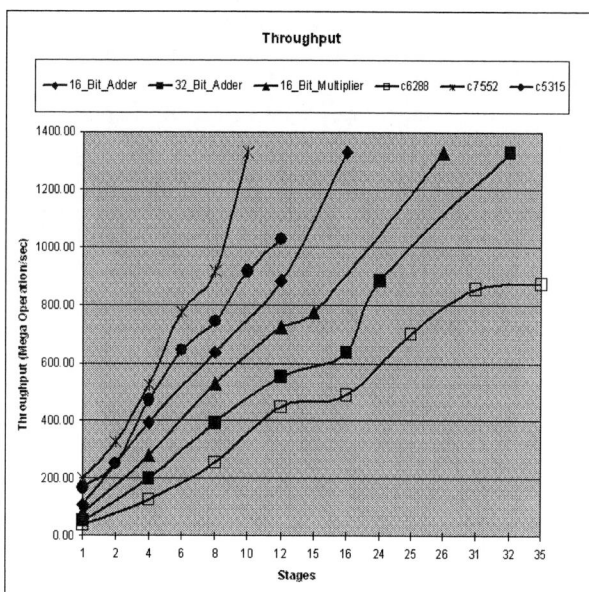

Figure 4. Circuit throughput vs. pipeline depth.

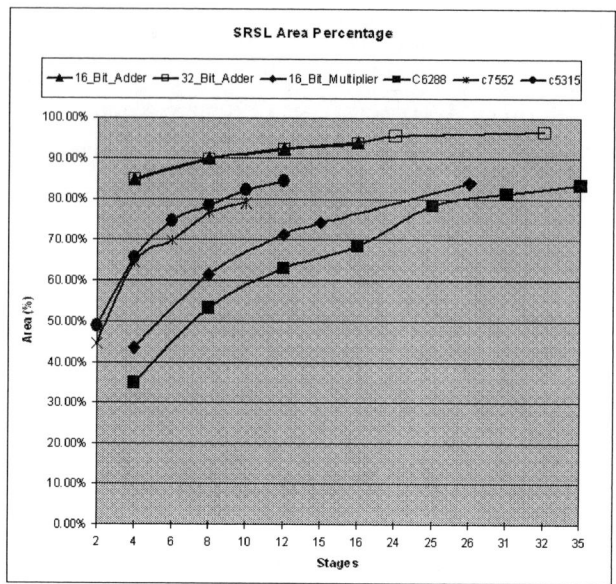

Figure 5. Circuit area vs. pipeline depth.

For example, the throughput of C6288 goes from 39.44 Mega-operations/sec in its non-pipelined version to 875.66 Mega-operations/sec in its 35-stage SRSL pipeline. This increase represents 22.2 fold in throughput improvement. This would make sense in deep circuits where there is ample room to add more stages with smaller delays to the pipeline to increase its throughput. While the throughput increases as more stages are added to the pipeline, it is obvious that the rate of throughput increases is not the same for all circuits. It seems that shallow circuits, such as C7552 and C5315, display the fastest throughput increase contrary to deep circuits such as C6288 and 32_Bit_Adder. By examining stage delays in equal depth pipelines, the delay of a single stage is usually higher in deep circuits than the delay of a single stage in shallow circuits. As a result, the throughput will be higher in shallow circuits as opposed to deep circuits for the same pipeline depth. Furthermore, it is obvious that the maximum possible pipeline depth will be higher in deep circuits than in shallow circuits.

Figure 5 shows the SRSL area as a percentage of the total area of a pipeline for each circuit across different pipeline depths. It is meant by SRSL area the area occupied by SRSL circuitry consisting of latches, delay buffers, AND and NOR gates. It is clear that the area of each circuit increases as the circuit is partitioned into a deeper pipeline. However, the largest increases in areas tend to occur in deeper circuits partitioned into deeper pipelines such as C6288. Furthermore, it is clear from the figure that the area occupied by SRSL circuitry tends to be smaller in general for large and deep circuits than for large and shallow circuits. In any case, small circuits tend to experience high SRSL areas

regardless of pipeline depth. This shows that the tradeoff of throughput gain vs. SRSL area is beneficial for deep SRSL pipelines and costly in shallower SRSL pipelines. Partitioning small and deep circuits requires relatively larger SRSL areas to support their SRSL pipelines. The increase in area cost can be offset in throughput gains only when large and deep datapaths are converted into deep pipelines. It can be concluded that the SRSL pipelining technique proposed in this paper is highly suitable for coarse-grain datapaths.

6. CONCLUSION

In this paper, a methodology to synthesize SRSL pipelines has been presented. The synthesis of SRSL pipelines is formulated as an IP problem subject to area and timing constraints for which a two-phase algorithm is proposed and applied on six circuits. Experiments reveal that SRSL pipelining is highly suitable for coarse-grain datapaths where these pipelines can yield throughput beyond the 1 GHz mark for deep circuits

7. REFERENCES

[1] C. P. Sotiriou, "Implementing asynchronous circuits using a conventional EDA tool-flow," *Design Automation Conference*, New Orleans, LA, 2002, pp. 415-418.

[2] A. Ejnioui and A. Alsharqawi, "Pipeline Design Based on Self-Resetting Stage Logic," *IEEE Computer Society Annual Symposium on VLSI*, 2004, pp. 254-257.

[3] G. DeMicheli, *Synthesis and optimization of digital circuits*: McGraw-Hill, 1994.

An Efficient Hardware Implementation of a Self-Adaptable Equalizer for WCDMA Downlink UMTS Standard

Romualdo Begale Prudêncio, Leandro Soares Indrusiak, Manfred Glesner

Microelectronic Systems Institute (MES)
Darmstadt University of Technology (TUD)
Karlstr. 15 – 64283 Darmstadt, Germany
rbegale@yahoo.com.br, <lsi, glesner>@mes.tu-darmstadt.de

Abstract

This paper proposes an adaptable receiver architecture for the WCDMA downlink UMTS standard. The architecture is aimed to support the receiver in such a way that it can self-adapt to different channel conditions. Architectural optimizations aiming to a more efficient implementation are presented and the advantages of the self-adaptable behavior of the receiver under a Raleigh channel are shown.

1. Introduction

In a CDMA communication system, users share the same channel by means of a spread spectrum modulation. In the UMTS standard, each user is designed a unique code sequence and the spread is carried out by directed multiplication of the code by the user data (DS-CDMA). At the receiver side, by knowing the code sequence of a given user it is possible to decode the received signal and recover the original data.

The separation between the users is granted by the orthogonal codes used to spread the original signal. In an ideal wireless channel where the only disturbance is the Additive White Gaussian Noise (AWGN), the receiver can recover the original data by simply correlating the received signal with the code. The others users will appear to the receiver as noise.

However, the actual wireless transmission channel is better characterized by a multipath Rayleigh fading channel. In a multipath propagation environment the receiver receives several delayed copies of the same signal, so the received signal is sometimes intensified or weakened. Although the use of orthogonal codes minimizes the multiple access interference (MAI), that is, interference induced by other users in the network, the presence of multipath destroy this orthogonality and MAI increases significantly.

In [1], Hooli addressed the suppression of intra-cell interference with channel equalizers. Several kinds of linear equalizers are discussed and compared. It has been shown that linear channel equalizers provide significant performance improvements with an acceptable increase in complexity, in particular in the cases of frequency selective channels and high data rate services. However, the use of equalizers does not have any significant impact on the overall network performance when the terminal is receiving only some control information every now and then, as for example in idle mode, or when the terminal is in a cell with a light traffic load. Hence some savings can be achieved in terminal battery consumption if a rake receiver with a lower complexity is used in such situations. In other words, it is desirable that the terminal can dynamically select between the rake receiver and the equalizer.

This paper presents a receiver structure that improves the WCDMA downlink FDD UTRA performance and can self-adapt to the channel conditions. The adaptation process is based on a choice between two receivers: a conventional rake receiver using maximal ratio combiner, as proposed by [2], and the same rake but using an adaptive linear minimum mean square error to perform the signal equalization and interference suppression. The decision variable is the correlation between the CPICH symbol and the received CPICH symbol.

2. Background on WCDMA

In the wideband CDMA (WCDMA) option of the FDD (frequency division duplex) mode of the 3GPP UMTS proposal for cellular wireless communication, both uplink and downlink use direct sequence CDMA (DS-CDMA) communication. Data at the transmitter side is spread by multiplying a fixed pseudo-random sequence, often called spreading sequence, in a DS-CDMA.

There are two main codes used in WCDMA: the spreading codes and the scrambling codes. The further, also called channelization codes, are orthogonal codes with real values based on orthogonal variable spreading factor (OVSF) technique. The codes are fully orthogonal, i.e., they do not interfere with each other, only if the codes are time synchronized. Thus, channelization codes can separate the transmissions from a single source. Therefore, in the downlink it can separate different users within one cell/sector. Scrambling codes, on the other hand, reduces inter-base-station interference.

In the downlink of the UMTS terrestrial radio access using WCDMA, payload and control data destined to a certain user are conveyed by means of one or eventually more so-called dedicated physical channels (DPCHs). In parallel to

---

0-7695-2533-4/06 $20.00 © 2006 IEEE

the DPCHs, a so-called common pilot channel (CPICH) is transmitted a signal carrying known symbols spread with the all-one OVSF sequence of length 256.

In DL WCDMA (FDD Mode), the data transmission is organized in frames of 10ms, each divided into 15 slots. Each slots has of length 2560 chips, where the chip rate is 3.84 Mchips per second. Figure 1 shows the frame structure of a typical downlink (DL) channel [6].

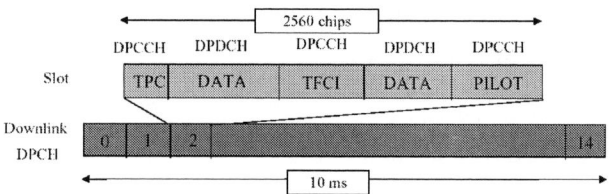

Fig. 1. Frame structure for the downlink DPCH

3. Receiver Structure

The proposed structure is based on the fact that while a rake receiver with MRC is the optimum receiver in Rayleigh fading channels with additive white Gaussian noise [3], the adaptive LMMSE (Linear Minimal Mean Square Error) equalizer is effective against digital interferers. Actually the MMSE linear equalizer is capable to minimize the MSE no matter what kind of noise may be present, but it suffers from the convergence problem of the weight in a fading channel.

The receiver depicted on figure 2 is proposed in such way that it takes advantage of the rake receiver and LMMSE characteristics. Another advantage is that the equalizer works at symbol rate instead of chip rate.

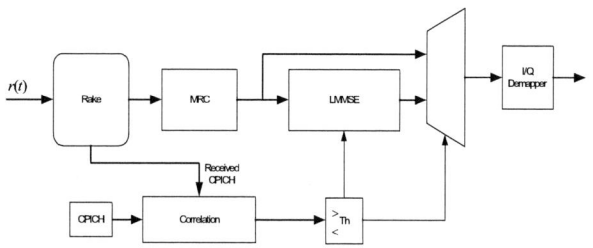

Fig. 2. Proposed receiver structure

To address the dynamic behavior of the receiver, specifically its self-adaptation, it is necessary to measure or obtain information about the channel quality. Both the short term mean squared error (MSE) and the projection of the adaptive filter to the space spanned by the signature sequence of the desired user were addressed in [4]. They conclude however that MSE is not robust since its works only in extreme situation (too bad or very good). The second approach is

not applicable here since we want to switch one receiver off to save power while in [4] the adaptive filter must work continually to provide the weight to the projection.

In our approach, we propose a simple correlation between the CPICH symbol and the despreaded received CPICH symbol, as will be detailed in subsection 4.1. If the correlation is high, it means that the number of users is low so the LMMSE can be switched off. When the number of users gets high, the correlation goes down, and if it is below a predefined threshold it triggers the receiver to switch to the equalizer.

4. System Architecture

4.1 Rake Receiver

The structure of a conventional rake receiver is depicted in Figure 3. Its functionality starts with the descrambling process, where the received signals are multiplied sample-wise by the scrambling code and delayed versions of the scrambling code. It is important to notice that delaying the input signal and multiplying it by the code has the same effect. The delays are determined by a path searcher prior to descrambling. Each delay corresponds to a separate multipath that will eventually be combined by the rake receiver.

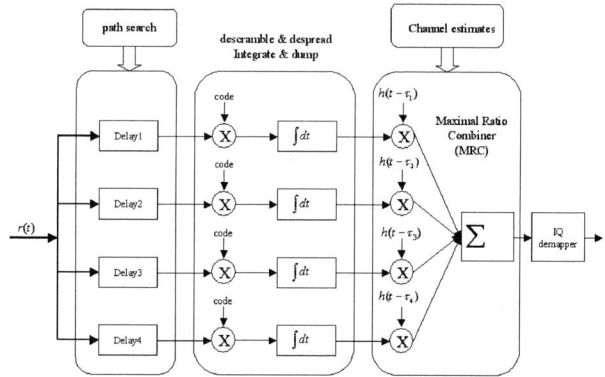

Fig. 3. Rake receiver combined with MRC

The descrambled data of each path is then multiplied by the spreading code and afterwards it is integrated over one symbol period, giving one complex sample output per quadrature phase-shift keying (QPSK) symbol. This process is carried out for all the paths that will be combined by the rake receiver. The same symbols obtained via different paths are then combined together using the corresponding channel information using MRC. The combined outputs are then sent to a demodulator to decide on the transmitted bits.

The objective of the channel estimation block is to estimate the channel response (phase and amplitude) $h(t,\tau i)$ for every identified path. Once this information is known, it can be used for combining each path of the received signal.

In the downlink of the WCDMA system, a common pilot channel (CPICH) is transmitted with 10% of the total transmitter power, so it is the best way to estimate the channel response. All the mobile terminals inside a cell receive this channel. CPICH is transmitted with a constant spreading factor of 256, which correspond to 30Kbps, and a spreading code of all ones. This results that there are 10 symbols per slot of CPICH. All the symbols of the CPICH are 1+j.

At the receiver end, the CPICH symbols are then used for channel estimation. The advantage of using CPICH to estimate the channel response is that all the data inside a frame can be used for channel estimation as opposed to only a few symbols in the DPCCH/DPDCH (DPCH). Furthermore, it is transmitted with a higher power than the time multiplexed pilot channel, resulting that it has better reception at the mobile terminal.

For each independent path, three steps must be taken in order to obtain the channel estimate (Figure 4). First the CPICH modulation is removed through multiplication of the CPICH data by its conjugate. In this case, the conjugate is 1-j since all the CPICH symbols are 1+j. This produces a channel estimate that is noisy due to the presence of AWGN and multiple user interferences.

Next, to overcome distortion caused by the random noise, the despread CPICH is averaged by a moving window average with duration of 6 pilot symbols. Then the obtained averaged channel estimates are either decimated or interpolated to match the data rate of the CPICH to the data rate of the DPCH. The process of interpolation is done by a simple zero-order hold (i.e. a simple repeater). This procedure works well if the channel is assumed to be stable for the symbol duration.

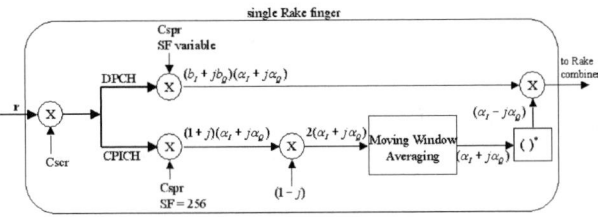

Fig. 4. Channel estimation

4.2 Adaptive LMMSE multiuser detector

The LMMSE is basically composed of two parts, the linear filter and the weights update. The basic equations are defined in (1) [5].

$$
\begin{aligned}
y_I(n) &= \hat{\mathbf{w}}_I^T(n)\mathbf{u}_I(n) + \hat{\mathbf{w}}_Q^T(n)\mathbf{u}_Q(n) \\
y_Q(n) &= \hat{\mathbf{w}}_I^T(n)\mathbf{u}_Q(n) - \hat{\mathbf{w}}_Q^T(n)\mathbf{u}_I(n) \\
e_I(n) &= b_I(n) - y_I(n) \\
e_Q(n) &= b_Q(n) - y_Q(n) \\
\hat{\mathbf{w}}_I(n+1) &= \hat{\mathbf{w}}_I(n) + \mu\left[e_I(n)\mathbf{u}_I(n) + e_Q(n)\mathbf{u}_Q(n)\right] \\
\hat{\mathbf{w}}_Q(n+1) &= \hat{\mathbf{w}}_Q(n) + \mu\left[e_I(n)\mathbf{u}_Q(n) - e_Q(n)\mathbf{u}_I(n)\right]
\end{aligned} \tag{1}
$$

The first two equations are the filter equations and the last two are the weights update (WUD). As we can see the weight update needs four real multipliers to execute the multiplication of the error by the input signal ($e(n)\mathbf{u}(n)$). Applying the equation identities shown in (2),

$$
\begin{aligned}
x &= (ac - bd) = a(c-d) + d(a-b) \\
y &= (ad + bc) = b(c+d) + d(a-b)
\end{aligned} \tag{2}
$$

we can reduce to three multipliers. Then the weight update equations become

$$
\begin{aligned}
\hat{\mathbf{w}}_I(n+1) &= \hat{\mathbf{w}}_I(n) + \mu\left[\left(e_I(n) + e_Q(n)\right)\mathbf{u}_I(n) - \left(\mathbf{u}_I(n) - \mathbf{u}_Q(n)\right)e_Q(n)\right] \\
\hat{\mathbf{w}}_Q(n+1) &= \hat{\mathbf{w}}_Q(n) + \mu\left[\left(e_I(n) - e_Q(n)\right)\mathbf{u}_Q(n) - \left(\mathbf{u}_I(n) - \mathbf{u}_Q(n)\right)e_Q(n)\right]
\end{aligned} \tag{3}
$$

Next we analyse the filters equations. For each output component, in-phase y_I and quadrature y_Q, we need two filters to produce them, that is $\hat{\mathbf{w}}_I^T(n)\mathbf{u}_I(n)$ and $\hat{\mathbf{w}}_Q^T(n)\mathbf{u}_Q(n)$ for y_I, and $\hat{\mathbf{w}}_I^T(n)\mathbf{u}_Q(n)$ and $\hat{\mathbf{w}}_Q^T(n)\mathbf{u}_I(n)$ for y_Q. But here we can follow the same reasoning that we used before to reduce the multipliers and reduce the number of filters. This procedure will result in the following filters equations,

$$
\begin{aligned}
y_I(n) &= \left(\hat{\mathbf{w}}_I^T(n) + \hat{\mathbf{w}}_Q^T(n)\right)\mathbf{u}_I(n) - \hat{\mathbf{w}}_Q^T(n)\left(\mathbf{u}_I(n) - \mathbf{u}_Q(n)\right) \\
y_Q(n) &= \left(\hat{\mathbf{w}}_I^T(n) - \hat{\mathbf{w}}_Q^T(n)\right)\mathbf{u}_Q(n) - \hat{\mathbf{w}}_Q^T(n)\left(\mathbf{u}_I(n) - \mathbf{u}_Q(n)\right)
\end{aligned} \tag{4}
$$

Continuing the analysis, we can conclude that now we need just three filters, because the filter $\hat{\mathbf{w}}_Q^T(n)\left(\mathbf{u}_I(n) - \mathbf{u}_Q(n)\right)$ is common to both signal components. Then we have reduced the total number of multiplier from 4N to 3N, where N is the number of taps filter. If we take into account the weight update equations, the total reduction is from (4+4)N to (3+3)N. However, now three delay lines are needed instead of two as before. The complete circuit is shown in Figure 5, while Figure 6 depicts the module for weights update.

5. Implementation and Performance Results

The system architecture presented in the previous section was first designed with Ptolemy II, an actor-based modelling and simulation framework [7]. It allows the integration of other simulators and handles heterogeneous compositions of blocks using distinct Models of Computation (MoCs), making it possible to describe both the designed system and the complex testbenches which are needed to model its usage environment. More details on this design methodology can be found elsewhere [8].

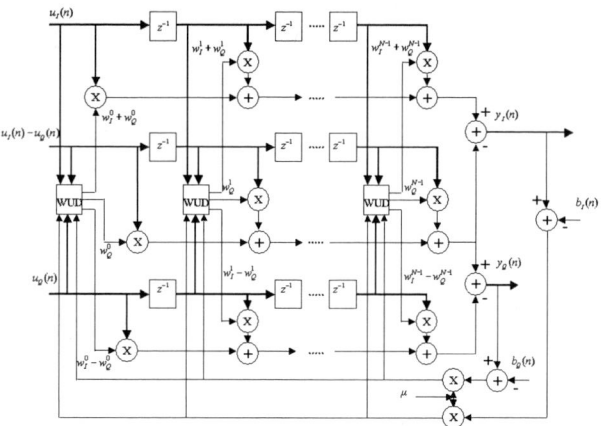

Fig. 5. Linear adaptive MMSE

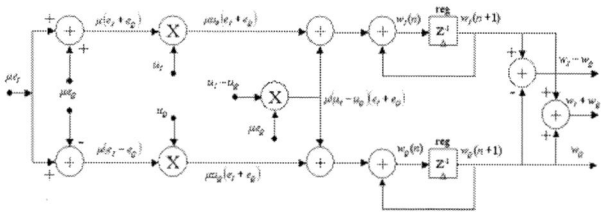

Fig. 6. WUD – Weights Update module

The following step was a refinement into a fixed-point cycle-accurate model, which was done in JHDL [9]. This model was exported as an EDIF 2.0 netlist and was implemented on a Virtex XCV 800 FPGA using Xilinx ISE. Table 1 shows the implementation results. Just for sake of comparison, the complete receiver was split in its three main components: the rake, the MRC and the LMMSE equalizer. One consideration is that for the power analysis it was consider the symbol rate equal to 30kHz, equivalent to spreading factor of 256 and a 4 taps equalizer.

The rake is the smaller circuit, as it has no multipliers, but it has the highest power consumption. This is due the higher clock in comparison with the others, as it works at chip rate while the others works at symbol rate. As expected, the LMMSE equalizer is the biggest circuit due the number of multipliers and consumes three times more power than the MRC, justifying the approach presented here that it should only be used when channel conditions require it. Another important result to be analysed is the maximum operation

frequency. Because of its complexity, the LMMSE limits the overall operation frequency up to 15.672MHz. However this is not critical, since the operation frequency required by the standard in this case is of 3.84MHz.

Table 1: FPGA implementation results

	Rake	MRC	Equalizer	Receiver
Number of Slices	772 out of 9408 [8%]	1464 out of 9408 [15%]	2557 out of 9408 [27%]	4793 out of 9408 [50%]
Connection Delay	1.821 ns	1.995 ns	1.944 ns	1.992 ns
Minimum Period	23.388 ns	6.273 ns	63.808 ns	63.808 ns
Maximum Frequency	42.757 MHz	159.413 MHz	15.672 MHz	15.672 MHz
Power	75 mW	12 mW	35 mW	115 mW

To validate the proposed approach, the performance of LMMSE adaptive equalizer and rake/MRC receiver with maximal ratio combiner are compared under the influence of a Rayleigh channel. The transmitted signal contains a variable number of users using spreading factor (SF) from 64 to 512.

Figure 7 shows the bit error rate for a receiver in the vehicular channel with velocity of 60 km/h, spreading factor of 64 and 15 users. A single correlator receiver is also shown just to give an idea about the performance of the other receivers. The MRC and the LMMSE present very good improvement in performance, as expected. The LMMSE equalizer provides better performance than the MRC already in a SNR of 4 dB. Figure 8 shows the variation of the bit error rate (BER) in function of the number of users. In this analysis the signal-to-noise ratio (SNR) is 10dB, SF 64, and terminal velocity of 60Km. It shows that the performance of the LMMSE is better with the increase of the number of users. With 10 users the advantage can already be noticed, but the difference is more prominent with 15 users or more, so that would be a potential threshold for the switching between the rake/MRC and the LMMSE. A more detailed analysis is left for future work, taking into consideration the user increase and decrease rates within a cell so that an optimum switching point can be found.

6. Conclusions and Future Work

This paper presented a receiver structure that can adapt itself according to the channel conditions. Two well-known receivers were combined to provide a way to overcome the channel impairment while keeping a reasonable balance in power consumption and hardware complexity. The first is the rake receiver using maximal ratio combiner, which is the optimal solution for wideband wireless systems. It takes advantage from the multipath channel characteristics to get

0-7695-2533-4/06 $20.00 © 2006 IEEE

time diversity. The second is an adaptive linear minimum mean square error multi-user detector (LMMSE), used to overcome the loss of user orthogonality introduced by a fading channel. The LMMSE is sub optimum in the sense that it does not eliminate the complete interference caused by others users, but it is still a good solution because it has low hardware complexity in comparison with the optimum solution, and is more suitable for a mobile terminal. Furthermore, a number optimizations were presented in section 4.2 aiming to reduce the number of multipliers and achieving a more efficient hardware implementation.

From the analyses in Section 5 we conclude that LMMSE has better performance than the MRC when the number of users increase, and that the correlation criteria can be used to measure the number of users. This advantage is followed by a reasonable increase of chip area and power consumption, justifying the use of a multi-user detection solution in a mobile terminal.

Current results are based on the implementation of both receiver structures in hardware, but future work will address the design of a single structure aiming towards the possibility of the dynamic reconfiguration of the receiver, as well as the reconfiguration control circuitry that takes into account both the variations of the channel conditions and the overhead time for the reconfiguration.

References

[1] Hooli, Kari: Equalization in WCDMA Terminals. Academic Dissertation, Oulu University Press, 2003, ISBN 951-42-7183-1.

[2] 3GPP: Third Generation Partnership Project, http://www.3gpp.org.

[3] Proakis, John G.: Digital Communications. MacGraw-Hill, 4. ed, 2000.

[4] Latva-aho, Matti; Juntti, Markku; Oppermann, Ian: Reconfigurable Adaptive Rake Receiver for Wideband CDMA Systems, IEEE 1998, ISBN: 07803-4320-4.

[5] Haykin, Simon: Adaptive Filter Theory, Fourth Edition. Prentice Hall, 2002, ISBN 0-13-090126-1.

[6] Holma, Harry; Toskala Antti: WCDMA for UMTS. John Wiley & Sons, Ltd, 2001.

[7] Indrusiak, L.S.; Prudêncio, R. B.; Glesner, M. Modeling and Prototyping of Communication Systems using Java: a Case Study. In: Proceedings of 16th IEEE International Workshop on Rapid System Prototyping (RSP), 2005, Montreal, Canada.

[8] Brooks, C.; Lee, E. A. et al: Heterogeneous Concurrent Modeling and Design in Java. Technical Memorandum UCB/ERL M04/27, University of California, Berkeley.

[9] Bellows, P.; Hutchings, B. L.: JHDL - An HDL for Reconfigurable Systems. In: Proceedings of FCCM '98, Napa Valley, 1998.

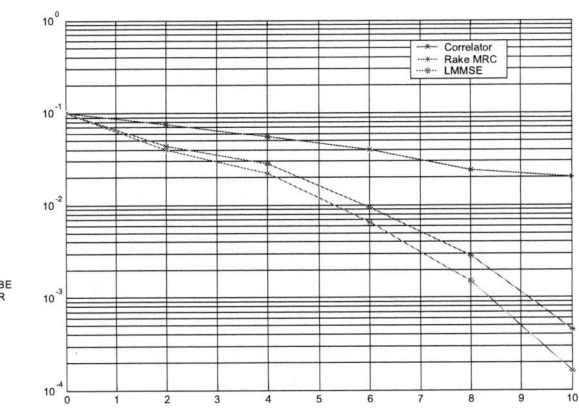

Fig. 7. BER versus SNR for receiver in a vehicular channel with 60Km/h, spreading factor 64, and 15 users

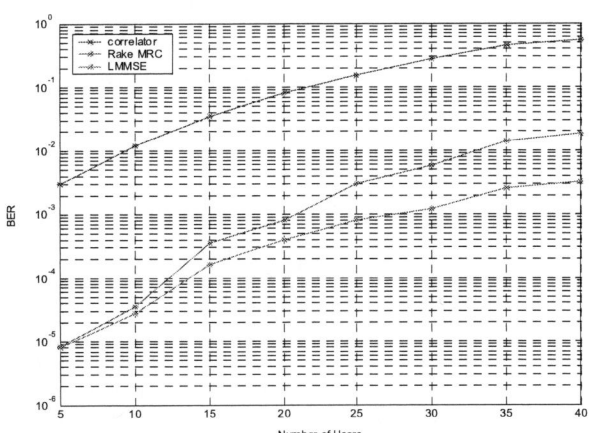

Fig. 8. BER versus Number of Users for receiver in a vehicular channel with 60Km/h, spreading factor 64 and SNR 10dB

Reconfigurable Systems Integration

Autonomous Realization of Boeing/JPL Sensor Electronics based on Reconfigurable System-on-Chip Technology

Evangelos F. Stefatos, Tughrul Arslan
School of Engineering & Electronics
The University of Edinburgh,
King's Buildings, Mayfield Road,
EH9 3JL, Scotland, UK,
Evangelos.Stefatos@ee.ed.ac.uk

Didier Keymeulen, Ian Ferguson
Jet Propulsion Laboratory
California Institute of Technology
4800 Oak Grove Drive
PASADENA, CA 91109
Didier.Keymeulen@jpl.nasa.gov

Abstract

This paper presents a reconfigurable architecture that is specifically tailored for the autonomous realization of the Boeing/JPL gyroscope and its interface circuitry. The overall architecture consists of a reconfigurable hardware substrate that suits the implementation of transposed form FIR filters and a custom non-stochastic search algorithm, which is in charge of autonomously providing the best binary configuration on the adaptive hardware. Preliminary simulation results show that our architecture is able to cope with single hard errors that occur on the reconfigurable substrate, while power analysis demonstrates that the proposed methodology is capable of implementing filters that consume almost 4.5 times less power, compared with industrial application specific reconfigurable fabrics.

1. Introduction

Digital finite impulse response (FIR) filters constitute irreplaceable hardware components in the crushing majority of electronic devices. Due to the nature of numerous applications (channel equalization, adaptive noise cancellation, fault-tolerance etc.), reconfigurability has become an important issue. Previous research works have introduced numerous reconfigurable FIR filter architectures [1], [2]. However, although these are able to change on demand the specification of the filter, they are vulnerable to the presence of hardware faults because the granularity of critical components (multiplier and accumulation block) within the filter is not adequate for exploiting redundancy. On the contrary with the previous research works, the authors in [3] have presented a multiplier-less reconfigurable architecture, which is also capable to compensate for single event hard errors

(SHEs). However, the injection of faults has been constrained only on the multiplier-block and not on the accumulative unit. This paper presents a novel architecture, which is uniquely suited for the implementation of reconfigurable linear-phase transposed form FIR filters. Moreover, a custom non-stochastic search algorithm based on a genetic algorithm (GA) has been implemented for efficiently providing the binary configuration on the reconfigurable hardware. The scope of this architecture aims at accomplishing the filtering tasks of the Boeing/JPL micromachined gyroscope [4]. Therefore, due to the nature of the targeted application, the architecture employs techniques in both physical and algorithmic level for reducing the power dissipation of the realized filters and efficiently mapping alternative configuration solutions on the reconfigurable design in order to cope with hardware failures. Figure 1 depicts a high-level schematic, which presents the functionality of the Boeing/JPL gyroscope. Alike common vibratory sensors, it employs feedback compensation to achieve harmonic excitation of selected modes, disturbance rejection and tuning of the sensor's dynamics. These tasks are employed by two control loops named drive and rebalance loop, respectively. Finally, there is a demodulation stage that estimates the angular rotation rate of the gyroscope by demodulating the sense rebalance signal with respect to a measurement of the drive loop response. It can be seen in [4] that different kinds of FIR filters including lowpass, passband and allpass filters are employed for accomplishing the electronics associated with the gyroscope's filtering and control tasks. In this paper a representative example of a high-order lowpass filter is given. The rest of this paper is organized as follows. Section 2 initially presents the reconfigurable hardware topology of our architecture and later on introduces the architecture of the employed search algorithm, which employs several techniques in order to

0-7695-2533-4/06 $20.00 © 2006 IEEE

realize filters with minimum hardware resources. Subsequently, section 3 demonstrates the simulation results for the realization of a high-order lowpass filter and describes how our architecture adapts in order to cope with SHEs. Finally, section 4 presents the synthesis results and the power analysis of our reconfigurable hardware design, through the paradigm of a 41-tap linear phase lowpass filter.

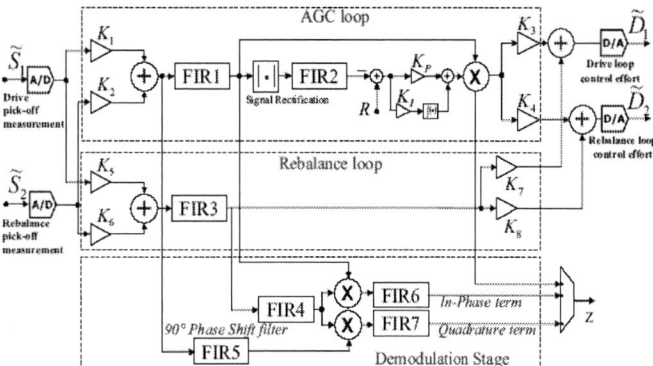

Figure 1. High-level schematic of sensor's electronics

2. Overall architecture

Our overall architecture consists of a reconfigurable hardware topology that aims at accomplishing the filtering tasks of the Boeing/JPL gyroscope and a custom search algorithm, which provides the binary configuration on the hardware. Both the hardware topology and the GA are combined in order to implement filters with minimum hardware resources. Figure 2 depicts the block diagram of the overall architecture. It can be seen that the GA during the evolution of the filters, considers the impulse response H(z), which is considered as the response of a discrete-time system to a unit-impulse u(n). The impulse response is represented by the coefficients of the evolved filter and is compared with the ideal impulse response (set of coefficients). Finally, after the implementation of the targeted filter, the GA provides the representative binary configuration and some additional control signals, which concern the accumulation unit of the filter.

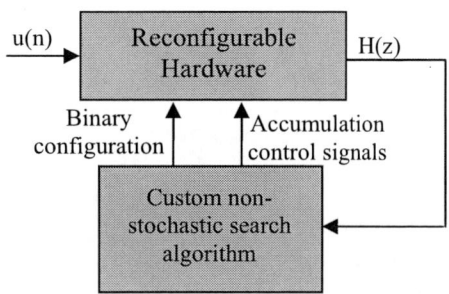

Figure 2. Block diagram of overall architecture

2.1. Reconfigurable hardware topology

The topology of the hardware has been specifically selected for the implementation of the transformed form of FIR filters. This technique has the advantage that only half coefficients have to be evolved and therefore the search algorithm needs considerably less time in order to obtain the binary configuration. In addition to this, our design consists of two parts named multiplier and accumulation block. Figure 3 shows the interface between these two parts. It must be mentioned here that the multiplier block does not employ a dedicated multiplier but instead it uses the primitive operation filtering (POF) technique for realizing the targeted coefficients. It has been proved in [5] that POF technique presents significant improvement in terms of silicon area and consequently power consumption, compared with canonical-signed-digit (CSD) algorithm [6]. Moreover, POF increases the granularity of the multiplier block and hence redundancy can be exploiting more efficiently. Based on this technique each coefficient in our architecture can be implemented with different combinations of addition/subtraction (A/S) and left/right (L/R) shift operations. Moreover, each coefficient may need different number of vertices in order to be implemented. A main target of this architecture is to minimize the number of the employed A/S units, as they consume considerably more power, compared with shifters and multiplexers. Hence, each vertex is able to be connected with either one of the three previous adjacent vertices in order to re-use former partial sums that facilitate the realization of subsequent coefficients. Figure 3 depicts that the first coefficient needs two vertices and therefore only the second tap is connected with the accumulation block. This mechanism is controlled by the GA and enables our architecture to be very flexible when system's adaptation is required or faults occur on hardware components either on the multiplier or accumulation block. Based on the folded transposed form of FIR filters, the multiplication of each data sample with each coefficient must be performed at the same time. Our design is a pipelined architecture and the multiplier block has been implemented so that each tap needs 6 clock cycles. Therefore, since each coefficient may need different number of vertices, each vertex must incorporate delay of variable clock cycles in order to cope with synchronization problems. Figure 4 shows that each vertex includes a programmable delay of 4 clock cycles. Hence, if for example a coefficient employs two vertices, the delay on these two must be adjusted to give a sum of 6. In addition to this, the A/S units includes a two's compliment module that determines whether addition or subtraction will be performed. Moreover, our architecture has been designed to realize either positive/negative and odd/even integer coefficients. Therefore, both L/R shift operations are needed. In figure 4 the upper shifter is capable to either perform binary multiplication or division

0-7695-2533-4/06 $20.00 © 2006 IEEE

by 256. On the other hand, the lower shifter performs binary multiplication by two or it just multiplies its input by one. This technique secures that any odd or even integer number can be realized. It must be mentioned here that decimal coefficient sets have first to be transformed into integer ones. On the contrary with the A/S units, the shifters are asynchronous circuits and their outputs feed the two inputs of each A/S unit, as it is depicted in figure 4. From the graph at the bottom end of figure 4, it can be deduced that when the input of the FIR filter is selected to feed the A/S units, this must be consistent in time with the A/S unit. Finally, figure 5 illustrates the internal design of the A/S units, which are attached on the accumulation block. These differentiate with respect to the A/S units of the multiplier block. More specifically, the control signal of the multiplexer is controlled by an adaptive mechanism, which collaborates with the search algorithm and determines whether the respective vertex on the multiplier block corresponds to a tap or it just constitutes an intermediate vertex. In the later case the horizontal input of the adder instantaneously feeds the output and neither addition or subtraction operation is performed.

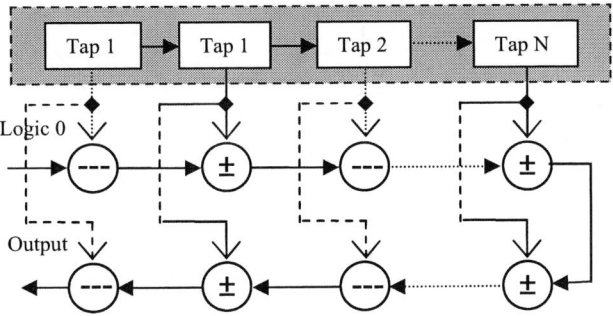

Figure 3. Reconfigurable hardware topology

2.2. Architecture of custom search algorithm

The GA, which has been employed to provide the binary configuration on the reconfigurable hardware substrate, composes a custom design that aims at realizing FIR filters with the minimum possible adder-cost and speeding-up the time that is needed for the filter implementation. This aim has been achieved by following several techniques in both reconfigurable hardware (each vertex re-uses former partial sums) and GA. From the algorithmic perspective, genetic operations take place only on the parts of the chromosome that correspond to the currently realized coefficient and the three previous vertices. According to this approach the search algorithm gradually realizes each coefficient individually and not the entire coefficient set at the same time. Hence, at any instance the search area of the GA is minimized and consequently the realization time of each coefficient is reduced. Figure 6 depicts the encoding approach of each

chromosome and shows how the GA assigns the coefficients to vertices. From the graph it can be deduced that the first coefficient utilizes two vertices, while the second realized coefficient employs the output of the second vertex plus two additional ones. During the realization of the second vertex, the GA performs genetic operations on vertices 1 to 4, since vertex 4 can utilize the output of each one of the three previous vertices. Concerning the third coefficient, the GA initially assigns to it one additional vertex (vertex 5) and it waits to see whether the coefficient 5 will be realized by utilizing vertex 5 and the partial sums of the vertices 2, 3 and 4. If the GA is not able to realize the third coefficient within a certain number of generations, then it assigns an additional vertex (vertex 6). Based on this approach, the GA ensures that the binary configuration does not employ redundant vertices for realizing a given coefficient set. Consequently, this has a significant impact on the power, which is consumed by the implemented FIR filter. Finally, the objective function of the GA is given as the summation of the ratios between the ideal and the evolved coefficients (C_{ideal} and $C_{evolved}$) as it is illustrated in the formula below. It can also be deduced that the maximum fitness-score for each coefficient is 1.

$$\text{Fitness} = \begin{cases} C_{ideal} / C_{evolved} & \text{if } C_{evolved} >= C_{ideal} \\ C_{evolved} / C_{ideal} & \text{if } C_{evolved} < C_{ideal} \end{cases}$$

Moreover, only when the fitness-score is equal to 1 (first coefficient has been evolved), the GA considers the fitness-score of the second coefficient. This is another point that our GA differentiates from previous research works [3], [7], in which all the sub fitness-scores of the different coefficients are considered from the start of the evolution, even if none of these has been precisely realized. The later methodology has been proved to create hardware dependencies between coefficients, which are not adjacent within the coefficient set and thus, a small change in the binary configuration that improves a certain coefficient it may negatively affect other non-adjacent coefficients. Therefore, this effect introduces significant delay in the convergence process of the GA or even worse forces the GA to get stuck on sub-optimal solutions. Finally, the GA collaborates with an adaptive mechanism, which controls the control-bits of the A/S units in the accumulation block. More specifically, this controller initially configures the A/S unit that must operate as accumulation node, considering that the realization of a specific coefficient needs one vertex. Subsequently, it considers the fitness-score, which can be obtained with this specific configuration and if for a certain number of generations the fitness-score does not achieve an acceptable threshold, then the controller assigns an additional vertex in the realization of the coefficient by turning-off the previous A/S unit and making accumulation node the next A/S unit. More details are given in section 3.

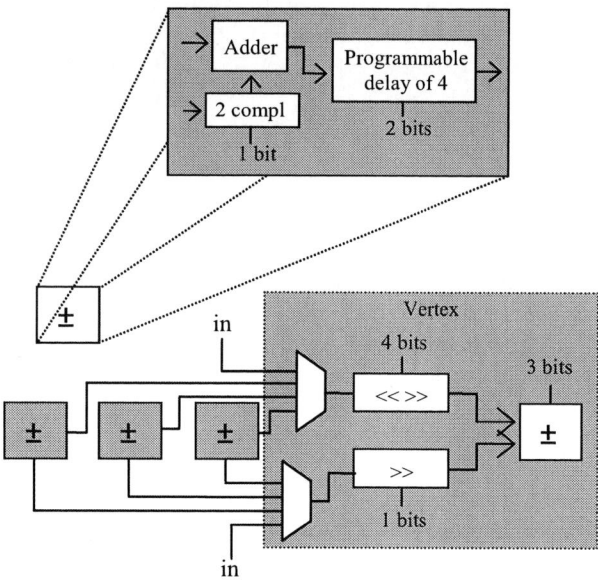

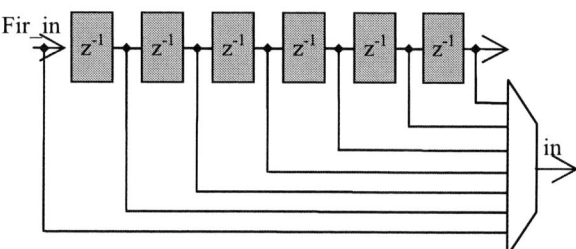

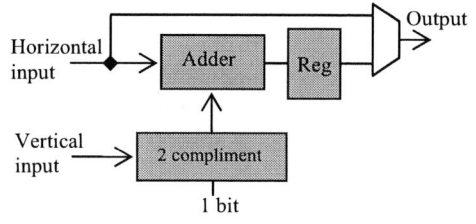

Figure 4. Interconnection between vertices within taps

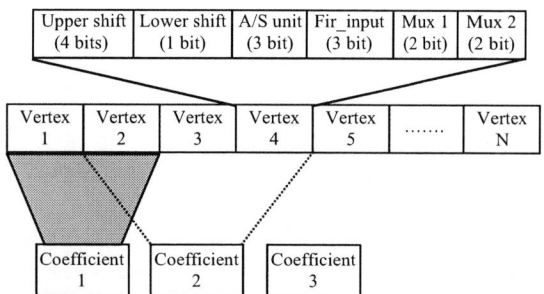

Figure 5. A/S unit in the accumulation block

Figure 6. Encoding scheme for search algorithm

3. Simulation results

The capability of our architecture to accomplish the electronics of the Boeing/JPL gyroscope is demonstrated through a 41-tap lowpass filter. Based on the targeted specification the passband stops at 0.2 rad/s, the stopband starts at 0.284 rad/s and there is 50db attenuation in the stopband. Initially, the realization of the filter is performed in an error free environment. Subsequently, different scenarios of SHEs are injected either in the multiplier and accumulation unit of the realized filter.

3.1. Error-free simulation of FIR filter

Due to the folded transposed form that our architecture supports, the GA has to realize only 21 coefficients. According to the simulation results the filter needs 45 A/S units in order to accomplish this specific coefficient set. This implies that on average the realization of each coefficient takes 2.14 vertices, which is quite impressive, considering that the coefficient set consists of 16 bits numbers. Finally, figure 7 depicts how precisely the evolved filter matches the specification. The green graph represents the frequency response of the evolved filter, while the blue one represents the ideal one, as it was generated in MatLab.

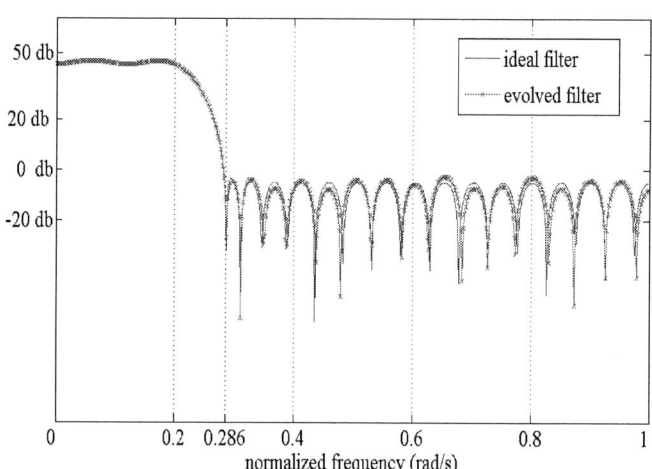

Figure 7. Comparison of the evolved and ideal filter

3.2. Adaptive mechanism for fault robustness

The proposed architecture employs an adaptive mechanism in order to recover from the occurrence of faults. Preliminary results show that our reconfigurable design is capable to cope with SHEs, which occur either in user (registers) or configuration memory. According to this mechanism, there is a control unit embedded in the search algorithm that when the fitness-score of the GA cannot reach a certain threshold, it changes the control bits of the multiplexers that exist inside the A/S units in the accumulation block after a number of generations.

0-7695-2533-4/06 $20.00 © 2006 IEEE   88

In the first scenario of faults injection, SHEs occur on the user memory of the multiplier or accumulation block. Figure 8 depicts how the controller modifies (lower graph) the configuration in the accumulation block and assigns an extra vertex-block in the currently realized filter. It can also be seen that a vertex-block consists of a vertex (multiplier block) and two A/S units (accumulation block). According to this specific scenario, the filter initially employs three vertices in order to realize a coefficient. However, because there are faults (modeled as stuck-at faults) in the third vertex-block, the fitness-score cannot overcome a satisfactory threshold and thus the controller changes the configuration and assigns an additional vertex block for the coefficient realization. Thus, after the reconfiguration only the A/S units in the fourth vertex-block accumulate and vertex 3 is isolated because vertex 4 selects only vertices 1 and 2 in order to realize the coefficient.

Moreover, the architecture is able to recover even when faults occur on the configuration memory. A representative example is given in figure 9, where the control bit of the multiplexer in the third A/S unit in the accumulation block has stuck-at logic 1, while the controller has assigned it to be logic 0 because the currently realized coefficient needs two vertices. However, due to the injection of a fault in the A/S unit of the third vertex-block, there are two A/S units that accumulate for the same tap and furthermore the additional A/S unit introduces an extra delay of one clock cycle that affects the behaviour of the filter. Therefore, the architecture has to take two steps in order to recover the functionality of the filter. Firstly, the three vertices (multiplier block) have to realize the coefficient within 5 clock cycles instead of six in order the architecture to compensate for the introduction of the additional delay in the accumulation block. Secondly, the second vertex has to be appropriately configured by the GA in order to output logic 0. Thus, the second vertex will be isolated from the realization process of the coefficient and the accumulation will be performed in the third vertex-block.

4. Synthesis and power analysis

For the purpose of this paper, we synthesized a design, which consists of 50 vertices. For the synthesis we employed Synopsys synthesis tools and UMC 0.13μm technology cell library. Moreover, since in this example we deal with 16 bits data samples and coefficient sets, we designed our design with 32 bits bus, in order to avoid overflow. Table 1 shows the synthesis results, which concern the die area that the multiplier and accumulation block of the filter occupies, respectively.

Subsequently, the power analysis of the 41-tap lowpass filter has been carried out using Synopsys Prime-Power tool. Moreover, interconnect parasitics were taken

into account for achieving a more realistic result. Figure 10 depicts the power, which is consumed by the simulated filter (0.693mW/MHz). Furthermore, the operational voltage of the design is 1.2 Volt. This in-turn implies that the evolved filter pulls current equal to 0.577 mA/MHz. This value is outstanding compared with ATMEL series FPGAs [8], which consume 2.62 mA/MHz in order to realize a symmetrical 32-tap filter. Finally, the power analysis of the simulated netlist showed that the 69% of the total power is consumed by the multiplier block and the rest 31% by the accumulation block.

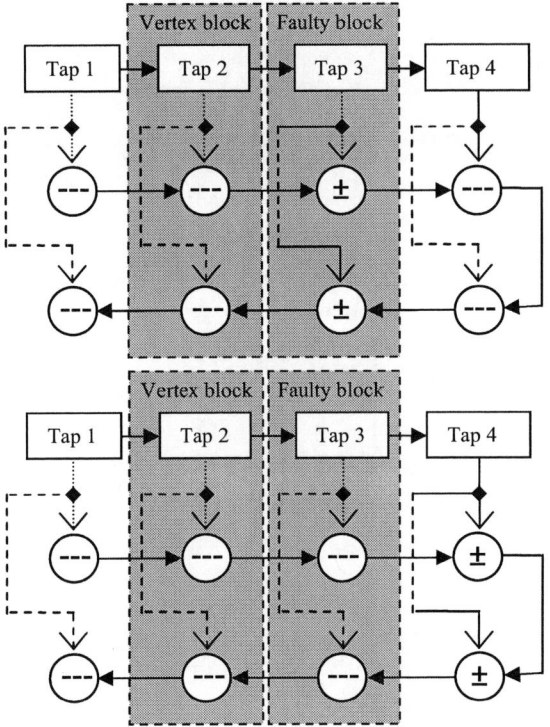

Figure 8. Recovery from SHE in user memory

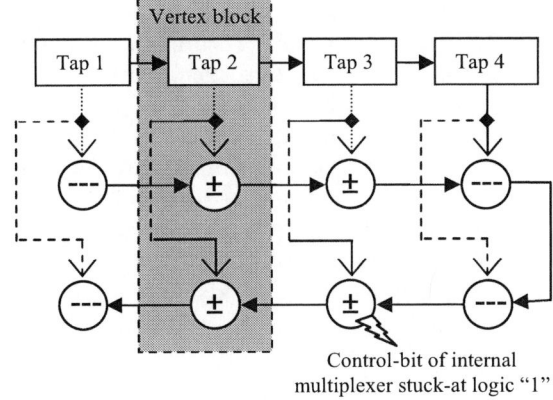

Figure 9. Recovery from SHE in configuration memory

Table 1. Area results of our design (50vertices)

32-bit Bus Design	Area (mm^2)
Multiplier block	0.964
Accumulation block	0.401
Total design	1.365

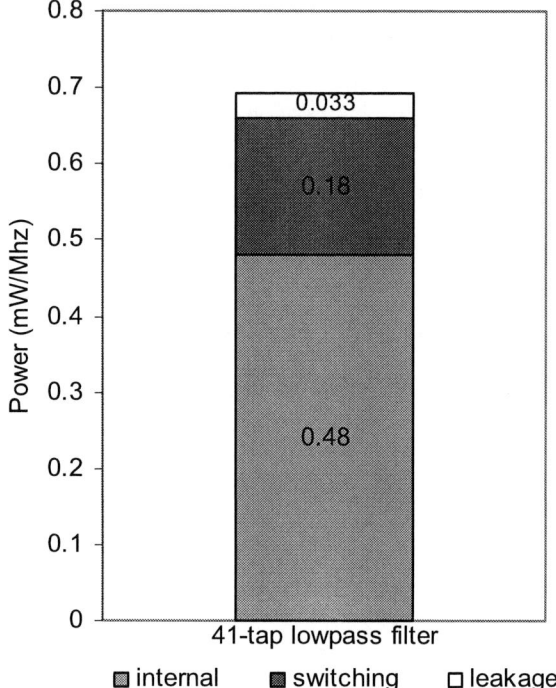

Figure 10. Power analysis of the 41-tap lowpass filter

5. Conclusions

This paper has presented an adaptive architecture, which is specifically tailored for the implementation of linear-phase FIR filters within Boeing/JPL gyroscope's control and interface circuitry. In conjunction with a custom non-stochastic search algorithm that has been specially designed for the reconfigurable hardware substrate, this architecture forms an autonomous low power fault-tolerant SoC solution. Several techniques have been employed both in physical and algorithmic level to reduce the power consumption of the realized filters, since battery autonomy is very crucial issue in space applications. Power analysis of a 41-tap lowpass filter netlist demonstrates that our hardware architecture consumes 4.5 less power than ATMEL series FPGAs, which target similar applications. Finally, preliminary simulation results show that our architecture is capable of compensating with SHEs that occur either on the user or configuration memory of our reconfigurable fabric.

6. References

[1]. Kuan-Hung Chen, Tzi-Dar Chiueh, "Design and implementation of a reconfigurable FIR filter", Circuits and Systems, 2003. ISCAS 2003. Proceedings of the 2003 International Symposium on Volume 4, 25-28 May 2003 Page(s): IV-205-IV-208.

[2]. Bruce, H., Veljanovski, R., Owall, V., Singh, J., "Power optimization of a reconfigurable FIR-filter", Signal Processing Systems, 2004. SIPS 2004. IEEE Workshop on 2004 Page(s): 321-324.

[3]. Hounsell, B. L., Arslan, T., "Evolutionary design and adaptation of digital filters within an embedded fault tolerant hardware platform", Evolvable Hardware, 2001. Proceedings. The Third NASA/DoD Workshop on 12-14 July 2001 Page(s): 127-135.

[4]. Yen-Cheng Chen, M'Closkey, R. T., Tran, T. A., Blaes, B., "A control and signal processing integrated circuit for the JPL-Boeing micromachined gyroscopes" Control Systems Technology, IEEE Transcactions on Volume 13, Issue 2, Mar 2005 Page(s): 286-300.

[5]. Bull D. R., Horrocks D. H., "Primitive operator digital filters", IEE Procedings. -G, 1991, Page(s). 401-412.

[6]. Avizienis A., "Signed-digit number representation for fast parallel arithmetic'', IRE Trans. Electon. Comput. 1961, 10, Page(s): 389-400.

[7]. Stefatos E. F., Arslan, T., Keymeulen D., Ferguson, I., "An EHW architecture for the design of unconstrained low-power FIR filters for sensor control using custom-reconfigurable technology'', Evolvable Hardware, 2005. Proceedings. 2005 NASA/DoD Conference on 29-1 June 2005, Page(s): 147-153.

[8].www.atmel.com/dyn/products/app_notes.asp?family_id=623 &part_id=2073.

Defect-Aware Design Paradigm for Reconfigurable Architectures

Rahul Jain, Anindita Mukherjee, Kolin Paul
Indian Institute of Technology, Delhi
{jvl042433, jvl042434, kolin}@cse.iitd.ernet.in

Abstract

With advances in process technology, the feature sizes are decreasing, which leads to higher defect densities. More sophisticated techniques, at increased costs are required to avoid defects. If nanotechnology based fabrications are applied, the yield may even go down to zero, as avoiding defects during fabrication will not be a feasible option. Hence, future architectures have to be defect-tolerant. Most of the current defect-tolerance schemes introduce redundancy in architecture to combat defects. Alternatively we can introduce defect tolerance in the design-flow.

In this paper we analyze the bottlenecks faced by current design-methodologies while addressing defect tolerance. We study the performance of present Place and Route tools on a defective fabric in terms of area and critical delay penalty, and explore routing aware placement in this context. We have proposed a new cost function, CA-RISA for improving the performance in a defect-aware environment.

1. Introduction and Motivation

VLSI technology progress for finer dimensions and larger chip area has lead to more complex processes and introduction of new and more complex material systems [3]. Due to higher defect densities, and complicated fabrication techniques the cost of manufacturing has increased. The increase in *defect/fault-sourcing complexity factor*[2], has lead to more devices being defective, hence reduced yields.

The problem of lower yield can be overcome by incorporating defect tolerance or fault tolerance. This will relax the stringent constraints of the manufacturing process. We can then leverage cheap but defect-prone technologies. In other words, there is a trade-off between process technology complexity in terms of yield, and post-fabrication complexity in terms of defect-tolerance. This is the design-space dimension which we will explore in this paper.

The issue of defect-tolerance assumes greater importance as we near the lithographic-limits. This

brings us to the era of molecular electronics. One alternative is the chemically assembled electronic nanotechnology (CAEN)[1]. The CAEN-based fabric can easily result in dense, highly regular, homogeneous structures. Reconfigurable architectures would then be a natural target for these technologies. It is expected that CAEN-based fabrics will have around 10% defect-density. It will be impossible to throw away a defective chip and we have to live with the defects. This makes the defect tolerance important for these devices.

Reconfigurable devices e.g FPGA's can provide a solution for defect tolerance in both the scenarios, present lithographic fabrics as well as future nanoFabrics. FPGA like, reconfigurable devices, enable programming the device after manufacture, thereby providing a means to incorporate information about defects in the fabric and build the circuit around the defects. This provides us with a method to have defect-tolerance with these fabrics, at an increased configuration cost, that is, area and time penalty.

Researchers have previously tried to build systems around defective chips by interconnecting many boards together *Teramac*[7]. But in present scenario the number of devices on a chip has exponentially increased. Hence, large designs can be realized on a single board itself. Also area penalty becomes less significant, due to high device densities and *free* Silicon resources, that is, we are using Silicon which would have been thrown anyway.

Most approaches for defect tolerance mainly concentrate on introducing hard-ware redundancy to overcome faults/defects. But this slows down the circuit, requires additional resources, increases the circuit complexity, and can work only for a certain threshold of defects.

This paper introduces a new paradigm for incorporating defect-tolerance. Unlike earlier approaches which incorporate redundancy in architectures, we propose a defect-aware design-flow for reconfigurable architectures. We analyze how the current place-and-route methodologies adapt to a defect-aware design paradigm, and try to quantify the relative performance of different algorithms after incorporation of defect awareness. We also try to

0-7695-2533-4/06 $20.00 © 2006 IEEE

explore how the cost functions associated with place-and-route can be changed, and propose a new cost function, CA-RISA (Congestion Aware RISA) and obtain an increase in performance of the P&R tools.

2. Defect-Aware Design Flow

The Design flow is modified such that the place-n-route tools now expect a *defect map*, and the design is compiled accordingly. Defect maps are discussed in detatil in section 3.3

We use VPR [13] as the FPGA place and route infrastructure. The VPR infrastructure was changed to incorporate defect-awareness. Following were some observations made during the course of the experiments:

1. The main bottleneck in P&R for defective FPGAs is defective interconnects, rather than defective devices.

2. Different defect distributions give different results. The defect distribution on a die is clustered[14]. The defects map randomly generated have to follow this distribution for correct analysis.

3. Different P&R approaches behave differently. It is found that congestion aware placement gave best results in section.

4. Based on the above observations we propose a new cost-function CA-RISA which gives us better performance in terms of time and delay penalty.

The above observations have been discussed in detail in section3.

3. Experimental Setup

VPR[13] is used as the FPGA place and route infrastructure. The VPR infrastructure was changed as follows to incorporate defect-awareness. We can extract the defect information for the whole fabric, by mapping each component of the FPGA on the defect map. The defective components are assigned zero capacities; hence they cannot be used as place and route resources.

3.1. Motivational Example: CLB defects vs Routing Resource defects

This experiment gives us a perspective on how the place and route algorithm behaves according to defects densities in clbs and routing resources.

The experiments are run at uniform random defect distribution for CLBs and routing resources. In Table 1 CLB defects indicate the percentage defects for clbs, where as RR defects indicate the percentage defects of the routing resources. VPR uses the Pathfinder algorithm. The 4th column shows us the number of

congested resources. For a circuit that has been routed this is 0. Circuits that cannot be routed, encounter congestions 1-10 show that as the percentage of defective routing resources fall below a certain threshold, the circuit becomes routable. This result is as expected, but we see unusual behavior in 6-10. We can see that for 6-10, against a given percentage of RR defects, the number of congestion faults decrease with increasing percentage of CLB defects. In fact, at a defective CLB percentage of 40 the design becomes routable again. This could be because at this stage there are lots of defective CLBs, so in principle the placed components are placed far from each other, and congestion is less intense.

Table 1				
	% CLB defects	% RR defects	congested RR	Routing success
1	10	50	193	No
2	10	40	29	No
3	10	35	11	No
4	10	30	3	No
5	10	27	0	Yes
6	47	35	0	Yes
7	40	35	1	No
8	30	35	1	No
9	20	35	3	No
10	10	35	11	No

Generally in FPGAs, logic utilization of up to 100% can be achieved but the routing resource utilization is usually low [9]. The results in Table 1 indicate that the feasibility of compilation is mainly dependent on routing resource defects. The CLB defects don't have much bearing on placement, as long as there are enough number of defect-free CLBs in the fabric. Whereas, defects in the routing-resources increase the congestion, which leads to routing failure. After a certain threshold of defective routing-resources, the design is no longer routable.

3.2. Base Cases

All the experiments have been run on 7 MCNC benchmarks with varying sizes. Table 2 shows their base area and critical delay for defect-free successful compilation.

Table 2: Base Cost			
Bench	CLB	T_Crit	Total Area
gcd	220	5.70e-08	34596
scf	418	5.57e-08	78400
table3	480	7.57e-08	148225
table5	485	8.07e-08	161604
apex1	700	8.55e-08	220900
ex5p	1064	1.10e-07	422500
alu4	1522	1.12e-07	454276

3.3. Defect Maps

The proposed design solution will expect a *defect-map* for each defective design. The test-phase is conducted only once for each fabric, which can gives us information about which CLBs and routing resources are defective. The compilation methodologies take the map as an input, and incorporate defect-tolerance. The defect-map is introduced after technology mapping and is an input to the place and route tools. We can extract the defect information for the whole fabric, by mapping each component of the FPGA on to a defect map. The defective components are assigned zero capacity. Hence they cannot be used as place and route resources.

It has been observed that the defects tend to cluster and do not follow a uniform distribution over the chip. For creating the defect maps we have used the simulation model presented by C.H.Stapper in [14]. The modeling involves 3 steps. The first step is to generate a random symmetric cluster. The model takes a chip as a grid array with the x axis representing real line and y axis representing imaginary line. The origin is assumed to be at the centre of the chip. The occurrence or absence of a fault is determined by assigning a probability to each grid point. This is done with a symmetrical bivariate Gaussian distribution that is described mathematically as

$$P(x,y) = C \exp \{ (x^2 + y^2) / 2\sigma^2 \}$$

where C is a constant, σ is the standard deviation, x is the real part of the coordinates, and y is the imaginary part. Both C and σ are simple scalars. The value of σ typically is 0.3 or less, thus assuring that the 3σ limits of the distribution fall within the map boundaries. The value of C sets the density of the simulated faults, and depends on the grid area.

Since the defect clusters are rarely symmetrical so a second step of cluster shaping is done. In this step the real and imaginary coordinates are divided by random numbers between 0 and 1. We have not taken scratches into account since they are rare. Finally the clusters are randomly rotated and displaced.

The above method gives a defect map on a chip die. To map it onto a FPGA fabric we assume each cell of the defect map grid equal to the interconnect width. This is equivalent to dividing the whole FPGA fabric into a array of cells of side equal to the interconnect width. This implies that a interconnect is now a 1-D array of cells according to its position in the fabric, while the CLB is a 2-D array of cells. For simplicity we have taken the same width for both the horizontal and vertical tracks. The defect map on the above grid is found and if a defect maps onto any of the CLB's or interconnects, they are assumed to be defective.

3.4. VPR placement cost functions

The following experiment was run to find out which placement and routing strategy adapted best to defect-aware environment.

VPR uses simulated annealing for placement, and Path_finder algorithm for routing. There are several cost functions which can be used. Table 3 gives the performances of different cost functions for placement. VPR implements 4 types of cost function, namely, Bounding Box with Linear Cost function (BB_L), Bounding Box with Non-Linear Cost function (BB_NL), net_timing_driven (Net) and path_timing_driven (Path).

Table 3						
Bench	nx ny	chan wd	T_Crit (10^{-8} s)			
			BB_NL	BB_L	Net	path
gcd	15	6	5.70	5.49	NA	NA
scf	23	11	5.57	NA	NA	NA
table3	27	11	7.58	8.32	6.28	6.0
table5	22	11	8.07	8.14	NA	NA
apex1	21	7	8.55	NA	NA	NA

Table 3 shows the comparison of the 4 cost functions at minimum number of CLB's and Channel Width returned by BB_NL. We find that other algorithms fail to do the routing for some benchmarks. But the nonlinear, bounding box algorithm by far gave us the best results. This is an implementation of congestion aware cost-function RISA[12].

0-7695-2533-4/06 $20.00 © 2006 IEEE

RISA works on the principles of balancing demand and supply over an array of regions. In order to make the routing congestion visible to a metric, the analysis is performed over an array of N x N regions. The metric mainly tries to balance the demand with the supply in a region.

The reason RISA performs better with respect to other routing algorithm in the defect-aware scenario, is because it is congestion aware. We know from pervious analysis that introduction of defects mainly causes pressure on routing resources. RISA takes routability into account, during placement, hence, it eases this pressure and gives lesser number of congestions.

3.5. Analysis of the current design methodology

Our proposition is to use a defective fabric, at the cost of reduction in its functionality. Among the choices, RISA is the most defect-tolerant. Hence we have used it for the analysis. Table 4 has results generated for a set of 7 MCNC benchmarks of different sizes. We see that to port a design on a defective fabric, we incur area as well as critical delay penalty.

For calculating the penalties we use, results calculated in Table 2 as the base case.

Table 4					
Bench	No.of CLB's	Defect (%)		Penalty (%)	
		clb	int	Time	Area
gcd	220	2.5	14.2	52.5	192
scf	418	1.3	7.4	39.6	382
table3	480	1.9	11.0	20.1	251
table5	485	3.7	12.6	-4.4	246
apex1	700	2.4	15.8	2.5	369
ex5p	1064	2.8	15.5	9.2	462
alu4	1522	3.2	17.6	12.6	610

From Table 4, we see that the design flow can tolerate 10-15% defects in routing resources, with some area and time penalty. Note that we incur more than 200% area penalty for larger designs. But the significance of area penalty is comparatively less, because the Silicon resources are getting relatively cheaper with progress in technology, moreover we are using defective fabrics, which would have been otherwise thrown away.

The time penalty is less than the area penalty. We even get a improvement in time in the first case. The time penalty does not follow a trend. This observation indicates, that we can get our designs working, with

low critical path degradation for defective FPGAs. As the technologies allow more devices on a die, the area penalty will not cause significant setbacks to the feasibility of defect-tolerance.

3.6 Proposed Cost Function

The proposed cost function tries to balance the demand and supply in the NxN regions dividing the FPGA[12]. Following is the way demand and supply is defined in RISA:

Demand: Expression for the routing demand on a region Ri (W,L), due to a net n_k (net b-box X,Y) with an overlap of (w,l) is,

$$D_k^{i,horiz} = q \times \frac{w \times l}{Y \times L} ; D_k^{i,vert} = q \times \frac{w \times l}{X \times W}$$

If each channel is modeled as a region, then for a FPGA architecture we can take, W=L=w=l=1. Then the total routing demand on a routing-element is the sum of each net's routing [11]. However in VPR the FPGA is divided into regions which are coarser grained than just one channel. In our experiments the FPGA has been divided into 4x4 regions.

Supply: Supply indicates the sum of channel capacities over a region. In VPR, the capacities only take channel width into account.

Following expression give the supply in RISA.

$$_v S_{ij} = T_v / N \quad , \quad _h S_{ij} = T_h / N$$

Where, $_h S_{ij}, _v S_{ij}$ are indicative of supplies of the available horizontal and vertical tracks, that is, routing resources in a region.

In our approach we want this to reflect the reduced routing capacities in the NxN regions, that divide the fabric. Hence we change, the supply function as follows:

$$_h S_{ij} = T_h / N - W * \sum_i \sum_j N_{hdef} ,$$

$$_v S_{ij} = T_v / N - W * \sum_i \sum_j N_{vdef}$$

Where N_{hdef}, N_{vdef} are the number of defects in the horizontal and vertical tracks, within the region i,j respectively. W here is the weighting factor. This is the factor by which we penalize the defects. Higher value of W would assign high penalties to defective resources. High values of W, would aggressively reduce the supplies of regions with more defects, as seen by the cost function. Hence higher values of W, push the placement away from regions having high defects (where the clustering takes place). Thus the pressure of routing in these regions ease, and the routing may succeed.

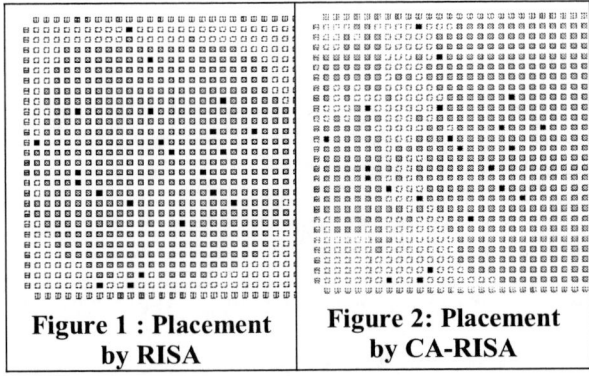

| **Figure 1 : Placement by RISA** | **Figure 2: Placement by CA-RISA** |

Figure 1, shows the placement where for W = 0. This is equivalent to the original RISA cost function, as can be seen. Where as Figure 2 shows the placement for CA-RISA, with W=10.

In this case the cluster lies in the middle of the die. Due to simulated annealing, the placement tools would try to place all the blocks in the middle of the fabric. We can see that the first placement solution, can hardly avoid the defect cluster. But when we run the same netlist, with high values of W, then the tool starts avoiding the defective regions, and we are able to route the design.

3.7. Performance Evaluation of Proposed Cost Function

The Table 5 and Table 6 shows the results obtained for the benchmarks for the placements with CA-RISA.

Table 5 : Least Time Penalty				
Bench	**%Def**		**% Penalty**		**% Imp**
	clb	int	Time	Area	
gcd	2.5	14.2	45.6	192	4.50
scf	1.3	7.4	8.6	382	22.15
table3	1.9	11.0	1.1	251	15.79
table5	1.8	12.0	-6.6	248	2.33
apex1	2.4	15.8	-3.1	369	5.45
ex5p	2.8	15.5	7.3	462	1.8
alu4	3.2	17.6	4.3	610	8.0

Table 5 shows the performance of CA-RISA when we keep the area same for a particular benchmark, as required for routability by RISA. Time and Area penalty, show the penalty incurred by CA-RISA as compared to a defect-free place-n-route solution. The last column shows the, improvement in critical delay obtained by CA-RISA, when compared to RISA. We

see a improvement in critical delay, keeping the same area.

Table 6 : Least Area Penalty				
Bench	**%Def**		**% Penalty**		**%Imp**
	clb	int	Time	Area	
gcd	2.5	14.2	45.6	192	0
scf	2.7	9.8	30.1	341	8.59
table3	4.0	12.1	14.6	180	20.44
table5	2.9	13.9	15.6	200	13.42
apex1	1.5	9.8	24.1	340	6.19
ex5p	2.3	14.9	26.5	437	5.8
alu4	2.8	18.3	16.5	553	10.4

Table 6 shows the performance of CA_RISA, when we try to achieve minimum area. Time and Area penalty columns, show the penalty incurred by CA-RISA as compared to a defect-free place-n-route solution. The last column shows the, percentage improvement in area of CA-RISA as compared against RISA, computed for minimum area. CA-RISA gives better results for area also.

Table 7				
Bench	**nx,ny**	**chan_wd**	**Congestion**	
			RISA	CA-RISA
gcd	17	12	0	0
scf	26	17	0	0
table3	26	16	578	0
	25	22	0	0
	25	19	373	0
	24	18	280	10
table5	25	21	692	0
	26	20	738	0
	25	23	0	0
	26	22	0	0
apex1	30	26	591	0
	31	25	669	2
	31	26	0	0
ex5p	35	33	0	0
	34	33	1466	0
	35	32	1324	34
alu4	43	32	0	0
	42	31	578	0

While comparing time penalty columns of Table 6 and Table 4, we observe that for the last three benchmarks, we get an improvement in time also with CA-RISA. Hence, we can conclude that CA-RISA gives us several pareto-optimal solutions, with respect to time and area. It is superior to RISA for defect-aware design-flow.

The Table 7 shows how effective CA-RISA is in reducing the congestion. For most cases in which RISA placement encountered congestion, our cost function produced a routable placement, or brought down the congestion considerably. This clearly shows the effectiveness of our cost function.

4. Future Work

The proposed methodology requires the circuit to be placed and routed for each chip separately. With increasing device density this might not be a feasible option, so integration of the defect maps into the synthesis flow should be as late as possible. One option is to have a initial placement solution for a non-defective FPGA. This solution can now be incrementally processed to map onto a defective fabric. This might reduce the place-and-route time considerably.

4. Conclusion

The present design methodology for defective FPGA has been explored. The bottlenecks faced by these have been studied. We have come up with a cost function CA-RISA, which reduces congestions.

This work does not provide a complete solution to the problem of using defective reconfigurable architectures, but provides a motivation for a defect-aware design paradigm. It tries to highlight the bottlenecks of the present methodology, and provides pointers in the direction of better solutions for defect-tolerance.

10. References

[1] S. C. Goldstein and M. Budiu, "NanoFabrics: Spatial Computing Using Molecular Electronics", *in Proceedings of the 28th Annual International Symposium on Computer Architecture*(ISCA 2001), pp. 178–191, July 2001.

[2] Milton Godwin, Manuela Huber, Richard Jarvis, and Fred Lakhani, "Examining upcoming yield enhancement challenges in the 2001 roadmap", *Micro Magazine*, 2002 Issue.

[3] S.M.Sze, *"VLSI Technology"*, Tata McGraw-Hill, 2nd edition.

[4] Mishra, M. and Goldstein, S.C., "Defect Tolerance at the End of the Roadmap", in *Proceedings of International Test Conference*, Sept. 30-Oct. 2, 2003 (ITC 2003), Volume 1, pp:1201 – 1210.

[5] R. Amerson, R. Carter, B. Culbertson, P. Kuekes, and G. Snider, "Teramac–configurable custom computing", in *Proceedings of IEEE Workshop on FPGAs for Custom Computing Machines*, 1995, pp: 32-38

[6] J.R. Heath, P.J. Kuekes, G. Snider, and R.S. Williams, "A Defect-Tolerant Computer Architecture: Opportunities for NanoTechnology", *Science*, 280, 1716 (1998)

[7] B. Culbertson, R. Amerson, R. Carter, P. Kuekes, G. Snider, "Defect tolerance on the Teramac Custom Computer", *Proceedings of the 1997 IEEE Symposium on FPGA's for Custom Computing Machines* (FCCM 97), pp. 116-123

[8] Aman Gayasen, N. Vijaykrishnan, M. J. Irwin, "Exploring Technology Alternatives for Nano-Scale FPGA Interconnects", *Proceedings of the 42nd annual conference on Design automation*, (DAC 05), pp. 921-926

[9] Andre DeHon, "Design of Programmable Interconnect for Sublithographic Programmable Logic Arrays", in *Proceedings of the International Symposium on Field-Programmable Gate Arrays* (FPGA2005), pp. 127-137.

[10] Amit Singh, Malgorzata Marek-Sadowska "FPGA interconnect planning, International", Workshop on System-Level Interconnect Prediction, *Proceedings of the 2002 international workshop on System-level interconnect prediction*, pp: 23 – 30

[11] Parivallal Kannan, Shankar Balachandran, and Dinesh Bhatia, "On Metrics for Comparing Routability Estimation Methods for FPGAs", *Proceedings of the 39th conference on Design automation Conference*, June 2002, pp. 70-75.

[12] Chih-Liang Eric Cheng, "RISA: accurate and efficient placement routability modeling", *Proceedings of the 1994 IEEE/ACM international conference on Computer-aided design*, pp. 690-695

[13] V. Betz and J. Rose, "VPR: A New Packing, Placement and Routing Tool for FPGA Research," *International Workshop on Field Programmable Logic and Applications*, 1997.

[14] C. H. Stapper, "Simulation of spatial fault distributions for integrated circuit yield estimations," *IEEE Trans. Computer-Aided Design*, vol. 8, pp. 1314-1318, Dec. 1989.

0-7695-2533-4/06 $20.00 © 2006 IEEE

New 2-Dimensional Partial Dynamic Reconfiguration Techniques for Real-time Adaptive Microelectronic Circuits

Michael Hübner, Christian Schuck, Matthias Kühnle, Jürgen Becker

Institut für Technik der Informationsverarbeitung – ITIV
Universität Karlsruhe (TH), Germany
http://www.itiv.uni-karlsruhe.de/
{huebner, schuck, kuehnle, becker}@itiv.uni-karlsruhe.de

Abstract

Short time-to-market pressure, high cost and risks and power consumption are keywords in development of microelectronic solutions for embedded systems as well as for universal and application tailored processor architectures. Modularity and flexibility while design-time, e.g. for System-on-Chip (SoC) component design, is not sufficient if the possibility of run-time reconfiguration of novel architectures has to be considered. Here, exploitation of real-time and on-demand reconfiguration of silicon area personalized on suitable granularities demonstrates high situation adaptivity and perspectives for next generation microelectronics. This paper discusses our implemented, synthesized and tested on-demand and partial reconfiguration approaches for fine-grain (Xilinx Virtex FPGAs) data paths. This includes also very new dynamic and partial reconfiguration for 2D placement and routing adaptation for today's fine-grain Xilinx FPGAs.

Keywords: On-Demand & Partial Reconfiguration, 2D FPGA placement, Dynamic Online Routing & Mapping

1 Introduction

The possibility to perform dynamical and partial hardware reconfiguration of FPGAs increases their flexibility and ability of run-time adaptation. This feature is provided by for example Xilinx Virtex II FPGAs, which can be dynamically reprogrammed using the so called ICAP interface. By reconfiguring parts of the chip's architecture on-demand, the application bitstreams that are currently not needed can be outsourced on a off-chip memory, for example on a FLASH memory, to reduce the necessary chip area. Since very often not all functionality is needed at the same time, this would make it possible to realize more functionality on one chip or to choose a smaller FPGA. An example of a system which uses this feature is presented in [10]. This feature is used in new approaches by outsourcing configuration data which makes it possible to use FPGAs with smaller configuration memory and consequently smaller chip size. Thus, it is possible to save costs and reduce power consumption since not actually used modules of a complete system do not allocate configuration memory and corresponding power consuming hardware [1]. Introducing the paradigms of [3] into a reconfigurable system and exploiting the adaptivity while run-time, opens a wide spectrum for power and performance aware designs. Systems-on-Chip is the very promising way to overcome problems with traditional used technologies. Designing chips with the features of processors, reconfigurable architectures and even micromechanical parts can help to shrink the size of devices and optimise the performance and power consumption. The unique solutions for systems with the goal of enhancement of human living are adaptive systems. Embedded systems have to realize the requirements of the application and adapting themselves to this task. Concurrently the system does this adaption with respect to the power consumption. The demands of the application changes while run-time initiated e.g. by variable requested quality of service. Like organic life, systems have to adapt abutted to this behaviour. One example is Network on Chips. These networks connect parts in a Chip which have several tasks. A new challenge is to design these networks adaptive to optimise data rates and power. Adaptivity and networks have to be seen as parallel tasks to be optimised in embedded systems of the future. As mentioned above, the state of the art approaches for run-time reconfigurable systems offer the possibility of substituting hardware modules while other parts of the system stay in operation. Unfortunately the architecture of Xilinx Virtex FPGAs only provides the substitution of complete columns with the logic parts (configurable logic blocks) and the routing resources. This leads to designs, where fixed rectangular shaped modules of hardware, were substituted. The size of these modules normally is related to the maximum required area of the biggest hardware module. This means, the biggest hardware module sets the size of the partial reconfigurable areas. Considering this, enables the allocation of every configurable area for each hardware module. Figure 1 shows the hardware reconfigurable system described in [2], designed using the tools described in [4][5]. The four reconfigurable modules have exactly the same size of resources on the FPGA. Certainly this approach is adequate for a variety of applications as described in [2]. The lack of this approach is that the resources of the FPGA cannot be utilized in an optimal manner since not all hardware modules exploit all

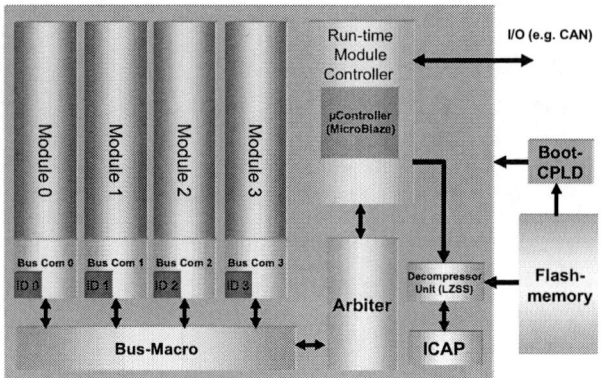

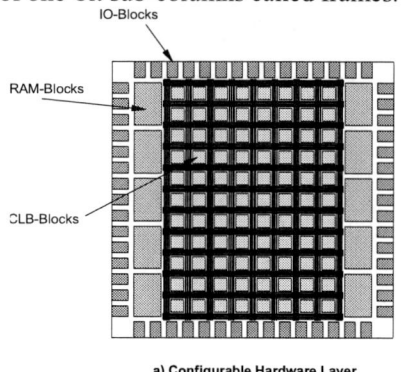

Figure 1. Dynamic Reconfigurable System

logic which is within their area. Therefore a new approach was developed which allows the placement of rectangular hardware modules with a variable size. This paper presents an approach for two-dimensional placement with Xilinx Virtex-II FPGAs. This solution is based on parametrizable pre-routed hardware blocks, stored within an external memory. A further step for this solution is an integrated on-line routing method for individual connection of hardware blocks while run-time. Also this approach will be presented and described in detail in the next chapters. This paper is organized in the following manner: section 2 describes the new approach for 2D partial HW reconfiguration. In section 3, the read-modify-writeback technique is described, and section 4 and 5 further presents the implemented reconfigurable modules as well as the interconnect and routing technique. Finally, the implemented system and performed tests are described in section 6, and the paper is concluded in section 7.

2 New 2D Partial FPGA Reconfiguration

Due to the high performance and parallelism of FPGAs, this technology is now more and more common in the design and integration of system-on-chips, rather than exploited only for rapid prototyping purposes. With the new possibility of dynamical and partial reconfiguration, the performance and flexibility of an embedded system can be even further increased. Using this feature, functionality on the FPGA can be replaced on-demand during run-time. However, the design of dynamical and partial reconfigurable systems is very complicated and time-consuming. The designer must be careful since the misplacement of one signal line can lead to destruction of the FPGA during reconfiguration. In order to simplify the design process and improve the security, new methods and tools must be developed in order to perform the design process on a higher abstraction level. Especially by introducing the new method of 2D dynamic and partial reconfiguration necessitates the development of basic functionalities for implementation into existing systems. These basic functionalities enable a transparent usage of the features in reconfigurable architectures. Basically a Virtex FPGA consists of two layers and additional configuration and control logic which handles the configuration bitstream

loading and the distribution of the configuration data on dedicated positions of the second layer. The first layer contains the reconfigurable hardware like logic-blocks (CLBs), RAM-blocks, I/O-blocks and configurable wiring resources. All blocks of the same type are aligned into columns (except the I/O-blocks), see Figure 2. The second layer which matches to the first layer is organized into columns as well. Each column whose width depends on the covered block-columns from the first layer consists of a set of one-bit sub-columns called frames.

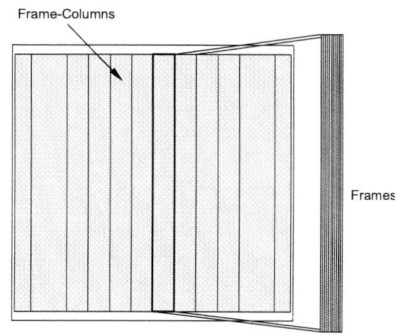

Figure 2 Layered FPGA structure (a,b)

A frame is the smallest piece of reconfiguration information that can be written on an FPGA and each frame contains fractions of the configuration information needed to configure the blocks assigned to the column. So if one block in a column is to be reconfigured all other blocks in the same column have to be rewritten as well. This restriction leads to that partial reconfiguration of functions can only be done on groups of consecutive columns. It is possible to write the frames in a random manner of order, but the configuration files are normally structured in an ordered way.

Virtex-II FPGAs provide no possibility to overcome the circumstance to write configuration data in complete columns. A work-around is, to exploit the possibility of the read-modify and writeback method. Virtex-II FPGAs includes an internal configuration access port (ICAP) which enables to read-back the actual configuration data from the internal memory. After reading back the data, a software tool is used to modify the configuration and write it back via the same configuration port. This method is described more in detail in the next section.

0-7695-2533-4/06 $20.00 © 2006 IEEE

3 Read-Modify-Writeback

Figure 3 shows a schematic view of the modular reconfigurable system. As a control element the soft-core processor MicroBlaze or the hard-IP PowerPC 405 in a Virtex-II Pro architecture can be used. The system is connected via a Flash controller to an external Flash memory. The controller enables both read or write access to the memory device. Another device integrated into the FPGA is the Hardware-ICAP module which provides the access to the FPGA internal ICAP-interface for reading or writing configuration data to or from the devices' configuration memory. This Xilinx IP-Core is described more in detail in [7]. The software drivers provide several functions to access the configuration memory. As a buffer, an internal block-RAM is used to store the data, which has been read from the device. The software to handle the data has access to this block-RAM for further processing while run-time. First implementation uses the connection via RS232 (Uart-IP) to read the configuration data from an external PC. For this purpose, the software running on the internal processor only provides the GUI to handle external commands for accessing the hardware ICAP IP-core. As shown in Figure 3, rectangular shaped modules are connected via a user-IP interface to the controlling element. The size of these modules and their interfaces are described in the next section.

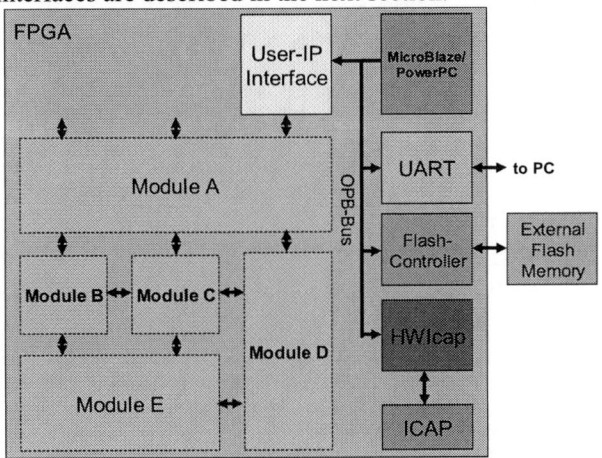

Figure 3 Dynamic Reconfigurable Modular System

After receiving the configuration data with the PC, JBits was used to modify the bitstream and write it back to the device. This first approach was developed for debugging purpose. To establish the data-base of modules, the complete content of the FPGA is read for extracting the relevant data of the modules. This is achieved by using the Java based functions of JBits. After extraction of the modules' configuration data, the Flash is programmed with the different configurations of the modules. Important here is, that the stored data does not contain any placement information. This is the initialization phase of the system. While run-time, it is possible to place the stored module to an area by including the position information. The process to place a module, starts by reading back the columns of the area, which is dedicated for the placement. The data which were read-back contains also the configuration information of the upper and lower areas as described in section 2. Now the stored module data were merged into this bitstream. Only the positions of the bitstream concerning the module which has to be placed were substituted. Therefore, configuration outside of the dedicated area is not affected.

The implemented system enables to position a module variable to a vertical position. This is due to the fact that the Virtex-II architecture is homogeneous in vertical direction. Positioning a module to any horizontal coordinate is not possible. The cause here is the heterogeneous architecture which includes block-RAM and multiplier blocks. Routing resources crossing these blocks are different to those which can be used in an area without BRAM elements. However, placement of modules to an area with same horizontal conditions is possible, which is described in section 5.

4 Reconfigurable Modules and Interconnect

One important issue in the design of such a system is the connection between the included components. Using a shared medium bus would in this matter not be sufficient enough to handle the internal/external communication. A network optimised for system-on-chip would be more suitable. In an adaptable system however, it must be considered that different system functionality may require a different network topology or performance, and that this must be adaptive during run-time to the demand of the current application. Therefore a mesh of different but standardized routers were developed. To enable a standardized communication and the possibility to connect to neighbour modules, a set of basic elements were developed.

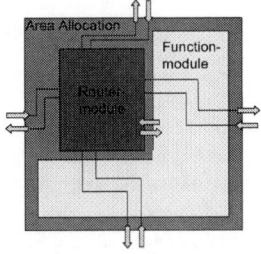

Figure 4 Basic 1x1 Reconfiguration Module

Figure 3 shows the basic elements for the presented reconfigurable system. From a basic cell, called 1x1 cell (module C in figure 3 for example), all other modules (cells) were derived. Figure 4 shows a schematic overview of a 1x1 basic module. Different sizes of the modules are necessary because the integrated applications within the cells can have different complexity. The communication interfaces, in Figure 3 visualized as arrows, fit to other modules since the distances between the connection points are equal. While writing back a module to the FPGA, the connection is established by positioning the cell in neighbourhood to other modules which provide exactly the same points for connection.

0-7695-2533-4/06 $20.00 © 2006 IEEE

These interfaces are based on the LUT-Based communication primitives described in [8] and have a bit width of 8 bits in both input and output direction.

The logic content of a module is a simple router, enabling the connection to all available interfaces. The implemented methodology is packed based. The in- and output ports are unidirectional and parameterized in their bit-width, which is set to 8 data bits and one control bit in this case. The router contains several control units, which will be described in the following.

There are two independent communication processes for receiving and sending. Due to synchronization, each of these processes is equipped with a timer module and contain acknowledge and control signals. These two processes communicate with the network through a port select module to avoid multiple sourcing.

The port select module recognizes the states of the communication processes. Using this information, it connects the intern data with the in and output ports of the router. Additionally, a FIFO is implemented. It stores incoming data as long as the router is in a busy state, and processes it afterwards. The router is busy as long as data cannot be routed or is still stored in the FIFO (see figure 5).

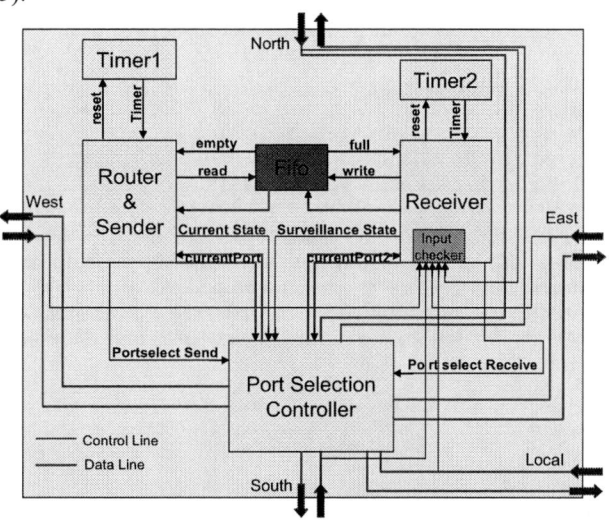

Figure 5 Schematic of Router Module

The most important task of the router is to find a proper way to the destination address, which is given in the header. Therefore the FPGA is divided matrix-like into basis squares of all the same size. Hence, the partitioning of the FPGA looks like a chessboard. The IDs of each square is then coded and separated in vertical and horizontal coordinates. In this particular case a 4x4 Matrix is implemented. The routing algorithm then calculates the distances of either dimension and compares them. When having chosen the output ports, the router tries to contact the neighbour router for a couple of times. If there is no answer, a reroute signal is set to '1', which tells the read and route state to look for another way to the destination address. Figure 6 shows a system with 8 1x1 cells in the FPGA editor. No functions are included, only the routers and corresponding Function Connect Units.

There is however implementation space allocated for the functionalities around each cell. Traditional computer

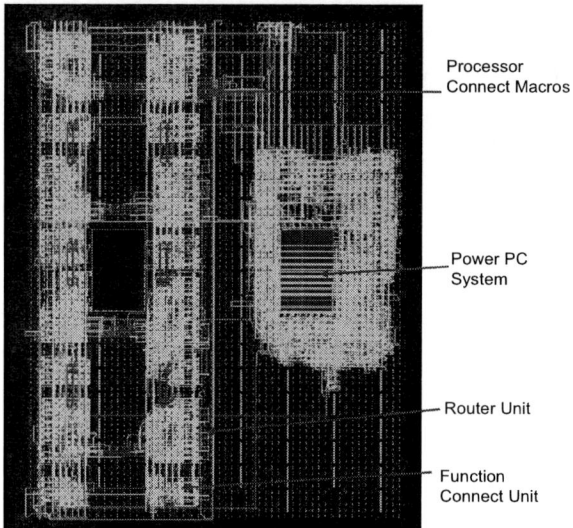

Figure 6 FPGA Editor: 8 1x1 Module System

architectures with system bus concept suffer from the fact that the bottleneck caused by the bus architecture and arbitration, even when highly optimized, increases by including more and more processors of a SoC, connected to the bus. Flexible, programmable and adaptive Networks-on-Chip (NoC) approaches solve this problem. As an extension of the design standard of today's multiprocessor-SoCs (MPSoCs), in future these NoCs has to be adaptive in all these disciplines while run-time and is therefore a challenge for future architecture design [9].

5   Online FPGA Routing

To overcome the difficulties described in section 3, a new approach for online-routing was developed. The implemented system allows connecting placed modules to a LUT-based communication primitive by modifying the bitstream.

To enable the feature of online routing the reconfigurable area of the design was divided into vertical configuration slots similar to the approach described in [2]. The key property of such a slot is that the underlying FPGA architecture is homogenous in both vertical and horizontal direction and that it is identical for each slot. This allows rectangular shaped modules stored in a flash repository to be placed into any of the configuration slots regardless of its vertical position within the slot. It is also possible to place multiple smaller modules, that don't use the complete slot height, on top of each other into one configuration slot, as long as the sum of the module heights does not exceed the slot height. The modules in the repository are self contained with respect to their used routing and CLB resources. In order to communicate with the outer world each module has a LUT-based communication interface in the lower right corner of the module. Via its interface the module can be dynamically connected during runtime to a LUT-based communication primitive. The communication primitives are located

0-7695-2533-4/06 $20.00 © 2006 IEEE   100

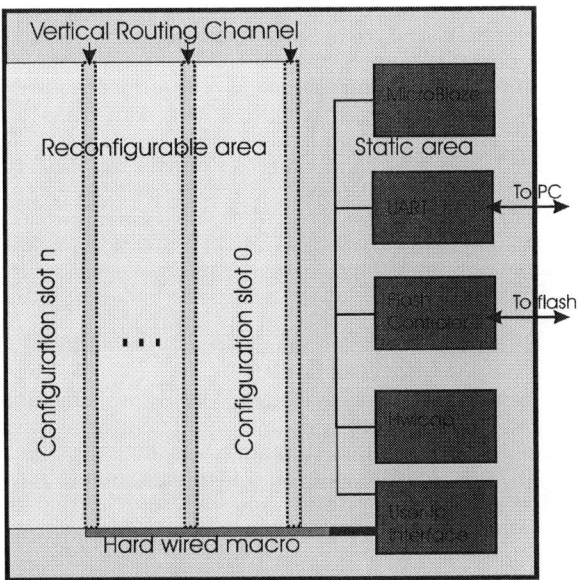

**Figure 7   2D Partial FPGA Reconfiguration
Hardware Template**

adjacent to each configuration slot in so called "vertical routing channels", which can be seen in Figure 7. Each slot has its own routing channel attached to it which covers the whole height of the slot. This ensures that the modules have access to the communication primitives, no matter on which vertical position they are placed within the slot. A hard wired macro, which extends to the static area, is placed on the bottom of the reconfigurable area in horizontal direction, to hook up to all vertical communication primitives and route the signals to the static part of the design, e.g. a general-purpose I/O (GP-IO) module of the MicroBlaze soft-core processor. The area on top and on the bottom of the reconfigurable area belongs to the static design and is predominantly used to access the IO-pins at the top and at the bottom of the FPGA.

The schematic in Figure 7 shows the static part of the design which is loaded into the FPGA as the initial configuration. The reconfigurable area, the slots and the vertical routing channels at the left are completely empty. All configurable elements are loaded (placed) into this area form a C-based configuration management system, running on the MicroBlaze processor [6] at runtime. The online routing is based on only two types of communication primitives, which can be seen in Figure 8 A. Type-I primitives route the signals from the previous to the next Type-I or Type-II primitive in a vertical direction (up/down) along the routing channel. Type-II primitives do the same as Type-I primitives but in addition they also route the signals to the left in horizontal direction. The height of a Type-I primitives is one CLB-row. The width of Type-I and Type-II primitives and the height of a Type-II primitive depends on the number of signals to route. Per CLB-row/column height/width it is possible to route 8 signals.

When the configuration management system is initialised all vertical routing channels are filled up

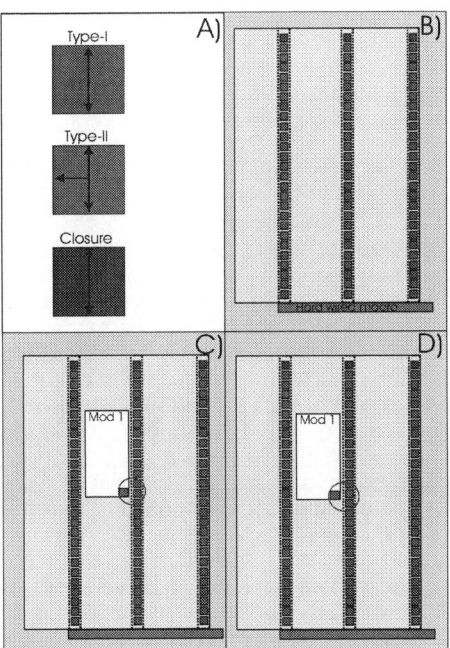

Figure 8   Online FPGA Routing Approach

with Type-I primitives and a special closure primitive on top of each channel, see Figure 8 B .It ensures that the dangling inputs of the last Type-I primitive have a defined low-level. As with the modules the routing primitives are stored in the flash repository and can be placed into the routing channels just as the modules can be placed into the slots.

After a module is placed into one of the slots Figure 8 C, it needs to be connected to the routing primitives. This is done by just replacing the Type-I primitive, which is adjacent to the modules interface, with a Type-II primitive. As stated before the Type-II primitives also routes the signals to the left and therefore it hooks up to the modules interface. The module is connected as described in Figure 8 D. This is the same for each module regardless of its position within the slots.

Unrouting a module is done in the opposite way. The Type-II primitive, which was connected to the module interface, is replace by a Type-I and the area where the module was placed is overwritten with a blank configuration. Routing and unrouting a module by swapping Type-I with Type-II primitives and vice versa does not effect the other modules as the primitives are designed in a way that they feature glitch less switching of the Virtex-II configuration logic.

6   FPGA Implementation and Tests

The 1x 1 basic cells as they are described in section 4 have been modelled in VHDL for implementation on a Virtex II-Pro board from Xilinx. For each cell a number of 220 CLBs has been allocated, from which the router unit takes up 62 and the function connect unit 28. This allocation leaves 130 CLBs for the functionality in each 1x1 basic module. The system is connected to a Power PC processor kernel, which manages the complete system

0-7695-2533-4/06 $20.00 © 2006 IEEE

control. This control includes the communication with the system environment, the reconfiguration management and the on-chip resource management. The Power PC can however be replaced by for example a MicroBlaze, for implementation on FPGAs without integrated processor kernels.

On-line routing can be performed by a MicroBlaze/Power-PC- system. All the algorithms for the reconfiguration control were realized in C and executed on a MicroBlaze soft-core processor. The complete code size is currently 16 KB. The code has however not yet been optimised, and further reductions of the code size are assumed. The reconfiguration time of a module increases due to the on-line routing process. The execution of the algorithm which calculates the positions of the CLBs in the module is more time consuming compared to the reading and writing of configuration files over the ICAP interface. For example, a module with the size of 15 CLBs in 5 columns, will take 1.5 seconds to reconfigure. This time includes fetching the module description from the FLASH memory, the reading and writing of the configuration file to the ICAP interface, and the time for calculating the vertical positions of the CLBs. Also, the time for connecting the module in its position is included. As mentioned above, the calculations of the CLB positions in all columns of a module, is very time consuming. This part can however also be optimised in order to further reduce the reconfiguration time significantly.

Future work will be to provide a fully automated online routing and placement while run-time without pre-routed modules in a repository. The goal is to store netlists as a database for the router. Integration of the complete routing engine compared with an intelligent run-time system enables to develop a high adaptive and optimised dynamic and partial reconfigurable system.

7 Conclusions

The paper discussed new perspectives and corresponding approaches for today's microelectronic embedded and processor solutions incorporating on-demand reconfigurable datapath allocations on suitable granularities in real-time. This was demonstrated and verified by our approaches and techniques implemented, synthesized and tested. The focus of the paper has demonstrated substantially new dynamic and partial reconfiguration techniques for 2D FPGA placement and routing adaptation for today's fine-grain Xilinx devices. State of the art solutions of partial and dynamic reconfigurable systems provide the substitution, of fixed rectangular shaped blocks of hardware modules while other parts stay in operation. The circumstance caused by the FPGA architecture and its reconfiguration mechanism, forces to substitute parts of a design in complete columns. Therefore, the reconfiguration of variable rectangular shaped hardware modules is a challenging task since configuration of hardware in the same column, doesn't has to be affected while this process. The approach introduced here demonstrated variable two-dimensional placement of hardware modules to Xilinx Virtex-II FPGAs, which offers an optimized utilization of such kind of devices, in comparison to the traditional partial reconfiguration approach with fixed geometry column-based topologies of module slots.

In summary, the transparent and easy programmable integration of fine grain reconfigurable data paths into today's and future microelectronic solutions for embedded systems and for universal as well as application-tailored processor architectures demonstrates very beneficial perspectives for short time-to-market pressure, for low cost and power consumption, and for additional options in fault tolerance and risk management. Flexibility and adaptivity which is essential in future systems can only be provided by both hardware and software reconfigurable systems. Fixed architectures like general purpose processors or ASICs are not sufficient to adapt in performance and necessary hardware structures. Run-time reconfigurable, power efficient and high performance computation resources fills the gap between pure software systems and fixed hardware architectures. Adaptivity in both software and hardware enable the flexible migration of tasks to the required hardware and its performance.

8 References

[1] J. Becker, M. Hübner, M. Ullmann: "Power Estimation and Power Measurement of Xilinx Virtex FPGAs: Trade-offs and Limitations", SBCCI03, Sao Paulo, Sep. 03

[2] J. Becker, M. Hübner, M. Ullmann: "Real-Time Dynamically Run-Time Reconfiguration for Power-/Cost-optimized Virtex FPGA Realizations", VLSI03, Darmstadt, Sept. 03

[3] L. Benini, G. De Micheli: "Networks on Chip: A New Paradigm for Systems on Chip Design", Date 02, March 3~7, Paris France

[4] http://www.xilinx.com/ise/design_tools/

[5] http://www.xilinx.com/ise/embedded/edk.htm

[6] http://www.xilinx.com/ipcenter/processor_central/microblaze/literature.htm

[7] B. Blodget, S. McMillan: "A lightweight approach for embedded reconfiguration of FPGAs", DATE´03, Munich Germany

[8] M. Huebner, T. Becker, J. Becker "Real-Time LUT-Based Network Topologies for Dynamic and Partial FPGA Self-Reconfiguration", SBCCI04, Brasil

[9] S. Leibson, J. Kim. "Configurable Processors: A New Era in Chip Design," Computer, vol. 38, no. 7, July 2005

[10] M. Ullmann, M. Huebner, B. Grimm, J. Becker: "An FPGA Run-Time System for Dynamical On-Demand Reconfiguration", RAW04, Santa Fee

0-7695-2533-4/06 $20.00 © 2006 IEEE

A "Soft++" eFPGA Physical Design Approach with Case Studies in 180nm and 90nm

Victor Aken'Ova, Resve Saleh

Department of Electrical and Computer Engineering, University of British Columbia
Vancouver, British Columbia, Canada
(vaken, res)@ece.ubc.ca

Abstract

Our recent work in embedded FPGAs has been focused on a *soft* IP approach where programmable fabrics are described at the RTL level and implemented using the ASIC digital flow and *generic* standard cells. Early results showed significant penalties in area, delay, and power overhead. However, using *tactical* standard cells and a *structured physical design* approach within such a flow, we were able to obtain large savings in area and delay. We defined this new approach as *soft++* eFPGA. This paper provides details of the physical design flow, with particular emphasis on floor-planning, interconnect-planning, and clock tree synthesis. The advantages of our approach in handling larger circuits are demonstrated on a set of realistic benchmark circuits implemented in 180nm and 90nm CMOS process technology.

1. Introduction

Growing design complexity and cost has forced designers to build programmability into System-on-Chip (SoC) designs to reduce the number of costly chip re-spins, and amortize IC costs over several derivatives. Programmability in the form of embedded field programmable gate array (eFPGA) cores is one of a handful of design solutions that has emerged to meet this challenge. An eFPGA fabric is suitable for implementing small or medium-sized logic functions such as a local accelerator for a processor core, or a block that requires updating as an industry standard evolves.

Despite the potential benefits of such an approach in SoC design, its widespread use has been hampered by overhead typically associated with the use of programmable logic architectures. Our recent work [01,02,08] has focused on a new method that uses the digital ASIC design flow for synthesis of *on-demand* eFPGA fabrics. Further details of this so-called "soft" eFPGA approach are contained in Section 2.

This paper describes details of the "soft++" eFPGA physical design flow and contrasts it with the original "soft" eFPGA physical design flow. The "soft++" name was derived from two improvements: the use of tactical standard cells for layout and the use of a structured layout. This paper demonstrates that the use of a structured fabric is critical to the success of the ASIC flow-based synthesis of eFPGA fabrics.

Since prior "soft" programmable logic architectures [01,02]

did *not* have a regular structure that could be effectively exploited during physical layout, a mismatch existed between the actual layout produced by ASIC back-end tools and the physical model assumed by FPGA CAD tools. This caused the FPGA tools to make some incorrect decisions during FPGA place and route that would impact speed negatively. For example, if assumptions made by FPGA tools about the RC delays of interconnects do not correspond to the actual layout, the tool may inadvertently select a slow resource over a fast one. This issue was investigated in [07], and a CAD solution that extracts and then back-annotates RCs from layout into the FPGA tools was devised. This approach improved timing but increased CAD runtime significantly. Our work aims to resolve such mismatches through structured floor-planning and interconnect-planning.

A related issue is the layout of tactical standard cells. An approach described in [08] cut overall eFPGA area by 2.4X, and delay by 1.7X. In this paper, additional details of cell layout techniques that produced such gains are presented. Furthermore, models that accurately predict cell layout area are presented. Lastly we present area results for eFPGA fabrics in 180nm and 90nm based on the soft++ eFPGA flow. These results are presented for "real-world" logic circuits including encryption circuits, I/O protocols and coprocessor circuits that may be suitable for use within a given SoC design.

2. Background

Two research groups [01,02] have suggested the use of standard cells to implement programmable hardware in SoC designs. Figure 1 shows the details of one such eFPGA flow. It involves describing the FPGA architecture in RTL, and then synthesizing the architecture with standard cell gates to produce the desired fabric. The dashed regions relate to the eFPGA architecture exploration phase, the shaded regions relate to the physical design aspects of the flow, the dashed regions on the right relate to the FPGA CAD rerun flow (explained in Section 3) and the rest is the Front-end flow. The only difference between a "soft" eFPGA and an off-the-shelf programmable logic fabric (from a commercial eFPGA vendor) is its use of standard cell logic gates rather than custom designed gates and interconnect, dominated by pass transistor logic.

Although it uses the well-established ASIC flow, a major

0-7695-2533-4/06 $20.00 © 2006 IEEE   103

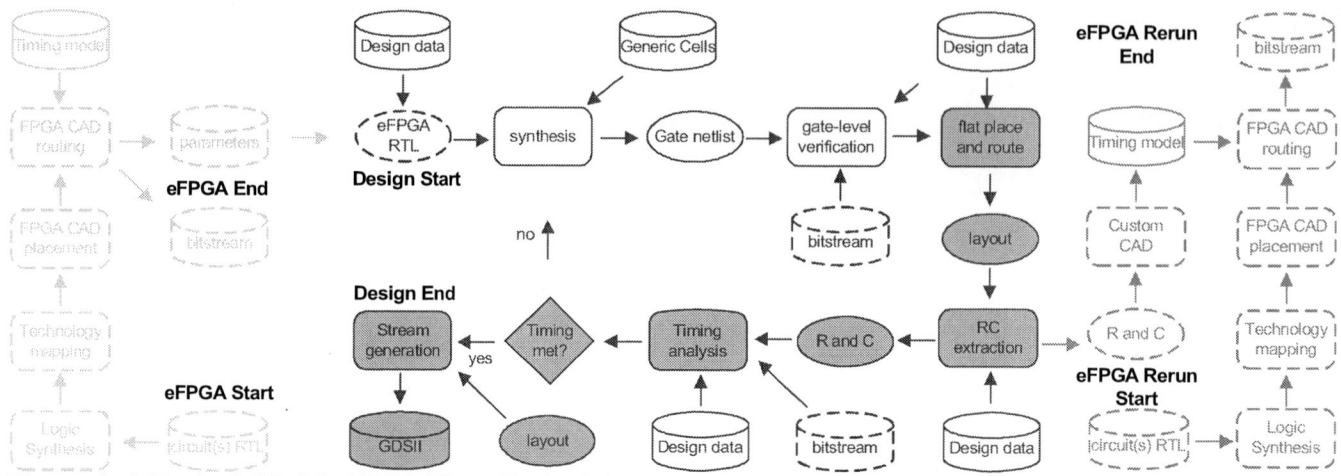

Figure 1: A Typical "Soft" eFPGA Flow Methodology based on an *Unstructured* Physical Design Strategy

drawback of the "soft" approach is the large area, delay, and power overhead since standard cells were not originally designed with programmable logic in mind [01,02,08]. To illustrate the problem, note that an FPGA uses of high-fanin multiplexers, and SRAM cells that *hold* configuration bits [03,05]. When implementing wide fan-in multiplexers with CMOS logic gates, it results in a large area, delay, and power overhead [02,08] compared to corresponding pass transistor logic. Similarly, flip-flops (that are much larger than SRAM cells) are used to store configuration bits in "soft" eFPGAs since SRAM cells are not available in commercial standard cell libraries. This adds only to area and power overhead. Configuration flip-flops do not impact delay because they are only used to hold bits. Based on these arguments, the reasons for the significant overheads in "soft" eFPGA fabrics should now be apparent.

In [08], we showed that the use of tactical standard cells in a typical semi-custom ASIC flow reduces area and delay in "soft" eFPGAs by 58% and 40%, respectively. Such fabrics are estimated to be between 1.6X to 2.8X the area of full-custom versions. More importantly, it was found that these fabrics have only 1.1X higher delay than full-custom versions. Overall, the results in [08] show that programmable logic fabrics created using tactical standard cells can significantly reduce the overhead associated with using eFPGAs in SoCs.

3. Details of Our Approach

Our physical design (PD) approach for "soft++" eFPGAs is based on a hierarchical flow. This approach to physical design has the potential to reduce CAD runtimes [02], improve layout quality and increase overall design efficiency. However, in large ASICs, a major difficulty of the hierarchical flow approach is timing budget allocation for heterogeneous blocks in the design. This can take several iterations of both design repartitioning and slack reallocation to ensure a design meets timing after placement and routing. In eFPGA fabrics, this difficulty is less likely if the architecture is based on a repeated, regular, and homogeneous tile structure. This approach is well-known in stand-alone FPGAs but only recently introduced in soft eFPGAs. In such a case, all the building blocks of the FPGA are identical and so their layout slack constraints should be more or less the same. Hence, a single "master" tile is constrained for the appropriate delay during PD and all the other tiles (clones) inherit its attributes.

Our adapted "soft++" design flow is shown in Figure 2 and is based on the Cadence™ hierarchical design flow for SoC Encounter™. The shaded regions are enhancements relative to the flow in Figure 1. In Figure 2, the flow begins with architecture exploration in the FPGA CAD flow (similar to Figure 1) to determine the "best" parameters for gate-level synthesis of the eFPGA fabric. In our work, the best set of architecture parameters are the ones that result in the smallest area-delay product. The ubiquitous island-style FPGA architecture [03,05] was selected for this work because it has a highly-regular, tile-based structure that is well-suited to hierarchical physical design. Previous "soft" programmable logic architectures [01,02] did not impose a regular structure, and so using a hierarchical layout approach would have been difficult. Instead, a flat layout approach was used [01,02], but this leads to problems that are mentioned in sections to follow.

Gate-level synthesis is an important aspect of the flow in Figure 2 because it is tightly-coupled with the physical design approach used here. The first step in our physical design flow is floor-planning and this includes interconnect-aware tile placement, partition definition, and specification of "keep-out" regions. Floor-planning is followed by power planning, and pin placement for master tiles and the top-level fabric itself. This is followed by cell placement within tile masters as well as clock tree planning and synthesis for tile "masters" and the top-level design. Finally, all the selected tile masters and the top-level design are routed using a standard ASIC router.

0-7695-2533-4/06 $20.00 © 2006 IEEE   104

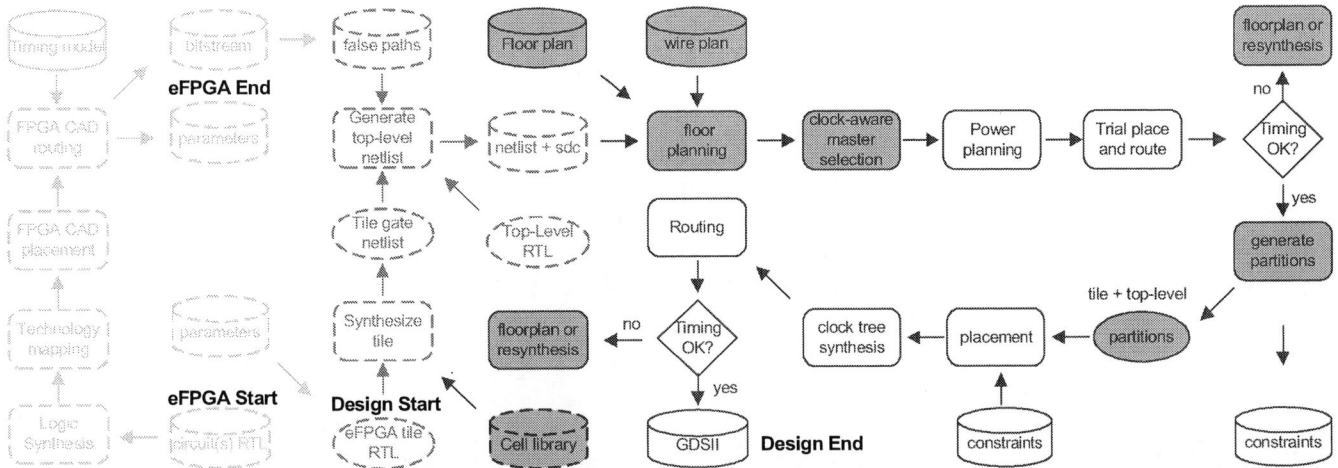

Figure 2 Typical "Soft++" eFPGA Flow Methodology based on a highly *Structured* Physical Design Strategy

Each master's attributes (placement and routing information) is subsequently propagated to their respective "clones". Global placement and routing data is also restored to the design top-level. Specific details are provided in the following sections.

3.1 Architecture

Since the selection of an appropriate set of parameters for the architecture is key to our physical design approach, an overview of the island-style architecture parameters used in all of our experiments is presented here. In Figure 3, the architecture is shown as a 2 x 2 array of tiles (islands) surrounded by "W" vertical and horizontal routing tracks. The highlighted tile in Figure 3 is comprised of mainly a configurable logic block (CLB) and switch block (SBK). (Later in Figure 8, we show the logic and routing details of a single tile.) For this work, a parameterized architecture was described in VHDL including a parameterized configuration state machine, and row and column address decoders for programming. Some key architecture parameters are: the array size ($D_X * D_Y$), lookup table (LUT) input size "K", the number of LUTs per tile "N", and the routing channel width "W".

3.2 Gate Synthesis

Successful "cloning" of the island-style architecture during physical design relies on a similar approach during gate-level synthesis. Figure 2 shows a bottom-up gate-level synthesis approach. A "master" tile is first synthesized using parameters from the FPGA CAD flow and other user-defined constraints. Then, during top-level synthesis, the master tile gate-level netlist and the top-level fabric RTL are used to generate the eFPGA gate netlist and constraint file. At the top-level, synthesis is reduced to simply writing out a gate netlist of the eFPGA because a "don't-touch" flag is set on the master tile netlist. Figure 2 shows false timing path constraints are applied during top-level gate "synthesis". This information is *not* used to constrain logic synthesis but is simply used to

generate the timing constraint file for physical design. The false timing paths constraints are generated using specialized CAD scripts, and essentially generate timing exceptions that sensitize *only* the paths needed to implement a particular circuit on the final eFPGA layout. In our case, the bitstream for the user circuit with the most stringent slack requirements is used for this task.

Finally, our synthesis strategy ensures all tile instantiations within the top-level netlist are clones of the master design since they inherit all its attributes. This netlist of *non-unique* tiles and its associated timing file are used for physical design.

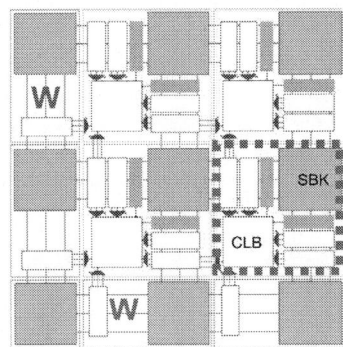

Figure 3: Top-level view of a 2 x 2 Island-style Array

3.3 Floor Planning

During floorplanning all the non-unique tile blocks are arranged in a 2-D array grid as commonly used in island-style architectures. Interconnect planning is also an essential part of the floorplanning stage because wire lengths must be kept as short as possible. In island-style architectures, interconnect planning is relatively easy because of the very regular structure of the architecture. For example, in such an architecture, it is known in advance which of the W channel routing tracks (wires) are intended to be long or short from architecture experiments using tools like VPR [03].

0-7695-2533-4/06 $20.00 © 2006 IEEE

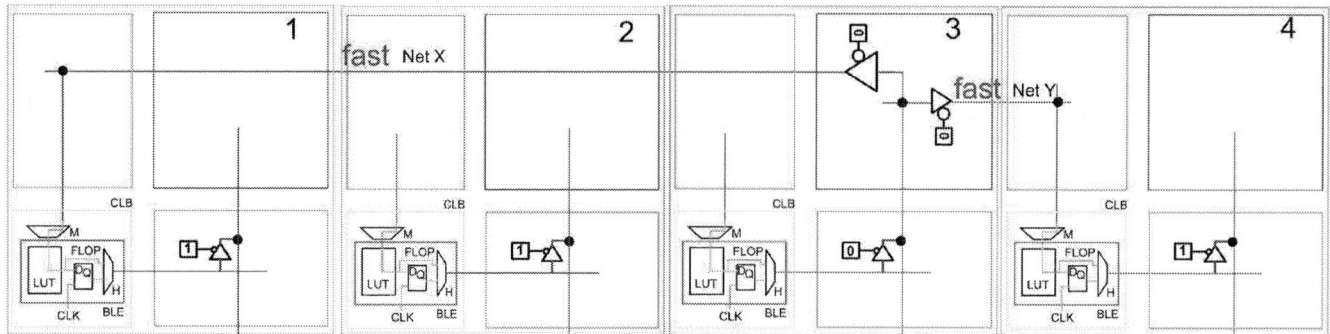

Figure 4: Expected layout of a row in a standard programmable logic fabric model assumed by FPGA CAD

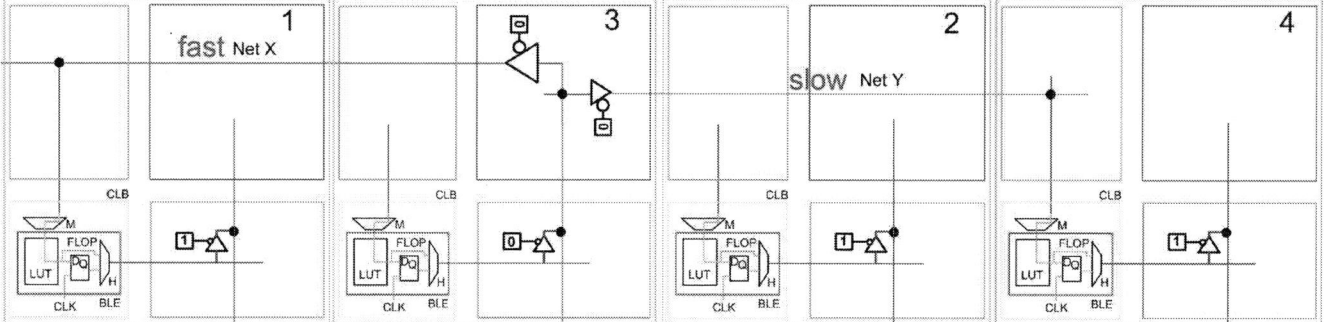

Figure 5: Possible layout in an eFPGA when wire-aware placement is not enforced during physical design

Therefore, a floorplan and interconnect template (see Figure 2) that captures such information is needed during place-and-route (P&R) in VPR.

Figures 4 and 5 illustrate why this is important. In Figure 4, the buffer that drives a long wire (net X) should be selected during P&R so that net X is a sufficiently fast connection. Also, the buffer on net Y is used because it drives a shorter wire. However, if the layout does not use tiles and is otherwise unstructured, then the components in the tiles can be placed anywhere in the layout, and the wire lengths and their associated delays are unknown by VPR during P&R. We represent this *conceptually* by swapping tiles 2 and 3 in Figure 5. Then, net X becomes a much smaller load driven by a disproportionately large buffer. Net Y, on the other hand, is now a much slower connection because a small buffer now drives a much larger load.

The flow of Figure 1 has this problem because the use of tiles and regular tile placement is not enforced. As a result, FPGA CAD tools may incorrectly assign timing critical signals to slow connections *that otherwise should have been fast*. This is due to significant timing gaps between FPGA CAD tool timing reports and actual post-layout timing results. Figure 6 (*section of an actual layout*) further illustrates the problem with unstructured layouts. The highlighted cells are parts of a single tile but, as shown, its cells are fairly widely scattered and this makes it difficult to predict wire lengths and delays.

One solution to this problem is to back-annotate the FPGA CAD tool with extracted RC delay information from layout [07]. A CAD-based flow was created for this purpose [07]. This is the CAD RERUN part of Figure 1. The "RC-aware" bitstream generated after layout RC annotation is used to reprogram the eFPGA. The results was an improvement in critical path timing, but CAD runtime was much higher [07].

Using the interconnect-aware flow in Figure 2, the same timing improvements as the CAD-based flow described above were achieved. In [08], structured layout (interconnect-aware) experiments using small benchmarks as those in [07] *yielded no timing improvement*. Hence, the conclusion was that CAD re-characterization was the only way to regain slack. Although some timing improvements were expected, it was realized that such gains can only be seen in *much larger fabrics* [08] than those used in CAD rerun experiments in [07]. That is, the true benefits of the soft++ approach are only seen if the benchmarks circuits are very large.

Additional experiments were used to gain further insight into the delay results [08]. In Figure 7, the bars represent delay for each cluster size N, and the line graph represents delay scaling relative to the previous cluster size. The graph shows delay increase with increasing cluster size (the delay "saturates" for cluster sizes 4 and 5, and 6 through 10). To investigate further, the buffer types encircled in Figure 8 were resized. After downsizing these buffers (by as much as 4 levels of drive-strength), static timing analysis (STA) on fabrics created *without structure* showed an increase in delay while results for *structured* layouts showed no change in

0-7695-2533-4/06 $20.00 © 2006 IEEE   106

delay.

 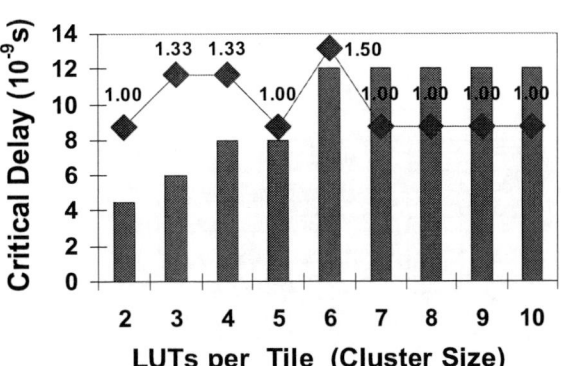 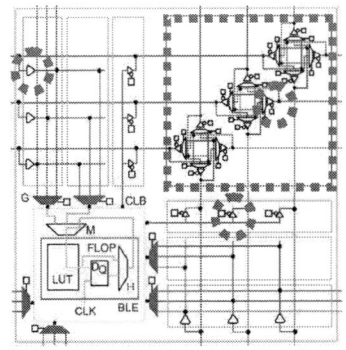

Figure 6: Unstructured Layout   Figure 7: Cell "saturation effect" in eFPGA Tile   Figure 8: A Island-style Tile

This suggests smaller drivers within the library can be used in the soft++ approach compared to the original soft approach. Therefore, we could reduce the size of the buffers in the tile without suffering a delay penalty. This is a somewhat inadvertent finding of the soft++ approach. In large fabrics (greater than 8 x 8 array), experiments showed delay improvement due to a *structured layout* approach increased from 3% to 7%. Larger savings are generally expected in architectures with longer channel tracks. However, our parameterized eFPGA "generator" currently supports short channel wires [03] only. Downsizing the buffers in structured layouts also saves area (roughly 4%).

To summarize, structured floorplanning and regular wire templates in the soft++ flow saves CAD runtime, and reduces area, delay and power. In addition, it eliminates timing gaps due to RC mismatches between VPR and the physical layout. This avoids the need for compute-intensive CAD iterations where the layout information about wire lengths and delays in the eFPGA layout are back annotated into FPGA P&R tools.

3.4 Clock Synthesis

Clock tree synthesis (CTS) typically occurs after placement and before detailed routing. In a hierarchical design flow, CTS is first performed on the master tile and then at the top-level. A macro-model with detailed specifications (e.g., phase, skew, capacitive loading) of the *pre*-route clock tree within a tile master is then used to constrain CTS at the top-level. Thus, CTS is a two-step process that reduces runtime (roughly 7X [08]) and CAD effort significantly relative to the "flat" CTS approach used in the soft flow of Figure 1. Furthermore, the quality of the clock tree (*post*-route) in the two-step approach of the soft++ flow is significantly better than in the soft flow. Specifically, clock *skew for the soft++ approach was as much as 15% less than the soft flow*. Similarly, the maximum clock buffer transition times were 20% better in the soft++ flow. The total clock tree area overhead in the soft++ flow was only about 2.5% higher than the soft flow. In both cases (soft and soft++), the same clock tree network constraints were applied.

Finally, the choice of a master tile is important relative to clock synthesis. In our case, we have chosen tiles farthest from the global clock input so as to model worst -case conditions.

3.5 Cell Area Reduction Techniques

Efficient physical design of tactical standard cells for programmable fabrics is important because an NMOS-dominanted logic style (such as pass transistor logic) is the best way to implement wide fan-in multiplexers in such architectures. Furthermore, implementing this logic style in standard cells can lead to underutilization of cell area since only the lower half of a standard cell is typically reserved for NMOS transistors. Our solution to this problem, as described in [08], uses p-well cutouts in the upper region of a standard cell to allow for denser layout of NMOS-dominated logic. A smaller n-well region is still needed for buffer implementation. Adjacent cell "guard-banding" is used to guarantee that layout design rules are satisfied with neighboring cells [08,09].

An analysis showed that our p-well cutout approach results in about 25% area savings for multiplexers (muxes) that have 16 or fewer inputs. However, for larger muxes, the p-well cutouts are insufficient. In this case, double height cells **and** the p-well cutout strategy were needed to achieve high layout densities. To illustrate how area grows with input size, Eqns. (3.1) and (3.2) empirically model the layout area of LUT muxes and generic pass tree muxes, respectively. Here, s_I is the select input width, and n_I is the number of inputs. We used best-fit curves to obtain these equations based on data from actual cell layout in TSMC 180nm CMOS process technology.

$$Area = 10.4e^{0.7s_I} \quad (LUT\ muxes) \qquad (3.1)$$

$$Area = 5.3n_I + 6.1 \quad (pass\ tree\ muxes) \qquad (3.2)$$

The two equations indicate that the growth of the mux area is exponential for LUTs and linear for the pass trees.

0-7695-2533-4/06 $20.00 © 2006 IEEE   107

Table 1: Table showing *Soft*++ eFPGA area results for 8 realistic logic circuits in 180nm and 90nm CMOS

Application Circuits	Size of Fabric(Soft)	Size of Fabric (Soft++)	Core Area 180nm (um^2) (Soft++)	Core Area 90nm (um^2) (Soft++)	90nmTile Area (um^2) (Soft++)	eFPGA Areas Ratios (Soft++ vs. Hard)
FHK	6 x 6	6 x 6	1.74E+06	4.24E+05	1.09E+04	0.1
Cordic CLA	7 x 7	7 x 7	4.48E+06	1.09E+06	2.07E+04	0.4
I^{2}C_master	7 x 7	7 x 7	4.34E+06	1.06E+06	2.01E+04	0.3
Cordic RCA	12 x 12	12 x 12	4.64E+06	1.13E+06	7.29E+03	0.4
UART*	-	14 x 14	1.73E+07	4.21E+06	2.05E+04	1.4
SPI	-	18 x 18	2.37E+07	5.78E+06	1.70E+04	1.8
Twofish*	-	30 x 30	2.58E+07	6.29E+06	6.72E+03	1.3
Rijndael*	-	33 x 33	2.05E+07	5.00E+06	4.45E+03	1.0

4.0 Design Results

In this section, we provide experimental area and delay results for realistic circuits that are suitable candidates for implementation as eFPGAs [01,02,06,07]. Technology files in 180nm (TSMC) and 90nm (ST Microelectronics) CMOS were used. VPR 4.30 [03] was the FPGA CAD tool in the flow. Timing results are based on STA experiments using Primetime™. Details of the timing flow method are provided in [08,09].

Table 1 provides a list of the circuits including Rijndael and Twofish encryption algorithms, coprocessor-type blocks such as cordic cores used in signal processing, and I/O standards such as I^{2}C, SPI and UART. Circuits with "*" were modified to exclude their RAM buffers. All of the circuits listed except FHK were obtained from either [10] or [11].

In addition, we include a proprietary core from an industrial standard Bluetooth baseband design. FHK is the baseband Frequency Hopping Kernel and it is interesting because recent changes to Bluetooth have called for adaptive frequency hopping (AFH) as a way to improve quality of service (QoS). A programmable fabric for the FHK block would provide the needed adapability as the standard changes.

Based on the first two columns, note that many of these designs could not be handled by the original soft eFPGA approach. Also, soft++ area results in Table 1 are about 0.4X of soft area. These results demonstrate that the soft++ approach is suitable for generating both small and medium size fabrics. The actual areas for 180nm and 90nm are given in the next few columns. With the reduced area consumption at 90nm, eFPGAs become an attractive option [06]. The last column shows soft++ eFPGAs can compete with hard eFPGAs in terms of area for certain cores. To model hard eFPGA area, we assume the same hard eFPGA library in [06]. The results show that for smaller circuits, the soft++ approach results in significant area savings compared to the hard approach. However, for larger circuits hard eFPGAs tend to fare better.

5.0 Summary

This paper described, in detail, the design flow and benefits of using a tile-based layout approach in soft eFPGAs and imposing structure on the final layout. It also demonstrated the importance of floorplanning, and interconnect planning. The original soft and new soft++ eFPGA were compared on realistic circuits. With the approaches described in this paper, the soft++ approach to eFPGA design using the ASIC flow becomes a viable alternative to hard eFPGA fabrics. Also, the results indicate that the area consumption of eFPGAs can be tolerated at the 90nm (whereas this has not been the case in 180nm), and perhaps moves the industry one step closer to widespread use of programmable logic fabrics in SoC design.

6.0 Acknowledgements

Funding was provided by PMC-Sierra and NSERC (Grant No. STPGP 257684). We also thank Guy Lemieux for help on FPGA architecture issues, and Pedram Sameni for tactical cell library development. The Canadian Microelectronics Corporation provided access to Bluetooth cores from Thales.

7.0 Bibliography

[01] S. Phillips, S. Hauck "Automatic Layout of Domain specific Reconfigurable Systems for System-on-Chip" FPGA Feb. 2002.

[02] J. Wu et al, "SoC Implementation Issues for Synthesizable Embedded Programmable Logic Cores", IEEE CICC, Sept.2003

[03] V. Betz, J. Rose, A. Marquardt, *Architecture and CAD for Deep Submicron FPGAs*, Kluwer Publishers, 1999.

[04] D. Chinnery, K. Keutzer, *Closing the Gap between ASIC and Custom*, Kluwer Academic Publishers, 2002.

[05] G. Lemieux et al "Directional and single wire drivers in FPGA Interconnect" IEEE FPT, Brisbane, December, 2004.

[06] P. S. Zuchowski et. al "A Hybrid ASIC and FPGA Architecture" IEEE International Conference on CAD San Jose 2002

[07] J. Wu "Implementation Considerations for Soft Programmable Logic Cores Master's (MASc) Thesis, October, 2004

[08] V. Aken'Ova "Bridging the Gap between Soft and Hard eFPGA Design", Master's (MASc) Thesis, March 10, 2005

[09] V. Aken'Ova, et. al "An Improved *Soft* eFPGA Design and Implementation Strategy" IEEE CICC, San Jose, Sept., 2005

[10] M. Holland, S. Hauck "Automatic Creation of Domain Specific Reconfigurable CPLDs for SoC", FPL, August, 2005

[11] Opencores IP http://www.opencores.org/browse.cgi/by_category

QUKU: A Two-Level Reconfigurable Architecture

Sunil Shukla[1], Neil W. Bergmann[2], Jürgen Becker[3]

[1,2] ITEE, University of Queensland, Australia, [1,3] ITIV, Universität Karlsruhe, Germany

{sunil, n.bergmann}@itee.uq.edu.au, becker@itiv.uni-karlsruhe.de

Abstract

FPGAs have been used for prototyping of ASICs, for low-volume ASIC replacement and for systems requiring in-field hardware upgrades. However, the potential to use dynamic reconfiguration to adapt FPGA operation to changing application requirements has been hampered by slow reconfiguration times, and poor CAD tool support. In this paper, a new architecture, QUKU (pronounced cuckoo), is described which uses a coarse-grained reconfigurable PE array (CGRA) overlaid on an FPGA. The low-speed reconfigurability of the FPGA is used to optimize the CGRA for different applications, whilst the high-speed CGRA reconfiguration is used within an application for operator re-use. An FIR filter kernel has been implemented on QUKU and is shown to have performance which bridges the gap between softcore CPUs and custom FPGA filter circuits.

1. Introduction

Traditionally, FPGAs have been thought of as a prototyping device for small scale digital systems. However current generation mega-gate FPGA chips have changed this perception. The geometric growth of logic density in FPGAs now makes them a viable alternative to ASIC implementations for many circuits. DSP applications, such as software defined radio, are very computationally intensive, requiring high communications bandwidth as well as high processing speed to achieve high throughput. Generally, ASICs are seen as the only low-power solution which can meet this demand for fast computational efficiency and high I/O bandwidth. However, the rapidly spiraling NRE (Non-Recurrent Engineering) costs of ASICs are a strong motivation for some sort of general purpose computing chip for mobile applications with better power efficiency than microprocessors. Recently, coarse grained reconfigurable architectures (CGRA)have been developed to bridge the gap between power-hungry microprocessors and single-purpose ASICs.

Hartenstein [1] gives a good overview of various CGRAs that have been developed or are being developed in various universities and industry labs. Coarse-grained arrays have the advantages of power-efficiency for their intended application domain, and also they are designed with efficient implementation of dynamic reconfiguration supported in hardware and in design software. Furthermore, this dynamic reconfiguration can be very fast, since configuration codes are very short. On the downside, they are not generally available as off-the-shelf commercial chips, and their particular processor mix usually ties them to one particular application domain. Indeed, CGRAs are often used as a co-processor core within an ASIC to extend their re-usability.

On the other hand, FPGAs are widely available as commercial chips, and their large consumer base ensures excellent tracking of the performance and price benefits of semiconductor technology improvements. However, poor hardware and software support for dynamic reconfiguration, and large configuration file sizes make dynamic reconfiguration difficult and inefficient, so most FPGAs are used with a single configuration during normal operation.

We have developed an architecture, QUKU [2] which is a merger of these two technologies: CGRAs and FPGAs. QUKU is a coarse-grained PE matrix overlaid on a conventional FPGA. The aim is to develop a system which is based on commercially available and affordable technologies, but at the same time provides active support for fast and efficient dynamic reconfiguration. This implementation will enable us to use the same platform for applications which are a mix of control flow and computationally intensive applications.

Our system is unique in that it provides two levels of application-specific reconfigurability. Within a single execution of an application, the system works as a CGRA of processing elements. The operation of each PE, and the interconnections between PEs can be

---

0-7695-2533-4/06 $20.00 © 2006 IEEE

reconfigured on a cycle-by-cycle basis, giving maximum reuse of arithmetic and logical operators without long reconfiguration delays. Between different applications, the CGRA can be reconfigured at the FPGA level to provide a different CGRA which is optimized for that application.

This approach allows applications which have larger hardware requirements than are available on a single FPGA to efficiently reuse operators through fast CGRA reconfiguration on a cycle-by-cycle basis. However, the CGRA does not suffer from the normal CGRA problem of trying to identify the optimal mix of operators for all present and future applications to be used on that array. Rather, the structure of the CGRA can be periodically re-optimized for each new application at a cost of several milliseconds of FPGA reconfiguration time.

A commonly perceived obstacle to the use of FPGAs to provide algorithm speedup is the difficulty of designing custom hardware for each new algorithm. The QUKU architecture moves the design problem from a complex hardware design problem to a simpler problem of programming a PE array. The programmer sees a homogeneous array of function-rich PEs. QUKU compiles these PEs to a heterogeneous array, with each PE optimized for just the range of operations it requires to implement for one application set.

It is our conjecture that such a PE array has the potential to provide area and power efficiencies not normally associated with FPGAs, and this paper outlines some of our initial investigations into this potential.

In this paper, we target the application of FIR filtering which is widely used in DSP applications. The application is mapped on QUKU.

The next section gives a short overview of FIR filters followed by an in depth explanation of mapping process on QUKU. Section 3 describes the FIR implementation on three different architectures, QUKU, custom FPGA, and an FPGA embedded processor based design. All the three architectures were implemented on the same FPGA platform. Section 4 concludes the paper and provides the comparison of performances on these architectures.

2. FIR filter implementation on QUKU

2.1 FIR Background

Digital filters are quite commonly used in DSP. They are used to modify the signal attributes in the time or frequency domain. The output of an FIR filter of order L is given by convolving the input time series $X(n)$ with the impulse response of the filter .

$$Y(n) = X(n) * C(n) = \sum_{k=0}^{L-1} X(k)C(n-k)$$

Equation 1

Where $C(0)$ through $C(L-1)$ are the filter coefficients, also known as the impulse response of the FIR filter. There are two representations of an FIR filter, the direct form and the transposed form. The direct form of an FIR filter is shown in fig. 1.

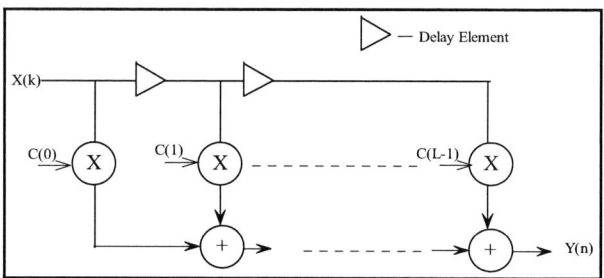

Figure 1 Direct form of FIR filter

The transposed structure is often used and has the advantage that the adders are pipelined [5]. But the transposed form requires the storage of partial sum either in registers or in memory. We will be implementing the direct form as it doesn't require any partial sum storage even for very large filter structure.

3. Mapping the application on QUKU

Once a set of applications is identified, the optimal PE mix is implemented on FPGA. Then the design steps, as shown in fig. 2 and described in subsequent sub-sections, are to be followed.

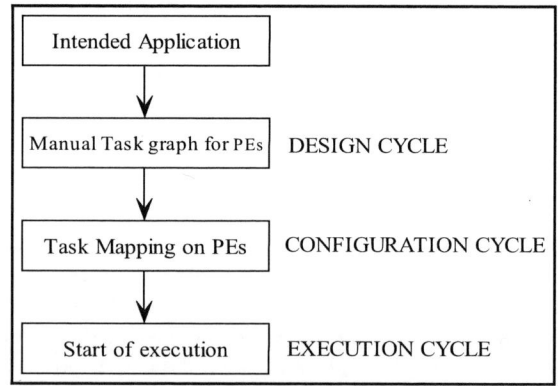

Figure 2 QUKU design flow

0-7695-2533-4/06 $20.00 © 2006 IEEE   110

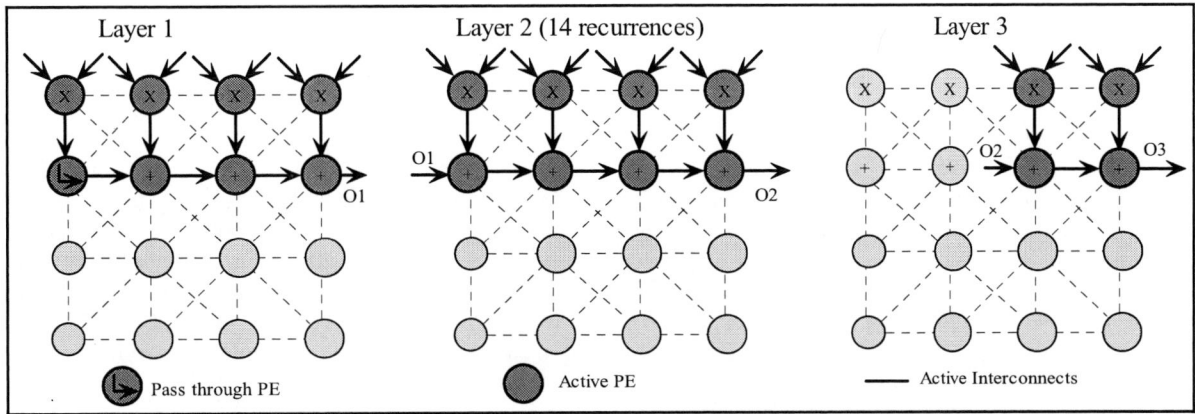

Figure 3 Temporal & Spatial task graph for a 62 tap FIR filter on QUKU

2.2.1. Design Cycle

A task graph describes the mapping of the application kernel onto the architecture both in the spatial and temporal domains. To demonstrate the task graph design for an FIR filter application, we have chosen a general PE layout where the first row PEs are multipliers and the succeeding rows PE are general purpose ALUs. This is a very general setup irrespective of the type of application. Although there may be PE layouts which can perform the task of FIR filtering in a better way, we have chosen this layout to demonstrate the mapping process. The application efficiency depends upon the task graph. In section 3.1, we will show that by choosing a different task mapping scheme, application execution is accelerated by a margin of 4.

To map an application, it is unfolded, in terms of its operations, in time and space. Quite possibly, a kernel size is bigger than the array size. In that case, multiple configuration contexts have to be loaded. An N tap filter requires N multiplications and $N-1$ additions. Hence on QUKU, a filter of tap length up to 4 requires just 1 configuration context. Filters having 5 to 8 taps require 2 configuration contexts and for any FIR filter size greater than 8, either 2 or 3 configuration contexts may be required depending upon the order of FIR filter. In case the order of FIR filter is a multiple of 4, 2 configuration contexts are sufficient else, 3 configuration contexts are required. Fig. 3 shows the different configuration layers for a 62 tap FIR filter. Each configuration layer can be executed for multiple cycles. Currently we have limited the recurrence of a configuration layer to 1024 as a compromise between hardware used and the application requirement. For the configuration shown in fig. 3, layer 1 is executed once then layer 2 is executed 18 times and layer 3 is executed once. We can program the above

configuration for a maximum FIR tap length of 4104 *(4+4*1024+4)*. However this is not the maximum size that can be supported in QUKU. The maximum tap processing can be upgraded to 16416 *[4X(4+4*1024+4)]* by choosing a different mapping scheme as described in section 3.1.

2.2.2. Configuration Cycle

From the task graph, configuration codes are generated for each PE. The configuration code includes 16 bit configuration word and 10 bit configuration recurrence counter value. PEs are addressed from left-to-right and top-to-bottom. In case a PE has same configuration throughout all the layers, it is just configured once with the exact value of configuration recurrence counter. The configuration code for the task graph in fig. 3 is summed up in table 1 (all values in Hex).

Table 1 Configuration code for task graph of fig. 4

Config word	Explanation
800000FF	PE 0-7 reset indication
03480001	Multiplier configuration
8000000F	Valid for PE 0-3
02540001	PE 4 configured as a pass through element
80000010	
04540001	PE 5-7 configured to add the O/P of previous adder and O/P of their multiplier
800000E0	
0414000E	PE 4-7 configured to add the O/P of previous adder and their result
800000F0	
04940001	PE 6 configured to add the O/P of previous stage, O2 and O/P of mult 2
00000040	
04140001	PE 7 configured to add the O/P of PE 6 and O/P of mult 3 and indicate final result
00000080	

QUKU configuration controller has the capability to address multiple PEs in one cycle. This results in

0-7695-2533-4/06 $20.00 © 2006 IEEE   111

simultaneous programming of PEs having the same configuration. In Table 1, same configuration words are grouped. The first entry indicates the address of PEs on which the application is mapped. Those PEs are reset before new configuration is loaded. After that, each odd numbered location contains PE 16 bit configuration word and 10 bit recurrence counter value and the consecutive even numbered location contains the PE address for which the configuration applies. The total configuration size for 62 tap filter is *6X16* bits for PE configuration and *6X10* bits for configuration recurrence counter. The complete configuration requires 15 clock cycles. Table 2 compares the configuration bit size for FIR filtering operation for Montium processor [6], developed by University of Twente and QUKU.

Table 2 Configuration bit size comparison

	FIR 5	FIR 20
QUKU	120 bits; 10 cycles	146 bits; 12 cycles
Montium	1968 bits; 123 cycles	4320 bits; 270 cycles

2.2.3. Execution Cycle

Once the configuration cycle is over, the configuration controller indicates to the local controller unit (LCU) of the PEs to start execution. Since the architecture is transport triggered [3], there is no need for cycle accurate synchronization. A PE starts processing once both the inputs are available and output FIFO is not full.

3. Performance Analysis

In this section, we will compare the implementation and performance of QUKU, custom FPGA and microblaze processor based solutions.

3.1. QUKU Performance

The mapping shown in fig. 3 is intended to demonstrate the design flow for any general purpose application. It does not represent a very optimal mapping scheme for FIR filter. It is mostly a sequential operation in the way that unless a result is available from the previous adder, the next adder cannot proceed. This factor reduces the speed of operation. The total processing time for this configuration is given by equation 3. FIR filtering is a convolution operation where an input time series is convolved with the impulse response of the filter. An optimum way would

be to break the FIR taps into the number of available multiplier PEs, say L. Then L number of sub-FIR are generated out of the tap length N where each sub-FIR unit processes *floor(N/L)* taps. This is shown in fig. 4 where the FIR is broken down into 4 sub-FIR operations and then the result from each sub-FIR is added up to get the final result. The rectangular shaded area represents a sub-FIR. It runs for a programmed number of cycles and then passes on the result to the triangular shaded area. The triangular shaded area is executed just once when each rectangular shaded area is ready with the result. The total processing time can be calculated according to equation 2.

$$T_{\Sigma} = T_{mult} + T_{add} * (N/4 + 2) ....(a)$$
$$T_{\Sigma} = T_{mult} + T_{add} * (floor(N/4) + 1 + 2) ...(b)$$

Equation 2

Eq. 2(a) represents the case where N is a multiple of 4. Eq. 2(b) represents the case where N is not a multiple of 4. The excess tap points *(rem (N/4))* are distributed among the 4 sub-filters. Using this distributed FIR technique, we can effectively increase the maximum number of taps that can be processed. Distributive mapping on QUKU gives up to four-fold increase in the maximum FIR length processing and there is almost four-fold increase in the performance.

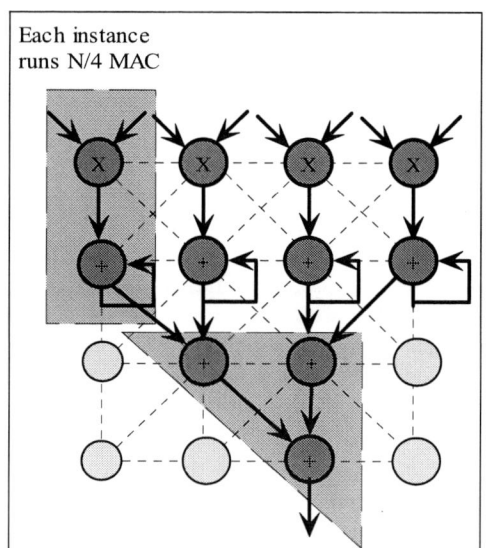

Figure 4 Distributed implementation of FIR filter

3.1.1. Multi-channel FIR filter on QUKU

Multi-channel filtering finds widespread application in wireless communication employing

0-7695-2533-4/06 $20.00 © 2006 IEEE   112

phase (I) and quadrate (Q) channels that need to be filtered separately but using the same set of filter coefficients.

In QUKU, multi-channel FIR filtering can be implemented for up to 8 channels. Each rectangular block in fig. 5 can accept input from a data channel. For the multi-channel filter application the throughput is fixed for the configuration shown in fig. 5, irrespective of the number of channels. The processing time for individual channel is given by:

$$T_{channel} = T_{mult} + T_{add} * N$$

Equation 3

Where, N is the order of individual filter. N can be different for different channels. We have conveniently programmed the different channels for different filter coefficients. This implementation can find place in adaptive filtering application where filter coefficients are required to be changed on the fly.

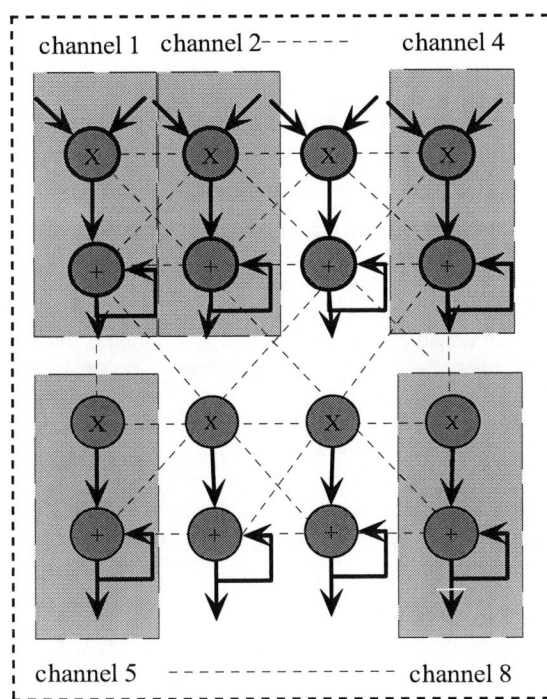

Figure 5 Multi-channel FIR implementation

3.2. Custom FPGA Implementation

The Xilinx Virtex series of FPGAs are essentially a multi-grain architecture. They provide hard macros and primitives of commonly used modules. Customization can be done using the Coregen [4] CAD Tool. It also provides options for automatic generation of FIR

modules. User can specify the throughput requirement and data sample rate. To meet a specified throughput, Coregen adds more DSP MAC units. For the multi-channel implementation throughput is equally divided among the channels. A limitation of Xilinx multi-channel FIR implementation is that it accepts only one set of filter coefficients for all the channels. Moreover, all the channels are required to be of same order.

3.3. FIR implementation on Microblaze

For comparison, the Microblaze [4] soft processor core has been chosen to implement FIR filter core. A Xilinx ML401 development board with a Microblaze based system running uClinux OS [7] was used.

Microblaze is based on Harvard architecture with separate interfaces for data and instruction. Microblaze uses pipelined instruction execution. The pipeline is divided into three phases: fetch, decode and execute. Each phase may take one or more clock cycles. If a phase takes more than one clock cycle then the pipeline is stalled until that phase is over. The instruction execution pipeline is shown in fig. 6. Fixed point multiplication in microblaze takes 3 clocks and fixed point addition requires 1 clock cycle.

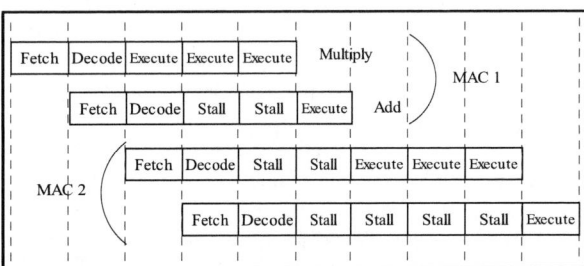

Figure 6 Microblaze instruction pipelining for FIR filtering

As evident from fig. 6, for execution of multi-cycle instructions, the whole pipeline is stalled until the execution cycle is over. The first MAC operation takes 6 cycles including fetch, decode and execute phases. The subsequent MAC operations require 4 additional cycles to complete all the three phases. The total processing time for FIR filtering can be given by:

$$T_{\Sigma} = 6 + (N-1) * 4$$

Equation 4

The performance deteriorated further when floating point representation is used. As a floating point multiplication and addition requires 6 cycles each for execution as compared to 3 and 1 cycle respectively for fixed point format. That indicates almost three fold deterioration in the performance with floating point representation as compared to fixed point representation.

4. Result Comparison and Conclusion

In section 3 we described in detail the FIR filter implementation using QUKU, custom FPGA and Microblaze soft processor. Figure 7 shows the processing time and area cost of the three different implementations with increasing FIR taps.

The custom FIR implementation has the best performance in terms of number of clock cycles, assuming that the hardware is continually expanded to maintain the same throughput with increasing filter size.

QUKU provides intermediate performance between Microblaze and the custom FIR filter, with approximately an 8-times speed improvement over the Microblaze.

QUKU has significant advantage in terms of configuration bit size. As shown in Table 2, the QUKU's configuration overhead is just about 13% of that of Montium processor.

Analysis of QUKU is still very much in progress. Area, area-speed product, power and energy comparisons are required to fully characterise QUKU performance in comparison to other approaches, and these are underway. This first set of experiments is mostly to confirm the QUKU design flow, and to establish ball-park performance speedups over soft-core CPUs. We are yet to develop a Microblaze based soft processor interface to QUKU for system and software control.

5. Acknowledgement

This project is proudly supported by the International Science Linkages programme established under the Australian Government's innovation statement, *Backing Australia's Ability*.

6. References

[1] Reiner Hartenstein, "Coarse Grain Reconfigurable Architectures" Proceedings of the 2001 conference on Asia South Pacific design automation

[2] Sunil Shukla, Neil Bergmann, Jürgen Becker, "APEX – A Coarse Grained Reconfigurable Overlay for FPGA", Proceedings of the IFIP VLSI SoC conference 2005, pp. 581-585

[3] H. Corporaal, J.A.A.J. Janssen and M.L.C.H. Arnold, "Computation in the context of Transport Triggered Architectures" *International Journal of Parallel Programming*, vol. 28(4), August 2000, pp. 401–427

[4] Xilinx products and services, online at http://www.xilinx.com

[5] Uwe Meyer-Baese, *Digital Signal Processing with Field Programmable Gate Arrays*" Springer-Verlag 2001

[6] Paul M. Heysters, "Coarse Grained Reconfigurable Processors" Ph.D. Dissertation, University of Twente

[7] J.A. Williams and N.W. Bergmann, "Embedded Linux as a platform for dynamically self-reconfiguring systems-on-chip", International Conference on Engineering of Reconfigurable Systems and Algorithms, Las Vegas USA (2004)

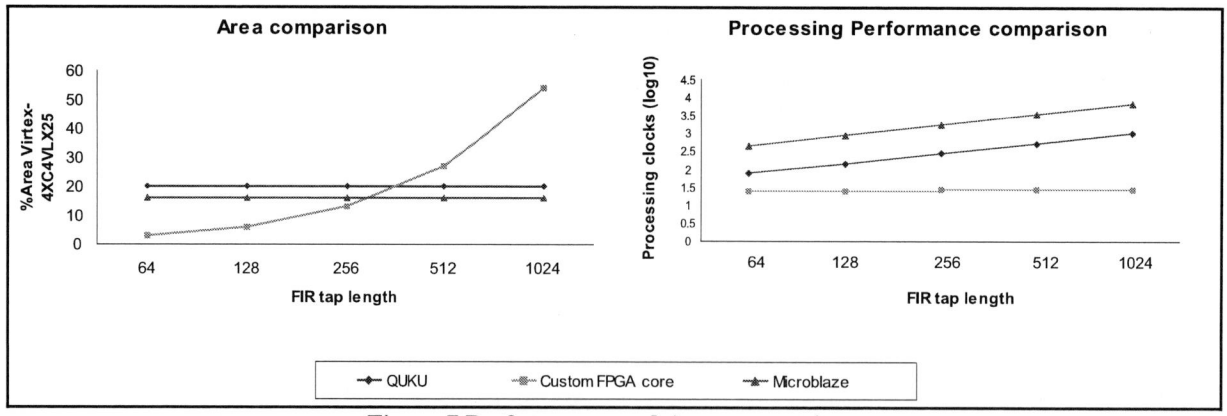

Figure 7 Performance and Area comparison

0-7695-2533-4/06 $20.00 © 2006 IEEE   114

Mixed-Signal Design and Analysis

Space-Saving Layout for Passive Components

Päivi H. Karjalainen and Pekka Heino

Tampere University of Technology, Institute of Electronics,
P.O.Box 692, FIN-33101 Tampere, Finland
Tel: +358 3 3115 2104, Fax: +358 3 3115 3394,
E-mail: paivi.karjalainen@tut.fi, pekka.heino@tut.fi

Abstract

The large number of passive components in mobile electronics devices require a large area. In this study, the passive components on the test chip are processed using a commercial CMOS process. The layout area is reduced by superimposing the on-wafer passive components of the basic inductor-capacitor and inductor-resistor circuits. The resonant frequency of the LC circuit using the stacked components matches well with the calculated value of the reference components. The effect of the parasitic components between the stacked passive components is found negligible in the operating frequency range.

1. Introduction

The part count of the passive components in portable electronics devices may be as high as 90 % of the total part count [1]. The transistor count is higher than that of the passive components, but they are integrated on a few chips. The passive components, instead, are either discrete passives, grouped as integrated passive devices (IPDs) or integral passives built into a substrate or interposer [1]. Thus, significant space saving for the whole device can be achieved by increasing the number of integrated passive components. On the other hand, the technical and economic factors determine the viability of the integration [1].

The quality of the analog circuits is highly dependent on the process variations, which can be minimized by interdigitation and a common-centroid layout [2]. These layouts are effective when certain components need to be matched to each other, but they do not improve the quality of an individual component. The characteristics of the components can be improved by appropriate selection of materials, e.g. by selecting high-k materials for the gate [3], and by using layout such as patterned ground shield [4].

The economic factor in the case of integrated components involves both the process costs themselves and the space the components need on the substrate. The unit price for an integrated component still remains high when the number of integrated components is low. The issue is device-specific and needs to be solved for each separate case. However, from the miniaturization point of view, the area the integrated components need is smaller than that of the discrete components. The trend towards miniaturization has succesfully driven IC technology. Transistor size has decreased to submicron scale, but the passive components are still quite large, due to the lack of suitable miniaturization methods. Fortunately, modern multi-metal CMOS processes enable novel layouts.

2. Motivations for the Study

In this study, space-saving layouts for integrated inductor-resistor (LR) and inductor-capacitor (LC) circuits are proposed. The effect of the unconventional layout on component characteristics is discussed.

A starting point for this study was to design the passive circuits using a standard process without any additional process steps. Another important aspect was the possible applications of the circuits such as radio receiver front-end [5], where a large number of inductors and capacitors are needed. New electronics modules such as System-in-Package (SiP) require small-area high-quality components, due to which area reduction emerged as the main point of the presented circuit layouts.

Also the quality aspect was taken into account. The quality factor (Q) of an inductor can be increased by thickening the inductor wire itself or by stacking multiple metal wires together [6]. In this study, the

0-7695-2533-4/06 $20.00 © 2006 IEEE

latter method is used and the processed 3.5-turn inductors consist of two stacked metal coils. The patterned ground shields processed on an nwell layer are added under the inductors to reduce the substrate eddy current. The dummy capacitors surround the capacitor matrice to ensure equal circumstances for all unit capacitors.

SiP is suitable especially for small, portable applications [7], because of which the operating frequency range of the inductors was designed to meet the requirements of the standards IEEE 802.11a and 802.11b. In line with these standards, Bluetooth [8] is operating at the frequency of 2.45 GHz and WLAN [9] at the frequencies of 2.45 GHz and 5 GHz.

2.1. Specification of the Measured Components

The passives are processed using an AMS 0.35 μm 4 metal 2 poly CMOS process. The circuits consist of two parallel connected passives, inductor-capacitor or inductor-resistor pairs. The single passives are used as references for the passive circuits.

The line width, the spacing between the lines, and the inner diameter of the inductors are 16 μm, 2.5 μm, and 75.5 μm, respectively. The inductor coils consist of via-connected third and fourth metal layers. The second metal is used as the output wire. The capacitor is a polysilicon-insulator-polysilicon (PIP) capacitor and the resistors are made of nwell and polysilicon. The characteristics of the components are collected in Table I.

Table 1. Characteristics of components

	Value	Type
Inductor, L	2.0 nH	M3+M4
Capacitor, C	3.7 pF	PIP
Resistor, R1	13.8 kΩ	nwell
Resistor R2	11.0 kΩ	high resistive polysilicon

The reference components lie in the normal way separately between the two rows of measurement pads. In contrast, in the passive circuits (LC, LR1, and LR2) two passive components lie one on the other. The layout of the processed passive circuits is presented in Fig. 1. For clarity reasons, the wiring between the components and the parts of components (i.e. capacitor matrix) is left out. The patterned ground shields are processed under the inductors, but they are not shown in Fig. 1. The input and output wires of the inductor are connected to the top and bottom layers of the capacitor, respectively. The resistor is connected to the

inductor in parallel. The hatched line of the inductor represents the output wire.

The total width of the adjacent parallel-connected inductor and capacitor would be approximately 320 μm, when the spacing between the components is 40μm. The length is defined by the inductor and it stays constant approximately 200 μm. The layout area of the proposed superimposed passive components is (320 -200) μm x 200 μm ≈ 24000 μm² smaller than that of the conventional layout.

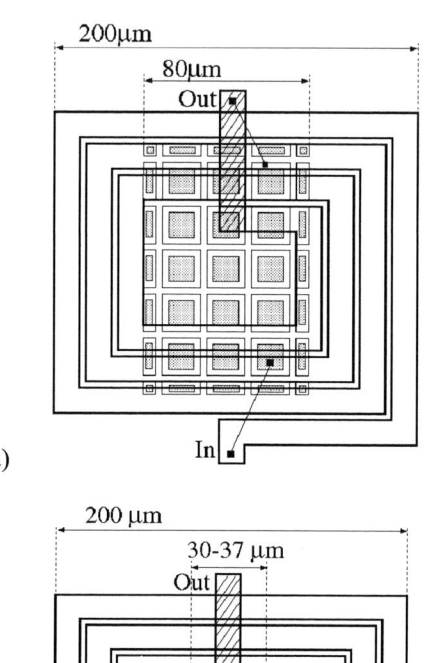

a)

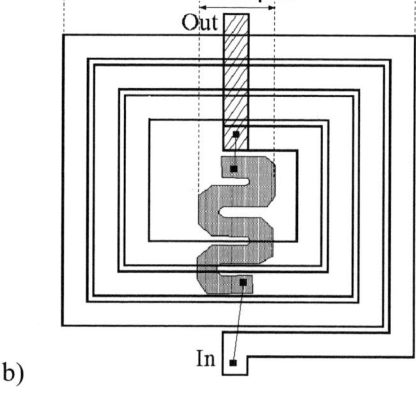

b)

Fig. 1. Layouts of the a) LC and b) LR circuits

The inductor coils are formed by the two highest metal layers and the second-lowest metal layer is used as the output wire. This form takes advantage of the better characteristics of the higher metal layers compared to the lower ones, which affects positively the operation of an inductor. The resistors and capacitor are processed using the lower IC layers. Thus, the inductors lie physically higher than the other components. A basic sectional view of the LC circuit is shown in Fig. 2. The black areas between the metal layers represent vias. In reality, the vias are spread throughout the metal area. The dashed areas between the polysilicon layers represent the insulator layer of the capacitor. The other insulator layers are left out for clarity reasons. Both Figs. 1 and 2 are drawn out of scale.

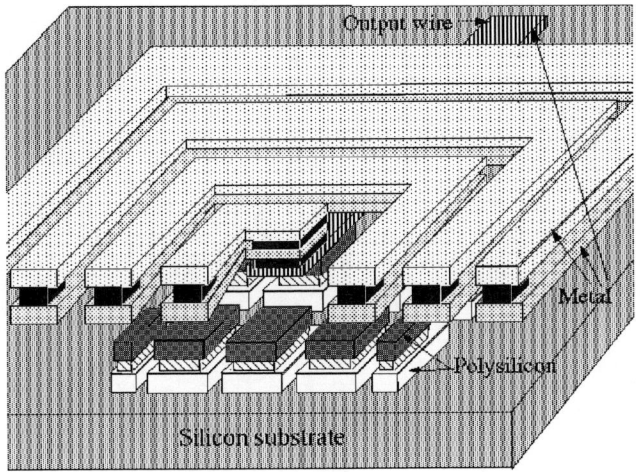

Fig. 2. Basic sectional view of LC circuit

3. Experimental Procedure and Results

The on-wafer testing for the passive components and circuits was performed with an HP810C network analyzer and Cascade Microtech coplanar ground-signal-ground (GSG) RF probes. Two-port S-parameters were measured at the frequency range of 45 MHz - 25 GHz. Open and short de-embedding structures were used to remove the parasitic components of the pads from the measurement data.

The quality factor (Q) defines the relation between stored and dissipated energies. It becomes zero at the resonant frequency of the component. At the resonant frequency the energy is stored in a different energy form than at lower frequencies. E.g. the energy of an inductor is stored in the electric field instead of in the magnetic field at frequencies above the resonant frequency. The differential Q value is defined as [10]

$$Q_D = \frac{Im(Z_D)}{Re(Z_D)} \, , \qquad (1)$$

where the differential impedance Z_D is defined as $Z_D = Z_{11} - Z_{12} - Z_{21} + Z_{22}$.

Also the measured values for inductance (L) and capacitance (C) are defined from differential impedance as

$$L = \frac{Im(Z_D)}{2\pi f} \quad \text{and} \quad C = \frac{-1}{Im(Z_D) \cdot 2\pi f} \, , \qquad (2)$$

Since the inductance and capacitance values are calculated as shown above, they include the frequency-dependent variations of the component characteristics. Thus, they are not the plain frequency-independent inductance or capacitance.

Furthermore, it is worth noting that some parasitic losses are neglected in the differential impedance [11] and that the capacitive coupling will cause over-evaluation of the inductance value at high frequencies [12].

3.1. Inductor-Resistor Circuits

The measurement results for the resistor-inductor circuits are presented in Fig. 3 and they are compared with the plain inductor L. The resistor connected in parallel does not change the inductance value at the operating frequency range, but it naturally causes losses and weakens the Q value. The differences in inductance and Q values between the plain inductor and LR circuits are at the frequency of 4 GHz 10 pH and 0.8, respectively. The reduction in Q value is almost equal for both resistors used. The nwell resistor R1 has a higher resistance value and it causes lower parasitic capacitances between the resistor and the inductor metals than the polysilicon resistor R2. The losses reduce the resonant frequency. The resonant frequencies of the circuits LR1 and LR2 are 500 MHz and 700 MHz lower than that of the reference inductor L.

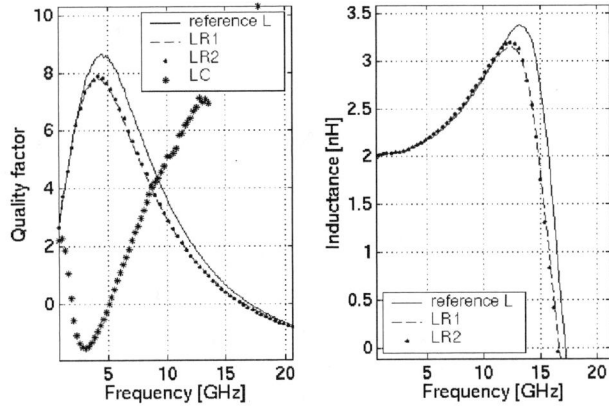

Fig. 3. Quality factor and inductance of circuits

The substrate losses, such as the eddy current losses, are almost equal for both circuits, since the peak values of Q are almost equal for both circuits. The eddy currents are reduced by using the patterned ground shield (PGS) under the inductors. In the case of the nwell resistor, a partial shield is used due to processing rules. On the grounds of the measurement results, the nwell resistor can be used as a part of the patterned ground shield for protection of the eddy current. The basic layout of the patterned ground shield in the case of LR1 and LR2 is shown in Fig. 4. The dashed line represents the part of the shield which was left out in the case of the circuit LR1.

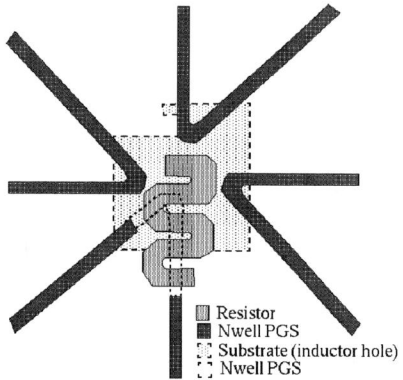

Fig. 4. Basic layout of patterned ground shield (PGS)

3.2 Inductor-Capacitor Circuit

The Q and capacitance values of the LC circuit and capacitor C are shown in Fig. 5a. The resonant frequency of the LC circuit is shown in Fig. 5b. The resonant frequency of an LC circuit is defined as

$$f = \frac{1}{2\pi\sqrt{LC}} \quad , \tag{3}$$

where L and C are inductance and capacitance, respectively.

The measured value for the resonant frequency of the LC circuit is 1.85 GHz. The measured inductance and capacitance values of the reference components L and C at the frequency of 1.0 GHz are 2.0 nH and 3.75 pF, respectively. As calculated with Eq. (3), the resonant frequency of the LC circuit should be 1.83 GHz. This matches quite well with the measured resonant frequency presented in Fig. 5b. Thus, it can

be said that the superimposed inductor and capacitor have the same characteristics as the separately processed components.

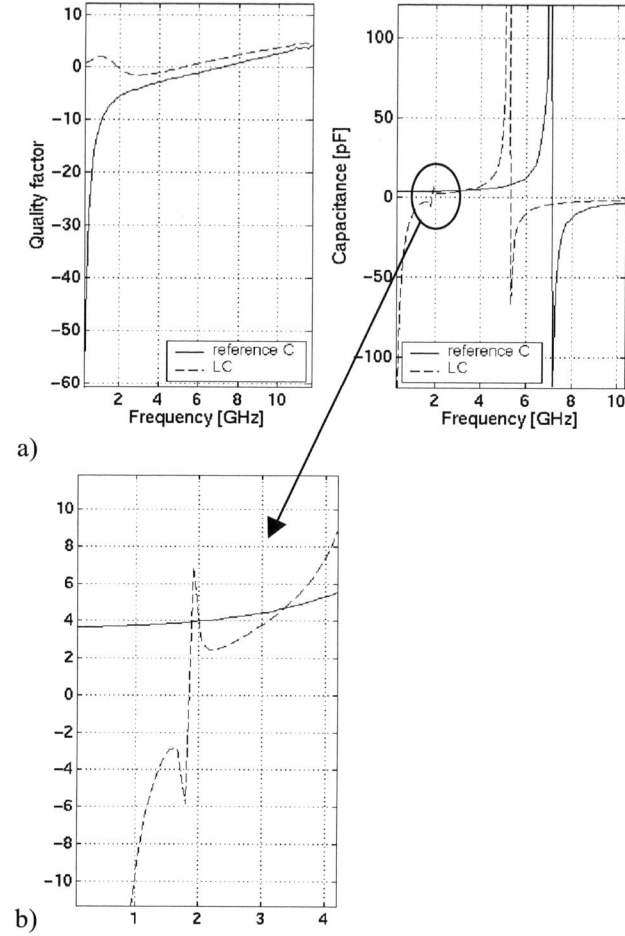

Fig. 5. a) Quality factor and capacitance of circuit. b) Resonant frequency of the LC circuit.

3.3 Error Factors of Superimposed Components

The parasitic capacitors are formed between different CMOS layers. The magnitude of the capacitances is higher between metal and polysilicon layers than between metal and nwell due to shorter vertical distance. Thus, the LC and LR2 circuits have higher parasitic capacitances compared to the LR1 circuit, which can be seen as lower resonant frequency of the circuits. Moreover, the stacked area of the capacitor and inductor is larger than that of the inductor and resistor. The resistive and capacitive parasitic effects are formed also by the wiring of the unit capacitors and the interconnections between the inductor and the other passives (resistor or capacitor). Since the metal layers have shorter vertical distances, the capacitances between the metal layers become more significant. It may be a problem especially in complex passive circuits, where a lot of wiring is needed under the inductor.

In the presented layouts, the length of the wiring is kept as short as possible. On the basis of the measurements, the parasitics are found to be less significant, since the measurement results of the single components and the passive circuits match quite well.

The slightly unstable values of the integrated passive components restrict their use in analog circuits. The maximum variations in the reference inductor and capacitor values at the frequency range of 45 MHz - 1.8 GHz were 0.12 nH and 0.25 pF, respectively. They correspond to 6 % and 7 % margins. The components were affected by some external factors, but their tolerances are likely to rise in actual circuitry due to increased wiring, the number of vias, and interference such as proximity effect. Moreover, the matching of different passive components to each other is quite difficult, whereas the similar components can be effectively matched using common-centroid or interdigitated layout.

4. Conclusions

A space-saving layout for passive circuits is proposed. The on-wafer passive components for two-component circuits LC, LR1, and LR2 are processed one on the other. It was found that their layouts have little significance for the characteristics of the circuits. The resonant frequency of the LC circuit is almost equal to the calculated value using the single components. Thus, the parasitic components formed between the stacked passive components have little significance and the proposed layout can be tested also in more complex designs.

Acknowledgement
The authors would like to thank Austria Micro Systems for processing the chips and Hannu Hakojärvi and Mikko Kantanen at the Technical Research Centre of Finland, Espoo, Finland for measuring the chips. P. Karjalainen would like to thank the Graduate School of Tampere University of Technology, Finnish Cultural Foundation, Ulla Tuominen Foundation, and Tuula and Yrjö Neuvo Foundation for financial support.

References

[1] S. K. Pienimaa and N. I Mart, "High-Density Packaging for Mobile Terminals", *IEEE Trans. on Adv. Packag.*, vol. 27, Aug. 2004, pp. 467-475.

[2] A. Hastings, *The Art of Analog Layout*, Prentice Hall, 2000.

[3] G. Ribes, J. Mitard, M. Denais, S. Bruyere, F. Monsieur, C. Parthasarathy, E. Vincent, and G. Ghibaudo, "Review on High-k Dielectrics Reliability Issues", *IEEE Trans. on Devices and Materials Reliability*, vol. 5 , March 2005, pp. 5-19.

[4] C. P. Yue and S. S. Wong, "On-Chip Spiral Inductors with Patterned Ground Shields for Si-Based RF IC's", *IEEE J. of Solid-State Circuits*, vol. 33, May 1998, pp. 743-752.

[5] F. Sabouri-S., C. Christensen, and T. Larsen, "A Single-Chip GaAs MMIC Image-Rejection Front-End for Digital European Cordless Telecommunications", *IEEE Trans. on Microwave Theory and Techniques*, vol. 48, Aug. 2000, pp.1318-1325.

[6] C. P. Yue and S. S. Wong, "Physical Modeling of Spiral Inductors on Silicon", *IEEE Trans. on Electron Devices*, vol. 47, March 2000, pp. 560 -568.

[7] K.-F. Becker, E. Jung, A. Ostmann, T. Braun *et al.* "Stackable System-on-Packages with Integrated Components", *IEEE Trans. on Advanced Packaging*, vol. 27, May 2004, pp.268-277.

[8] T. W. Rondeau, M. F. D'Souza, and D. G. Sweeney, "Residential Microwave Oven Interference on Bluetooth Data Performance", *IEEE Trans. on Consumer Electronics*, vol. 50, August 2004, pp. 856-863.

[9] G. C. T. Leung and H. C. Luong, "A 1-V 5.2-GHz CMOS Synthesizer for WLAN Applications", *IEEE J. of Solid-State Circuits*, vol. 39, Nov. 2004, pp. 1873-1882.

[10] G. J. Carchon, X. Sun, and W De Raedt, "High-Q Above-IC Inductors and Transmission Lines - Comparison to Cu Back-End Performance", in *Proc. Electronic Components and Technology 2004*, vol. 1, June 2004, pp.1118-1123.

[11] V. Ermolov, T. Lindström, M. Olsson, M. Read *et al.* "Microreplicated RF Toroidal Inductor," *IEEE Trans. on Microwave Theory and Techniques*, vol. 52, Jan. 2004, pp. 29-37.

[12] H. Feng, G. Jelodin, K. Gong, R. Zhan *et al.* "Super Compact RFIC Inductors in 0.18 um CMOS with Copper Interconnects", in *Proc. of IEEE Microwave Symposium Digest*, vol. 1, June 2002, pp.553-556.

A Novel Low Power Multilevel Current Mode Interconnect System

Supreet Joshi, Dinesh Sharma
Department of Electrical Engineering
Indian Institute of Technology, Bombay, India
{supreet, dinesh}@ee.iitb.ac.in

Abstract

We propose circuits for low power, high throughput multilevel current mode signaling using 2 bit simultaneous data transfer. A novel design of the receiver for very low line voltage swings is discussed. The technique involves matching the receiver impedance to the line impedance thereby reducing the ringing on the wire. Simulation results show upto 50% reduction in latency and upto 100 times reduction in power over voltage mode buffer insertion techniques. We also show that the delays through this system are largely independent of the interconnect lengths. Data rates of upto 1Gb/s have been obtained. A power consumption model is derived for the system which matches the simulation results to within 5%.

1. Introduction

Traditionally, voltage mode repeaters along the interconnect have been used to reduce the delays in signal transmission. However, there is a limit to the performance improvement that can be obtained with repeaters in deep submicron designs in terms of power and delay [5], [10]. Current mode signaling has been explored as an alternative for data transmission over interconnects in [1], [2], [11] and related works. It has already led to improved speed in SRAM CMOS circuits [8]. Multilevel current mode signaling [3] allows increased on-chip communication bandwidth and reduced interconnect area and power consumption as well as being more process and noise tolerant than voltage mode operation.

Power consumption on the interconnect can be reduced by reducing voltage swings on the line. However, in a voltage mode scenario, this means that the signals need to be amplified back, which consumes power and leads to a tradeoff between the circuit power loss and the interconnect power loss [9]. In a current mode situation, the swings can be independently controlled leading to extremely low power consumption in the wire and reduced wire delays. By using our impedance control technique in the receiver circuit, both

power and delay in the wire can be reduced.

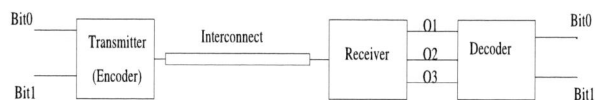

Two Bit Four Level Current Interconnect System

Figure 1. Block Diagram

Table 1. The encoding and decoding scheme

Tx0	Tx1	Line Current	O1	O2	O3	Rx0	Rx1
1	1	0	1	1	1	1	1
1	0	I	0	1	1	1	0
0	1	2I	0	0	1	0	1
0	0	3I	0	0	0	0	0

2. The Four Level Current Mode System

Figure 1 shows the block diagram of the quaternary current mode signaling system that we have developed. The driver circuit generates four distinct current levels from the two voltage levels, which then propagate along the interconnect. The line current is then compared against reference currents in the receiver and the three current comparator outputs are decoded back to the digital bits by the decoder. The receiver can control the voltage swing on the line by controlling the impedance seen by the interconnect. Table 1 shows the encoding, current comparison and decoding schemes where 'I' represents the unit current level.

2.1. The Driver circuit

Figure 2 shows the driver circuit we used to convert 2 voltage bits to 4 current levels. A small reference current is generated in M1-M2 arm. The M4-M6 arm is designed to carry twice the amount of current in the M3-M5 arm by appropriate sizing. The analog sum of the currents is then

0-7695-2533-4/06 $20.00 © 2006 IEEE   122

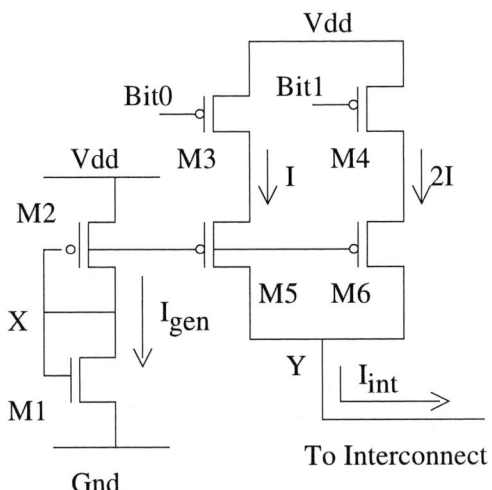

Figure 2. The Driver Circuit

put on the wire. When M3 and M4 are both cut off (both voltage bits 1), no current flows in the wire. When both bits are 0, M3 and M4 are in deep triode and M1-M5, M1-M6 form a current scaling structure - so the reference current can be scaled to obtain 'I' and '2I' in the two arms, and an interconnect current of '3I'. When Bit0 is 0 and Bit1 is 1, M3 is in deep triode while M4 is cut off - so the M3-M5 arm carries 'I' while no current flows in the M4-M6 arm, leading to an interconnect current of 'I'. Similarly, when Bit0 is 1 and Bit1 is 0, a current of '2I' flows in the interconnect. The choice of the current 'I' is determined by the tradeoff between power considerations and system latency as well as the resolution of the current comparator (for accurate decoding) in the receiver.

2.2. The Receiver circuit

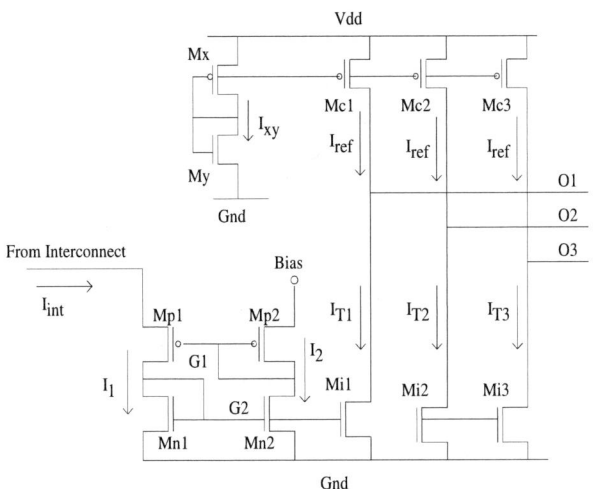

Figure 3. The Receiver Circuit

Figure 3 shows the circuit diagram of our receiver. The transistors Mx and My provide a small reference current that is scaled up and used for comparison with the scaled values of line current. There are two basic operations of the receiver:

1. Impedance control: The transistors Mp1, Mp2, Mn1, Mn2 are used to control the input resistance of the receiver. This technique has been used before to develop high injection efficiency readout circuit for infra red detectors [13], [4] but its use for controlling input resistance of receiver on current mode interconnects is novel. The input resistance of this circuit can be shown to be:

$$R_{in} = \frac{(1 - \Gamma_{gm})}{g_{mp1}}; \text{ where } \Gamma_{gm} = \frac{g_{mn2}/g_{mn1}}{g_{mp2}/g_{mp1}} \quad (1)$$

This means that by sizing the 4 transistors appropriately, we can match the impedance of the receiver to the line impedance. This prevents any reflections and ringing, thereby reducing power lost in charging and discharging the huge wire capacitance. The source voltages of Mp1 and Mp2 are forced to be identical due to the current feedback structure, and so the voltage at the receiver end of the interconnect can be held at $V_{dd}/2$ by fixing the bias voltage at source of Mp2.

2. Current comparison: The 3 arms Mc1-Mi1, Mc2-Mi2, and Mc3-Mi3 form the standard current comparator [12]. A single fixed reference current I_{ref} is replicated in transistors Mc1, Mc2 and Mc3 by scaling I_{xy} (see Figure 3). Scaled versions of the interconnect current I_{T1}, I_{T2}, I_{T3} attempt to flow through transistors Mi1, Mi2 and Mi3 respectively.

In any single arm, the lower current of the 2 transistors flows through, forcing the other into triode region. Hence, the common drain voltage comes close to the source voltage of the transistor supposed to carry higher current. For example, if $I_{T1} > I_{ref}$, then I_{ref} flows in the Mc1-Mi1 arm with Mi1 in deep triode region. The common drain is close to ground in this case. Similarly, if $I_{T1} < I_{ref}$, Mc1 is in deep triode with I_{T1} flowing in the arm and causing the common drain to be close to V_{dd}. This is the basis of current comparison operation.

The sizes of Mc1, Mc2 and Mc3 are identical, but those of Mi1, Mi2 and Mi3 need to be chosen to ensure logical correctness. Let the sizes of Mi1, Mi2 and Mi3 be such that currents I_{T1}, I_{T2} and I_{T3} are k_1, k_2 and k_3 times the interconnect current I_{int} respectively. For logically correct operation of the comparator (see Table 1) the following inequalities must hold (where 'I' is the unit current): $0 < I_{ref} < k_1 I$, $k_2 I < I_{ref} < 2k_2 I$

0-7695-2533-4/06 $20.00 © 2006 IEEE

and $2k_3I < I_{ref} < 3k_3I$. The sizes of the transistors Mi1, Mi2 and Mi3 are obtained from these relations.

2.3. The Decoder circuit

The decoder takes the outputs of the current comparator and converts them back to voltage bits. The decoding logic is shown below :

$$Rx0 = O3 \cdot (O1 + O2\,') \qquad (2)$$

$$Rx1 = O2 \qquad (3)$$

$Rx0$ has been implemented through simple CMOS logic (Figure 4), while $Rx1$ requires no additional logic.

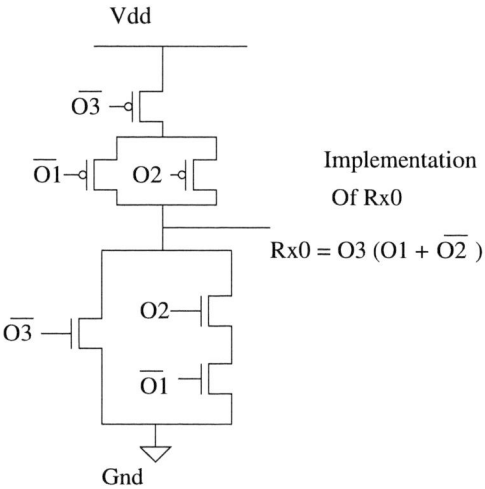

Figure 4. The Decoder circuit

3. Power Dissipation Equations

In this section, we obtain the equations governing the power loss in this system. The major component is the static loss due to constant current flow in the driver, the interconnect and the receiver. The dynamic power is very small because the large wire capacitances charge and discharge to only very small voltage swings.

3.1. The Driver circuit

The reference current generation consumes static power $= I_{gen} \cdot V_{dd}$ at all times. However, when both bits are 1, this is the only power loss in the driver as both arms carry no current. In the worst case, when both bits are 0 and both arms conduct, the power loss in M3 and M4 is very small as they operate in the deep triode region. The static power loss is given by $3I \cdot (V_{dd} - V_y)$, where V_y is the interconnect voltage at the driver end. Since the receiver end is held at $V_{dd}/2$, $V_y = V_{dd}/2 + 3I \cdot R_{int}$, where R_{int} is the total

interconnect resistance. In general, the total static loss in the driver is given by:

$$P_{dr} = I_{gen} \cdot V_{dd} + I_{int} \cdot (V_{dd}/2 - I_{int} \cdot R_{int}) \qquad (4)$$

3.2. The Receiver circuit

The static loss due to reference current generation is given by $I_{xy} \cdot V_{dd}$. In the impedance control section, the two arms carry equal current and since the sources of Mp1 and Mp2 are at approximately the same voltage $V_{dd}/2$, the power loss is given by $I_{int} \cdot V_{dd}$. In the comparator section, the current flowing in each arm is different for different cases :

1. When the wire current is '3I', all NMOS transistors (Mi1, Mi2 and Mi3) are operating in triode region, while all PMOS transistors (Mc1, Mc2 and Mc3) are in saturation. The current in each arm is I_{ref}. So, the static power loss is $3I_{ref} \cdot V_{dd}$.

2. When the wire current is '2I', Mi1, Mi2 and Mc3 are operating in triode region (because $I_{ref} > 2k_3I$). Arms 1 and 2 conduct I_{ref} while Arm 3 conducts $2k_3I$. Hence, the power loss is $(2I_{ref} + 2k_3I) \cdot V_{dd}$.

3. When the wire current is 'I', Mi1, Mc2 and Mc3 are operating in triode region ($I_{ref} > k_2I$ and $I_{ref} > k_3I$). The currents in the three arms are I_{ref}, k_2I and k_3I respectively. The power loss, then is given by $(I_{ref} + k_2I + k_3I) \cdot V_{dd}$.

4. When there is no current through the wire, Mc1, Mc2 and Mc3 operate at edge between triode and cutoff regions while Mi1, Mi2 and Mi3 are cutoff. No current flows in any arm and hence, there is no static power loss.

3.3. The Decoder circuit

Since $Rx0$ has been implemented in CMOS style, there is no static loss in the decoder, while $Rx1$ does not require any additional logic (see equations (2) and (3)). The only power loss is due to switching, which is very small compared to the static loss.

3.4. The Interconnect

There are two parts to the interconnect power loss - one, the static resistive losses which can be approximated by $I_{int}{}^2 R_{int}$, where R_{int} is the total interconnect resistance; and two, the dynamic power loss arising from capacitive charging, expression for which has been obtained in [1]. It is to be noted though, that the receiver design ensures that the voltage swing on the wire is almost negligible, and hence even for very long interconnects, the dynamic power loss is very small.

0-7695-2533-4/06 $20.00 © 2006 IEEE   124

These simple analytical expressions, which ignore the capacitive losses in the wire and the switching loss in the CMOS decoder, can be used to estimate the power loss in the system. Simulations will be shown to confirm that these assumptions hold true and that these equations are very accurate.

4. Simulation Methodology

Simulations were carried out assuming a 0.18μ TSMC process using Synopsys HSPICE and AvanWaves tools. The interconnect was modeled as a 5 section π RC network. The values of R and C were obtained from the TSMC 0.18μ process models for metal layer M3. The width of the wire was 5λ and the spacing between wires was 4λ. The wire resistance per unit length used was $0.1778\text{K}\Omega/\text{mm}$, while the wire capacitance per unit length used was 0.19pF/mm. These values are consistent with TSMC 0.18μ process. Nominal Vdd of 1.8V and operating temperature of 25°C were used for simulations. The power measurement feature of HSPICE was used to estimate the power dissipation in the system for comparison with the theoretical results.

5. Analysis of System Latency

In this section, we describe the factors affecting the latency of this system based on simulation observations. The fact that the large interconnect capacitance is never charged or discharged to large voltage swings, coupled with the fact that current switching is faster contributes to the extremely low system latency.

The rise time of the received bits is also lower because current rises much faster (almost quadratically with voltage) than the transmitted voltage bits, and over a much shorter time. For example, when the voltage input to the driver circuit goes from Vdd to 0 over a time of t_r (see Figure 2), the current starts increasing quadratically when the voltage reaches $Vdd - V_{Tp}$, and increases linearly when the gate voltage is close to 0 (when the transistor goes into triode region). The current comparator starts responding as soon as the corresponding scaled current (see Figure 3) exceeds the reference current, which happens midway through the current transition, thereby causing the comparator output to start switching much before the transmitted voltage has even completed its switching. The decoder adds a CMOS gate equivalent delay to the system.

Another issue to be noted is that different transitions lead to different latencies. For a 4-level signaling system, there are 12 different transitions, and since each of them involve different current levels, the delays and power loss will all be different.

6. Simulation Results

Figure 5 shows the operation of this signaling system, for a wire length of 1mm and a data rate of 1Gb/s and a unit

current level of $14.5\mu\text{A}$. The 4 current levels correspond to the 4 combinations of the two voltage bits. The lower rise and fall times of the received signals, and different latencies for different transitions can be seen clearly.

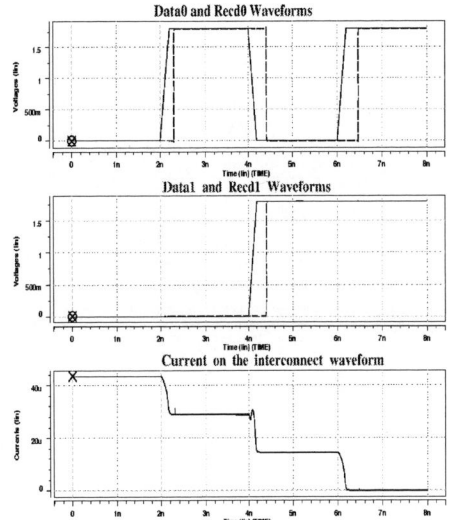

Figure 5. Waveforms of transmitted and received voltages and wire current

Figure 6 shows the observed average and maximum latency as a function of the interconnect length as well as a comparison with voltage mode delay values obtained from the pre-mid-post buffer strategy [7], [6]. It can be seen that our system has latency upto 50% lower than the voltage mode delays at wire lengths greater than 2mm and this advantage increases at higher wire lengths, indicating the suitability of this system for global and semiglobal interconnects.

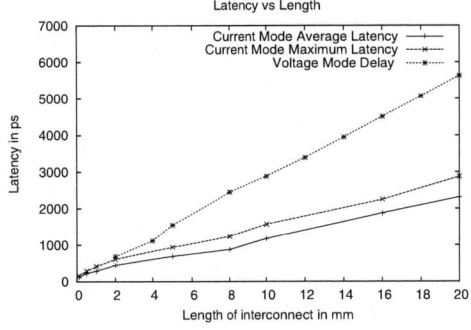

Figure 6. Current mode latency and Voltage mode delays vs. Interconnect length

0-7695-2533-4/06 $20.00 © 2006 IEEE

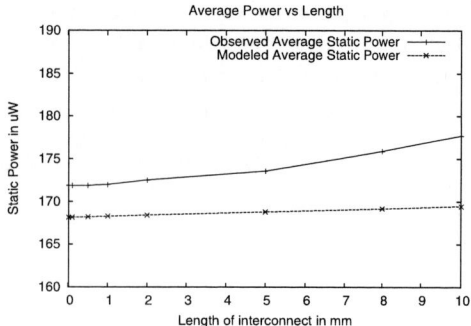

Figure 7. Observed and modeled average power vs. Interconnect length

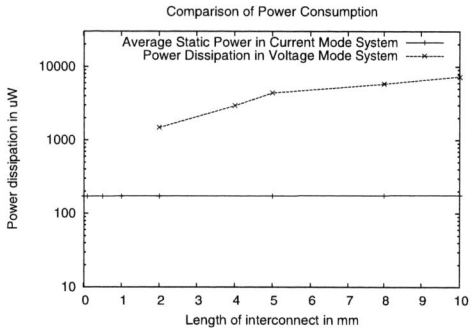

Figure 8. Comparison of power loss in current mode system with voltage mode system

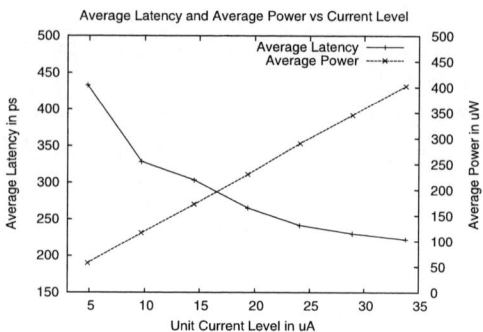

Figure 9. Average latency and Average power vs. Unit current level

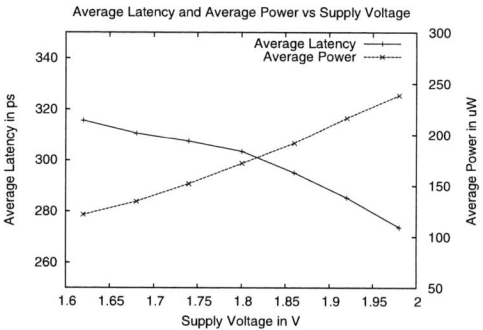

Figure 10. Average latency and Average power vs. Supply voltage

Figure 7 shows the observed average power consumption and the average power expected from our model as a function of the interconnect length at a unit current level of $14.5\mu A$ and data rate of 1Gb/s. The deviation is less than 3% from the expected for interconnect lengths upto 2mm, and remains less than 5% even upto 10mm. Since our model ignores capacitive losses, this indicates that the reduced wire ringing has reduced the wire capacitive losses to negligible quantities.

Figure 8 shows the comparison of the average power dissipation in our system with the power loss in the voltage mode system using pre-mid-post buffering strategy described in [7]. The gains vary from $10\times$ for semiglobal wires to almost $100\times$ for global wires. This gain is primarily due to the low static power dissipation and the absence of capacitive losses in the wire.

Figure 9 shows the average latency and average power loss as a function of the unit current level for an interconnect length of 1mm and data rate of 1Gb/s. Increased current leads to lower latency due to faster current comparator response. As expected from our model, power is approximately a linear function of the unit current level. This leads

to a latency-power tradeoff as lower latency can be obtained at the cost of higher power loss.

A worst case variation analysis of voltage, temperature and process parameters was carried out for interconnect length of 1mm and data rate of 1Gb/s with a unit current level of $14.5\mu A$. Figure 10 and Figure 11 show the variation of average latency and power with supply voltage and temperature respectively. A 10% variation in the power supply led to 7% variation in latency and 10% variation in the interconnect voltage swing from the nominal values at 1.8V. A $0^{o}C$ to $70^{o}C$ temperature variation led to variation of less than 3% in latency, 2% in power and 8% in interconnect voltage swing from the nominal values at $25^{o}C$. Current mode circuits are known to be process-tolerant and this was confirmed by simulations on all the 4 process corners. Latency, power and interconnect voltage swing measurements showed less than 1% variation across process corners. This demonstrates the robustness of our impedance control technique and indicates that the system performance is consistent across process, voltage and temperature variations.

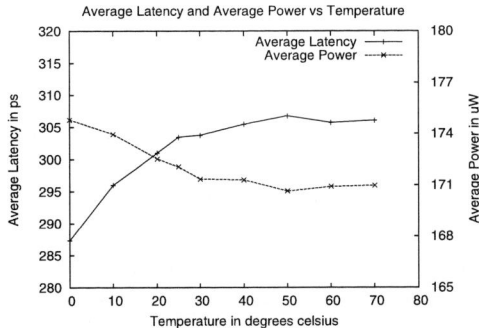

Figure 11. Average latency and Average power vs. Temperature

7. Discussion

The contribution of interconnect to the system latency is very small since no capacitive charging or discharging is required. Further, latency and power are largely independent of the wire length, which makes this system attractive for global interconnects over traditional voltage mode techniques. Figures 6 and 7 illustrate this point.

The circuit performance under scaling is expected to improve as the circuits become faster. In comparison, voltage mode buffering technique is expected to suffer from increasing leakage power loss and increasing area at lower technology nodes. Hence, the wire lengths at which this technique will perform better than voltage mode techniques will reduce, making it viable for even semi-global interconnects. Further, the absence of "ringing" on the wire leads to very low electrostatic coupling with nearby wires, thereby reducing noise injection. While wire inductance does not play a significant role at $0.18\mu m$ technology, its effects on this system at lower nodes will need to be studied.

8. Conclusion

In this paper, we presented a low power, low latency, high throughput multilevel current mode signaling system which is suitable for global and semiglobal interconnects. Accurate, yet simple analytical expressions which estimate the power in our system were derived. The system provides significant gains in latency and power over voltage mode buffer insertion techniques, without compromising on throughput. The circuits were also found to be very robust in presence of temperature, process and supply voltage variations. A power-latency tradeoff was identified which can be used to optimise the system for latency or power. The system offers clear advantages over existing interconnect systems at lower technology nodes in terms of latency and power, and is very economical in terms of area and bandwidth utilization.

Acknowledgment

The authors would like to thank Ms. Vani Prasad, Research scholar in Dept. of Electrical Engineering, IIT Bombay for her constructive suggestions and useful discussions throughout the duration of this effort.

References

[1] R. Bashirullah, W. Liu, and R. K. Cavin III. Current Mode Signaling in Deep Submicrometer Global Interconnects. *IEEE Transaction on VLSI Systems*, 11, No.3:406–417, June 2003.

[2] I. Dhaou, M. Ismail, and H. Tenhunen. Current Mode, Low Power, On-Chip Signaling in Deep Sub-micron CMOS Technology. *IEEE Transactions on Circuits And Systems*, 50, No.3:397–406, March 2001.

[3] H. C. Kirsch and E. Ku, Multiple-bit current-mode data bus, U.S. Patent 6 184 714, Feb, 2001.

[4] H. Kulah and T. Akin. A Current Mirroring Integration Based Readout Circuit for High Performance Infrared FPA Applications. *IEEE Transactions on Circuts And Systems*, 50, No.4:181–186, April 2003.

[5] D. Liu and C. Svensson. Power Consumption Estimation in CMOS VLSI Chips. *IEEE Journal Of Solid State Circuits*, 29, No.6:663–670, June 1994.

[6] V. Prasad. Interconnect Aware VLSI Design. Technical Report, Dept. of Electrical Engineering, IIT Bombay, 2005.

[7] V. Prasad and M. P. Desai. Interconnect Delay Minimization Using a Novel Pre-Mid-Post Buffer Strategy. *Proceedings of 16th International Conference on VLSI Design*, pages 417–422, June 2003.

[8] E. Seevinck, P. van Beers, and H. Ontrop. Current Mode Techniques for High Speed VLSI Circuits with Application to Current Sense Amplifier for CMOS SRAMs. *IEEE Journal Of Solid State Circuits*, 26:525–536, April 1991.

[9] C. Svensson. Optimum Voltage Swing on On-Chip and Off-Chip Interconnect. *IEEE Journal Of Solid State Circuits*, 29, No.6:663–670, June 1994.

[10] D. Sylvester and K. Kuetzer. Getting To The Bottom Of Deep Submicron II: The Global Wiring Paradigm. *Proceedings of International Symposium on Physical Design*, pages 193–200, April 1999.

[11] V. Venkatraman and W. Burleson. Robust Multi-Level Current-Mode On-Chip Interconnect Signaling in the Presence of Process Variations. *Proceedings of Sixth International Symposium on Quality Electronic Design*, pages 522–527, March 2005.

[12] K. Wayne Current. Current Mode CMOS Multiple Valued Logic Circuits. *IEEE Journal Of Solid State Circuits*, 36, No.7:1108–1112, July 2001.

[13] N. Yoon, B. Kim, H. C. Lee, and C.-K. Kim. High Injection Efficiency Readout Circuit for Low Resistance Infrared Detector. *IEE Electronic Letters*, 35, No.18:1507–1508, September 1999.

0-7695-2533-4/06 $20.00 © 2006 IEEE

The Design of Analog Front-End Circuitry for 1X HD-DVD PRML Read Channel

Sheng-Jang Lin
Industrial Technology Research Institute ,Taiwan, R.O.C.
shengjang@itri.org.tw

I-Shun Chen
Industrial Technology Research Institute ,Taiwan, R.O.C.
Shiunger@itri.org.tw

Bo-Wei Chen
Industrial Technology Research Institute ,Taiwan, R.O.C.
BWChen@itri.org.tw

Feng-Hsiang Lo
Industrial Technology Research Institute ,Taiwan, R.O.C.
fhLo@itri.org.tw

Abstract

In this paper, the design techniques and considerations for each building block required for analog signal processing in HD-DVD PRML read channel are presented and the procedures of analog signal processing are also described. The Analog Front-End Circuitry (AFE) includes the circuits of RF Summer、Attenuator、Equalizer、AGC and ADC. The Equalizer is constructed by seven-pole two-zero 0.05 degree equiripple linear phase Gm-C filter. It has a cutoff frequency (fc) tunable between 8 and 39MHz and it is also able to provide up to 12dB of boost at fc. The constant group delay bandwidth of the filter is 1.65 fc. And the AGC circuit which uses the exponential type of VGA can has nearly constant settling time within 10us and it has 1Vpp constant amplitude output. Behind the AGC is the flash ADC, it has a resolution of 6 bit and 300MHz conversion rate and it is enough to provide the digital data required for the digital part which uses the method of partial response maximum likelihood (PRML). The design was made using TSMC 0.35um 2P4M mixed-signal CMOS process. The AFE consumes 520 mW from a single 3.3v power supply, and occupies an area of 12.8_{mm^2}.

1. Introduction

With the increasing demand of high data storage, the high density DVD (HD-DVD) specifications have been instituted by DVD Forum in Japan. The new generation of optical drive system which uses blue-ray techniques can has 20GB storage volume in an optical disk. However, the received signal from the disk suffers from severer inter-symbol interference (ISI) not only due to the smaller laser spot supported by the blue-ray pickup head but also because of the smaller pits、lands and track pitch in a blue-ray disk [1]. Therefore, in order to fit the bit error rate required for HD-DVD, the traditional read channel which uses slicer behind the AGC but not the PRML system behind the ADC can not has data jitter larger than 7%. Unlike using the slicer method, the read channel of HD-DVD which uses PRML method can allow the SNR to be smaller, that is, a larger jitter tolerance.

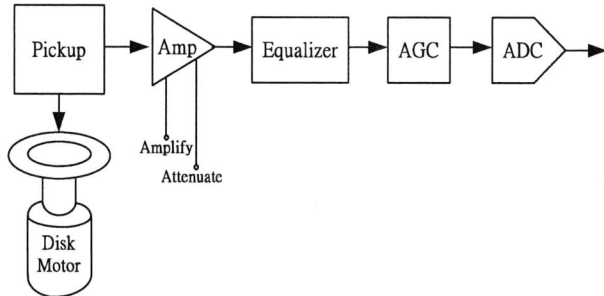

Figure 1. Analog Front-End Circuitry

Owning to the low pass characteristic of the optical channel, the RF signal is decayed in high frequency .The analog equalizer used here can boot the high frequency signal and thus reduce the impact of ISI which will induce jitter in read channel [1]. Besides, in PRML read channel applications, if a significant amount of equalization is performed digitally, the quantization noise generated by the ADC can be amplified by the digital equalization filter. This, then, results in an increased resolution requirement for the ADC in order to reduce the quantization noise contribution [2]. Thus, if the analog equalizer is used, it can perform some or all of the equalization required to match the amplified pulse shape and the number of tap in digital FIR filter can be reduced as well and it also can allow the ADC to has less resolution requirement [3]. The AGC circuit here can be used to suppress the amplitude modulation which is caused by the low frequency noise from the servo and the disk. Furthermore, if the AFE (as shown in Figure 1) is used in traditional read channel, that is, the AGC is connected to the slicer, then, the data jitter after the slicer will be reduced [1]. The characteristics of the partial-response system for the optical channel in HD-DVD can approximate to PR(1,2 ,2,2,1) which has 9 levels and thus the requirement of the resolution of ADC must be at least 6 bit [4].

This paper contains four sections. Section 2 describes the design of all building blocks in the AFE and Section 3 is the simulation results and the conclusions are made in Section 4.

0-7695-2533-4/06 $20.00 © 2006 IEEE

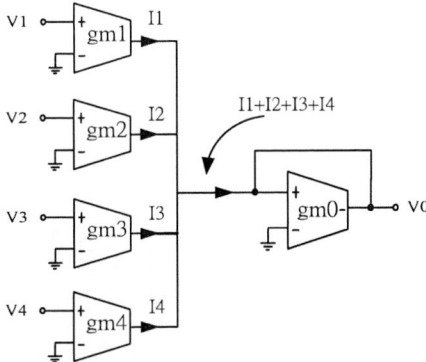

Figure 2. The Gm cell based amplifier

2. The AFE Design

2.1 The Design of RF Summer and Attenuator

The RF Summer and Attenuator are combined into one building block which performs the functions of summing and attenuation (as shown in Figure 2). And the bandwidth of the amp must choose to be moderate for the group delay and noise considerations. It realizes the function as below [5]:

$$V_o = \frac{g_{m1}}{g_{m0}}V_1 + \frac{g_{m2}}{g_{m0}}V_2 + \frac{g_{m3}}{g_{m0}}V_3 + \frac{g_{m4}}{g_{m0}}V_4$$

2.2 The Filter Architecture and Design Considerations

There are two kinds of filters which have constant group delay. One is the Bessel-Thomson filter, the other is the equiripple delay filter. For the same delay and the same degree, the equiripple delay filter has a wider constant-delay bandwidth and a little sharper cutoff than Bessel-Thomson filter. Thus, the equiripple delay responses are generally preferable over Bessel-Thomson responses. In this AFE design , the 7th $0.05°$ equiripple delay filter with two asymmetric zeros is used (as shown in Figure 3). Using two asymmetric zeros gives a more flexible approach to provide separate control for left and right plane real zeros. Because this could correct for the group delay slopes in the band of interest due to imperfect filter pole placement or help equalize somewhat asymmetrical pulses from the media [2]. The transfer function of the filter is given by :

$$H(s) = \frac{\frac{g_{m1}}{g_{m2}}}{\frac{sC_{m1}}{g_{m2}}+1} \cdot \frac{sC_{11}\frac{g_{15}}{g_{12}g_{14}} - \frac{g_{11}}{g_{14}}}{s^2C_{11}C_{12}\frac{1}{g_{12}g_{14}} + sC_{11}\frac{g_{13}}{g_{12}g_{14}}+1}$$

$$\cdot \frac{sC_{21}\frac{g_{25}}{g_{22}g_{24}} + \frac{g_{21}}{g_{24}}}{s^2C_{21}C_{22}\frac{1}{g_{22}g_{24}} + sC_{21}\frac{g_{23}}{g_{22}g_{24}}+1}$$

$$\cdot \frac{\frac{g_{31}}{g_{34}}}{s^2C_{31}C_{32}\frac{1}{g_{32}g_{34}} + sC_{31}\frac{g_{33}}{g_{32}g_{34}}+1}$$

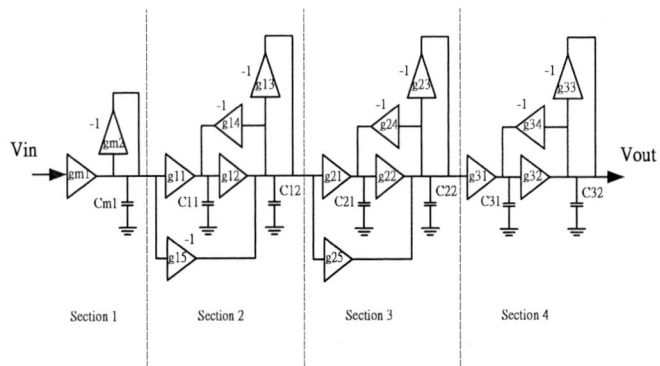

Figure 3. A Gm-C modular implementation of the seven-pole two asymmetric zero filter

Table I
Normalized Pole Frequency and Pole
Quality Factor for 0.05 Degree Equiripple
Linear Phase Filter

	pole frequency	pole Q
Section 1	0.86133	—
Section 2	1.14762	0.68110
Section 3	1.71796	1.11409
Section 4	2.31740	2.02290

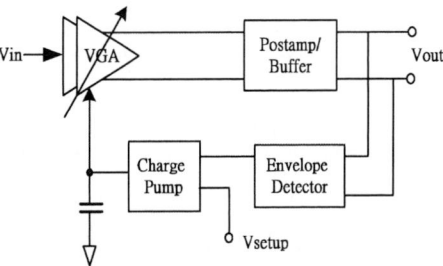

Figure 4. The AGC Architecture

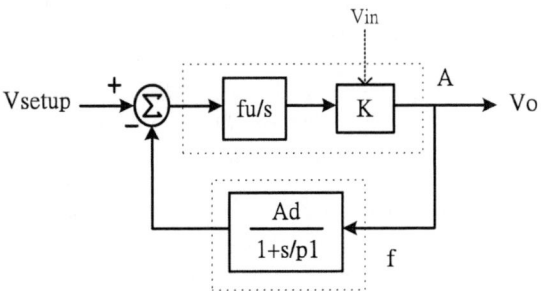

Figure 5. A simplified model used for AGC loop analysis

Thus the center frequency and the quality factor are given as follow:

$$\omega_k = \sqrt{\frac{g_{k2}g_{k4}}{C_{k1}C_{k2}}} \quad , \quad Q_k = \sqrt{\frac{C_{k1}}{C_{k2}}} \cdot \frac{\sqrt{g_{k2}g_{k4}}}{g_{k3}}$$

0-7695-2533-4/06 $20.00 © 2006 IEEE   129

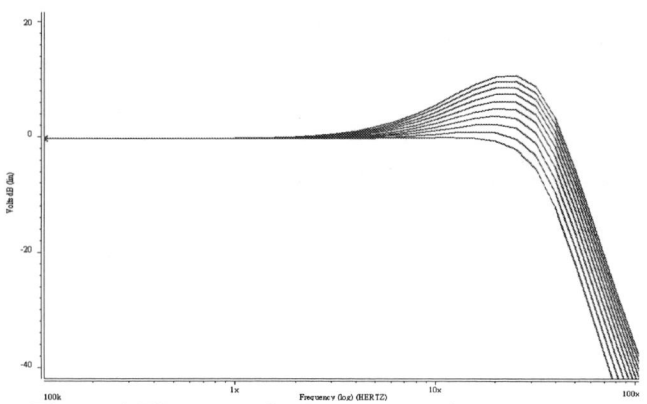

Figure 6. Filter magnitude response with boost range 3~12dB at fc=16.2MHz

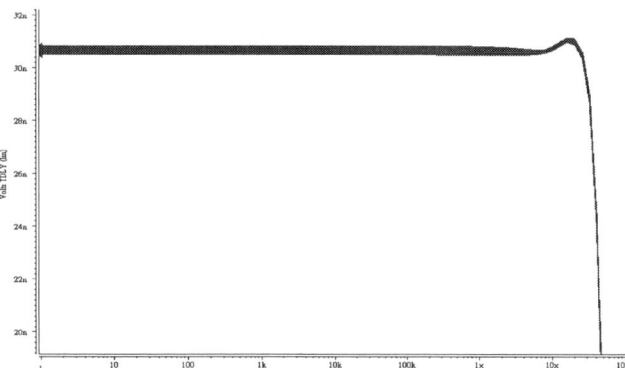

Figure 7. Constant group delay with boost range 3~12dB at fc=16.2MHz

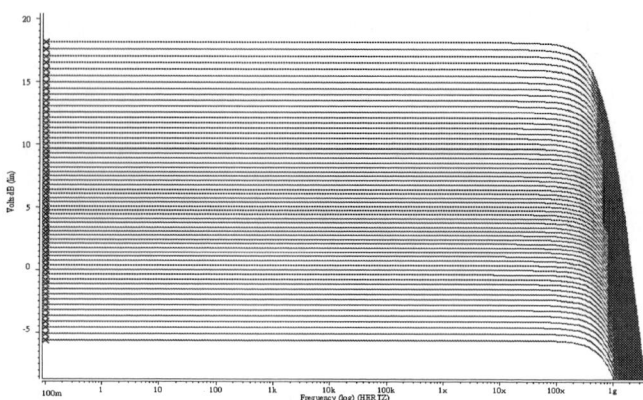

Figure 8. VGA gain with linear in dB

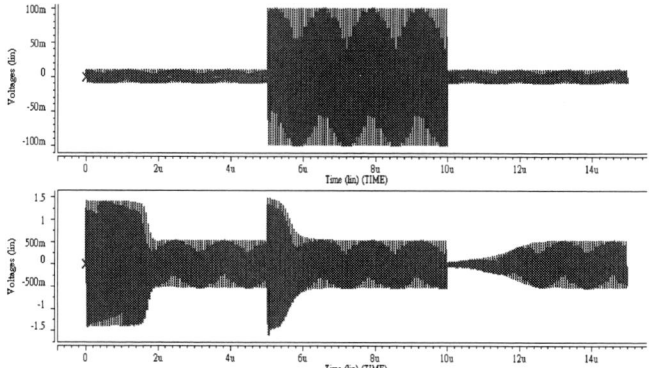

Figure 9. AGC output with nearly constant settling time

The ratios of capacitance and gm should be moderate because small ratios can be realized more accurately than large ones and for dynamic range optimization, we chose

$$gm3 = gm2, \quad \text{for} \quad Q < 1 \quad (\text{section2})$$

$$gm3 = gm1, \quad \text{for} \quad Q > 1 \quad (\text{section 3 and 4})$$

And the normalized pole frequencies and Q's of the filter sections realizing the 0.05° equiripple linear phase filter are given in Table I. However, due to the parasitic capacitors in each node of the filter, it will cause errors in ω and Q and therefore impact the performance of the filter. Consequently, proper predistortion and layout techniques should be made and precise process insensitive capacitor ratios can be obtained. Apart from the transconductor's excess phase caused by the parasitics, another important error is the integrator finite output impedance. This nonideality will impact the values of ω and Q as well. However, a design using cascade structures and large gate length devices can easily minimize the effect [2][5]. Besides, a master-slave tuning circuit can be used to correct the error of gm due to process and temperature variations [6].

2.3 The Building Blocks of AGC and Design Considerations

An automatic gain control loop (as shown in Figure 4),

composed of a variable gain amplifier、post amplifier、 envelope detector and integrator, can set the output amplitude constant. However, the RF signal of the optical drive system is the series combination of signal which has unequal period and amplitude. And it can not be set constant at the output of AGC because of the variation of the amplitude is too fast to let the AGC to settle. Hence the use of AGC here is just to suppress the amplitude modulation caused by low frequency noise. The AGC circuit is usually designed to has a constant settling time property which permits the AGC loop's bandwidth to be maximized for fast signal acquisition while maintaining stability over all operating conditions. [7]. In order to has a constant settling time, the VGA gain is designed to has pseudo-exponential gain control function which is realized by a source-coupled pair with diode-connected loads and the gain function is given as follow [7][8]:

$$gain_{n\ stage} = (\frac{g_{m,input}}{g_{m,load}})^n = K \cdot \left(\frac{1+x}{1-x}\right)^{n/2} \approx K \cdot e^{nx}$$

Higher order exponential function and larger dynamic control range can be achieved by cascading gain stages using the same amplifier cell with the same gain control characteristic. The AGC loop can be simplified as shown in Figure 5 and hence the loop transfer function is shown as follow [9]:

$$H(s) = \frac{V_o}{V_{setup}}$$

$$= \frac{A}{1 + A \cdot f}$$

$$= \frac{fu \cdot K \cdot (s + pl)}{s^2 + pl \cdot s + Ad \cdot fu \cdot pl \cdot K}$$

$$= \frac{fu \cdot K \cdot (s + pl)}{s^2 + \frac{\omega_n}{Q} \cdot s + \omega_n^2}$$

where K is the forward path gain and pl is the pole of low pass filter which is embedded in envelope detector and fu is the unity-gain frequency of integrator and ωn is the loop bandwidth of AGC. The loop bandwidth of AGC should be set properly. Because if the loop bandwidth is narrow, the tracking agility of AGC will degrade. On the other hand, if the loop bandwidth is large, the coupled inband noise will influence the output level.

2.4 The Design Considerations of Flash ADC

The characteristics of Flash ADC are high speed and easy to implement. But the price is large power consumption and large chip area. Besides, some attentions should be paid, such as large input capacitance and the problems of comparator offset voltage and resistor mismatch, because they will limit the speed and resolution of the Flash ADC.

3. The Simulation Results

Because the maximum frequency of RF signal is about 16.2MHz in 1x HD-DVD read channel so that the bandwidth of RF Summer and Attenuator must be at last 70MHz for group delay consideration and so does the bandwidth of VGA and Post amplifier in AGC. Besides, due to the bit error rate required for HD-DVD, the SNR of read channel must be at last 20dB [4]. Therefore, the SNR of each building block in AFE must be as large as possible and can not be smaller than 20 dB. In order to has better SNR, many considerations should be made to circuits such as noise、harmonic distortion and the problem of power supply rejection, etc. In this AFE design, the simulation results are summarized in table II、III、IV and V. Figure 6 and Figure 7 are the simulation results of equalizer. The constant group delay bandwidth of equalizer is about 1.65fc (with boost range 0~12 dB and within ±1.5ns delay variation) and the total harmonic distortion (THD) is below 1% (for 20~200mvpp input, maximum bandwidth setting, and no boost). And Figure 8 and Figure 9 show the simulation results of VGA and AGC. The order of exponential VGA is 5. The input dynamic range of AGC can reach 20dB and the total harmonic distortion of AGC is less than 2% (at 1Vpp constant amplitude output). The DNL and INL of ADC and the layout placement of each building block in AFE are also shown in Figure 10 and Figure 11 respectively. With a single 3.3v power supply, the AFE consumes 520 mW and occupies an area of 12.8_{mm^2}.

TableII
Post Simulation of RF Summer & Attenuator

Amplifier gain:	1~1/16 (V/V)
Input dynamic range:	20~200 mVpp
Bandwidth @ 1PF loading:	> 75MHz
Total harmonic distortion:	< 0.5%
Power consumption:	16 mW

TableIII
Post Simulation of Equalizer

Filter type:	0.05° equal ripple delay filter
Filter order:	7
Input dynamic range:	20~200 mVpp
F-3dB (fc)	16.2MHz
Tunable range:	8~39MHz
Boost range:	0~12dB
Constant group delay bandwidth:	> 1.65fc
Total harmonic distortion:	< 1%
Power consumption:	40 mW

TableIV
Post Simulation of AGC

VGA order:	5
Input dynamic range:	20~200 mVpp
Output dynamic range:	0.6~1.4Vpp
Forward gain bandwidth:	> 200MHz
Settling time:	< 10u sec
Total harmonic distortion:	< 2%
Power consumption:	56 mW

TableV
Post Simulation of Flash ADC

Resolution:	6 bit
Conversion rate:	300MHz max
DNL:	-0.1LSB ~ 0.1LSB
INL :	-0.1LSB ~ 0.1LSB
Power consumption:	350mW

4. Conclusions

The AFE for use in 1x HD-DVD PRML read channel has been designed and fabricated in 0.35um CMOS process. The required building block and design considerations for analog signal processing have also been described. With the techniques developed in this paper, the AFE achieves signal-processing and satisfies the HD-DVD specifications.

5. References

[1] Sorin G. Stan, " THE CD-ROM DRIVE A Brief System Description ", Philips Optical Storage, Optical Recording Development Laboratory, Eindhoven, The Netherlands,1998.

[2] Iuri Mehr and David R. Welland, " A CMOS Continuous-Time Gm-C Filter for PRML Read Channel Applications at 150Mb/s and Beyond ", IEEE J. Solid-State

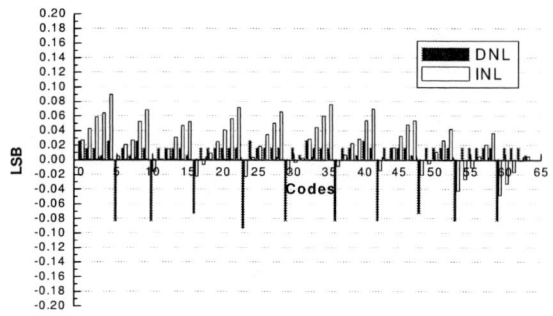

Figure 10. The DNL and INL of ADC

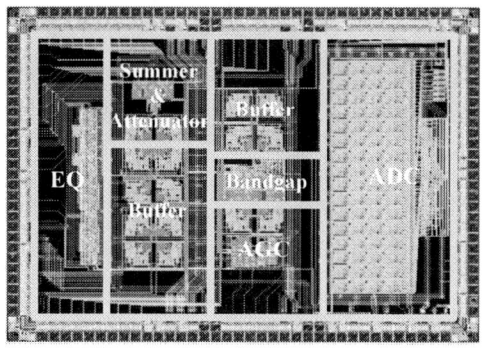

Figure 11. Layout Implementation of AFE

499~513April 1997.

[3] Gregory T. Uehara and Paul R. Gray, " Parallelism in Analog and Digital PRML Magnetic Disk Read Channel Equalizer ", IEEE Trans. On Magnetics, vol . 31 pp. 1174~1179, March 1995.

[4] Tai-Liang Lin, " The AFE System Specifications and Behavior Model Simulations in 1X HD-DVD PRML Read Channel ", Opto-Electronics & System Laboratories, Industrial Technology Research Institute, Taiwan, 2004.

[5] Rolf Schaumann and Mac E. Van Valkenburg, " Design of Analog Filters ", New York Oxford, 2001.

[6] Jose Silva-Martinez, Joseph Adut, Jose Miguel Rocha-Perez, Moises Robinson and Shahriar Rokhsaz, " A 60-mW 200- MHz Continuous-Time Seventh-Order Linear Phase Filter With On-Chip Automatic Tuning System ", IEEE J. Solid-State Circuits, vol. 38 pp. 216~225, February, 2003.

[7] John M. Khoury, " On the Design of Constant Settling Time AGC Circuits ", IEEE Trans. On Circuits and Systems – II : Analog and Digital Signal Processing, vol. 45 pp. 283~294,March, 1998.

[8] Po-Chiun Huang, Li-Yu Chiou and Chorng-Kuang Wang, "A 3.3-V CMOS Wideband Exponential Control Variable-Gain-Amplifier ", IEEE Int. Symp. On Circuits and Systems, pp. 285~288, June, 1998.

[9] Chorng-Kuang Wang and Po-Chiun Huang, " An Automatic Gain Control Architecture for SONET OC-3 VLSI ", IEEE Trans. On Circuits and Systems – II : Analog and Digital Signal Processing, vol 44. pp. 779~783, September, 1997.

Adaptive Signal Processing in Mixed-Signal VLSI with Anti-Hebbian Learning

Miguel Figueroa, Esteban Matamala and Gonzalo Carvajal
Department of Electrical Engineering
Universidad de Concepción
Concepción, Chile
Email: mfigueroa@die.udec.cl

Seth Bridges
Computer Science & Engineering
University of Washington
Seattle, WA 98195-2350, USA
Email: seth@cs.washington.edu

Abstract

We describe analog and mixed-signal primitives for implementing adaptive signal-processing algorithms in VLSI based on anti-Hebbian learning. Both on-chip calibration techniques and the adaptive nature of the algorithms allow us to compensate for the effects of device mismatch. We use our primitives to implement a linear filter trained with the Least-Mean Squares (LMS) algorithm and an adaptive decorrelation network that improves the convergence of LMS. When applied to an adaptive Code-Division Multiple-Access (CDMA) despreading application, our system, without the need for power control, achieves more than 100x improvement in the bit-error ratio in the presence of high interference between users. Our 64-tap linear filter uses $0.25mm^2$ of die area and dissipates $200\mu W$ in a $0.35\mu m$ CMOS process.

1 Introduction

A challenging aspect in the design of portable electronic systems is their need to operate under unknown environmental conditions such as interference, noise, and varying input statistics. Adaptive signal-processing techniques, often in the form of adaptive filters and neural networks, effectively optimize system performance under these conditions because they model the varying statistics of the environment by adapting a set of internal weights to optimize a goal function.

Portable systems also face severe restrictions in cost, power dissipation, and die area. In such cases, implementing adaptive signal-processing algorithms on an embedded processor is often infeasible. The computationally-intensive nature of these algorithms means that even custom digital VLSI solutions can be prohibitively large and power-hungry. For problems that require moderate arithmetic resolution, analog and mixed-signal circuits provide an attractive tradeoff in die area and power dissipation; however,

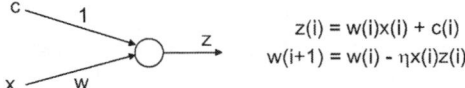

Figure 1. Anti-Hebbian synapse.

these circuits are plagued by problems such as charge leakage, signal offsets, device mismatch, and noise sensitivity.

In this paper, we present a set of analog and mixed-signal primitives that compensates for these problems through on-chip calibration and dynamic adaptation. Using these primitives, we implement adaptive LMS filters based on neural networks trained using anti-Hebbian learning [1] and we show that the same primitives can form decorrelating networks to improve filter convergence in the presence of correlated inputs. Finally, we show an application of our system to improve adaptive CDMA despreading without power control.

2 Anti-Hebbian Learning

Adaptive signal-processing algorithms that update their weights based on correlations between local signals are particularly amenable to hardware implementations because of their local and regular communication structure. Particularly, *anti-Hebbian* learning rules [1] minimize output energy by subtracting an update proportional to the correlation between the input and output. Fig. 1 illustrates this concept on a single-input neural network that computes the synaptic function $z(i) = w(i)x(i) + c(i)$. Using a stochastic gradient descent to minimize the variance of the output, $E[z^2]$, yields the anti-Hebbian learning rule $w(i+1) = w(i) - \eta x(i)z(i)$, where η is a constant learning rate. The independent input $c(i)$ prevents the trivial solution where $w = 0$. The function generalizes naturally to multiple inputs $x_j(i)$.

Despite its simplicity, anti-Hebbian learning is widely used in signal processing. In fact, the learning rule converges when $w(i)x(i)$ is the best approximation to $-c(i)$ in

0-7695-2533-4/06 $20.00 © 2006 IEEE   133

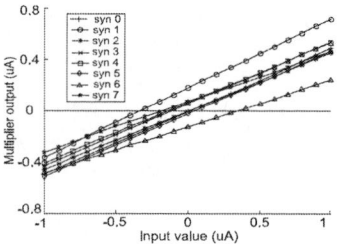

(a) Multiplier output vs. input value.

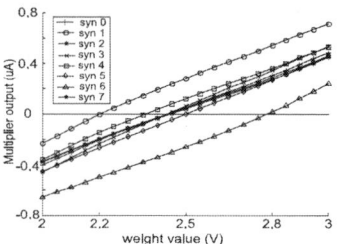

(b) Multiplier output vs. weight value.

Figure 2. Multiplier output for 8 synapses.

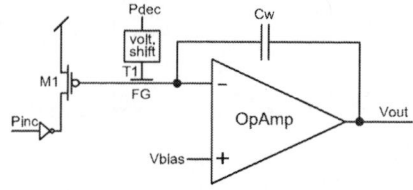

(a) Memory cell with linear updates.

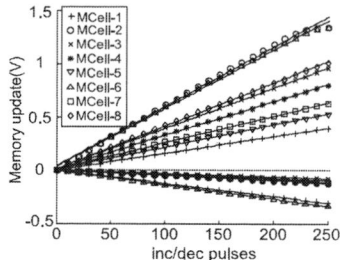

(b) Updates in eight memory cells.

Figure 3. A simple PDM analog memory cell.

the sense of the mean-squared value of the error $z(i)$. This is indeed the formulation of the well-known Least-Mean Squares (LMS) algorithm, commonly found in adaptive filtering applications. Another property of this algorithm is that it converges when the output has extracted all the information of $-c(i)$ available from the input $x(i)$ such that the correlation between x and z is minimal. Therefore, anti-Hebbian learning is also used to adaptively decorrelate signals in applications such as dimensionality reduction and blind source separation [2].

3 VLSI Blocks for Anti-Hebbian Learning

To build anti-Hebbian learning networks in VLSI, we need multipliers and adders to implement the forward-path synaptic computation $\sum_{j=1}^{N} x_j w_j$, distributed on-chip memory to store the weights, and memory-update mechanisms to implement the learning rule $\Delta w_j = -\eta z x_j$. This section, based on previous analyses [3] and our own systems simulations [4], describes how the design of these blocks allows us to correct for the effects of device mismatch. Unless otherwise noted, the data shown in this paper was measured from an implementation of these primitives in a $0.35\mu m$ CMOS process.

3.1 Forward-Path Multipliers

Multipliers impose the strongest requirements on die area and power dissipation in VLSI signal-processing systems. To achieve compact and low-power systems, we use

analog multipliers with current outputs in the forward-path computations and we sum the output currents from each synapse on common wires to form the neuron output.

A systems analysis of our filter design shows that the residual error is sensitive to multiplier linearity with respect to the synaptic input x, but relatively robust to their linearity with respect to the weight w, which is automatically compensated by the adaptive algorithm; also, adaptation compensates for weight offsets in the multipliers, provided the weight range is large enough to absorb the offset.

Based on our analysis, we minimize the convergence time and residual error of our filter by using a Gilbert-style multiplier with differential current inputs for x and a differential voltage representation for w. We used long transistors and above-threshold operation to maximize the linearity of x, but favored larger range in w. Fig. 2 shows the transfer function of eight different multipliers on a single chip.

3.2 Weight Storage and Updates

Using analog multipliers in the forward path requires we provide on-chip analog weight storage. Because the performance of the learning algorithm depends directly on the accuracy of the stored weights and update rules, conventional VLSI capacitors are inadequate: Charge leakage requires continuous updates, preventing the open-loop operation common in applications such as adaptive filters. VLSI capacitors are sensitive to charge injection, degrading performance when used with digital pulse-based updates which provide accurate and compact learning rules [5].

Instead, we use *synapse transistors* [6] to store and update our analog weights. These devices use charge on a

0-7695-2533-4/06 $20.00 © 2006 IEEE

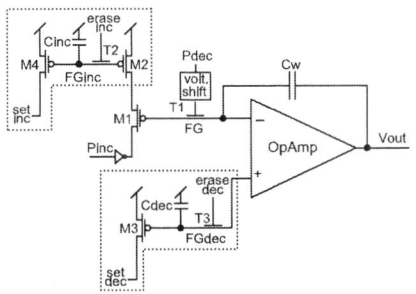
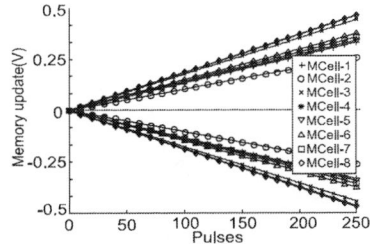
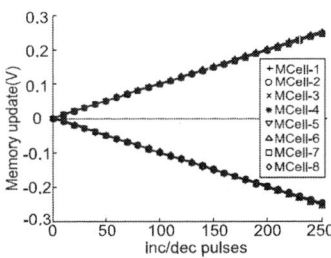

(a) Memory cell with on-chip calibration. (b) Symmetric updates in 8 memory cells. (c) Uniform updates in 8 memory cells.

Figure 4. PDM cell with on-chip calibration.

floating gate to provide compact and accurate nonvolatile analog storage. Fowler-Nordheim tunneling adds charge to the floating gate and hot-electron injection removes charge and both mechanisms accurately update the stored value during normal device operation [6]; however, the dynamics of injection and tunneling are highly nonlinear with respect to their control variables (gate, drain and tunneling-junction voltages), which leads to exponential learning rules that do not enable anti-Hebbian learning. This section describes a memory cell based on synapse transistors that supports accurate, linear updates.

Fig. 3(a) shows a pulse-density modulated (PDM) memory cell with linear updates. A negative-feedback loop around an operational amplifier pins the floating-gate voltage FG at V_{bias}. Fixed-width, fixed-amplitude pulses on P_{inc} and P_{dec} trigger electron tunneling and injection, respectively. Because all the control voltages are fixed, the magnitude of the charge updates depends linearly on the density of the update pulses. The charge is integrated on the feedback capacitor C_w, causing a linear update in the output voltage V_{out}. Fig. 3(b) shows the transfer function of eight PDM memory cells on a chip. The integral nonlinearity (INL) of most cells is less than 0.1%, corresponding to a linearity of more than 10 bits.

Fig. 3(b) also highlights an important problem: device mismatch makes it impossible to achieve symmetric updates within a single synapse or equal updates across different synapses using only a single reference voltage V_{bias}. Without symmetric updates, the residual error increases significantly, while unequal updates across synapses result in slower convergence [4]. Fig. 4(a) shows an improved design which adds two local degrees of freedom to the cell. First, we use an additional floating gate FG_{dec} to locally set the reference voltage at each memory cell. By tunneling and injecting to this floating gate, we can vary the voltage at FG, thus changing the relative strengths of tunneling and injection at P_{dec} and P_{inc} (increasing the voltage at FG weakens tunneling and strengthens injection and vice versa) until we

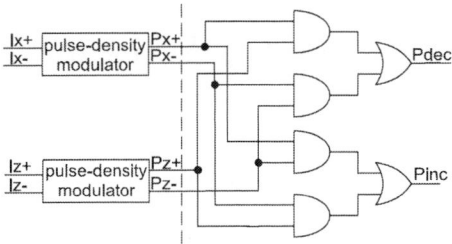

Figure 5. Anti-Hebbian update circuit.

achieve local symmetry. Second, setting the voltage at the floating gate FG_{inc} controls the source current through M1, limiting injection due to pulses on P_{inc} without affecting the tunneling rate due to pulses on P_{dec}. This added control allows us to achieve symmetric updates and equalize them for different cells across the chip. Fig. 4(b) shows the symmetric updates achieved tuning only FG_{dec}, while Fig. 4(c) shows uniform symmetric updates using both floating gates. The mismatch across cells is now less than 0.25% of their dynamic range (better than 8-bit matching).

3.3 Anti-Hebbian Learning Rules

Fig. 5 shows a block diagram of the digital circuit that implements the anti-Hebbian rules at each synapse. Pulse-density modulators [7] (off-chip in our current implementation) transform the synaptic inputs and neuron outputs into pairs of fixed-width digital pulses (P_x^+, P_x^-; P_z^+, P_z^-). The value of the input is represented as the difference between the density (frequency) of the pulses. We use a single modulator for each input or output in the system. We implement the anti-Hebbian learning rules as:

$$P_{inc} = (P_x^+ \& P_z^-) \mid (P_x^- \& P_z^+)$$
$$P_{dec} = (P_x^+ \& P_z^+) \mid (P_x^- \& P_z^-)$$

If the pulse streams are asynchronous and sparse, then the weight updates implemented by the expressions above con-

0-7695-2533-4/06 $20.00 © 2006 IEEE

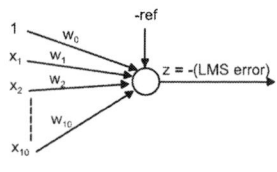

(a) LMS filter.

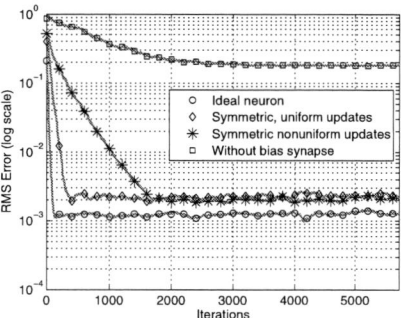

(b) Normalized LMS filter performance.

Figure 6. Filter with LMS adaptation.

verge to the negative correlation between x and z, effectively implementing an anti-Hebbian learning rule.

4   A Linear Filter with LMS Adaptation

Using the primitives described in Section 3, we implemented a 10-input LMS adaptive filter in a 0.35μm CMOS process. Fig. 6(a) shows the architecture of the filter as a neural network adapting with anti-Hebbian learning rules. If the inputs to the neuron are drawn from a tapped-delay line (such as the barrel-shifter analog line described in [8]), then the neuron implements an adaptive FIR filter. Because input offsets in the forward-path multipliers translate into nonzero-mean inputs and degrade the performance of the filter [4], we use a *bias* synapse w_0 with a constant input to compensate for the aggregated value of all the input offsets in the neuron [1]. We train the bias synapse using the same anti-Hebbian learning rule we use to train the other the synapses in the neuron.

Fig. 6(b) shows the evolution of the RMS value of the output error compared to an ideal mathematical implementation of the filter. We normalize the current output to its full-scale value (20μA differential) to ease the comparison to the ideal filter. We drew the inputs from a uniform random distribution and trained our filter using a reference generated by a non-adaptive, mathematically ideal filter with the same inputs. We added Gaussian noise to the reference, which resulted in a 60dB signal-to-noise ratio (SNR). Trained with this reference, the ideal LMS filter achieves a normalized RMS error of 10^{-3}. The output reference has

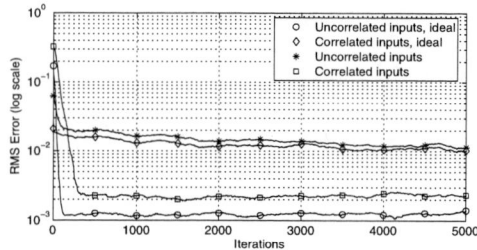

Figure 7. Effect of input correlation.

unity variance, therefore this RMS error is equivalent to a digital resolution of 10 bits. Calibrating the memory cells for symmetric and uniform updates as depicted in Fig. 4(c), the residual error reaches a RMS value of 2×10^{-3}, corresponding to a 9-bit output resolution. The performance of the circuit is mainly limited by the linearity of the forward-path multipliers. Fig. 6(b) also shows the performance of the filter with the memory cells calibrated for local symmetry only. In this case, the residual error is the same, but the convergence time is about four times longer because the learning rate η must be adjusted to stabilize the synapse with the fastest updates [4] (we tune the learning rule globally changing the gain of the pulse modulators). Without a bias synapse, the multiplier offsets limit the performance of the filter to a RMS error of 2×10^{-1} (2 bits).

In the experiment described above, the filter inputs were uncorrelated. Unfortunately, many real-world signals are generated by distributed sensors such as antenna arrays, and show a significant correlation between them. This correlation severely degrades the performance of the LMS algorithm, as shown in Fig. 7 for both the hardware and ideal filter. In this experiment, we mixed the inputs to achieve an average cross-correlation of 0.75. Algorithms such as Recursive Least Squares (RLS) keep an estimate of the inverse of the input correlation matrix and use it to improve the performance of the adaptation, but their computational structure is a poor match for custom VLSI implementations. In the next section, we explore an alternative approach.

5   A Triangular Decorrelating Network

As described in Section 2, we can train an anti-Hebbian synapse to decorrelate two signals. We can use this property to build a decorrelation network and use it as a preprocessing stage to the LMS filter. Fig. 8(a) shows the architecture of the system, based on the direct form of the triangular decorrelating filter described in [1]. Synapse a_{21} decorrelates y_2 from x_1, which is equal to y_1. Synapses a_{31} and a_{32} decorrelate y_3 from x_1 and x_2. Because y_1 and y_2 are linear combinations of x_1 and x_2, y_3 is decorrelated from them as well. As a result, the triangular network computes a linear combination of its inputs that minimizes the corre-

0-7695-2533-4/06 $20.00 © 2006 IEEE

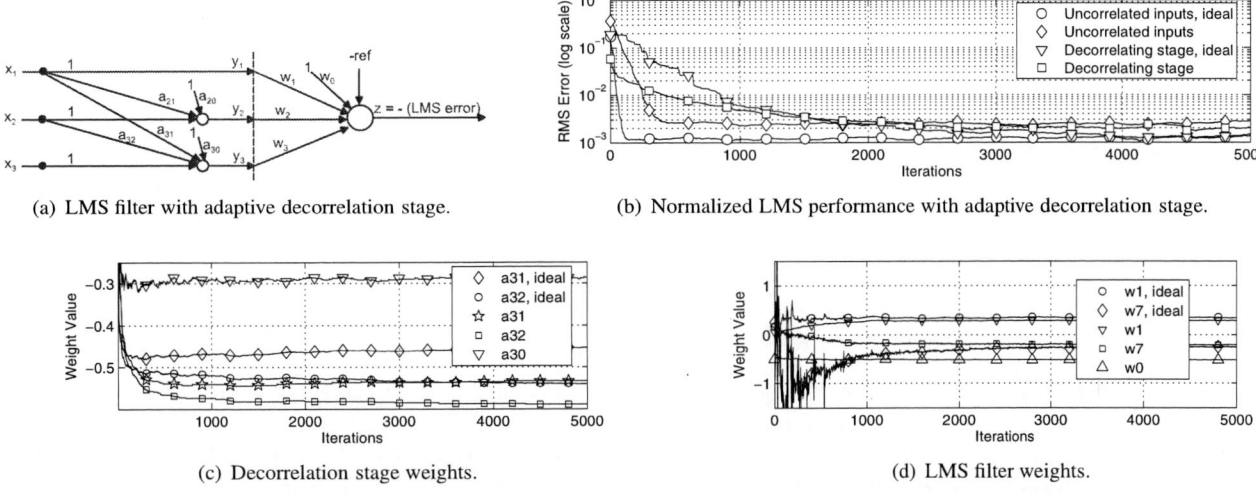

(a) LMS filter with adaptive decorrelation stage.

(b) Normalized LMS performance with adaptive decorrelation stage.

(c) Decorrelation stage weights.

(d) LMS filter weights.

Figure 8. Filter with adaptive decorrelation.

lation between the outputs. The LMS filter can now operate on these outputs and achieve better performance.

We tested a 10-input decorrelating filter using the architecture described above, training it with the same correlated inputs discussed in Section 4. Fig. 8(b) compares its performance to a filter trained with uncorrelated inputs. We include a mathematically ideal implementation for comparison purposes. Even though the decorrelation stage and the LMS filter learn concurrently, the residual error reaches the same value as the LMS filter with uncorrelated inputs after less than 2000 iterations, which corresponds to about four times the convergence time of the filter with uncorrelated inputs. This is a substantial improvement over the results shown in Fig. 7. In fact, the VLSI filter converges faster than its mathematically-ideal counterpart. This is partly because the ideal filter converges to a smaller error, but also because the discrete updates of the VLSI filter damp the variance of the weights in the decorrelation network, allowing better LMS performance.

Figs. 8(c) and Fig. 8(d) depict the evolution of selected weights in the decorrelation network and the LMS filter. We observe that the bias weights (a_{30} and w_0) converge to nonzero values to compensate for the multiplier offsets. We also see that the weights converge to different values than their ideal counterparts, to compensate for offsets in the memory cells and analog multipliers. Fig. 8(c) shows that it takes about 1500 iterations for the weights in the decorrelation network to approach their final value, at which point the cross-correlation of the LMS filter inputs is low enough to achieve good performance.

6 Application: Adaptive CDMA despreading

Direct Sequence Code Division Multiple Access (DS-CDMA) is a spread-spectrum technique widely used in wireless data communications. In this scheme, each bit transmitted by a user is encoded using a sequence of shorter binary values (chips), called the *user signature*. Ideally, the cross-correlation between signatures is zero, which makes it possible to detect a single user's message by simply correlating the received signal with the desired user's signature, thus canceling the contribution of other users and recovering only the bit stream of interest. In practice, finite correlation between signatures, unequal power among users and multipath fading causes interference between users, which lead to high error rates, low channel utilization and/or expensive power-control techniques.

An alternative to traditional CDMA detection is to use an adaptive filter to compute an optimal user signature using a training sequence. Because of multiple-user interference and different signal power, the optimal signature is often not the one used to encode the original bit stream (and it is, in fact, not binary). In this section, we show an application of the filters presented in Sections 4 and 5 to adaptively recover user data in a CDMA system. Our filter supports 64 inputs and occupies a die area of 0.25mm^2 in a $0.35\mu m$ CMOS process, dissipating $200\mu W$ of total power.

We configured a CDMA system with 16 simultaneous users and 64-chip signatures. The worst-case effective cross-correlation between user signatures is 0.4 and the worst power ratio between user signals is 10. Fig. 9(a) shows the estimation errors as a result of decoding a 2048-bit message with the traditional method for the user with the

0-7695-2533-4/06 $20.00 © 2006 IEEE

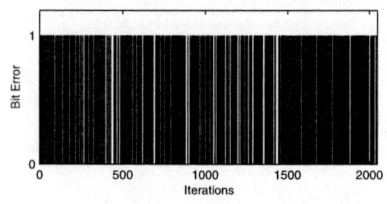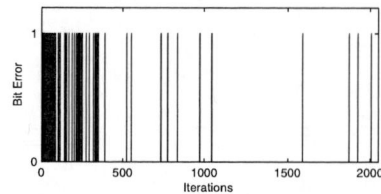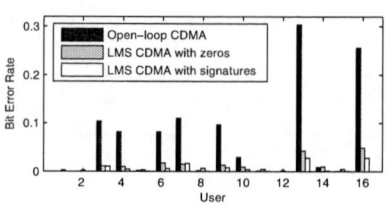

(a) Open-loop BER for user 13.　　(b) BER for user 13 with adaptive detection.　　(c) BER for 16 users with adaptive detection.

Figure 9. Application to CDMA despreading.

lowest power. The bit-error rate (BER) is 0.3 (30% of the bits were incorrectly detected). We used our adaptive filter to detect the same user's message, using an off-chip comparator to discriminate between bits, and using only the first 700 bits as a training sequence. The performance is greatly improved, reaching a BER of 0.05 including the incorrect estimations during training. The BER after training is only 0.01. Experiments published in [4] show that the output resolution of the 64-input filter is equivalent to 10 bits.

Fig. 9(c) shows the BER for all 16 users traditional CDMA detection and our adaptive filter (including training). In the adaptive case, we obtain slightly better performance if we initialize the weights to the original user signature instead of zeros. Considering only the detection after training, our filter improves the BER performance by a factor between 25 and 114, depending on the user. Other hardware implementations of CDMA detection [9] use less power and area than our filter, but they rely fundamentally on nonadaptive, strictly binary user signatures. Thus, these implementations can only operate in open loop using a traditional detection scheme.

7　Conclusions

We have described analog and mixed-signal primitives for adaptive signal processing in CMOS VLSI using anti-Hebbian learning. Analog-weight storage using synapse transistors enables the use of analog hardware while maintaining accurate weight updates. The adaptive nature of the algorithms compensates for some analog-hardware nonidealities such as gain and offset mismatches. We include on-chip calibration circuitry to tune those features not compensated by the adaptation, such as asymmetric and nonuniform learning rates. We demonstrated the effectiveness of our approach building a 10-input filter that adapts with a pulse-based implementation of the LMS algorithm, achieving an output resolution of 9 bits. We also used anti-Hebbian learning to implement an adaptive decorrelating stage for the LMS filter, greatly improving convergence time and residual error (by more than an order of magni-

tude) in the presence of correlated inputs. Finally, we built a 64-input LMS filter to perform adaptive DS-CDMA despreading, improving the BER over traditional detection by a factor of more than 100 in the presence of user interference and unequal signal power. The circuit die area is $0.25 mm^2$ and its power dissipation is $200 \mu W$ in a $0.35 \mu m$ CMOS process.

Acknowledgments

This work was financed in part by a FONDECYT grant No. 1040617. We fabricated our chips through MOSIS.

References

[1] F. Palmieri, J. Zhu, and C. Chang, "Anti-Hebbian Learning in Topologically Constrained Linear Networks: A Tutorial," *IEEE Transactions on Neural Networks*, vol. 4, no. 5, pp. 748–761, 1993.

[2] A. Hyvärinen and E. Oja, "Independent Component Analysis: Algorithms and Applications," *Neural Networks*, vol. 13, no. 3, pp. 411–430, 2000.

[3] B. K. Dolenko and H. C. Card, "Tolerance to Analog Hardware of On-Chip Learning in Backpropagation Networks," *IEEE Transactions on Neural Networks*, vol. 6, no. 5, pp. 1045–1052, 1995.

[4] M. Figueroa, S. Bridges, and C. Diorio, "On-chip compensation of device-mismatch effects in analog VLSI neural networks," in *Advances in Neural Information Processing Systems 17*. Cambridge, MA: MIT Press, 2005.

[5] Y. Hirai and K. Nishizawa, "Hardware Implementation of a PCA Learning Network by an Asynchronous PDM Digital Circuit," in *Proceedings of the IEEE-INNS-ENNS International Joint Conference on Neural Networks (IJCNN)*, vol. 2, 2000, pp. 65–70.

[6] C. Diorio, P. Hasler, B. Minch, and C. Mead, "A Complementary Pair of Four-Terminal Silicon Synapses," *Analog Integrated Circuits and Signal Processing*, vol. 13, no. 1/2, pp. 153–166, 1997.

[7] C. Mead, *Analog VLSI and Neural Systems*. Reading, MA: Addison-Wesley, 1989.

[8] M. Q. Le, P. J. Hurst, and J. P. Keane, "An Adaptive Analog Noise-Predictive Decision-Feedback Equalizer," in *IEEE Symposium on VLSI Circuits*, D. Scott and M. Yamashina, Eds. Honolulu, Hawaii, USA: IEEE Solid-State Circuits Society, 2000, pp. 216–217.

[9] T. Yamasaki, T. Fukuda, and T. Shibata, "A Floating-Gate-MOS-Based Low-Power CDMA Matched Filter Employing Capacitance Disconnection Technique," in *IEEE Symposium on VLSI Circuits*, Kyoto, Japan, 2003, pp. 267–270.

0-7695-2533-4/06 $20.00 © 2006 IEEE

Test and Verification

Verification of Scheduling in High-level Synthesis

C Karfa   C Mandal   D Sarkar   S R Pentakota
Department of Computer Sc & Engg
Indian Institute of Technology, Kharagpur
WB 721302, INDIA
{ckarfa, chitta, ds}@iitkgp.ac.in, satya@ti.com

Chris Reade
Kingston Business School
Kingston University
England KT2 7LB, UK
Chris.Reade@king.ac.uk

Abstract

This paper describes a formal method for checking the equivalence between two descriptions of the target system, one before and the other after scheduling. The descriptions are represented as finite state machines with data paths (FSMD). The basic principle is to show that any computation of one FSMD is covered by a computation on the other, a computation being characterized by a concatenation of paths in the FSMD. These notions are formalized in the paper. The method is strong enough to accommodate merging of the segments in the original behaviour by the typical scheduler such as DLS, a feature common in scheduling. The method also works for limited arithmetic transformations. Although the proposed method is found to have a non-polynomial worst case complexity, many non-trivial examples encounter a low polynomial order of complexity. The technique is illustrated with an example.

1 Introduction

High-level synthesis is the process of generating the register transfer level (RTL) design from the behavioural description. The synthesis process consists of several interdependent sub-tasks such as, specification, compilation, scheduling, allocation and binding. The operations in the behavioural description are assigned time steps through the scheduling process. Input to the scheduling phase is a control data flow graph (CDFG)[3]. While a CDFG is better suited to scheduling algorithms, an FSMD is a more appropriate model for verification. We therefore construct FSMDs from the CDFGs before and after scheduling. In the process of scheduling, operations are often moved across basic block boundaries for various optimizations. In general several transformations may be made to improve the performance of a design. For example, path based scheduling techniques [7] perform several such non-trivial path based transformations. Hence, it is important to ensure that the

scheduling process preserves the behaviour of the original specification, irrespective of the scheduling technique that is used. The objective of this work is to check that the design descriptions before and after scheduling, as represented by FSMDs, are computationally equivalent.

The equivalence problem of FSMDs (EPFSMD) is the same as the equivalence problem of flowchart schemas[4, 6] which is undecidable and not even partially decidable[6]. However, since the final targeted hardware has only a finite datapath, the restricted problem can be reduced to the equivalence problem of FSM models (EPFSM) which is decidable. Unfortunately, an FSMD with an n-bit datapath results in a number of states of the order of 2^{kn}, where k is the number of storage elements of n bits. The value of *kn* easily exceeds several hundreds. Thus, deciding EPFSMD with a finite datapath by reducing them to EPFSM is of little use in practice. On the other hand specialized analytical treatments, such as the work described here, may aid in revealing problems in the working of the algorithm which may never use the finiteness in producing the output which is to be checked. In this case the equivalence checking algorithm would identify paths that are not matched up, which could be particularly helpful in fixing the scheduling algorithm. This benefit would normally be lost by trying to reduce a finite EPFSMD to EPFSM.

Most of the algorithms proposed in the literature can successfully verify the basic block based scheduling but apparently fail to verify when structure of the scheduled FSMD differs from the input FSMD due to path based transformation. In this paper, we propose a scheduling verification method which is strong enough to work even when the basic path structure is changed by the scheduler. This method formally establishes equivalence between the FSMDs before and after scheduling.

This paper is organized as follows. In section 2, FSMDs and the notions of computations on FSMDs and the equivalence of FSMDs are defined. The verification method is described in section 3. the complexity of the proposed method is treated in section 4. An example has been treated in sec-

0-7695-2533-4/06 $20.00 © 2006 IEEE

tion 5 to illustrate the working of the algorithm. Some experimental results have been given in section 6. The paper is concluded in section 7.

2 FSMDs and their Equivalence

2.1 FSMDs

An FSMD (*finite state machine with data-path*) is a universal specification model, proposed by Gajski in [2], that can represent all hardware designs. The model is used in the present work with the addition of a reset state, for encoding the designs to be verified. The FSMD is defined as an ordered tuple $\langle Q, q_0, I, V, O, f, h \rangle$, where

1. $Q = \{q_0, q_1, q_2, \ldots q_n\}$ is the finite set of control states,

2. $q_0 \in Q$ is the reset state,

3. I is the set of primary input signals and Σ_I is the input alphabet,

4. V is the set of storage variables and Σ is the set of all data storage states or simply, data states,

5. O is the set of primary output signals and Σ_O is the output alphabet,

6. $f : Q \times S \rightarrow Q$, is the state transition function and

7. $h : Q \times S \rightarrow U$, is the update function of the output and the storage variables, where U and S are as defined below.

   (a) $U = \{x \Leftarrow e | x \in O \cup V \text{ and } e \in E\}$ represents a set of storage or output assignments, from variables (storage or output) or expressions constructed over (input or storage) variables. Thus, $E = \{g(x, y, z, \ldots) | x, y, z, \ldots \in I \cup V\}$ represents a set of arithmetic expressions over the set $I \cup V$.

   (b) $S = \{R(a, b) | a, b \in E \text{ and } R \text{ is any arithmetic relation}\}$ represents a set of status signals as a result of comparisons $(=, \neq, >, \geq, <, \leq)$ between two expressions from the set E.

It may be noted that we have not introduced final states in the FSMD model as we assume that the systems work in an infinite outer loop.

2.2 Walks and Transformations along a Walk

A (finite) *walk* α from q_i to q_j, where $q_i, q_j \in Q$, is a finite transition sequence of states of the form $\langle q_i = q_1 \xrightarrow[c_1]{} q_2 \xrightarrow[c_2]{}, \ldots, \xrightarrow[c_{n-1}]{} q_n = q_j \rangle$ such that $\forall l, 1 \leq l \leq n-1, \exists c_l \in S$ such that $f(q_l, c_l) = q_{l+1}$, and $q_k, 1 \leq k \leq n-1$, are all distinct.

The state q_n may be identical to q_1. The *condition of execution of the walk* $\alpha = \langle q_{l_0} \xrightarrow[c_0]{} q_{l_1} \xrightarrow[c_1]{} q_{l_2} \cdots \xrightarrow[c_{k-1}]{} q_{l_k} \rangle$, R_α, is a logical expression over the variables in V such that R_α is satisfied by the (initial) data state at q_{l_0} iff the walk α is traversed.

We assume that inputs and outputs occur through named ports. The ith input from port P is a value represented as P_i. Thus if some variable v stores input from port P (for the ith time along a walk), it is equivalent to the assignment $v \Leftarrow P_i$.

The simple data transformation of a walk α over V (s_α): It is an ordered tuple $\langle e_i \rangle$ of algebraic expressions over the variables in V and the inputs in I such that the expression e_i represents the value of the variable v_i after the execution of the walk in terms of the initial data state (i.e., the values of the variables at the initial control state) of the walk.

Taking into account outputs that may occur in a walk, the data transformation r_α of a walk α over V is the tuple $\langle s_\alpha, O_\alpha \rangle$, where the output list $O_\alpha = [OUT(P_{i_1}, e_1), OUT(P_{i_2}, e_2), \ldots]$. For every expression e output to port P along the walk α, there is an $OUT(P, e)$ in the list, in the order in which the outputs occurred.

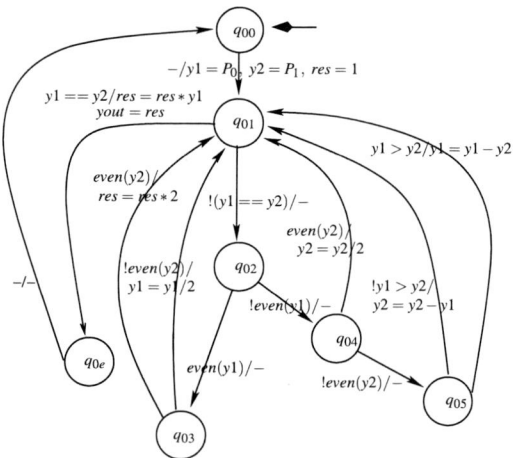

Figure 1. M_0, the FSMD of GCD before scheduling

Computation of the condition of execution R_α can be by *backward* substitution or by *forward* substitution. The former is more readily described and is based on the following rule: If a predicate $c(y)$ is true after execution of $y \leftarrow g(y)$, then the predicate $c(g(y))$ must have been true before the execution of the statement [6]. The transformation s_α is found indirectly using the same principle. The forward substitution method of finding R_α is based on symbolic execution.

0-7695-2533-4/06 $20.00 © 2006 IEEE   142

2.3 Characterization of Walks and their Concatenations

The characteristic formula τ_α of a walk α with initial storage and input variables as $\bar{v}$, final variables as $\bar{v}_f$ and outputs along the walk as O is $\tau_\alpha(\bar{v}, \bar{v}_f, O) = R_\alpha(\bar{v}) \wedge (\bar{v}_f = s_\alpha(\bar{v})) \wedge (O = O_\alpha(\bar{v}))$, where s_α is the data transformation and O_α output list in the walk α.

Let $\tau_\alpha(\bar{v}, \bar{v}_f, O) : R_\alpha(\bar{v}) \wedge (\bar{v}_f = s_\alpha(\bar{v})) \wedge (O = O_\alpha(\bar{v}))$ be the characteristic formula of the walk α and $\tau_\beta(\bar{v}, \bar{v}_f, O) : R_\beta(\bar{v}) \wedge (\bar{v}_f = s_\beta(\bar{v})) \wedge (O = O_\beta(\bar{v}))$ be the characteristic formula of the walk β. The characteristic formula for the concatenated walk $\alpha\beta$ is $\tau_{\alpha\beta}(\bar{v}, \bar{v}_f, O) = \exists \bar{v}_\alpha \exists O_1 \exists O_2 (\tau_\alpha(\bar{v}, \bar{v}_\alpha, O_1) \wedge \tau_\beta(\bar{v}_\alpha, \bar{v}_f, O_2)) = R_\alpha(\bar{v}) \wedge R_\beta(s_\alpha(\bar{v})) \wedge (\bar{v}_f = s_\beta(s_\alpha(\bar{v}))) \wedge (O = O_\alpha(\bar{v})O_\beta(s_\alpha(\bar{v})))$. O is the concatenated output list of $O_\alpha(\bar{v})$ and $O_\beta(s_\alpha(\bar{v}))$. The detail of incrementing the input indices on each port in the formulas for β to start after the last index of the corresponding port in α has been omitted for notational clarity.

2.4 Computations on FSMDs and their Path Covers

A computation of an FSMD is a finite walk from the reset state q_0 back to itself without having any intermediary occurrence of q_0 (as a new computation starts from the reset state). A computation c of an FSMD M may be characterized as $\tau_c(\bar{v}_i, \bar{v}_f, O) : R_c(\bar{v}_i) \wedge (\bar{v}_f = s_c(\bar{v}_i)) \wedge (O = O_c(\bar{v}_i))$, where $\bar{v}_i$ is the vector of initial input and data state with which the computation is started, R_c is a satisfiable condition over the domain of I and V, s_c is a function over this domain to the co-domain of values over V and O_c is the concatenation of the output lists resulting from output operations along c.

Two computations c_1 and c_2 having the characteristic formulae τ_{c_1} and τ_{c_2}, respectively, are said to be equivalent if $R_{c_1} = R_{c_2}$, $r_{c_1} = r_{c_2}$. The computational equivalence of two walks p_1 and p_2 is denoted as $p_1 \simeq p_2$. Equivalence checking of walks, therefore, consists in establishing the computational equivalence of the respective conditions of execution and the respective data transformations.

A finite set of paths[1] $P = \{p_0, p_1, p_2, \ldots, p_k\}$ is said to cover an FSMD M if any computation c of M can be looked upon as a concatenation of paths from P. P is said to be a *finite path cover* of the FSMD M.

2.5 Arithmetic Expressions and their Normalization

Since the condition of execution and the data transformation of a walk involve the whole of integer arithmetic, checking equivalence of walks reduces to the validity problem of first order logic; the latter is undecidable because a canonical form does not exist for integer arithmetic. Instead, in this work we use the following normal form adapted from [5, 8].

Every formula is converted into the conjunctive normal form; every conjunct, therefore, is a disjunction of literals where a literal is an atomic formula (atom) or its negation. An atom is a boolean variable or an arithmetic relation of the form $S \ r \ 0$, where S is a normalized sum, $r \in \{\leq, \geq, =, \neq\}$. The relation $>$ ($<$) can be reduced to $\geq$ ($\leq$) over integers. A normalized sum is a sum of terms with at least one constant term; each term is a product of primaries with a non-zero constant primary; each primary is a storage variable, an input variable or an output variable or of the form $abs(s)$, $mod(s_1, s_2)$, $exp(s_1.s_2)$ or $div(s_1, s_2)$, where s, s_1, and s_2 are normalized sums. Any normalized sum is arranged by lexicographic ordering of its constituent subexpressions from the bottom-most level. The common subexpressions in a sum are collected. Thus, $x^2 + 3x + 4z + 7x$ is simplified to $x^2 + 10x + 4z + 0$. A relational literal is reduced by a common constant factor, if any, and the literal is accordingly simplified. For example, $3x^2 + 9xy + 6z + 7 \geq 0$ is simplified to $x^2 + 3xy + 2z + 2 \geq 0$, where $\lfloor 7/3 \rfloor = 2$. A conjunct $C = l_1 \vee l_2 \vee \ldots \vee l_n$ is first expressed as $\neg(\neg l_1 \wedge \neg l_2 \wedge \ldots \wedge \neg l_n)$ and then literals are deleted by the rule "if $(l \Rightarrow l')$ then $l \wedge l' \equiv l$." C reduces to *true* if $\neg l_i \Rightarrow l_j$ for $1 \leq i, j \leq n$. Symmetry of $\{=, \neq\}$, reflexivity of $\{\leq, \geq, =\}$ and irreflexivity of $\{\neq\}$ are accounted for by the above transformations. The above normal form may be shown to be canonical for multivariate polynomials.

2.6 Equivalence of FSMDs

Let M_0 be the FSMD representation of the CDFG given as the input to the scheduler and M_1 be the FSMD of the scheduled behaviour. Our main goal is to verify whether M_0 behaves exactly as M_1. This means that for all possible input sequences, M_0 and M_1 produce the same sequences of output values and eventually, when the respective reset states are re-visited, they are visited with the same storage element values. In other words, for every computation from the reset state back to itself of one FSMD, there exists an equivalent computation from the reset state back to itself in the other FSMD and vice-versa.

Thus two FSMDs M_0 and M_1 are said to be computationally equivalent if for any computation c_0 of M_0, there exists a computation c_1 of M_1 such that c_0 and c_1 are computationally equivalent and vice-versa.

---

[1] A path is a walk in which all the states (nodes) are distinct. A cycle is like a path where the first and last nodes are identical but all other nodes are distinct. Here we allow our paths to be cycles also.

The following theorem, stated without proof, is key to our algorithm for checking the equivalence of two FSMDs.

Theorem 1 *Two FSMDs M_0 and M_1 are computationally equivalent if there exists a finite cover $P_0 = \{p_{00}, p_{01}, \ldots, p_{0l}\}$ of M_0 for which there exists a set $P_1^0 = \{p_{10}^0, p_{11}^0, \ldots, p_{1l}^0\}$ of paths of M_1 such that $p_{0i} \simeq p_{1i}^0$, $0 \leq i \leq l$ and vice-versa.*

The following (inductive) notion of *corresponding states* will be used in the algorithm to be presented. Let $M_0 = \langle Q_0, q_{00}, I, V_0, O, f_0, h_0 \rangle$ and $M_1 = \langle Q_1, q_{10}, I, V_1, O, f_1, h_1 \rangle$ be the two FSMDs having identical input and output sets, I and O, respectively, and $q_{0i}, q_{0k} \in Q_0$ and $q_{1j}, q_{1l} \in Q_1$.

- The respective reset states q_{00}, q_{10} are corresponding states.

- If $q_{0i} \in Q_0$ and $q_{1j} \in Q_1$ are corresponding states and there exist $q_{0k} \in Q_0$ and $q_{1l} \in Q_1$ such that, for some path α from q_{0i} to q_{0k} in M_0, there exists a path β from q_{1j} to q_{1l} in M_1 such that $\alpha \simeq \beta$, then q_{0k} and q_{1l} are corresponding states.

3 Verification Method

The above theorem, therefore, suggests a verification method which consists of the following steps:

1. Construct the set P_0 of paths of M_0 so that P_0 covers M_0. Let $P_0 = \{p_{00}, p_{01}, \cdots, p_{0k}\}$.

2. Show that $\forall p_{0i} \in P_0$, *there exists a path* p_{1j} *of* M_1 such that $p_{0i} \simeq p_{1j}$.

3. Repeat steps 1 and 2 with M_0 and M_1 interchanged.

Owing to the presence of loops it is difficult to find a path cover of the whole computation comprising only finite paths. So any computation is split into paths by putting *cutpoints* at various places in the FSMD so that each loop is cut in at least one cutpoint. The set of all paths from a cutpoint to another cutpoint without having any intermediary cutpoint is a path cover of the FSMD. The method of decomposing an FSMD by putting cutpoints is identical to the Floyd-Hoare's method of program verification [1, 5]. We choose the cutpoints in any FSMD as follows.

1. The reset state is chosen.

2. Any state with more than one outward transitions is also chosen.

Obviously, cutpoints chosen by the above rules cut each loop of the FSMD in at least one cutpoint, because each

internal loop has an exit point (ensured by our notion of computation in §2).

In the following we propose one method which combines the first two steps listed above into one. More specifically, the method constructs a path cover of M_0 and also finds its equivalent path set in M_1 hand-in-hand. An initial set of cutpoints is chosen for M_0 as described above. The reset state of M_0 is always a cutpoint of M_0 and the reset states of M_0 and M_1 is the initial pair of *corresponding states*. Starting from a corresponding state q_{0i} of M_0 the algorithm traverses *all* the paths leading out of q_{0i} to the next cutpoint in M_0 and for each path it tries to find a corresponding equivalent path in M_1. On success the end points of the two paths may be recorded as a new pair of corresponding points. Otherwise, the path in M_0 is extended in all possible ways (without re-entering loops) and again matching paths are sought in M_1 for each extension. When all possibilities are exhausted without finding a match the algorithm reports a failure. The algorithm continues until all pairs of corresponding points are processed. The following pseudocode describes this process more precisely.

3.1 Verification Algorithm

Step 1: Insert cutpoints in M_0 by the following rules.

- the start state is a cutpoint,

- any state with more than one outward transition is a cutpoint.

Step 2:

```
/*Main data stores:
  η: Set of corresponding nodes
  P₀: path cover of M₀
  P₁⁰: paths in M₁ with matching paths in P₀
Working data stores:
  F: list of paths of M₀ starting with nodes having
  corresponding nodes but ending with nodes whose
  corresponding nodes have not yet been found
  P: Working list of corresponding nodes from which
  paths will be examined */
F := [ ] ; P₀ := [ ] ; P₁⁰ := [ ] ;
η := {⟨q₀₀,q₁₀⟩} ;
P := {⟨q₀₀,q₁₀⟩} ;
while ( P is not empty || F is not empty )
{ // main loop continues till termination
  if ( F is empty )
  // new paths starting from entries in P to be examined
  { ⟨q₀ᵢ,q₁ⱼ⟩ := deQ P ;
    put in F all the paths from q₀ᵢ to its successor
    cutpoints (in M₀) ;
  } else // now work on the un-matched path frontier
```

```
{ β := deQ F ; // endPtNd ( β ) is un-matched!
  if ( ( α = findEquivalentPath ( β , q₁ⱼ ) ) != NULL )
  { if ( ! ⟨ endPtNd(β), endPtNd(α)⟩ ∈ η)
      enQ ( P, ⟨ endPtNd(β), endPtNd(α) ⟩ );
      // new paths will start from here
    η := η ∪ {⟨ endPtNd(β), endPtNd(α) ⟩ ;
    P₀ := P₀ ∪ {β} ; P₁⁰ := P₁ ∪ {α} ;
  } else // no match
  { // so continue along all paths through successors
    if ( the path is marked NOT_EXTENDIBLE ) fail ;
    tF := all the paths obtained by concatenating to β
    all the paths from endPtNd (β) to all the successor
    cutpoints of endPtNd ( β ) ;
    if ( endPtNd of any member of tF is a node of the
      same path other than its start node )
      fail;
    if ( endPtNd of any member of tF is same
      as its start node ‖ the reset state)
      mark the path as NOT_EXTENDIBLE ;
    F := append ( tF, F ) ;
  } // else-if
} // else-if
} // end while
```
$\beta := \text{deQ F}$ is referenced above.

Step 3: Identify the cutpoints in M_1

Step 4: Repeat the same procedure as described in Step 2 with the roles of M_0 and M_1 interchanged.

Step 5: If it succeeds for both Step 2 and Step 4 then report *M_0 and M_1 are computationally equivalent.* Otherwise report a failure.

The functions used are specified as follows.

- $findEquivalentPath(\beta, q_{1j})$: It tries to find a path α in M_1 so that $R_\alpha = R_\beta$ and $r_\alpha = r_\beta$. If such an α exist then this function returns α, otherwise a NULL path.

- $endPtNd(\beta)$: returns the state where the path β terminates.

4 Algorithm Complexity

The complexity of the algorithm is determined in step 2. Let there be up to n control states (nodes) in M_0 or M_1. This is an upper bound on the number of cutpoints in M_0 or M_1. The complexity of the function $findEquivalent(\beta, q_{1j})$ is proportional to the number of paths from q_{1j} to be examined, for a given β. If there are up to k parallel edges between any two states, then up to $O(kn)$ paths of length $O(n)$ need to be examined. We avoid the k^n possibilities of branching here because the paths are examined with respect to β; only the k possibilities at each node of the path need to

be examined for only a single viable choice. Thus the time complexity of $findEquivalent$ is $O(kn^2)$. This is not a tight upper bound, so we also consider the lower bound which is $\Omega(1)$. This represents the case when we obtain straight matches without any exploration.

While finding the equivalent paths in M_1 corresponding to those in M_0, in the best case we only need to consider a path from one cutpoint to the next. However, path extensions may be required. The maximum length of any path in M_0 after repeated path extensions is $O(n)$. Starting from a cutpoint, considering repeated path extensions and an edge multiplicity of k, up to $O(k^{n-1})$, if $k \neq 1$ or $O(n^2)$, if $k = 1$, paths of M_0 may need to be examined. Note that if $k \neq 1$, then along a path of n nodes there are k path segments of length one, k^2 path segments of length two, etc. The total number of such paths is $k + k^2 + k^3 + \ldots + k^{n-1} = k\frac{k^{n-1}-1}{k-1}$, which is $O(k^{n-1})$. Since paths may start from any cutpoint, $O(nk^{n-1})$, if $k \neq 1$ or $O(n^3)$, if $k = 1$, paths of M_0 may need to be examined. The total number of pairs of the form $\langle \beta, q_{1j} \rangle$ that need to be examined is $O(n^2k^{n-1})$, if $k \neq 1$ or $O(n^4)$, if $k = 1$. Thus, the overall worst case complexity of step 2 is $O(n^4k^n)$ if $k \neq 1$ or $O(n^6)$, if $k = 1$ and $\Omega(n)$. In practice this is often not hit.

5 An Example

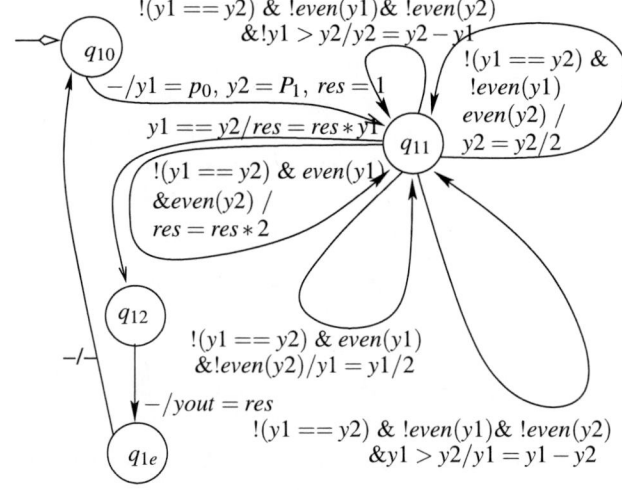

Figure 2. M_1, **the FSMD of GCD after scheduling**

The flow of step 2 of the proposed algorithm applied to the GCD example (fig 1 and fig 2) is briefly discussed here. The algorithm first considers the path $\langle q_{00} \rightarrow q_{01} \rangle$. The findEquivalentPath function successfully finds an equivalent path which is $\langle q_{10} \rightarrow q_{11} \rangle$. So, $\langle q_{01}, q_{11} \rangle$ are the corresponding pair of states. The path $\langle q_{01} \rightarrow q_{0e} \rightarrow q_{00} \rangle$ has to be considered next. Its equivalent path is $\langle q_{11} \rightarrow$

0-7695-2533-4/06 $20.00 © 2006 IEEE   145

$q_{12} \rightarrow q_{1e} \rightarrow q_{10}\rangle$. For the path $\langle q_{01} \rightarrow q_{02}\rangle$, there is no equivalent path in M_1. So, this path needs to be concatenated with its successor paths. These concatenated paths are $\beta_1 = \langle q_{01} \rightarrow q_{02} \rightarrow q_{03}\rangle$ and $\beta_2 = \langle q_{01} \rightarrow q_{02} \rightarrow q_{04}\rangle\}$. β_1 is again concatenated with its successor paths as it has no equivalent path in M_1. The concatenated paths have the same sequence of states $(q_{01} \rightarrow q_{02} \rightarrow q_{03} \rightarrow q_{01})$ with different conditions of execution and data transformations. Similarly, β_2 is also concatenated with its successor paths. These concatenated paths are $\beta_{21} = \langle q_{01} \rightarrow q_{02} \rightarrow q_{04} \rightarrow q_{05}\rangle\}$ and $\beta_{22} = \langle q_{01} \rightarrow q_{02} \rightarrow q_{04} \rightarrow q_{01}\rangle\}$. Path β_{21} will be concatenated with two of its successor paths next. Now each of these concatenated paths has an equivalent path in M_1. These will be explored one by one in the following iterations. It may be noted that Step 2 takes only 18 iterations which is much better than the worst case complexity of $6^6 = 46656$ (here $k = 1, n = 6$).

6 Experimental Results

The proposed algorithm has been implemented in 'C' and has been run for some standard high-level synthesis benchmarks as shown in table 1. These have been run on an Intel Pentium 4, 1.70 MHz, 256MB RAM machine. The number of states, number of paths explored in each FSMD M_0 and M_1, number of consecutive path segments merged by the scheduler and the CPU time are tabulated for each benchmark example. It is evident from table that execution time is sensitive on number of paths explored. It also may be noted from the table that run time of this algorithm is less sensitive on the number of states in the FSMDs. For example, in table 1, the run times of EWF and DCT are small compared to GCD and MODN even though EWF and DCT have greater number of states. These examples also suggest that the upper bound is not necessarily hit for practical scheduling verification cases.

7 Conclusions

Advances in VLSI technology have enabled its deployment into complex circuits. Synthesis flow of such circuits comprises various phases where each phase performs the task algorithmically providing for ingenious interventions of experts. The gap between the original behaviour and the finally synthesized circuits is too wide to be analyzed by any reasoning mechanism. The validation tasks, therefore, must be planned to go hand in hand with each phase of synthesis. The present work concerns itself with the validation of the scheduling phase. Both the behaviours prior to and after scheduling have been modeled as FSMDs. The validation task has been treated as an equivalence problem of FSMDs.

The method presented in this paper has been proved to be sound, completeness being ruled out by the fact that the

Name	#state in FSMD		#path in cover		#path extn	CPU time
	M_0	M_1	M_0	M_1		in ms
DIFFEQ	4	12	3	3	0	2.442
EWF	4	35	1	1	0	1.820
GCD	7	4	11	7	3	3.976
DCT	3	29	1	1	0	1.754
TLC	7	8	13	14	2	4.196
MODN	6	7	8	12	2	4.324
PERFECT	9	6	7	5	2	4.028

Table 1. Results for different high-level synthesis benchmarks

equivalence problem of FSMDs has been reported to be not even partially decidable. The method is strong enough to accommodate merging of the segments in the original behaviour by the typical scheduler such as, DLS [7]. It is also able to handle arithmetic transformations and expected to handle simple code motion. Similar methods reported in the literature have been found to fail under such situations. Although the proposed method is found to have a non-polynomial worst case complexity primarily because of presence of parallel edges between the same pair of states, the best case complexity is found to be linear. The initial experiments show that the algorithm is usable for practical equivalence checking cases of scheduling.

References

[1] R. W. Floyd. Assigning meaning to programs. In J. T. Schwartz, editor, *Proceedings the 19th Symposium on Applied Mathematics*, pages 19–32, Providence, R.I., 1967. American Mathematical Society. Mathematical Aspects of Computer Science.

[2] D. Gajski and L. Ramachandran. Introduction to high-level synthesis. *IEEE transactions on Design and Test of Computers*, pages 44–54, 1994.

[3] D. D. Gajski, N. D. Dutt, A. C. Wu, and S. Y. Lin. *High-Level Synthesis: Introduction to Chip and System Design*. Kluwer Academic Publishers, 1992.

[4] W. E. Howden. *Functional program testing and analysis*. McGraw-Hill, New York, 1987.

[5] J. C. King. Program correctness: On inductive assertion methods. *IEEE Trans. on Software Engineering*, SE-6(5):465–479, 1980.

[6] Z. Manna. *Mathematical Theory of Computation*. McGraw-Hill Kogakusha, Tokyo, 1974.

[7] M. Rahmouni and A. A. Jerraya. Formulation and evaluation of scheduling techniques for control flow graphs. In *Proceedings of EuroDAC'95*, pages 386–391, Brighton, 18-22 September 1995.

[8] D. Sarkar and S. C. De Sarkar. Some inference rules for integer arithmetic for verification of flowchart programs on integers. *IEEE Trans. Softw. Eng.*, 15(1):1–9, 1989.

0-7695-2533-4/06 $20.00 © 2006 IEEE

An Efficient Wrapper Scan Chain Configuration Method for Network-on-Chip Testing

Ming Li, Wen-Ben Jone, Qing-An Zeng
Department of Electrical & Computer Engineering and Computer Science
University of Cincinnati, Cincinnati OH 45221, USA
Email: {lim0, wjone, qzeng}@ececs.uc.edu

Abstract

Network-on-Chip (NoC) is the new paradigm for core-based system design. Reuse of the on-chip communication network for testing cores in a NoC-based system is critical to reduce test cost for this new architecture. However, many new challenging issues come up correspondingly. In this paper, we propose a test data transportation method with multiple data flit formats and a novel scan chain configuration method to maximize the utilization of the on-chip network channel without adding too much hardware overhead. Experimental results on ITC'02 benchmarks show that the new wrapper scan chain configuration method (with the aid of multiple data flit formats) leads to substantial reduction in network channel waste, and thus results in a significant overall reduction of time and energy for testing the entire system. The test wrapper architecture that supports the new method of test data transportation is very simple, and has been verified by VHDL simulation.

1. Introduction

To provide high bandwidth and low latency, the Network-on-Chip (NoC) design paradigm has recently been proposed as an alternative to traditional broadcast and shared-bus architectures for core-based systems [1] [2]. This new paradigm relies on a packet-switching network implemented on the chip to provide high performance interconnection to embedded cores. Testing the embedded cores in a system that contains a NoC poses considerable challenges. Reuse of the existing on-chip communication resources, such as routers and channels, is critical to avoid additional area overhead [3]. However, reusing the NoC resources efficiently is challenging because the design of routers and channels in on-chip networks is optimized for communication in mission-mode, not for test. For example, there may be a mismatch between the available network channel width and the core scan chain width (which is usually equal to the Test Access Mechanism (TAM) width for traditional SoC architectures [4][5]), and this can adversely affect test efficiency and test cost.

In this paper, we propose a new test data transportation method using *multiple data flit formats* (MDFF). With this method, a data flit[1] can contain multiple bits for each wrapper scan chain, instead of only 1 bit/chain in traditional test

application methods. Also, the data flits for a core can have different formats to adapt with the number of unfilled scan chains, for maximum utilization of network channels. A heuristic wrapper scan chain configuration method is developed based on the concept of MDFF, to reduce both the test application time and the waste of data flits for testing cores in a NoC. The performance of the configuration method is proved by the simulation results on the ITC'02 benchmark set.

2. Background

For traditional SoC, test stimuli and test results are transported through the TAM circuit. Hence, research efforts for SoC testing are mostly focused on the co-optimization of test wrapper and TAM [4][5][6]. Functional I/Os and internal scan chains of an embedded core are generally configured to a designated number of balanced wrapper scan chains whose lengths are as equal as possible. The number of wrapper scan chains is determined by some optimization algorithms (e.g., some ILP algorithms), which take both test wrapper and TAM into consideration, and try to minimize the overall test time and area overhead.

With the introduction of NoC, valuable works have been done for embedded core testing based on this new architecture. In [3], two methods of using an on-chip network for the test of core-based systems were introduced, and advantages in reducing test time, area overhead and pin-overhead were shown by experimental results. In [7], the authors extended the results of a previous on-chip network research to a test scheduling algorithm with power constraints considered. In this algorithm, scheduling was based on every single packet and the test pipeline for a core can be interrupted. Since this is not applicable for the non-preemptive case, an improved test scheduling algorithm with BIST and precedence constrains was further proposed in [8]. For all these test scheduling algorithms, each packet only contains a single bit for each wrapper scan chain of the embedded core. Since this may result in a huge waste of packets, a further improved test scheduling algorithm has been proposed in [9]. The algorithm allows a packet to contain multiple bits for each wrapper scan chain. To support test under this new

---

1 There are different definitions of a 'flit' for NoC. In this paper, a flit means the data unit that can be sent within one clock cycle. A packet is composed of multiple flits, called head flit, body flit, and tail flit by function. 'Data flits' in this paper are in fact the body flits that contain in-band data in stead of control information. Further, the width of a data flit indicates the effective bits in a data flit for in-band data.

0-7695-2533-4/06 $20.00 © 2006 IEEE   147

packet format, the authors proposed to use on-chip clocking to speed up the test data transfer for certain cores by faster clocks, and use slower clocks for other cores to limit the power consumption. An algorithm was presented to determine the clock rate distribution among the cores.

A network architecture using the star-connected on-chip network was proposed in [10]. The authors implemented an example NoC and analyzed the core access time and the communication throughput. It was shown that this architecture can result in reduction of test time due to the high bandwidth and smaller area overhead due to the network reuse. Another work [11] evaluated the impact of reusing processors to test a NoC-based system. The results demonstrated that using available processors as source/sink for test can achieve better test parallelism, and hence can reduce the system test time without additional area and test pins.

3. Motivation

Most embedded cores use scan test to verify the functional and structural correctness for their random logic circuits. Scan chain configurations in SoC architectures have been fully researched, and good results have been successfully accomplished. However, the objective of scan testing in a NoC is different from that in a SoC, so the detailed configuration method is also different. The major difference comes from the following two points.

1) In a traditional SoC architecture, the width of TAM directly affects the cost of test, so each embedded core can allow only very few wrapper scan chains. The scan chain configuration has to be limited by this requirement, and the test application time for a single core has to sacrifice. However, it is no longer a bottleneck in a NoC where there is no common TAM. Instead, test patterns and output responses are transferred using the existing on-chip communication network. Each embedded core is equipped with a wrapper (i.e., network interface) for mission-mode operations to serve all I/Os of the core. These connections to all I/Os can be used in test mode as a test access port, and the number of wrapper scan chains only has a limit with the network channel width; in most common cases, it is much smaller than the network channel width. So, the minimum test application time for each core can be easily satisfied as long as we limit the length of the longest internal scan chain as the maximum length of all wrapper scan chains.

2) In a traditional SoC, wrapper scan chains are configured as balanced (i.e., equal length) as possible, and all bits of each test pattern are scanned into the scan chains simultaneously. Since the assigned TAM channel width is the same as the number of wrapper scan chains, this can minimize the waste of the channel bandwidth. However, it is not the case in a NoC. The network channel structure is fixed and designed for mission-mode operations in a NoC, so there may be a mismatch between the network channel width and the number of wrapper scan chains in a core logic. This problem will not affect the test time for a single core, but the waste of the network channel bandwidth will result in extra network traffic, and thus has a great effect on the total test time for the entire chip.

4. MDFF Test Data Application

During scan test, all scan chains involved need to be filled with test patterns. To minimize the test pattern ap-

plication time, data for different scan chains are scanned in simultaneously. In a NoC architecture, communication is in the format of packet-switching, and each packet is composed of a series of data flits, where the width of a data flit is equal to the network channel width. Since the width of the network channel may be more than the number of wrapper scan chains, each data flit can contain several bits of test data for each wrapper scan chain instead of only one bit for each. The simplest case is that there is only one data flit format. However, wrapper scan chains may have different lengths, and some of them may finish before others. In this case, after short chains are finished, their input channels are wasted. If there are significant variations between the lengths of different wrapper scan chains, the waste can be excessive. To reduce the waste of network channel due to the variation in wrapper scan chain lengths, instead of using only one data flit format, we propose to use multiple data flit formats for test data application. That is, when some wrapper scan chains finish their test applications, we will re-assign a new data flit format such that the remaining wrapper scan chains can fully use the network channel.

For example, if there are four scan chains in an embedded core of a NoC, and their lengths are 160, 160, 80, and 80 respectively. Assume each data flit can transfer 32 bits of data. By using only one fixed format of data flits to apply test patterns to the core, each data flit will include eight bits of data for each scan chain (Fig. 1(a)). After 10 flits are transferred, both shorter scan chains are finished, and the network channel for them will be totally wasted. If variable data flit formats are allowed, after transferring 10 data flits, we can use a new flit format such that each flit carries 16 bits for each of both longer chains (Fig. 1(b)). Thus, no network channel will be wasted, and we can use fewer data flits (15 flits) than the single format option (20 flits) to finish the test data application.

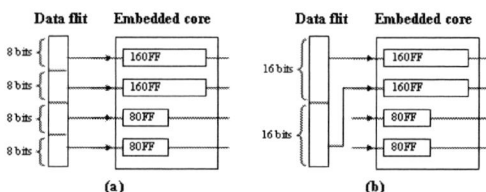

Figure 1. (a)Data flit format with each flit containing 8 bits/chain; (b)Data flit format with each flit containing 16 bits/chain.

Assume the width of each data flit is N, the total number of wrapper scan chains is m, and $m \leq N$. Based on the idea of multi-format test data transportation, we can calculate the total number of data flits required to finish the transfer of a test pattern as discussed below.

- At first, each flit includes data for all m chains.
  - Each chain has $x_0 = \lfloor \frac{N}{m} \rfloor$ bits where $\lfloor x \rfloor$ is the floor of x.
  - Since we have $l_0 \leq l_1 \leq \ldots \leq l_m$ (l_i denotes the length of wrapper scan chain i), the shortest chain with length l_0 will finish first and the number of data flits it takes is
  $k_0 = \lceil \frac{l_0}{x_0} \rceil$, where $\lceil y \rceil$ is the ceiling of y.
- After the shortest chain with length l_0 is finished, there are at most $m - 1$ chains remaining to be filled.

0-7695-2533-4/06 $20.00 © 2006 IEEE   148

– Each chain has $x_1 = \lfloor \frac{N}{m-1} \rfloor$ bits in each flit.

– Since we have $l_1 \leq ... \leq l_m$, the scan chain with length l_1 will finish first and this takes k_1 flits with

$$k_1 = \lceil \frac{(l_1 - k_0 * x_0)^+}{x_1} \rceil.$$

Note that $(z)^+$ returns z if z is larger than 0, otherwise, it returns 0.

- When the chain with length l_{i-1} is finished, there are at most $m - i$ chains remaining to be filled.

– Each chain has $x_i = \lfloor \frac{N}{m-i} \rfloor$ bits in each flit.

– To finish shifting the chain with length l_i, we need another k_i flits where

$$k_i = \lceil \frac{(l_i - \sum_{j=0}^{i-1} k_j * x_j)^+}{x_i} \rceil.$$

- For the longest chain to finish, totally k flits are required. We have

$$
\begin{aligned}
K &= k_0 + ... + k_{m-1} \\
&= k_0 + \sum_{i=1}^{m-1} \lceil \frac{(l_i - \sum_{j=0}^{i-1} k_j * x_j)^+}{x_i} \rceil \quad (1).
\end{aligned}
$$

Consequently, it takes totally K data flits to complete the application of a single test pattern.

Let us use a core with five wrapper scan chains whose lengths equal 300, 300, 300, 120, and 90 as an example. Again, we assume the size of each data flit is 32 bits. So, we have $N = 32$ and $m = 5$ in this example. At first, each data flit will carry $\lfloor 32/5 \rfloor = 6$ bits for each chain. It takes $90/6 = 15$ flits to finish the shortest chain. After this, each data flit will carry $32/4 = 8$ bits for the remaining 4 chains. It will take $\lceil (120 - 6 * 15)/8 \rceil = 4$ flits to finish the second shortest chain. Then, each data flit will carry $\lfloor 32/3 \rfloor = 10$ bits for each remaining chain, and it will take $\lceil (300 - 8 * 4 - 6 * 15)/10 \rceil = 18$ flits to finish the 3 chains left to scan. So, the total number of data flits required to finished a test pattern application is $(15 + 4 + 18) = 37$. If only one data flit format is used, however, the total number of flits required will be $300/6 = 50$. Thus, by using 3 data flit formats, 26% of data flits can be saved in this example. This is very worthwhile if the hardware overhead for implementing multi-format data flits is tolerable.

5. Wrapper Scan Chain Configuration

5.1. Guidelines for Scan Chain Configuration

Assume there are two possible wrapper scan chain configurations for a given embedded core. Configuration C_1 has M_1 equal length wrapper scan chains and each chain has length L. Configuration C_2 has M_2 wrapper scan chains and the longest chain in C_2 has length equal to L. Further, assume C_1 (C_2) requires N_1 (N_2) data flits to apply a test pattern to the core. Again, N is the bit-width of each data flit.

Lemma 1: For configurations C_1 and C_2 with $N \bmod M_1 \neq 0$, $N \bmod M_2 = 0$, where M_2 is greater than M_1 and is the smallest integer factor of N between M_1 and N. We have $N_2 \leq N_1$. *Proof:* [12].

For example, let us consider an embedded core with two wrapper scan chain configurations, C_1 and C_2. Here, C_1 has 17 chains with each equal to 160 bits. C_2 has 16 chains with each equal to 160 bits, and another set of 16 chains with each equal to 10 bits. Assume each data flit can carry 32 bits of data. The number of data flits required in the first configuration is $160/1 = 160$, while that in the second configuration is $10/1 + (160 - 10)/2 = 10 + 75 = 85$. The number of data flits saved by the second configuration is $(160 - 85)/160 = 46.875\%$.

The reason behind this lemma is that, under equal length wrapper scan chain configuration, if the number of (final) wrapper scan chains is not an integer factor of the network channel width, there is fixed waste for each data flit, which is equal to $N - \lfloor \frac{N}{M_1} \rfloor * M_1 = N \bmod M_1$. Different values of M_1 result in different values of flit waste, and the relationship between the waste of data flits and the number of wrapper scan chains can be shown in Fig. 2 (we still assume that each data flit carries 32 bits of data).

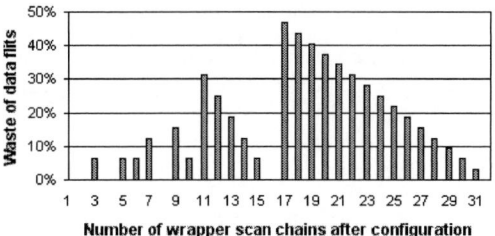

Figure 2. Relation between the waste of data flits and the number of wrapper scan chains.

So, in this case, the idea of equal length configuration is no longer the optimal solution as in the case of traditional SoC testing, instead, reconfiguring all scan chains in a core to M_2 wrapper scan chains may achieve some improvement. The best improvement can be estimated by the value of M_1, and it occurs when $M_1 = \frac{N}{2} + 1$, and the improvement can be up to $\frac{(N/2-1)}{N}$, as shown in the above example.

Assume l_0 and l_0' are the shortest chains in C_1 and C_2 respectively, and l_i (l_i') is the length of the i_{th} wrapper scan chain in C_1 (C_2).

Lemma 2: If C_1 and C_2 have the same number of wrapper scan chains (i.e., $M_1 = M_2 = M$, and $N \bmod M = 0$), and $l_0 \leq l_0'$, $l_i = l_i'$ for all $M/2 \leq i < M - 1$, then we have $N2 \leq N1$. C_2 requires the minimum number of data flits when $l_0 = l_1 = ... = l_{M/2-1}$. *Proof:* [12].

For example, consider three different configurations C_1, C_2 and C_3 with $M = 4$, $N = 32$. The lengths of all wrapper scan chains in C_1 are 1600, 1600, 1200, and 400, respectively. Those in C_2 are 1600, 1600, 1000, and 600, respectively. Finally, we have the wrapper scan chain lengths in C_3 equal 1600, 1600, 800, and 800. The number of data flits required to apply a test patter by C_1 can be derived by $400/8 + (1200 - 400)/10 + (1600 - 1200)/16 = 50 + 80 + 25 = 155$. The number of data flits required by C_2 is $600/8 + (1000 - 600)/10 + \lceil (1600 - 1000)/16 \rceil = 75 + 40 + 38 = 153$, and that by C_3 is $800/8 + (1600 - 800)/16 = 100 + 50 = 150$. Obviously, C_3 (C_1) requires the smallest (largest) number of data flits.

It can be observed from Lemma 2 that an embedded core will require a smaller number of data flits if: (1) the embedded core can be configured as S_1 long chains with equal length (S_1 is an integer factor of N) plus S_2 shorter chains ($S_1 + S_2$ is the minimal integer factor of N that is greater than S_1), and (2) the length of the shortest chain can be maximized. The best case occurs when the shorter S_2 chains can be configured to be of equal length.

Combining Lemma1 and Lemma2 and extending them to a general case, we can find that: the optimum configuration should result in *factorial wrapper scan chain groups* (FSCG) with balanced-length chains within each group, as shown in Fig. 3. Each group $FSCG_{i+1}$ is composed of sorted wrapper scan chains with index $F_i + 1$ to F_{i+1}, where F_i and F_{i+1} are consecutive integer factors of N. For example, when $N = 32$, we have $F_1 = 2$, $F_2 = 4$, $F_3 = 8$, $F_4 = 16$, and $F_5 = 32$. We assign $F_0 = 0$ by default. Thus, the wrapper scan chains in $FSCG_1$ is indexed from $F_0 + 1$ to F_1, which is from wrapper scan chain 1 to wrapper scan chain 2 (Fig. 3). The maximum wrapper scan chain length of group $FSCG_{i+1}$ is no greater than that of group $FSCG_i$. The reason for this configuration to be optimum is evident. With this configuration, there will always be y wrapper scan chains in a data flit where y is an integer factor of N, so there will be no channel loss for test data transportation.

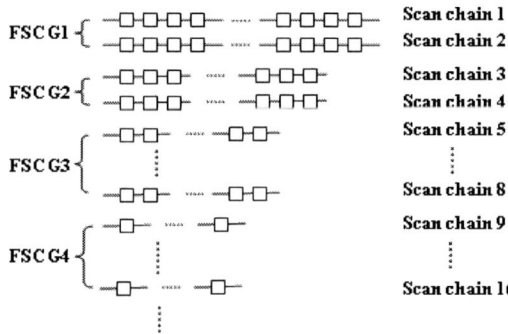

Figure 3. Factorial Scan Chain Groups with N=32.

5.2. A Heuristic Scan Chain Configuration Algorithm

In this section, we will introduce a heuristic wrapper scan chain configuration algorithm. The objective is to minimize the total number of data flits to apply a test pattern. Wrapper scan chain configuration has to obey a length constraint, i.e., the maximum length of all wrapper scan chains cannot be larger than the length of the longest internal scan chain of the core. The performance measure of a configuration is the ratio of waste in data flits, which is given by $W = (K - K_{ideal})/K_{ideal}$. Note that K is the minimum number of data flits required to transfer a test pattern based on a specific configuration, and can be calculated using equation (1). Further, K_{ideal} is the theoretical minimum number of data flits to apply a test pattern, and can be calculated using equation (2),

$$K_{ideal} = \lceil \frac{\sum_{i=0}^{m-1} l_i + N_{io}}{N} \rceil \quad (2)$$

where l_i is the length of each internal scan chain, m is the total number of internal scan chains, N_{io} is the number of all functional I/Os, and N is the width of the network channel, i.e., the bit-width of each data flit.

Based on the MDFF test data transfer protocol and two lemmas introduced above, three guidelines are used during the wrapper scan chain configuration process:

1. Try to configure the internal scan chains and functional I/Os into M wrapper scan chains where M is an integer factor of N, since there will be no channel loss in this case (refer to lemma 1).

2. When wrapper scan chains are sorted by length, try to balance the chains within each FSCG group (refer to lemma 2).

3. Attach as many functional I/Os together as possible, since this will result in less hardware complexity.

Three configuration strategies are used in the algorithm:

- Equal-length configuration — Assign internal scan chains or functional I/Os to wrapper scan chains and try to balance the lengths of all wrapper scan chains. This is based on guideline 2.

- Shortest-first configuration — Assign functional I/Os to existing wrapper scan chains with the shortest wrapper scan chain first. This is based on guideline 3.

- Best-fit configuration — Assign internal scan chains or functional I/Os to wrapper scan chains to allow more chains with the maximum length. This will also help to reduce the hardware complexity for configuration.

The wrapper scan chain configuration algorithm is given below.

Step 1:

- Sort all internal scan chains by length in a decreasing order.

- Apply the best-fit strategy to all internal scan chains and get M wrapper scan chains.

Step 2:

- Find the maximum integer M_1 which is no greater than M, and is an integer factor of N.

- Divide the longest M_1 wrapper scan chains into FSCG groups. If within any group, there is a chain whose length difference with the maximum length in the same group is larger than a given limit δ, fill it to the maximum length of the group using functional I/Os by the shortest-first configuration strategy. If there is no I/O left, stop. (note: since connecting a very small number of I/Os to wrapper scan chains will result in hardware complexity, if not necessary, we only connect I/Os to wrapper scan chains when more than δ I/Os can be grouped together.)

- If $M_1 = M$, and all remaining I/Os can be used to balance existing wrapper scan chains within each FSCG group without breaking the length constrain, do it using the shortest-first configuration strategy and stop. Otherwise, continue to step 3.

Step 3:

- Find a minimum integer M_2 which is larger than M and is an integer factor of N. We configure the shortest $M - M_1$ wrapper scan chains and all remaining functional I/Os into a new FSCG group with $M_2 - M_1$ wrapper scan chains, with a constraint that the maximum wrapper scan chain length of this new FSCG group is no larger than that of the closest FSCG group.

- If not all chains in this new FSCG group can be filled to the group's maximum length, try two options: 1) the equal-length configuration strategy and 2) the best-fit configuration strategy, and choose one of them based on the measure of data flit waste and the configuration complexity.

- Otherwise, fill all wrapper scan chains of this FSCG group to the group's maximum length. For extra I/Os, if they can be used to balance wrapper scan chains within each FSCG group without breaking the length constrain, do it using the shortest-first strategy. Otherwise, configure them to new FSCG groups by following the same method as step 3.

Let us use an example to show the procedure of the proposed algorithm. Given a core with 6 internal chains where the chain lengths are 400, 398, 250, 200, 190, and 150 respectively. There are also totally 168 functional I/Os around this core. The length constrains is that the longest wrapper scan chain length cannot exceed 400. Assume each data flit can contain 32 bits of data (i.e., $N = 32$). Ideally, the minimum number of data flits required to apply one test pattern is $\lceil (400 + 398 + 250 + 200 + 190 + 150 + 168)/32 \rceil = 55$.

Now, we begin to do wrapper scan chain configuration. At step 1, six internal scan chains are first sorted and then configured using the best-fit strategy with the length constraint. The longest two scan chains are each assigned to a wrapper scan chain, the third one (250) and the shortest one (150) are connected together and assigned to one wrapper scan chain, while the remaining two chains are also connected and assigned to another wrapper scan chain. So, now we have four wrapper scan chains with lengths: 400, 400, 398, and 390 respectively.

Assume $\delta = 5$, and we continue to step 2. In this example, the maximum integer factor of N (32) which is no larger than M (4) is 4, i.e., $M_1 = 4$. So, we first divide these four chains to two FSCG groups and try to balance chains within each group. The first FSCG group has two chains with full length, so nothing can be done further. For the second FSCG group, the difference between the shorter one and the longer one is 8, which is larger than δ, so we will assign eight I/Os to the shorter one and fill it to length 398 (based on the shortest-first strategy). Even we fill all these four chains to the maximum length, there are still I/Os left. So, we continue to step 3 without any more configuration to these four chains.

The minimum integer factor of N which is larger than M is 8 in this example, i.e., $M_2 = 8$. Now, we need to assign the remaining 160 (168-8) I/Os to a new FSCG group with 4 ($M_2 - M_1$) new wrapper scan chains. Try 2 options. With option 1, all 160 I/Os are assigned to 4 chains each with length 40. By option 2, all 160 I/Os are assigned to a single wrapper scan chain. Let us check the performance measure. The number of data flits required to apply a single test pattern with option 1 is $40/4 + (400 - 40)/8 = 55$. The total number of data flits required by option 2 is $\lceil 160/6 \rceil + \lceil \frac{400 - 6*\lceil 160/6 \rceil}{8} \rceil = 57$. Although the second option may result in less hardware configuration complexity,

the waste of data flits by option 1 is smaller than that by option 2. So we will choose option 1 as the optimal configuration.

In the above algorithm, we only considered the case where $M \leq N$. If $M > N$, we can divide the wrapper scan chains to 2 parts. Firstly, try to fill the longest N chains to maximum length, then recursively apply the algorithm to configure the remaining $N - M$ wrapper scan chains.

6. Experimental Results

We have applied our heuristic configuration algorithm to 40 cores from ITC'02 SoC benchmark set [13]. Since traditional SoC test wrapper optimization algorithms prefer to configure internal scan chains and functional I/Os to balanced wrapper scan chains, we also applied this equal-length configuration strategy to the benchmark set for comparison. In this algorithm (used for comparison), we first sort the internal scan chains, and then use the best-fit strategy to configure them into wrapper scan chains with the length constraint. After that, we use the best-fit strategy to assign functional I/Os to fill existing wrapper scan chains to maximum length, without any δ limit as in our heuristic algorithm. If there are extra I/Os left, we further use the best-fit strategy to assign them to new wrapper scan chains with the length constraint. The results obtained by these two configuration algorithms are shown in Fig. 4.

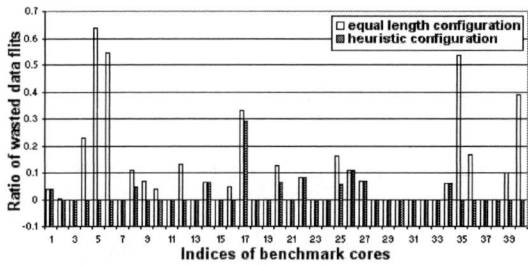

Figure 4. Comparison of data flit waste between equal length configuration and our algorithm.

As we can see, among 40 benchmark cores, 20 of them can be solved by using the simple equal-length configuration method without any waste, but the other 20 of them have significant waste. For the 20 remaining cores, our configuration algorithm can reduce the data flit waste to 0 for 14 cores, and have slight improvement for the other 6 cores, as shown in Fig. 4. For the cores that have significant improvement using our configuration, only 2 or 3 data flit formats are required, so the hardware overhead is quite small.

For a core with special internal scan chain structure, there is very little space for any configuration method to reduce its data flit waste without increasing the testing time of the core. But, when testing a NoC-based system, although the test time for a single core is increased, the total test time for the system can also be reduced using a good scheduling method based on the concept of MDFF to interleave the test data application of different cores and test them in parallel. Assume the clock frequency of the on-chip network is the same as test clock frequencies of the embedded cores. The basic concept is: for data flits containing multiple bits (x bits) of each wrapper scan chain, we only need to send one such flit to the core every x clock cycles, and the other ($x - 1$) cycles during the x cycles period can be utilized

0-7695-2533-4/06 $20.00 © 2006 IEEE

to send other data flits to other cores. The details of parallel core testing based on MDFF test data application protocol are presented in [12].

7. Test Wrapper Architecture for MDFF Test Data Application

The test wrapper architecture for MDFF is slightly more complex than that for a single data flit format application. The sketch of the architecture is shown in Fig. 5. The test wrapper is located between the Network Interface(NI) unit and the embedded core as shown in Fig. 5(a). In normal mode, data from the network will firstly pass the network interface (for synchronization, error check and etc.), go through the decoder and finally reach the I/O pins of the embedded core. In test mode, data from network interface will directly go to the test wrapper (instead of the decoder), and then scanned into the wrapper scan chains of the embedded core.

The test wrapper shown in Fig. 5(b) is composed of a load-shift register, a test pattern distributor, a controller and several counters. The load-shift register will load a data flit from the network (when the control signal 'load' equals '1'), and will operate as a shift register to pass different bits to the test pattern distributor (when control signal 'shift' equals '1'). The test pattern distributor is used to select different bits from a data flit to the wrapper scan chains of the embedded core. The controller receives input signals from NI and counters, and sets control signals to load-shift-registers, test pattern distributor, counters and embedded core for testing.

This architecture is suitable for test data transportation where each data flit contains one or more bit of information for each wrapper scan chain. The only difference between single data flit format and multiple data flit formats is the number of counters and internal complexity of the controller. As shown in our experimental results, using 2 or 3 data flit formats can significantly reduce the waste of data flits in most cases. The small number of data flits formats ensures tolerable hardware overhead. The functional correctness of this architecture has been verified using VHDL simulation.

8. Conclusions and Future Work

In this paper, we have analyzed the test problem coming up with wrapper scan chain configuration for embedded cores in a NoC architecture. Although the final objective of wrapper scan chain configuration is to reduce the total test cost for the entire chip, the problem is very different between NoC and traditional SoC architectures. We have identified their differences, proposed a new test data transportation concept with the aid of multiple data flit formats, and have summarized rules to guide the optimal wrapper scan chain configuration search process for the NoC architecture. The algorithm is effective and the results show significant improvement on the data flit waste. The test wrapper architecture that supports the idea of multi-format test pattern application is also introduced. It is simple and does not add too much overhead to the system. By combining the MDFF concept with a delicate test scheduling algorithm, the test of different cores of a NoC-based system can be performed in parallel, and the test time of the whole system can be minimized due to the full utilization of network

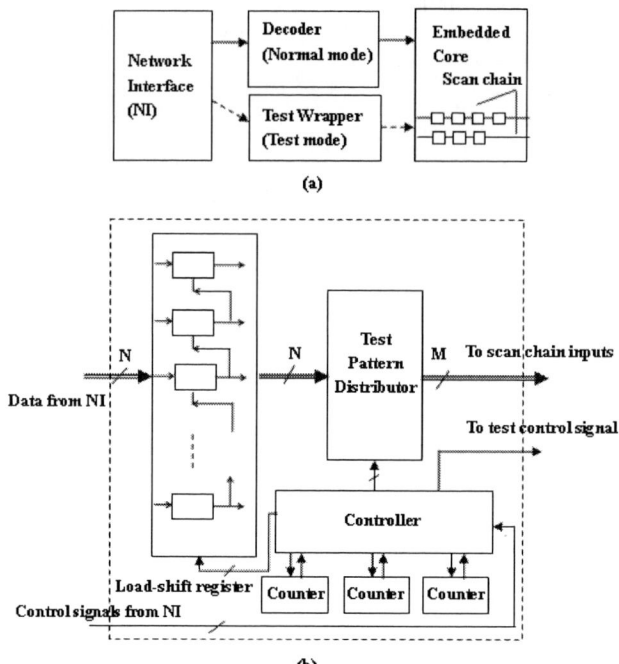

Figure 5. (a)Location of the test wrapper; (b)Test wrapper architecture.

channels. To develop a complete test scheduling algorithm for the entire system is our future work.

References

[1] W. J. Dally and B. Towles. Route Packets, Not Wires: On-Chip Interconnection Networks. In *proc. of Design Automation Conf.*, pages 684–689, 2001.

[2] L. Benini and G. D. Micheli. Networks on chips: a new SoC paradigm. *IEEE Computer*, 35:70–78, Jan 2002.

[3] E. Cota, M. Kreutz, C.A. Zeferino, L. Carro, M. Lubaszewski, and A. Susin. The impact of NoC reuse on the testing of core-based systems. In *proc. of IEEE VLSI Test Symp.*, pages 128–133, 2003.

[4] V. Iyengar, K. Chakrabarty, and E.J. Marinissen. Test wrapper and test access mechanism co-optimization for system-on-chip. In *proc. of International Test Conference*, pages 1023 – 1032, Oct 2001.

[5] V. Iyengar, K. Chakrabarty, and E.J. Marinissen. Test access mechanism optimization, test scheduling, and tester data volume reduction for system-on-chip. *IEEE tranc. on Computers*, pages 1619 – 1632, Dec 2003.

[6] M. Nourani and C. Papachristou. An ILP formulation to optimize test access mechanism in system-on-chip testing. In *proc. of International Test Conference*, pages 902–1000, 2000.

[7] E. Cota, L. Carro, F. Wagner, and M. Lubaszewski. Power-aware NoC reuse on the testing of core-based systems. In *proc. of International Test Conference*, pages 612–621, 2003.

[8] C. Liu, E. Cota, H. Sharif, and D.K. Pradhan. Test scheduling for network-on-chip with BIST and precedence constraints. In *proc. of International Test Conference*, pages 1369–1378, 2004.

[9] C. Liu, V. Iyengar, J. Shi, and E. Cota. Power-aware test scheduling in network-on-chip using variable-rate on-chip clocking. In *proc. of IEEE VLSI Test Symp.*, pages 349–354, 2005.

[10] J.S. Kim, M.S. Hwang, S. Roh, J.Y. Lee, K. Lee, S.J. Lee, and H.J. Yoo. On-chip network based embedded core testing. In *proc. of IEEE International SoC Conference*, pages 223 – 226, 2004.

[11] A.M. Amory, M. Lubaszewski, F.G. Moraes, and E.I. Moren. Test time reduction reusing multiple processors in a network-on-chip based architecture. In *proc. of Design, Automation and Test in Europe*, pages 62 – 63, 2005.

[12] M. Li, W.B. Jone, and Q.A. Zeng. An efficient wrapper scan chain configuration method for network-on-chip testing (in preparation).

[13] http://www.extra.research.philips.com/itc02socbenchm. ITC'02 SoC test benchmarks (online).

0-7695-2533-4/06 $20.00 © 2006 IEEE

An Efficient Data-Independent Technique for Compressing Test Vectors in Systems-on-a-Chip

Xiaoyu Ruan and Rajendra Katti

Department of Electrical and Computer Engineering

North Dakota State University, Fargo, ND 58105-5285

E-mail:{xiaoyu.ruan,rajendra.katti}@ndsu.edu

Abstract— **We present an efficient approach, namely, pattern run-length (PRL) coding, for reducing the volume of test vectors that must be stored in automatic test equipment (ATE) and transferred to each core in a system-on-a-chip (SOC) during manufacturing test. The need for compressing test data is due to the bandwidth bottleneck between the ATE and the SOC. In our new coding scheme, the test vectors for the SOC are stored in compressed form in the ATE memory and transferred to the chip. An embedded processor is employed to perform decompression. The decompressed test set is then applied to the scan chains of each core-under-test. Pattern run-length coding works by compressing consecutive patterns in an innovative manner. The proposed compression is data-independent. The program for decompression is very small and simple, thereby allowing fast and high throughput to minimize test time. Experimental results for ISCAS-89 benchmarks show that for almost all of the circuits our new technique results in much better compression ratios than former methods.**

I. INTRODUCTION

As the complexity of today's SOC circuits increases, testing becomes a challenging task due to several limitations, including large test data sets, long test application time, high power consumption, and limited bandwidth of ATEs [1]. Test data compression is used to speed up the interaction between ATE and SOC during the test. A typical SOC consists of several intellectual property (IP) cores that are to be tested by a large number of precomputed data. The deterministic test vector needs to be transferred from the ATE to the chip, and then applied to the cores. Test data is usually compressed since ATEs have limited bandwidth and memory, and the test vectors are often large. By compression, both transmission time and memory requirement can be reduced. Compressed test data are decompressed by the decoder in the SOC and then applied to IP cores.

Many techniques for test data compression have been introduced in the literature. Scan chain architectures for core-based designs are proposed in [6]. Compression and decompression of scan vectors using cyclical decompressors and run-length coding is described in [3]. An approach based on Burrows-Wheeler transform is investigated in [2]. Statistical coding techniques, such as run-length coding and Golomb coding, have also been studied in [9], [10], [3], and [11]. The use of variable-length Huffman coding on compressing test data is presented in [5]. Later, Jas *et al.* proposed a method that encodes only binary patterns with high frequencies, called selective Huffman coding [4]. Dictionary-based compression applied to test data is reported in [8], [21], and [20]. The construction of frequency-directed run-length (FDR) coding is described in [13]. The FDR coding compresses a so-called "difference vector" by encoding the run-lengths of 0's. The authors in [14] analyzed the concept of entropy and suggested applying arithmetic coding to compressed test data. In [12], Balakrishnan and Touba presented a novel compression scheme based on matrix operations. Nourani and Tehranipour proposed a methodology combining run-length and Huffman coding in [7]. Recently Tehranipoor *et al.* demonstrated a new compression technique that uses exactly nine codewords (9C) [18].

The existing test data compression approaches can be classified as data-independent and data-dependent. The former implies that the on-chip decoder or decompression program can be adopted for *any* test set. The 9C [18], FDR [13], VIHC [5], and MTC [19] methods belong to this type. In contrast, the decoder of a data-dependent technique is designed for decompressing a specific test vector. All dictionary-based methods such as LZW [21] and LZ77 [20] are data-dependent. They often have difficulties in terms of size and organization for compression and require large on-chip memory. Hence data-independency is a preferable property of a compression scheme.

The automatic test pattern generator (ATPG) provides a test cube with a large number of unspecified bits (also referred to as don't-cares). Since the ATPG fills don't-cares randomly, the chance of detecting nonmodeled faults increases, and the fault coverage increases too [18]. Therefore it is an advantage to have at least a portion of don't-cares remain unchanged. We notice that except 9C [18], all other compression schemes in the literature replace *all* unspecified bits with zeros and ones in a certain manner in order to achieve better compression.

In this paper we consider software-based decompression. Fig. 1 shows the approach being used. The SOC consists of an embedded processor, memory and a set of IP cores. Usually the IP cores come from different vendors along with a test set. The embedded processor is first tested, after which the IP cores are tested. In software-based decompression, the compressed test is fetched by the processor from the ATE and decoded using a program stored in the processor memory. A buffer (a set of memory locations) may be necessary if the rate at which

0-7695-2533-4/06 $20.00 © 2006 IEEE

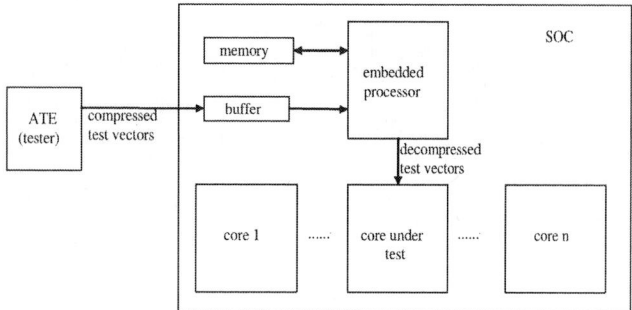

Fig. 1. SOC Test Architecture.

the ATE is supplying data is not the same as the rate at which the processor is supplying test vectors to the IP core under test. The buffer size is determined by the ATE speed and the compression program speed.

We propose an efficient coding scheme for reducing the volume of test data in SOCs. The main strategy is to compress consecutive patterns that are equal or complementary. Four major properties make our approach advantageous:

1) It is data-independent.
2) It leaves a portion of unspecified bits in original test cubes unchanged.
3) It results in better compression performance than most known techniques for most benchmark circuits.
4) The decompression algorithm is very simple, hence it can be easily implemented by an embedded processor in the SOC with a small number of instructions.

The remainder of the paper is organized as follows. The next section demonstrates the code construction. Section III develops programs for decompression. Experimental evaluation for the six largest ISCAS-89 benchmark circuits [15] is reported in Section IV. Comparisons with other methods are presented. Finally Section V concludes the paper.

II. PATTERN RUN-LENGTH CODING

In this section we introduce the new technique, which can be described with two different styles. They are respectively detailed in the following subsections.

A. Fix-Length Pattern Run-Length (PRL) Coding

The compression is performed by simply retaining one copy of a pattern and using "10" or "11" for specifying whether the other patterns are equal to or complement of the retained pattern. A "0" is used to terminate a round. For example, the vector $101x$ $101x$ $10xx$ $010x$ $01xx$ $x1xx$ would be compressed into $101x$ 10 10 11 11 11 0. In this compressed sequence the first 4 bits specify the retained pattern $101x$. The next "10" specifies the fact that the next 4 bits are equal to the retained pattern. Therefore the ten bits (1010111111) after the first 4 bits $101x$ specify the vector $101x$ $101x$ $010x$ $010x$ $010x$. This is the same as the original vector if some of the don't-cares (x's) assume the right values. A compressed test vector will consist of a k-bit retained pattern

followed by 10's and 11's followed by a 0 and another k-bit retained pattern and so on.

Let the number of bits in the whole test vector be len. The length of a pattern, k, is set to a *fixed* value. We first pad ($k - len \bmod k$) don't-cares at the end of the vector so that k divides ($len + k - len \bmod k$). The ($len + k - len \bmod k$) bits are then split into ($\frac{len - len \bmod k}{k} + 1$) patterns. Initially the first k bits are set to be the retained pattern, which will be referred to as "common pattern" afterwards. We compare it with the next k bits and see if they can be equal/complementary or not. If they can, then we find their common pattern and update the first pattern accordingly. Notice that the common pattern is always the first pattern in that round. We continue operating like this until it is impossible to match a k-bit pattern with the common pattern. The unmatchable pattern is then set to be the new common pattern and a new round of encoding starts. Our aim is to generate as many equal or complementary patterns as possible in a round since this increases compression. Unspecified bits provide flexibility that can be used to achieve this goal.

For example, let $k = 7$. We examine the following sequence

$$001xxxx \; xx11xxx \; xx0x0xx \; xxxxxx0 \; 1xx1xxx \; 10 \cdots$$

For clarity neighboring patterns are separated by a blank. The common pattern is initialized to $001xxxx$. Compare the first two patterns

$$001xxxx$$
$$xx11xxx$$

We may make them equal by replacing the leftmost x of the upper row by "1" and the leftmost two bits of the lower row by "0", resulting in

$$0011xxx$$
$$0011xxx$$

The common pattern becomes $0011xxx$. Compare it with the next k bits

$$0011xxx$$
$$xx0x0xx$$

They are complementary if proper values are assigned to some x's, as shown below.

$$00111xx$$
$$11000xx$$

The updated common pattern $00111xx$ is compared with $xxxxxx0$.

$$00111xx$$
$$xxxxxx0$$

We have two options at this point. If the rightmost bit of the upper row is replaced by "0", then $xxxxxx0$ is equal to the updated common pattern $00111x0$. Otherwise, if the rightmost bit of the common pattern assumes "1", then $xxxxxx0$ is complimentary to $00111x1$. Since the common pattern will impact on future pattern matching, the decision we make may affect the compression performance. However, simulations

0-7695-2533-4/06 $20.00 © 2006 IEEE   154

show that this influence is very small. For simplicity of encoding, the two patterns are set to be equal by default whenever this condition occurs. Hence for this example the updated common pattern is $00111x0$.

The next k bits are $1xx1xxx$. No matter how we play with the unspecified bits they cannot match $00111x0$. Thus in this round we encode $001xxxx$ $xx11xxx$ $xx0x0xx$ $xxxxxx0$. A round is a set of consecutive k-bit patterns that are equal/complementary. Notice that $1xx1xxx$ will be the initial common pattern of the next round.

We now construct the compressed sequence. The first m bits of the compressed sequence are simply the binary expansion of k. They are sent only once and k is fixed for the whole test cube. It is found from empirical observations that the best k for most test cubes is unlikely to exceed 128, therefore m can be set to 7. Since m is very small compared to the total length of compressed data, it lowers the compression by just a negligible amount. The compressed code of a round consists of two parts: the final common pattern and a list of $p - 1$ codewords indicating if a k-bit pattern is equal to or complementary of the common pattern. Here p is the number of patterns in the round. Since the first pattern is always equal to the common pattern, $p - 1$ indicator codewords are enough. The indicator codeword is arranged as follows: "10" indicates a k-bit pattern that is equal to the common pattern; "11" indicates a k-bit pattern that is complementary to the common pattern; and "0" means here is the end of a round. For the aforementioned example, the compressed code for $001xxxxxx11xxxxx0x0xxxxxxxx0$ is $00111x01011100$. The leftmost 7 bits $00111x0$ specify the common pattern, i.e., the first 7 bits of the original test vector. The following "10" suggests the second pattern is equal to $00111x0$. Next two bits ("11") imply that the third pattern is bitwise opposite to $00111x0$, i.e., $11000x1$. Similarly next "10" implies $00111x0$. Finally the "0" ends this round and indicates that the next 7 bits of the compressed sequence will be the common pattern of a new round.

The decompression process is straightforward. The decoder computes k from the first m bits of the compressed sequence. This is done only once since k is fixed for the entire test set. In a round, the decoder stores the first k bits in memory and outputs them. Then a 3-state finite state machine (FSM) is used. If the next two bits are "10" then the decoder outputs the contents in the memory sequentially; if the next two bits are "11" then the contents in the memory are bit-wise exclusive-ored with 1 and the results are output. This procedure is performed until a "0" is encountered, and a new round begins. The decompression algorithm will be discussed in detail in Section III. For our trivial example, compressed data $00111x01011100$ is decoded into $00111x000111x011000x100111x0$, which is identical to the original sequence $001xxxxxx11xxxxx0x0xxxxxxxx0$ if some x's assume appropriate values. Notice that some x's remain unspecified after decompression.

The pseudo code for the compression process is given in Algorithm 1. Symbol "$\|$" means "concatenated with". Function $length(vector)$ returns the number of bits in $vector$. The k-bit vector pc and puc stand for "common pattern" and

---

Algorithm 1 Data-Independent Fixed-Length Pattern Run-Length Coding

Input: Test vector $t\_vec[i]$ with $0 \leq i < len$ and pattern size k.

Output: Compressed sequence out_seq.

$out\_seq \leftarrow m$-bit binary expansion of k

$t\_vec \leftarrow t\_vec\|x\cdots x$ where the number of x is $k - length(t\_vec) \bmod k$

$cp \leftarrow t\_vec[0]t\_vec[1]\cdots t\_vec[k-2]t\_vec[k-1]$

$ind\_vec \leftarrow \emptyset$

for $1 \leq n \leq (length(t\_vec) - length(t\_vec) \bmod k)/k$ **do**

  $puc \leftarrow t\_vec[nk]t\_vec[nk+1]\cdots t\_vec[nk+k-2]t\_vec[nk+k-1]$

  for $1 \leq i \leq k$ **do**

    if $cp[i] = puc[i] \neq x$ **then**

      $ind[i] = 0$

    else if $cp[i] \neq puc[i]$ and $cp[i] \neq x$ and $puc[i] \neq x$ **then**

      $ind[i] = 1$

    else

      $ind[i] = x$

    end if

  end for

  if $ind[i_1] \neq x$ and $ind[i_2] \neq x$ and $ind[i_1] \neq ind[i_2]$ for $1 \leq i_1 \neq i_2 \leq k$ **then**

    $out\_seq \leftarrow out\_seq\|cp\|ind\_vec\|0$

    $cp \leftarrow puc$

    $ind\_vec \leftarrow \emptyset$

  else

    if there exists i_0 such that $1 \leq i_0 \leq k$ and $ind[i_0] \neq x$ **then**

      $xor \leftarrow ind[i_0]$

    else

      $xor \leftarrow 0$

    end if

    $cp[i] \leftarrow puc[i] \oplus xor$ for $1 \leq i \leq k$ with $cp[i] = x$ and $puc[i] \neq x$

    $ind\_vec \leftarrow ind\_vec\|1\|xor$

  end if

end for

return(out_seq)

"pattern under consideration" respectively. Symbol "$\emptyset$" is used to denote an empty vector. It is easy to see that the complexity of Algorithm 1 is $O(n)$.

We now discuss the choice of block length. For a deterministic test vector, different values of k give different compression ratios (defined in Section IV). Experimental results for two ISCAS-98 benchmarks [15] s5378 and s13207 are displayed in Fig. 2. The lower curve is for s5378. The maximum compression ratio is found to be 54.9592%. This happens when $k = 16$. For s13207, compression ratios remain almost the same for k ranging from 20 to 90. The maximum 83.1120% is achieved when $k = 40$.

0-7695-2533-4/06 $20.00 © 2006 IEEE

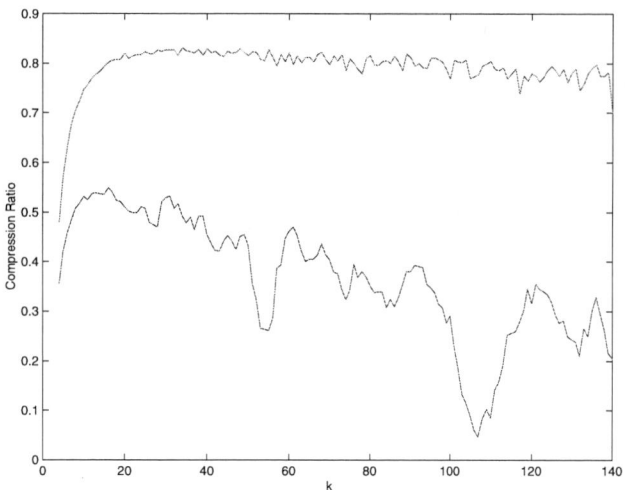

Fig. 2. Relationship between k and compression ratio for s5378 and s13207.

B. Variable-Length PRL (vPRL) Coding

Generally speaking, the earlier compression techniques that adopt variable-length strategy [11] [16] [5] result in better compression ratios than the fixed-length ones [10] [4]. The fixed-length PRL coding applies a fixed value of k to the entire test vector. This makes the decompression very simple, yet the compression ratio is pretty satisfactory. To achieve further improvement, we consider varying k.

In this style, each round chooses its own block length. Since the block length is not constant for the entire test set, it is not enough to indicate k only once by the very first m bits of the compressed sequence as in the fixed-length case. Instead the value of k_j must be specified at the beginning (i.e., immediately before the k_j-bit common pattern) of the compressed code of the jth round. For example, recall that uncompressed sequence

$$001xxxx \ xx11xxx \ xx0x0xx \ xxxxxx0$$

is encoded to $00111x01011100$ in fixed-length PRL coding. For vPRL coding, block length $k = 7$ has to be included in the compressed sequence

$$111 \ 00111x01011100.$$

The first 3 bits "111" are the binary expansion of 7. The decoder determines the size of the common pattern according to this information.

Now a critical problem follows. How to split the original test set into rounds in order to maximize the compression ratio? Suppose there are N rounds in total and R bits are used to denote block lengths, then $k_j \in G = \{i : 3 \le i \le 2^R - 1\}$ for $1 \le j \le N-1$ and $1 \le k_N \le 2^R - 1$. That is,

$$len = \sum_{j=1}^{N} k_j \cdot p_j$$

where p_j represents the number of k_j-bit patterns in the jth round. We set $k_j \ge 3$ for all rounds except the Nth one

because, from the construction of fixed-length PRL coding it is easy to see that 3 is the minimum block length that results in compression rather than expansion. The pattern size of the last round is allowed to assume any integer value ranging from 1 to $2^R - 1$ so no don't-cares need to be concatenated. We thereafter have $|G| = 2^R - 3$ candidates for rounds 1 to $N-1$. The size of total search space is $(2^R - 3)^{(N-1)}$. The complexity of the encoding algorithm for finding the optimal splitting is $O(2^{R \cdot N})$, which is exponential. Obviously the encoder is incapable of performing the exhaustive search in reasonable time. Thus it is impractical to figure out the best splitting that results in the maximum compression ratio.

Alternatively, we employ a suboptimal approach. The main idea is to decide the termination of the jth round only based on its own compression ratio, but not on that of the other $N-1$ rounds. For round j we test all the $|G|$ block length candidates and choose the one that results in the maximum compression. Notice that the starting points (the first bit of a round in original test vector) of the $|G|$ trials are the same, while the end points (the last bit of a round in original test vector) can be different. We call this strategy "local optimization". The size of search space reduces to $(2^R - 3) \cdot (N - 1)$, corresponding to complexity $O(2^R \cdot N)$.

As an example, again consider compressing

$$001xxxxxx11xxxxx0x0xxxxxxxx01xx1xxx10\cdots$$

In case more than one round of encoding is required we shall analyze only the first round. Let $R = 3$ and hence $G = \{3, 4, 5, 6, 7\}$. The compression ratios for the 5 candidates are listed in the sixth column of Table I. The first column gives the value of block length k. The portion of the original test vector that is compressed by this round is shown in the second column. Sizes of sequences in column 2 are recorded in the third column. The compressed sequences are illustrated in column 4 and their lengths are shown in column 5. The first 3 bits of a compressed sequence are the binary representation of corresponding k. The rightmost column gives the original sequence of the next round, i.e., removing the sequence of the second column from $001xxxxxx11xxxxx0x0xxxxxxxx01xx1xxx10\cdots$. This table implies that $k = 5$ gives the best compression ratio. Therefore $001xxxxxx11xxxxx0x0xxxxxxxx01xx1xxx$ is coded as 101001011011101011110 in this round. Next round of encoding starts from uncompressed sequence $10\cdots$. Block lengths $\{3, 4, 5, 6, 7\}$ will be again respectively applied to $10\cdots$ before the best one is selected.

The code construction is described below. The very first n bits of the compressed sequence are the binary expansion of R used for all rounds of the entire test vector. Experiments indicate that for most benchmarks optimal R is unlikely to exceed 15 hence we can set $n = 4$. These n bits are sent only once so their impact on compression can be ignored. What follows are the compressed sequences of the N rounds as explained above.

The decoder computes R from the first n bits of compressed data. For decompressing the jth round, the decoder reads the

TABLE I

USING DIFFERENT BLOCK LENGTH FOR ENCODING $001xxxxxx11xxxxx0x0xxxxxxxx01xx1xxx10\cdots$

k	original sequence of current round	length (bits)	compressed sequence of current round	length (bits)	CR	original sequence of next round
7	$001xxxxxx11xxxxx0x0xxxxxxxx0$	28	11100111x01011100	17	0.3929	$1xx1x\cdots$
6	$001xxxxxx11xxxxx0x0xxxxxx$	24	11000111x1011100	16	0.3333	$xxx01\cdots$
5	$001xxxxxx11xxxxx0x0xxxxxxxx01xx1xxx$	35	101001011011101011110	21	0.4000	$10\cdots$
4	$001xxxxx$	8	100001x100	10	−0.25	$x11xx\cdots$
3	$001xxxxxx11xxxxx0x0xxxxxxxx$	27	01100110100110101010100	23	0.1481	$01xx1\cdots$

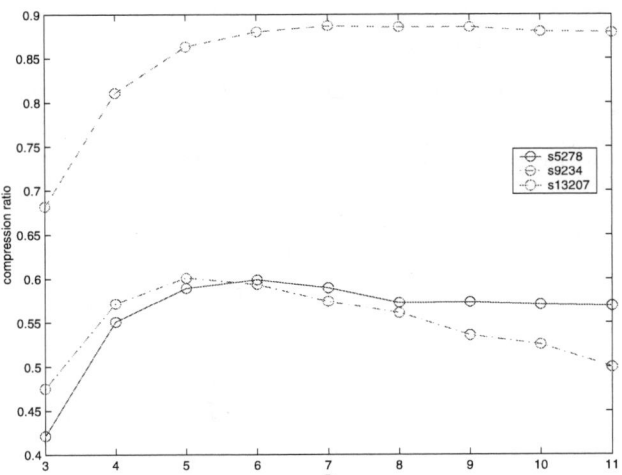

Fig. 3. Relationship between R and compression ratio for s5378, s9234, and s13207.

first R bits and calculates block length k_j used for this round. It then stores the next k_j bits in memory and outputs them. The rest of the decoding procedure is identical to fixed-length PRL coding.

The value of R determines the compression performance. Small R leads to small $|G|$. Consequently the encoder has to choose k from a small number of block length candidates, which may lower the efficiency. On the other hand, recall that $R \cdot N$ bits are added to the entire compressed sequence. Furthermore, $(2^R - 1)$-bit memory is required by the decoder for storing common patterns. These two factors limit the value of R. Fig. 3 shows the compression ratios resulting from R ranging from 3 to 11 for s5378, s9234 (another ISCAS-98 benchmark circuit [15]), and s13207. Best R for each case is given in the sixth column of Table II in Section IV.

III. DECOMPRESSION ALGORITHMS

The decompression for the fixed-length PRL coding is realized by Algorithm 2. Vector cp stands for "common pattern". The first m bits of the compressed sequence are used to specify k, which will be adopted for decoding the entire $CompLen$-bit compressed vector. Notice that m is fixed and known by both compressor and decompressor ends.

The decoding algorithm for vPRL coding can be developed

Algorithm 2 Decompression for Data-Independent Fixed-Length Pattern Run-Length Coding

Input: Compressed sequence $c[i]$ with $0 \leq i < CompLen$.
Output: Decompressed test vector $d[q]$ with $0 \leq q < len$.
  $q \leftarrow 0$
  $k \leftarrow$ decimal value of $c[m-1]c[m-2]\cdots c[0]$
  $i \leftarrow m$
  while $i < CompLen$ **do**
    $d[q+j] \leftarrow c[i+j]$ for $0 \leq j \leq k-1$
    $cp[j] \leftarrow c[i+j]$ for $0 \leq j \leq k-1$
    $q = q + k$
    $i = i + k$
    while $c[i] = 1$ **do**
      $d[q+j] \leftarrow cp[j] \oplus c[i+1]$ for $0 \leq j < k$
      $q = q + k$
      $i = i + 2$
    end while
    $i = i + 1$
  end while
  return(d)

---

by slightly modifying Algorithm 2. The only additional function is to decode k_j according to its R-bit binary form before recognizing cp.

IV. EXPERIMENTAL RESULTS

The proposed coding scheme is applied to the MinTest [17] test data sets for the six largest ISCAS-98 benchmarks [15]. Most contributions on test data compression use the same circuits for experiments.

The compression ratio is computed as

$$CR = \frac{\text{size of original data} - \text{size of compressed data}}{\text{size of original data}}.$$

Comparisons of the fixed-length PRL and vPRL coding with other data-independent methods, including V9C [18], FDR [13], VIHC [5], and MTC [19], are reported in Table II. The sizes of the original test data are listed in the second column. Columns 4 and 6 give the corresponding k and R for the fixed-length PRL and vPRL coding, respectively. The CR values are given in percentage (%). The best results in each row are written in boldface. Our vPRL coding significantly outperforms other techniques in most cases.

0-7695-2533-4/06 $20.00 © 2006 IEEE   157

TABLE II

COMPARING CRs OF FIXED-LENGTH PRL AND vPRL CODING WITH OTHER DATA-INDEPENDENT TECHNIQUES

circuit	original size (bits)	PRL	k	vPRL	R	V9C [18]	FDR [13]	VIHC [5]	MTC [19]
s5378	23754	54.97	16	59.88	6	55.75	50.77	51.52	38.49
s9234	39273	54.03	11	60.09	5	54.77	44.96	54.84	39.06
s13207	165200	83.11	40	88.67	7	85.20	80.23	83.21	77.00
s15850	76986	66.54	20	74.28	6	71.26	65.83	60.68	59.32
s38417	164736	53.25	13	57.07	6	62.89	60.55	54.51	55.42
s38584	199104	66.89	18	74.17	6	69.11	61.13	56.97	56.63
Avg.		63.13		69.03		66.49	60.58	60.28	54.32

TABLE III

DECOMPRESSION PROGRAM SIZES

	PRL	vPRL
# of instructions	83	130
# of bytes	332	520

Decompression algorithms are implemented in C language and cross-complied for the ARM7 architecture. Table III measures the sizes of the two programs. Since all instructions on ARM7 systems are of 32 bits, the number of bytes is always 4 times the number of instructions. It is understandable that the fixed-length PRL decoding requires fewer instructions than the vPRL decoding does. Overall, the program sizes are small and suitable for on-chip implementation.

V. CONCLUSION

Compressing test data is necessary due to the bandwidth bottleneck between ATE and SOC. We have proposed an innovative data-independent method for the compression of test data for IP cores in SOCs, called pattern run-length coding. This compression scheme is a modified run-length coding that encodes consecutive equal/complementary patterns. The software-based decompression algorithm is small and requires very few processor instructions. Experiments for the six largest ISCAS-98 benchmarks [15] show that our method outperforms most known techniques.

ACKNOWLEDGMENT

This work has been supported in part by the National Science Foundation under Grant CCR-0429523. The authors would like to thank Prof. K. Chakrabarty and his students for providing the MinTest test cubes.

REFERENCES

[1] R. Chandramouli and S. Pateras, "Testing systems on a chip," *IEEE Spectrum*, pp. 42-47, Nov. 1996.

[2] M. Ichihara, D. S. Ha, and T. Yamaguchi, "COMPACT: a hybrid method for compressing test data," in *Proc. VLSI Test Symp.*, pp. 418-423, 1998.

[3] A. Jas and N. A. Touba, "Test vector decompression via cyclical scan chains and its application to testing core-based designs," in *Proc. Int'l Test Conf.*, pp. 458-464, 1998.

[4] A. Jas, J. Ghosh-Dastidar, M. Ng, and N. A. Touba, "An efficient test vector compression scheme using selective Huffman coding," *IEEE Trans. Compter-Aided Design of Integrated Circuits and Systems*, vol. 22, no. 6, pp. 797-805, June 2003.

[5] P. Gonciari, B. M. Al-Hashimi, and N. Nicolici, "Improving compression ratio, are overhead, and test application time for system-on-a-chip test data compression/decompression," in *Proc. Design Automation Test Euro.*, pp. 604-611, 2002.

[6] J. Aerts and E. J. Marinissen, "Scan chain design for test time reduction in core-based ICs," in *Proc. Int'l Test Conf.*, pp. 448-457, 1998.

[7] M. Nourani and M. H. Tehranipour, "RL-Huffman encoding for test compression and power reduction in scan applications," *ACM Trans. Design Automation of Electronics Systems*, vol. 10, no. 1, pp. 91-115, Jan. 2005.

[8] L. Li, K. Chakrabarty, and N. A. Touba, "Test data compression using dictionaries with selective entries and fixed-length indices," *ACM Trans. Design Automation of Electronics Systems*, vol. 8, no. 4, pp. 470-490, Oct. 2003.

[9] V. Iyengar, K. Chakrabarty, and B. T. Murray, "Deterministic built-in-self-testing of sequential circuits using precomputed test sets," *J. Electronic Testing: Theory and Applications (JETTA)*, vol. 15, pp. 97-114, Aug./Oct. 1999.

[10] A. Jas, J. Ghosh-Dastidar, and N. A. Touba, "Scan vector compression/decompression using statistical coding," in *Proc. IEEE VLSI Test Symp.*, pp. 114-120, 1999.

[11] A. Chandra and K. Chakrabarty, "Test data compression for system-on-a chip using Golomb codes," in *Proc. IEEE VLSI Test Symp.*, pp. 113-120, 2000.

[12] K. Balakrishnan and N. A. Touba, "Matrix-based software test data decompression for system-on-a-chip," *J. Systems Architecture*, vol. 50, pp. 247-256, 2004.

[13] A. Chandra and K. Chakrabarty, "A unified approach to reduce SoC test data volume, scan power, and testing time," *IEEE Trans. Computer-Aided Design Integr. Circuits Syst.*, vol. 22, no. 3, pp. 352 3 63, Mar. 2003.

[14] H. Hashempour and F. Lombardi, "Application of arithmetic coding to compression of VLSI test data," *IEEE Trans. Computers*, vol. 54, no. 9, pp. 1166-1177, Sept. 2005.

[15] F. Brglez, D. Bryan, and K. Kozminski, "Combinational profiles of sequential benchmark circuits," in *Proc. IEEE Int'l Symp. Circuits and Systems*, pp. 1929-1934, 1989.

[16] A. Chandra and K. Chakrabarty, "Test data compression and test recouce partitioning for system-on-a-chip using frequency-directed run-length (FDR) codes," *IEEE Trans. Computers*, vol. 52, no. 8, pp. 1076-1088, Aug. 2003.

[17] I. Hamzaoglu and J. H. Patel, "Test set compaction algorithms for combinational circuits," in *Proc. Int'l Conf. Computer-Aided Design*, pp. 283-289, 1998.

[18] M. Tehranipoor, M. Nourani, and K. Chakrabarty, "Nine-Coded Compression Techinque for Testing Embedded Cores in SoCs," *IEEE Trans. VLSI Systems*, vol. 13, no. 6, pp. 719-731, Jun. 2005.

[19] P. Rosinger, P. Gonciari, B. Al-Hashimi, and N. Nicolici, "Simultaneous reduction in volume of test data and power dissipation for system-on-a-chip," *Electron. Lett.*, vol. 37, no. 24, pp. 1434-1436, 2001.

[20] F. Wolff and C. Papachristou, "Multiscan-based test compression and hardware decompression using LZ77," in *Proc. Int. Test Conf. (ITC'02)*, 2002, pp. 331-339.

[21] M. Knieser, F. Wolff, C. Papachristou, D. Weyer, and D. McIntyre, "A technique for high ratio LZW compression," in *Proc. Design Automation Test in Europe (DATE'03)*, 2003, pp. 116-121.

0-7695-2533-4/06 $20.00 © 2006 IEEE

Methods for Run-time Failure Recognition and Recovery in dynamic and partial Reconfigurable Systems Based on Xilinx Virtex-II Pro FPGAs

Katarina Paulsson, Michael Hübner, Markus Jung, Jürgen Becker
Universitaet Karlsruhe (TH), Germany
http://www.itiv.uni-karlsruhe.de/
{paulsson, huebner, jung, becker}@itiv.uni-karlsruhe.de

Abstract

The rapid development of hardware/software and microelectronic technology enables the realization of more complex systems with new characteristics. These characteristics could lead to further advances in electronic measurement-, control- and regulation systems.

The industrial demands of future electronic systems rely on systems to be fault-tolerant, since the complexity will increase to the point where it is impossible to detect all errors during the design phase.

The ability for a system to recover from a failure requires that incorrect system operation can be detected and analysed during run-time. To achieve this, methods for performing tests of functionalities and components dynamically must be incorporated in the system behaviour during the design phase.

This paper presents methods for efficient on-line failure detection, integrated in a reconfigurable system for execution and test of multiple automotive inner cabin functions. These methods also allow a certain degree of failure recovery, and even make it possible for a system to heal itself from more advanced faults. By exploiting the ability of dynamic and partial hardware reconfiguration, the monitoring can also be performed with less hardware overhead since the monitoring functionalities are configured only when they are required.

Keywords: Automotive, Organic Computing, Reconfigurable Architectures, FPGA, Fault Tolerance

1. Introduction

The advances in reconfigurable hardware now allow system to be dynamic and partial reconfigurable. This brings many new possibilities to the design of adaptive systems, since the system resources can be configured and executed when they are required. When they are no longer required, their chip resources are available for other functions.

In [1], a dynamic and partial reconfigurable system for an automotive application is described. This system replaces the required number of microcontrollers with one Xilinx Virtex-II FPGA, which can be configured on-demand with the required functionalities. The functions being executed are automotive inner-cabin functions.

In modern automotive electronic systems, some of the systems components can be diagnosed by monitoring functionalities. The hardware components required to enable this diagnosis must be provided for statically. Also, redundancy is achieved for certain safety functionalities by integrating not only one but even two extra microcontrollers to ensure correct operation. This method, referred to as triple module redundancy, naturally causes an overhead in hardware but is in some cases necessary.

In future adaptive electronic systems, the complexity caused by the many integrated system components will require that the system is able to detect failures and recover from them by itself and during run-time. Due to the complexity of these systems, it will not be possible to detect all errors during the design phase. In order not to further increase the complexity, failure detection should be done with as little hardware overhead as possible.

Typical errors in this sense could for example be errors in the original design or errors caused by radiation in FPGAs, which is further described in [3].

Also, future systems that can be categorized in the field of Organic Computing [10] must have the ability of Self- healing. Self-healing is one of the self-X characteristics of Organic Computing Systems, among Self- organising, self-optimising and Self-protection. Self-healing is an important characteristic to assure system emergence, which means that the system can overcome failures and disturbances without interrupting its operation.

This paper describes methods for integrating monitoring functionalities in a system with as little hardware overhead as possible. For this purpose, we use the automotive application which was mentioned above as an example application, and exploits the dynamic and partial hardware reconfiguration to perform the necessary tests without wasting expensive chip area required for implementation of a static test function.

The paper is organised in the following way: section 2 gives a short overview of the example application. Section 3 presents the different suggested methods for

0-7695-2533-4/06 $20.00 © 2006 IEEE

detecting failures and what can be done to recover from them and in section 4 the actual implementation and results of the monitoring functionalities are described. Finally the further steps in this area are discussed in section 5.

2. System Overview

The work presented in this paper is based on a dynamic and partial hardware reconfigurable system which was designed for an automotive application. Figure 1 presents an overview of the system. The functionalities that are being executed in the system are inner cabin functions such as the window control, seat control and lightening control. Normally, it is required that each function is executed on separate ECU´s (electronic control units). Due to the high number of electronic control functions in modern cars, the increasing number of microcontrollers is becoming a difficult matter regarding the design and test of such systems.

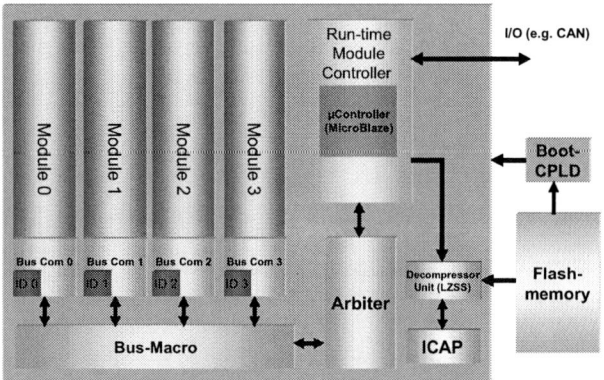

Figure 1. Hardware Reconfigurable System

The system presented in figure 1 however can manage the parallel execution of up to four functionalities and also provides a platform for up to 15 inner cabin functions, by exploiting hardware reconfiguration.

During system operation, a function is configured on the chip whenever it is required. If a function is currently not needed, its resources are available for a different function. In the system described here, the reconfigurable area must be divided into different slots for the different functionalities. This is due to the fact that Xilinx Virtex-II FPGA´s, which were used for implementation of this system, only allows partial reconfiguration of the chip in complete columns. This means that the FPFA must be reconfigured in complete columns from the top of the chip to the bottom. For this reason, the reconfigurable area is divided in slots of equal sizes. This allows the execution of any inner cabin function in any of the different slots. The restriction on performing

reconfiguration in complete columns has been resolved lately by exploiting the read-modify-write mechanism, which is further described in [6].

To manage the complete system a control module is required. For this purpose a Xilinx MicroBlaze soft-core processor [7] is implemented in the static area of the FPGA, and is executing a small run-time system. The control module then controls the communication between the system and the system environment, keeps a record of which slots that are configured with which function and also manages the context load/save mechanism. This mechanism is required to load a function the variables defining its last state after reconfiguration, so that it can start executing in the correct state. Also, before replacing one function with another one, the current state of the function to be replaced must be stored by saving its state variables.

For a more detailed description of the system and its internal communication structure, refer to [1] and [9].

3. Failure Detection and Recovery

Both of the failure detection methods that are described here exploit dynamic and partial hardware reconfiguration. One advantage thereby is the reduction of hardware overhead. Earlier work in this area also exploits reconfiguration of FPGAs, for example in [2], where so called "scrubbing" is investigated. Using this method, the bit streams are continuously configured on the FPGA, without reading the bit streams from the FPGA before to detect possible errors. This method is not suitable for our example application, since it causes a lot of overhead in reconfiguration.

Another interesting method is referred to as built-in-self-test (BIST) and is also further investigated in [4], where all resources of an FPGA are tested to detect failures. This method was designed for detecting permanent errors which occur during system life-time. The FPGA is then divided in different squares, and some of them operate normally while some of them are performing tests. These squares are moving continuously on the FPGA in order to test the complete FPGA. In the work presented in this paper, we do not perform tests on the entire FPGA, but on the functions which are being executed. With our methods we can detect errors that have caused incorrect behaviour and that are caused by for example radiation or bit flips that occurred during reconfiguration. Testing the complete FPGA continuously would interfere with the system operation in our specific application.

The first method which was investigated in this work is an on-chip hardware test bench. A suitable test function is then modelled in a hardware description language for integration in the FPGA-system. Here two possibilities were considered. Either the function can be connected to a

0-7695-2533-4/06 $20.00 © 2006 IEEE   160

module for configuration in the same slot as the function during run-time, see figure 2.

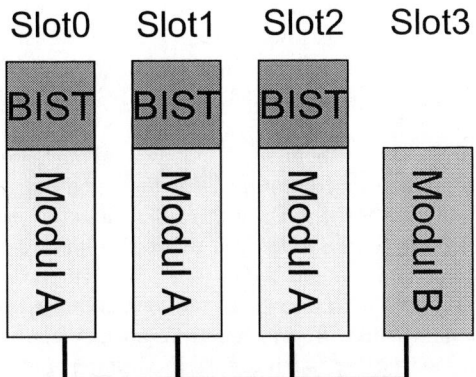

Figure 2. Test bench directly connected to the module

This way the test function would be configured together with the function under test in every reconfiguration.

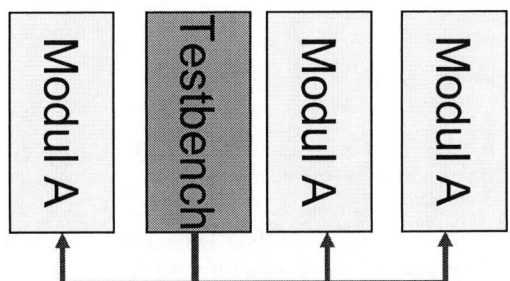

Figure 3. Hardware test bench in separate slot

It is however not required that the function can be tested continuously, and therefore it is also not necessary to configure the test function each time the module itself is configured. That would cause extra overhead in hardware. Therefore, we choose to implement the test function as a separate module for configuration in an available slot whenever a test session is required, see figure 3. For this purpose, the hardware test bench was designed with the same kind of interface as the other modules, for direct connection with the system bus. This way, the same test bench can be used for testing different configurations of the same function. For example, if three window control functions are required and configured, then only one test function is needed to test all of them. Dynamic and partial hardware reconfiguration also allows the test function to be configured whenever there is available chip area. For example, the run-time module can detect when a function was inactive for a certain amount of time, and then choose to exploit the hardware in this slot for testing the other functions. Of course, one important criteria of such a test is that it can be performed without interrupting system operation. Therefore the time required for testing one function must be restricted.

It is also possible to implement the test bench in software in the control module. When integrating the monitoring function in the control module in this specific application system, the tests will require more time. This because of the interface between the control module and the system bus, described in section 4, which will cause extra delay in the communication between the test function and the function under test. One advantage with this kind of test is that it can be performed even if there is no available slot in the reconfigurable area.

A different method for testing on chip functions is to exploit Triple Module Redundancy. Since the FPGA can be configured dynamically, the extra redundancy caused by this method does not need to be provided for statically and we therefore refer to this method as Dynamic Triple Module Redundancy, DTMR. When performing DTMR, two "copies" of the function under test is configured on the chip. These two extra configurations are then loaded with the context data defining the current state of the function. All commands to the specific function are delivered to all three configurations of the function, and the responses produced by the three implementations can then be compared to see if they are the same. A voter implemented in either hardware or software can be used to compare and analyse the results, as well as suggesting possible actions, see figure 4. A failure can be detected if for example two of the slots are producing the same responses, but the third one is delivering a different one. Then actions can be taken to locate the failure and overcome it.

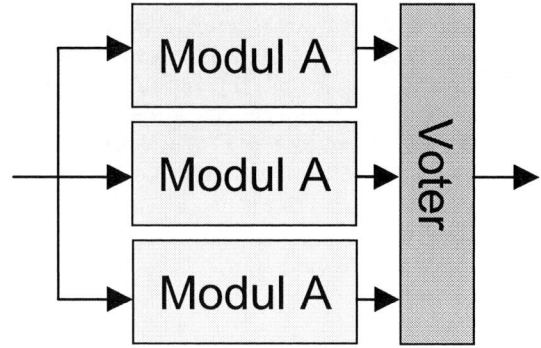

Figure 4. Triple module redundancy

After identifying an error using an on-chip test bench or DTMR, the reason for the failure must be identified. By reconfigure the slot where the error is detected with the same function, it can be detected if the failure occurred during the earlier reconfiguration. If the slot is still delivering incorrect responses, then the slot can be reconfigured with a different mapping of the same function in order to avoid a malfunctioned transistor or a similar hardware problem. If the new configuration is still not functioning properly, then the specific function can be configured in a different slot. In case of a real-time condition, the last suggested action can be performed earlier in the failure recovery. Finally, the faulty slot can be configured with a completely different function of a different size and mapping to see if this overcomes the problem. If not, the slot can be blocked for configuration of any function.

This way, failures due to reconfiguration or malfunctioning hardware components (e.g. transistors) can be detected. The method based on an on-chip test bench can also be used for testing external components, or for monitoring the internal communication interface. For the specific application in this work, the monitoring functionality is only configured whenever a slot is free for reconfiguration. The tests must be performed as quickly as possible to avoid interference with normal system operation.

4. Implementation and Test

The system described here for integration of test mechanisms is implemented on a Xilinx Virtex-II Pro device [5][7][8], which offers two PowerPC 405 cores. Since only one of the Power PC's is required to control the system and to provide an interface to the outside world, the second Power PC is left unused. Gpio instances were used to connect the processor to the arbiter which controls the multiplexed bus system for the communication between the PowerPC and the slots. See figure 5 for a schematic of the complete system. The slots contain hardware models (Finite State Machines, FSM) of automotive inner cabin functions, and can be dynamically reconfigured. The multiplexed bus system also offers the possibility for the slots to communicate which each other. This is important when implementing an on-chip testbench, a so called built-in-self-test (BIST) module, for testing a cabin function which is located in a neighbouring slot. Whenever a slot is available, this slot can be used for configuration of a test module. As the BIST circuit is a FSM itself and resides alone in one slot, it is referred to as the diagnosis module. Figure 6 shows the implemented system as it looks like in the Xilinx FPGA Editor. There the four red boxes represent the different slots with different functions already configured. As mentioned before, such a system would normally not be reconfigurable without destroying the FPGA, due to the lines crossing the different reconfigurable areas. However, in [6] a method which enables a safe reconfiguration of such a system is presented.

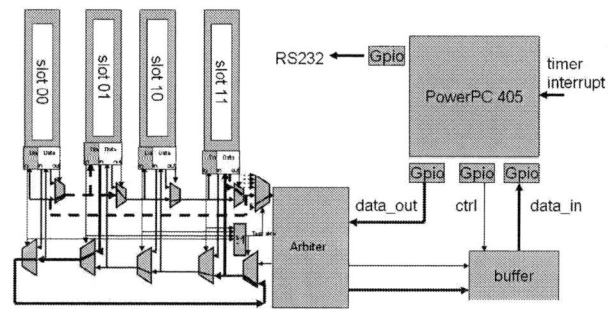

Figure 5. Local Cluster Node

In case of tests with an on-chip test-bench, the diagnosis module (DM) offers stimuli (a test scenario) and analyses the output of the function-under-test to detect errors or confirm correct behaviour. The DM was realized for the window regulator, and can be instantiated in any available slot. The control module sends a message to the DM to start the test and also gives information about where the function-under-test is placed. Then the DM connects to the target slot and starts the test procedure.

Figure 6. System in the FPGA Editor

After connecting to the target function-under-test a "context save" is performed to store the origin state of the target function before testing. Then the window regulator is initialised by the DM with the information that a test is to be performed. A diagnosis procedure can be done in approximately 0.5ms with a clock frequency of 33 MHz. This means that a diagnosis of the automotive function can be carried out within a normal state transition of 4ms, which is the time basis of the implemented automotive applications. In the test scenario, the window function moves up and down with either manual or automatic commands, checking motor movement, scaling and termination control. The expected output is compared to the monitored output in the different stages and the results are stored in diagnosis vector. After the test scenario an additional "context save" is done to get additional information about counter value and scaling. After the test scenario is performed, the original state is stored by performing a "context load" operation. The control module can now claim for the diagnosis results as well as the "context save" data from the DM.

The PowerPC runs at 100 MHz. Its peripherals and the other parts run at 33 MHz and the complete system occupies 4334 slices. The exemplary FSM used in these tests is a window regulator that occupies 254 slices. The corresponding diagnosis module needs 619 slices itself.

This work also covers the implementation of a software test bench for integration in the run-time system. As earlier mentioned, this method slows the test down, since all communication with the function under test must occur over the internal communication bus. Therefore, a test performed in software took approximately 4-5 seconds. Figure 7 presents a comparison between a test bench in hardware and one implemented in software.

	HW Diagnose (onChip Testbench)	SW Diagnose
Time Duration	0.5 ms	4-5 s
Resources	619 slices	320 LOC
Dynamic configurable	yes	no
Interference with system operation	no	yes

Figure 7. Comparison, HW/SW Test benches

As described in section 3, another approach for fault tolerance is triple module redundancy (TMR). For single fault assumption fault masking can be provided. The control module compares the outputs of the three different configurations and if one differs from the other two, the control module will forward the correct result and mask the incorrect.

In the above described system, four slots are available which can be configured with any hardware function and also with three configurations of the same function for performing dynamic TMR. The control module can configure the TMR system on demand or when it detects that enough slots are available. For this purpose the middleware in the control module must contain functionalities such as message triplication, TMR message arrival storage and a watchdog for freezed modules, which do no longer response to incoming commands. See figure 8 for an overview of a possible SW partitioning in a TMR system.

0-7695-2533-4/06 $20.00 © 2006 IEEE

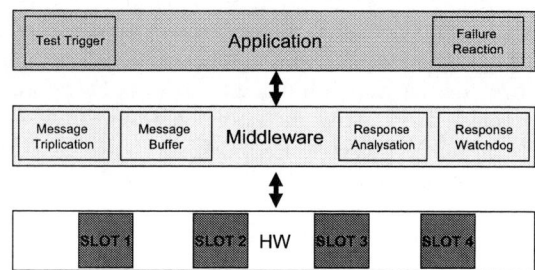

Figure 8. SW Partitioning in a TMR System

The described system was implemented on a Xilinx ML310 board. Debugging was performed by using the RS232 interface. One configured function out of three was corrupted to produce wrong outputs to see if this could be detected by the control module. The implemented system reacted correctly by detecting the failures and masking them.

The build-in-self-test approach was tested in a similar way, one function-under-test was corrupted in several different ways to see if these faults could be detected by the diagnose module. The system reacted as expected and was able to detect the incorrect functional behaviour.

5. Conclusions and Future Work

This paper presents novel implementations of different methods for detecting system failures during system operation. The suggested methods have been designed and tested for a specific automotive application, but are general enough to be integrated in many other different applications.

One of the advantages when using reconfigurable hardware is that the extra hardware required for the test function need not to be occupied constantly, but only when it is required or when enough chip area is available. The suggested test benches can be implemented in hardware or software, whatever is required. However, our tests showed that the hardware test benches speeded up the test time remarkably. Software test benches could be more suitable in cases when the size of the hardware test bench is too large.

Regarding failure recovery, this paper presents some first possibilities. There are however other methods to be investigated. If necessary, complete functions could for example be transferred to a different FPGA platform for execution of the function there. If the last state of the function is lost, the function can be further initialised. This requires however that it is also possible to initialise the component which is controlled by the specific function.

The methods presented in this paper were designed to test inner cabin functions, but they are also well suited for testing other components such as the internal bus structure. They can also be extended for testing external components.

6. References

[1] M. Ullmann, M. Huebner, B. Grimm, J. Becker: "An FPGA Run-Time System for Dynamical On-Demand Reconfiguration", RAW04, Santa Fee

[2] Lima Kastensmidt F., et al.

On the Optimal Design of Triple Modular Redundancy Logic for SRAM-based FPGAs Design, Automation and Test in Europe Conference and Exhibition (Date'05)

[3] Lima Kastensmidt F., Neuberger G., Carro L., Reis R.

Designing and Testing Fault-Tolerant Techniques for SRAM-based FPGAs

[4] Abramovici M., Stroud C. E., Emmert J. M.

Online BIST and BIST-Based Diagnosis of FPGA Logic Blocks, IEEE Transactions on very large scale integration systems, December 2004

[5] Virtex-II Pro Virtex-II Pro X FPGA User Guide, UG012 (v4.0) 23 March 2005

[6] J. Becker, M. Hübner, K. Paulsson, A. Thomas: "Dynamic Reconfiguration On-Demand: Real-time Adaptivity in Next Generation Microelectronics", ReCoSoc2005, Montpellier, France

[7] http://www.xilinx.com/ise/embedded/edk_docs.htm

[8] http://www.xilinx.com/support/sw_manuals/xilinx6/index.htm

[9] M. Huebner, T. Becker, J. Becker: "Real-time LUT-based Network Topologies for dynamic and partial FPGA Self-Reconfiguration", SBCCI04, Porto de Galinhas, Brasil

[10] www.organic-computing.de/spp

Low Power System Design

Design and Analysis of a Low Power VLIW DSP Core

Chan-Hao Chang
Industrial Technology Research Institute
chanhao@itri.org.tw

Diana Marculescu
Carnegie Mellon University
dianam@ece.cmu.edu

Abstract

Power consumption has been the primary issue in processor design, with various power reduction strategies being adopted from system-level to circuit-level. In order to develop a power efficient system, architecture design, compiler optimization, as well as user evaluation must be employed in a unified framework. This paper presents an architecture-level power/performance simulator for a VLIW DSP processor core. Relying on parameterized power models and cycle accurate simulation, it provides fast and accurate power estimation for architecture exploration. Furthermore, the proposed modeling methodology can be used with minimal changes in the evaluation of other VLIW processor cores or for characterizing the efficiency of compiler-driven power efficient transformations.

1. Introduction

Power-related issues have been the primary concerns for battery operated devices, such as communication systems and multimedia systems. A portable device needs to rely on a long battery life to enable the convenience leaded by mobility. To reduce the energy dissipation, processor design is especially important in low power/energy computing. Some well-known power reduction techniques such as resource scaling [2] and clock gating have been adopted for low power architectures, while, in recent years, voltage scaling and multiple voltages [5] have been extensively considered to save the unnecessary energy dissipation. In order to optimize architectures for power budgets and meet performance requirements, designers have to investigate the trade-off between various low power architectures and, by doing so, rely on tools that need to provide fast, but sufficiently accurate estimates for performance, as well as power consumption. Frequent iterations among RTL design, synthesis, gate-level simulation, as well as power analysis is prone to lengthening the design cycle and possibly violating

time-to-market deadlines. In addition to architecture exploration, compiler designers need to rely on a rapid cycle, accurate framework for compile-time optimizations. For a VLIW digital signal processor (DSP) which runs statically scheduled instructions, the scheduler needs to perform instruction ordering, resource scheduling and insert specialized power down instructions to reduce power consumption. Therefore, a framework with fast power estimation characteristics has to be adopted in support of such compiler optimizations. Moreover, user evaluation tools are always essential for processors. Evaluating the power consumption for battery operated applications is especially important.

In this paper, we present an architecture-level power estimation simulator, which could be employed in exploring architecture tradeoffs, compiler optimizations as well as user evaluations of a given architecture. The power estimation engine of this simulator is based on parameterized power models and cycle accurate simulation, while power reduction strategies such as voltage scaling and clock gating are considered to provide various design alternatives. Furthermore, this simulator is also configurable to deal with various instruction set architectures (ISAs), while scalable power models can be employed to estimate the power consumption of soft cores, which could be synthesized according to specific applications.

2. Prior work

In recent years, several simulation-based power analysis engines at instruction-level, RT-level and architecture-level have been proposed. Some of the recent power modeling techniques which play an important role on the accuracy of simulators are described in the sequel.

Power analysis at RT-level based on an application driven methodology [7] has been proposed. Instead of using macro modeling, this work relies on program profiling to extract the parameters from applications. For instruction-level, the power model is derived through the simulation of various instruction sets

0-7695-2533-4/06 $20.00 © 2006 IEEE

[11,14]. However, in general, if the full spectrum of transitions between instructions is considered, the complexity of instruction-level power model increases exponentially with the number of instructions. Moreover, power models derived for specific instruction set architectures do not provide flexibility for varying different architecture parameters.

More recently, some ideas related to architecture-level power estimation have been presented. For example, Wattch [4] and SimplePower [12] provide cycle accurate framework to collect switching activities and employ unit power models for power estimation. Their power models are configured by width and input transitions, but this may not be sufficient to be used for various microarchitectures since the load capacitance may need to be adjusted according to the delay budget used for optimizing individual modules. In power modeling and pre-characterization, several techniques have been introduced, such as RT-level Power Modeling [8,1,17], structure-oriented technique [13] and Dual Bit Type model [9]. In contrast with using the random input assumption, the structure-oriented method considers all possible input transitions and further reduces the complexity of state transition graph (STG) by finding the minimum number of compatible pattern set. The work proposed in the Dual Bit Type model organizes an operand into sign region and uniform region due to their different transition behavior.

3. Framework

This proposed simulator is a cycle accurate design exploration environment based on a VLIW infrastructure and cycle-by-cycle simulation. In order to support multiple instruction issue processing, we treat each instruction way as an independent cluster to provide the flexibility for various architectures, with different number of instruction ways. Moreover, the architecture parameters in this simulator can be adjusted according to architecture specifications. For instance, examples of parameterized units include: the processing elements (PEs) or functional units, memory organization, or the number of pipeline stages. The flexible infrastructure enables designers to explore various design architectures.

The organization of simulator is illustrated in Figure 1. The core of the VLIW simulation engine is the pipeline operation composed of seven stages (where each stage could be skipped or extended to multi-cycle as well). The pipeline operation is implemented in software, so as to accomplish the work of each stage. In each stage, ways are executed sequentially, although the simulator emulates the actual parallel execution

taking place in hardware. Therefore, hazards occurring in shared register files and data memory are correctly detected in this simulator. For instance, there are two instructions issued at the same cycle for loading and storing data to the same memory address respectively, with the load instruction being completed prior to the store instruction. During the pipeline operation, the simulator calculates the transition probabilities of operands and uses pre-characterized power models for obtaining cumulative power numbers. In addition, the switching activity of pipeline registers is also accounted for. In order to simplify the usage of simulator and offer complete cycle accurate information for analysis and debugging, the loader in this simulator is responsible for program loading, while the debugger allows user to trace the state of operation through user interface. To launch the simulator, ISA parameters and benchmarks must be specified to configure memory organization, instruction path and corresponding power models. The operating conditions for frequency, voltage and synthesis frequency can also be selected independently for further exploration. In addition, the simulator can be used to figure out the optimal operation frequency/voltage point according to a given performance constraint. The simulator relies on fast simulation speed, which is more than 1000 times the simulation time of commercial tools. With such design exploration speed-up enabling various ISA evaluations, the extensive architecture analysis covers most useful strategies of power reduction.

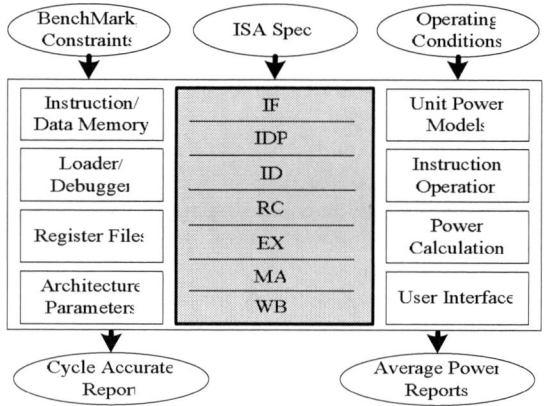

Figure 1. Block diagram of proposed simulator.

4. Power modeling

Our power modeling methodology is based on using Hamming distance between two consecutive input operands for determining switching activity and on the pre-characterization of main RT-level modules. Hamming distance has been extensively applied in dynamic power analysis [6], but it just examines a portion of input vectors since many identical top bits caused by sign extension, and multi-function units increase the complexity of modeling [15]. Therefore, we analyze processing elements further to establish practical and parameterized power models. The parameters for the proposed parameterized power models are the bitwidth of PE, transition probability of input operands, operation voltage, as well as delay budget allocated to each PE. To truly support various architectures, the power models must be configurable for different timing constraints. Since most soft cores can be synthesized according to the characteristics of the target application, this simulator can also be used for soft processor evaluation as described next.

4.1. Parameterized power models

We sort RT-level modules into two classes, combinational and array based components. Combinational components include PEs, fetch, dispatcher and decoder. We extract their power models based on input transition characterization. As shown in Figure 2, we illustrate our analysis methodology through a DesignWare adder [18]. The curves represent different widths for the adder, and each point is derived by simulating 1000 input patterns characterized by a specific transition probability. In fact, each curve could be a look-up table characterizing the power cost for certain bitwidths of the adder. For components that can have variable bitwidths (such as adders, etc.), we employ one curve and the ratio of average power (Figure 3) to determine the power tables for various bitwidths. To be able to reuse these power models, we consider the following factors. In an integrated system, the same type of components may be used to implement units which are characterized by different timing constraints. Furthermore, different architectures and user-defined synthesis speed targets for soft cores may also affect the average load capacitance according to the imposed timing constraints. Therefore, we analyze the influence of timing constraints on load capacitance. In Figure 4, we show the relation between average load capacitance and synthesis-driven timing constraint. The analysis provides not only the effect on the time budget of components but also the limit on the minimum time budget. With this practical information, the simulator could find the maximum operation frequency according to ISA specifications, and our power models could be scaled for various delay budgets.

Most processors provide specific registers for accumulating results. For example, the 40 bit-registers in a 32-bit processor provide 256 accumulation operations without precision penalty. However, a 40-bit adder must be employed to perform the operation.

But the smaller utilization causes that the extra 8 bits to be generally the sign extension (which is all zero or all one). For a two-operand component, we consider the switching activity of most significant bits (MSB's) in three situations: one sign bit transition, two sign bit transitions, and no sign bit transition. In other words, we separate the operands into MSB's and uniform part, and establish the power table for each situation.

The array based components triggered with clock signal, include pipeline registers, memory and register files. We consider the power consumption contributed by clock signal in each array based power model. For instance, a single register with synchronous reset signal is analyzed to estimate the power consumption of pipeline registers. The power model of single register is composed of three coefficients, leakage power, dynamic power and idle power. The dynamic power represents the average power that a register latches different value every rising edge of clock signal, in contrast with dynamic power, the idle power consider only clock swing and no switching on latched bit. The power consumption when there is no switching on input and clock signals determines leakage power. For register files, the models are established in terms of the data switching on cells and read-ports. The power consumption of a register file could be modeled in the following equation:

$$P_{rf} = P_{idle} + \sum_{1 \le i \le n} P_{wd} + N_{cell,i} P_{cb} + \sum_{1 \le i \le n} P_{rd} + N_{r,i} P_{rb}$$

where P_{idle} is the idle power, P_{wd} is the average power of write-decoder while one write-port is active, $N_{cell,i}$ denotes the number of changed bits by a write-port, P_{cb} is the average power associated with one changed cell bit, P_{rd} is the average power of read-decoder while one read-port is active, $N_{r,i}$ is the Hamming distance of two successive vector on a read-port and P_{rb} is the average power associated with one changed bit on a read-port.

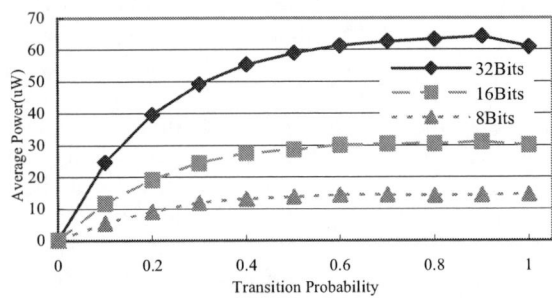

Figure 2. Average power as a function of input transition probability of Synopsys DesignWare adder.

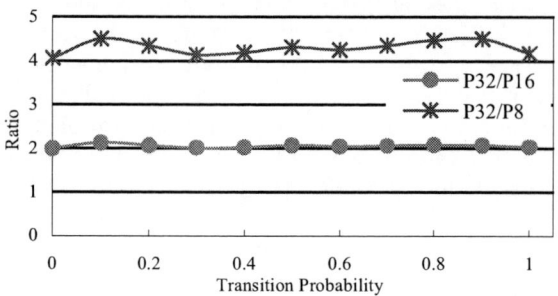

Figure 3. Ratio between the average power of 32-bit, 16-bit and 8-bit adder as a function of input transition probability.

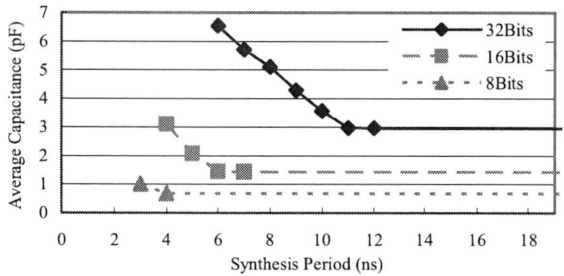

Figure 4. Average capacitance as a function of synthesis specified timing constraint.

4.2. Multi-function components

As they need to rely on resource sharing, existing processors typically use multi-function processing elements to perform various types of operations, while targeting full hardware utilization. In the case of multi-function processing elements, each operation mode may be characterized by different power values, especially when the control signals have transitions. Therefore, using a single set of switched capacitance models may be not sufficient to accurately model a multi-function component. We analyze input operands as well as control signals. The change in control signals makes a multi-function component switch the function from previous instruction to next one. For the purpose of computing switching activities, we treat both directions of transition between two types of functions as one transition. Figure 5 illustrates the average power for DesignWare Adder-Subtractor. The curves represent successive addition, successive subtraction and the switching between addition and subtraction. The average power numbers of addition and subtraction are quite close, but the curve add/sub which represents the switching between addition and subtraction contributes more power consumption. In this case, we establish two power tables for the Adder-Subtractor, one is the mean of curve add and curve sub, and the other one is exactly the curve add/sub. In fact, some functions of a multi-function component could

be incorporated into one type according to the power analysis. Doing so significantly decreases the size of power look-up tables.

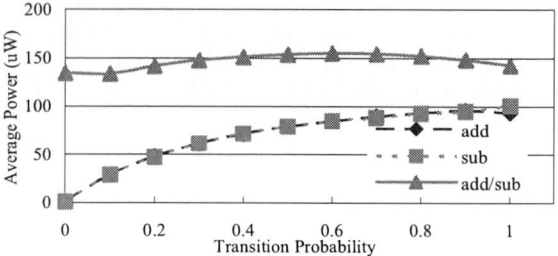

Figure 5. Average power as a function of input transition probability of multi-function component.

5. Experimental results

In this section, we describe how the simulator has been completely validated. Our experimental architecture is the PAC VLIW DSP v1.0 (Table 1) explored by Industrial Technology Research Institute, and the benchmark suite includes a series of DSP algorithm kernels including BDTI benchmarks [3]. The validation has been performed for all modules and the entire processor in terms of detailed power distribution, as well as the influence of synthesis timing constraints. The experiments are performed on a SparcII 500Mhz machine running SunOS 5.9 operating system. ModelSim SE 5.8b and Synposys PrimePower 2002.05 are employed on gate-level simulation and power analysis respectively. Our simulator is more than 1000 times faster than the combination of PrimePower and ModelSim. Table 2 shows the comparison of average power between our simulator and the gate-level power analysis flow for entire DSP core. The DSP core is synthesized at 100MHz. The error is within 10% in all cases. In addition, we extract the power numbers of each component type. Figure 6 illustrates the percentage of register files which consists of accumulation registers, address registers as well as data registers. The average power consumption is about 28% of entire DSP core. We illustrate also processing elements in Figure 7. The power consumption depends on the active elements and input vectors. In Figure 8, we show the power consumption of pipeline registers.

For a soft core, the timing constraints provided for synthesis could be specified according to the target applications. Moreover, the effects of tight timing constraint can be reflected on the critical path. We demonstrate the modeling of MAC element which is on the most critical path of PAC VLIW DSP under different synthesis frequencies in Table 3. The target frequency for synthesis to be used in the simulator is set to 66Mhz, 83MHz and 100MHz. We validate these

0-7695-2533-4/06 $20.00 © 2006 IEEE

benchmarks which have MAC element related instructions. As shown in Table 3, the results show a reasonable accuracy for different timing constraints.

TABLE 1. Configuration of the target architecture

VLIW Processor Core	
Fetch width	> 4 instructions/cycle
Decode width	5 instructions/cycle
Issue width	5 instructions/cycle
Instruction buffer	768 bits
Data path	2 cluster, 1 scalar unit
Functional units	2 ALUs/cluster
	2 MACs/cluster
	1 MUL/cluster
	2 Shifters/cluster
Memory Hierarchy	
Instruction mem	256 KB, 2 ports
Data mem	256 KB, 4 banks
Max. Load/Store	5x32 bits/cycle

TABLE 2. Overall power consumption (TSMC 0.13um)

Benchmark	Cycles	Sim. (uW)	Prime. (uW)	Error (%)
dct	77	10505	11024	-4.7
rand1	80	7162	7360	-2.7
rand2	89	15748	16718	-5.8
fft64	2293	10018	10936	-8.4
bkfir	1779	9602	10655	-9.9
conv	3365	10526	10728	-1.9
iirn	4816	10619	10696	-0.7
fir2d	29964	10553	10663	-1.0
comult	280	10676	10714	-0.3
dot	338	10414	10572	-1.5
m1x3	1609	9826	10106	-2.8
comud	499	10256	10642	-3.6

TABLE 3. Results for various synthesis frequencies

Bench-mark	66MHz(uW)		83MHz(uW)		100MHz(uW)	
	Sim.	Prime.	Sim.	Prime.	Sim.	Prime.
dct	241	213	265	248	338	348
rand1	260	265	286	246	364	345
rand2	817	944	899	1035	1144	1139
fft64	112	108	123	121	157	141
bkfir	321	311	353	356	449	479
conv	8.9	9.8	11.4	11.2	13.9	13.3

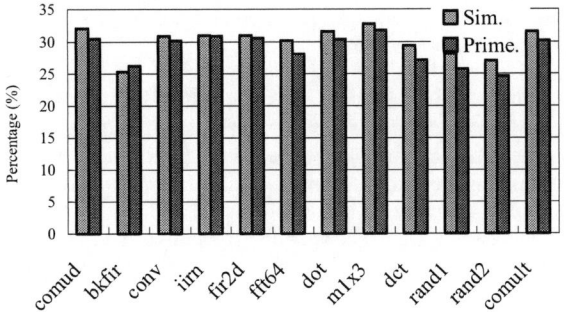

Figure 6. Percentage of register files power from total power budget.

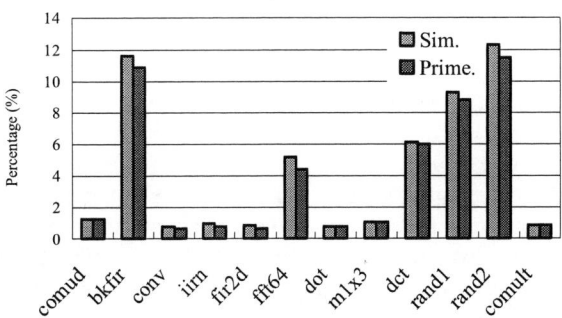

Figure 7. Percentage of processing elements power from total power budget.

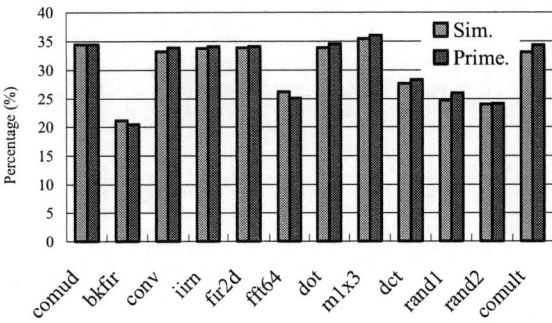

Figure 8. Percentage of pipeline registers power from total power budget.

5.1. Case study of clock gating

We have also performed a detailed analysis of the power distribution across various datapath modules (Figure 6-8). As seen in Figure8, approximately 35% of the total power consumption is contributed by pipeline registers, while register files consume 30% of the total power budget (Figure 6).

For benchmarks that have fewer parallel instructions because of longer data dependency chains, many resources turn out to be redundant and consume power in multi-ways processors. Clock gating is a well-known technique for pipeline power reduction [16,10] especially for multi-issued processors. We employ this technique in our simulator to investigate the influence of clock gating on pipeline registers and register files (Figure 9). The three levels we consider are cluster base (CB), register file (RF) as well as pipeline register (PR). CB is a coarse-grain clock gating, the entire data path is isolated by clusters and one scalar unit, the clock signal of specific cluster is gated while there is no valid instruction traveling in it. RF focuses on register files. Each cluster has one address register file, one accumulation register file and two data register files. We detect the instructions in read-operand stages and write-back stages in order to determine the idleness of register file. Regarding PR,

0-7695-2533-4/06 $20.00 © 2006 IEEE

each instruction in data path is analyzed to determine if there will be any valid instructions to be transferred through pipeline registers. For instance, the clock signal of the pipeline register between execution stage and memory-access stage is gated while the instruction in execution stage is invalid. In addition, the corresponding leakage power is considered when clock gating is used. This simulator provides valuable data through fast simulation. The estimates illustrate the percentages of power saving and the comparison of different strategies. Given the efficiency with which they have been obtained, they could determine alternatives of architecture optimizations without frequent gate-level analysis flow.

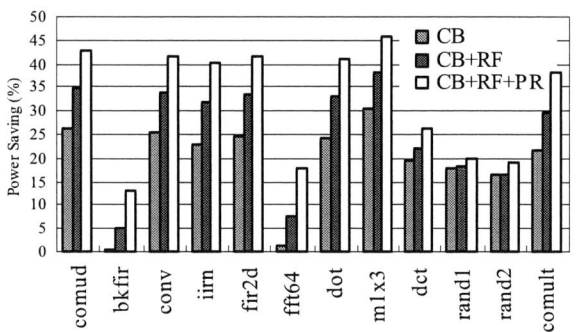

Figure 9. Power saving due to clock gating.

6. Conclusion and future works

The processor design is getting more complex, with energy consumption being now the primary design issue for portable devices. Iterating between synthesis, gate-level simulation and power analysis is very expensive and may lead to time-to-market violations. We have presented a simulator, which is at the heart of an architecture-level power/performance estimation framework providing fast simulation and reasonable accuracy. For user evaluation, our simulator relies on the parameterized power models to enable hard core and soft core power/performance evaluation.

Future work for this simulator will include power analysis for the memory subsystem. In general, the memories in processors are macro blocks, thus our memory models will be established according to the characteristic of memory macros. Moreover, our power models will be augmented to include information about more aggressive technologies (90 and 65nm).

7. References

[1] A. Bogliolo, R. Corgnati, E. Macii and M. Poncino, "Parameterized RTL Power Models for Soft Macros," *IEEE Transactions Very Large Scale Integration Systems*, vol. 9, pp. 880-887, Dec. 2001.

[2] A. Iyer and D. Marculescu, "Power Aware Microarchitecture Resource Scaling," *Design Automation and Test in Europe Conference*, pp. 190-196, Mar. 2001.

[3] Berkeley Design Technology Inc., "Evaluating DSP Processor Performance," http://www.bdti.com.

[4] D. Brooks, V. Tiwari and M. Martonosi, "Wattch: A Framework for Architecture-Level Power Analysis and Optimizations," *Proceedings of the 27th International Symposium Computer Architecture*, pp. 83-94, Jun. 2000.

[5] D. Marculescu, "Power Efficient Processors Using Multiple Supply Voltages," *Workshop on Compilers and Operating Systems for Low Power*, Oct. 2000.

[6] G. Jochens, L. Kruse, E. Schmidt and W. Nebel, "A New Parameterizable Power Macro-Model for Datapath Components," *Design, Automation and Test in Europe Conference and Exhibition*, pp. 29-36, Mar. 1999.

[7] H. Cheng-Ta and M. Pedram, "Profile-Driven Program Synthesis for Evaluation of System Power Dissipation," *Proceedings of the 34th Design Automation Conference*, pp. 576-581, Jun. 1997.

[8] J. Costa, J. Monteiro, L. M. Silveira and S. Devadas, "A Probabilistic Approach for RT-Level Power Modeling," *Proceedings of 6th IEEE International Conference Electronics, Circuits and Systems*, vol. 2, pp. 911-914, 1999.

[9] J. Y. Lin, W. Z. Shen and J. Y. Jou, "A Power Modeling and Characterization Method for Macrocells Using Structure Information," *IEEE/ACM International Conference Computer-Aided Design*, pp. 502-506, Nov. 1997.

[10] L. Hai, S. Bhunia, Y. Chen, T. N. Vijaykumar and K. Roy, "Deterministic Clock Gating for Microprocessor Power Reduction," *The Ninth International Symposium on High-Performance Computer Architecture*, pp. 113-122, Feb. 2003.

[11] M. Sami, D. Sciuto, C. Silvano and V. Zaccaria, "An Instruction-Level Energy Model for Embedded VLIW Architectures," *Computer-Aided Design of Integrated Circuits and Systems*, vol. 21, pp. 998-1010, 2002.

[12] N. Vijaykrishnan, M. Kandemir, M. J. Irwin, H. S. Kim and W. Ye, "Energy-driven integrated hardware-software optimizations using SimplePower," *Proceedings of the 27th International Symposium Computer Architecture*, pp. 95-106, Jun. 2000.

[13] P. E.Landman and J. M. Rabaey, "Activity-Sensitive Architectural Power Analysis," *IEEE Transactions Computer-Aided Design of Integrated Circuits and Systems*, vol. 15, pp. 571-587, Jun. 1996.

[14] P. Stanley-Marbell and M. S. Hsiao, "Fast, Flexible, Cycle Accurate Energy Estimation," *International Symposium Low Power Electronics and Design*, pp. 141-146, Aug. 2001.

[15] S. Haga, N. Reeves, R. Barua and D. Marculescu, "Dynamic Functional Unit Assignment for Low Power," *Design, Automation and Test*, pp. 1052-1057, 2003.

[16] S. Manne, A. Klauser, and D. Grunwald, "Pipeline Gating: Speculation Control for Energy Reduction," *Computer Architecture*, pp. 132-141, Jul. 1998.

[17] S. Theoharis, G. Theodoridis, P. Merakos and C. Goutis, "Accurate data path models for fast RT-level power estimation," *IEE Proceedings Computers and Digital Techniques*, vol. 147, pp. 209-214, Jul. 2000.

[18] Synopsys, Inc., http://www.synopsys.com.

0-7695-2533-4/06 $20.00 © 2006 IEEE   172

High-Performance Noise-Robust Asynchronous Circuits[†]

Pankaj Golani and Peter A. Beerel

Department of Electrical Engineering-Systems

University of Southern California, Los Angeles, CA 90089

{pgolani,pabeerel}@@usc.edu

Abstract—

This paper presents the development of a prototype high-performance asynchronous standard-cell library based on the static single-track full buffer family. It focuses on the design choices and challenges that mitigate sensitiveness to noise, including transistor sizing and wire spacing rules. Post-layout simulation results demonstrate its robustness to noise while achieving a peak cycle time of 5.7 FO4 delays.

Index Terms—Asynchronous, library, robustness to noise, worst-case crosstalk environment, cycle time, overlap period.

I. INTRODUCTION

Driven by overwhelming design-time constraints, standard-cell based synchronous design styles supported by mature CAD design tools and a largely automated flow dominate the ASSP and ASIC market places. As device feature sizes shrink and process variability increases, however, the reliance on a global clock becomes increasingly difficult, yielding far-from-optimal solutions. Because standard-cell designs use very conservative circuit families and are often over-designed to accommodate worst-case variations, the performance and power gap between full-custom and standard-cell designs continuously widens [10]. Our research demonstrates that it is possible to dramatically narrow this gap using asynchronous techniques by showing that conventional standard-cell techniques with an asynchronous cell library can produce very high-performance power-efficient circuits [1].

Previously, we created a single-track full-buffer STFB standard-cell library in TSMC 0.25µm technology and a associated back-end place and route flow [2]. The library enables chip design with measured frequencies of over 1.2 GHz [2][4], 4X faster than the typical synchronous ASICs. In addition, the circuits are 50% smaller, are over 2X faster, and consume a fraction of the power per operation than the quasi-delay-insensitive asynchronous templates of Caltech and Fulcrum Microsystems [14]. The basic concern of the STFB circuit family is that the communication wires can be tri-stated for some period of time with only a small staticizer fighting leakage and crosstalk noise. While effective for 0.25µm, the

STFB technology noise margin in deeper submicron technologies may be too low.

The new challenges addressed in this paper are the development of a prototype high-performance *static* STFB library (SSTFB) for the IBM 180nm technology that guarantees the communication wires are always actively driven and is more robust to process variability and crosstalk noise.

The paper is organized as follows. Section II provides background on STFB and the SSTFB templates along with crosstalk noise background. Section III presents the factors that impact crosstalk of our SSTFB circuits and the crosstalk model we adopted to evaluate our design tradeoffs. Section IV presents the trade-off analysis of the various sub-components of the SSTFB cell templates. Section V presents the layout methodology and post-layout simulation results, followed by conclusions presented in Section VI.

II. BACKGROUND

This section begins by describing various implementations of asynchronous channels. It then describes the single track full buffer (STFB) template and the proposed static STFB designed for higher robustness in deeper submicron technologies.

A. Asynchronous channels

An asynchronous channel is a bundle of wires and a protocol to communicate data across the wires from one pipeline stage (the sender) to another one (the receiver). Figure 1 shows a 1-of-N single track asynchronous channel

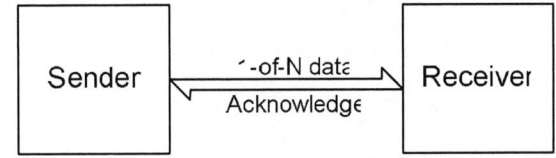

Figure 1. A 1-of-N single track asynchronous channel

In this type of channel, the receiver detects the presence of the token, as in the 1-of-N channel, but is also responsible for consuming it (by resetting all the wires). The sender detects that the token was consumed before sending another token. Notice that validity of the data is embedded in the data itself. Consequently, this type of channel has much less timing assumptions than the single-rail-encoded bundled-data channel

---

[†] This work was partially supported by NSF ITR Award No. CCR-00-86036.

0-7695-2533-4/06 $20.00 © 2006 IEEE

that rely on a separate request line.

Related designs include that from Berkel et al. [6] who proposed single-track handshake circuits to control medium-grain bundled-data pipelines. Sutherland et al. [12] later developed faster single-track GasP circuits to control fine-grain bundled-data pipelines. Nyström [13] also proposed a dual-rail (1-of-2) single-track template based on self-resetting pulsed-logic circuits like GasP but which requires significantly more transistors and is significantly slower.

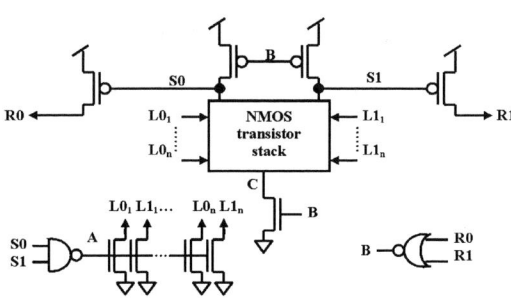

Figure 2. Typical STFB transistor level diagram

STFB templates

Figure 2 shows a typical STFB cell's block diagram. When there is no token in the right channel (R) (the channel is empty), the Right environment Completion Detection block (RCD) asserts the "B" signal, enabling the processing of a next token. In this case, when the next token arrives at the left channel (L) it is processed lowering the state signal "S", which creates an output token to the right channel (R) and causes the State Completion Detection block (SCD) to assert "A", removing the token from the left channel through the Reset block. The presence of the output token on the right channel resets the "B" signal which activates the two PMOS transistors at the top of the N-stack, restoring "S", and deactivates the NMOS transistor at the bottom of the N-stack, as shown in Figure 2, disabling the stage from firing while the output channel is busy.

The cycle time of the STFB template can be as low as 6 transitions with a forward latency of 2 transitions. This implies that the peak pipeline throughput can be achieved with just three stages per token, which allows high pipeline occupancy and the implementation of high performance small rings. The full-buffer characteristic of STFB stage refers to each stage capacity of holding one token. The STFB template is very flexible and can be expended to different functionalities including logic gates (e.g., AND, OR, XOR) gates, functional elements (e.g., half and full-adders), and non-linear pipeline templates (e.g., forks, joins, splits and merges) [2][3][4].

B. Static STFB template

Because the communication wires can be tri-stated for some period of time with only a small staticizer fighting leakage and crosstalk noise, the STFB technology noise margin in deeper submicron technologies may be too low. In particular, a cross-coupling noise event on a long-wire when it is tri-stated can either create a new token or remove a token from the system,

causing system failure (often in the form of a deadlock). Moreover, in smaller geometries leakage currents may become so high that the staticizers would need to be up-sized to the point that they cannot be easily over-powered.

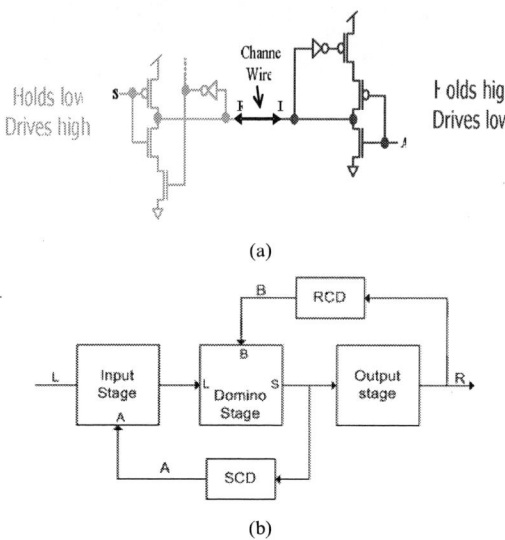

(a)

(b)

Figure 3. (a) Static STFB driver circuitry which improves robustness to noise. (b) Typical SSTFB block diagram

For this reason, Ferretti et al. [2] proposed a new STFB family called static STFB in which the channel wires are always actively driven with the modified driver circuits illustrated in Figure 3(a). Notice that after the sender drives the line high, the receiver is responsible for actively keeping the line high until it wants to drive it low. After the receiver drives the line low, the sender is responsible for actively keeping the line low until it wants to drive it high. The line is always statically driven and no fight with staticizers exists. This means that the hold circuitry can be sized to what ever strength is needed and a tradeoff between performance/power/area and robustness to noise is created. Moreover, the inverters in the hold circuitry can be skewed such that they turn on early creating overlap between the driving and hold logic. This overlap avoids the channel wire being in a tri-state condition thus making the circuit family more robust to noise. The overlap helps ensure that the channel wires are always driven close to the power supplies further increasing noise margins.

With this circuit technology there are six ways to combat noise and process variations 1) increase the size of the hold transistors 2) increase the minimum separation between wires in the place and route flow 3) decrease the maximum allowable length of any route 4) shield long communication wires 5) skew the hold inverters to create more time overlap between the driving and hold transistors and 6) reduce the power supply. This paper analyzes many of these trade-offs.

III. MODELING OF WORST-CASE CROSSTALK ENVIRONMENT

In this section we discuss various factors that affect cross talk and the distributed RC model that we adopted as our model of the worst case crosstalk environment. The quantity

0-7695-2533-4/06 $20.00 © 2006 IEEE   174

of noise can be measured in many ways including peak noise voltage or area under the noise curve. In this paper we chose the former.

A. Factors affecting worst-case crosstalk

To properly model and quantify the worst case crosstalk noise on a static STFB interconnect we first study the impact of number of factors on the severity of crosstalk effects.

1) Number of aggressors: In order to model the worst-case noise environment we studied the impact of noise on a quiet victim as we increase the number of aggressors around it. Figure (a) illustrates that the crosstalk noise setup. In particular, as the number of aggressors increases, the coupling capacitance per victim and aggressor pair is proportionally decreased. Note that senders are assumed to be on the left and all wires are modeled with a Π model, the victim being a multi-stage model with resistance between each coupling capacitor. Figure 4(b) shows the simulation results which indicate that as we increase the number of aggressors, the voltage bump asymptotically increases. This asymptotic behavior can be partially attributed to the fact that the resistance related to the aggressors is increasingly being dominated by the resistance of their driver while the total coupling capacitance remains constant.

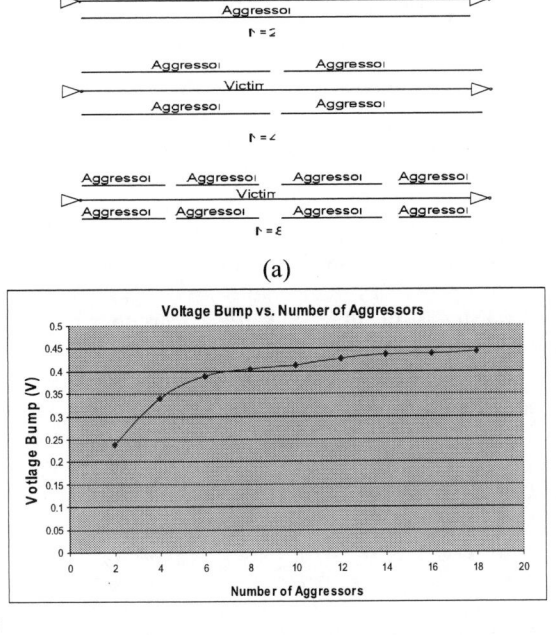

(a)

(b)

Figure 4. (a) Various interconnect topologies (b) its effect on the crosstalk noise on the victim.

2) Phase alignment: To determine worst-case noise, the phase alignment of aggressors switching times should maximize the induced noise on the victim. We analyzed a quiet victim with two capacitive coupled switching neighbor nets (aggressors). We changed the input arrival times at the aggressors and analyzed the magnitude of the noise pulse on the victim. The results, as shown in Figure 5, show that the maximum crosstalk noise occurs when the aggressors are aligned. We attribute this to the fact that the two aggressors

had equal drive strength and note that when the drive strengths are different the maximum-noise alignment occurs with a phase difference.

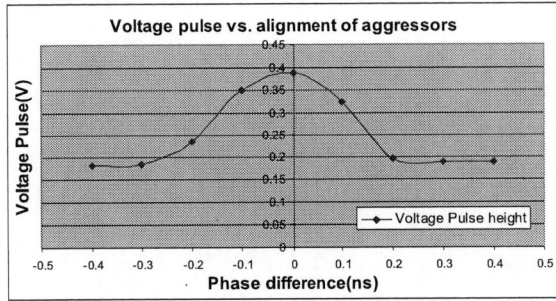

Figure 5. Impact of phase difference on crosstalk noise.

3) Location of drivers: To further explore the impact of the location of the coupling on the magnitude of the noise, we experimented with two nets one long and one short with equal driver strengths. We selected the longer net as the victim and altered the location of the coupling to the shorter aggressor net as shown in Figure 6(a). The magnitude of the coupling capacitance was fixed. Here as we increase x, the location of the aggressor sender driver moves towards the victim receiver. As shown in Figure 6(b), as the coupling location moves towards the victim receiver, the crosstalk noise at the receiver modestly increases. This can be attributed to the decreasing shielding effect of the resistance between the coupling point and receiver.

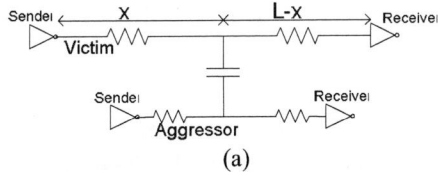

(a)

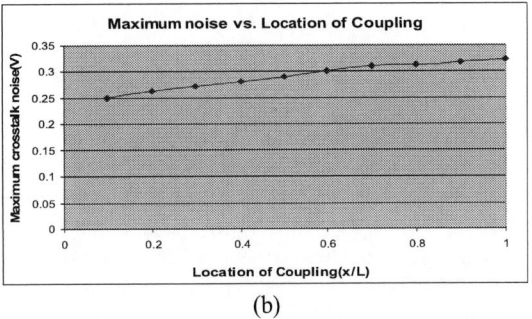

(b)

Figure 6. (a) Set up to study the impact of location of coupling on victim net (b) Noise at the receiver vs. location of coupling

B. Worst-case crosstalk model

Our goal is to select a worst-case crosstalk model that is relatively easy to understand but still captures all factors discussed in Section II-A. We adopt the distributed circuit model illustrated in Figure 7. Unlike simpler lumped and Π circuit model, it can model the impact of location of coupling which have particular important to bi-directional nets in sub-nanometer wires.

We assume that for a long victim there will be at most 8 aggressors near it. In Figure 7 we show only 4 aggressors but the coupling capacitances for each aggressor-victim pair (Cm)

has been doubled thus in effect modeling 8 aggressors. We gave all aggressors identical drive strength and thus aligned their input arrival times to maximize crosstalk. This distributed model also captures the realistic impact of the location of coupling factor by spreading the aggressors over the entire wire length.

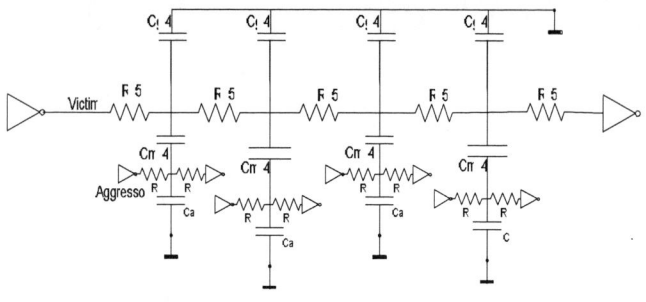

C_t = wire to ground capacitance for a 400 um long wire R = Line resistance for 400um long wire
C_i = wire to ground capacitance for 100 um long wire R = Line resistance for 100 um long wire
$Cm = 2$ couling capacitance between two 400 um long wire

Figure 7. RC network used to model the noise environment

IV. SPEED ROBUSTNESS TRADE-OFFS

In this section we describe the sizing strategies that we adopted for major blocks such as input drivers, output drivers and domino logic stage. We also present the trade-offs we analyzed in order to achieve noise robustness along with desired speed.

A. Input and Output drivers

The output driver is a sub-cell referred to here as SSTFB2_OUT and is utilized in all SSTFB cells [2]. It utilizes three PMOS transistors utilized to restore the state input ("S") high, as illustrated in Figure 8. If the output channel is empty, the "B" signal is high, "R" is low, and "NR" is high. During this time, M6 alone fights leakage and holds "S" high. At the same time, M2 and M3 which act as a keeper holds "R" low. When "S" is driven low, the output driver PMOS transistor M1 drives the output "R" high, which makes the inverter INV_HI drive "NR" low, deactivating M3 and activating M4. The RCD will also make the "B" signal fall activating M5.

The input stage is a sub-cell (SSTFB2_INP) and is utilized in all SSTFB cells and is shown in Figure 8. The initial empty channel condition is that "A" is low, M8 is on, M9 is off. When L is driven high by the left environment, the inverter INV_LO turns M9 on, M9 and M8 continues to hold the input L high. The SCD will turn "A" high, activating M7 and deactivating M8 thus resetting the input.

B. Overlap protocol

As IC process technology continues to scale, on-chip variations in process parameters plays an important role. In particular, the switching point of the inverter (INV_LO) in Figure can be different from the switching point of the output driver of the previous stage. This can cause the output driver of the previous stage to turn off before INV_LO can turn the keeper on, leading to a tri-stated wire during the transition. In order to avoid tri-stating the wires, the two inverters in the output driver stage and input driver stage INV_HI and INV_LO are skewed to implement an overlap protocol. In particular, INV_LO is skewed to lower the switching threshold voltage and INV_HI is skewed to higher the switching threshold voltage.

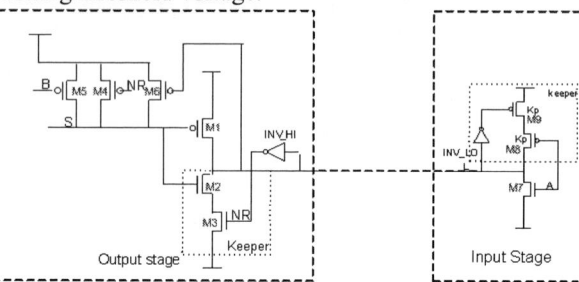

Figure 8. Implementation of the input and output drivers that yield an overlap protocol

C. Domino Logic stage

Although the IBM 0.18μm process allows somewhat smaller transistors (minimum width 0.4μm), we choose the minimum NMOS transistor width of the domino logic to be 0.5μm and minimum PMOS width to be 1μm. Also we made sure that the strength of the N-stack is almost twice of the minimum size NMOS. For example, for a 2-transistor logic stack, the width of each transistor is 2μm.

D. Speed vs spacing vs noise trade-offs

Crosstalk noise in a circuit can have two affects. Switching activities on the neighboring aggressors can cause an unwanted transition on a quiet victim. In addition, simultaneous switching of the aggressor and victim can either speed up or slow down the transition on the victim [8][9]. The slow down of the transition can cause significant trouble in synchronous design as it may lead to catastrophic setup time violation on the D flip flops, but in asynchronous design this will only lead to a slow-down in performance. We therefore focus on the crosstalk effects on a quiet victim as this can more likely lead to functional errors. Our analysis is divided into two cases:

The first case is when the victim is at logic state 1 and there is a transition on aggressors from 1→0. The fall transition on the aggressors will create a voltage pulse on the victim. As we increase the driver size (Wn) on the aggressors keeping the victim drivers and keeper sizes constant the crosstalk noise (voltage pulse height) will increase. Here we have a tradeoff between the speed of the falling transition and the amount of crosstalk noise for different spacing between the aggressor and victim. As observed in Figure , stronger drivers cause more crosstalk noise but faster aggressor cycle times.

0-7695-2533-4/06 $20.00 © 2006 IEEE

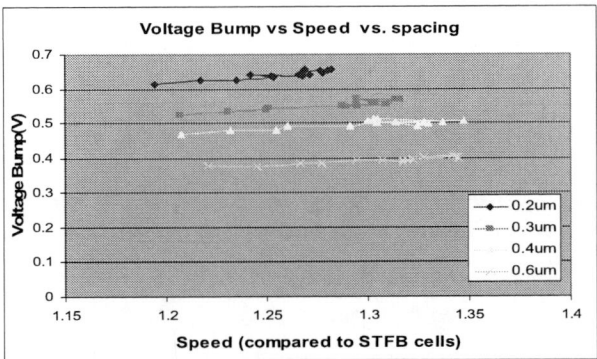

Figure 9. Analysis of voltage pulse height vs maximum driver size measured as aggressor cycle time with different wire spacing rules.

The second case is when the victim is at logic state 0 and there is a transition on aggressor from $0 \rightarrow 1$. The rise transition on the aggressors will create a voltage pulse on the victim. As we increase the driver size (Wp) on the aggressors keeping the victim drivers and keeper sizes constant, the voltage pulse height increases. Here we have a tradeoff between the speed of the rising transition and the amount of crosstalk noise for different spacing between the aggressor and victim. As observed in Figure 10, stronger drivers cause larger voltage pulse height but faster transition speeds.

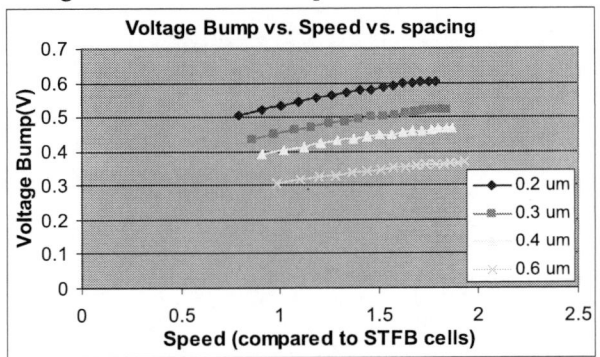

Figure 10. Tradeoff between speed and the voltage pulse height with different spacing

E. Overlap period vs. Keeper sizes trade-offs

The keeper in the input and output drivers are responsible to hold the wire at its current value during idle periods. The keeper in the input stage is responsible to hold the wire high, while the keeper in the output block is responsible to hold the wire low. We designed the library cells such that the maximum wire load they can drive is 400μm long. Consequently, the keeper should be strong enough to hold a 400μm long wire in a worst-case crosstalk environment. The stronger the keeper, the more robust the design is with respect to noise when the victim line is quiet. But as we increase the keeper size, the *overlap period,* defined as the time difference between the output of INV_LO (INV_HI) goes low (high) 50% and S0 (A) of the previous stage (next stage) starts to go high (low) 50%, decreases, thus leading to the possibility that the wire floats, making the transition time susceptible to noise. The tradeoff between keeper sizes and amount of overlap is shown in Figure 11.

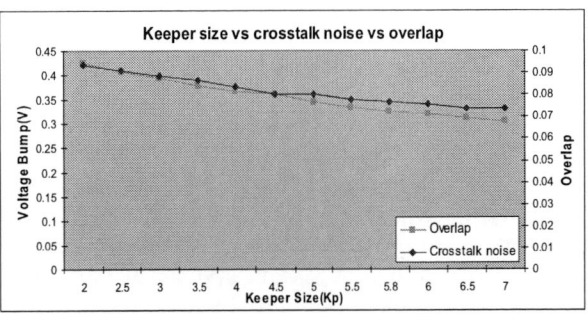

Figure 11. Keeper size vs voltage bump vs overlap

V. LAYOUT & LIBRARY EVALUATION

Our prototype SSTFB standard cell library contains 14 cells along with a variety of sub-cells used to simplify the development of new library cells. For each cell we currently have four views: *functional* views contain the behavioral description of the cell in Verilog HDL, *schematic* views contain the transistor level implementation of the cell, *layout* view, and finally its *symbol*. For proper functioning of the cells in noisy environment we put a constraint that worst-case crosstalk noise cannot exceed 0.4V. Also for longer wires greater than 100μm we decided to have spacing of 3 times of minimum spacing allowed, similar to what we have seen in commercial libraries. So from the trade-off curves in Figures 8 and 9 we picked a point at which speed is 1.2 times the speed of STFB cells.

A. Layout

The entire library was laid out using Cadence Virtuoso Layout Editor. Hierarchy was used extensively throughout the library to save time and reduce errors. The layouts thus are a combination of sub-cells and cell specific transistors. The cells have been designed for abutment and have a fixed height of 6.04μm with a minimum vertical and horizontal pitch of 0.56um. The cell widths are variable and are a multiple of the horizontal pitch. This accommodates 8 horizontal routing tracks over the cell for internal routing. The ground and power rails are 0.8μm wide. Metal layers 1, 2 and 3 were used for the internal routing leaving layers 4, 5 and 6 primarily for global routing and the power grid. For some of the complex logic cells such as the adder cells, metal 4 was used sparingly. Substrate and n-well contacts are placed on the ground and power rails respectively

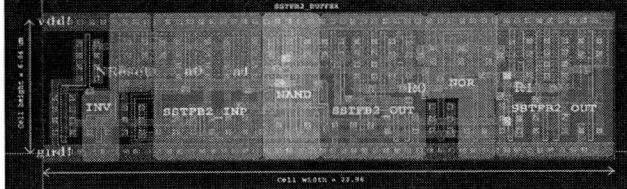

Figure 12 SSTFB buffer cell layout, dimensions, and cell hierarchy

B. Performance of a 7-stage linear pipeline

To evaluate the performance of this library we analyzed the performance of 7-stage linear pipeline. The pipeline structure we simulated consisted of 5 buffers, a bit generator and a bucket as shown in Figure 13.

0-7695-2533-4/06 $20.00 © 2006 IEEE   177

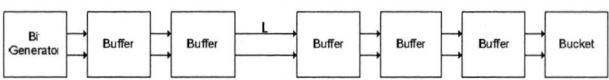

Figure 13. 7-stage linear pipeline

The throughput of this pipeline is dictated by the critical cycle time arising from the one buffer driving another buffer with dual-rail interconnect wires of length L. The pipelines worked with L over 1mm even though transistor sizing of these cells targeted 400µm. The throughput of the pipeline varies from 2.27GHz for L=10µm, 1.92GHz for L=400µm, and 1.69GHz for L=900µm. Given that a FO4 delay in IBM 0.18µm technology is 77.6ps, the performance of SSTFB pipelines thus range from 5.7 to 7.6 FO4 delays.

To analyze the effect of dynamic voltage scaling on the performance and robustness to noise of the circuit we measured the performance and power consumption of this pipeline for L = 400µm as a function of supply voltage as shown in Figure 14.

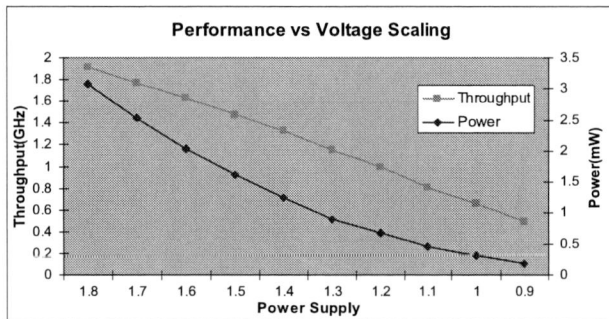

Figure 14. Throughput vs. Voltage scaling

For further verification we simulated the pipeline for L=400 µm in noisy conditions by introducing an aggressor pipeline coupled to the long wire as shown in Figure 7. We first swept the start time of the aggressor pipeline while keeping the victim pipeline quiescent. Both the victim and aggressor pipelines functioned as expected in all simulations with a worst-case voltage bump on the victim lines of 0.32V. We then swept the start time of the aggressor pipeline with the victim pipeline running at peak throughput. Due to crosstalk noise, the pipeline performance of the victim pipeline varied from 2GHz to 1.45GHz. One can contribute this robustness to noise to the fact that the overlap period in all cases was no less then 80ps, meaning that the wire is never in the floating state.

Moreover, as we reduce the voltage supply from nominal 1.8V to 0.9V the circuit still operates properly and the voltage bump reduces from 0.32V to 0.2V.

VI. CONCLUSIONS

This paper introduces a prototype SSTFB standard-cell library which facilitates a conventional back-end flow for ultra-high-performance asynchronous blocks. The library was designed in IBM 0.18u process and various trade-offs for mitigating the crosstalk noise were identified and analyzed for optimum transistor sizing and wire spacing rules. Post layouts simulations show that the templates can operate at a peak throughput of over 1.7 GHz. Also voltage scaling to 66% of V_{dd} yields significant decreases in noise, quadratic savings in

power, and over 1GHz buffer cycle times.

To illustrate applications of this standard cell library we are currently implementing a high-performance asynchronous Turbo decoder using SSTFB standard cells. Preliminary results show that, for the same throughput, the asynchronous turbo decoder is actually 2-3 times smaller then its synchronous counterpart because it needs far less parallelism. Moreover, preliminary estimates suggest it has comparable power consumption.

REFERENCES

[1] P. A. Beerel. Asynchronous Circuits: An Increasingly Practical Design Solution, in *ISQED'02* March. 2002.

[2] M. Ferretti. Single-track Asynchronous Pipeline Template, Ph.D. Thesis, University of Southern California, Aug, 2004.

[3] M. Ferretti, R. O. Ozdag, P. A. Beerel. High Performance Asynchronous ASIC Back-End Design Flow Using Single-Track Full-Buffer Standard Cells. *ASYNC'04*, April, 2004.

[4] M. Ferretti and P. A. Beerel. Single-Track Asynchronous Pipeline Templates using 1-of-N Encoding, *DATE'02*, Mar. 2002.

[5] P. Golani and P. A. Beerel. Back-Annotation in High-Speed Asynchronous Design, *PATMOS*, Sep. 2005.

[6] R. O. Ozdag and P. A. Beerel, High Speed QDI Asynchronous Pipelines, *ASYNC'02*, April 2002.

[7] K. van Berkel, and A. Bink, Single-Track Handshake Signaling with Application to Micropipelines and Handshake Circuits, *ASYNC'96*, pp: 122–133.

[8] F. Dantu and L.T. Pileggi, Calculating Worst-Case Gate Delays Due to Dominant Capacitance Coupling, *DAC'97*, June 1997

[9] K. L. Shepard, Design Methodologies for Noise in Digital Integrated Circuits. *DAC'98*, June 1998.

[10] D. G. Chinnery and K. Keutzer, Closing the Gap between ASIC and Custom: Tools and Techniques for High-Performance ASIC Design, Kluwer Academic Publishers, ISBN 1-4020-7113-2, May 2002.

[11] R. O. Ozdag. Template-Based Asynchronous Circuit Design, Ph.D. Thesis, University of Southern California, Nov, 2003.

[12] I. E. Sutherland, and S. Fairbanks. GasP: a Minimal FIFO Control, *ASYNC'01*, March 2001.

[13] M. Nyström and A. J. Martin. Asynchronous Pulse Logic. Boston: Kluwer Academic Publishers, 2001.

[14] U. Cummings, A. Lines, and A. Martin. An Asynchronous Pipelined Lattice Structure Filters, *ASYNC'94*, March 1994.

A Low Power Lookup Technique for Multi-Hashing Network Applications

Ilhan Kaya and Taskin Kocak
School of Electrical Engineering and Computer Science
University of Central Florida
Orlando, FL 32816, U.S.A
{ikaya, tkocak}@cs.ucf.edu

Abstract

Many network security applications require large virus signature sets to be maintained, retrieved, and compared against the network streams. Software applications frequently fail to identify so many signatures through comparisons at very high network speeds. Bloom filters are one of the main multi-hashing schemes utilized in hardware to support this level of security. Nevertheless Bloom filters consume significant power to store, retrieve and lookup virus signatures owing to many hash function computations required to index to the memory. We present a novel lookup technique and architecture to decrease the power consumption of multi-hashing schemes, predominantly Bloom filters, in hardware. The theoretical analysis has shown that power gain achieved through new lookup technique can go up to 90%. Simulation results with three different classes of the hash functions embedded into the Bloom filter have indicated that power consumption of the Bloom filters can be considerably decreased by employing the low power lookup technique.

1 Introduction

Many network security applications make use of multi-hashing schemes. Either they require functionalities offered by hash tables or hash functions. Not only the software applications but also some hardware systems depend upon the properties of a high performing multi-hashing scheme. Such a multi-hashing scheme generally appears in the form of a Bloom filter [3]. Bloom filters have been used for many network applications like resource routing [6], string matching [1, 7], and packet filtering [2]. They are also used to improve lookup operations in hash tables [8]. A hardware system, consisting of Bloom filters to detect malignant content, is described in [7]. A detailed survey of Bloom filters for networking applications can be found in [4].

Although Bloom filters have found wide spread usage in

networking applications, they are not conservative in terms of power. A network intrusion detection system (NIDS) consists of 4 Bloom filter engines can dissipate up to 5 W.To decrease the power consumption of Bloom filters, we propose a new lookup technique which basically makes use of less number of hash function computations to determine the maliciousness of the network stream. The architecture to implement this new lookup technique in Bloom filters is presented in this paper. Furthermore, a comparative power analysis of the Bloom filter architecture which realizes the new lookup operation is given. A mathematical analysis carried out in this paper clearly states the efficiency of the new lookup technique in terms of power.

In spite of the importance of the hashing techniques in Bloom filters, hashing analysis of the Bloom filters is somewhat overlooked sofar. This paper also presents hardware architectures for implementation of the different classes of the hash functions utilized in programming and lookup operations of Bloom filters. The simulation results with the different hashing functions in varying configurations of the Bloom filters is also discussed.

2 Low power lookup technique

Before describing the low power lookup technique for multi-hashing schemes, it is important to consider what a multi-hashing scheme stands for. In this paper, we use a Bloom filter as a multi-hashing scheme for network applications. A Bloom filter is a data structure that stores a given set of signatures, by first computing multiple hash functions on each of the members of the set, and then it queries the database for a given input string, by again computing many hash functions of the input. The first operation is called *programming* of the Bloom filter, and the second operation is *lookup*. A block diagram of a typical Bloom filter is illustrated in Fig. 1. Given a string X, which is a member of the signature set, a Bloom filter computes k many hash values on the input X by using k different hash functions. Then it uses these hash values as index to the *m-bit* long lookup

0-7695-2533-4/06 $20.00 © 2006 IEEE

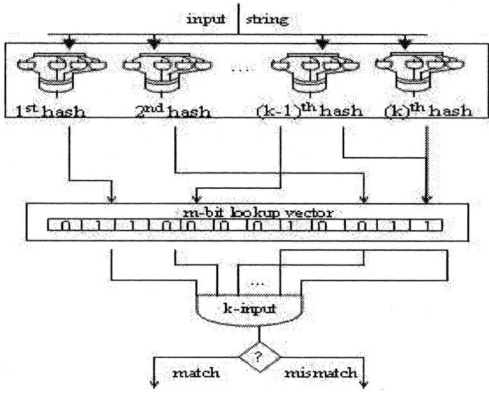

Figure 1. Block diagram of a typical Bloom filter

vector. It sets the bits corresponding to the index given by the hash values computed. It repeats this procedure for each member of the signature set.

For an input string Y, Bloom filter computes k many hash values by utilizing the same hash functions used in programming of the bloom filter. Bloom filter looks up the bit values located on the offsets (computed hash values) on the bit vector. if it finds any bit unset at those addresses, it declares the input string to be a nonmember of the signature set, which is called a *mismatch*. Otherwise, it finds all the bits are set, it concludes that input string may be a member of the signature set with a certain probability (*false positive probability*), which is called a *match*.

A Bloom filter never produces *false negatives*, which means if it decides that an input is a nonmember, input certainly does not belong to the signature set. However, it may produce false positives. It may conclude that the input is a member of the signature set, although in reality the input may not be a member of the set. Following the analysis of [7], the false positive probability f is calculated by,

$$f = \left(1 - e^{\frac{-nk}{m}}\right)^k \qquad (1)$$

where n is the number of signatures programmed into the bloom filter, m is the length of the lookup vector, and k is the number of the hash functions used to implement the Bloom filter. In order to minimize the false positive probability, the value of m must be quite larger than n. For a fixed value of $\frac{m}{n}$, k must be large enough such that f gets minimized. Since the number of hash functions in Bloom filters is large to reduce the false positive probability, it is intuitive that their total power consumptions are large. During the programming phase of the Bloom filter, not much can be done to reduce the power consumption, otherwise Bloom filter will produce many false positives. However, while performing lookups over the Bloom filter, the num-

ber of hash functions used to produce a decision can be reduced significantly. This is because Bloom filter never makes false negatives, and it is enough to find a zero on the *m-bit* long lookup vector to conclude that there is a *mismatch*. We call this type of lookup operation as *low power lookup technique*. The architecture to support such a lookup operation for a multi-hashing scheme is illustrated in Fig. 2.

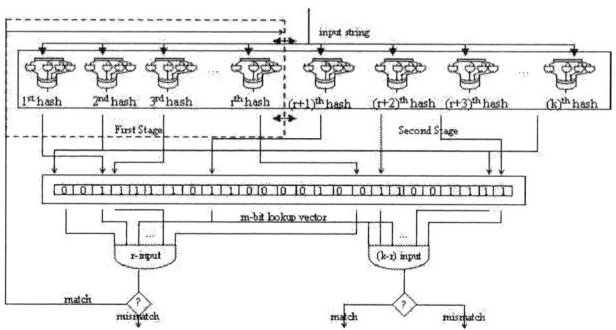

Figure 2. Block diagram of a two stage Bloom filter supporting low power lookup

At the core of the proposed architecture supporting low power lookup technique lies the division of the hash functions into two groups. These two groups are clearly identifiable on the Fig. 2. The first stage of hash functions always computes the hash values. By contrast, the second stage of hash functions only compute the hash values if in the first stage there is a match between the input and the signature sought. The result produced at the end of the first stage will be used as a select signal to start computing the second stage of hash functions. In the worst case, the new lookup operation will make use of all the hash functions in both groups, nonetheless most of the time the first group of hash functions will be enough to make a decision, which is a *mismatch*, claiming input is free of malicious content. Instead of computing k many hash functions, now it is enough to compute r many indices. This results in computational power savings.

3 A comparative power analysis

In this section, a theoretical approach is followed to analyze and compare the power consumptions of the different lookup operations available through two Bloom filter architectures presented in Fig. 1, and Fig. 2 respectively. A single Bloom filter shown in Fig. 1 uses k many hash functions in order to make a decision on the input given. Hence, the power consumption of a Bloom filter when performing a regular lookup operation is a summation of the power consumptions of each of the hash function computations, P_{H_i}, plus the power consumed accessing the memory for each hash value computed, P_Q, plus the power consumed by an AND gate.

$$P_{BF_{regular}} = \sum_{i=1}^{k} (P_{H_i} + P_Q) + P_{AND} \qquad (2)$$

Power consumption of an AND gate is ignored hereafter, since it is minimal compared to the power used by the hash functions. We assume that the power required to query m-bit vector is approximately constant for each index calculated by any of the hash functions. The power consumption equation for a single Bloom filter simply becomes the total power used up by the hash functions and the power consumed by querying the m-bit vector for each hash value calculated.

$$P_{BF_{regular}} = \sum_{i=1}^{k} (P_{H_i} + P_Q) \qquad (3)$$

In order to compare the power consumption of a regular lookup operation to that of the low power lookup proposed, we use 16-bit implementation of hash functions. For comparison reasons, we do not consider different class of hash function implementations till the next section. We assume all of the hash functions implemented are from the universal class of hash functions called H_3 [5]. Hence, all of the k many hash functions are of type 16-bit H_3 class of hash functions, so Equ. 3 becomes

$$P_{BF_{regular}} = \sum_{i=1}^{k} (P_{H_{i(H_{16})}} + P_Q) = k.(P_{H_{16}} + P_Q) \quad (4)$$

To derive the power consumption of the new lookup operation proposed, we follow an mathematical analysis similar to the analysis done in [9].Let us first derive the probability of match in the first stage. The probability that a bit is still unset after all the signatures are programmed into the the Bloom filter by using k-many independent hash functions is α.

$$\alpha = \left(1 - \frac{1}{m}\right)^{kn} \approx e^{\frac{-kn}{m}} \, (for \, large \, m) \qquad (5)$$

where $\frac{1}{m}$ represents any one of the m bits set by a single hash function operating on a single signature. Then $\left(1 - \frac{1}{m}\right)$ is the probability that the bit is unset after a single hash value computation with a single signature. For it to remain unset, it should not be set by any of the k-many hash functions each operating on all of the n-many signatures in the signature set. Consequently, the probability that any one of the bits is set is

$$(1 - \alpha) \approx 1 - e^{\frac{-kn}{m}} \qquad (6)$$

In order for the first stage to produce a match, the bits indexed by all r of the independent random hash functions should be set. So the match probability of the first stage is, represented as p,

$$p = \prod_{i=1}^{r} (1 - \alpha) = (1 - \alpha)^r \approx (1 - e^{\frac{-kn}{m}})^r \qquad (7)$$

The mismatch probability of the first stage is simply $1-p$,

$$1 - (1 - e^{\frac{-kn}{m}})^r \qquad (8)$$

With a probability of $(1-p)$ the first stage of the hash functions in the Bloom filter will produce a mismatch when performing a lookup operation. Otherwise, the first stage produces a match, then the second stage is used to compare the input with the signature sought as it is suggested by the architecture proposed. Therefore the power consumption of a Bloom filter shown in Fig. 2 is given by

$$\begin{aligned}
P_{BF_{lowpower}} &= P_{1st-stage} + P\{match\} \times P_{2nd-stage} \\
P_{BF_{lowpower}} &= \sum_{i=1}^{r} (P_{H_i} + P_Q) + p \times \sum_{j=r+1}^{k} (P_{H_j} + P_Q) \\
&\quad + P_{AND} \qquad (9)
\end{aligned}$$

As we stated previously, for comparison purposes, $P_{H_{i,j}}$ are of type 16-bit H_3 class of universal hash functions. By substituting Equ. 7 into Equ. 9 power consumption of a Bloom filter shown on Fig. 2 becomes

$$\begin{aligned}
P_{BF_{lowpower}} &= \sum_{i=1}^{r} (P_{H_{i(H_{16})}} + P_Q) \\
&\quad + (1 - e^{\frac{-kn}{m}})^r \times \sum_{j=r+1}^{k} (P_{H_{j(H_{16})}} + P_Q) \\
&= r.(P_{H_{16}} + P_Q) + \\
&\quad (1 - e^{\frac{-kn}{m}})^r (k - r)(P_{H_{16}} + P_Q) \quad (10)
\end{aligned}$$

The power saving ratio, PSR, in a single Bloom filter implemented based on the architectures presented functioning on two different lookup techniques can be calculated as

$$PSR = \frac{(P_{BF_{regular}} - P_{BF_{lowpower}})}{P_{BF_{regular}}} \qquad (11)$$

By substituting Equ. 4 and Equ. 10 into Equ. 11, the average power saving ratio, PSR, is found out to be

$$PSR = \frac{k - r + (r - k)\left(1 - e^{\frac{-kn}{m}}\right)^r}{k} \qquad (12)$$

0-7695-2533-4/06 $20.00 © 2006 IEEE

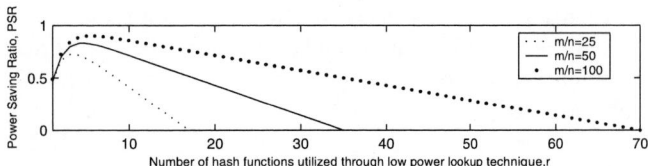

Figure 3. Power saving ratio w.r.t. number of hash functions in the first stage

For different values of the number of bits allocated to per signature, $\frac{m}{n}$, power savings over the number of hash functions utilized in the first stage are illustrated in Fig. 3.

As it is seen in the Fig. 3, the number of bits per signature, $\frac{m}{n}$, increases, the amount of power conserved in the system increases. In other words, the power saving ratio becomes larger as $\frac{m}{n}$ increases. This is because, as $\frac{m}{n}$ increases, although probability of mismatch in the first stage stays the same for all configurations for a fixed value of r (See Equ. 8), the number of hash functions deployed in the first stage becomes a smaller portion of the overall hash functions deployed in each configuration. For a fixed value of r, $\frac{r}{k}$ decreases. This explains the reduction in power consumption. The less are the number of hash functions utilized through low power lookup technique, the more the power is saved. Another observation from Fig. 3 is that as the number of hash functions utilized through low power lookup technique increases, the power saving ratio, *PSR*, first increases to an optimum *PSR* value, thereafter it drops gradually.

4 Practical hashing functions utilized in Bloom filters

In this section, we will analyze the effects of utilizing different hashing functions in Bloom filters. Performance of different hash functions in hardware are investigated in [10]. We utilized three different types of hash functions in Bloom filters to examine the effects of them on the performance of low power lookup technique and possible power savings on multi-hashing schemes.

4.1 H_3 class of universal hash functions

Universal class of hash functions are first introduced by Carter et al. [5]. They defined a special class of hash functions and called them as class H_3. The definition is as follows. Given any string X, consisting of b bits, X = $<x_1, x_2, x_3, \ldots, x_b>$
i^{th} hash function over the string X is defined as

$$h_i(x) = d_{i1} \bullet x_1 \oplus d_{i2} \bullet x_2 \oplus d_{i3} \bullet x_3 \oplus \ldots d_{ib} \bullet x_b \quad (13)$$

where d_{ij}'s are random coefficients uniformly distributed between *1* to size of the lookup vector, m, and x_k is the kth bit of the input string. $\bullet$ is a bit by bit AND operation, and $\oplus$ is a logical exclusive OR (XOR) operation. A block diagram of the H_3 class of hash functions implemented is given in Fig. 4.

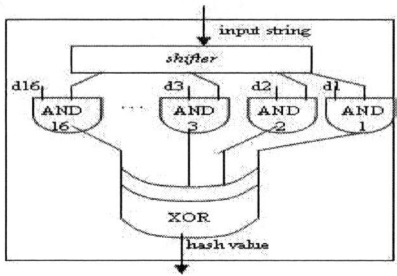

Figure 4. A block diagram of a H_3 class of universal hash function

Input is shifted one bit left till 16 bits are handled. Each bit is logically AND-ed with the random number. At the end, all AND results XOR-ed together to get a hash value. This type of hash functions are linear transformations, as a result they distribute the index values randomly. Implementation of these type of hash function requires 16 2-input AND gates and a single 16-input XOR gate for a 16 bit signature. They produce key values as the same size of the input. Pseudocode to implement H_3 class of hash functions is given below.

Pseudocode 1 A H_3 class of universal hash function

```
for each signature:
    - generate as many random numbers
    as the bits in the signature
    - left shift the signature to
    get to the specified bit
    - AND each shifted signature
    with the random number
    - XOR all the results of AND's
```

4.2 Bit extraction hashing functions

These type of hashing functions consists of selecting j bits out of b bits of the signature. Depending on the selection fashion of these bits out of input signature, they are classified as *regular* and *randomized* bit extraction hash functions. Since regular bit extraction hashing functions are constrained in number by the input length, we have used randomized bit extraction hash functions. Definition of a randomized bit extraction hashing function is as

0-7695-2533-4/06 $20.00 © 2006 IEEE

follows. Given any string X, consisting of b bits, $X = <x_1, x_2, x_3, \ldots, x_b>$

i^{th} hash function over the string X is defined as

$$h_i(x) = <x_{l_1}, x_{l_2}, x_{l_3}, \ldots, x_{l_j}> \qquad (14)$$

where l_j's are random bit positions uniformly distributed between one to size of the input signature in bits, b, and x_{l_j} is the input bit located at l_j. A block diagram of randomized bit extraction hash functions implemented is illustrated in Fig. 5.

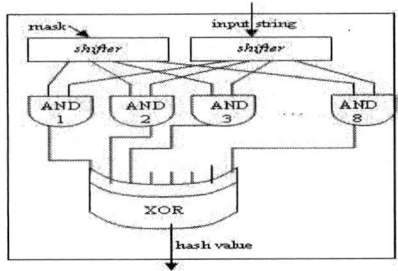

Figure 5. A block diagram of a bit extraction hash function

Implementation of these types of hash functions requires 8 2-input AND gates and a single 8-input XOR gate for a 16 bit signature. Shifter is necessary to left shift the bits in input as specified by random number, l_j. These type of hash functions produce key values shorter in bits than the size of the signature. They distribute keys randomly since the bit positions to extract the bits based on random numbers. Pseudocode to simulate this hash function is given below.

Pseudocode 2 A bit extraction hash function

```
for each signature:
    - generate as many random numbers
    as the bits in the indices
    - right shift  the signature to
    get to random bit position
    -adjust the bit at random position to
    the correct position at index
    by left or right shifting
    - XOR all the results of shifting
```

4.3 Hashing functions from exclusive OR method

These types of hash functions partition the b bit long input signature into j bits of segments. The segments are XOR-ed to get the hash value. The segments can be formed either in a regular manner or randomly like bit extraction

hash functions. Since we want to have random indices, we have used random segment forming hash functions. The definition of the hashing functions from XOR method is as follows. Given any string X, consisting of b bits, $X = <x_1, x_2, x_3, \ldots, x_b>$

i^{th} hash function over the string X is defined as

$$h_i(x) = (x_{s_1} \oplus x_{s_2})(x_{s_3} \oplus x_{s_4}) \ldots, (x_{s_{j-1}} \oplus x_{s_j}) \quad (15)$$

where s_j's are the uniformly distributed random bit positions in input string. x_{s_j} are the bits at the position specified by s_j. There are two segments of length j-bits are formed and XOR-ed. Fig. 6 illustrates a block diagram of a hash function from XOR method.

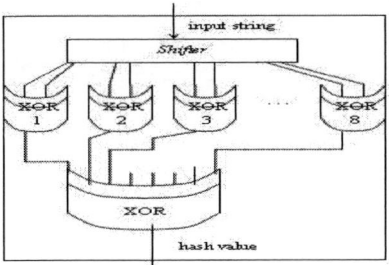

Figure 6. A block diagram of a hash function using the XOR method

Implementation of these types of hash functions requires a shifter to get to the bit at the random position, plus 8 2-input XOR gates, and a 8-input XOR gate. The length of the resulting hash value is smaller in bits than the input. However they map the inputs to the hash values in a completely random manner due to the random selection of bits from input. Following is a pseudocode to implement these type of hash functions.

Pseudocode 3 A hash function using the XOR method

```
for each signature:
    - generate twice as many random numbers
    as the bits in the indices
    - right shift  the signature to get the
    random bit positions for two segments
    - XOR the bits at each segment
    - right shift the XOR result to get
    correct position
```

---

5 Simulations

We have simulated the low power lookup technique presented in Fig. 2 with three different hash functions described in the previous section by using our custom-written

C program. In the simulation, parameters such as the size of the signature set, the type of signatures, m-bit long lookup vector size are varied to observe the power saving ratio. The length of the signatures are 16 bits. Indices, however, depend on the type of hash function utilized since both bit extraction and hash functions using the XOR method produce 8 bit hash values when the signature length is 16 bits. Table 1 depicts the results of the simulations illustrating large power gains through low power lookup operation.

Table 1. Power Savings through low power lookup operation for different hash functions

Simulation number	hash function type	number of signatures	lookup vector length	Power Saving (%)
Simul. 1	H_3	101	65536	97
Simul. 2	H_3	1000	65536	82
Simul. 3	EXT	6	256	88
Simul. 4	HXOR	6	256	85
Simul. 5	EXT	10	256	71
Simul. 6	EXT	20	256	60

The most important observation from the simulation results presented on Table 1 is that the low power lookup operation indeed provides significant power savings. The type of hash functions does not effect the power savings ratio as long as the bits allocated to the per signature, $\frac{m}{n}$, stays constant. However the type of the hash function affects the number of signatures that can be programmed, n, and size of the lookup vector, m. This is an expected result, since bit extraction and hash functions from XOR method produces 8-bit long indices whilst H_3 type of universal hash functions produce 16-bit indices. Consequently, the size of the lookup vector is limited by 256 or 65536 respectively. Hence, the type of the hash functions determines the size of the lookup vector m. As the number of signatures programmed in Bloom filter, n, increases, the number of hash function computations required to generate a decision on the maliciousness of the input rises. As a result, the power gain that is achieved in the Bloom filter through new lookup technique drops.

6 Conclusions

In this paper, we have proposed a low power lookup technique for multi-hashing schemes. Furthermore, an architecture supporting this new low power lookup technique for Bloom filters is described. Mathematical power analysis as well as simulations are carried out to show the effectiveness of the proposed method. In addition to that, three different types of hash functions are examined to observe the effects on the power savings and the operation of the multi-hashing scheme. The simulations performed revealed that the new lookup technique drastically decreases the power consumption of the Bloom filter. Simulations have also shown that the type of the hash function utilized in Bloom filter does not largely affect the power savings by low power lookup technique. However the type of the hash functions determines the size of the lookup vector, which in turn affects the number of allowed signatures programmed in to Bloom filter.

References

[1] M. Attig, S. Dharmapurikar, and J.L. Lockwood. "Implementation Results of Bloom Filters for String Matching", *Proc. of IEEE Symp. on Field-Programmable Custom Computing Machines*, pp. 322-323, 2004.

[2] M. Attig, and J.L. Lockwood. "SIFT: Snort Intrusion Filter for TCP", *IEEE Symp. on High-Performance Interconnects*, Stanford, CA, 2005.

[3] B. Bloom, "Space/Time Trade-Offs in Hash Coding with Allowable Errors", *Commun. ACM*, vol. 13, no. 7, pp. 422-426, July 1970.

[4] A. Broder and M. Mitzenmacher, "Network Applications of Bloom Filters: A Survey", *Internet Mathematics*, vol. 1, no. 4, pp. 485-509, July 2003.

[5] J. L. Carter and M. Wegman, "Universal classes of hash functions", *Journal of Computer and System Sciences*, vol. 18, pp. 143-154, 1978.

[6] S. Czerwinski, B. Y. Zhao, T. Hodes, A. D. Joseph, and R. Katz. "An Architecture for a Secure Service Discovery Service", *Proc. ACM/IEEE Int'l Conf. on Mobile Computing and Networking*, pp. 24-35, 1999.

[7] S. Dharmapurikar, P. Krishnamurthy, T.S. Sproull, and J. W. Lockwood, "Deep Packet Inspection Using Parallel Bloom Filters", *IEEE Micro*, vol. 24, no. 1, pp. 52-61, 2004.

[8] S. Dharmapurikar, H. Song, J. Turner, and J. W. Lockwood, "Fast Hash Table Lookup Using Extended bloom Filter: An Aid to Network Processing", *ACM/SIGCOMM*, pp. 181-192. Philadelphia, 2005.

[9] M. Mitzenmacher, "Compressed Bloom filters", *IEEE/ACM Transactions on Networking*, vol. 10, no. 5, pp. 604-612, October, 2002.

[10] M. Ramakrishna, E. Fu, and E. Bahcekapili, "Efficient Hardware Hashing Functions for High Performance Computers", *IEEE Trans. on Computers*, vol. 48, no. 12, pp. 1378-1381, 1997.

0-7695-2533-4/06 $20.00 © 2006 IEEE

A Low Power Pipelined Maximum Likelihood Detector for 4x4 QPSK MIMO Wireless Communication Systems

J. H. Han[1], A.T. Erdogan[1,2], T. Arslan[1,2]

[1]The University of Edinburgh, School of Engineering and Electronics
Edinburgh, EH9 3JL, UK
[2]Institute of System Level Integration, The ALBA campus
Livingston, EH54 EG, UK
j.han@ed.ac.uk, Ahmet.Erdogan@ee.ed.ac.uk, Tughrul.Arslan@ee.ed.ac.uk

Abstract

The authors present a maximum likelihood (ML) detector for multiple-input multiple-output (MIMO) wireless communication systems. The ML detector has been specifically designed to reduce the implementation complexity without significant degradation in bit error rate (BER) performance. In order to identify the optimized fixed-point representation, the ML detector has been simulated with various representations for the received data. The computation process of the channel matrix and constellation symbols in ML detector is simplified by using normalized symbols. Simulation results are provided showing 42% saving in area usage and 68% saving in power consumption compared to a conventional architecture.

1. Introduction

High throughput is one of the key issues in wireless communication systems. Multiple-input multiple-output (MIMO) systems provide a breakthrough for achieving high data rates for wireless communication systems such as 3GPP, WiMax, and WLAN [1]. Since the introduction of a simple space-time diversity technique in [2], many researchers have studied MIMO systems in order to improve their performance in terms of bit error rate (BER) and capacity. Various algorithms for MIMO channel detection have been proposed in literature to reduce their complexity for practical applications. Maximum likelihood (ML) algorithm can provide the best BER performance for MIMO systems, while the number of searching steps increases exponentially with the number of receiver antenna (M) and the modulation method (Q), which is given by 2^{MQ}.

Several methods are suggested to reduce the number of searching steps in the literature. Sphere decoder method [3] can reduce the searching steps without significant performance degradation as compared to ML method. VBLAST method with nulling and cancellation [4] has significantly reduced the searching steps with a coast of less BER performance than ML method.

Although the number of searching steps exponentially increases with the number of antennas and constellation, ML method can still provide lower complexity implementations compared to other methods mentioned above. For example, ML does not need complex matrix computations such as matrix inversion and decomposition [5], which are necessary in Sphere decoder and VBLAST methods.

In this paper, the authors present a fixed-point ML detector implementation for a 4x4 QPSK MIMO system. The ML detector has been implemented with various quantization levels for received symbols in order to investigate its BER performance and identify the optimum quantization level. In practical digital signal processing applications, fixed-point implementations are commonly used since they directly contribute to area and power savings. Moreover, a detection algorithm has been suggested in order to reduce the complexity of the ML detector by reducing the number of multiplications. The implemented ML detector has also been compared with the conventional and other methods in the literature in terms of BER performance, area usage and power consumption.

2. ML detection in MIMO systems

Fig. 1 shows a simple diversity diagram of a MIMO system where multiple antennas are used at transmitter

0-7695-2533-4/06 $20.00 © 2006 IEEE

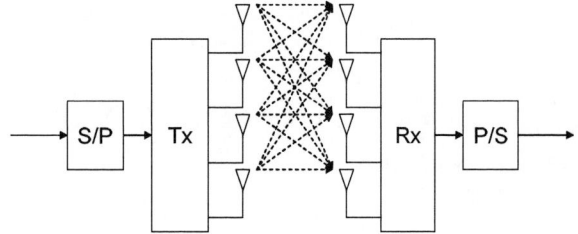

Figure 1. A simple diagram of a MIMO system

and receiver. The MIMO system can be represented as follows :

$$y = Hx + n \qquad (1)$$

where $y = [y_1, \ldots, y_M]^T$ is the vector of received symbols, H is the M (the number of Tx antenna) x N (the number of Rx antenna) channel matrix, $x = [x_1, \ldots, x_N]^T$ is the vector of transmitted symbols, and n is additive white Gaussian noise (AWGN). The transmitted symbol vector, x, is represented by a modulated symbol, such as QPSK and QAM. In this paper, QPSK modulation, $s \in \{1+i, 1-i, -1-i, -1+i\}/\sqrt{2}$, which are generated by following Gray coding method, is used and H is assumed to be known at receiver.

The transmitted symbols, x, can be obtained by calculating a minimum Euclidian distance from the received symbols, y, the channel matrix, H, and the modulation symbols, $s = [s_0, \ldots, s_N]^T$. Therefore, the ML detection algorithm can be represented with the equation below:

$$\arg\min \|y - Hs\|^2 \qquad (2)$$

A straightforward approach to solve Eq. (2) is an exhaustive search. However, the corresponding computational complexity grows exponentially with the number of antennas and the number of bits per symbol in the constellations. For example, for a 4x4 MIMO

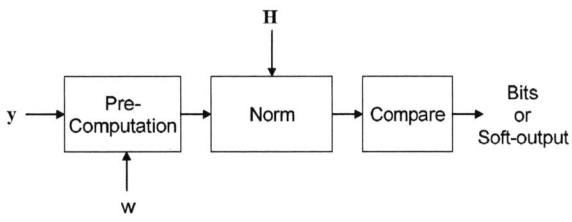

Figure 2. A block diagram of a ML detector

system with QPSK modulation, the required number of searching steps is 256, which is regarded as a limit for the use of ML algorithm for MIMO systems.

3. Fixed-point implementation

Fig. 2 illustrates the main blocks in a ML detector implementation. The received symbols, y, and the channel matrix, H, are represented in a fixed-point format $(t.p)$ where t is the total number of bits, and p is the number of precision bits. The optimization of the number of total and precision bits is crucial in hardware implementations for saving area and power consumption.

Before describing the simulation results of fixed-point ML detector implementation, Eq (2) is slightly modified in order to reduce the matrix computations for H and s, as shown below:

$$\|y - Hs\|^2 = \|y - H(2^{-1/2} \times s')\|^2 = 2^{-1} \times \|y' - Hs'\|^2 \quad (3)$$

$$\arg\min \|y' - Hs'\|^2 \qquad (4)$$

where s' is $[s_0', \ldots, s_N']^T$, in which $s' \in \{1+i, 1-i, -1-i, -1+i\}$. Therefore, the matrix computation of Hs' can be implemented without using any multipliers. However, the computation of y' will require multiplying y with a parameter, $w (= \sqrt{2} \times 2^p)$, which is determined by the number of precision bits, p, and the modulation method for y. Although, Eq. (4) requires multiplications for computing y', it can still lead to less complexity when compared to Eq. (2), where the computation of Hs requires more multiplications compared to the multiplications required for obtaining y'. Moreover, in our implementation, there is no need for a memory block for storing the multiplication results, Hs, as in [6].

Fig. 3 illustrates BER performance with different fixed-point representations for E_b/N_0 of 5, 10, 15, and 20. The numbers inside the brackets represent the number of total and precision bits $(t.p)$. For example, (8.5) denotes that the number of total and precision bits is 8 and 5, respectively. As can be seen, the BER performance gets better as the total number of bits increases. However, the performance is strongly affected by the number of precision bits. For example, $p=3$ provides the best BER when $t=6$. Therefore, we have simulated our ML detector with the best $(t.p)$ combinations. Fig. 4 shows the BER performance versus E_b/N_0 for (6.3), (7.4), (8.4), (9.5), and (10.5). As can be seen from Fig. 4, the BER performance does not change significantly until Eb/N$_0$ is 10. In overall, a fixed-point representation of (9.5) can provide good

0-7695-2533-4/06 $20.00 © 2006 IEEE

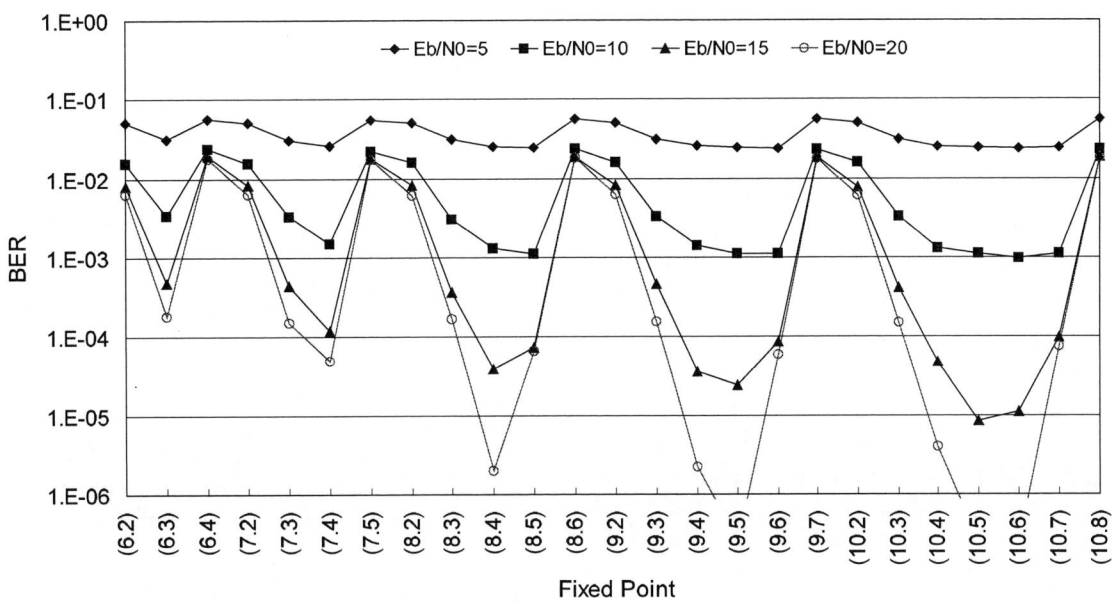

Figure 3. BER performance for different fixed-point values with Eb/N0=5, 10, 15, and 20

BER performance without any significant performance degradation.

4. Complexity reduction

The ML detector algorithm in Eq (2) represents a simple form for implementation. Although the complexity of the ML algorithm is less compared to other detector algorithms, implementing the ML detector still requires many multiplications and additions for computing the norm value. This results in

![Figure 4 plot: BER vs Eb/N0 with curves labeled (6,3), (7,4), (8,4), (9,5), (10,5)]

Figure 4. BER performance for different fixed-point implementations

increased area usage, power consumption, and critical path delay for practical implementations. In [7], a low complexity ML detector implementation is proposed where the required number of multiplications and additions are also provided. However, the detector proposed in [7] is not suitable for practical wireless communications systems since it can only be used for constellations s ∈ {1, i, -i, -1} with QPSK modulation. On the other hand, an approximation method for calculating the norm value has been introduced in [8], as shown below:

$$\|A+iB\|^2 \approx [\max(|A|,|B|)+0.5\min(|A|,|B|)]^2 \qquad (5)$$

This approximation can reduce the complexity of ML detector implementation. However, it still requires a number of multipliers and adders. To reduce the complexity further, in this paper, Eq. (4) is transformed as follows:

$$\arg\min\left[\text{Re}\left|\mathbf{y'-Hs'}\right| + \text{Im}\left|\mathbf{y'-Hs'}\right|\right] \qquad (6)$$

As can be seen, this equation does not require any multiplications, and hence it can reduce the implementation complexity of the ML detector much more compared to other methods suggested in literature. To verify its functionality, the BER performance of the ML detector based on Eq. (6) has been analyzed and compared with the BER performance achieved with the conventional method based on Eq. (2), as shown in Fig.

0-7695-2533-4/06 $20.00 © 2006 IEEE

187

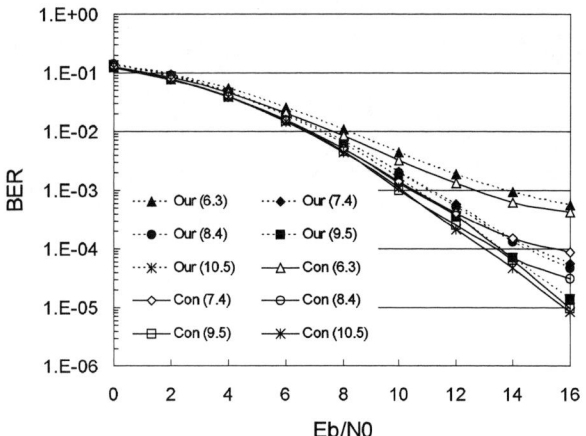

Figure 5. BER performance comparisons

5 where 'CON' and 'OUR' denote the results of ML detector implemented with Eq. (2) and Eq. (6), respectively. As can be seen, there is no significant degradation in BER performance between 'CON' and 'OUR'. Therefore, a less complex ML detector implementation can be achieved by using Eq. (6).

5. Implementation of maximum-likelihood detector

In previous sections, the ML detector has been investigated and simulated with various fixed-point representations to find the optimized quantization level for the received symbols, **y,** and the channel matrix, **H**. Fig. 6 illustrates a block diagram of the ML detector implemented in this paper for a 4x4 QPSK MIMO system. The architecture consists of five pipelined blocks which are named as pre-computation and norm1 (PCN1U), norm2 (N2U), norm3 (N3U), norm-summation (NSU), and decision (DEU). Here, the first three units are used to compute the real and imaginary

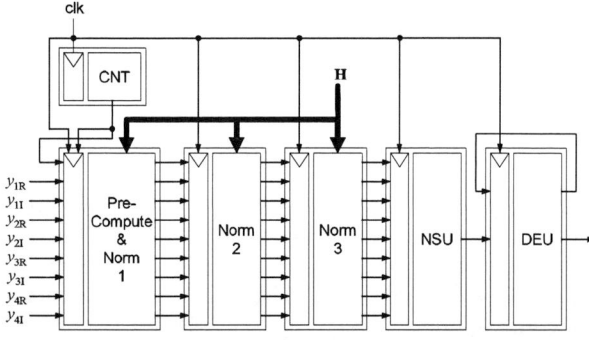

Figure 6. A block diagram of Maximum Likelihood detector implementation

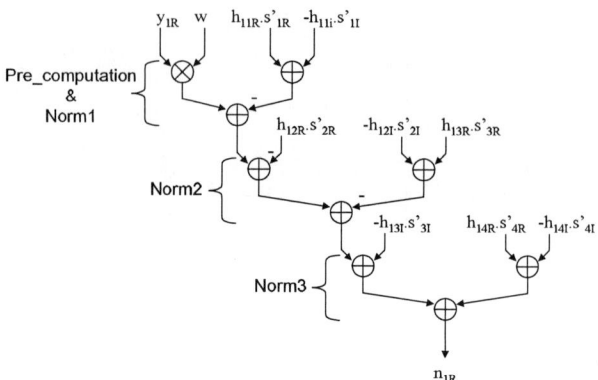

Figure 7. Pipelined data path for the computation of the norm value.

parts of **y'-Hs'** for each received symbol, as shown in Fig. 7. These are then summed by the NSU to generate the norm value. This process is repeated 256 times after which the minimum value is finally chosen by the decision unit (DEU). In Fig. 6, the counter (CNT) is used to provide all 256 symbol sequences.

In our implementation, the NSU block has been implemented with and without using multipliers in order to investigate the impact of multiplier elimination in terms of area usage and power consumption.

6. Simulation results

The ML detector has been implemented as Verilog HDL and synthesized to a 0.18um standard cell library with Synopsys DesignCompiler[TM]. RTL and gate level simulations have been performed using Cadence Verilog-XL[TM]. The results of the power consumption were obtained with Synopsys PowerCompiler[TM] based on gate level simulations with a clock speed of 100MHz.

As illustrated in Fig. 8 and 9, our ML detector has been implemented for different fixed-point representations for evaluating its area usage and power consumption. In these figures, 6m and 6 denote to 6 bits word length for **y** and **H** with and without multiplier for computing the *norm* value in the ML detector, respectively.

Clearly, the ML detector implementations without multiplier can save more area and power than with multiplier. The main contribution for the savings is due to NSU block. Particularly, power consumption savings is more significant than the area usage savings. For example, with a word length of 9-bits for **y** and **H**, the ML detector without multiplier can save 42% area usage and 68% power consumption compared to the ML detector with multiplier.

0-7695-2533-4/06 $20.00 © 2006 IEEE   188

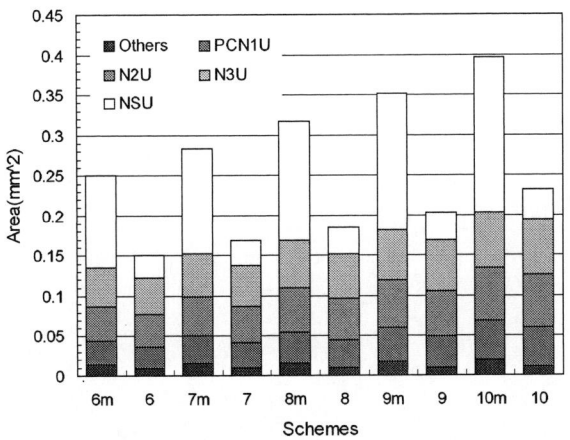

Figur 8. Area results of the ML detector with and without multiplier for different fixed-point implementations

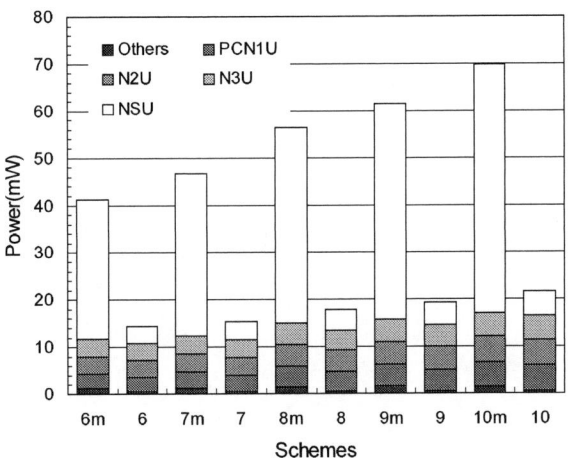

Figure 9. Power simulation results of the ML detector with and without multiplier for different fixed-point implementations

7. Conclusions

The authors have presented a fixed-point ML detector for 4x4 QPSK MIMO systems. The ML detector has been simulated with various quantization levels for input data in order to find an optimized fixed-point representation. The authors have also

proposed an implementation method to reduce the complexity of the matrix and norm value computations for the ML detector. It is shown that eliminating the multiplier in the ML detector implementation results in significant savings in area usage and power consumption. For example, with 9-bits word length for y and H, area and power savings are 42% and 68% respectively compared with conventional implementations based around the use of multipliers.

8. References

[1] G. Foschini and M. Gans, "On limits of wireless communications in fading environment when using multiple antennas", Wireless Personal Communications, vol. 6, pp. 311-335, 1998.

[2] S. M. Alamouti, "A simple transmit diversity technique for wireless communications", IEEE Journal on Selected Areas in Communications, vol. 16, no. 8, pp. 1451-1458, Oct. 1998.

[3] B. M. Hochwald and S. ten Brink, "Achieving near-capacity on a multiple-antenna channel", IEEE Trans. on Communications, vol. 51, no. 3, pp. 389-399, March 2003.

[4] P. W Wolniansky, G. J. Foschini, G. D. Golden, and R. A. Valenzuela, "V-BLAST: An architecture for realizing very high data rates over the rich-scattering wireless channel", International Symposium on Signals, Systems, and Electronics (ISSSE 98), pp.295-300, 29 Sept. – 2 Oct. 1998.

[5] L. M. Davis, "Scaled and decoupled Cholesky and QR decompositions with application to spherical MIMO detection", IEEE Wireless Communications and Networking (WCNC 2003), vol. 1, pp. 326-331, 16-20 March 2003.

[6] D. Garrett, L. Davis, S. ten Brink, B. Hochwald, and G. Knagge, "Silicon complexity for maximum likelihood MIMO detection using spherical decoding", IEEE Journal of Solid-State Circuits, vol. 39, no. 9, pp.1544-1552, Sept. 2004.

[7] A. Burg, N. Felber, and W. Fichtner, "A 50 Mbps 4x4 maximum likelihood decoder for multiple-input multiple-output systems with QPSK modulation", 10th IEEE Int. Conf. on Electronics, Circuit and Systems (ICECS 2003), vol. 1, pp. 332-335, 14-17 Dec. 2003.

[8] D. Garrett, G. K. Woodward, L. Davis, and C. Nicol, "A 28.8 Mb/s 4x4 MIMO 3G CDMA receiver for frequency selective channels", IEEE Journal of Solid-State Circuits, vol. 40, no. 1, pp. 320-330, Jan. 2005.

System-on-Chip

Optimal Periodical Memory Allocation for Logic-in-Memory Image Processors

Masanori Hariyama and Michitaka Kameyama

Graduate School of Information Sciences
Tohoku University
Aoba 6-6-05, Aramaki, Aoba, Sendai, 980-8579, Japan
{hariyama,kameyama}@ecei.tohoku.ac.jp

Yasuhiro Kobayashi

Oyama National College of Technology
Nakakuki 771, Oyama, Tochigi, 323-0806, Japan
y-kobayashi@oyama-ct.ac.jp

Abstract

One major issue in designing image processors is to design a memory system that supports parallel access with a simple interconnection network. This paper presents a design methodology for a logic-in-memory architecture where each of memory modules is connected to its dedicated processing element(PE). An efficient memory allocation to minimize the number of memory modules and PEs under a time constraint is proposed based on regularity.

1 Introduction

Highly-parallel image processors requires a complex interconnection network between memory modules and processing elements(PEs) for parallel memory access. The complex interconnection network causes significant overhead in delay and power in deep-submicron and more advanced technologies since the delay and the power of interconnection units are more dominant than those of logic units.

This paper presents a design methodology for a logic-in-memory architecture where each of memory modules is connected to its dedicated PE. The advantage of the logic-in-memory architecture is its simple interconnection network between memory modules and PEs. Its disadvantage is that it may require a large number of PEs when a large number of memory modules are required for parallel access.

To solve this problem, this paper also presents a memory allocation technique to enable parallel access with the minimum number of memory modules. From a practical point, a memory allocation should have a simple address function that calculates which memory module a pixel is stored in. Hence, we consider a periodical memory allocation where a memory allocation for a whole image is given by repeating a memory allocation for a partial image. A periodical memory allocation has a simple address function because of its regularity. The memory allocation problem is usually mapped onto the clique partitioning problem that is known as an NP-complete problem[1]. An efficient search method

is presented based on the regularity of a periodical memory allocation. As a result, an optimal memory allocation for a practical image size is found in much less than 1 second on Pentium4@2GHz.

The proposed method is used for a road-extraction VLSI for highly-safe vehicles and a stereo-matching VLSI[2]. In the former and latter cases, the number of memory modules is reduced to 91.7% and 19.4%, respectively.

2 Problem formulation

2.1 Target algorithms

We consider window-type algorithms. In these algorithms, the output/intermediate output depends on a small neighborhood of an input image, where the neighborhood size is fixed and given as a window as shown in Fig. 1. Algorithms of this type frequently appear in practical situations: spatial filter, morphology, and image matching, and so on. Moreover, they usually have high degree of parallelism, and are suited for VLSI implementations.

We use window-serial-and-pixel-parallel scheduling as shown in Fig. 2. In this scheduling, operations are performed in parallel with pixels in a window, whereas operations are performed in a serial manner with windows. Figure 2(a) shows the location of the window at each step. The thick line denotes a window. Figure 2(b) shows the scheduled data-flow graph(SDFG) corresponding to Fig. 2(a). A node in the SDFG denotes a operation. The labels **A** and **B** on operations denote the operation types of the nodes. There is an edge from node V_1 to node V_2 when the output of V_1 is used as an input of V_2. At Step 1 in Fig. 2, pixels:$(0, 0)$, $(0, 1)$, $(0, 2)$ and $(1, 1)$ are used as inputs of operations of type **A**. The pixels must be accessed in parallel since the **A**-type operations are performed in parallel. Their results are used as inputs of a **B**-type operation. As the location of the window changes, the input pixels change.

0-7695-2533-4/06 $20.00 © 2006 IEEE

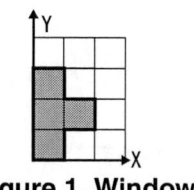

Figure 1. Window.

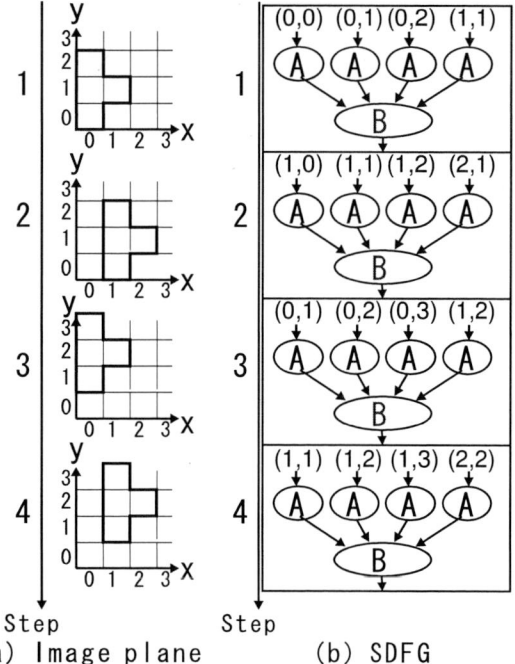

(a) Image plane   (b) SDFG

Figure 2. Data-flow graph of a pixel-parallel-and-window-serial schedule.

2.2 Target architecture

Figure 3 shows the basic structure of the logic-in-memory architecture. A single dedicated PE is connected to each memory modules. Pixels of input images are distributed among the memory module. The PEs performs operations of type **A** according to the window-serial-and-pixel-parallel scheduling shown in Fig. 2(b). Their outputs are used as inputs of the unit for type-**B** operations. In order to extend the architecture model, you can add inputs to PEs as required.

2.3 Periodical memory allocation

From a practical point, a memory allocation should have a simple address function to map the coordinates of a pixel onto the memory module number. If the address function is complex, its overhead in area and delay become serious. For example, a look-up table is required for address mapping of all pixels in the worst case. Therefore, we consider a periodical memory allocation where a memory allocation

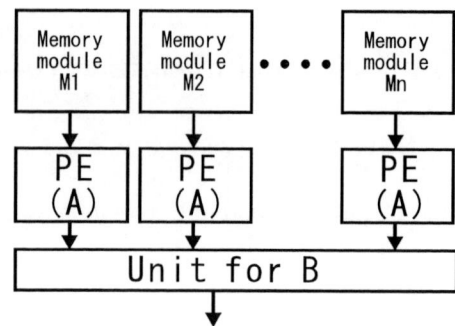

Figure 3. Logic-in-memory architecture.

for a whole image is given by repeating a memory allocation for a partial image. The periodical memory allocation has a simple address function because of its regularity. Let K be the number of memory modules. Let N_i be the number of pixels that are allocated to the memory module M_i for $1 \leq i \leq K$. Let (x_i^j, y_i^j) be the coordinates of a pixel that is allocated to the memory module M_i for $1 \leq j \leq N_i$. Then, we exactly define a periodical memory allocation as a memory allocation where the coordinates (x_i^j, y_i^j) of pixels allocated to M_i are expressed as

$$\begin{bmatrix} x_i^j \\ y_i^j \end{bmatrix} = s_i^j \begin{bmatrix} U_x \\ U_y \end{bmatrix} + t_i^j \begin{bmatrix} V_x \\ V_y \end{bmatrix} + \begin{bmatrix} x_i^0 \\ y_i^0 \end{bmatrix} \quad (1)$$

where $\mathbf{U} = [U_x\ U_y]^T$ and $\mathbf{V} = [V_x\ V_y]^T$ are vectors to represent periods(called period vectors); the variables s_i^j and t_i^j are integers; the coordinates (x_i^0, y_i^0) are those of the reference pixel allocated to M_i. Note that you can select an arbitrary pixel as the reference one from the ones allocated to the same memory module.

Fig. 4(a) shows an example of a periodical memory allocation for the SDFG shown in Fig. 2. The label on a pixel denotes the module number where the pixel is allocated. For example, the pixels: $(0,0),(0,1),(1,0)$ and $(1,1)$ are allocated to M_1, M_2, M_3 and M_4, respectively. From Fig. 4 (b), the coordinates of the pixels allocated to M_1 are given by

$$\begin{bmatrix} x_1^j \\ y_1^j \end{bmatrix} = s_1^j \begin{bmatrix} 1 \\ 2 \end{bmatrix} + t_1^j \begin{bmatrix} 2 \\ 0 \end{bmatrix} + \begin{bmatrix} 0 \\ 0 \end{bmatrix} \quad (2)$$

where the coordinates of the reference pixel and the period vectors are $[0\ 0]^T$, $\mathbf{U} = [1\ 2]^T$ and $\mathbf{V} = [2\ 0]^T$, respectively. Each of the pixels with label 1 in Fig. 4 (b) are given by $(s_1^j, t_1^j) = (0,0), (1,0), (2,0), (0,1), (1,1)$ or $(2,1)$. Figures 4-(c), 4-(d) and 4-(e) show the pixels allocated to M_2, M_3 and M_4, respectively. These figures show that the same period vectors as M_1 are used for M_2, M_3 and M_4. They also show that the coordinates of the reference pixels for M_2, M_3 and M_4 are $[0\ 1]^T$, $[1\ 0]^T$ and $[1\ 1]^T$, respectively. The memory allocation shown in Fig. 4 satisfies Eq.((1)), that is, the definition of a periodical memory allocation.

0-7695-2533-4/06 $20.00 © 2006 IEEE   194

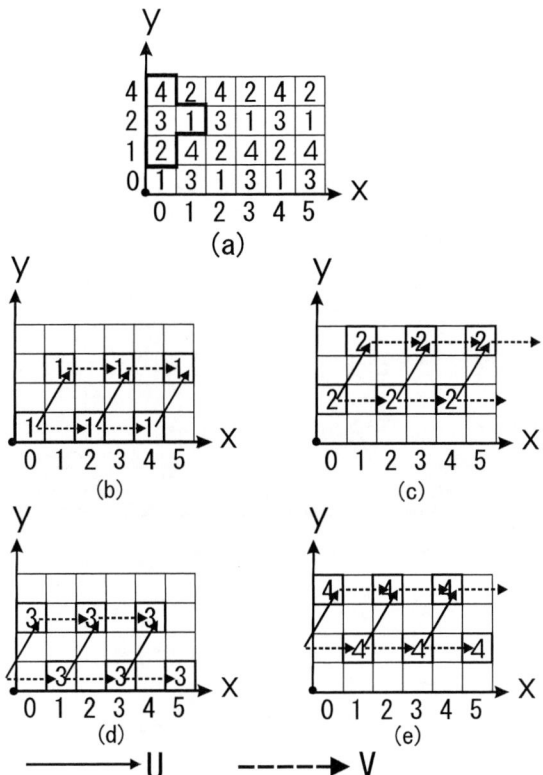

Figure 4. Example of a periodical memory allocation.

As mentioned at the beginning of this section, the advantage of the periodical memory allocation is its simple address function. For the periodical memory allocation shown in Fig. 4(a), the address function that maps the coordinates (x, y) of a pixel to the module number is given by

$$f(x, y) = (2x + y) \bmod 4 + 1 \qquad (3)$$

2.4   Optimal memory allocation

Let us minimize the hardware amount when the SDFG is specified. In the logic-in-memory architecture, the hardware amount is determined by a memory allocation task since the number of PEs is determined by the number of memory modules. The SDFG imposes the degree of parallelism in memory access. For example, in the SDFG shown in Fig. 2, the degree of parallelism in memory access is 4. From these observations, minimizing the hardware amount is formulated as finding a memory allocation that minimizes the number of memory modules under a specified degree of parallelism in memory access. Given a window, the optimal memory allocation is defined as the memory allocation that satisfies the following conditions:

C1: For an arbitrary location of the window, pixels in the window can be retrieved in parallel. In other words, the

pixels in the window are allocated to different memory modules.

C2: Each pixel is allocated to a single memory module. This condition ensures that the total memory capacity is minimized.

C3: The number of memory module is minimized. This condition ensures that the hardware amount is minimized in the logic-in-memory architecture.

C4: The memory allocation is a periodical one. This condition ensures that the hardware for the address function is small.

For example, Fig. 4(a) is the optimal memory allocation for the window shown in Fig. 1 from the following reasons. The condition C1 is satisfied since pixels in the window are allocated to different pixels for an arbitrary location of the window. The condition C2 is satisfied since each pixel has a single label. The condition C3 is satisfied since the number of memory module is 4 and is equal to the minimum number of memory modules required for parallel access, i.e. the number of pixels in the window. The condition C4 is satisfied since the memory allocation is periodical as mentioned in section 2.3.

2.5   Overview of search method

In the periodical memory allocation, the period vectors **U** and **V** determine which pixels are allocated to the same memory module as shown in Fig. 4. The period vectors make a parallelogram. Given the period vectors, the number of memory modules is estimated by the area of the parallelogram. The area S of the parallelogram is given by

$$S = |U_x \cdot V_y - U_y \cdot V_x|. \qquad (4)$$

For example shown in Fig. 4,

$$S = |1 \cdot 0 - 2 \cdot 2| = 4,$$

and S is exactly same as the number of memory modules. This is because the memory allocation for a whole image is given by repeating the one for the parallelogram, and the parallelogram must be filled with pixels allocated to different memory modules. From these observations, finding the optimal memory allocation is reduced to finding period vectors that make the minimum parallelogram still satisfying the parallel access condition.

Basically, the search for the optimal period vectors for $\mathbf{U}_{min}$ and $\mathbf{V}_{min}$ is performed by repeating following steps for **U** and **V** within possible coordinates of pixels.

Step 1 Check the parallel access condition, that is, whether the window includes the pixels allocated to the same memory module.

Step 2 If the parallel access condition is satisfied, calculate the area S by (4). Otherwise, go to Step 1.

0-7695-2533-4/06 $20.00 © 2006 IEEE   195

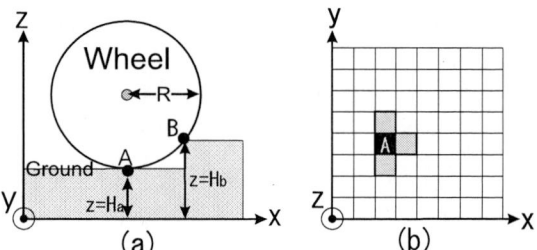

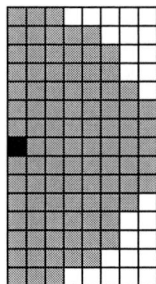

Figure 5. Road extraction.

Figure 6. Window for road extraction .

Step 3 If S is smaller than current minimum area S_{min}, then $S_{min} \leftarrow S$, $\mathbf{U}_{min} \leftarrow \mathbf{U}$, $\mathbf{V}_{min} \leftarrow \mathbf{V}$.

To reduce the computational amount, we can limit the reference pixel to $[0\ 0]^T$ as shown in Fig. 4(b). Moreover, we can limit the search space using the branch-and-bound method such that

$$S = |U_x \cdot V_y - U_y \cdot V_x| < S_{min},$$

where S_{min} is the current minimum area of the parallelogram made by the current period vectors.

3 Design examples

3.1 Road extraction VLSI

Let us design a road extraction VLSI for highly-safe vehicles. The "road" is defined as a set of regions that a vehicle can climb over. We assume that the 3-D information of the land surface is obtained from pre-processing such as stereo vision. Given the 3-D coordinates of the land surface and the size of the wheel of a vehicle, let us examine a 3-D pixel to see if it is a 3-D pixel on the "road". Let R be the radius of the wheel. As shown in Fig. 5, let A be the contact between the wheel and the land. Let B be an arbitrary 3-D pixel within the radius from A. The 3-D pixel A is a part of the "road" if the vertical interval $|H_b - H_a|$ is less than acceptable range H_{max}. This condition is given by

$$-H_{max} \leq H_b - H_a \leq H_{max} \qquad (5)$$

where H_a and H_b are the z-coordinates of A and B, respectively. Considering the traveling direction of the vehicle,

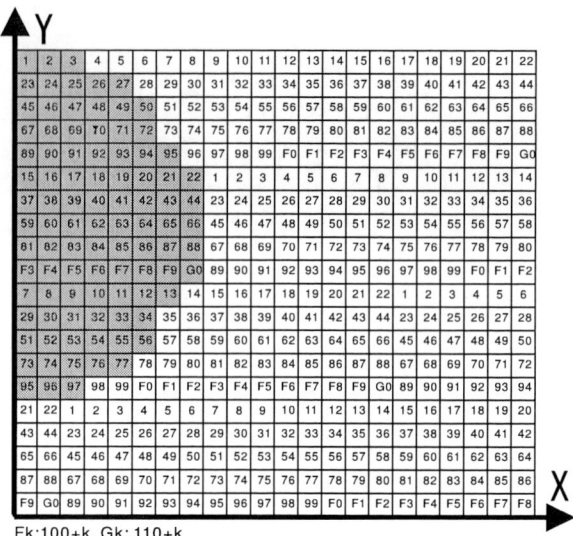

Fk:100+k, Gk: 110+k

Figure 7. Optimal allocation for Fig. 6.

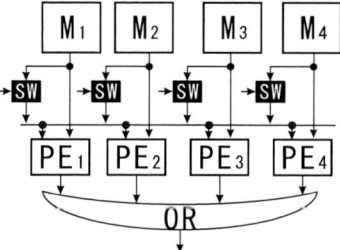

Figure 8. Logic-in-memory architecture for road extraction.

Eq. (5) is checked for B within a front half circle centered on A as shown in Fig. 6.The black pixel and gray pixels denote A and B, respectively.

Let us find the optimal memory allocation for the window shown in Fig. 6. Suppose that the area for road extraction is $10[m] \times 50[m]$ ahead and that the area is divided into a set of squares of $5[cm] \times 5[cm]$. Then, the 3-D information of the land can be considered as an image of 200×1000 pixels. Figure 7 shows the optimal memory allocation. It shows the result for a partial image since the space is limited. The labels F_k and G_k mean memory modules M_{100+k} and M_{110+k}, respectively. The period vectors are $\mathbf{U} = [14\ 5]^T$ and $\mathbf{V} = [6\ 10]^T$. The number of memory modules is 110. The search time for the optimal allocation is 40ms on a PC(Pentium4@2GHz, 2G-byte main memory, OS:Window XP). Figures 8 and 9 show the logic-in-memory architecture for road extraction and the block diagram of the PE, respectively. Note that the interconnection between memory modules and PEs is extended for road extraction. The road extraction VLSI is designed in a 0.5μm CMOS technology. Its performance is 1200 times higher than that of Pentium4@2GHz.

Let us compare the proposed method with the conventional approach from the point of the number of memory

0-7695-2533-4/06 $20.00 © 2006 IEEE   196

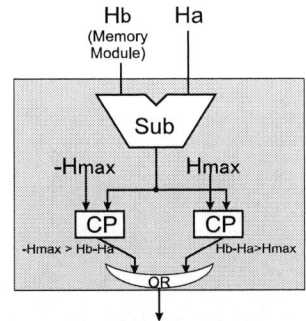

Figure 9. Block diagram of the PE.

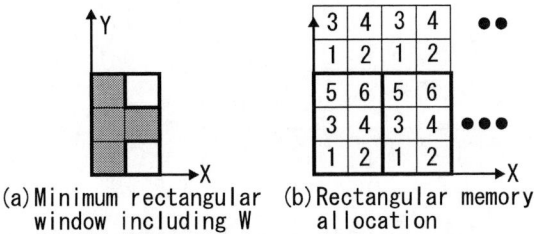

(a) Minimum rectangular window including W

(b) Rectangular memory allocation

Figure 10. Rectangular memory allocation.

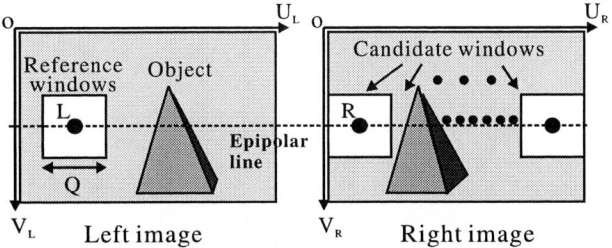

Figure 11. Search for a corresponding point.

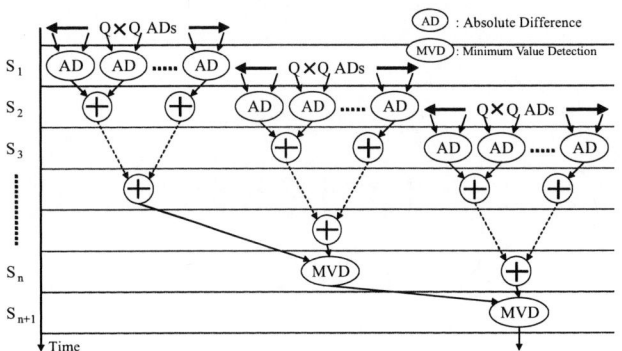

Figure 12. Data flow graph of stereo matching using multi-resolution images.

modules. For the window-type image processing, one efficient memory allocation method is a rectangular memory allocation[3] as shown in Fig. 10. A minimum bounding rectangle approximates a given window(Fig. 10(a)). The optimal memory allocation for the rectangle window is obtained by the rectangular memory allocation(Fig. 10(b)). The number of memory modules is $K \times L$ for a rectangle window of size $K \times L$. The window shown in Fig. 6 is approximated by a rectangle window of size 8×15. The rectangular memory allocation requires $8 \times 15 (= 120)$ memory modules. As a result, the number of memory modules is reduced to 91.7%($110/120 \times 100$).

3.2 Stereo matching VLSI processor using multi-resolution images

A coarse-to-fine approach using multi-resolution images is one of important techniques in image processing such as pattern matching. This section describes the optimal memory allocation for stereo matching using multi-resolution images. The method proposed in Section 2 is extended to the case with several different windows in the following.

Stereo matching is one efficient method to obtain 3-D information of real scene. It uses two images taken from two different cameras at the same time. After correspondence between the two images is established, the 3-D information of the scene is computed based on the triangular method. The correspondence search is time-consuming. Figures 11 and 12 show the correspondence search and the DFG, respectively. Figure 14 shows the logic-in-memory architecture for stereo matching. As a similarity measure between a reference window and a candidate window, a sum of absolute differences(SAD) is used. Given a reference window,

SADs are computed for all the possible candidate windows. The candidate window with the minimum SAD is determined to be the corresponding pixel.

Multi-resolution images are used to reduce the computational amount. Given sampling periods $SR = SR_{max}$, SR_{max-1},..., 2, 1, reduced images are made by sampling the original images every SR pixels. Beginning with the lowest-resolution image ($SR = SR_{max}$), the resolution is iteratively increased until $SR = 1$. The possible location of candidate windows at a higher resolution are limited by using the matching result at a lower resolution. Hence, the computational amount is reduced[2].

Figure 13 shows the windows at multi-resolution images for window size $Q = 3$. As shown in Fig. 13(a), the window of 3×3 at $SR = 1$ corresponds to the window of 3×3 on the original image. This is because the image at $SR = 1$ is exactly same as the original image. As shown in Fig. 13(b), the window of 3×3 at $SR = 2$ corresponds to the window of 5×5 on the original image. Note that only 3×3 gray pixels must be accessed in parallel. In order to find the optimal memory allocation for such different windows, we make a single window by the union of the windows of $Q \times Q$ at $SR = SR_{max}, SR_{max-1}, ..., 2, 1$ on the original image.

Figure 15 shows the resulting optimal memory allocation for $Q = 3$ and $SR_{max} = 3$. The period vectors are $\mathbf{U} = [5\ 0]^T$ and $\mathbf{V} = [0\ 5]^T$. The number of memory modules is 25. Figures 15 (a), (b) and (c) corresponds to SR=1,

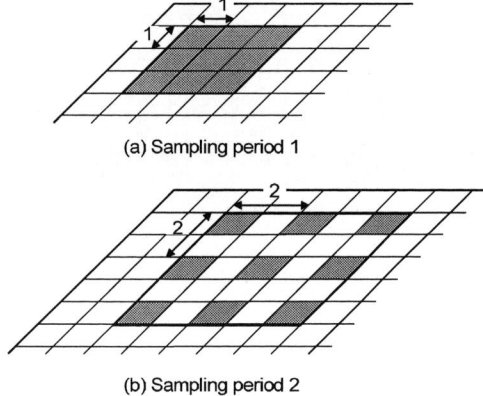

(a) Sampling period 1

(b) Sampling period 2

Figure 13. Window.

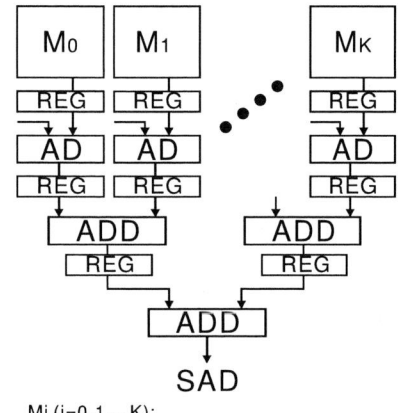

M_i (i=0,1,..,K):
Memory modules for a candidate image

Figure 14. Logic-in-memory architecture for stereo matching.

2 and 3, respectively. The label (x, y) and $k(= 1, 2, \cdots 25)$ on a pixel denote the coordinates of the pixel on the original image and the memory module number where the pixel is allocated, respectively. You can see that all the pixels in a 3×3 window of an arbitrary location are distributed between different memory modules at each sampling period. In other words, parallel memory access for a 3×3 window is achieved at each sampling period. Table 1 shows comparison between the proposed and the rectangular memory allocation for $Q = 4$ and $SR_{max} = 8$, *i.e.* a more practical parameters. The number of memory modules (AD units) are reduced to 19.4%. The search time is $30ms$ on the PC.

4 Conclusion

This paper presents an interconnection-aware design methodology for image processors. A key to success is a optimal memory allocation for a logic-in-memory architecture. The method is also useful for FPGA implementations where interconnection overhead in delay is significant large.

(0,0) 1	(1,0) 2	(2,0) 3	(3,0) 4	(4,0) 5	(5,0) 1	(6,0) 2	(7,0) 3	(8,0) 4	(9,0) 5
(0,1) 6	(1,1) 7	(2,1) 8	(3,1) 9	(4,1) 10	(5,1) 6	(6,1) 7	(7,1) 8	(8,1) 9	(9,1) 10
(0,2) 11	(1,2) 12	(2,2) 13	(3,2) 14	(4,2) 15	(5,2) 11	(6,2) 12	(7,2) 13	(8,2) 14	(9,2) 15
(0,3) 16	(1,3) 17	(2,3) 18	(3,3) 19	(4,3) 20	(5,3) 16	(6,3) 17	(7,3) 18	(8,3) 19	(9,3) 20
(0,4) 21	(1,4) 22	(2,4) 23	(3,4) 24	(4,4) 25	(5,4) 21	(6,4) 22	(7,4) 23	(8,4) 24	(9,4) 25
(0,5) 1	(1,5) 2	(2,5) 3	(3,5) 4	(4,5) 5	(5,5) 1	(6,5) 2	(7,5) 3	(8,5) 4	(9,5) 5
(0,6) 6	(1,6) 7	(2,6) 8	(3,6) 9	(4,6) 10	(5,6) 6	(6,6) 7	(7,6) 8	(8,6) 9	(9,6) 10
(0,7) 11	(1,7) 12	(2,7) 13	(3,7) 14	(4,7) 15	(5,7) 11	(6,7) 12	(7,7) 13	(8,7) 14	(9,7) 15
(0,8) 16	(1,8) 17	(2,8) 18	(3,8) 19	(4,8) 20	(5,8) 16	(6,8) 17	(7,8) 18	(8,8) 19	(9,8) 20
(0,9) 21	(1,9) 22	(2,9) 23	(3,9) 24	(4,9) 25	(5,9) 21	(6,9) 22	(7,9) 23	(8,9) 24	(9,9) 25

(a) Original image(SR=1)

(0,0) 1	(2,0) 3	(4,0) 5	(6,0) 2	(8,0) 4
(0,2) 11	(2,2) 13	(4,2) 15	(6,2) 12	(8,2) 14
(0,4) 21	(2,4) 23	(4,4) 25	(6,4) 22	(8,4) 24
(0,6) 6	(2,6) 8	(4,6) 10	(6,6) 7	(8,6) 9
(0,8) 16	(2,8) 18	(4,8) 20	(6,8) 17	(8,8) 19

(b) Reduced image (SR=2)

(0,0) 1	(3,0) 4	(6,0) 2	(9,0) 5
(0,3) 16	(3,3) 19	(6,3) 17	(9,3) 20
(0,6) 6	(3,6) 9	(6,6) 7	(9,6) 10
(0,9) 21	(3,9) 24	(6,9) 22	(9,9) 25

(c) Reduced image (SR=3)

Figure 15. Optimal memory allocation for multi-resolution images.

Table 1. Comparison between the proposed method and the rectangular allocation.

	Proposed	Rectangular Allocation
Number of modules	121	625
Number of AD units	121	625
Number of adders	123	624

Acknowledgment

This work was supported in part by industrial technology research grant program from New Energy and Industrial Technology Development Organization (NEDO) of Japan.

References

[1] D. Gajski, N. Dutt, A. Wu, and S. Lin, "HIGH-LEVEL SYNTHESIS -Introduction to Chip and System Design," Kluwer Academic Publishers, pp.277-278(1992).

[2] M. Hariyama, H. Sasaki, and M. Kameyama, "Architecture of a Stereo Matching VLSI Processor Based on Hierarchically Parallel Memory Access", IEICE Trans. Info. and Syst., Vol. E88-D, No.7,pp.1486-1491(2005).

[3] M. Hariyama, S. Lee, and M. Kameyama, "Highly-Parallel Stereo Vision VLSI Processor Based on an Optimal Parallel Memory Access Scheme", IEICE Trans. Electron., Vol.E84-C, No.5, pp.382-389(2003).

Globally Asynchronous Locally Synchronous Wrapper Circuit based on Clock Gating

Esmail Amini   Mehrdad Najibi   Hossein Pedram

IT and Computer Engineering Department, Amirkabir University of Technology, Tehran, IRAN.
{ es_amini, najibi, pedram@ce.aut.ac.ir }

Abstract

In this paper we propose an asynchronous wrapper with new asynchronous communication port controllers and reliable clock generation scheme for locally synchronous modules. This is achieved by utilizing clock gating idea within GALS wrappers which makes the use of reliable and robust off-chip clock generator possible for locally synchronous modules. In addition to clock robustness, the clock generator part becomes totally synchronous. To validate the proposed solution, we employed the wrapper circuit in Viterbi error detection and correction circuit. The synthesis results show that our GALS approach gains 44%~48% performance improvement in contrast to pausible clock GALS wrappers.

1. Introduction

By technology improvements and increasing complexity of VLSI systems, pure synchronous methodology doesn't look much promising to handle all design requirements. As a well-known example with increasing die size and clock frequency the clock skew problem is getting much troublesome. On the other hand, power consumption is getting even more important design consideration and large clock distribution network is one of the major sources of power consumption [1].

Asynchronous methodology is shown to be a good candidate solution which can solve nearly all the problems arising form the clock. While Asynchronous circuits potentially can have better performance, lower power consumption and more robustness, they can be larger due to handshaking overheads and they also suffer from design complexity and lack of automated CAD tools.

Globally Asynchronous Locally Synchronous or in abbreviate GALS methodology [2] is an intermediate solution that combines benefits of both synchronous and asynchronous methodologies.

The general architecture of pausible clock based GALS module is shown in figure 1. In GALS Systems,

synchronous modules are surrounded by asynchronous wrappers. In other words, GALS modules communicate with each other inside an asynchronous medium. A GALS module consists of a Locally Synchronous (LS) module, a clock generator and asynchronous port controllers. Asynchronous port controllers intervene between synchronous and asynchronous environments. Local clock generator is used to generate LS module clock signal.

Most contemporary GALS methodologies use on-chip clock generators to feed locally synchronous (LS) modules. Most of the proposed on-chip clock generators for GALS systems are based on pausible clocking which is based on on-chip ring oscillators [2][3][4][5][6]. Since on-chip ring oscillators is quite dependent to process variations, the generated clock signal will be unstable [7][8]. Therefore we should consider a marginal penalty in clock frequency and in result lower performance will be expected.

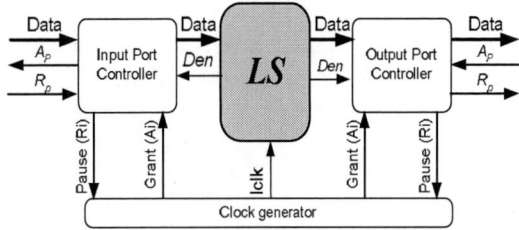

Figure 1- Pausible clock based GALS wrapper

When LS module needs data, it must inform the related port controller by toggling *Den* signal at the falling edge of the current clock pulse. Then port controller generates a request signal (R_i) for clock generator circuit which is responsible for synchronizing the LS module with incoming asynchronous data and stops the local clock. Afterward clock generator replies by an acknowledge signal (A_i) and stretches the low phase of the clock if needed. This prevents metastablity during asynchronous communication. The *Den* signal is activated on the falling edge of clock because using left edge of the

0-7695-2533-4/06 $20.00 © 2006 IEEE   199

clock leaves the controller of the LS module unchanged [8]. After data communication, the clock will be released. These sequences of events are shown in figure 2.

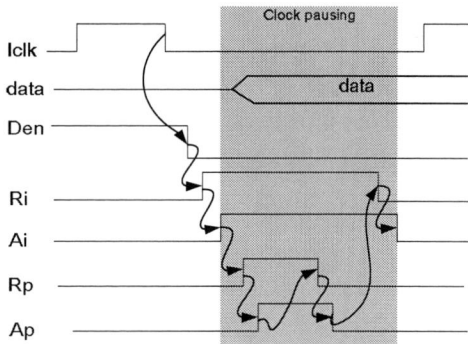

Figure 2- Timing diagram of clock pausing

The STG diagrams of pausible clock based GALS port controllers could be found in [4][8].

The circuit in figure 3 is a typical implementation of pausible clock generator [4].

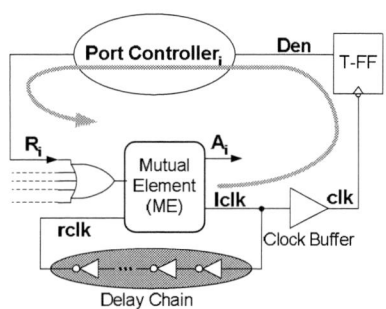

Figure 3- Pausible clock generation

The OR gate in figure 3 has N input for N port controllers inside wrapper. The Mutual Exclusion (ME) element in figure 3, arbitrates between R_i and rclk signals. If incoming data arrives too close to sampling clock edge, either the clock edge or data transfer is shifted to a later point in time [4]. It must be also taken into consideration that when the clock generator stops the clock, there may be still clock edges in the buffer tree and therefore, the delay of the chain must be longer than the delay of clock buffer and tree [8]. Also the delay of the port controller (T_{pc}) and delay of the OR gate (T_{or}) has much effect on LS modules frequency.

At this article a new method for generating local clock pulse is introduced that is based on clock gating. The clock gating is also used as a technique to reduce the power consumption in totally synchronous systems. However in totally synchronous systems, gating the clock would produce problems itself because it can worsen the clock skew problem. Because of the

simplifying skew nature of GALS methodology, clock gating is suitable in designing GALS systems. We applied clock gating idea in GALS wrapper circuit and compared it with pausible clock method introduced in [4]. There are few works done related to the clock gating the only notable work on using of clock gating method in GALS systems without any details and results has mentioned in [9].

The general architecture of the GALS wrapper circuit based on clock gating is shown in figure 4. In proposed wrapper circuit, each locally synchronous module has a local clock which is obtained by gating separated external clock (eclk) signal regarding to requests which comes from port controllers. When locally synchronous module enters data communication phase, it informs the related port controller that it needs data communication. Afterward port controller will generate a gate request for clock generator and the external clock will be gated. After handshake completion, the clock will be released.

Gated clock based GALS scheme has no major impact on the interface of LS module as it is clear by comparing Figure 1 and Figure 4. Thus the same LS modules can be used in both pausible and gated clock based GALS systems.

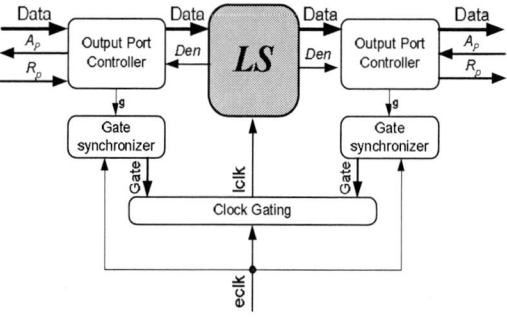

Figure 4- Gated clock based GALS wrapper

Because no handshaking is required between asynchronous port controllers and local clock generation circuit, the port controllers are simpler than pausible clock based port controllers. While simpler port controller will able us to utilize faster LS modules, it requires more considerations; Since there is no acknowledge signal which indicates proper gating of the clock, it is needed for the synthesis tool to guarantee that the gate request always reaches the clock generator with a reliable delay margin. It can be done by enforcing some timing constraints on the path from Den to the gate signal within each GALS module. In section 2 we introduced wrapper circuit in more details. Viterbi Error Detection and Correction GALS implementation as our benchmark is introduced in section 3. In section 4 we have shown the results of

0-7695-2533-4/06 $20.00 © 2006 IEEE

implemented circuits and final conclusion of the outcomes of this research is provided in Section 5.

2. Wrapper implementation

As stated before, each galsified block consists of a synchronous module, a clock generator and port controllers to communicate with asynchronous environment. Various parts of GALS wrapper shown in figure 4 is introduced in the following subsections.

2-1. Clock generation

The gated clock based clock generation circuit in GALS wrapper is as simple as usual clock gating as shown in figure 5. There is no asynchronous element in clock generator circuit. This simplifies implementing clock generator circuit.

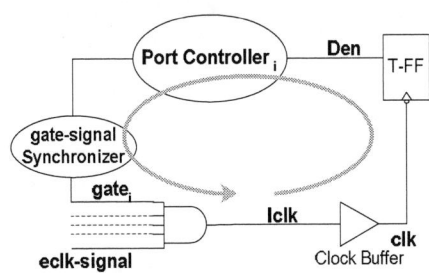

Figure 5- Clock gating in gated clock GALS wrapper

During the activity phase of LS module all of the gate-signals are high. When LS module enters data communication phase, one or more of the gate-signals go low and clock will be gated.

The period of the local clock is a multiple of external clock signal period because the gating and releasing of the clock is done when external clock signal is in transition.

As pausible clock based GALS scheme, the overall probability of stopping the clock relates to both the number of port controllers (*gate* signals) and *gate* signals overlapping. Also the propagation delay from *eclk-signal* to *gate* signal has to be smaller than the low phase of *eclk-signal*. This path consists of clock tree buffers, a T-flip flop, the port controller, gate signal synchronizer circuit and an AND gate as shown in figure 5. The AND gate has N+1 input where N is the number of port controllers inside wrapper. So by growing N, it takes more time to produce *gate* signal.

The larger the clock tree latency, the less time is available for generating *gate* signal. This scenario exists in pausible clock generation too. So for comparing two methods we will ignore the clock tree latency.

2-2. Port controllers

Each port controller is activated by Den signal which is generated by LS module. When LS module needs data

for next clock cycle,it activates Den signal at the coming negative edge of the current clock pulse like pausible clock based GALS scheme.

After Den signal activation, the port controller starts its work. At the first step, external clock will be gated to prevent metastablity during asynchronous data exchange. The flow of signal events to gate the eclk-signal is shown in figure 6.

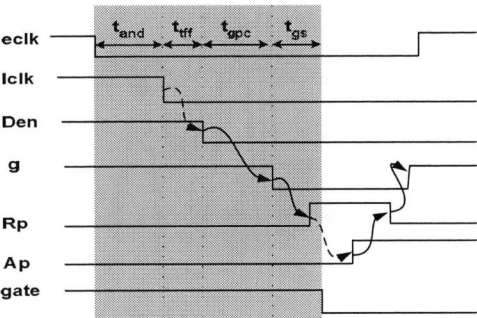

Figure 6- Timing diagram of external clock gating

In figure 6 the meaning of indicated times are:
-   t_{and}: delay of the AND gate in figure 5.
-   t_{tff}: the delay of T-flip flop to toggle Den-signal.
-   t_{pc}: delay of port controller
-   t_{gs}: delay of gate synchronizer circuit

The *gate* signal must be generated before the rising edge of *eclk-signal* to prevent metastablity problem during asynchronous communication and ensure that incorrect data not being processed by LS module. Thus the gray rectangle in figure 6 should not exceed the incoming rise edge of eclk-signal. In the other words we must guarantee $t_{and}+t_{tff}+t_{gpc}+t_{gs}<t_{eclk}/2$.

The STG diagram of the asynchronous port controller in clock gating wrapper circuit is mentioned in figure7. Petrify [10] is used for synthesis of the port controllers and the synthesis results for port controllers for tsmc018 library cells are shown in figure 8.

2-3. Gate synchronizer

Gating the eclk-signal must not be done later than the next positive clock edge. As stated before, this can be guaranteed by enforcing timing constraints on GALS modules during synthesis process.

Releasing the clock in the high phase of external clock will decrease the clock period (figure 9) so it must be avoided. To avoid this, the gate synchronizer circuit is used to generate the final gate-signal from port controller's output (g-signal) and eclk (figure 10). This prevents any spike or clock period reduction.

In the circuit in figure 10, gating will be done as soon as g-signal goes high. It is done by lowering the gate-signal so the external clock will be gated.

Input port controller STG diagram

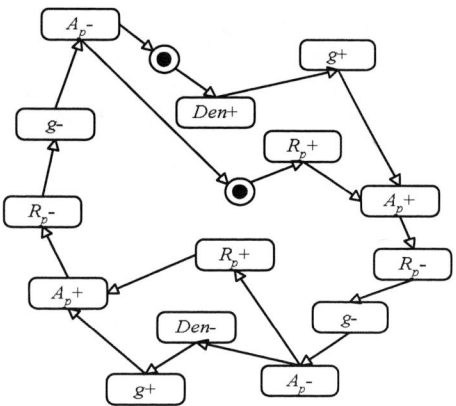

Output port controller STG diagram

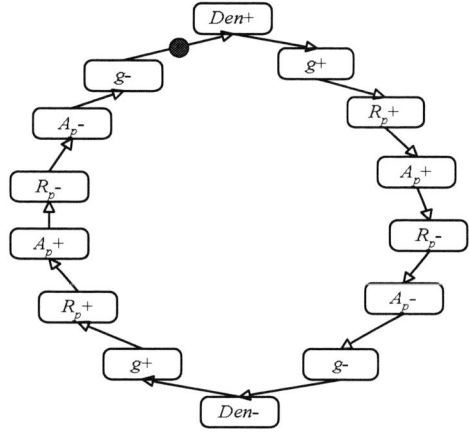

Figure 7- STG diagram of the port controllers

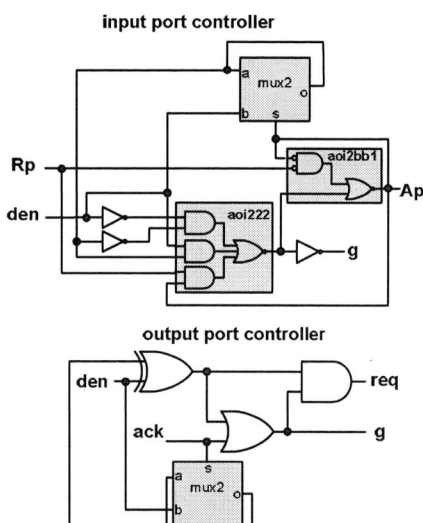

Figure 8- Gated clock based Port controllers

The time for the gate signal to stop the eclk signal is depicted in figure 10, by solid Line (T3—T5—T8).

The clock will be released when both g-signal and eclk-signal are in low state. It takes the normal time to release the eclk dedicated by dashed line in figure10 (T2—T6—T7).

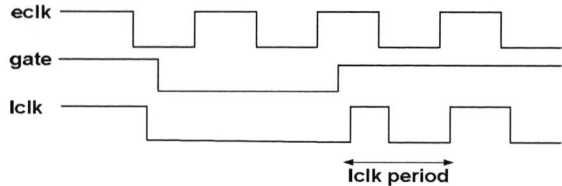

Figure 9- Invalid time for clock releasing

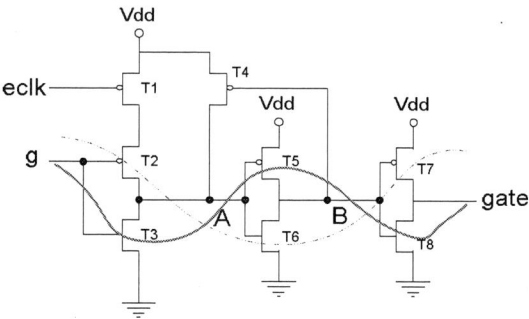

Figure 10- Gate-signal synchronizer

The releasing scenario is as following:
Suppose that the external clock is gated. Thus the value of A-signal is low. Depending on eclk-signal value, one of the following situations can occur:

- *eclk-signal* is high: in this situation A is not connected directly to Vdd or GND until the *eclk-signal* changes its value; A remains low and in result *gate-signal* will remain low. When the *eclk-signal* becomes low, *A-signal* goes high and clock will be released. During this state transistor T4 will keep the A's value.

- *eclk-signal* is low: In this situation *A-signal* goes high, in result *gate-signal* goes high and clock will be released. While clock is in released mode, transistor T4 will charge A's value.

- *eclk-signal* is in transition mode: In this situation transistor T4 will prevent node A from getting charged to an intermediate value because regarding to inverter N1, *B-signal* could have one of the following values:
  - The first inverter gate (T5 and T6) assumes *A-signal* is low. So *B-signal* is high, afterward *gate-signal* goes low and clock remains gated.
  - The first inverter assumes *A-signal* is high. So *B-signal* goes low and *gate-signal* becomes high meanwhile T4 charges *A-signal* to high.
  - Depend on A's value, both T5 and T6 may be ON. Thus *B-signal* would be between low and high. This state is unstable because transistor T4

0-7695-2533-4/06 $20.00 © 2006 IEEE   202

will charge *A-signal*. So clock will be released eventually. In this situation, the releasing path of *eclk-signal* is T2—T5—T4—T6—T7.

This analysis shows that the clock will be gated and released safely in all situations, and we do not face the metastablity problem. This is mainly correct because the local clock will be always in its low phase during asynchronous data communication.

3. GALS Implementation of Viterbi Circuit

As of benchmarking, Viterbi error detection and correction circuit is used. In Viterbi (2,1,6) circuit for each cycle, a 2-bit input stream is entered and 1-bit output is generated. For each input stream, the cost of each branch metric (BM) cost is calculated by branch metric unit (BMU). After calculating BM, the total path metric (PM) cost is calculated by path metric unit (PMU). We keep the PM costs in Survivor Path Storage. The Trace Back Unit reconstructs the true path from the data in Trace Back Unit and sends the result to shift register. Shift Register gets the 32-bit input vector and generates 32 single bit output signals. The details of the implementation of the Viterbi could be found in [11]. The Structure of Viterbi GALS Implementation is shown in figure 11.

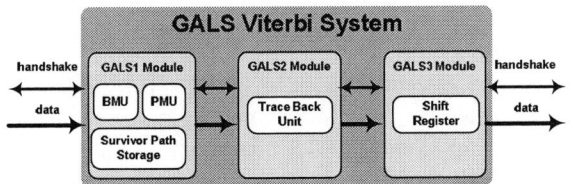

Figure 11- Viterbi GALS block diagram

The bottleneck of Viterbi implementation is the Trace Back Unit. Fortunately Trace Back Unit is not a busy module and is off in most of times. If we encapsulate BMU, PMU and Survivor Path Storage by a single GALS wrapper, Trace Back Unit in a separate GALS wrapper and shift register in another wrapper, we could expect power saving and performance improvement in contrast to totally synchronous implementation.

4. Results

For synthesizing port controllers in both pausible and gated clock based GALS schemes, we must prevent synthesis tool from optimizing port controllers, because Petrify resulted circuits must be instantiated exactly in wrapper circuit to guarantee speed independent nature of generated circuits. After acquiring timing reports of synthesis tool, we simulated synthesized port controllers. The delays of asynchronous port controllers for both pausible clock based wrapper circuit and gated clock based wrapper circuit, synthesized in 0.18μ technology are shown in

Table 1. To estimate the delay of gate-signal synchronizer circuit, we simulated the gate synchronizer circuit in 0.18μ technology. GM1, GM2 and GM3 modules in Viterbi GALS system (figure 11) each need one input port controller and one output port controller for asynchronous data communication. By the way, the AND gate in figure 5 has 3 inputs.

Table 1- pausible clock based wrapper port delays

Wrapper component		Delay
Pausible clock based wrapper	Input port latency (T_{ppc}+ T_{or})	210
	Output port latency(T_{ppc}+T_{or})	410
Gated clock based wrapper	Input port latency (Tgpc)	112
	Output port latency (Tgpc)	200
	gate-signal synchronizer clock gating (Tgs)	90 ps
	AND gate delay (Tand)	90 ps

In pausible clock based GALS wrapper, the time it takes to pause the clock is 0.41 ns. In gated clock GALS wrapper, the time it takes to gate the clock is 0.2 ns. By using of gated clock based GALS wrapper, the minimum low phase duration of clock-signal is $t_{and}+t_{gpc}+t_{gs}=0.38$ ns. So in gated clock GALS Systems, LS modules can operate in higher frequencies.

After synthesizing all Viterbi modules in 0.18μ technology for the totally synchronous circuit the maximum clock frequency was reported as 166 MHz. The Viterbi GALS implementation results for locally synchronous modules mentioned as follow:

Table 2- GALS Viterbi Implementation Results

Module	clock Frequency	Activation
GM1 LS module	256 MHz	32
GM2 LS module	153 MHz	1
GM3 LS module	555 MHz	32

The LS modules can operate up to the frequencies shown in Table1. Note that these results are valid both for clock pausing and clock gating schemes, so the analysis of both systems can be done in the same way; If we ignore the environment delays due to the special structure of the circuit, pausing or gating of the clock-signal and also releasing it occurs before the rising edge of clock pulse of Viterbi GALS modules. In other words no clock pausing or gating will occur in case of the dominance of GM1 module in Viterbi GALS system because of its operational duration. Thus the frequency of the implemented Viterbi GALS system is 256 MHz. So the maximum frequency of GM1 module is 54% higher than totally synchronous Viterbi circuit frequency. By the way, implementing GALS methodology has performance yield of 0.54 in Viterbi error detection and correction system.

There are some performance variations between pausible and gated clock method regarding to the difference in their clock generator behavior.

In pausible clock based GALS scheme we generate the clock signal from chain of inverters and the delay of the chain specifies the generated clock frequency (figure 3). To calculate the number of inverters that can guarantee the critical path timing requirements of the circuit for pausible clock scheme, we have to use fast model for chained inverters, while using typical model for the circuits themselves. The clock gated system can take advantage of eliminating these margins and can result in notable performance improvement. The fallowing typical analysis shows the order of the possible performance improvement in 0.18μ technology. The chain of inverters post synthesis results can be seen in table 3 for both fast and typical delay models.

Table 3- Modules frequencies in process modes

Circuit		GM1	GM2	GM3
Pausible clock based Wrapper	*Fast mode frequency*	254 MHz	150 MHz	554 MHz
	Typical mode frequency	177 MHz	103 MHz	385 MHz
Gated clock based wrapper		256 MHz	153 MHz	555 MHz

By using gated clock scheme, an external clock can be used with least amount of frequency variation. The LS modules in gated clock based GALS scheme have a performance improvement of 44%~48% in contrast to LS modules in pausible clock based GALS scheme as can be computed from table 3. This performance improvement as stated before is due to the elimination of the extra clock margins required for pausible clocking. So in gated clock based GALS scheme 44%~48% performance can be achieved, moving from fast to typical operation models in 0.18μ technology.

5. Conclusion and Future Work

GALS methodology smoothes clock skew problems and utilizes the benefits of both synchronous and asynchronous design. While using on chip pausible clock generators in GALS wrapper circuits, leads to a tradeoff between robustness and performance, we have shown that both performance and robustness can be improved using the gated clock scheme.

Gated clock GALS systems combines the benefits of GALS design and clock gating idea. The new GALS wrapper circuit has an external clock input that is gated whenever asynchronous communication is needed. Better stability and robustness in our clock generator circuit is the direct consequence of using an external more robust, stable and reliable clock generator. By using more stable clock signal in gated clock scheme, we would have in turn improvements in performance. Also converting the pausible clock GALS

implementations to gated clock GALS implementation, is as simple as replacing port controllers and clock generator with proposed circuits and the LS modules need no change.

Meanwhile gated clock GALS systems if can be implemented as Multi FPGA can present more benefits in prototyping and less design time, it requires the asynchronous port controllers to be mapped into FPGA LUT's which is our future target.

6. References

[1] A. Hemani, T. Meincke, et al. Lowering power consumption in clock by using globally asynchronous, locally synchronous design style. *In Proc. ACM/IEEE Design Automation Conference*, pp. 873-878, June 1999.

[2] D.M. Chapiro. Globally-Asynchronous Locally-Synchronous Systems. *PhD thesis*, Stanford University, 1984.

[3] David S. Bormann and Peter Y.K. Cheung. Asynchronous wrapper for heterogeneous systems. *In Proc.International Conf. Computer Design (ICCD)*, October 1997.

[4] J. Muttersbach, T. Villiger, and W. Fichtner. Practical Design of Globally-Asynchronous Locally-Synchronous Systems. *In Proc. International Symposium on Advanced Research in Asynchronous Circuits and Systems*, April 2000.

[5] K.Y. Yun and A. E. Dooply, Pausible clocking based heterogeneous systems. *IEEE Transactions on VLSI Systems*, December 1999.

[6] S. Moore, G. Taylor, R. Mullins and P. Robinson. Point to point GALS interconnect. *In Proc. International Symposium on Advanced Research in Asynchronous Circuits and Systems*, pages 69-75, April 2002.

[7] Charles L. Seitz. System timing. In Carver A. Mead and Lynn A. Conway, editors, Introduction to VLSI Systems, chapter 7. Addison-Wesley, 1980.

[8] K. Saleh, M. Najibi, M. Naderi, H. Pedram, M. Sedighi, A Novel Clock Generation Scheme for Globally Asynchronous Locally Synchronous Systems: An FPGA-Validated Approach, *Proceedings of the 15th GLSVLSI ,Chicago*, April 2005.

[9] M. Singh, M. Theobald. Generalized Latency-Insensitive Systems for GALS Architectures. Proc. of the Workshop on Formal Methods for Globally Asynchronous Locally Synchronous (GALS) Architecture (FMGALS-03), held in conjunction with the 12th International Formal Methods. Europe Symposium, Pisa, Italy, September 2003.

[10] J. Cortadella, M. Kishinevsky, A. Kondratyev, L. Lavagno, and A. Yakovlev. Petrify: a tool for manipulating concurrent specifications and synthesis of asynchronous controllers. In *IEICE Trans. on Information and Systems*, pp. 315-325. March 1997.

[11] K.Saleh. Viterbi Hardware Implementation.Amirkabir University of Technology. Internal Report. June 2004.

0-7695-2533-4/06 $20.00 © 2006 IEEE

Connection-oriented Multicasting in Wormhole-switched Networks on Chip

Zhonghai Lu, Bei Yin and Axel Jantsch
Laboratory of Electronics and Computer Systems
Royal Institute of Technology, Sweden
{zhonghai,axel}@imit.kth.se, {beiy}@kth.se

Abstract

Network-on-Chip (NoC) proposes networks to replace buses as a scalable global communication interconnect for future SoC designs. However, a bus is very efficient in broadcasting. As the system size scales up to explore the chip capacity, broadcasting in NoCs must be efficiently supported. This paper presents a novel multicast scheme in wormhole-switched NoCs. By this scheme, a multicast procedure consists of establishment, communication and release phase. A multicast group can request to reserve virtual channels during establishment and has priority on arbitration of link bandwidth. This multicasting method has been effectively implemented in a mesh network with deadlock freedom. Our experiments show that the multicast technique improves throughput, and does not exhibit significant impact on unicast performance in a network with mixed unicast and multicast traffic if the network is not saturated.

1 Introduction

As the technology steadily scales, chip design is increasingly becoming communication-bound. Network-on-Chip [1, 5, 10] addresses the design challenges by proposing networks to replace buses as a scalable global communication platform. In a NoC, heterogeneous resources such as processors, DSPs, FPGAs/ASICs, and memories are interconnected by switches. These resources communicate by routing packets instead of using dedicated wires.

Buses (a single bus, segmented or crossbar-type buses and a hierarchy of buses) do not scale well with the system size in bandwidth and clocking frequency. However, a bus is very efficient in broadcasting since all clients are directly connected to it. A network allows many more concurrent transactions, but it does not directly support multicast. As there exists a variety of SoC applications, many applications necessitates to support multicast in the case of passing global states, managing and configuring the network, and implementing cache coherency protocols etc. Particularly, real-time constrained, throughput-oriented embedded applications for multi-media processing will demand

an efficient means to implement multicast. One crucial aspect for supporting multicast in SoCs is Quality-of-Service (QoS), which means that the performance of multicast traffic should be predictable. Implementing multicast by sending multiple unicast messages is neither efficient nor scalable. In addition, in a network with a mixture of unicast and multicast traffic, multicast traffic should not degrade the performance of unicast traffic since multicast traffic takes only a portion of the total network traffic.

In this paper we present a connection-oriented multicast scheme in wormhole-switched networks on chip. Wormhole switching [3] is a network flow control mechanism that allocates buffers and physical channels (PCs) to flits instead of packets. A packet is encapsulated into one or more flits. A flit, the smallest unit on which flow control is performed, can advance once buffering in the next hop is available to hold the flit. This results in that the flits of a packet are delivered in a pipeline fashion. In order to make an efficient use of link bandwidth, wormhole switching can employ *virtual channels* (VCs or lanes) to enhance throughput [2]. Because of these advantages, namely, *better performance, smaller buffering requirement* and *greater throughput*, wormhole switching with lanes is being advocated for on-chip networks [5, 8].

By our multicast scheme, multicasting consists of three phases: *group setup, communication,* and *group release.* A multicast is realized by sending a single copy of multicast packets to multicast group members along a pre-established path. This results in low packet overhead for multicasting. During the setup phase, multicasting can be aware of QoS in the sense that a multicast group may request to reserve VCs, and enjoy a higher priority against unicast packets for link bandwidth arbitration. Although the three-phase (setup, transmission, and release) communication has been used for establishing virtual-circuit communication to support QoS in store-and-forward packet-switched networks, applying the technique to implement multicast in a wormhole-switched network on chip is the novel aspect of this paper. Moreover, we shall look at how much impact multicast traffic will exert on unicast traffic, and the performance tradeoff between *multicast without VC reservation* and *multicast with VC reservation.*

0-7695-2533-4/06 $20.00 © 2006 IEEE

2 Related Work

Multicasting in wormhole-switched networks has been extensively studied in parallel machines in order to support collective communications such as barrier synchronization, reduction and global combining [6, 9]. Multicast can be achieved via software or hardware approach. With software implementation of multicast, a multicast operation is implemented by sending a separate copy of the messages from the source node to every destination or to a subset of destinations, each of which in turn forwards the message to one or more other destinations in a multicast tree.

As the software approach is not efficient enough, hardware support of multicast communication is proposed. With the *tree-based multicast* [9], the destination set is partitioned at the source, and separate copies of the message are transmitted. A message may be replicated at intermediate nodes and forwarded to disjoint subsets of destinations in the tree. This scheme does not perform well to be deadlock free unless messages are very short because the entire tree is blocked if any of its branches are blocked. A solution is to forbid branching at intermediate nodes, leading to a multicast path pattern, called *path-based multicast* [6]. In order to reduce the length of the multicast path, the set of destination nodes may be divided into multiple disjoint subsets. A copy of the source message is sent across several multicast paths, each path for each subset of the destination nodes. In this scheme, multicasting is realized by sending multi-destination messages. The header of multi-destination messages must carry the addresses of all the destination nodes. As the header is an overhead, the message latency is increased and the effective network bandwidth is reduced. Besides, multicast traffic does not reserve network resources, equally competing with unicast traffic for buffers and link bandwidth, resulting in no QoS for multicasting.

By the traditional multicast schemes, group formation and multicast communication are not decoupled. The multicast-packet overhead is high and there is no QoS concern. In our multicast scheme, there is an explicit multicast group setup phase. After a group is set up, multi-destination messages carry only the group identity number not the addresses of all the destination nodes. In addition, a multicast group can be aware of QoS by reserving lanes for performance enhancement.

For a circuit-switched network on chip, a multicasting scheme using global traffic information is proposed in [7]. This scheme is difficult to scale to a large system size since it relies on the global network state. Connection-oriented communication has been proposed in the Mango [1] and Æthereal [5] NoCs to achieve QoS for unicasting. In Mango, connections are created using asynchronous/clockless circuitry. By reserving link bandwidth, Æthereal builds connections to provide a virtual contention-less path from sources to destinations. We use the connection-oriented technique to realize QoS-aware multicasting in a best-effort network. As stated, the contention-free route in the Æthereal NoC [5] and the looped containers [10] in the Nostrum NoC can be used to realize multicasting. But no concrete results are released so far.

3 The Multicast Scheme

3.1 Unicast in wormhole networks

Figure 1 sketches an input-buffering wormhole switch with lanes. It employs credit-based link-level flow control to coordinate packet delivery between switches.

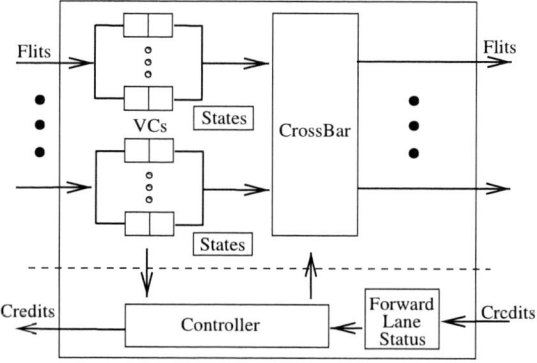

Figure 1: An input-buffering wormhole switch

A packet is segmented into flits, which are then delivered in the network. After the segmentation, a packet is typically composed of a head flit, a tail flit and body flit(s). A single-flit packet is also possible. A packet passes the switch through four states: *routing, lane allocation, flit scheduling*, and *switch arbitration*. In the routing state, the routing logic determines the routing path the packet advances. Routing is performed only when the head flit of a packet becomes the earliest-come flit in the lane. This means that if flits of a previous packet still stay in the lane, the routing will not be performed. Only when the earlier-coming flits are switched out, the head flit becomes the earliest-come flit. Then routing is performed, and the packet path and output physical channel are determined. In the state of lane allocation, the lane allocator *associates* the lane the packet occupies with an available lane in the next hop on its routing path, i.e., to make a *lane-to-lane* association. Note that it is not necessarily required that there is an empty buffer in the lane in order for the lane to be associated or allocated. A lane-to-lane association fails when all requested lanes in the next hop are already associated to other lanes in directly connected switches. If the lane-to-lane association succeeds, the packet enters into the scheduling state. If there is a buffer available in the associated lane, the lane

0-7695-2533-4/06 $20.00 © 2006 IEEE

enters into the switch arbitration. The first level of arbitration is performed on the lanes sharing the same physical channel. The second level of arbitration is for the crossbar traversal to output physical channels. If the lane wins the two levels of arbitration, the earliest-come flit in the lane is switched out. Otherwise, the lane returns back to the scheduling state. Once the tail flit is switched out, the lane-to-lane association is released, thus the allocated lane is available to be used by other packets. Credits are passed between adjacent switches in order to keep track of the statues of downstream/forward lanes, such as if a lane is free, and a count of available buffers in the lane.

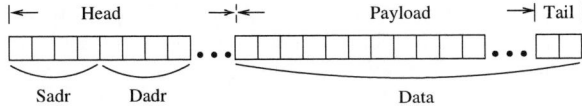

Figure 2: Unicast packet format

Figure 2 shows a typical unicast packet format, which consists of a head, a payload and a tail. The head and tail are the overhead for transmitting the payload. The head typically consists of routing and sequencing information. Basic routing information includes source address (sadr) and destination address (dadr). A switch uses the destination address to perform routing and switches the packet to the right output physical channel (PC). When the packet is split into flits, each flit contains a flit type field to identify if it is a head (H), body (B), tail (T) flit, or a single-flit packet.

3.2 The multicasting protocol

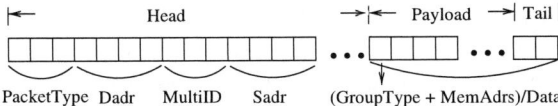

Figure 3: Multicast packet format

In order to support multicasting, we expand the packet format into that shown in Figure 3. We explain the packet fields as follows:

- **PacketType** indicates the purpose of a packet. It has six options, namely, *unicast, multicast setup, multicast setup response, multicast data, multicast group release* and *multicast group release acknowledgment*.

- **Dadr**: the destination address. In the case of multicast, it is the address of the next group member.

- **MultiID**: the multicast group identity number, which is unique for each multicast group to be established.

- **Sadr**: the source address. In the case of multicast, it is the address of the node that initiates the multicast setup. This node is called *group master*.

- **GroupType**: the type of the multicast group. It is used to inform the switches whether the multicast group will reserve a lane or not. It occupies only one bit. One group is allowed to reserve only one lane in a switch since the number of lanes is limited.

- **MemAdrs**: the multicast members' addresses. The order of the address list specifies the multicast path [1]. Specifically, the node with the first address in the list will be reached first and then second and so on. Upon reaching the node with address **Dadr**, the next member address in the list will replace the **Dadr** field.

Fields **GroupType** and **MemAdrs** are only needed for multicast setup packets, **MultiID** for multicast packets. A response packet for multicast group setup or release is handled as a unicast packet. By our scheme, a multicast group can be established and released dynamically. A multicasting procedure consists of three phases explained as follows:

1. *Group establishment*: First the group master sends a setup packet, which passes downstream to all the group member nodes along the predetermined path as indicated by **MemAdrs**. When the setup packet reaches a node, the switch records the multicast information and reserves resources according to the group type. This record will be used later to transmit multicast data. If the setup packet reaches the last group member, a setup response packet will be sent back to the group master to acknowledge the success. If the setup fails in a node, for example, due to lane unavailability, a response packet will be sent back to the master from the current node.

2. *Multicast communication*: After a successful setup, the master can send multicast data packets. The packets carrying **MultiID** will be transmitted along the same path and the same VCs which the setup packet used before. When a data packet reaches a destination node in the group, its payload is replicated and the packet is forwarded to the next member. In this way, all the members will receive the packet. The group members can also send multicast data packets, but only to the members in the downstream since a multicasting path is simplex and thus only simplex communication is allowed.

3. *Group release*: A group can only be released by its master by sending a release packet to its members. When the release packet reaches a node, the multicast record in the switch and the reserved lane will be freed after all on-going group transactions complete. Upon reaching the last member, a release acknowledgment is sent back to the master.

---

[1]In our approach, the multicast setup path can be diverse, and it shall follow the path-based schemes. But this is not the focus of the paper.

3.3 Multicast implementation

3.3.1 Extending the unicast switch

We have implemented the unicast switch in VHDL according to the model shown in Figure 1. The implementation consists of a data path and a control path. The data path is concerned with the flit movement through the crossbar and virtual channel. The control path realizes the functionality of the controller. For flit ejection, we implement a p-sink model to reduce cost [8]. By this model, a switch uses p flit sinks to eject flits, where p is typically equal to the number of PCs per switch. These p sinks are shared by the $p \cdot v$ lanes, where v is the number of VCs per PC.

Based on the unicast switch model, we have implemented the multicast scheme. The resultant switch supports both unicast and multicast. The data path is maintained the same as unicast while the control path is complicated. Specifically, the controller is extended to distinguish different packet types and perform actions according to the protocol. The switch must record the multicast information for each group passing it. The record of a multicast group includes {**MultiID**, **GroupType**, **Sadr**, **VCID**, **VCID downstream**, **output PC**, **next member adr.**}, where **VCID** is the identity number of the lane a multicast packet passes in the current switch; **VCID downstream** is the lane allocated downstream; **output PC** is the output physical channel the packet is to be switched out; **GroupType** indicates if the group reserves a lane or not. The current implementation arbitrates link bandwidth in favor of multicast traffic.

3.3.2 Deadlock avoidance

Deadlock is catastrophic to a network. It happens when a packet waits for an event that cannot happen. For example, a group of packets are unable to make progress because of waiting on one another to release buffers or channels. Forbidding such a cyclic resource dependency is a sufficient condition to design a deadlock-free network. Deadlock is related to many factors such as the network topology, flow control scheme, communication protocol and so on. Restricting the routing choice and adding buffer classes are the basic ways to deal with it.

Our multicast scheme has been implemented in a 2D mesh network employing dimension-order XY routing, which is proven to be deadlock free for unicast traffic on meshes. We constrain that a multicast path follows XY routing. This removes cyclic dependencies involving the two dimensions. However, care must be taken when planning a multicast path. Resulting from an improper path, a multicast packet may involve the turn from Y to X. As shown in Figure 4, a cycle is formed if the group is organized as $A \rightarrow B \rightarrow C \rightarrow D$. To avoid such a cycle, a group path must be organized so that no turn from Y to X

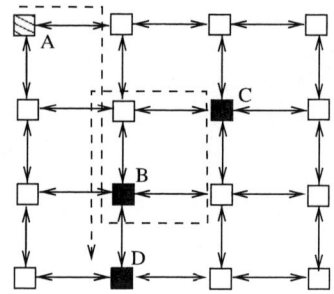

Figure 4: A cycle in a group

is resultant. For example, the group may be organized as $A \rightarrow C \rightarrow B \rightarrow D$. This simplification avoids deadlock at the expense of restrictions on planning groups and paths.

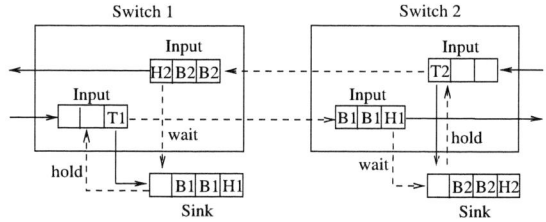

Figure 5: Deadlock while sinking and forwarding

Since a multicast packet has to sink locally at a member node and meanwhile to be forwarded downstream, the allocation of both a local sink and a lane in the next hop must be successful. In order not to introduce extra buffers and complicate the control, we decide to sink a flit when both the allocation conditions are true. This means that we perform sinking and forwarding on a multicast flit simultaneously. This may lead to deadlock with the p-sink ejection. As illustrated in Figure 5, two four-flit multicast packets pass two adjacent switches. While sinking and forwarding, both packets hold the sink the other waits for. The dashed lines in Figure 5 shows the wait-for graph [3], which forms a dependency cycle. Two solutions may be used to remove the cycle. One is to make the lane size long enough to hold an entire packet. The other is to ensure that the actual number p of sinks is larger than the number of multicast groups passing a switch, and a multicast packet does not wait for the availability of a particular sink. This guarantees that there is at least one sink available to break a possible dependency cycle due to the exhaustion of sink resources.

When a multicast setup fails, a negative response (nack) packet will be sent from the failing switch back to the group master, which in turn will send a release packet to the network. This may create a dependent loop (A positive response does not cause a loop). By adopting the technique of the credit-based end-to-end flow control and separate buffer classes in [4], we are certain that this never causes deadlock.

4 Experiments

The purposes of our experiments are to (1) compare multicasting with unicasting multiple packets; (2) investigate the impact of multicast traffic on unicast traffic in a mixed unicast-multicast network; (3) evaluate the multicast scheme with/without lane reservation.

Using the multicast-supported wormhole model, we construct a 1×7 and 4×4 mesh. The networks operate synchronously. With the switch model, it takes 5 cycles for a head flit and 3 cycles for other flits to pass through a switch. Each switch has the same configuration parameters as follows: the number of VCs per PC is four for the 1×7 mesh and six for the 4×4 mesh; the depth of a VC is two, which is the minimal number in order to pipeline flits; the number of sinks is eight for the 1×7 mesh and 24 for the 4×4 mesh. Four-flit packets are injected into the network synchronously at a constant rate. Each node has a workload of 1000 packets. Simulations terminate when anyone of the nodes completes transmission. Latency of a packet is recorded from the instant that the packet is queued in the source FIFO to that the packet is ejected from the network. Network load is the average percentage of active links through the simulation cycles. Throughput is defined as the number of packets received per cycle per node.

4.1 Multicast vs. unicast

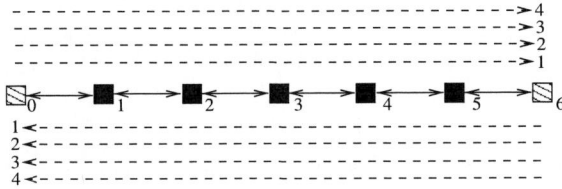

Figure 6: Traffic scenario in the 7-ary 1-mesh

As shown in Figure 6, we use the seven-node array in this set of experiments. Either unicast or multicast packets exist in the network. In the case of pure unicast, sending packets to multiple destinations is implemented by unicasting. In the experiments, node 0 and 6 send packets to the other six nodes randomly. In the multicast setting, four groups per direction are created, and the multicast groups do not reserve lanes. Node 0 and 6 are the group masters, which send multicast packets to the four groups alternatively.

Figure 7 depicts the results. With the same injection rate, the network is more loaded with the multicast, since multicast packets are delivered to all other nodes until reaching the other end while a unicast packet is sent to a particular node. The latency is worse with the multicast packets, this is due to the co-allocation of lane and sink for a multicast packet. However, the throughput of the multicast case is six

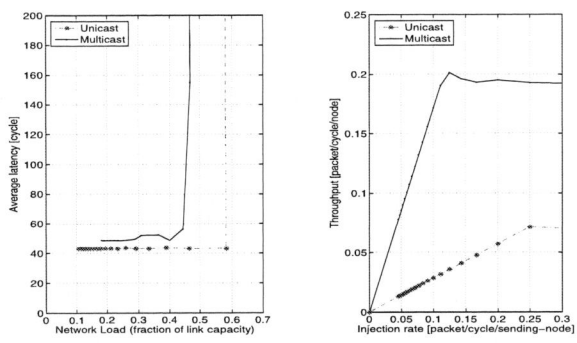

Figure 7: Multicast vs. unicast performance

times as much as that of the unicast case before the multicast network reaches saturation.

4.2 Multicast vs. mixed traffic

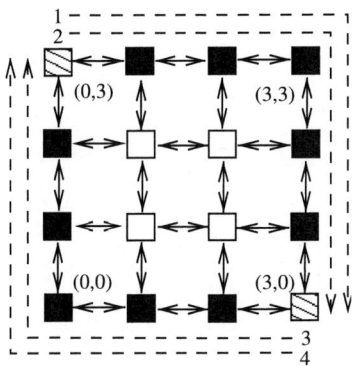

Figure 8: Traffic scenario in the mesh

The 4×4 mesh is used, and two traffic scenarios are created for the following experiments. One is *purely unicast traffic*. All network nodes send unicast packets to random destinations except themselves. The other is a *mixed unicast-multicast* scenario where four multicast groups start establishment upon simulation starts (meanwhile, unicast traffic is also injected.). As illustrated in Figure 8, four multicast groups are set up. From node (0, 3) to node (3, 0), group 1 and 2 are built following +X to -Y; from node (3, 0) to node (0, 3), group 3 and 4 are built following -X to +Y. Only the group masters send multicast packets to their members. They send one multicast packet every four packets. If a multicast packet is sent to n members, the amount of traffic is counted as n packets. The resultant multicast traffic takes 16.2% percent of the total network traffic.

We consider two cases. **Case 1**: the multicast groups do not reserve VCs (lanes); **Case 2**: the multicast groups reserve VCs. Figure 9 and Figure 10 draw the network per-

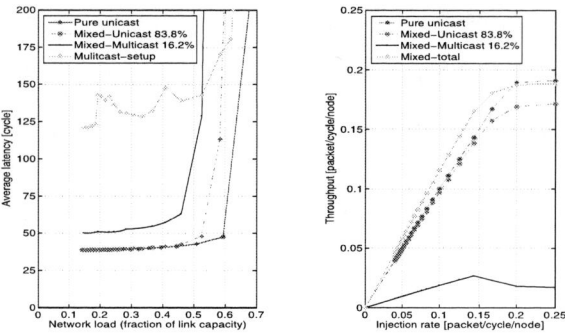

Figure 9: Performance without VC reservation

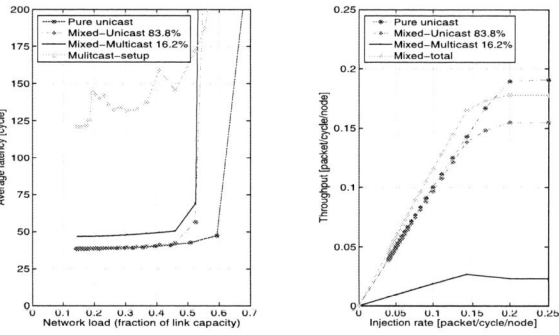

Figure 10: Performance with VC reservation

formance with the two scenarios for case 1 and case 2, respectively. In comparison with the purely unicast traffic, the unicast in the mixed traffic scenario performs equivalently to each other when the network load is below 0.45 for both cases. This means that the multicast traffic does not degrade the unicast performance if the network is not overloaded. The throughput is higher with the mixed traffic if the network is not saturated, since a multicast uses link bandwidth more efficiently. As shown in Figure 10, if a group reserves lanes, the average latency of the multicast traffic is improved 4.6 cycles on average if the network operates below load 0.45. However, due to lane reservation, the network saturation throughput is decreased by 4.2% from 0.185 to 0.177 packet/cycle/node.

As can be observed in Figure 9 and 10, the average group setup latency is 3-4 times as much as the average latency of unicast packets even when the network is not overloaded, since a group setup takes at least a round-trip time. This overhead suggests that our multicast scheme is beneficial to send block data where the amount of multicast traffic is high, if a multicast group is to be established dynamically. If a group is set up statically during the system warm-up phase, this overhead may be ameliorated.

5 Conclusions

We have presented our multicasting scheme in wormhole-switched networks on chip. With this scheme, multicasting starts after a multicast group is established. During establishment, a multicast group can reserve virtual channels. Our experimental results suggest that multicasting is beneficial in throughput. In addition, in a network with mixed unicast and multicast traffic, the multicast traffic does not show negative impact on the performance of unicast traffic if the network is not saturated. With the lane reservation, the latency of multicast traffic can be improved at the expense of slightly decreased throughput.

In future work, we aim at designing a synthesizable wormhole switch supporting the multicast scheme in order to obtain its cost overhead in area and speed penalty. The optimization of the controller will be essential for enhancing the hardware speed. Another direction is to use the QoS-aware multicast communication to emulate traditional buses. This could potentially address the problem of large amount of legacy code written for buses.

References

[1] T. Bjerregaard and J. Sparso. A router architecture for connection-oriented service guarantees in the MANGO clockless network-on-chip. In *Proceedings of the Design, Automation and Test in Europe Conference*, volume 2, pages 1226–1231, 2005.

[2] W. J. Dally. Virtual-channel flow control. *IEEE Transactions on Parallel and Distributed Systems*, 3(2):194–204, March 1992.

[3] W. J. Dally and B. Towles. *Principles and Practices of Interconnection Networks*. Morgan Kaufman Publishers, 2004.

[4] B. Gebremichael, F. Vaandrager, M. Zhang, K. Goossens, E. Rijpkema, and A. Rădulescu. Deadlock prevention in the Æthereal protocol. In D. Borrione and W. Paul, editors, *Proc. Working Conference on Correct Hardware Design and Verification Methods (CHARME)*, volume 3725 of *Lecture Notes in Computer Science (LNCS)*, pages 345–348, Oct. 2005.

[5] K. Goossens, J. Dielissen, and A. Rădulescu. The Æthereal network on chip: Concepts, architectures, and implementations. *IEEE Design and Test of Computers*, 22(5):21–31, Sept-Oct 2005.

[6] X. Lin, P. K. McKinley, and L. M. Ni. Deadlock-free multicast wormhole routing in 2-d mesh multicomputers. *IEEE Transactions on Parallel and Distributed Systems*, 5(8):793–804, 1994.

[7] J. Liu, L.-R. Zheng, and H. Tenhunen. Interconnect intellectual property for network-on-chip. *Journal of System Architectures, Special issue on networks on chip*, 50(2):65–79, February 2004.

[8] Z. Lu and A. Jantsch. Flit ejection in on-chip wormhole-switched networks with virtual channels. In *Proceedings of the IEEE Norchip Conference*, November 2004.

[9] M. P. Malumbres, J. Duato, and J. Torrellas. An efficient implementation of tree-based multicast routing for distributed shared-memory multiprocessors. In *Proceedings of the 8th IEEE Symposium on Parallel and Distributed Processing*, 1996.

[10] M. Millberg, E. Nilsson, R. Thid, and A. Jantsch. Guaranteed bandwidth using looped containers in temporally disjoint networks within the Nostrum network on chip. In *Proceedings of the Design Automation and Test in Europe Conference*, 2004.

0-7695-2533-4/06 $20.00 © 2006 IEEE

A Virtual Channel Network-on-Chip for GT and BE traffic

Nikolay Kavaldjiev, Gerard J. M. Smit, Pierre G. Jansen, Pascal T. Wolkotte
Department of EEMCS, University of Twente, the Netherlands
{n.k.kavaldjiev, g.j.m.smit, p.t.wolkotte, p.g.jansen}@utwente.nl

Abstract[*]

This paper presents an on-chip network for a run-time reconfigurable System-on-Chip. The network uses packet-switching with virtual channels. It can provide guaranteed services as well as best effort services. The guaranteed services are based on virtual channel allocation, in contrast to other on-chip networks where guarantees are provided by time-division multiplexing. The network is particularly suitable for systems in which the traffic is dominated by streams.

We model the data traffic in the system and simulate the behaviour of the network with this model. The results show that the network is capable of handling the system traffic and can provide the required guarantees.

1 Introduction

Advances in silicon technology bring, among others, two problems that chip designers have to face – a high design complexity and a signal integrity problem [1],[2],[3]. The design complexity problem is the concern that the complexity of a system that fits on a single chip is getting so high that the time needed to design a completely new system using the current design methods and tools is becoming impractical long. For that reason it is foreseen that future System-on-Chip (SoC) will be based mostly on pre-designed IP blocks relying on extensive IP reuse. To be practical, such a design methodology needs to be complemented with a unified and simple solution for interconnecting and integrating IP blocks in a system. Currently on-chip buses offer such a solution, but since the bus bandwidth does not scale with the number of IP cores on the chip it will soon become a system bottleneck.

The second problem, the signal integrity problem, is due to the fact that with the technology scaling transistors get smaller and faster while wires get thinner and slower. Wire delay becomes proportional to the

wire length and a few long wires on a chip can degrade the performance of the entire chip. Thus, the on-chip interconnects become a limiting factor for SoC performance and their physical parameters must be taken into account at an early design stage. The current approach of ad-hoc on-chip wiring becomes questionable especially for the global (inter IP) on-chip interconnects, which are longer and slower.

A possible approach for coping with the two problems is to use a Network-on-Chip (NoC) [2][3], which is a light weight communication network built on the chip. This network replaces the traditional on-chip bus and provides a scalable high-bandwidth solution for interconnecting the increasing number of IP blocks. On the other hand, when a regular network topology is used, the global on-chip wires are short and well structured which helps coping with the signal integrity problem.

The performance of a network strongly depends on the characteristics of the data traffic generated by the system. Therefore a NoC cannot be evaluated separately, but has to be considered in the context of its operational environment. In this paper we present a network-on-chip solution for a run-time reconfigurable SoC used in mobile multimedia devices. The dynamic nature of such a system requires a flexible NoC solution. Since many of the system applications have real-time requirements, the system and the network have to be predictable. Network predictability is achieved by providing guaranteed network services. The proposed network provides guaranteed throughput (GT) and best effort (BE) services.

We discuss the system and the applications running on it and construct a model of the data traffic in such system. The model is then used to evaluate the proposed network. We perform simulations of the network behaviour the results of which show that the network can handle the system traffic and can provide the guarantees requested by the applications

The paper is organized as follows: Section 2 gives an overview of the SoC where the proposed network is used. Section 3 presents the network-on-chip and explains how guaranteed services are provided. Section

---

[*] This research is supported by the research program of the Dutch organisation for Scientific Research NWO (project number 612.064.103) and the EU-FP6 project 4S (IST 001908)

0-7695-2533-4/06 $20.00 © 2006 IEEE

4 presents simulation results showing that the network can handle the system traffic and can provide the required traffic guarantees. Section 5 presents related work.

2 System overview

The network proposed in this paper is intended for Multiprocessor System-on-Chip (MPSoC) for mobile multimedia devices. In such a system streaming applications dominate the demand for computation and communication capacity and high performance has to be achieved with a limited power budget. The operational environment requires for a dynamic system capable to adapt it self at run-time.

The MPSoC consists of a number of processing elements (PEs) arranged in a matrix and connected in a grid by a NoC. A PE consists of a processor with its own code and data memory that can operate independently. Several PEs are general purpose processors shared between the control oriented tasks running in the system. Most of the PEs are domain specific – simplified processors designed to perform fast and efficiently the computationally intensive algorithms in certain application domains, such as baseband processing, audio/video processing, etc. An example of a domain specific PE is the processing tile proposed by Heysters et al [4]. For 0.13 μm technology the area of the tile is 2 mm². On a typical chip of 14x14 mm near 100 such PEs can be fitted. The next technology generations will allow hundreds of domain specific PEs to be fitted on a single chip.

The system exploits the coarse grain parallelisms in streaming applications. The streaming applications are partitioned in a set of *processes* to run on separate PEs. The data exchanged between the processes are transported on *communication channels* handled by the NoC. After partitioning, the streaming applications typically have a structure like that shown in Figure 1. Such a structure is observed for the applications in the domain of base-band processing for wireless communications, like HiperLAN/2, UMTS, Bluetooth and DRM [5][6]. A similar observation for media processing applications is made by Dally et al [7]. Two parts can be distinguished in the structure in Figure 1 - a *processing part* and a *control part*.

The processing part does the actual stream processing. It has a simple pipeline structure. The data from the stream flow through the pipe and each process applies some transformation on it. The processes constructing the pipe (denoted as P_1, P_2, ..., P_n)

typically implement DSP algorithms, e.g. FFT, DCT, FIR etc. These algorithms are small, computationally intensive and specific for the application domain. Therefore, they are suitable to run on domain specific PEs. The communications between the processes in the pipeline carry the main data stream and therefore require high throughput. Because the pipe usually works under real-time constraints, the communications there are also real-time, hence requite GT services. Therefore the communication traffic generated by the processing part of the applications is high throughput GT traffic.

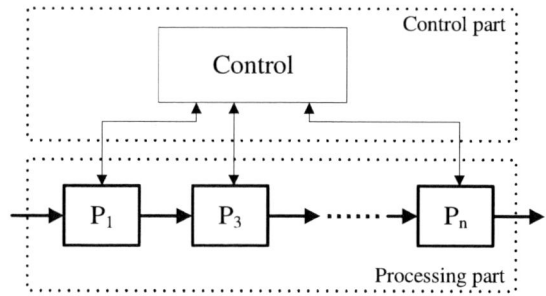

Figure 1 Generalized structure of a streaming application

The control part of the application comprises all the application tasks dealing with the application organisation and control, run-time adaptation and reconfiguration. Because of the reactive nature of these tasks they run not so often, require lighter computation and therefore are more suitable to run on shared general purpose PEs. The data traffic between the control and the processing part consists of infrequently exchanged short control messages, like event notifications, status and parameters exchange. These are not real-time communications, hence require BE services. Therefore the communication traffic generated by the control part is low throughput BE traffic. As a rough estimate for the applications we consider (base band processing), we observe that 90% of the traffic is GT traffic and 10% is BE traffic.

In summary, the traffic in the stream-processing system for mobile multimedia devices consists of: i) 90% high throughput GT traffic coming from streaming data, ii) 10% low throughput BE traffic coming from short control messages.

The system is dynamic and applications are started and stopped at run-time. When a new application is started, PEs are allocated to its processes and network resources are allocated to its communication channels are. The system is centralized – a central authority, which we call *Configuration Manager* (CM), starts and

0-7695-2533-4/06 $20.00 © 2006 IEEE   212

stops applications and manages the system resources (PEs, network, etc.) at run-time. The CM runs as a high priority task on one of the general purpose processors. By sending control messages on the network to PEs the CM can control and configure each PE separately while the rest of the system is running.

The configuration messages sent by the CM to the PEs in the system also contribute to the BE traffic in the system. The size of these configuration messages depends on the configuration space of the PEs. For example the total configuration space of the processing tile proposed by Heysters et al [4] is 2.6KB. Although this size is not small, configuration messages are generated infrequently, only when a new application is started (in period from seconds to hours), and therefore do not contribute significantly to the system traffic - our estimate is for less then 0.1% of the system traffic.

3   Network

Here we present the on-chip network we propose for interconnecting the PEs in the system described in the previous section. It is a packet switching virtual channel network that provides GT as well as BE services.

Each PE in the system is equipped with a network router which it uses for inter processor communications. The routers are connected in a mesh by full-duplex channels build of two unidirectional channels – one in each direction. We refer to the unidirectional channels as *physical channels*. The network is a virtual channel network [8]. In a virtual channel network there are several *virtual channels* (VCs) on each physical channel and data are transported on the VCs. The physical channel is time shared between its VCs on a cycle-by-cycle basis in a round-robin fashion. Cycles are only used by a VC when data is transmitted on it; the idle VCs do not consume cycles. The VCs are separately buffered in FIFOs at the router inputs. In our network, each physical channel is shared between 4 VCs, this number being motivated by the trade-off between performance and buffer area of a virtual channel router studied by Dally [8].

The physical channel is shared in a round-robin fashion, which means that the VCs equally share the bandwidth of the physical channel. The bandwidth is shared only between the VCs that currently transport data; the VCs that are idle do not consume bandwidth. If on a physical channel of bandwidth b there are v VCs currently transmitting data, then each of these v VCs is guaranteed throughput of

$$TH_{\min} = \frac{b}{v} \qquad (1)$$

This is the worst case throughput for the v VCs; whatever traffic load is applied to the v virtual channels their throughput will not go below $TH_{\min}$. Since in our network the number of VCs per network channel is 4, the values that $TH_{\min}$ can take are b, $b/2$, $b/3$ or $b/4$.

In traditional virtual channel networks VCs are allocated to packets dynamically by the routers [8]. The number of VCs on a network channel that are currently used depends on the current traffic and cannot be determined. Therefore, no throughput bounds can be given for a VC. In contrast, in our network VC are statically allocated to communication channels. The allocation is done centrally by the CM and the number of occupied VCs on a network channel is known. Therefore, we can give a lower bound on the VCs throughput.

The network provides GT services on a connection basis in the following way. A route is found over VCs between the source and the destination of a GT communication. The chain of VCs on which the route traverses the network is called a *connection*. The VCs in the chain are reserved and used only by the connection. The minimal throughput of the connection is determined by the VC with minimal throughput bound. We can a guarantee a lower throughput bound to a VC, therefore we can guarantee a lower throughput bound to a connection. To guarantee a minimal throughput bound of TH_R to a connection, according to (1) the following must hold for all the VCs in the connection

$$v \leq \left\lfloor \frac{b}{TH_R} \right\rfloor \qquad (2)$$

Given a lower bound TH_{min} on the connection throughput, the maximal latency T_{max} of a message of length L bits traversing the connection (in a wormhole manner) is [16]

$$T_{\max} = N * t_r + \frac{L}{TH_{\min}} \qquad (3)$$

where, N is the hop count or the number of network channels traversed by the message, and t_r is the delay of the message per router or the time that the head of the message spends waiting in a router. In our design t_r is at most v clock cycles (v has the same meaning as in equ. (1)). Let T_c be the network clock period and w the network channel width. Taking into account equation (1) and that $b=w/T_c$, then equation (3) is transformed to

$$T_{\max} = \left(N + \frac{L}{w}\right) * v * T_c \qquad (4)$$

This is the maximal latency that a message of length L bits, travelling distance N hops experiences on a GT channel of $TH_{min}=b/v$.

Our network uses source routing - a network address specifies not only the destination PE, but also the route to be followed. The route is described as e sequence of physical channels and their VCs to be taken. The destination address is carried by the packet header. When a header is send by the source PE, it is handled by the network and transported to the destination PE following the route described by the destination address. Along its way in the network the header allocates the VCs it traverses. The data form the packet body follows the allocated VCs and reaches the destination PE without need for additional control information. At the end, the packet tail releases the VCs. When an application is started the CM reserves VCs for the connections required by the application and gives the connections description to the PEs in the form of network addresses. When the application starts its PEs send packet headers and open the GT connections. When the application is stopped packet tails are sent and the connections are closed.

The BE services are provided by using shared VCs. The VCs allocated by the CM to a BE communication channel can be shared between sever BE communication channels. Sharing entails packet blocking, which makes the throughput prediction difficult. Therefore no guarantees are given. The BE data are sent on packet-by-packet basis. Since the VCs for BE connections are statically allocated and shared, a behaviour similar to that of a wormhole network should be expected. The wormhole networks do not perform well for intensive traffic consisting of long packets. But as we saw in the previous section, the BE traffic in our system is of low intensity and consists mainly of short messages (several bytes), therefore no performance problems are expected.

A network router has been implemented and synthesised for 0.13 μm technology. The router has an area of 0.18 mm^2 and can operate at a maximal frequency of 500 MHz [9]. Almost half of the router area is consumed by the buffers. More implementation details can be found in Kavaldjiev et al [10] and results for the energy consumption are presented by Wolkotte et al [11].

4 Simulation

To validate the network a cycle-accurate simulation is performed. A mesh network of size 6-by-6 is simulated using a traffic model with the characteristics presented in Section 2.

4.1 Setup

The communication traffic pattern used in the simulation is derived as follows. The nodes of a directed graph with ring topology are scattered over the nodes of a 6-by-6 mesh network. The graph contains 36 nodes and each graph node is placed on a separate network node. The edges between the placed graph nodes define the communication channels between the network nodes. The ring communication pattern is representative for the GT traffic of the streaming applications, because a large ring graph can be seen as a serial connection of many short pipeline graphs. We consider random scattering as worst case strategy for mapping applications on the system. In a real system the application mapping will be done such that the communication locality is maximized, which leads to better network conditions.

Each node in the network generates both types of traffic, GT and BE, transmitted on two separate communication channels following the ring communication pattern (a communication channel in the ring graph is substituted by one GT and one BE channels). Although the ring pattern is not the most realistic traffic pattern for BE traffic, we use it because the purpose of the BE traffic in this simulation is only to disturb the GT traffic and to create heavy traffic conditions. We shall see that what ever the BE traffic load is the GT traffic cannot be disturbed. During the simulation the GT traffic is kept constant while the BE traffic is gradually increased to the point of network saturation. During the simulation statistics are collected for the message latencies. The aim is to show that the latency guarantees given by the network to the GT traffic are not violated by any traffic condition.

The GT traffic generated by a network node has the characteristics of the traffic generated by a HiperLAN/2 receiver [5] – a typical high throughput baseband processing application. Every 4 μs a new message of size 256 Bytes is generated, which equals 512 Mbit/s average throughput per node or 18.4Gbit/s total aggregated throughput for the 36 nodes in the system. The BE traffic generated by a node consists of packets with 10 Bytes payload. The packet generation period is

0-7695-2533-4/06 $20.00 © 2006 IEEE

gradually reduced in order to increase the BE traffic intensity.

We guarantee that GT message latency is at most 1/3 of the message generating period 4 μs or <1.3 μs. In this way in a real system a PE will spent at most 1/3 of the time transmitting messages, at most 1/3 of the time receiving messages and the rest of the time will be for message processing. On 16 bit network channels ($w = 16$) and network clock period $T_c = 3$ ns latency <1.3 μs can be guaranteed by requesting GT connections of throughput TH_R $b/3$ ($v=3$). The maximal message distance in a 6-by-6 mesh network is $N_{max}=10$ hops. The length of the GT messages is $L=8*256B=2048$ bits. According to equation (4) the maximal GT message latency is 414 clock cycles or 1.242 μs. To provide GT connections of throughput $b/3$, the routing is done such that at most 3 VC are used per network channel ($v=3$). Therefore, with this simulations setup we expect no GT message latency to exceed the given latency bound of 414 clock cycles.

4.2 Simulation and results

Figure 2 presents the simulation results. The graph shows how the latency of the GT and BE messages depends on the offered BE load. The offered BE load is given per PE as all the PEs generate equal amount of data. For the GT packets the mean and the maximal latency are given. The horizontal line represents the 414-cycle latency guarantee for the GT packets.

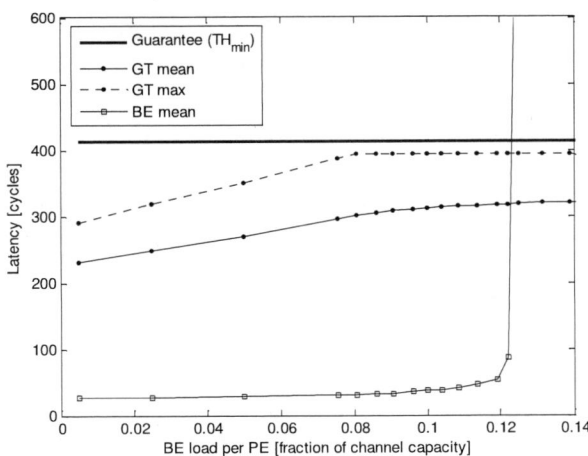

Figure 2 Message delay of the GT and BE traffic vs. BE load for 6-by-6 network

When the offered BE load is low the latency of the GT packets is smaller than the guaranteed latency. The reason is that the GT traffic utilizes the bandwidth unused by the BE traffic. The latency of the GT packets is higher than the latency for the BE traffic because the GT packets are larger (256B) than the BE packets (10B). With the increase of the BE load the latency of the GT traffic increases too and at some point it saturates. Further increase of the BE load increases the GT mean latency but the GT maximum latency does not increase and never exceeds the guaranteed latency. The GT maximum latency never reaches the latency bound, because the guarantee given by equation (1) is for worst case conditions when all v VCs constantly transmit data, while in our simulation setup the GT channels transmit data only 1/3 of the time. Thus, even beyond the point of network saturation for the BE traffic there is no GT packet that experience latency higher than 414 cycles – the packet latency is bounded according to the given guarantees.

The GT traffic offered to the network per PE is 512 Mbit/s or 0.09 of the channel capacity $w/T_c=5.3$Gbit/s. In Section 2, the BE traffic in baseband processing applications is estimated to be 10% of the total traffic while the remaining 90% is GT traffic. Thus, the intensity of the BE traffic expected in a real system per PE is about 57 Mbit/s or 0.01 of the channel capacity, which means that the network will operate in the very left part of the graph. According the simulation results the network saturates when the BE load per PE reaches about 0.12 of the channel capacity or 640 Mbit/s - more then ten times the expected BE load.

5 Related work

The RAW processor is a parallel architecture that exploits the applications instruction level parallelism [12]. For interconnecting its processing components it incorporates two types of networks dynamic and static that handle the different classes of traffic. The dynamic network is a dimension-ordered wormhole network, while the static network implements time-division multiplexing. In our solution both types of traffic are handled by a single virtual channel network.

ETHEREAL is a packet switching NoC solution based on time-division multiplexing (TDM) [13]. It provides both guaranteed and best-effort services and is targeted at general multimedia SoCs.

aSOC is a framework for on-chip communications in heterogeneous tiled architectures [14]. The proposed network implements a kind of advanced time-division multiplexing that can handle efficiently more complex communication patterns than the traditional TDM do. Instead of a simple timetable, each router there has a sequencer that allows more irregular switching behaviour. TDM (as used in aSOC and Ethereal) requires recomputation of a schedule for all the communications in the network even when only one communication link changes, which makes the system rather static. We avoid this by using virtual channels instead of TDM. We can add links incrementally without affecting the performance of already allocated links.

Wolkotte et al [9] proposes a circuit switching network which benefits small area and low energy consumption, while providing more flexibility than the traditional circuit switching solutions. Each network channel is divided into four lanes. Switching can be done at different granularity – from single lane to a whole channel. A network interface hides the real channel size from the application. Disadvantage of the circuit switching solution is that it does not support channel sharing and BE traffic. An additional network is required for configuring the switches and for carrying the BE traffic.

The SPIN network is a wormhole network that uses adaptive routing and is based on a fat tree topology [15]. It is targeted at general multiprocessor systems. While good in performance the topology used for this network is not natural for plane layout. Thus it is not clear whether it can help in structuring the global on-chip wiring.

6 Conclusion

This paper presented a network-on-chip for a run-time reconfigurable multi-processor system-on-chip. The system is used in mobile multimedia devices where most of the intensive applications are stream-processing applications. A model of the data traffic in the system is constructed and the network is simulated with this traffic model. Considering the communication locality, the simulated network conditions are worst case conditions. Nevertheless, the network manages to transport the system traffic and is able to provide the requested guaranteed services. The maximum message latency in the network never exceeds the guaranteed latency bound.

The proposed network is suitable for carrying the traffic in a real-time stream-processing system. The network provides guaranteed as well as best effort services. Data steams requiring guaranteed services are efficiently handled by network connections which reserve network resources, while in the same time best effort traffic is handled by allowing network resource

sharing. Thus the same network handles both types of traffic. The network requires minimum configuration which is done partially and only where and when it is required. Therefore the network is suitable for dynamic systems where fast reconfiguration is required.

References

[1] T. Whitney and G. Neville-Neil, "SoC: Software, Hardware, Nightmare, Bliss", *ACM Queue*, vol. 1, April 2003, pp. 24-31.

[2] W. Dally and B. Towles, "Route Packets, Not Wires: On-Chip Interconnection Networks", *DAC*, June 2001, pp. 684-689.

[3] L. Benini, G. De Micheli, "Networks on Chips: A New SoC Paradigm.", *IEEE Computer*, vol. 35, January 2002, pp. 70-78.

[4] P. Heysters., G. Smit, E. Molenkamp, "A Flexible and Energy-Efficient Coarse-Grained Reconfigurable Architecture for Mobile Systems", *Journal of Supercomputing*, vol. 26, November 2003, pp. 283-308

[5] G. Rauwerda, P Heysters, G. Smit, "Mapping Wireless Communication Algorithms onto a Reconfigurable Architecture", *Journal of Supercomputing*, vol. 30, December 2004, pp 263-282.

[6] P. Wolkotte, G. Smit, L. Smit, "Partitioning of a DRM receiver", *Proceedings of the 9th International OFDM-Workshop*, September 2004, pp. 299-304.

[7] W. Dally, P. Hanrahan, M. Erez, T. Knight, F. Labonté, J. Ahn, N. Jayasena, U. Kapasi, A. Das, J. Gummaraju, I. Buck, "Merrimac: Supercomputing with Streams", *Proceedings of the ACM/IEEE Supercomputing Conference (SC2003)*, November 2003, pp. 35.

[8] W. Dally, "Virtual-channel flow control", *IEEE Transactions on Parallel and Distributed systems*, vol. 3, March, 1992, pp. 194-205.

[9] P. Wolkotte, G. Smit, G. Rauwerda, L. Smit, "An Energy-Efficient Reconfigurable Circuit Switched Network-on-Chip", *Proceedings of the 12th Reconfigurable Architectures Workshop (RAW 2005)*, April 2005, pp. 155-161.

[10] N. Kavaldjiev, G. Smit, P. Jansen, "A Virtual Channel Router for On-chip Networks", *Proceedings of the IEEE International SOC Conference*, September 2004, pp. 289-293.

[11] P. Wolkotte, G. Smit, N. Kavaldjiev, J. E. Becker, J. Becker, "Energy Model of Networks-on-Chip and a Bus", *Proceedings of the International Symposium on System-on-Chip*, Tampere, Finland, November 2005, pp. 82-85.

[12] M. Taylor, W. Lee, Saman Amarasinghe, and Anant Agarwal, "Scalar Operand Networks", *IEEE Transactions on Parallel and Distributed Systems*, vol. 16, February 2005, pp. 145-162.

[13] E. Rijpkema, K. Goossens, A. Radulescu, J. Dielissen, J. van Meerbergen, P. Wielage, E. Waterlander, "Trade Offs in the Design of a Router with Both Guaranteed and Best-Effort Services for Networks on Chip", *Proceedings of Design, Automation and Test Conference in Europe*, March 2003, pp. 350-355.

[14] J. Liang, A. Laffely, S. Srinivasan, and R. Tessier, "An Architecture and Compiler for Scalable On-Chip Communication", *IEEE Transactions on VLSI Systems*, vol. 12, July 2004, pp. 711-726.

[15] A. Andriahantenaina, H. Charlery, A. Greiner, L. Mortiez, C. Zeferino, "SPIN: a Scalable, Packet Switched, On-Chip Micro-network", *Proceedings of the Design Automation and Test in Europe Conference (DATE'2003) Embedded Software Forum*, March 2003, pp. 70-73.

[16] W. Dally, B. Towles, "Principles and Practices of Interconnection Networks", Morgan Kaufmann, 2003.

Delay-Insensitive On-Chip Communication Link using Low-Swing Simultaneous Bidirectional Signaling

Ethiopia Nigussie, Juha Plosila, and Jouni Isoaho
Communication Systems Laboratory
Department of Information Technology, University of Turku
Turku, Finland
{ethnig, juplos, jisoaho}@utu.fi

Abstract

In this paper we present the circuit implementation of a new asynchronous delay-insensitive on-chip link structure, where two modules placed on the opposite sides of the link can exchange data simultaneously. Unlike the conventional delay-insensitive dual-rail link which requires 2N + 1 interconnects to transfer N-data bit, N +1 interconnects are required in this design. As two transceivers can access simultaneously the same physical interconnect the number of required interconnects halves compared to bidirectional transfer based on two separate unidirectional dual-rail links.This makes the link cost effective for future SoC. The transceiver circuits are designed using multiple-valued current-mode logic, linear summation is implemented by wiring without active devices simplifying the resulting circuitry. By using 110mV voltage swing the power consumption of the link is 8.32mW for 689ps propagation delay and 5mm interconnect length. Some of the potential application areas of this link are between locally clocked modules in GALS system, between routers of NoC nodes, and in adaptive and reconfigurable system where feedback information is crucial. The circuit is designed and simulated using Cadence Analog Spectre with 0.13um CMOS technology.

1. Introduction

System-on-Chip (SoC) designers face a major challenge in achieving the required functionality, performance and testability whilst minimizing design cost and time to market. The key to achieving this goal is a design methodology that allows component reuse. These design methodologies rely upon the use of a standardised interconnection interface for connecting component blocks together to form an on-chip system. This system-level interconnect encounters delay, power consumption, and signal integrity problems as feature sizes advances towards nanometer regime and multi-gigahertz frequencies.

Due to the challenge to overcome clock skew and switching noise in synchronous CMOS design, asynchronous circuit design becomes an increasingly practical alternative. Some potential advantages of asynchronous circuits over synchronous circuits are higher speed, lower power dissipation, and higher modularity [1-2]. Asynchronous circuits are also free from the problem of clock skew and have relatively low electromagnetic interference because of their distributed switching activities in time. The advantage of Multiple-Valued Current-Mode (MVCM) logic over conventional CMOS logic style include circuit simplicity, higher speed due to much smaller signal swing and frequency independent power dissipation, which results in lower power dissipation in multi-GHz frequencies. In addition, the steady current source of MVCM reduces the power supply fluctuations and noise. This property and the small voltage swing on the interconnect help to minimize crosstalk.

Asynchronous simultaneously bidirectional delay-insensitive link is presented in this paper. It uses MVCM signaling which combines the benefits of multiple-valued logic and current-mode operation. These make the link a good candidate for a future high speed, energy efficient and noise tolerant on-chip link.

2. Signaling protocol

The signal transmission system used in CMOS circuits can be broadly classified into two categories:

0-7695-2533-4/06 $20.00 © 2006 IEEE

voltage-mode and current-mode signaling. In voltage-mode the voltage has to swing rail-to-rail over the entire length of the interconnect. This leads to large transient currents consuming more power, larger delay and it also generates power-supply noise [7]. The optimal repeater insertion technique [4]–[6] used in voltage-mode signaling, was developed to reduce the wire delay and improve performance of lengthy global interconnections. However, with the increase in number and density of interconnects with technology, the number of repeaters necessary would increase manifold, presenting significant overhead in terms of power and area. Hence, there is a need for a new interconnection scheme that reduces wire delay, its variation and increases noise immunity. Unlike voltage-mode signaling, current-mode signaling allows voltage-swing to be reduced without separate voltage references and isolates the received signal from power supply noise [7]. It also translates into increased bandwidth performance [8], decreased delay and power dissipation and higher noise immunity. For these reasons, current-mode signaling technique is a better alternative to voltage-mode technique, for the future multi-GHz, noise-prone SoC chips.

In this link current-mode signaling is designed using multiple-valued logic. This enables to encode the data together with request using different current levels. In addition this logic allows linear current summation and subtraction just by wiring without active devices. Therefore the combination of multiple-valued logic with current-mode signaling makes the link circuit simple, propagates the signal faster, dissipates less power and isolates the signal from power-supply noise.

Communication on the link and between modules and transceivers follows 2-phase signaling protocol. This signaling protocol is a good alternative to 4-phase signaling for long interconnects because each communication across the link requires only two communication actions as opposed to four. This saves energy and time by eliminating the return to zero phase of four-phase signaling.

In asynchronous data encoding, dual-rail encoding is preferable over single-rail (bundled-data) encoding due to its delay-insensitive property. This property makes the data transfer very robust, because the sender and receiver can communicate reliably regardless of delays in the wire, in a delay-insensitive manner. But at the same time it has wiring cost because it requires 2N wires to encode N-bit data. In the conventional single-rail encoding data signals use normal Boolean levels to encode data and separate request and acknowledgement

wires are required [3]. In this encoding data should be stable before the data validity signal is activated. If the data interconnect has larger delay than the request one, the communication can be erroneous due to the arrival of request before the stable data reaches to the receiver. This timing constraint makes the communication delay-sensitive. But in this link special single-rail encoding which has delay-insensitive property is designed. This is achieved by using different current values to encode data and request together on one interconnect. Since the communication protocol is two-phase signaling there are four possible combinations of request and data as shown in Figure 1. These four different combinations are represented using four different current values of encoder current output as shown in Figure 3.

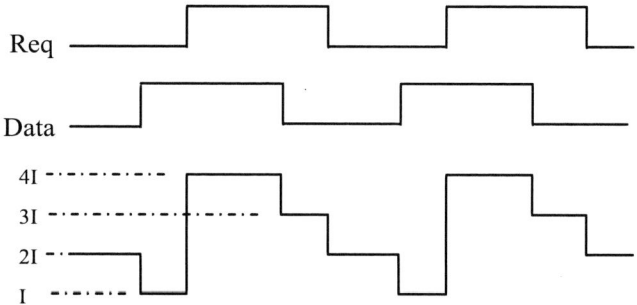

Figure 1. Encoding of Request and Data together using current-values

3. Link architecture

The proposed asynchronous link is capable of transferring data between two modules simultaneously in opposite directions. The block diagram of the link is shown in Figure 2, which has two transceivers placed opposite each other. The transceivers receive data from their own modules and output the data which is sent by the opposite module. The functions of transceiver blocks are explained as follows.

The encoder encodes the received data and request together into four different current levels. Then this encoded current is divided into two by the Current Subtractor unit and the pre-line. To restore to full current levels before it is fed to the line a current multiplier unit is used. The function of current Subtractor unit is to generate the current encoded by the opposite encoder. The decoder transforms the encoded current levels into its previously defined voltage levels after it takes output of Current Subtractor unit. In other words it decodes the current which is encoded by the other side of encoder into data and request. Since one acknowledgment signal is generated per N-bit data transfer, only one Current Adder unit is required per N-bit data transfer. This current adder sums at least the output of three different Current Subractor unit. For example, if there is 8-bit data transfer from both sides of modules. It takes first bit, last bit and 4th bit output of Current Subtractor and sum up together using the current adder. Then this sum sends to the acknowledgment line. The current adder which is placed in the other transceiver also performs the same thing. This means the output of the two current adders superimposed on each other on the acknowledgment interconnect. Then the current comparator compares the acknowledgments interconnect current with its own current adder output current. If the current of the acknowledgment interconnect is greater than the output current of the current adder then acknowledgment will be send to the module. This indicates the data sent by the module is received by the other module which is placed in the other side of the link. The current multiplier unit is designed from simple current mirror.

4. Design of link blocks

Both transceivers receive data and handshake signals from the module in voltage form then transform them into current representation and send to the interconnect. Then each transceiver changes data and handshake signals which are sent by the opposite module back to voltage form and make them ready for the module. The designs of link components are discussed below.

4.1. Encoder

The encoder bundles data and request together and change into four different current values because there are four possible combinations of data and request as shown in Figure 2. The encoder circuit is shown in Figure 3. Mn and Mp transistors serve as current source which generates constant current Is. Then Mp1, Mp2, Mp3 and Mp4 duplicate the current Is into four different values. The NMOS pass transistors are used to assign the current levels which are mapped to the combinations of data and request signals.

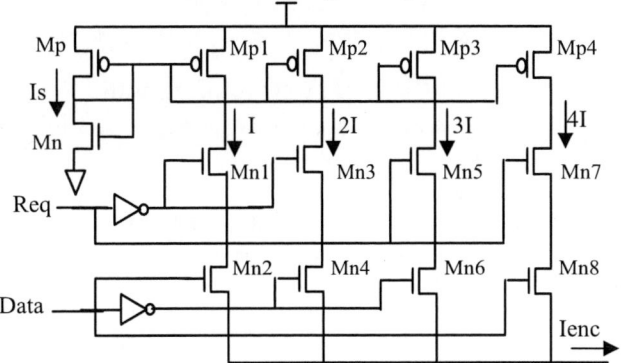

Figure 3. Encoder circuit

4.2. Current Subtractor

The Data + Req interconnect current is the sum of the two encoders current output, I_link = Ienc1 + Ienc2. The current subtractor shown in Figure 4 subtracts its side encoder current output from the

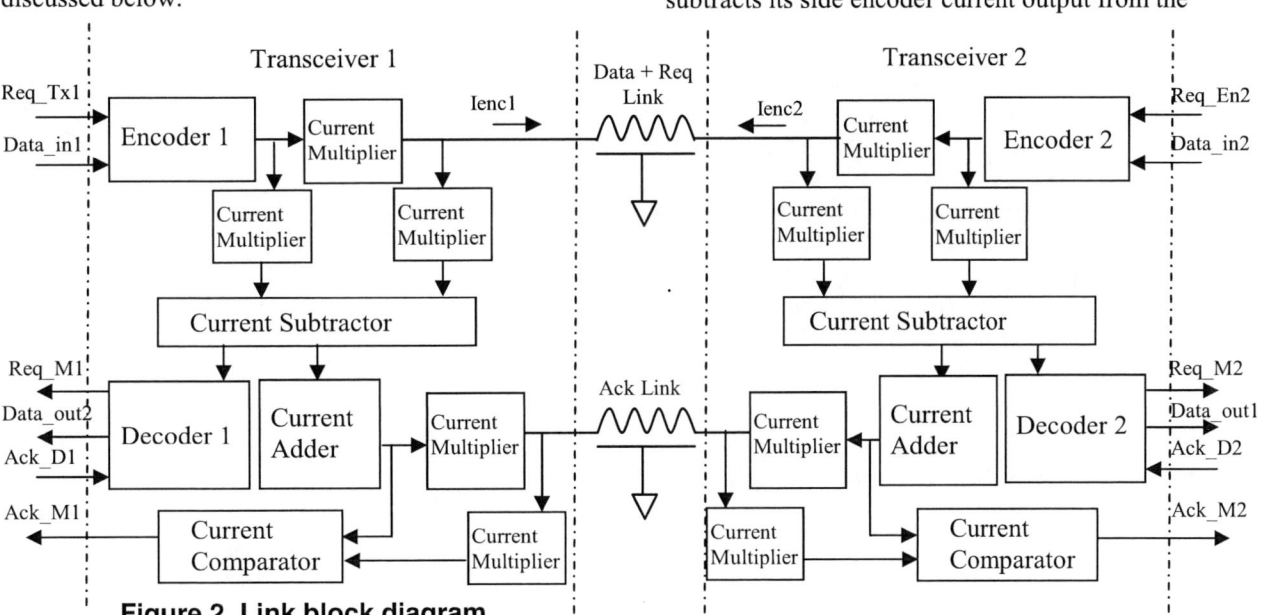

Figure 2. Link block diagram

0-7695-2533-4/06 $20.00 © 2006 IEEE   219

interconnect current. As shown in Figure 4 the input I_link current divides between Mn2 and Mn3. By using simple current mirror the current which can pass through Mn2 becomes Ienc1. Since I_link = Ienc1 + Ienc2, current through Mn3 becomes Ienc2. Thus the output of the current subtractor is the current encoded by the encoder which is placed on the opposite transceiver.

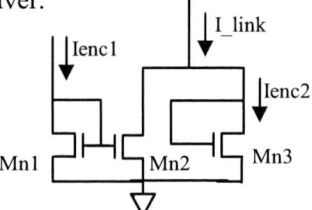

Figure 4. Current subtractor circuit

4.3. Decoder

The decoder consists of current comparator, XNOR and NAND gates. The XNOR gate is designed using Differential Cascode Voltage Switch Logic (DCVSL), which eliminates the static power consumption. The current comparator compares the input current with four different threshold currents as shown in Figure 5a. Transistors Mn and Mp used as current source and other four PMOS transistors which duplicate the Ith current into four different threshold currents. Its input current Ienc2 comes from the current subtractor output, the current which is sent by the other side of the encoder. This input current has four different current levels as shown in Table II. When the input current is I only V1 goes down because M1 consumes the entire 0.5I threshold current to drive the current. When the input current is 2I, V1 and V2 go down because M1 and M2 consume their threshold currents to drive the 2I current. The same principle applies for all voltage values shown in Table I. The data output which is sent by the other side of the module is XNOR of V2 and V4. And the request output is NAND of V3 and V4 as shown in Figure 5b. For example take the case where Ienc2 is 3I, then Data_out2 = V2 XNOR V4 = 0 XNOR 1= 0 and Req_M1= V3 NAND V4 = 0 NAND 1 = 1. This result can be cross checked from Figure 2, the current representation of 3I means Data = 0 and Req = 1.

Table 1. Decoding of encoded current to voltage

Ienc2	I	2I	3I	4I
V1	0	0	0	0
V2	1	0	0	0
V3	1	1	0	0
V4	1	1	1	0

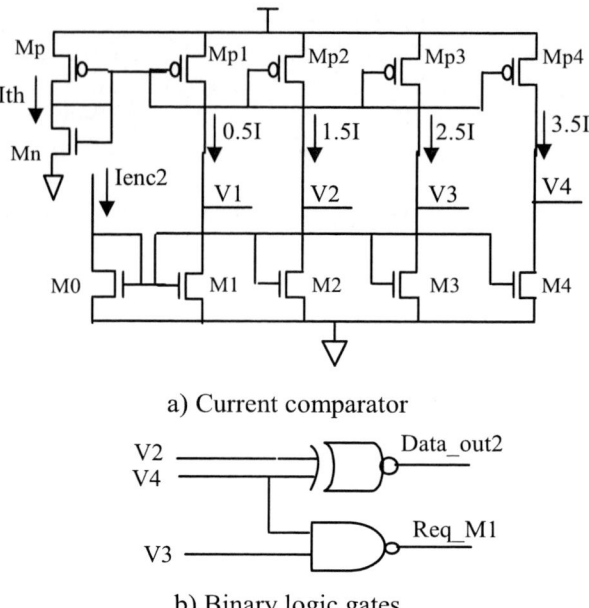

a) Current comparator

b) Binary logic gates

Figure 5. Decoder circuit

4.4. Current Adder and current Comparator

The purpose of these two blocks is to generate an acknowledgment signal. One acknowledgment signal is required for N-data bit transfer. Thus one current adder and current comparator is required for N-bit data transfer in each transceiver. Each of these current adders takes current output of three different Current Subtractor units, sum up together and send it to acknowledgment link. Then the current comparator compares the current on the acknowledgment link with its current adder output. When the acknowledgment link current is greater than the current adder output, the D-latch output, Ack_M1 becomes Req_Tx1. In other words acknowledgment signal follows request signal transition. The circuit which outputs acknowledgment signal is shown in Figure 6.

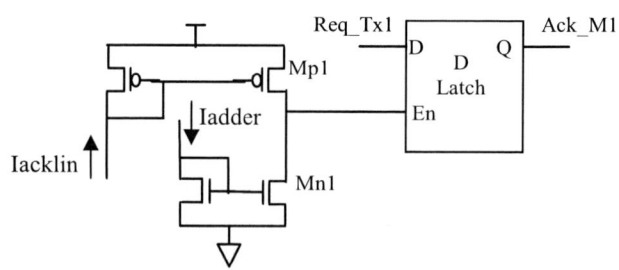

Figure 6. Circuit for generating acknowledgment signal to the module

5. Performance Comparison

The corresponding unidirectional link consists of only Encoder and Decoder. The number of transistors used for unidirectional and bidirectional link design is 44 and 112 respectively. The number of transistors used for bidirectional is greater than two times of the unidirectional, the remaining 24 transistors used in multiplying and subtracting the current. On-chip interconnects can be divided into three different types depending on their length [9]. Local interconnect less than 2mm, intermediate interconnect within 2 to 4mm and global interconnect greater than 4mm. The propagation of signal in unidirectional link is two times faster compared to bidirectional for local interconnects as shown in Figure 7. For intermediate interconnects both link has comparable signal propagation speed. But for global interconnects bidirectional link has faster signal propagation. In case of average power consumption of these links, bidirectional link consumes greater than four times of unidirectional as shown in Figure 8. Therefore the tradeoff for reducing the number of interconnects by half and becoming faster for global interconnects is power consumption. In terms of power supply noise both links isolate the signal from power supply noise.

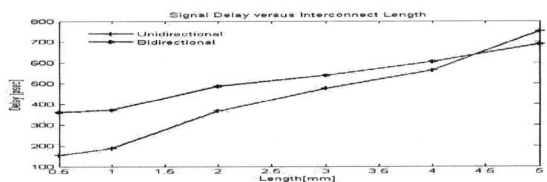

Figure 7. Signal Propagation Delay Versus Interconnect Length

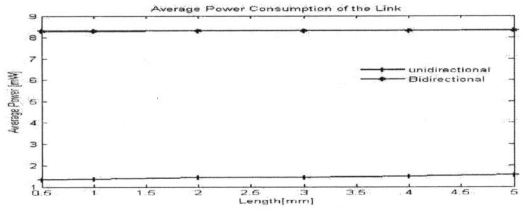

Figure 8. Average Power Consumption versus Interconnect length

5. Simulation of bidirectional link

The link between the two transceivers is designed using 10π RC model of interconnect. When the interconnect length varies from 0.5 to 5mm, the voltage swing on the link and average power dissipation varies

from 95 to 110mV and 8.30 to 8.32mW respectively. The effect of interconnect length variation on signal propagation delay is shown in Figure 8. Since the link is simultaneous bidirectional link, the sum of the two encoder current output on the link results in seven different current levels as shown in Table II. As expected the seven distinct current-levels on the link is shown in simulation waveforms Figure 9. Also waveforms of the two encoder's current output, the voltage-swing on the link and the two decoder's data and request outputs are shown.

Table 2. Sum of two encoders current output on the link

Ienc1 / Ienc2	I	2I	3I	4I
I	2I	3I	4I	5I
2I	3I	4I	5I	6I
3I	4I	5I	6I	7I
4I	5I	6I	7I	8I

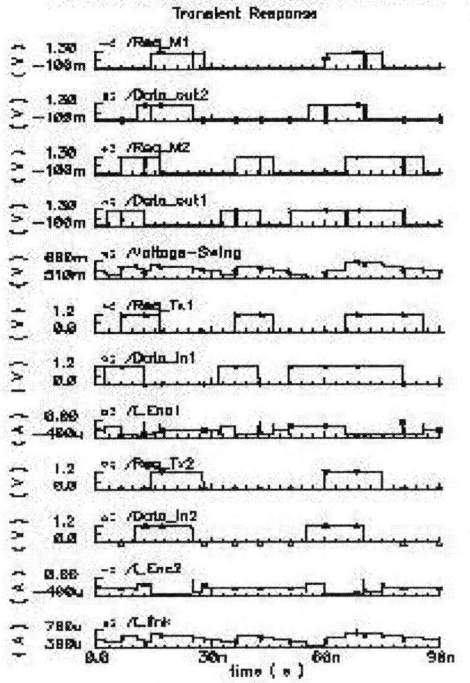

Figure 9. Simulation waveforms

6. Application of the Link

This link can be used in Globally Asynchronous Locally Synchronous System (GALS). In SoC clock distribution and alignment has become an increasingly challenging problem, consuming an increasing portion of resources such as wiring area, power, and design time. One of the solutions to this problem is to build SoC from several independently clocked subsystems which communicate each other through self-timed handshake signaling [3]. Such system enables flexible use of stoppable clocks providing automatic power down of idle system modules [10]. In addition to this, it makes easy to have a modular system. This simplifies the design process of a complex system enabling easy re-usage of different synchronous nodes. So this link can be used in GALS system between synchronously clocked local modules which has synchronous to asynchronous interface.

The other application area of this link is between two routers of Network-on-Chip (NoC). The design of larger more complex systems becomes an increasingly difficult task because of the many different issues related to the productivity, design reuse, technology and cost that have to be tackled simultaneously [11-12]. In this context Intellectual Property (IP) reuse becomes more and more important. However, building a system reusing existing IP blocks requires standardaised interfaces. One of a disciplined and scalable solution is offered by NoC [13]. Thus our link can be used between two routers which have one or more IP attached to it. Since the link is bidirectional, it is inherently reconfigurable in terms of information transfer direction. This means the router can route the information in opposite directions without any other effort.

It is also a good nominee for adaptive and reconfigurable link. Due to increasing number of noise sources and its level, technological parameter and environmental variations, communicating information reliably becomes difficult in future SoC. So to have reliable communication there should be some way of error detecting and correcting mechanism. To do this there should be feedback from the receiver side about the information transfer quality. This type of feedback communication can be done easily without any other effort using our bidirectional link.

7. Conclusions

In this work the realization of simultaneous bidirectional delay-insensitive asynchronous link using 2-phase handshake protocol and multiple-valued current mode scheme is presented. Unlike the conventional dual-rail delay-insensitive link which requires $2N +1$ interconnects $N/2 + 1$ interconnect is required for N-data bit transfer since both transceivers can send data simultaneously. The link circuit is composed of simple current mirrors, current comparators, and current subtractors. From circuit-level simulation results using CMOS 0.13um technology when the interconnect length varies from 0.5 to 5mm the power consumption of the link varies from 8.30 to 8.32mW. Also the voltage swing and signal propagation delay on the link varies from 95 to 110mV and from 360 to 689ps respectively. These results show that the effect of link length on power consumption and signal propagation delay is not as significant as the conventional voltage-mode link. Furthermore its signal propagation delay for 5mm length is 689psec which makes the link suitable for high speed data transfer on global on-chip interconnect. As performance comparison of unidirectional and bidirectional link indicates the cost of making the link simultaneously bidirectional is power consumption increase.

8. References

[1] M. Shams, et al, "Asynchronous Circuits", John Wiley's Encyclopedia of Electrical Engineering, pp. 716-725, 1999.

[2] Peter A. Beerel, "Asynchronous Circuits: An increasingly Practical Solution", Proc. ISQED, pp. 367-372, 18-21 March 2002.

[3] J. Sparso, and S. Furber, "Principles of Asynchronous Digital Design – A System Perspective", Kluwer Academic Publishers, Boston 2001.

[4] H. B. Bakoglu, Circuits, interconnections and Packaging for VLSI. Addison-Wesley, 1990.

[5] V. Adler, et al, "Repeater design to reduce delay and power in resistive interconnect," IEEE Trans on Circuits and Systems - II, vol. 45, no. 5, pp. 607–616, May 1998.

[6] D. Pamunuwa, H. Tenhunen, "Repeater insertion to minimise delay in coupled interconnects", VLSI Design, 3-7 Jan 2001, pp. 513-517.

[7] W. J. Dally and J. W. Poulton, Digital Systems Engineering. Cambridge University Press, 1998.

[8] R. Bashirullah, W. Liu, and R. K. Cavin, "current-mode signaling in deep submicrometer global interconnects", vol. 11, no. 3, pp. 406-417

[9] J. Nurmi, H. Tenhunen, J. Isoaho, and A. Jantsch, "Interconnect-Centric Design for Advanced SoC and NoC ", Kluwver Academic Publishers, 2004.

[10] J. Muttersbach, et al, "Practical Design of Globally-Asynchronous Locally-Synchronous Systems", Proc. ASYNC, April 2000.

[11] A. Jantsch, and H. Tenhunen, "Will Networks-on-Chip close the productivity gap?", in Networks on chip Kluwver Academic Publishers, 2003, pp. 3-18.

[12] L. Benin, and G. De Micheli, "Networks on Chips: A new SoC Paradign", Computer, 2002, 35, (1), pp. 70-78.

[13] Dally, W.J., and Towles, B.: 'Route packets, not wires: on-chip inteconnection networks'. Proc. 38th Annual Design Automation Conf. (ACM Press, 2001), pp. 684–689.

Nano Electronics

Nanowire Addressing in the Face of Uncertainty*

Eric Rachlin and John E. Savage
Department of Computer Science
Brown University

Abstract

Exploiting the high-potential of nanoscale architectures requires that they be controlled by CMOS technology. Such an interface, a decoder, must control many nanowires (NWs) with a small number of meso-scale wires (MWs). Multiple types of decoder have been proposed, each of which can be modelled as embedding resistive switches in NWs. In this paper we present a general model for NW decoders and use it to specify the criteria they must meet to function correctly and be fault-tolerant. To illustrate the power of our model, we derive the first bounds on the size of a fault-tolerant randomized contact decoder.

1 Introduction

Nanowire-based crossbars constitute a technology with the potential to extend Moore's Law beyond the limits of photolithography. A fundamental challenge in crossbar architectures is providing a reliable means of controlling individual NWs in each dimension with a small number of MWs. A circuit that provides this control is called a NW decoder. (See Figure 1.) These circuits address the general problem of controlling individual NWs. They are applicable to a wide range of architectures.

1.1 Nanowire Decoding

Four types of decoder have been proposed. All work on the principle that switchable resistances are created in NWs by the assembly process. The first two, the "randomized contact decoder" and the "randomized mask decoder," work with undifferentiated NWs, NWs that are all the same, except for possible manufacturing defects. Such NWs can be grown on chip using nanoimpring lithography [1, 2] and the superlattice nanowire pattern transfer (SNAP) method [3, 4]. The second two decoders, the "axially-doped NW decoder" and the "radially-doped NW decoder," work

---

*This research was funded in part by NSF Grant CCF-0403674.

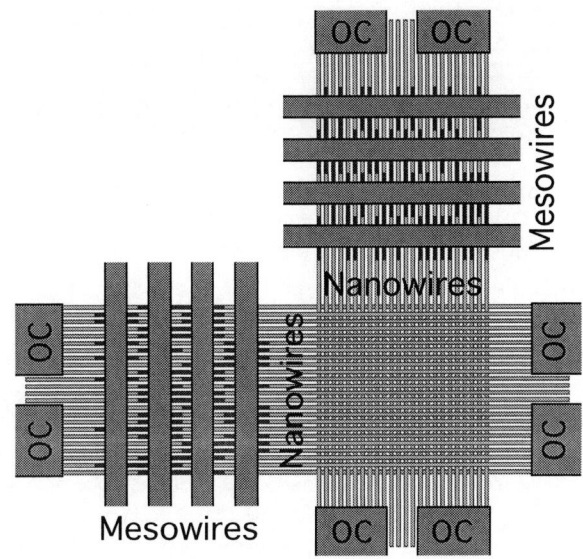

Figure 1. A crossbar with two parallel sets of NWs controlled by mesoscale address wires (MWs). FETs are defined at the intersection of lightly doped (dark) NW regions with MWs. Ohmic contacts (OCs) are made at ends of each set of NWs. Data is stored in the conductivity of molecular switches at crosspoints, intersections of orthogonal NWs.

with differentiated NWs. Different large batches of identical NWs are grown off chip, mixed together, and deposited on chip. In each case the goal is to produce NWs with sections whose resistance can be controlled by fields applied by MWs. The set of MW fields that cause a NW to be conducting is called its **codeword**.

The **randomized contact decoder** [5] scatters particles randomly between NWs and orthogonal MWs, thereby making switcheable resistances at the point of contact between NWs and MWs. (See Figure 2 (a).) A NW is conducting only when all its resistances are small. The **ran-**

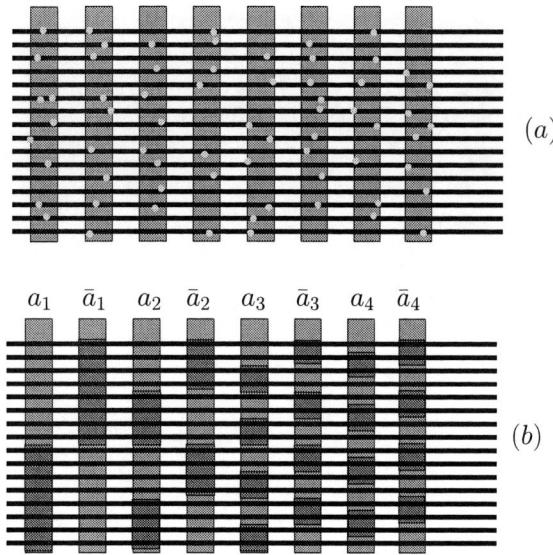

$a_1 \quad \bar{a}_1 \quad a_2 \quad \bar{a}_2 \quad a_3 \quad \bar{a}_3 \quad a_4 \quad \bar{a}_4$

Figure 2. (a) A randomized contact decoder and (b) a randomized mask decoder.

domized mask decoder [6, 7] interposes lithographically defined high-K dielectric regions, many of which are randomly placed, between MWs and NWs. A high-K dielectric that is adjacent to a MW intensifies its electric field, allowing the MWs to substantially increase the resistance of NWs exposed to the intensified field. (See Figure 2 (b).)

The third decoder, the **axially-encoded NW decoder**, controls axially differentiated NWs. As NWs grow through chemical vapor deposition, heavily and lightly doped regions are introduced along their axis. When a lightly doped NW region lies under a MW carrying an electric field, the NW's resistance is increased. (See Figure 1.) The fourth decoder, the **radially-encoded NW decoder**, operates on radially differentiated NWs that are formed by growing differentiably etcheable shells on lightly doped cores. To control these NWs, etching is done in lithographically defined regions to expose the cores of certain NWs to certain MWs, effectively simulating axial encodings.

Differentiated NWs have been assembled on chip using a fluid-based method [8], which aligns NWs in parallel through fluid flow but does not guarantee end-to-end alignment of NWs. Misalignment of axially encoded NW doped regions can lead to ambiguous control of NWs, a problem that doesn't arise with radially encoded NWs.

1.2 The General Decoder

We now present our **general** model for NW decoders. Small sets of NWs are controlled by "simple NW decoders" that are aggregated into "composite NW decoders." Since nanoscale manufacturing is stochastic, simple decoders become difficult to produce when N is large. For this reason, we use composite decoders.

Definition 1.1 *In a* **simple nanowire decoder***:*

1. *M large MWs control N much smaller NWs. The NWs are tightly packed and (at least partially) aligned, but are* not *in electrical contact with one another.*

2. *A pair of ohmic contacts applies a voltage across* all *NWs simultaneously. In the absence of MW control, all NWs conduct, effectively behaving like a single wire.*

3. *Each MW provides control over some subset of the NWs. A MW* **controls** *a NW if its resistance increases substantially when that MW carries a voltage.*

4. *When multiple MWs carry a voltage, the resistances induced in a NW are summed. The decoder* **addresses** *the NWs with low resistance.*

Definition 1.2 *A* **composite nanowire decoder** *uses multiple simple NW decoders, each associated with a pair of independently controllable ohmic contacts, to control groups of N NWs. All simple decoders share a single set of MWs, saving a substantial amount of space.*

To analyze these NW decoders, we require a model for MW control of NWs. We provide such a model in Section 2 and provide criteria that decoders must meet to function properly. In Section 3 we describe several classes of code that meet these criteria. Section 4 extends our model to cope with real-world manufacturing defects. To conclude, we provide a bound on the size of a fault-tolerant randomized contact decoder.

2 NW Codewords

Consider a simple NW decoder with M MW inputs and N NW outputs. Each MW provides control over a subset of NWs. In an **ideal decoder**, a NW is completely turned off (made nonconducting) by applying a field on a controlling MW and unaffected by a noncontrolling MW.

To describe an idealized NW decoder, we associate an M-bit vector (a **binary codeword**) with each NW. Let c^i be the codeword associated with NW n_i, $1 \le i \le N$, and let c^i_j ($1 \le j \le M$) be the jth bit of c^i. $c^i_j = 1$ (0) if and only if NW n_i is controlled (unaffected) by the jth MW.

Let a be the M-bit input (**activation pattern**) supplied to the decoder where $a_j = 1$ if and only if the jth MW carries a voltage. An activation pattern turns off NW n_i if and only if there exists a j such that $a_j = 1$ and $c^i_j = 1$. Equivalently, a turns off NW n_i if and only if $a \cdot c^i = \sum_{j=1}^{M} a_j c^i_j \ge 1$, where addition is over the integers.

2.1 Random Codeword Assignment

The four decoders described above assign codewords stochastically, a characteristic of all known NW decoders. Although some proposals allow portions of a NW's codewords may be assigned deterministically [9], a stochastic decoder is still needed.

The probability distribution governing the selection of NWs depends on the manufacturing process. This distribution is very important when evaluating decoding technologies. To illustrate this point, we compare the randomized contact and axial decoders when there is no misalignment.

1. In an ideal randomized contact decoder, codeword bits are generated independently. $c_j^i = 1$ with probability p, a parameter of the manufacturing process.

2. In an axial decoder codewords are assigned to NWs independently with some fixed probability. In an ideal axial decoder (which is not misaligned), $c_j^i = 1$ if NW n_i has a lightly-doped region under the jth MW. Otherwise $c_j^i = 0$. When NWs are deposited, a random subset of axial encodings is selected. Each encoding maps to a codeword. Codeword probabilities are a function of the corresponding encoding's relative concentrations in the original ensemble. Axial decoders are analyzed in [10].

We now show that ideal axial decoding is at least as good as ideal randomized contact decoding (or any decoding method which assigns bits independently at random).

Theorem 2.1 *An ideal axial decoder can simulate an ideal randomized contact decoder by assigning lightly- and heavily-doped regions to NW sections with equal probability while they are grown.*

Proof If all possible doping patterns are produced with the proper concentrations, bits in an axial decoder will be independent random variables, simulating codeword assignment in a randomized contact decoder. ∎

2.2 NW Addressability

We define the criteria codewords in a properly functioning decoder must meet using the binary codeword model.

1. Consider two NWs, n_a and n_b. If c^a contains a 1 in every position that c^b has a 1, it is impossible to turn off n_b without also turning off n_a. For this reason, we say c^b **implies** c^a, denoted $c^b \Rightarrow c^a$.

2. The **complement** of codeword c^i, denoted $\overline{c^i}$, is the NOT of M-bit vector c^i. If $c^b \nRightarrow c^a$, activation pattern $a = \overline{c^b}$ turns off NW n_a, but not n_b (since $\overline{c^b} \cdot c^b = 0$). If $c^b \nRightarrow c^a$ and $c^a \nRightarrow c^b$, NWs n_a and n_b are **independently controllable**.

3. A set S of NWs is **addressable** if there exists an activation pattern that turns off every NW not in S, and no NWs in S. A subset S' of S is addressable **with respect** to S if there exists an activation pattern that turns off every NW in $S - S'$ and no NWs in S'. A single NW, n_i, is addressable if the set $\{n_i\}$ is addressable. A codeword is addressable if the set of NWs with that codeword are addressable.

To understand the importance of codeword implication, we provide the following lemma.

Lemma 2.1 *A NW n_a with codeword c^a is addressable if and only if $c^i \nRightarrow c^a$ for all $i \neq a$. A codeword c^a is addressable if and only if $c^i \nRightarrow c^a$ for all $c^i \neq c^a$*

Proof If n_a (c^a) is addressable, an activation pattern, a, exists that turns off all NWs (codewords) other than n_a (c^a). None of these NWs (codeword) imply c^a. For the converse, recall that if $c_i \nRightarrow c_a$, $a = \overline{c^a}$ turns off n_i. ∎

Notice that if two NWs are addressable, they are necessarily independently controllable. Also, if all pairs of NWs are independently controllable, every NW is addressable.

2.3 Memory Decoders

NW decoders can be used to control a memory. (See Fig 1.). In a NW crossbar, a composite decoder is used along each dimension to select one or more crosspoints. For the memory to function properly, binary addresses must be mapped to disjoint sets of crosspoints.

A simple NW decoder which can address D disjoint sets of NWs is called a D-**address simple memory decoder**. Its codewords must meet the following requirement.

Lemma 2.2 *A decoder is a D-address simple memory decoder if and only if there exist D addressable codewords.*

Proof If D addressable codewords exist, the NWs with these codewords form D disjoint addressable subsets. For the converse, assume D disjoint addressable subsets of NWs exists. Each set must contain a codeword which no other codeword implies. Since each set is addressable, the D codewords are addressable and also distinct. ∎

In an N-address simple memory decoder all NWs, and hence all codewords, must be addressable. This implies that all pairs of NWs are independently controllable. In Section 3 we give the minimum length of such a code.

2.4 Circuit Decoders

A memory decoder must activate disjoint subsets of wires in order to control a memory efficiently. Now consider a decoder that acts as input to a circuit. If the circuit has D inputs, the decoder must provide control over a set S of D NWs. To supply the circuit with all 2^D possible inputs, the decoder must be able to address every subset of S with respect to S. We call such a decoder a D-**address circuit decoder** and give a condition on its codewords.

Definition 2.1 *Let S be a set of NWs that contains NW n_i. n_i is **uniquely controllable** with respect to S if there exists a j such that $c_j^i = 1$ and $c_j^a = 0$ for every other $n_a \in S$. Here a_j **uniquely controls** n_i, with respect to S.*

Lemma 2.3 *A decoder is a D-address circuit decoder if and only if there exists a set S of size D such that each NW in S is uniquely controllable with respect to S.*

Proof If every NW in S is uniquely controllable, there is an a_j which uniquely controls each $n_i \in S$. To turn off a subset of NW S', set $a_j = 1$ if and only if a_j uniquely controls a NW in S'. For the converse, assume each subset of S is addressable with respect to S. Since each set $S - \{n_i\}$ is addressable with respect to S, each NW, n_i, is uniquely controllable, with respect to S. ■

Our lemma does not assume a simple NW decoder. The condition on S still holds even if the decoder is composite.

To control N NW inputs to a circuit, we require at least N MWs, which presents an I/O challenge. It also presents a manufacturing challenge, as shown below.

In randomized-mask decoders, each MW controls multiple consecutive NWs. They cannot produce a set of adjacent, uniquely controllable NWs. Axial, radial and random-contact decoders assign codewords to NWs independently, and with fixed probability. In this case, the following holds.

Theorem 2.2 *Consider a NW decoder in which all N NWs are uniquely controllable using M MWs with probability $(1 - \epsilon)$. If NWs are assigned codewords independently and with fixed probability,*

$$M \geq e(1 - \epsilon)(N - 1)$$

Proof Since probabilities are fixed, the probability than NW n_i receives a particular codeword is independent of i. Thus $p_j = Prob(c_j^i = 1)$ is independent of i. It follows that the probability that the jth MW uniquely controls NW n_i is $Q(p_j) = p_j(1 - p_j)^{N-1}$. The probability that the jth MW controls one of the N NWs is exactly $NQ(p_j)$ because the N events are disjoint. $Q(p_j)$ is maximal when $p_j = 1/N$. So $NQ(p_j) \leq (1 - 1/N)^{N-1} = (N/(N - 1))(1 - 1/N)^N \leq e^{-1}N/(N - 1)$.

Since all N NWs are uniquely controllable with probability $(1 - \epsilon)$, the expected number of MWs that uniquely control some NW is at least $(1 - \epsilon)N$. The expected number of NWs controlled by M MWs is at most $Me^{-1}N/(N - 1)$. Thus, $M \geq e(1 - \epsilon)(N - 1)$. ■

3 Minimum Code Size

Even when codeword assignment is stochastic, some decoders (axial and radial, for example) provide substantial control over which codewords are generated. Future nanotechnology may even permit codewords to be deterministically programmed. In both cases, we wish to satisfy the conditions established by Lemmas 2.2 and 2.3 using minimum length codewords. In this section we present codes (sets of codewords) that satisfy this condition.

3.1 Codewords for Circuits

In a circuit decoder, Lemma 2.3 requires a set of D uniquely controllable NWs. Each wire must have a distinct codeword and each of these codewords must have a 1 in a unique position. The lemma implies that M, the number of bits in each codeword, is at least D. $M = D$ if and only if codewords are drawn from a $(1, D)$-hot code. A (k, M)-**hot code** [11] consists of codewords of length M in which each codeword has k 1s. Thus, a 1-hot code has a single 1 in each of its **M** codeword.

3.2 Codewords for Memories

In a memory decoder, Lemma 2.2 requires a set of D NWs with distinct codewords, none of which imply other codewords. Given M MWs, we now consider methods for generating length-M codewords in which every codeword is addressable.

First consider binary reflected codes, introduced in [10]. A **binary reflected code** (BRC) is code that contains all length-M codewords of the form $x\overline{x}$, where x is an arbitrary binary vector (and M is even).

In our previous work, we have found it convenient to use BRCs because each codeword directly corresponds to a binary sequence, x. (They also have the property that they are closed under cyclic shift, which models misalignment of axial codes.) BRCs and their subsets have also been used by others for the same reason [12, 13].

A BRC contains $2^{M/2}$ codewords. All pairs of codewords are independently addressable (since no codeword implies another codeword). Unfortunately, they are not optimal in that they use more MWs than some other codes with the same number of codewords.

The $(\lceil M/2 \rceil, M)$-hot codes are optimal in their use of MWs, as we show. A (k, M)-hot code contains $\binom{M}{k}$ codewords. As with BRCs, each codeword is addressable. (They are also closed under cyclic shift.) An $(\lceil M/2 \rceil, M)$-hot code has $\binom{M}{\lceil M/2 \rceil}$ addressable codewords, which is optimal.

Lemma 3.1 *Consider a set of C addressable codewords. The set consisting of the complement of each codeword also contains C addressable codewords.*

Proof All codewords are addressable if and only if for any pair of codewords, $c^a \not\Rightarrow c^b$ and $c^b \not\Rightarrow c^a$. If $c^a \not\Rightarrow c^b$ and $c^b \not\Rightarrow c^a$, then $\overline{c^a} \not\Rightarrow \overline{c^b}$ and $\overline{c^b} \not\Rightarrow \overline{c^a}$. ∎

Lemma 3.2 *Consider a set of C codewords which are independently addressable. If the minimum weight codeword has weight $w < \lfloor M/2 \rfloor$, there exists a code with C' codewords, such that $C' > C$ and all codewords have weight at least $w + 1$.*

Proof If $w < \lfloor M/2 \rfloor$, replace each w-weight codeword with each of the $M - w$ $(w + 1)$-weight codewords it implies. It can be show that this both increases the size of the code, and maintains the condition that all codewords are independently addressable. ∎

Lemma 3.3 *Consider a set of C codewords which are independently addressable. If all codewords have weight $\lfloor M/2 \rfloor$ or $\lceil M/2 \rceil$, there exists a code with $C' \geq C$ codewords such that all codewords have weight $\lceil M/2 \rceil$.*

Proof If M is odd, consider the same replacement described in the proof of Lemma 3.2. Now code size may remain unchanged. ∎

Theorem 3.1 *Given M MWs, there exist at most $\binom{M}{\lceil M/2 \rceil}$ addressable codewords.*

Proof Consider a code that maximizes the number of codewords. Lemma 3.1 states that the complement of this code also maximizes the number of codewords.

Lemma 3.2 implies that the code and its complement both have minimum weight codewords of weight at least $\lfloor M/2 \rfloor$. This means that all codewords in either code have weight $\lfloor M/2 \rfloor$ or $\lceil M/2 \rceil$.

Lemma 3.3 states that an equal size code exists where all codewords have weight $\lceil M/2 \rceil$. There are at most $\binom{M}{\lceil M/2 \rceil}$ such codewords. ∎

4 Tolerating Codeword Errors

Binary codewords describe the behavior of an ideal decoder. In ideal decoders, MWs provide all-or-nothing control over NWs. If NW on/off ratios are large, this is not an unreasonable assumption, although in practice it may be necessary to accommodate defects of various kinds. We do this by modifying our binary model slightly, assuming that some bits can be corrupted by an error. If bit c_j^i is in error, NW n_i behaves unpredictably with regard to MW a_j. To address codeword n_i, we must not activate MW a_j. To turn off n_i, we must activate an a_k for which c_k^i is not in error (and equal to 1). If MW a_j is activated, a MW a_k must also be activated.

Codewords in BRC and (k, M)-codes are addressable in the absence of errors. If codes tolerate d errors, we can accommodate the following real-world decoder defects.

1. After codewords are assigned, d bit flips occur. (This can model etching errors in radial decoders.)

2. d bits become partially corrupted, no longer behaving as 0s and 1s. (This can model misalignment, doping variation or poor separation between NWs and MWs.)

3. d transient errors in which a 1 becomes a 0 or is partially corrupted.

Unfortunately, transient errors in which a 0 becomes a 1 (and a NW is turned off incorrectly) cannot be tolerated because even one such error can prevent a corrupted NW from being addressed. Fortunately, these errors may be detectable by measuring the current produced by the decoder.

Two NWs, n_a and n_b are independently controllable *without errors if $c^a \not\Rightarrow c^b$ and $c^b \not\Rightarrow c^a$.*

Definition 4.1 *Two NWs are **d-independently controllable** if they remain independently addressable when up to d bit flips occur in their codewords. A set of NWs is **d-addressable** if every pair of NWs in the set is d-independently controllable.*

NWs which are d-independently controllable will be able to tolerate up to d errors. Let $|c^a - c^b\rangle$ be the number of is for which $c_i^a = 1$ but $c_i^b = 0$. The **balanced Hamming distance** between c^a and c^b is $2 \min \left(|c^a - c^b\rangle, |c^b - c^a\rangle \right)$.

Lemma 4.1 *If all pairs of codewords have a balanced Hamming distance of at least $2d + 2$, they are d-independently controllable. In this case, a total of d errors can be corrected. Also, up to $d/2$ errors in every codeword can be corrected.*

Lemma 4.2 *If codewords have a balanced Hamming distance of $2d + 2$, they have a normal Hamming distance of at least $2d + 2$. Two BRC codewords $x_1 \overline{x_1}$ and $x_2 \overline{x_2}$ have a balanced Hamming distance $2d + 2$ if and only if x_1 and x_2 have a normal Hamming distance $d + 1$.*

The last observation allows the x_i to be codewords from a standard error correcting code. The previous work on fault tolerant NW decoders assumes binary reflected codewords [12] and thus fails to define the more general notion of balanced Hamming distance. This work also does not accurately categorize which codewords are addressable. If errors cause $c^a \Rightarrow c^b$, the authors assume neither codeword is addressable. Lemmas 2.1 and 2.2 correct this assumption.

4.1 Randomized Contact Decoder

In this section we enforce the condition of balanced Hamming distance for the randomized contact decoder. We give an upper bound on M, the number of MWs required for all codewords in a randomized contact decoder to have a balanced Hamming distance of at least $2d+2$ with probability at least $1 - \epsilon$. When $d = 0$, the bound simply ensures that all NWs are addressable. A bound on M when $d = 0$ is cited in [5] but gives no mention of ϵ.

Theorem 4.1 *All N NWs in a randomized contact decoder (where $P(c_j^i = 1) = p$) are d-addressable with probability at least $1 - \epsilon$ when the number of MWs M satisfies*

$$M \geq \frac{\left(d + \sqrt{d^2 + 4\ln(N^2/\epsilon)}\right)^2}{(4p(1-p))}$$

The bound is $M \geq \ln(N/\sqrt{\epsilon})/(2p(1-p))$ when $d = 0$.

Proof Two NWs n_a and n_b are d-independently controllable if their balanced Hamming distance is at least $2(d+1)$. If NWs n_a and n_b are not d-independently controllable, then either $|c^a - c^b\rangle \leq d$ or $|c^b - c^a\rangle \leq d$. Since there are $N(N-1)$ such inequalities, all pairs of NWs are d-independently controllable with probability at least $1 - \epsilon$ when the following condition holds.

$$N(N-1)Pr(|c^a - c^b\rangle \leq d) \leq \epsilon$$

Let $x_j = 1$ denote the event $c_j^a = 0$ and $c_j^b = 1$. Then,

$$X = \sum_{i=1}^{M} x_j = |c^a - c^b\rangle$$

The probability that $x_j = 1$ is $q = p(1-p)$. The bound $P(X \leq d) \leq e^{-MqE(\theta)}$ is given in [10, Lemma A.3] when $d = \theta Mq$ and $E(\theta) = (1 - \theta + \theta \ln \theta)$. We must choose M so that $N^2 P(X \leq d) \leq \epsilon$. Since $(Mq)E(d/Mq) \geq Mq - d\ln(Mq)$, our condition on M holds when $Mq - d\ln Mq \geq \ln(N^2/\epsilon)$. Because $Mq \geq 1$, we can impose an even stronger condition and replace $\ln Mq$ by $\sqrt{Mq}$. This gives the inequality $Mq - d\sqrt{Mq} \geq \ln(N^2/\epsilon)$. This is a quadratic inequality satisfied when $4Mq \geq \left(d + \sqrt{d^2 + 4\ln(N^2/\epsilon)}\right)^2$. ∎

5 Conclusions

We introduce the binary codeword model, permitting a general discussion of NW decoders. We specify criteria decoders must satisfy and describe codes that meet these criteria. We also introduce balanced Hamming distance which characterizes fault-tolerant decoders. To illustrate the utility of our work, we derive a bound on the number of MWs required to create a fault-tolerant random contact decoder.

References

[1] S. Y. Chou, P. R. Krauss, and P. J. Renstrom. Imprint lithography with 25-nanometer resolution. *Science*, 272:85–87, 1996.

[2] W. Wu, G. Y. Jung, D. L. Olynick, J. Straznicky, Z. Li, X. Li, D. A. A. Ohlberg, Y. Chen, William M. Tong, S.-Y. Wang, J. A. Liddle, W. M. Tong, and R. Stanley Williams. One-kilobit cross-bar molecular memory circuits at 30nm half-pitch fabricated by nanoimprint lithography. *Applied Physics A*, 80:1173–1178, 2005.

[3] Nicholas A. Melosh, Akram Boukai, Frederic Diana, Brian Gerardot, Antonio Badolato, Pierre M. Petroff, and James R. Heath. Ultrahigh-density nanowire lattices and circuits. *Science*, 300:112–115, Apr. 4, 2003.

[4] E. Johnston-Halperin, R. Beckman, Y. Luo, N. Melosh, J. Green, and J.R. Heath. Fabrication of conducting silicon nanowire arrays. *J. Applied Physics Letters*, 96(10):5921–5923, 2004.

[5] R. S. Williams and P. J. Kuekes. Demultiplexer for a molecular wire crossbar network, US Patent Number 6,256,767, July 3, 2001.

[6] Eric Rachlin, John E Savage, and Benjamin Gojman. Analysis of a mask-based nanowire decoder. In *Procs 2005 Int. Symp. on VLSI*, Tampa, FL, May 11-12, 2005.

[7] Robert Beckman, Ezekiel Johnston-Halperin, Yi Luo, Jonathan E. Green, and James R. Heath. Bridging dimensions: Demultiplexing ultrahigh-density nanowire circuits. *Science*, 310:465–468, 2005.

[8] Y. Huang, X. Duan, Q. Wei, and C. M. Lieber. Directed assembly of one-dimensional nanostructures into functional networks. *Science*, 291:630–633, 2001.

[9] André Dehon. Deterministic addressing of nanoscale devices assembled at sublithographic pitches. *IEEE Transactions on Nanotechnology*, 4(6):681–687, 2005.

[10] Benjamin Gojman, Eric Rachlin, and John E. Savage. Evaluation of design strategies for stochastically assembled nanoarray memories. *J. Emerg. Technol. Comput. Syst.*, 1(2):73–108, 2005.

[11] André DeHon, Patrick Lincoln, and John E. Savage. Stochastic assembly of sublithographic nanoscale interfaces. *IEEE Transactions on Nanotechnology*, 2(3):165–174, 2003.

[12] Philip J Kuekes, Warren Robinett, Gabriel Seroussi, and R Stanley Williams. Defect-tolerant interconnect to nanoelectronic circuits. *Nanotechnology*, 16:869–882, 2005.

[13] Philip J Kuekes, Warren Robinett, and R Stanley Williams. Improved voltage margins using linear error correcting-codes in resistor-logic demultiplexers for nanoelectronics. *Nanotechnology*, 16:1419–1432, 2005.

[14] H. Chernoff. A measure of asymptotic efficiency for tests of a hypothesis based on a sum of observations. *Ann. Math. Stat.*, 23:493–507, 1960.

0-7695-2533-4/06 $20.00 © 2006 IEEE

Si Nanocrystal MOSFET with Silicon Nitride Tunnel Insulator for High-rate Random Number Generation

Ryuji Ohba, Daisuke Matsushita, Koichi Muraoka, Shinichi Yasuda, Tetsufumi Tanamoto, Ken Uchida and Shinobu Fujita

Advanced LSI Technology Laboratory, Toshiba Corporation,
8 Shinsugita-cho, Isogo-ku, Yokohama 235-8522 JAPAN
oba@amc.toshiba.co.jp

Abstract

It is shown that sub-0.1μm Si nanocrystal bulk MOSFET with thin SiN tunnel insulator is a very strong random noise source used in high-rate small-size random number generation circuit, which is required for cryptograph application in mobile network security. A fast random number generation rate of 0.12 MHz is demonstrated using Si nanocrystal MOSFET and a simple small circuit. It is suggested that a small-size random number generator with MHz generation rate, which is applicable for almost all security uses in our computer society, is possible by Si nanocrystal MOSFET device design.

1. Introduction

Random number generator (RNG) is requested for cryptographic application in network security systems. Recently, a small size RNG is more important for mobile network security. However, the present RNG using white noises are very large circuits [1,2]. Their large sizes are due to amplification of very small white noises. It is necessary to find out a strong noise source device, which can generate random numbers in combination with only small converting circuits.

For this purpose, we have proposed SOI Si nanocrystal MOFET with narrow-channel and thin tunnel oxide as a noise source device [3]. We generate high-quality random numbers at a generation rate 25 kbits/s using the SOI Si nanocrystal MOSFET in combination with simple small circuit (astable multi-vibrator and one-bit counter). However, the generation rate of 25 kbits/s is still short for many security uses. 1Mbits/s generation rate is requested, because it is applicable for almost all security systems.

Here, we propose a bulk Si nanocrystal sub-0.1μm MOSFET having thin tunnel SiN insulator as a new noise source device used in RNG. Based on the experimental results, we show that 200 times stronger noise is attained compared to the SOI Si nanocrystal MOSFET that enables 25 kbits/s generation rate. The 200 times noise enhancement is induced by substrate change from SOI to Bulk and scaling to sub-0.1μm region as well as by lower tunnel resistance of thin tunnel SiN and larger Si nanocrystal density. It is suggested that further noise enhancement is enough possible by device design.

We also demonstrate 0.12MHz excellent random number generation using aub-0.1μm Si nanocrystal MOSFET and a simple small circuit. It is concluded sub-0.1μm bulk Si nanocrystal MOSFET with extremely low tunnel resistance is a promising noise source device for MHz small-size RNG.

2. Experiments

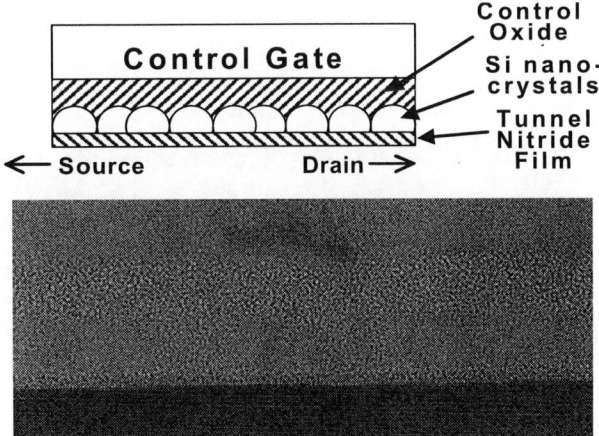

Fig.1 Cross sectional device structure.
(a) Schematic diagram. (b) TEM view.

0-7695-2533-4/06 $20.00 © 2006 IEEE

Sub-0.1μm Si nanocrystal MOSFETs with 0.15μm channel width were fabricated on bulk substrate based on 40 nm CMOS process [4]. To get large random noise in drain current, we use thin tunnel SiN and make Si nanocrystal density high. The tunnel SiN was about 1nm thick, and was formed by a low temperature plasma nitridation. So, it has a very low tunnel resistance. Thin tunnel oxide with 1nm thickness was also fabricated for comparison. Si nanocrystal diameter was typically 10nm, and the area density was about 10^{12} cm^{-2}. The Si nanocrystals can contact with each other. But, it is not essential in the noise source use. The drain current noise was measured at a constant voltage condition.

3. Results and Discussion

In this section, we show that 200 times stronger

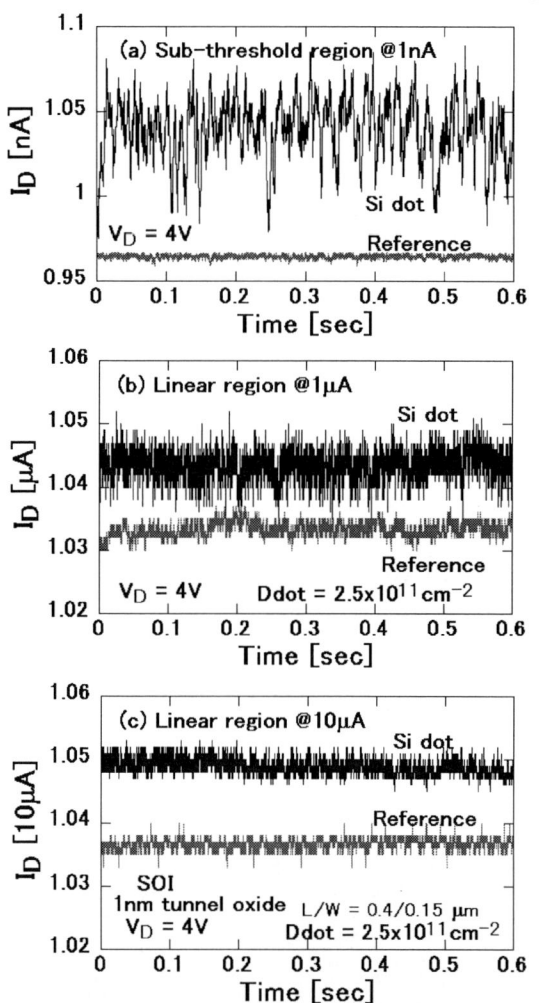

Fig.2 Random noise fluctuation in SOI Si nanocrystal MOSFET with 1nm tunnel oxide. "Reference" shows reference MOSFET without Si nanocrystals.

random noise is attained by device designs (substrate structure, Si nanocrystal density, tunnel insulator resistance and gate length scaling), and further improvement is enough possible since device parameters have still margins to improve.

3.1. Screening Effects

In RNG using SOI Si-nanocrystal MOSFET [3], we cannot obtain more than 25kbits/s RNG. One of the reasons is the screening effects by inversion electrons themselves. In order to show the screening effects, we examine gate voltage dependence of random noise.

A slow but large N/S (Noise/Signal) ratio random noise is found in sub-threshold region (Fig.2(a)). In a linear region at 1μA (Fig.2(b)), N/S ratio is smaller due to smaller gate voltage sensitivity in linear region. But, noise frequency is much higher, which will be due to many inversion electrons. We could get 25 kbits/s RNG using this frequent noise. However, we cannot attain stronger noise in a higher gate voltage. The random noise at 10 μA (Fig.2(c)) is almost the same as the reference MOSFET's noise without Si nanocrystals. This remarkable reduction of N/S ratio in random noise is due to the screening effects by too many inversion electrons. So as to resolve the screening effects, we have to change device structure.

3.2. Substrate Structure (SOI and Bulk)

First improvement comes from substrate structure.

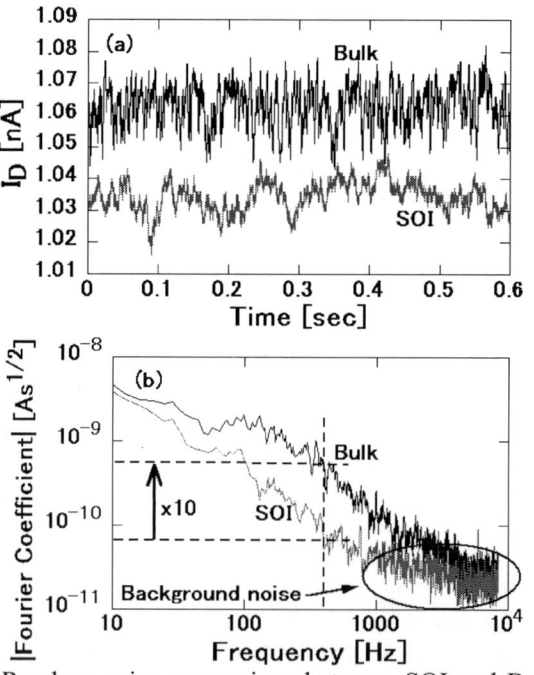

Fig.3 Random noise comparison between SOI and Bulk Si nanocrystal MOSFETs at all the same parameters. (a) Random noise. (b) Fourier characteristics.

0-7695-2533-4/06 $20.00 © 2006 IEEE

At the same condition, it is experimentally found that bulk structure shows 10 times larger random noise than SOI structure (Figs.3). The carrier supply from substrate will enhance injection/ejection in Si nanocrystals. We can obtain clear random noise at 10 µA in bulk structure (Fig.4), while random noise at 10 µA is too small in SOI structure (Fig.2(c)). The Fourier characteristics show that 10 times stronger noise is attained in bulk compared to SOI (Fig.5).

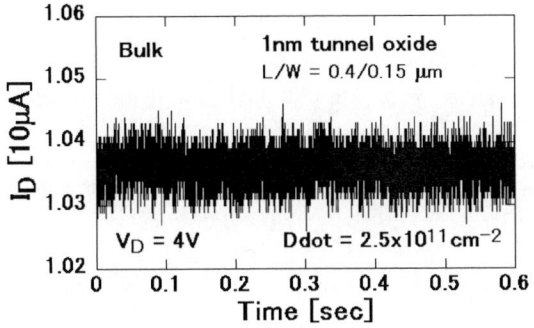

Fig.4 Random noise in a bulk Si nanocrystal MOSFET.

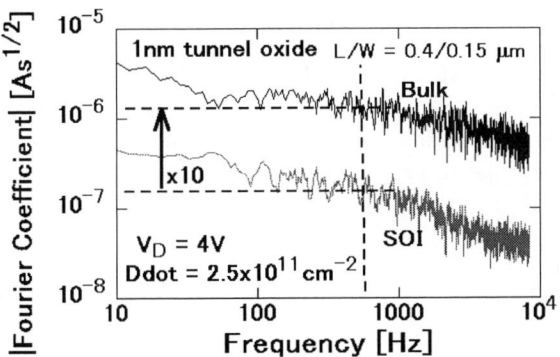

Fig.5 Fourier characteristics of random noise in Bulk Si-nanocrystal MOSFET (Fig.4) and in SOI Si-nanocrystal MOSFET (Fig.2).

3.3. Si Nanocrystal Density

Next, we consider noise enhancement by Si

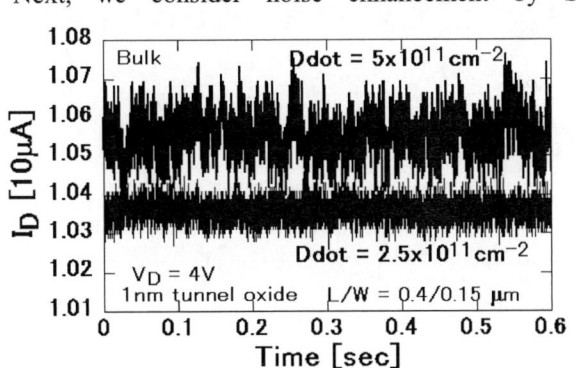

Fig.6 Random noise comparison between diffenent Si nanocrystal density in Bulk Si-nanocrystal MOSFETs.

nanocrystal density increase. Experimental results show that, by making Si nanocrystal density twice, random noise is not twice, but about $2^{1/2}$ times larger (Fig.6). In slow random noise, the noise strength is proportional to Si nanocrystal density [3]. However, in frequent random noise, it will be proportional to square root of Si nanocrystal density according to statistical law.

3.4. High-k Tunnel Insulator

Lower tunnel resistance is also advantageous for high-frequency noise. We fabricated bulk Si nanocrystal MOSFET with thin SiN tunnel layer and

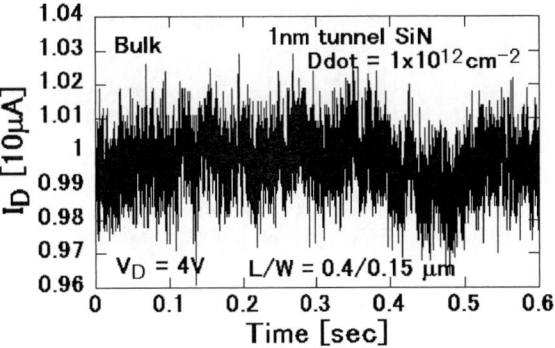

Fig.7 Random noise in a bulk Si nanocrystal MOSFET with thin tunnel SiN and dense Si nanocrystals.

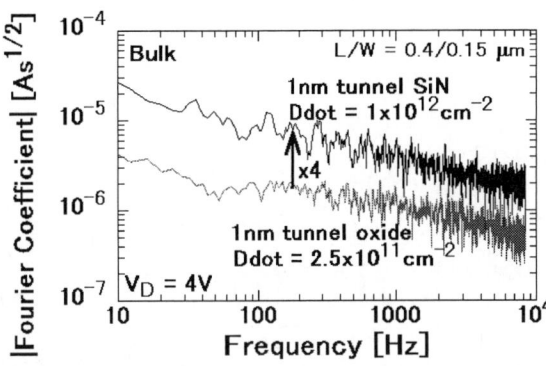

Fig.8 Fourier characteristics of random noise at 10µA. 2 times from Si nanocrystal density, and 2 times from tunnel resistance.

dense (10^{12} cm^{-2}) Si nanocrystals. The random noise at 10 µA (Fig. 7) is stronger than that in Fig.4. The Fourier characteristics show about 4 times stronger noise due to both tunnel SiN and higher nanocrystal density (Fig.8). 2 times improvement comes from square root law in Si nanocrystal density increase, and the other 2 times originates in lower tunnel resistance of thin tunnel SiN.

0-7695-2533-4/06 $20.00 © 2006 IEEE   233

3.4. Short gate length

The last improvement comes from gate length scaling. For 0.04μm Si nanocrystal MOSFET with thin tunnel SiN (Fig. 9) shows very large random noise. In comparison with Fig.7, stronger random noise is attained in shorter gate length. This is explained by reduction of screening effects in short gate length. At the same channel width 0.15μm, 10μA drain current flows at a smaller inversion electron density in shorter gate length because of low channel resistance in short gate length. As a result, screening effects reduces in short channel length at the same drain current, and 5 times larger random noise is attained in 0.04μm than 0.4μm (Fig.10).

Figure 11 summarizes the random noise enhancement in 0.04μm MOSFET with many Si nanocrystals and thin tunnel SiN, compared to SOI Si nanocrystal MOSFET which enables 25 kbits/s small circuit RNG. We note that 200 times enhancement is accomplished due to the device designs explained above.

Shorter gate length than 0.04μm is enough possible, because suppression of I_{OFF} is not always necessary in the noise source use. Lower tunnel resistance is also enough possible when we use high-k tunnel insulator other than oxide or nitride. Therefore, further enhancement of random noise is still possible.

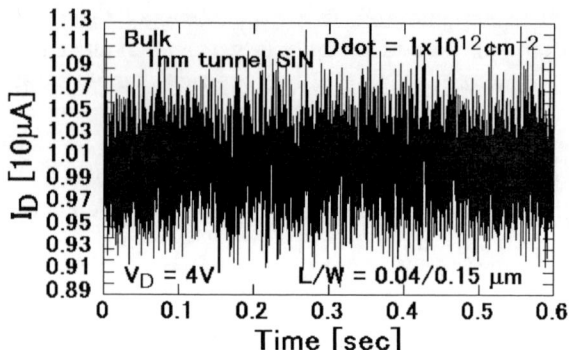

Fig.9 Random noise in 0.04μm bulk Si nanocrystal MOSFET with thin tunnel SiN and dense Si nanocrystals.

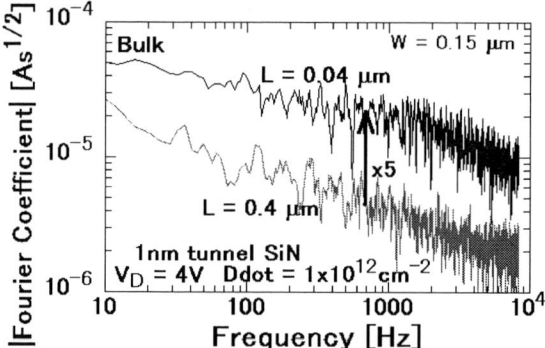

Fig.10 Fourier characteristics of random noise in a bulk Si nanocrystal MOSFET with thin tunnel SiN and dense Si nanocrystals for various gate lengths.

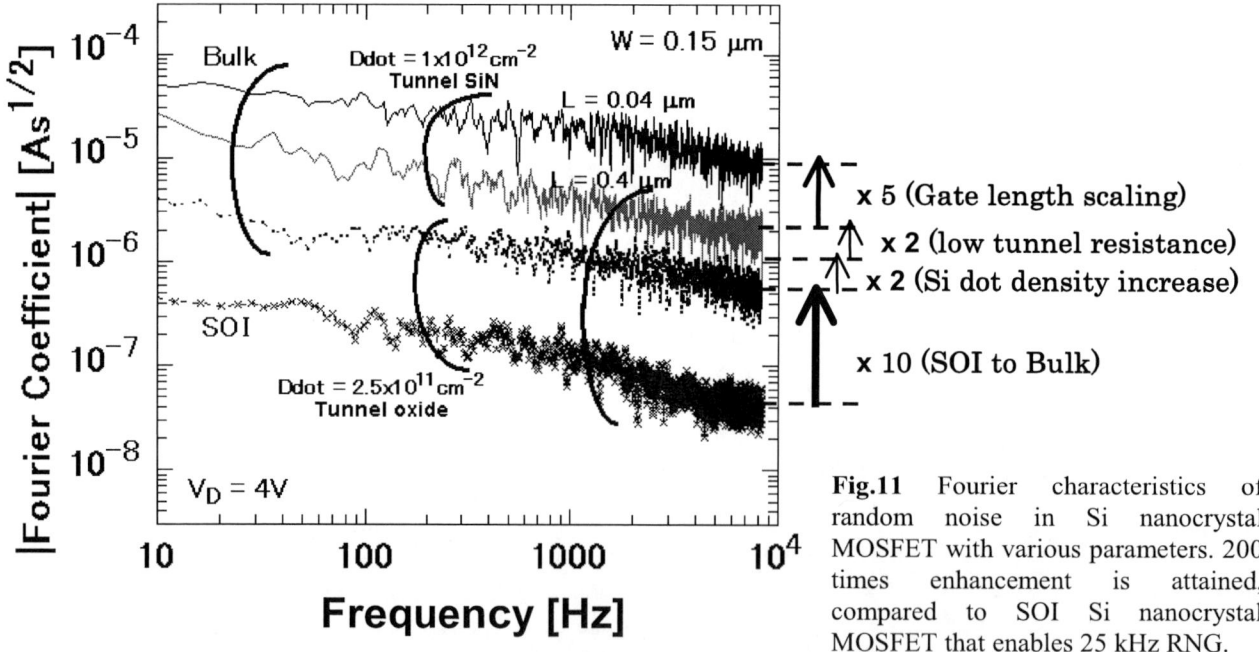

Fig.11 Fourier characteristics of random noise in Si nanocrystal MOSFET with various parameters. 200 times enhancement is attained, compared to SOI Si nanocrystal MOSFET that enables 25 kHz RNG.

0-7695-2533-4/06 $20.00 © 2006 IEEE

4. Random Number Generation

In this section, we demonstrate a random number generation by the Si nanocrystal MOSFET where the strongest random noise is attained in the last section. Only simple small converting circuit is used in the generation without amplification circuit.

4.1. Multi-vibrator Circuit

Random number sequence is generated in a converting circuit composed of astable multi-vibrator and one-bit counter [5,6] with the Si nanocrystal MOSFET (Fig.12(a)). Since the period of output pulse t_J is RC product ($t_J = R_B C_B$) in the circuit, it fluctuates

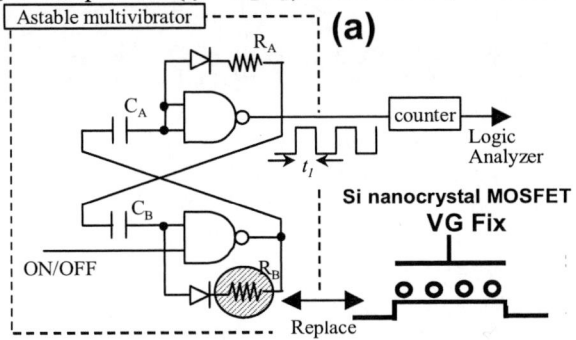

(a)

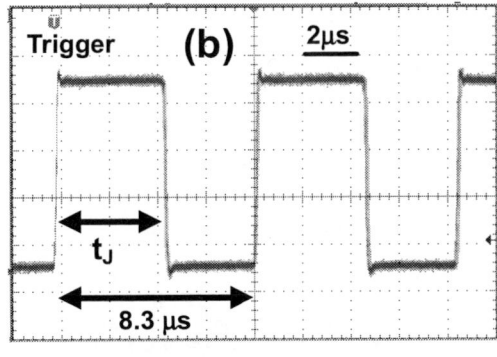

(b)

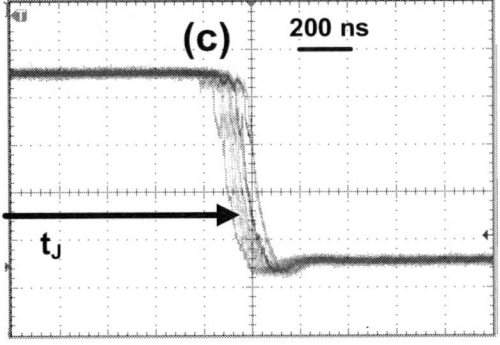

(c)

Fig.12 (a) Multi-vibrator circuit to generate random numbers. (b) 0.12 MHz output pulse. (c) 0.12 MHz output pulse magnified near pulse tail. Period of output pulse ($t_J = R_B C_B$) fluctuates due to ID fluctuation in Si nanocrystal MOSFET.

according to the ID fluctuation shown in the last section. Figure 12(b) shows the output pulse with 0.12 MHz frequency (8.3 µs period), and Fig.12(c) shows that each period t_J fluctuates.

The one-bit counter in the circuit has a clock which can generates a faster clock pulse, and converts each period t_J to 1-bit random number '0' or '1'. When the number of the faster clock pulse within each period t_J is an odd (/ even) number, one-bit number "0" (/ "1") is given. As a result, we can obtain a random number sequence with 0.12 Mbits/s generation rate.

4.2. Quality Analysis of Randomness

The quality of random numbers is checked by a statistical test FIPS140-2 [7,8], which is necessary for commercial uses. The 0.12 Mbits/s Si nanocrystal MOSFET RNG passed all the tests (Table1).

We also check the Si nanocrystal RNG by higher-level tests. Figure 13(a) shows the frequency test [8] to

Test		Requirement	Our RNG	
Monobit		9725–10275	9822	O
Poker test		2.16–46.17	14.3616	O
Long run test	"0"	1–26	13	O
	"1"		12	O
Length of run 1	"0"	2315–2685	2399	O
	"1"		2503	O
Length of run 2	"0"	1114–1386	1237	O
	"1"		1225	O
Length of run 3	"0"	527–723	598	O
	"1"		604	O
Length of run 4	"0"	240–384	338	O
	"1"		277	O
Length of run 5	"0"	103–209	178	O
	"1"		149	O
Length of run 6	"0"	103–209	181	O
	"1"		172	O

Table 1 Standard statistical test FIPS for 20000 random numbers. All requirements are satisfied.

check an independent uniform distribution between "0" and "1". Figure 13(b) shows serial test [8] to check a uniform distribution of pairs of successive numbers (as (0,0), (0.1), (1.0), (1.1)). Both Figs.13(a) and (b) show that our Si nanocrystal RNG shows a good randomness near ideal distributions.

Self-correlation plots for sequential 8-bit random numbers [8] is shown in Fig.14. This is a spectral test to check a full period in the random number sequence, and excellent randomness is attained safely.

Our results show that Si nanocrystal MOSFET RNG leads to a near-ideal randomness, which will be sufficient for commercial uses.

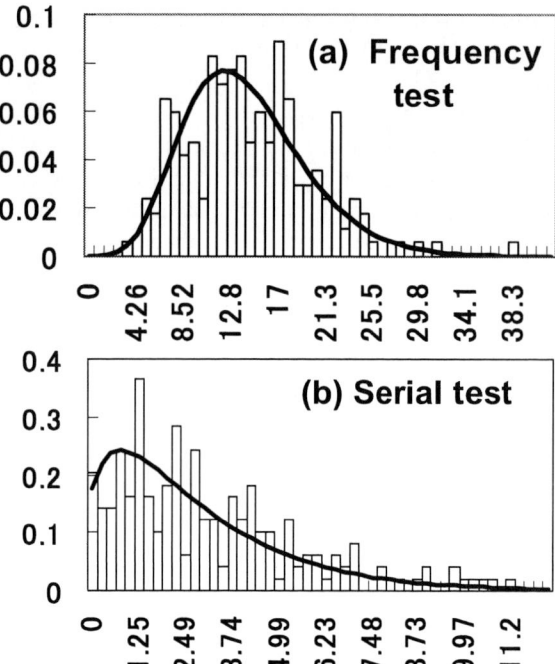

Fig.13 Higher level statistical tests for 60000 random numbers. (a) Frequency test. (b) Serial test. Bold lines show ideal χ^2 distributions for perfect randomness. Degrees of freedom in χ^2 distributions are (a) 15 and (b) 3. In frequency test, sequential 4 bits numbers of '0' and '1' are transformed to 0-15 numbers.

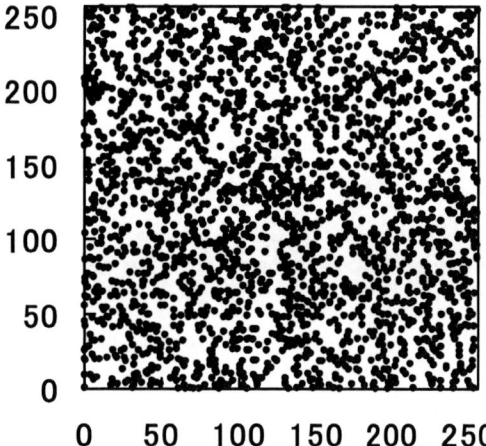

Fig.14 Self-correlation plots for sequential eight bit numbers for 60000 random numbers, generated at 0.12Mbits/s speed. Periodicity is not found.

4.3. Circuit design for higher generation rate

We have realized 0.12 Mbits/s generation rate using astable multi-vibrator and one-bit counter. However, this is very inefficient for fast generation rate, because this circuit reduces generation rate due to high resistance of noise source device. For faster generation rate, we can use smaller coupled capacitance, since output pulse period is RC product. Alternatively, we are developing differential amplifier type RNG circuit where the noise source resistance does not affect generation rate [9]. The combination with those improved converting circuits will be advantageous for MHz small-size RNG realization.

5. Conclusion

We have proposed sub-0.1μm bulk Si nanocrystal MOSFET with thin SiN tunnel insulator as a random noise source used in high-speed small-size RNG. We experimentally show that 200 times stronger random noise is attained, compared to SOI Si nanocrystal MOSFET that enables 25kbits/s small-size RNG. Further noise enhancement is still possible by device design, since tunnel resistance reduction and gate length scaling have enough margins to improve. We also showed 0.12MHz excellent random number generation using a simple small circuit. It is concluded sub-0.1μm bulk Si nanocrystal MOSFET with thin high-k tunnel insulator is a promising noise source device for small MHz RNG circuit.

Acknowledgements

We would like to thank K. Eguchi, E. Sekine and S. Inaba for their support. This work was supported in part by NICT (the National Institute of Information and Communications Technology of Japan).

References

[1]http://www.t-rs.co.jp/trs_new/trs-english/products/rmh.htm (Toshiba)
[2]http//www.intel.co..design.security rng/rng.htm:
[3] R. Ohba, et al., IEEE Technical Digest of IEDM 2003, pages 745-748.
[4] S. Inaba, et al, IEEE Technical Digest of IEDM 2001, pages 641-644..
[5] S. Yasuda, et al., IEEE J. Solid State Circuits, vol. 39, pp. 1375-1377, Aug. 2004.
[6] S. Fujita, et al., IEEE International Solid-State Circuits Conference, pp.294-295, 2004.
[7] Federal information processing standard publication FIPS PUB 140-2 (2001) May 25.
[8] D. Knuth, The art of Computer programming 3rd. ed. Vol.2 (Addi-son-Wesley.1998)
[9] S. Yasuda, et al., ESSCIRC, pp.399-402, 2005

Finite State Machine Implementation with Single-Electron Tunneling Technology

Jialin Mi and Chunhong Chen

Department of Electrical and Computer Engineering, University of Windsor, Ontario, Canada
cchen@uwindsor.ca

Abstract

In this paper we propose an implementation technique for finite state machines using single-electron tunneling (SET) technology towards ultra-low power dissipation. We take an example from radio-frequency identification (RFID) systems, and implement the state machine based on single-electron encoded logic (SEEL). The circuit is simulated by SIMON (a simulator for SET circuits), and is compared favorably with its CMOS counterpart.

1. Introduction

Finite state machine (FSM) is well understood as a typical module for most sequential circuit designs [1]. It can be described by a state transition graph, where each vertex represents a state and each directed edge represents the transition from one state to another. Implementation of an FSM involves (a) the logic design which includes the state assignment followed by combinational logic synthesis, and (b) the circuit realization with specific technology. Due to the increasing complexity and power dissipation of digital systems, a lot of research works have focused on FSM synthesis towards low power requirement [2-4]. As MOS transistors are expected to finally meet their physical fabrication limits in further power and/or area reduction, various new revolutionary device architectures have been recently proposed at nano-scale [6], with an increasing interest in Single Electron Tunneling (SET) devices as promising candidates for future ultra-low power integrated circuits [5-7].

Single-electron devices utilize one-electron-precision charge transfer based on the *Coulomb blockade* effect, and provide an alternative to implement digital logic. Single-electron transistor and single-electron memory are such an example [5, 7, 8]. The main advantages of single-electron devices are their fast and ultra-low power operation, simply because they use only a few electrons to accomplish logic or arithmetic operations. While the prospect of CMOS components being completely replaced by SET remains to be seen, more and more SET circuits are emerging.

In this paper, we present an FSM implementation using SET technology, and apply our approach to radio-frequency identification (RFID) protocols where the power dissipation is a critical concern. We simulate the proposed circuit and provide its power estimation to show its particular advantages over its CMOS counterpart.

2. SET Threshold Logic for FSM

2.1. SET junction

For single-electron transistors, a tunnel event is defined as the transport of a single electron through the tunnel junction. In the orthodox theory [9], we ignore the electron energy quantization inside the conductors, the time of electron tunneling through the barrier and coherent quantum processes consisting of several simultaneous tunneling events (i.e., co-tunneling). At a very low temperature, the voltage across a tunnel junction is the only factor to determine the possibility of a tunnel event. The critical voltage V_c of a tunnel junction is given by [7].

$$V_c = \frac{q_e}{2(C_e + C_j)} \qquad (1)$$

where C_j is the capacitance of the tunnel junction, C_e the equivalent capacitance viewed from the tunnel junction's perspective, and q_e the charge of an electron ($1.602 \times 10^{-19} C$). A tunnel event will occur through a junction only if the absolute voltage difference across the junction: $|V_j| > V_c$. Otherwise the circuit is in a stable state, meaning there will be no tunnel event. An electron tunneling through a junction in the quantum scale is a stochastic process. If we define P_{error} as the probability that the desired transport did not occur, the switching delay t_d can be expressed as [7]

$$t_d = \frac{-\ln(P_{error})q_e R_t}{|V_j| - V_c} \qquad (2)$$

where R_t is the tunnel resistance.

For logic design using SET devices, there exist two different architectures: CMOS/SET hybrid architecture and SET-only architecture. The former uses the tunnel junctions as switches to control ('open' or 'close') current through CMOS transistors. While this architecture can provide a high voltage gain and high driving ability from CMOS circuits, the ultra-low power feature of SET technology is partially sacrificed. In the latter architecture, the charge transport in SET transistors is scaled down to a few electrons or even one electron. Therefore, the Boolean logic '0' and '1' can be encoded as a net charge of zero or one electron charge, leading to a single-electron encoded logic (SEEL). As a result, much lower power can be expected in this SET-only architecture. However, its low driving ability may introduce extra difficulty in building up large logic circuits. In the following discussions, we will focus on the SET-only architecture.

2.2. SET threshold logic

0-7695-2533-4/06 $20.00 © 2006 IEEE

The most fundamental brick for digital circuits is a 'threshold' scheme, with which we can define differentiated signals into '0', '1' or others if needed. A generic SET threshold gate structure has been reported [7, 14, 15], and is depicted in Fig. 1. In this threshold gate, the critical voltage V_c acts as the 'threshold'. If the voltage drop across the tunnel junction is higher than the critical voltage, one electron tunnels through the junction. This can be specified as a 'high' output. Otherwise, the output is meant to be logic 'low'. Input voltages V^p and V^n, weighted by their input capacitors C^p and C^n respectively, are added across the tunnel junction (the superscripts p and n stand for positively and negatively weighted inputs, respectively). V_b is the bias voltage weighted by its input capacitance C_b.

This SET threshold gate, when weighted properly, can implement any two-dimension logic (such as AND, OR) gates. Also, with a number of threshold gates working together, more complex logic components, such as flip-flop, can be implemented. For example, an edge-triggered D-flip-flop is shown in Fig. 2, where an circle denotes a threshold logic gate, and the numbers appearing at the inputs of the gate represent the values of positive (+) or negative (−) weights. The detailed description of all gates is given in Table 1.

An FSM can be implemented at gate level using flip-flops and a combinational logic. With SET AND, OR gates and rising-edge triggered D-flip-flops, one can fully implement a given FSM with SET technology. In the next section, we will describe an FSM implementation example from RFID systems. However, the same method can be extended to the general FSM implementation problems. To improve the driving ability of the gates, the buffers/inverters are generally needed in the circuits. The buffer structure and its typical parameters will also be discussed in the next section.

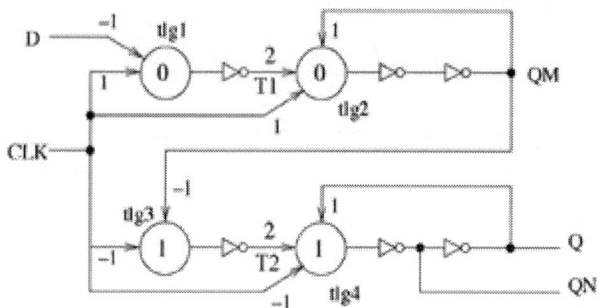

Figure 2. Positive edge-triggered D-flip-flop

3. An FSM Example from RFID Protocol

One of the most important characteristics of SET circuits is their ultra low power dissipation which is essential for some applications. RFID (radio frequency identification) is such an example. An RFID system [10] typically consists two parts: reader and tags. Tags are passive IC chips with unique IDs built in. Tag design depends on communication protocols to be used in the system. One popular protocol for this purpose is the *Binary-tree protocol* [11]. The implementation of this protocol using SET devices is shown in Fig. 3, where logic '1' = 16mV, logic '0' = 0V and, for all tunnel junctions, $R_t = 10^5 \ \Omega$, $C_t = 10^{-19}$ F. Table 3 shows all capacitance parameters of Fig. 3.

Table 1. Threshold logic gates in Fig. 2

Gate	Threshold logic
tlg1	sgn$\{CLK - D\}$
tlg2	sgn $\{CLK + 2T1 + QM\}$
tlg3	sgn $\{-CLK - QM - 1\}$
tlg4	sgn$\{-CLK + 2T2 + Q - 1\}$

4. Simulation and Analysis

4.1. Simulation

The SET circuit of Fig. 3 was simulated using SIMON (SIMulation Of Nanostructure) [12]. The simulation time step was set to be 0.1 *ms*. Fig. 4 shows a 4-bit ID identification procedure. The simulation starts with state S1 (i.e, *Q1* = '0', *Q0* = '1'). The state transitions are shown partially in Table 3. We see that in clock cycles #2 through #8, the state changes between S1 and S2, and ends up with S3. This process represents the successful identification of a 4-bit ID tag. A 'high' *DIFF* at clock cycle #13 puts the machine into state S0 which holds until a next 'high' *NULL* signal arrives.

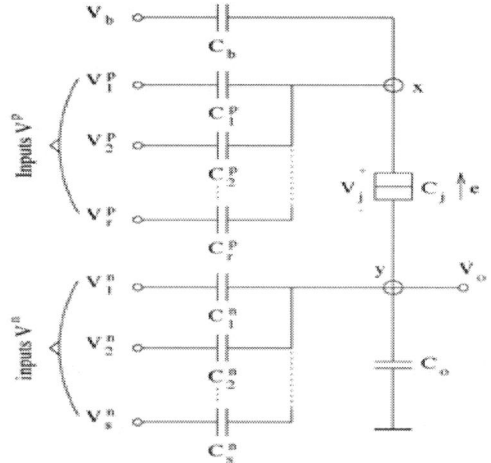

Figure 1.   SET threshold logic gate

0-7695-2533-4/06 $20.00 © 2006 IEEE   238

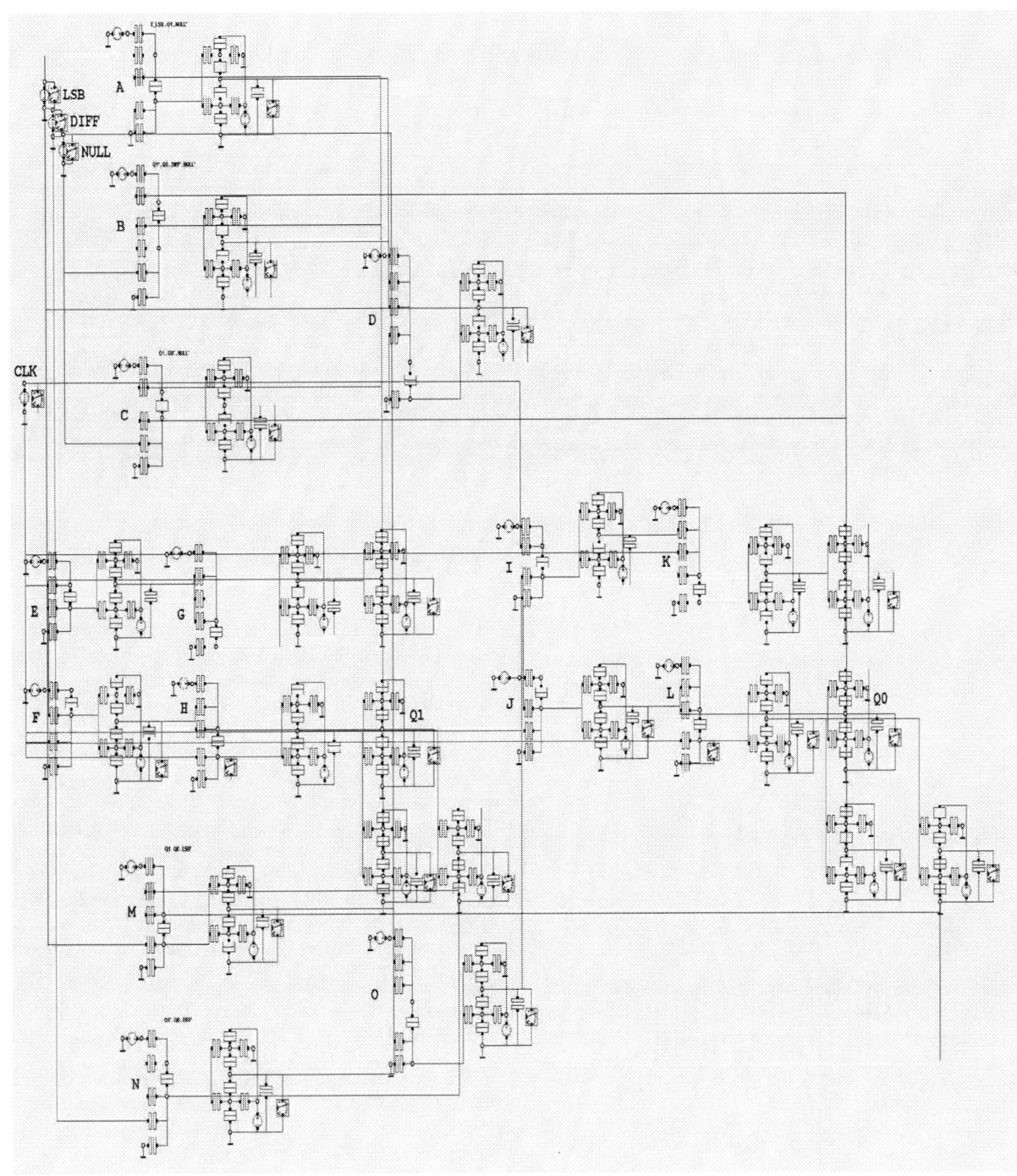

Figure 3. SET implementation of the state machine for Binary-tree protocol

Table 2. Capacitance parameters ($\times 10^{-19}$ F) for threshold gates in Fig. 3

Gate	C_b	C_o	C^p			C''		
A	117	80	5		5	5		
B	110	80		5		5	5	5
C	117	80		5		5	5	
D	117	80	5	5	5	-		
E	122	86		5		4		
F	132	80		-		5	5	
G	117	90	10	5	5			
H	131	86	13.3	5		4		
I	122	86		5		4		
J	132	80		-		5	5	
K	117	90	10	5	5	-		
L	131	86	13.2	5		4		
M	117	80	5	5		5		
N	117	80		5		5	5	
O	117	80	5	5		5		

Table 3. State transitions in Fig. 4

Clock Cycle	State	Q1	Q0	State transition Condition
#1	S2	1	1	DIFF = '0'
#2	S1	0	1	NULL = '1'
#3	S2	1	1	DIFF = '0'
#4	S1	0	1	LSB = '0'
#5	S2	1	1	DIFF = '0'
#6	S1	0	1	LSB = '0'
#7	S2	1	1	DIFF = '0'
#8	S3	1	0	LSB = '1'
#9	S3	1	0	NULL = '0'
#10	S1	0	1	NULL = '1'
#11	S2	1	1	DIFF = '0'
#12	S1	0	1	LSB= '0'
#13	S0	0	0	DIFF = '1'

4.2. Delay and Power Estimation

The critical path of the circuit in Fig. 3 starts from the *CLK*, and ends at the output of the two-level combinational logic. By using the estimation method from [7], the delay can be calculated as around $0.3 \mid ln(P_{error}) \mid ns$ which turns out to be 5.53 *ns* when $P_{error} = 10^{-8}$ is assumed.

In order to estimate the power dissipation of Fig. 3, we have to select a sequence of state transitions that occur during the tag identification process, because the switching energy varies for different states. If we have only one 4-bit ID tag in the reader's working range, the state machine will have the following sequence of state transitions: $S1 \rightarrow S2 \rightarrow S1 \rightarrow S2 \rightarrow S1 \rightarrow S2 \rightarrow S3$. With the power analysis method of [7], the total switching energy can be calculated to be about 702 *meV*. Assuming a clock frequency of 100 MHz, one can expect the total power consumption to be about 10 *pW*.

5. Conclusions

We have presented an FSM implementation using single electron encoded logic in SET technology with the goal of obtaining ultra-low power solution. We have taken the RFID binary-tree protocol as an example to show the details about implementation procedure together with its power and delay analysis. The simulation results have shown significant power savings, compared with the traditional CMOS circuits.

6. References

[1] G. D. Micheli, Synthesis and Optimization of Digital Circuits, McGraw-Hill, 1994.

[2] A. Iranli, P. Rezvani, and M. Pedram, "Low Power Synthesis of

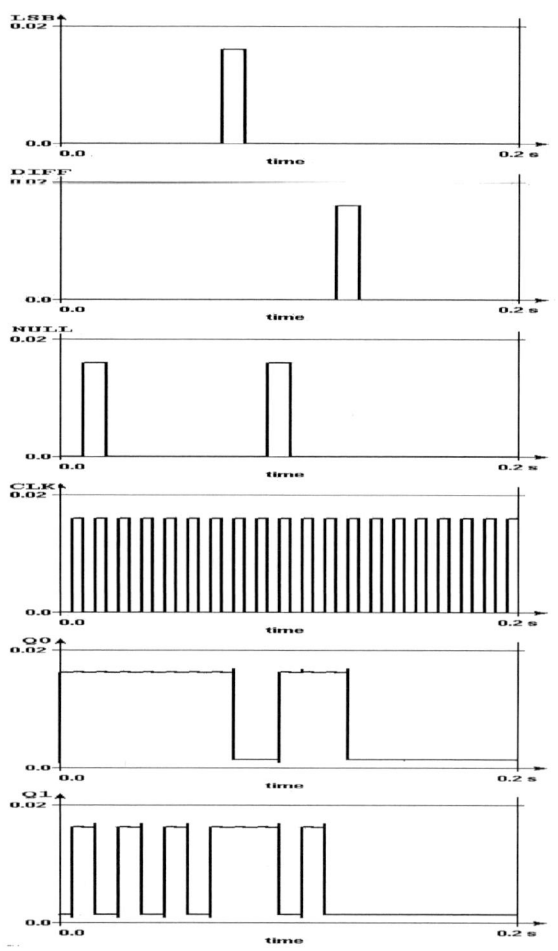

Figure 4. Simulation result of Fig. 3

Finite State Machines with Mixed D and T Flip-Flops," in Proceedings of 2003 Asia and South Pacific Design Automation Conference (ASPDAC), January 2003, pp. 803-808.

[3] S. Chattopadhyay and P. N. Reddy, "Finite State Machine State Assignment Targeting Low Power Consumption," IEE Proceedings – Computers and Digital Techniques, vol. 151, no. 1, pp. 61-70, January 2004.

[4] L. Benini, G. D. Micheli, and F. Vermeulen, "Finite State Machine Partitioning for Low Power," in Proceedings of the 1998 IEEE International Symposium on Circuits and Systems (ISCAS), vol. 2, June 1998, pp. 5-8.

[5] K. K. Likharev, "Single-Electron Devices and Their Applications," Proceedings of IEEE, vol. 87, no. 4, pp. 606-632, April 1999.

[6] A. M. Ionescu, M. Declercq, S. Mahapatra, K. Banerjee, J. Gautier, "Few Electron Devices: Towards Hybrid CMOS-SET Integrated Circuits," in Proceedings of ACM/IEEE Design Automation Conference (DAC), June 2002, pp. 88-93.

[7] C. Lageweg, S. Cotofana, and S. Vassiliadis, "Single Electron Encoded Latches and Flip-Flops," in Proceedings of the 4th IEEE Conference on Nanotechnology, August 2004, pp. 327-330.

[8] K. Yano, et al, "Single-Electron Memory for Giga-to-Tera Bit Storage," Proceedings of the IEEE, vol. 87, no. 4, pp. 633-651, April 1999.

[9] C. Wasshuber, Computational Single-Electronics, Springer-Verlag Wien New York, 2001.

[10] K. Finkenzeller, RFID Handbook, Jhon Wiley & Sons, 2003.

[11] F. Zhou, C. Chen, et al, "Evaluating and Optimizing Power Consumption of Anti-Collision Protocols for Applications in RFID system," in Proceedings of ACM/IEEE International Symposium on Low Power Electronics and Design (ISLPED), August 2004, pp. 357-362.

[12] C. Wasshuber, H. Kosina, and S. Selberherr, "SIMON - A Simulator for Single-Electron Tunnel Devices and Circuits," IEEE Transactions on Computer-Aided Design of Integrated Circuits and Systems, vol. 16, no. 9, pp. 937-944, September 1997.

[13] R. H. Klunder and J. Hoekstra, "Programmable Logic Using a SET Electron Box," in Proceedings of the 8th IEEE International Conference on Electronics, Circuits and Systems (ICECS), vol. 1, September 2001, pp. 185-188.

[14] C. Lageweg, S. Cotofana, and S. Vassiliadis, "Binary Addition Based on Single Electron Tunneling Devices," in Proceedings of the 4th IEEE Conference on Nanotechnology, August 2004, pp. 327-330.

[15] C. Lageweg, S. Cotofana, and S. Vasiliadis, "A Linear Threshold Gate Implementation in Single Electron Technology," in Proceedings of IEEE Computer Society Workshop on VLSI, April 2001, pp. 93-98.

PLAs in Quantum-dot Cellular Automata

Xiaobo Sharon Hu   Michael Crocker
Dept. of Comp. Sci. & Eng.
University of Notre Dame
Notre Dame, IN 46545, USA
Email: {shu, mcrocker}@nd.edu

Michael Niemier
College of Computing
Georgia Institute of Technology
Atlanta, GA 30332, USA
Email: mniemier@cc.gatech.edu

Minjun Yan   Gary Bernstein
Dept. of Elec. Eng.
University of Notre Dame
Notre Dame, IN 46545, USA
Email: {myan, gbernste}@nd.edu

Abstract— **Research in the fields of physics, chemistry and electronics has demonstrated that Quantum-dot Cellular Automata (QCA) is a viable alternative for nano-scale computing. However, little work on QCA has studied designing implementation-friendly programmable QCA circuits. This paper fills this gap by presenting a novel QCA-based Programmable Logic Array (PLA) structure. In addition to being compact, the proposed PLA structure exploits some unique properties of QCA cells to achieve ease of implementation, programming and defect detection. These features are indispensable to the successful adoption of any nano-scale circuits.**

I. INTRODUCTION

In order to sustain the remarkable growth rate in computing performance, device research is clearly needed. However, equally important are research efforts in computational models, circuits, and architectures for the devices. Not surprisingly, advances in one aspect do not necessarily correlate to advances in others. For example, quantum computing is a potentially powerful computational model that could easily solve several notoriously difficult problems (such as factoring large numbers) [36]. [2] shows that it is possible to create architectures that can handle the immense overhead of required error correction. However, to date, the largest number factored is 15 [21]. On the other hand, advances in nano-scale technologies and devices based on traditional diode or transistor models (i.e., using nanowires or carbon nanotubes) seem to indicate that fabricating these nano-scale circuits is markedly more plausible in the nearer-term [7], [35].

This paper focuses on another computational model, Quantum-dot Cellular Automata (QCA). First proposed in the early 1990s [32], QCA accomplishes logical operations and data movement via Coulombic interaction rather than electric current flow. It has been shown that QCA-based circuits could be clocked at an extremely high frequency (adiabatically at 1 THz [29] with cell switching times on the order of 10^{-12} to 10^{-15} seconds [30], [3]), potentially leading to circuits with densities that are multiple orders of magnitude beyond what end-of-the-curve CMOS can provide [22], and dissipate very little power [29]. Simple QCA circuits based on metal-dots have already been demonstrated [1]. More importantly, recent advances in DNA tiling and molecular self-assembly show great promise in implementing molecular QCA circuits [26], [18], [17]. As QCA-based systems could provide various "wins" over end-of-the-curve CMOS, an existing body of research on device physics has been joined by efforts that

explored the construction of traditional computer architectures [22], [24], provided simulators [33], considered how to test circuits [28]. Recent research has begun to examine the implications of physically implementing QCA-based circuits [5].

It is well recognized that realized molecular electronic circuits must be able to tolerate higher percentages of defects than current CMOS circuits. The task of ensuring that a system is still functional post-fabrication falls largely to the circuit designer or computer architect who, in many instances, have leveraged reconfigurable/reprogrammable structures. Largely for this reason, Programmable Logic Arrays (PLAs) have been studied by a number of research groups for different nanoscale technologies. For example, DeHon and Wilson proposed a nano-wire-based sublithographic PLA design [8]. Likharev and Strukov presented CMOL FPGA [27]. Hogg and Snider showed interesting algorithmic approaches to map logic functions to PLAs with defective crossbar switches. Though many papers have been published on designing QCA-based circuits, only a few have examined design issues with programmable structures that use QCA cells [10], [23], [14]. However, these designs either are difficult to implement or sacrifice performance.

In this paper, we present a novel PLA structure based on the QCA device architecture. It is fully re-programmable and this programmability is achieved by exploiting the fact that QCA logic can be bidirectional. Furthermore, the PLA structure allows defects to be readily detected and isolated, again by exploiting logic bidirectionality. The design is compact and easily extensible. Even with the consideration of near- to mid-term implementation constraints, the size of a QCA PLA is still very small. We have verified the functionality of our proposed design through simulations based on statistical mechanics for a 4-dot molecular QCA cell. Since correction operation of a QCA circuit is also dependent on the clocking fields controlling the QCA circuit, we have investigated the impact of utilizing CMOS technology to generate the necessary electric field distribution for our PLA structure. Several representative clocking wire layouts were studied and their resulting electric fields were simulated with a commercial Finite-Element based software [9].

II. BACKGROUND AND RELATED WORK

QCA represents information by encoding binary numbers into cells having a bi-stable charge configuration. A QCA cell

0-7695-2533-4/06 $20.00 © 2006 IEEE

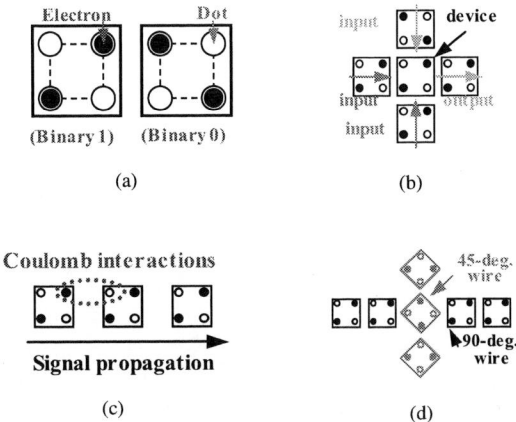

Fig. 1. (a) Two 4-dot QCA cells. (b) A majority gate made of 4 QCA cells. (c) A QCA wire. (d) Physically crossing two QCA wires.

can consist of 2 or 4 "charge containers" (i.e., quantum dots) and 1 or 2 excess charges, respectively. One configuration of charge represents a binary '1', while the other a binary '0' (Fig. 1(a)) [20]. Logical operations and data movement are accomplished via Coulomb interactions. QCA cells interact because the charge configuration of one cell alters the charge configuration of the next cell. Additionally, information transmission and processing is carried out by the same entities – QCA cells – rather than by separate devices and wires. As QCA cells are made smaller, interconnects shrink too.

Fig. 1(b)–Fig. 1(d) illustrate several basic QCA circuit elements [32], [22]. The majority gate (Fig. 1(b)) implements logic function $AB + BC + AC$. The output cell assumes the polarization of the majority of the 3 input cells [20]. By setting one of the majority gate inputs to logic '0' or '1', the gate reduces to an AND or OR gate, respectively. A QCA wire (Fig. 1(c)) is just a line of QCA cells; at the input end, one cell must be polarized to act as the driver for the wire. The cells do not need to be spaced exactly the same distance apart. QCA wires with different orientations (Fig. 1(d)) can cross in the plane without destructing the binary value on either wire. An inverter can also be built easily with QCA cells [32].

A "clock" structure is required to provide signal gain for a QCA circuit. The clock structure supplies a necessary electric field for groups of QCA cells to transition from a null state to a bistable, active state, and then back to the null state. A QCA cell can only change its charge configuration (from a logic '0' to '1' or vice versa) by going through a transition from the null state to the active state. Specifically, a negative electric field first pulls the charges in QCA cells "down" so that the cells are put in the null state, then a positive electric field pushes charge "up" to the active state, while a driver (a cell already in the active state) determines if a cell just turned on will be a 1 or 0. If the positive electric field is maintained, QCA cells will keep their original charge configurations.

For systems of molecular QCA cells (the target of this paper), the clock structure, i.e., the electrical field, can be generated using wires formed by conventional lithography [11], [29] on a silicon substrate to which QCA molecules can be attached [13]. The phases of a clock signal could take the form

of time-varying, repetitious voltages applied to such CMOS wires. Multiple wave fronts could exist simultaneously – even along the same QCA wire – resulting in inherent pipelining determined by the granularity of the clock structure [20], [24].

In a **molecular** implementation, a QCA device could be made from a single chemical molecule. A recent experiment demonstrates that applying reasonable electric fields can move a charge between two sites of a molecule engineered to function as a two-dot QCA cell [25]. The significance of this experiment is that it shows a self-assembled monolayer (or plane) of these molecules switching between a chemical representation of a binary 0 and 1. Four-dot QCA molecules have also been made [16] and promising I/O methodologies exist. Also, active research in DNA-scaffolding can provide viable substrates for the molecules. These and other advances in QCA implementation propel research in QCA-based circuits and architecture to an important position.

The work discussed here presents a novel programmable structure for QCA-based circuits. The design principle revealed here could be useful for developing other QCA-based programmable devices. The PLA structure is essentially implementation *independent* despite the possibilities of multiple QCA implementations (e.g., metal-QCA [1], [19] and magnetic-QCA [6], [4] have also been proposed).

III. PLA STRUCTURE

A. Re-programmable PLA Cells

We use a combination of AND and OR planes made of QCA cells to implement a PLA. AND or OR gates can be implemented by using a single majority gate. Our challenge is to make the PLA cells programmable so that they may function either as a logic gate or as a wire. Each PLA cell, no matter which plane it is in, is made to contain a programmable select bit, denoted as S. In the AND plane, each PLA cell consists of two majority gates: one behaves as an AND gate, while the other as an OR gate to aid programming. A QCA cell schematic of the AND PLA cell is shown in Fig. 2(a) and its equivalent logic representation is given in Fig. 2(b). The OR PLA cell is constructed by simply switching the positions of the AND and OR gates in Fig. 2(a) and Fig. 2(b).

We will use the AND cell as an example to illustrate how a PLA cell can be programmed to function as either a logic gate or a wire. The AND cell has the following functionality.

```
MintermOut = (LiteralIn) · (MintermIn), if S = 0.
```

```
MintermOut = (MintermIn), if S = 1.
```

We denote the mode of the PLA cell as "**logic**" if S = 0 and as "**wire**" otherwise. Thus during normal operation, the QCA cell corresponding to the select bit is used as an input to the OR gate. Initially, the mode of a PLA cell (i.e., the value of S) is undefined. To see how programming is achieved, it is interesting to note that a QCA-based majority gate is logically *bidirectional*, i.e., any one of the four boundary QCA cells can be treated as the output while the rest of the three as inputs.

0-7695-2533-4/06 $20.00 © 2006 IEEE   243

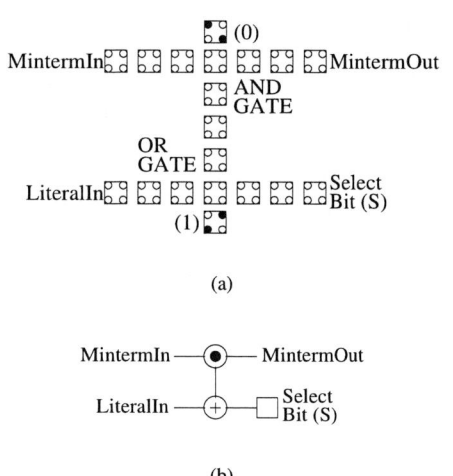

(a)

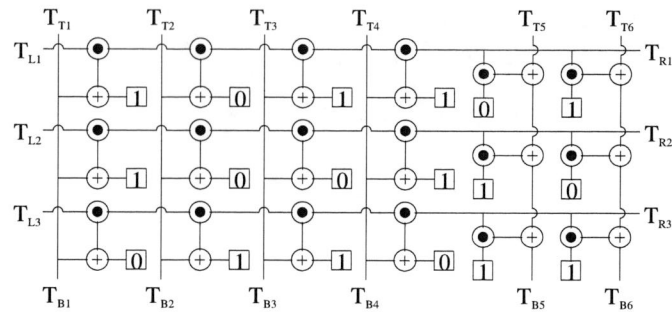

Fig. 3. A PLA array that implements a XOR and a OR function.

(b)

Fig. 2. (a) QCA representation of PLA cells in the AND plane. (b) Logic representation of PLA cells in the AND plane.

To program a cell, we treat the QCA cells corresponding to the select bit as the output of the OR gate. To drive a logical '1' or '0' to the select bit, we set MintermIn to '0' and LiteralIn to '1' or '0', respectively. The functionality and programming of OR cells are very similar except that '1' corresponds to **logic** mode and '0' corresponds to **wire** mode.

Readers may wonder how to differentiate QCA cells whose values need to be held (such as the QCA cells representing the select bit) from those whose values change when driver cells change their values. This can be achieved by judiciously controlling the clocking structure for the QCA cells. Recall from Section 2 that QCA cells can be in one of two states: a null state or a bistable, active state, depending on the specific electric field they are in. To allow a QCA-based logic gate to re-evaluate upon arrival of new input values, the electric field of the QCA cells associated with the logic gate needs to be altered (i.e., going from negative to positive). On the other hand, to hold the value of a set of QCA cells, simply supply a constant positive electric field. Therefore, to ensure that the PLA cell works properly, we must be able to individually control the electric field, i.e., having two separate clocking regions for the QCA cells of the select bit and for those of the logic gates. We will discuss this point further in Section 3.2.

B. The array structure

Based on the AND and OR cells introduced in the previous subsection, it is straightforward to construct a PLA. Fig. 3 depicts a logic representation of a PLA. The T_{XX}'s represent the terminal QCA cells at the PLA boundary. During normal operation, the inputs come from below, the minterm signals move from left to right, and the sum-of-minterm signals move from top to bottom. By setting $T_{L1} = T_{L2} = T_{L3} = 1$, $T_{T5} = T_{T6} = 0$, $T_{B1} = X$, $T_{B2} = X'$, $T_{B3} = Y$, $T_{B4} = Y'$, and leaving the rest of the terminals unset, it is easy to verify that the PLA performs the following two logic functions: $T_{B5} = X$ XOR Y, and $T_{B6} = X'$ OR Y'.

We have made a conscious effort to keep the number of wire crossings in a PLA to a minimum. This is due to the fact that implementing a wire crossing as shown in Fig. 1(d)

can be extremely challenging with near- to mid-term chemical synthesis processes envisioned by chemists [5]. One possible approach is to adopt the logic crossing idea proposed in [22]. The downside of employing the logic crossing idea is that it increases the number of QCA cells since as many as 12 majority gates are needed to achieve one logic crossing. Because our PLA structure only requires one crossing per cell, this increase is tolerable. (Additionally, our ongoing – and currently unpublished work – shows that a logical crossing could be mapped to a DNA substrate and would require less area than two metal1 wires would need to cross – at the 22 nm ITRS node [15]).

In order for a PLA array to operate properly, a carefully designed underlying clocking structure is required. For each PLA cell, two separate clocking regions are needed; one for the QCA cells representing the select bit and the other for the rest of the QCA cells. Let us consider the implications of this requirement. It is certainly possible to simply keep two clocking regions for every PLA cell. However, such a design may make the circuitry for generating the electric fields unnecessarily complicated. The reason is that each clocking region would need its own clock wire, which increases the demands on signal routing. Furthermore, since a PLA cell tends to be much smaller than the minimum metal wire in the end-of-the curve CMOS technology (if a crossing could be implemented with the structure shown in Fig. 1(d)), having a separate clocking region for the logic gates in each PLA cell would be somewhat wasteful. Individually controlled clocking regions would also increase the programming time and latency.

A much better approach is to treat the AND and OR planes as two large, alternate clocking regions. The entire AND plane, except the clocking regions for the select bits, is clocked by a larger clock wire. The same is true for the OR plane. The terminal QCA cells are grouped as follows: all T_{Li}'s in one group, all T_{Ri}'s in another group, all AND plane T_{Ti}'s and T_{Bi}'s in a group, and all OR plane T_{Ti}'s and T_{Bi}'s in a group. Such partitioning allows both efficient normal operation and programming. In normal operation, the clocking regions for the inputs (AND plane T_{Bi}'s) and outputs (OR plane T_{Bi}'s), together with those for the AND plane and OR plane can be driven by four 4-phase clock signals, each assuming a distinctive phase at any give time. We refer to this approach as the **global clocking** approach while the one in the previous paragraph as the local clocking approach. The

0-7695-2533-4/06 $20.00 © 2006 IEEE   244

normal operation for the global clocking approach proceeds as follows. In the first quarter of a cycle, the inputs are updated. Then the minterms in the AND plane are evaluated during the second quarter of the cycle. In the third quarter, the OR plane cells are switched based on the results of the AND plane. Finally, in the last quarter, the outputs are made available.

The global clocking approach is fast and easy to implement. Furthermore, the structure immediately leads to a pipelined PLA and can be readily used to implement finite state machines without requiring any additional registers. However, there is a drawback to this clocking approach. The number of QCA cells that can be driven reliably by a single clocking region is finite and depends on the number of QCA cells along the longest path in the region as well as the clock frequency [20], [22]. Based on the current estimate, one clocking region can easily drive anywhere between 10^3 to 10^5 QCA cells along the longest path. We feel that this upper bound on the PLA size is sufficient for many realistic designs, especially considering the impact of defects. If a PLA array is too large for the global clocking approach, we can partition the PLA cells into smaller regions so that the number of QCA cells in each clocking region is within the acceptable range. How to partition the PLA cells can greatly impact the complexity of the clocking wire layout (as we discussed for the case of the local clocking approach–an extreme partitioning). Furthermore, different partitioning may require different clock signal patterns. We plan to investigate various clocking region partitions in our future work if the needs become evident.

C. Programming and defect detection

The ability to reprogram a PLA is very powerful. It makes the PLA more versatile and it also helps defect tolerance. Unlike mask-programmable CMOS PLAs which can be programmed once at fabrication time, our QCA-based PLA can be reprogrammed over and over again for different applications. Furthermore, by judiciously programming the PLA cells into **logic** or **wire** mode, an entire PLA can be tested for defects after fabrication. Defective columns, rows, and cells can be avoided during the programming stage, allowing PLAs with defective QCA cells to remain useful.

Since putting a cell into either **logic** or **wire** mode can be done by setting the select bit of the cell to '0' or '1' (see Section 3.1), programming a PLA can thus be achieved by driving the desired select bit values to corresponding PLA cells. The fact that a number of PLA cells, such as those on the same column of the AND plane, share the same input requires that a proper programming sequence be followed.

A simple and efficient programming sequence is to program row by row for the AND plane first and then column by column for the OR plane. To program one row, say row i of the AND plane, the desired select bit value for each PLA cell j on this row is applied to the corresponding terminal T_{Bj} (or equivalently T_{Tj}). Terminal T_{Li} is set to '0' to ensure that the programmed bit makes it to S unchanged. After a row is programmed, the electric field for the select bits on this row must be kept adequately positive such that these values

would not change even when different input signals appear at the OR gates connected to the select bits. A column can be programmed in a similar way except that T_{Ri} is used to supply the needed S values and T_{Tj} (or equivalently T_{Tj}) is set to '1'.

Defect detection is a multi-step process that requires analysis of both the QCA wires and the PLA cells. Testing the QCA wires is straightforward for the AND plane columns, as no logic gates interfere with signal propagation from bottom to top. We can simply apply a logic '0' and '1' to each T_{Bi} and test if the same signal appears at T_{Ti}. To test the AND plane rows, the OR plane columns and rows, the logical gates must first be placed into **wire** mode in order for the input signal to pass through. Once in **wire** mode, the same test as used for the AND plane rows can be performed. Testing the rows and columns is a quick process as it can be done in parallel.

In order to test the PLA cells, only a single AND plane column or OR plane row can be tested at a time. Each PLA cell in that column (resp. row) is placed into **logic** mode while the rest of the cells are all put into **wire** mode. Then, we set T_{Li} (resp. T_{Ti}) to '0' (resp. '1'), drive a '0' and '1' alternately to T_{Bj} (resp. T_{Rj}), and monitor the values at terminal T_{Rj} (resp. T_{Bj}) in the AND (resp. OR) plane. By repeating this process, all cells in the PLA can be tested. After the testing procedure is completed, methods for mapping a specific function to a PLA with known defects can be used to determine the mode in which each functioning PLA cell should be in [12]. We summarize the entire testing procedure of our QCA-based PLA circuit below.

1. Check AND plane columns
2. Program *all* cells to **wire** mode
3. Check all rows and OR plane columns
4. Repeat the following steps for all AND plane columns and OR plane rows
5. Program *all* cells to **wire** mode
6. Program a single AND (resp. OR) plane column (resp. row) of cells to **logic** mode
7. Check row (resp. column) outputs for defects in individual cells

Special care must be taken when programming *all* cells into **wire** mode (step 2 and 5 above) in order to avoid being overly pessimistic. Suppose a column, say column 2 in Fig. 3, is found to be defective after step 1 above. Then it seems that we may not be able to program some (or even all) of the PLA cells along column 2 since we could not drive a logic '1' from T_{B2} all the way up. To alleviate this problem, we again make use of the logic bidirectionality property of QCA cells. Note that our goal is to put as many PLA cells in **wire** mode as possible. With the original programming method discussed at the beginning of this section, the select bit in an AND plane cell can only be set to '1' by a terminal at the bottom of the AND plane. To increase the probability of select bits in the AND plane receiving a logic '1', we propose to set *all* the terminals of the PLA to logic '1'. By doing so, we have introduced two additional routes for a logical '1' to reach a select bit, i.e., from the terminal at the top of the PLA array and from the AND gate of the corresponding PLA cell. This greatly reduces the impact of defective QCA cells and hence increases the probability of discovering operational QCA cells. The bottom line is that the yield of QCA-based PLAs could be increased. The detailed yield study will be presented in our future work.

IV. EXPERIMENTAL RESULTS

We have conducted several simulation-based experimental studies to validate the PLA design presented in the previous section. These experiments both verify the functionality of our PLA design through simulations based on a statistical mechanics model, and examine the electric field distribution resulting from the global clocking approach.

Based on the analytical model provided by [34], we have developed a statistical mechanics simulation tool for QCA circuits built with 4-dot QCA cells (Fig. 1(a)). This tool takes a pattern of QCA cells and determines the electron potential energy for all states. It then reports the lowest twenty energy states, including the global "ground" (i.e., the lowest energy) state.If the electron distribution in the global ground state matches the desired behavior of the QCA circuit, we can be confident that the design will work.However, If it does not match the global energy minimum, but is still one of the lower energy states, it may be a local minimum, and other simulation can determine if the design will still work in practice.

In order to verify the operation of our PLA design, we have tested our PLA cells using the statistical mechanical model. These tests considered the full truth table of possible inputs to the cells. Through careful layout and extension of the majority-gate intersections, we have found a PLA cell design that produces a global energy minimum for the desired behavior in each truth table combination, indicating that our PLA should function as intended.

Clocking region design plays a key role in ensuring the correct operation of the PLA. In our global clocking approach, one unique requirement is that the regions corresponding to the select bits must be able to supply a constant positive electric field even when the other regions' electric fields are changing. Hennessy and Lent have studied electric fields introduced by evenly spaced point charges [11]. Their main results demonstrated that desirable electric field distributions for a pipelined QCA circuit can be obtained by applying proper charges on infinitesimally small wires underneath the QCA circuit. The results should also apply to wires with finite sizes. However, to our best knowledge, no experimental results have been published regarding this, and there is no experimental study on the possibility of supplying the more demanding electric field distribution for our proposed PLA structure.

To validate that our global clocking approach indeed provides the desired electric field, we have conducted a number of experiments to investigate the electric field distribution from metal wires (such as those used in CMOS technology). Fig. 4(b) shows a sample layout of clocking regions resembling those that may be used for one plane of a PLA. The small square shapes (referred to as *select wires*) represent the clocking wires that supply the electric field for the QCA cells of the select bits. The long rectangular shapes (referred to as *logic wires*) correspond to the wires that supply the electric field for the columns/rows of the PLA logic cells. We have used different sizes for the wires as well as different separations between the wires. The minimum metal pitch is 40nm, in accordance with the 2004 ITRS for 2018 technology node [15]. A metal plane is added 20nm above the logic and select wire plane to act as the ground plane for directing electric fields [11]. It is envisioned that the QCA-based PLA would be sandwiched between the two metal planes. Such a structure has been demonstrated in [25]. We used silicon dioxide as the dielectric with relative permittivity of 3.9 and gold as the metal for the wires. (We note that only 2 layers of metal should be needed to build the clock structure.)

To study the electric field distribution, two distinct clocking signals are applied. All the select wires are driven by a constant positive voltage while the logic wires are driven by a quasi-adiabatic 4-phase clock signal [31]. This setup emulates the PLA in normal operation mode. (Note that we did not study the programming scenario as it is similar to the operations of pipelined circuits which has been studied in [11].) FEMLAB software [9] is used to obtain the electric field distribution at the plane where QCA cells reside. Since only the component of the electric field that is perpendicular affects the states of the QCA cells [11], we only examine the distribution of the vertical (i.e., the z-direction) electric field. We applied a 5 volt signal to select wires, and applied five different voltages, -5 volts, -2.5 volts, 0 volt, 2.5 volts, and 5 volts, to the quasiadiabatic signal (logic wires). The selection of 5 volts as the maximum voltage is based on the assumption that the intensity of 2 Mv/cm is sufficient to maintain the states of QCA cells. Based on conversations with chemists, this is a realistic assumption for molecular QCA cells.

The simulation results are summarized in Fig. 4(a)–Fig. 4(f). These figures depict the intensity of the vertical electric field 10nm above the logic/select wire plane. According to the intensity scale bar in Fig. 4(a), one can see that the vertical electric field above the logic wires changes from -2.5Mv/cm to 2.5Mv/cm while the vertical electric field above the select wires is unchanged. The field intensity above the select wires exceeds 2Mv/cm, which indicates that the QCA cells on top of these wires are in the active/bistable state [11]. At this state, the QCA cells cannot receive any inputs and maintain their previous logic values. This is precisely what we need in order to guarantee the correct functionality of our PLA structure. The field distributions remain essentially the same regardless of the sizes of the wires and distances between them.

V. CONCLUSIONS AND FUTURE WORK

We have presented a novel QCA-based PLA structure and showed how to detect defects and program around them by exploiting the bi-directionality property of QCA cells. We also demonstrated that our PLA design is implementation friendly in two regards. First, the clocking regions are regular and the electric field distribution is relatively simple. Therefore, the circuitry to provide the electric field can be designed and fabricated with traditional CMOS technology. This has been

0-7695-2533-4/06 $20.00 © 2006 IEEE

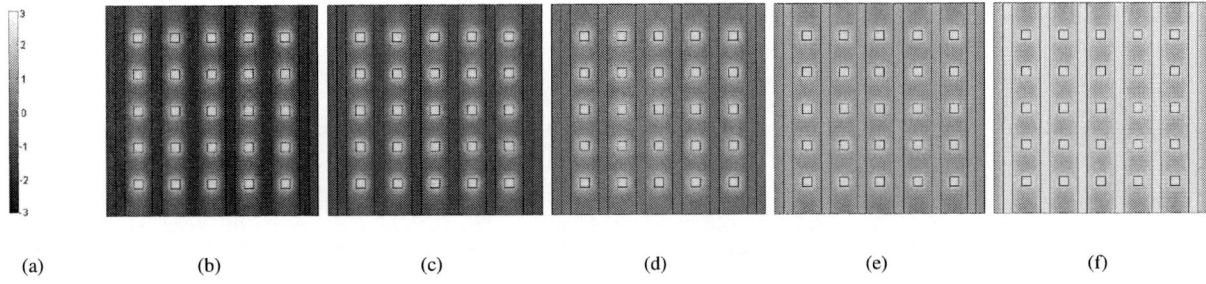

Fig. 4. Vertical electric field distribution at the plane of QCA cells. (a) shows the intensity scale bar from -3 Mv/cm to 3 Mv/cm, and (b)-(f) correspond to the five different voltage values on the logic (i.e., long) wires, -5 volts, -2.5 volts, 0 volt, 2.5 volts, and 5 volts, respectively.

validated via detailed electric field simulation. Second, the area of each PLA cell is small and requires only a single wire crossing. Wire crossings can be a major road block in the near- to mid-term molecular QCA fabrication. By having a single wire crossing, we achieve a minimal PLA cell size.

Our proposed PLA structure laid down some ground work for designing re-programmable QCA circuits. More detailed investigation for QCA-based PLA is still needed. For example, the actual layout of a PLA needs to be conducted to examine the impact of PLA cell size on its functionality. This should be studied together with the clocking wire layout. Another important aspect is to study the yield based on realistic QCA defect models. Designing reconfigurable/reprogrammable QCA-based circuits is still at its infancy. We expect that our work will initiate a much needed effort in this area.

REFERENCES

[1] Amlani I., Orlov A. O., Snider G., and Lent C.S., "Demonstration of a functional. quantum-dot cellular automata cell," *J. Vac. Sci. Technol. B*, 16, 1998, p. 3795-3799.

[2] S. Balensiefer, L. Kregor-Stickles, and M. Oskin. "An Evaluation Framework and Instruction Set Architecture for Ion-Trap based Quantum Microarchitectures," in the *Int. Symp. on Comp. Arch.*, 2005.

[3] P.F. Barbara, T.J. Meyer, and M.A. Ratner, "Contemporary Issues in Electron Transfer Research", *J. Phys. Chem.*, 1996, 100, p. 13148-13168.

[4] G.H. Bernstein, A. Imre, V. Metlushko, A. Orlov, L. Zhou, L. Ji, G. Csaba, and W. Porod, "Magnetic QCA systems," *Microelectronics Journal*, 36(2005), p. 619-624.

[5] A. Chaudhary, D.Z. Chen, X. Sharon Hu, M.T. Niemier, R. Ravichandran, and K. Whitton, "Eliminating Wire Crossings for Molecular Quantum-dot Cellular Automata Implementation," in *Proc. of Int. Conf. on Comp. Aided Des.*, p. 565-571, Nov. 6-10, 2005.

[6] R.P. Cowburn and M.E. Welland, "Room Temperature Magnetic Quantum Cellular Automata," *Science*, Vol. 287, Issue 5457, 1466-1468, 2000.

[7] A. DeHon, "Array-based Architectures for FET-based Nano-scale Electronics," *IEEE Nano.*, Vol. 2, No. 1 pp 23-32, March 2003.

[8] A. DeHon and M.J. Wilson, "Nano-wire based sublithographic Programmable Logic Arrays," published in *Proc. of Int. Symp. on FPGAs*, (February 22-24, 2004), pp. 123-132.

[9] FEMLAB 3.0 software by COMSOL, Inc., http://www.comsol.com.

[10] C.R. Graunke, D.I. Wheeler, D. Tougaw, and J.D. Will, "Implementation of a Crossbar Network Using Quantum-dot Cellular Automata," *IEEE Trans. on Nano*, Vol. 4, No. 4, July 2004.

[11] K. Hennessy, C.S. Lent, "Clocking of molecular quantum-dot cellular automata", *J. of Vac. Sci. & Tech. B*, 19, 5(Sep-Oct 2001), 1752-55.

[12] T. Hogg and G. Snider, "Defect-tolerant logic with Nanoscale crossbar circuits," HP Labs, http://www.hpl.hp.com/research/idl/papers/molecularAdder/.

[13] W. Hu, K. Sarveswaran, M. Lieberman, and G. H. Bernstein, "High resolution electron beam lithography and DNA nano-patterning for molecular QCA," *IEEE Trans. Nano.*, 2005, 4, 312-316.

[14] J. Huang, M. Momenzadeh, M.B. Tahoori, F. Lombardi: "Design and characterization of an and-or-inverter (AOI) gate for QCA implementation," *ACM GLVLSI Gymp. 2004*, 426-429.

[15] International Technology Roadmap for Semiconductors, 2004, http://www.itrs.net/Common/2004Update/2004Update.htm.

[16] Jiao J.Y., Long G.J., Grandjean F, Beatty A.M., Fehlner T.P. "Building blocks for the molecular expression of quantum cellular automata," *J. Am. Chem. Soc.*, 2003, 125, 7522-7523.

[17] T.H. Labean, S. Park, S.J. Ahn, and J.H. Reif, "Stepwise DNA self-assembly of fixed-size nanostructures," *Found. of Nanoscience, Self-assembled Arch., and Dev.*, pp. 179-181, 2005.

[18] J.D. Le, Y. Pinto, N.C. Seeman, K. Musier-Forsyth, T.A. Taton, and R.A. Kiehl, "DNA-Templated Self-Assembly of Metallic Nanocomponent Arrays on a Surface," *Nano Lett.*, 4 (12), 2343-2347, 2004.

[19] Lent C.S., Snider G.L., Bernstein G., Porod W., Orlov A., Lieberman M., Fehlner T., Niemier T.,and Kogge P. *Quantum-Dot Cellular Automata*. Chapter in Electron Transport in Quantum Dots, pp. 397-433, Johathan P. Bird (ed.) Kluwer Academic Publishers, 2003.

[20] C.S. Lent, P.D. Tougaw, "A Device Architecture for Computing with Quantum Dots", *Proc. of the IEEE*, Vol. 85, p. 541, 1997.

[21] M. Lieven, K. Vandersypen, M. Steffen, G. Breyta, C.S. Yannoni, M.H. Sherwood, and I.L. Chuang, "Experimental realization of Shor's quantum factoring algorithm using nuclear magnetic resonance", *Nature*, 414, 883887 (20 December 2001).

[22] M.T. Niemier, "The Effects of a New Technology on the Design, Organization, and Architectures of Computing Systems," *Thesis.*, 2004.

[23] Niemier, M.T. and Kogge, P.M. "The '4-Diamond' Circuit - A Minimally Complex Nano-scale Computational Building Block in QCA," In *Proc. of the IEEE Comp. Soc. Symp. on VLSI*, pp. 3-10, February 2004.

[24] M. Niemier and P. Kogge, "Exploring & exploiting wire-level pipelining in emerging technologies", *Int. Sym. of Comp. Arch.*, 2001, p.166-177.

[25] Qi H., Sharma S., Li Z.H., Snider G.L., Orlov A.O., Lent C.S., Fehlner T.P., "Molecular quantum cellular automata cells," *J. Am. Chem. Soc.* 2003, 125, 15250-15259.

[26] P.W.K. Rothemund, N. Papadakis, and E. Winfree, Algorithmic self-assembly of DNA Sierpinski triangles, *PLOS Biology*, 2004, 2041-2053.

[27] D.B. Strukov and K.K. Likharev, "CMOL FPGA: a reconfigurable architecture for hybrid digital circuits with two-terminal nanodevices," *Nanotechnology*, 16(2005) 888-900.

[28] M.B. Tahoori, J. Huang, M. Momenzadeh and F. Lombardi, "Defects and Faults in QCA at the nano-scale", *VLSI Test Sym.*, 2004.

[29] J. Timler and C.S. Lent, "Power gain and dissipation in quantum-dot cellular automata", *J. of App. Phys.*, 91, 2002, p.823-831.

[30] G.M. Tom and H. Taube, "The Mixed Valence State Based on μ-Cyanogen-bis(pentaammineruthenium)," *J. of Am. Chem. Soc.*, Vol. 97 p.5310-5311, 1975.

[31] G. Toth and C. S. Lent, "Quasiadiabatic switching for metal-island quantum-dot cellular automata", *J. Appl. Phys.*, 85, 2977, 1999.

[32] P.D. Tougaw, C.S. Lent, "Logical Devices Implemented Using Quantum Cellular Automata", *J. of App. Phys.*, Vol. 75, 1994, p. 1818.

[33] K. Walus, T. Dysart, G.A. Jullien and R.A. Budiman, "QCA Designer: A Rapid Design and Simulation Tool for Quantum-dot Cellular Automata", *IEEE Trans. on Nano.*, 3(1), 2004, pp. 26-31.

[34] Yuliang Wang and Marya Lieberman, "Thermodynamic behavior of molecular-scale quantum-dot automata (QCA) wires and logic devices," *IEEE Trans. on Nano.*, 2003.

[35] D. Whang, S. Jin, Y. Wu, and C.M. Lieber, "Large-scale hierarchical organization of nanowire arrays for integrated nanosystems," *Nanoletters*, Vol. 3, No. 9, pp. 1255-1259, September, 2003.

[36] C.P. Williams and S.H. Clearwater, *Explorations in Quantum Computing*, Springer-Verlang New York, Inc., New York, NY, 1997.

0-7695-2533-4/06 $20.00 © 2006 IEEE

Reconfigurable System Design and Technologies

Dynamic Hardware Multiplexing:
Improving Adaptability with a Run Time Reconfiguration Manager

P. Benoit, L. Torres, G. Sassatelli, M. Robert, G.
Cambon
LIRMM, UMR University of Montpellier 2-CNRS
C5506
Email: {name@lirmm.fr}

J. Becker
University of Karlsruhe
ITIV
Email: becker@itiv.uka.de

ABSTRACT

Dynamic reconfiguration provides interesting features offering hardware flexibility and adaptability. Unfortunately, the lack of programming tools to manage it has limited its use in current SoCs. This paper presents a method to abstract, at design-time, dynamic reconfiguration management. Dynamic Hardware Multiplexing is a generic principle based on a scheduler dedicated to reconfigurable resources management at run-time. Formal background, implementation, simulation results and validations are exposed to illustrate the contribution of this study.

1. Introduction

FPGA cores integration has become reality in embedded systems in a recent past. Several companies, eASIC[1] and M2000[2] for instance, offer dense customizable FPGA macros that can be used as dedicated logic and associated to a general purpose processor.

Despite dynamic reconfiguration capabilities, system designers use the reconfigurable elements as low cost prototyping hardware to make several iterations of the system before manufacturing it with a fixed configuration. Dynamic configuration could bring several obvious benefits such as multi-application handling, but there is a real lack of efficient programming tools or methods.

The objective of this paper is to present a technique to handle dynamic configuration of the hardware, *i.e.* allowing the application designer to abstract dynamic reconfiguration management constraints. This technique is based on a Run-Time scheduler [1]implementing an algorithm called Dynamic Hardware Multiplexing (DHM). Thanks to this approach, the configuration of each application is generated at design time. At run-time, the scheduler takes care of the placement and modifies the original configuration in order to adapt the computational resources to handle several applications or to implement different quality of service. The efficiency of our

---

[1] The term "scheduler" is used in this paper as "spatial scheduler" or "mapper"

---

approach is characterized by two metrics showing the increase in the resource usage and multi-application management. The suggested method has been designed and validated on a coarse grain reconfigurable architecture.

This paper is organized as follows: section 2 presents some related works, then the formal framework and characterization metrics are exposed; section 3 describes the Dynamic HW Multiplexing algorithms; section 4 then illustrates an implementation of DHM carried out on an existing coarse grain reconfigurable architecture; simulation results are exposed in section 5 and HW/SW DHM schedulers validations are discussed in section 6; finally, conclusions and perspectives are drawn.

2. State of the art

In the literature, dynamic reconfiguration management for multi-application purposes has been studied in [3, 4, 5, 6, 7] (non-exhaustive list). The different existing schemes target commercial fine grain reconfigurable devices. Proposed techniques generally provide a support for relocation or replication of tasks where those are handled as variable-size rectangle shapes that can be placed anywhere on the FPGA, or also as a fixed-size rectangle like in the approach proposed in [10,11]. The scheduler can be either conceptual (only an algorithm), a processor or dedicated logic implemented in the FPGA [8], or even an external block in [9].

Most of the papers present conceptual techniques and do not give neither many details on the implementation of the scheduler, nor simulation/validation results. Actually, fine grain reconfigurable elements require complex and time/ area consuming configuration schedulers. The goal of this research work is to propose a new simple method for automatic placement, especially on coarse grain reconfigurable architectures, to design the scheduler and analyze trade-offs of several hardware implementations.

3. Formal framework

In this article, we assume a heterogeneous SoC with processing cores and peripherals, including at least one general purpose processor (GPP) and at least one Reconfigurable Coprocessor (RC). Each task considered here can be mapped onto the system with a software program, on the GPP, <u>or</u> with a configuration program on the RC.

The underlying problem is the management of dynamic reconfiguration to achieve an improved adaptability and flexibility. We propose here a formal framework in order to define a generic solution to multi-application handling through dynamic reconfiguration.

We assume a set θ of applications (or independent tasks/threads) to be accelerated, $i.e.$ potentially assignable to the RC. The system is running applications from cycle 0 to cycle n. Each task $T_k \in \theta$ is randomly executed during $n_k \leq n$ cycles. The set Γ of <u>running tasks</u> corresponds to the set of tasks T_k handled by the accelerator <u>at each cycle</u>: ε is the set of <u>tasks executed by the RC</u>[2] ($\varepsilon \subseteq \theta$).

We consider the RC as a set of homogenous processing elements P, and a set of reserved memory channels C. Each task T_k is represented by a set of processing elements P_k and a set of memory channels C_k (a value set to 1 means that the resource is used, otherwise it is set to 0). We define the two following conditions to achieve run-time reconfiguration of the RC:

Condition 1. A task $T_k \in \theta$ can be mapped onto the RC if the number of required resources is currently available $i.e.$ $card(P_k) \leq card(P) - \sum_{Tr \in \Gamma} card\ (P_r)$ and $card(C_k) \leq card(C) - \sum_{Tr \in \Gamma} card\ (C_r)$

Condition 2. A task $T_k \in \theta$ is mapped onto the RC if $condition\ 1$ is verified and if the configurations of the running tasks are compatible with the configuration of T_k $i.e.$ $(\sum_{Tr \in \Gamma} P_r) \cap P_k = \varnothing$ and $(\sum_{Tr \in \Gamma} C_r) \cap C_k = \varnothing$

We can easily assume that mapping directly the configuration generated at design-time would lead to a configuration conflict (for instance, 2 configurations sharing some processing elements would create an incompatibility). In order to obtain a compatible solution, it is necessary to apply some modifications to the configuration. This can be done at design-time with hypothesis on application scenarios, but our approach is directly performed at run-time. Our purpose is to provide a configuration manager able to modify dynamically the initial configuration to fit to the available resources.

The so-called configuration manager is implementing a transformation process that allows generating a new configuration that has the same functionality. If $f(T_k)$ is the function implemented by T_k the function $transform()$ must verify $f(T_k)=f(transform(T_k))$, In other words, $transform()$ is able to generate a new configuration $\{P_k,\ C_k\}$, functionally equivalent to the configuration of T_k. This

---

[2] We assume that a task not assigned to the reconfigurable accelerator is executed by the microprocessor with its equivalent software program.

$transform()$ function depends on the geometry of the targeted architecture. We will give an example how to define it in the case study, section 5.

In order to measure the effect of this approach, we define the two following **metrics**. The <u>Multi-Tasking efficiency</u> MT_{eff} measures the percentage of applications admitted by the RC: $MT_{eff}(\%)= 100.\ card(\varepsilon)\ /\ card(\theta)$.

The <u>Processing efficiency</u> (P_{eff}) is measures the space and time utilization percentage ratio of the coprocessing resources as a function of the total number of cycles: $P_{eff}(n)= 100/(n.card(P)).\ \sum_{Tk \in \varepsilon} card(P_k).n_k$

4. Dynamic Hardware Multiplexing

The objective is now to provide an efficient method, first to ease the use of dynamic reconfiguration, second to use the reconfigurable resources more efficiently. The basic concept consists of abstracting run-time constraints thanks to a configuration manager associated to the RC.

At <u>design-time</u>, a static mapping of the application (or task) is performed. The configuration program is then encapsulated with a **header**. This header contains relevant information about the configuration: $e.g.$ the number of resources, memory bandwidth, required connections... At <u>run-time</u>, the configuration manager of the RC is interrupted by the CPU when an application needs to be accelerated. The interruption is formatted with the header of the task to be accelerated. This one is then analyzed by the configuration manager in order to define a new configuration matching the current constraints. Several cases are possible, and the scheduler has to take decision following the current restrictions.

The configuration manager is running a predefined program implementing a Dynamic Hardware Multiplexing (DHM) algorithm. DHM algorithm allows an adaptive spatial scheduling of tasks on a given reconfigurable architecture. It is based on two mapping procedures relying on a relocation of tasks (Simultaneous Multi-Task allocation algorithm), and duplication of tasks when possible (UnRolling allocation algorithm). The program running on the configuration manager is initially in a "wait for interrupt" mode. When an application running on a GPP has to be accelerated, the CPU generates an interruption, triggering the interruption handler $SMT\_allocation()$ of the configuration manager.

$SMT\_allocation$ (Figure 1) allows handling multiple tasks on a same RC thanks to a run-time relocation process. The initial configuration is simply modified by the $transform()$ function so that it matches the current configuration of the RC.

When the workload of the system is not important, it may be interesting to use the available resources in order to reduce the power consumption, or increase the throughput. This is performed by replicating the current running tasks, as long as the available resources allow it. The process *UnRolling_allocation* (*UR*, Figure2), tries to reallocate an instance of each running task. It is triggered every time an interruption has occurred and modified the state of the RC. Both conditions are tested so that an assignable task is selected. Then, transformations are applied in order to fit to the current configuration of the RC. Replicated tasks are stopped in priority when a new task request is ordered by the CPU. According to the current OS policy and priority levels, tasks might be killed for freeing resources or either left on the structure until completion.

```
1: process SMT_allocation()
2:    Γ←∅
3:    while System_on
4:       if New_Task_Starts then
5:          T_k←New_Task_Starts
6:          if Condition 1 then
7:             if Condition 2 then
8:                Assign(T_k)
9:                R←R ∪ T_k
11:            else
13:               T_k ←Transform(T_k)
12:               do
14:                  if Test 2 then
15:                     Assign(T_k)
16:                     R←R ∪ T_k
17:                     possible=false
18:                  end if
19:               while possible
20:            end if
21:         end if
22:      else if Task_stops
23:         T_k ←Task_stops
24:         Γ← Γ-{ T_k }
25:      else
26:         UR_allocation(Γ)
27:      end if
28: end while
29: end process
```

Figure 1. *Multi-application allocation (SMT)*

```
1: process UR_allocation(Γ)
2:    for each Ti∈R do
3:       UT_k ← T_k
4:       if Condition 1 then
5:          if Condition 2 then
6:             Assign(UT_k)
7:             R←R ∪ UT_k
8:          else
9:             do
10:               UT_k←Transform(UT_k)
11:               if Test 2 then
12:                  Assign(UT_k)
13:                  R←R ∪ UT_k
14:                  possible=false
15:               end if
16:            while possible
17:         end if
18:      end if
19: end for
20: end process
```

Figure 2. *Unrolling allocation algorithms for DHM (UR)*

5. Case Study

In order to design a DHM scheduler, it was necessary first to choose a test-architecture. It is generally admitted that it is less flexible but more efficient to use dedicated operators instead of LUTs (density, reconfiguration latency…). This is in this context that Systolic Ring [13] was designed. The proposed DHM algorithm has been therefore applied to this architecture.

The Systolic Ring is a customizable coarse grain reconfigurable model. It features a compact DSP-like coarse grain reconfigurable datapath, the *Dnode*, which is the building block of the architecture. The architecture is configured by a microinstruction code loaded from program memory to its configuration memory. The figure 3 describes this architecture.

With the formalism developed in section 3, the architecture is represented as $P=\{\{p_{0,0}, p_{0,1}, ..., p_{0,d-1}\}, \{p_{1,0}, p_{1,1}, ...,p_{1,d-1}\}, ..., \{p_{l-1,0}, p_{l-1,1}, ..., p_{l-1,d-1}\}\}$ and $C=\{c_0, c_1, ..., c_l\}$ with $dim\ P = l{\times}d$ and $dim\ C = l$. The total number of processing elements of the architecture is then given by $card\ P = l.d$.

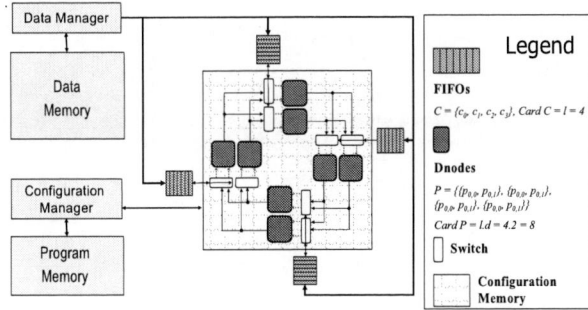

Figure 3. *Simplified overview of Systolic Ring: Dnodes are organized in l layers with d Dnodes per layer (here l=4 and d=2). A programmable switch component is used to interconnect the layers of processing elements and external memory (generally, l FIFO channels from the memory).*

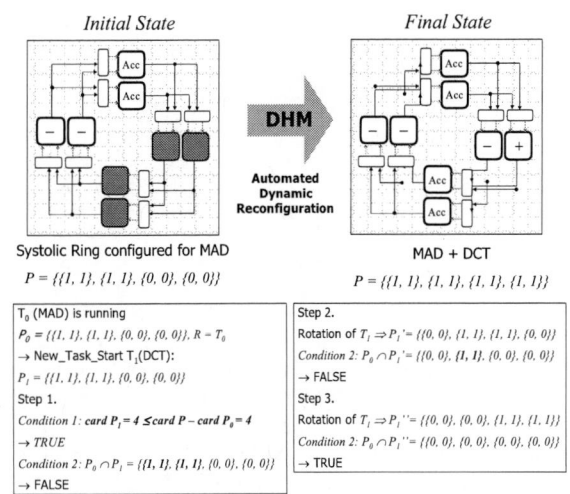

Figure 4. *Multi-threading of Minimum Absolute Difference (MAD) and Discrete Cosine Transform (DCT) thanks to a rotation based transformation scheme of the initial configuration of the task. Each task handled by the Systolic Ring is assumed to be designed so that $P_k \subseteq P$ and $C_k \subseteq C$. In this case, the tasks refer to validations made on the prototype of the Systolic Ring [13].*

0-7695-2533-4/06 $20.00 © 2006 IEEE   253

The proposed DHM algorithms can be applied to any reconfigurable architecture supporting run-time partial reconfiguration. The only architecture-dependent function is the transformation process applied to modify the configuration. Actually, it can be easily derived from the geometry of the targeted architecture. For instance, the Systolic Ring architecture is based on a homogenous ring topology where each processing element is able to implement the same set F of arithmetic and logic functions. Thus, a simple **rotation** (an example is depicted on figure 4) **of the configuration** following the dataflow direction (i) **is functionally-equivalent** i.e.:

$$\forall\ r,i \in [0..l\text{-}1]\ and\ j \in [0..d\text{-}1], f(p_{i,j})=f(p_{(i+r)\%l,\ j})$$
$$and\ f(c_i)=f(c_{(i+r)\%l})\ where\ f \in F$$

6. Results

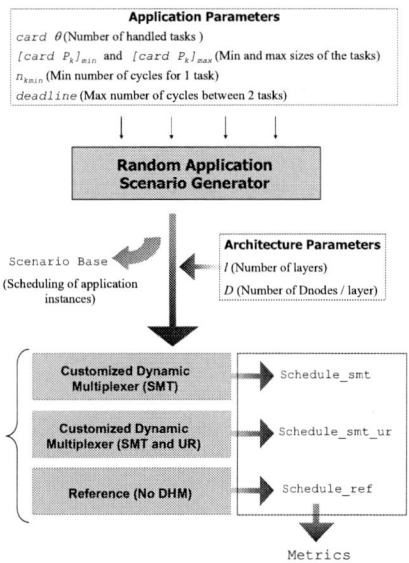

Figure 5. Simulation environment. *A set of application parameters is provided as an input to the simulator, relatively to the proposed formalism. These values are used to generate a base of random application scenarios. This set of threads is then scheduled following three different methods: DHM method ("SMT" with or without unrolling "UR" capabilities) and the static method (non-DHM) which is the "Reference". Architecture parameters are also available as an input to the scheduling algorithm (allowing simulating different sizes of the architecture).*

Figure 5 depicts the simulation environment developed in order to extract a significant number of situations to test the proposed algorithms. Several scripts and macros have been used to automate all the phases of the simulation process. For a given set of architecture parameters, 10 random application scenarios are generated in a given range of application parameters. Then, monitoring data are dumped into files thanks to an automated macro. These data are used to plot the different mapping and

scheduling obtained through the three methods. The resulting mapping and scheduling are compared (figure 6 and 7) on a single chronogram in order to depict the effects of the DHM method. The graphic on figure 8 shows the differences obtained on the Processing efficiency metric (Peff) for several scenarios[3].

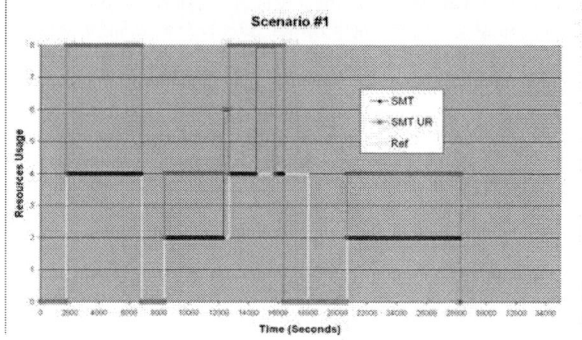

Figure 6. *Scenario number 1 mapped, scheduled and plotted*

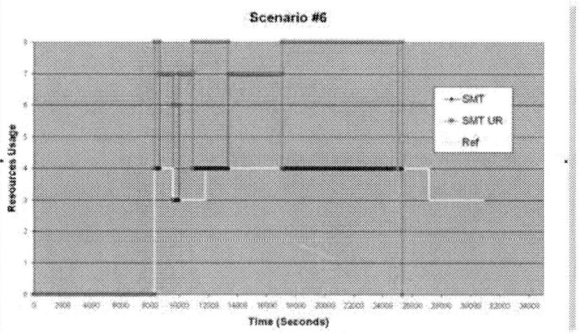

Figure 7. *Scenario number 6 mapped, scheduled and plotted The Y-axis represents the number of used resources at each cycle (there are 8 Dnodes in these examples).*

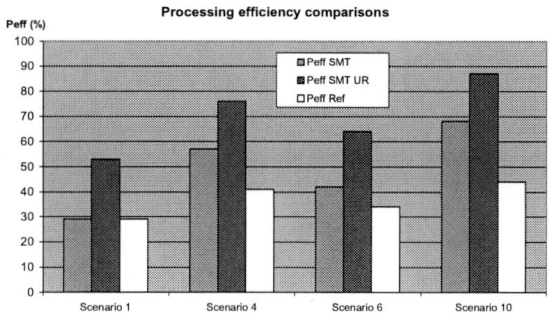

Figure 8. *Processing efficiency metric for 4 given scenarios (1, 4, 6, 10)*

---

[3] In the graphics depicted figure 6, 7 and 8, the parameters of the application scenario generator are the same: : *card θ=5, [card $P_k]_{min}$ =1, [card $P_k]_{max}$=5, n_{kmin}=1000s (100.10^9cycles@100 MHz) and deadline=10000s, architecture parameters: l=4 and d=2.*

0-7695-2533-4/06 $20.00 © 2006 IEEE   254

Two instances of the Systolic Ring have been tested: an *8 processing elements* version (*l=4, d=2*) and a *32 processing elements* instance (*l=8, d=4*). For each one, a variable number of tasks from *2 to 80* (*card $\theta \in [2..80]$*) have been implemented. For each set of tasks, all the parameters have been drawn lots: the *start* and *stop* cycles for each task (and consequently the duration n_k of T_k), the time interval between two tasks (with the possibility that a new task may be started before previous tasks' completion), the size and the shape of the task (*i.e.* the number of required processing elements *card P_k* and memory channels *card C_k*, the topologies). The DHM algorithm has been emulated for the Systolic Ring in a C++ program. The emulator takes as input a given generated scenario, and executes the DHM algorithms on the Systolic Ring. In order to increase the significance of statistics, a total of 300 hundred scenarios have been simulated through the emulator, for each set of tasks. Thus, the presented values in the following figures correspond to a mean value on 300 samples.

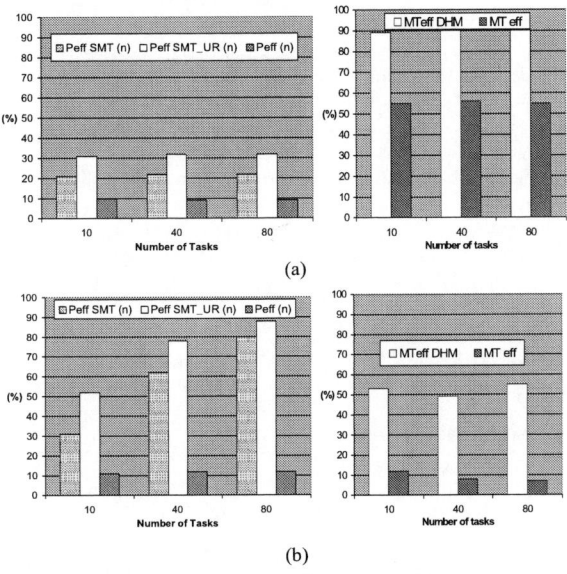

Figure 9. *Performance metrics[4] on various tasks sets for Systolic Ring l=8, d=4, with **8 processing elements**. The values depicted here correspond to random sets of 10, 40 and 80 tasks, with an low task load (a) and a high task workload (b).*

Figure 9 and 10 depict performances obtained by DHM. The metrics *Peff(n)* and *MTeff* are plotted for 2 instances of the Systolic Ring. For each one, two cases are represented: one depicts a low task workload (a) and the other one a high task worload (b), corresponding to

---

[4] For *Peff* comparisons, DHM is based on Simultaneous Multi-Tasking allocation (SMT), or SMT and Unrolling allocation (SMT_UR). This does not influence the *MTeff* results.

different task request contexts. As a performance reference, a non-DHM scheduling control has been also implemented and simulated (it does not allow simultaneous multi-tasking). This one is compared to the DHM implementation following the proposed metrics. One can notice also that for the *Peff* value, the version of DHM allowing only-SMT is compared to the one allowing also UnRolling allocation (SMT_UR).

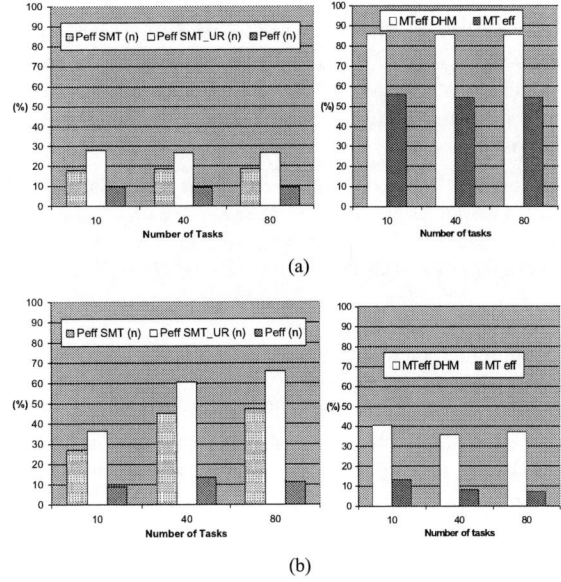

Figure 10. *Performance metrics[1] on various tasks sets for Systolic Ring l=8, d=4, with **32 processing elements**. As in figure 5, the values depicted correspond to tasks sets of 10, 40 and 80 tasks, with low task load (a) and a high task workload (b).*

In all simulated scenarios, the DHM algorithm has proved its capacities by increasing both *Peff* and *MTeff*. Not surprisingly, the *Peff* for low task workloads is lower than *Peff* for high task loads. But in both instances, the whole DHM algorithm (*SMT_UR*) allows improving about *15 to 20%* the initial the reference *Peff*. Moreover, the *MTeff* is about *85 to 90%* (*20 to 45%* higher than the reference). With a high task workload, *i.e.* most of tasks are started before previous tasks' completion; the DHM algorithm really shows great improvements in the processing efficiency, from 25 to 85%. Moreover, the *Peff* increases with the number of tasks to handle. However, the *MTeff* is then lower, about *40%* for the *32* processing elements instance and *50%* for the *8* one. The simulations also highlight that the algorithm provides better results on a smaller version of the RC. This is especially true when the task workload is high. Actually, this is a consequence of the interaction between the size of tasks and the number of processing elements: equivalent performances may be easily reached with a bigger set of tasks. This measure is very interesting to evaluate the trade-off on size of the RC and performance. Following the size of tasks and acceleration requests, these metrics could be used to define the size of the RC (*i.e.* the number of processing elements, …).

7. Validation

DHM is implemented with a dedicated scheduler allowing to abstract dynamic reconfiguration management from design-time. There are basically 2 ways to handle DHM algorithms: either by designing a dedicated sequencer (HW-DHM) or by programming a general purpose processor (SW-DHM). We have implemented these two approaches and we compare obtained results in the following section.

HW and SW-DHM controllers have been designed both for a Systolic Ring instance composed with 8 Dnodes. After synthesis, the designs have been validated by simulations. For HW-DHM, the whole algorithm takes less than 20cycles (0.2μs@100MHz) to dynamically modify and (re-)allocate a pre-defined configuration. The area overhead is about 0.04 mm² (in 0.13μm CMOS technology), which represents less than 10% of the Systolic Ring. For SW-DHM, a 16 bits RISC processor has an equivalent area and implies in the worst case up to 20000cycles (200μs@100MHz) to perform the same allocation task. These results are listed in Table 1.

***Table 1.** HW and SW-DHM results*

	HW-DHM Scheduler	SW-DHM Scheduler
Area (mm²)	0.041	0.043
Time (μs)	0.2	200
1/(A.T)	122	0.12

Not surprisingly, these results clearly show the superiority of the dedicated approach with a global performance (1/AT) 1000 times better than the SW-DHM. However, the HW-DHM needs to be redesigned for each different instance of the reconfigurable architecture, while the SW-DHM can be easily changed by simply reprogramming the processor. In both cases, we can consider acceptable times in terms of latency introduced in the dynamic reconfiguration process.

8. Conclusion and perspectives

We have proposed a generic method to address the problematic of dynamic reconfiguration management. This method implements Dynamic Hardware Multiplexing algorithms allowing an abstraction, at design-time, of the dynamic reconfiguration management. The DHM algorithm aims at exploiting more efficiently the processing resources thanks to an adaptive on-line scheduling technique. A case study on the Systolic Ring architecture has proven the feasibility of the proposed approach. The simulation results show an improved processing efficiency and dynamic reconfiguration is

directly and automatically managed by the hardware. The HW-scheduler offers a performance/area trade-off clearly better, while the SW-DHM approach is attractive from a flexibility point of view.

A "fine-grain" implementation could also be considered. This technique is currently ported to other reconfigurable devices in order to widen the application spectrum of DHM. At the present time, we strongly believe that a SW-DHM could be the lowest-cost solution thanks to its portability, from a given architecture to another one, and its flexibility in dynamic reconfiguration management. The background motivation stands in its system integration, as a "smart-middleware" with a full OS compatibility.

References

[1] http://www.easic.com/

[2] http://www.m2000.fr/

[3] J. Teich, S. Fekete, and J. Schepers, "Optimisation of dynamic Hardware reconfigurations", The Journal of Supercomputing, 19(1): pp. 57–75, May 2000

[4] K. Bazargan, R. Kastner, and. M. Sarrafzadeh, "Fast template placement for reconfigurable computing systems", IEEE Design and Test of Computers 17(1), pp. 68–83, 2000.

[5] C. Steiger, H. Walder, and M. Platzner, "Operating Systems for Reconfigurable Embedded Platforms: Online Scheduling of Real-time Tasks", IEEE Transactions on Computers, 53(11): 1392–1407, November 2004.

[6] O. Diessel, H. El Gindy, M. Middendorf, H. Schmeck, and B. Schmidt, "Dynamic scheduling of tasks on partially reconfigurable FPGAs", IEE Proceedings – Computers and Digital Techniques, 147(3): 181–188, May 2000.

[7] H. Simmler, L. Levinson, and R. Manner, "Multitasking on FPGA coprocessors", In FPL'2000, (LNCS 1896), pp. 121–130, 2000.

[8] E. Carvahlo, N. Calazans, F. Moraes and D. Mesquita, "Reconfiguration Control for Dynamically Reconfigurable Systems", DCIS 2004, pp.405-409, 2004.

[9] E. Caspi, M. Chu, R. Huang, N. Weaver, J. Yeh, J. Wawrzynek, and A. DeHon, "Stream Computations Organized for Reconfigurable Execution (SCORE)", FPL'2000, LNCS 1896, pp. 605-614, 2000.

[10] M. Huebner, M. Ullmann, L. Braunn, A. Klausmann, and J. Becker, "Scalable Application-Dependent Network on Chip Adaptativity for Dynamical Reconfigurable Real-Time Systems", FPL'2004, LNCS 3203, pp. 1037-1041, 2004

[11] M. Ullman, M. Hübner, B. Grimm and J. Becker, " On-demand FPGA Run-Time System for Dynamical Reconfiguration with Adaptive Priorities", J. Becker, M. Platzner, S. Vernalde (Eds): FPL 2004, LNCS 3203, pp. 454-463, Springer-Verlag, 2004

[12] D. Robinson and P. Lysaght. Modeling and Synthesis of Configuration Controllers for Dynamically Reconfigurable Logic Systems using the DCS CAD Framework. In FPL'99, Lecture Notes in Computer Science v. 1673. UK. 1999.

[13] G. Sassatelli, L. Torres, P. Benoit, T. Gil, G. Diou, G. Cambon, J. Galy, *Highly Scalable Dynamically Reconfigurable SystolicRing-Architecture for DSP applications*, IEEE DATE'02 , France, 2002

Regular Routing Architecture for a LUT-based MPGA

Francisco-Javier Veredas[†][‡], Michael Scheppler[†], Bumei Zhai[†][‡], Hans-Joerg Pfleiderer[‡]

[†]Infineon Technologies AG, D-81699,Munich, Germany

[‡]Microelectronics Department, University of Ulm, D-89081, Ulm, Germany

Abstract

Mask Programmable Gate Arrays (MPGAs) are an attractive solution to reduce design cost and turnaround time in ultra-deep submicron technologies. Several design methodologies have been proposed in the recent years for converting an evaluated Field-Programmable Gate-Array (FPGA) prototype design into an MPGA. In this paper, we investigate a predefined regular routing architecture of an MPGA. The routing architecture is easily scalable. A simple model for the MPGA interconnect is presented which facilitates static timing analysis. We explain the difference of this interconnect with the FPGA interconnect. The resulting MPGA is implemented in 130nm. Circuit level simulations show that our model is accurate in terms of delay. The study presents tradeoffs with the placement and routing to reach timing closure. A special MPGA routing tool is used. The study shows that high number of tracks in the MPGA is area prohibitive, but with better timing closure.

1 Introduction

The increased complexity of FPGAs and the decreased per unit prices make them an alternative to low-end ASICs. On the other hand FPGAs are not optimum for higher volumes and for limited power budgets. MPGAs have the potential to step between these two alternatives because they offer increased flexibility at moderate cost of non-recurring engineering (e.g. mask cost) and fast turnaround time.

A methodology to design MPGAs into a system consists of prototyping the system with an FPGA framework and then converting the design to an MPGA for cost reduction. Existing conversions use for the substitute a target architecture with elementary cell structures because this offers high density. The downside of this approach is the need of re-synthesis and complete place and route, i.e. the portion of customization masks is high. So, for an MPGA the decision is to use a predefined master architecture and one question is how many customization masks are required. Another question is how timing integrity after the conversion can be facilitated. To avoid re-synthesis an MPGA that preserves the gate-level structure of an FPGA has been presented recently [2]. The MPGA presented in [2] re-adapts semi-custom tools for place and route. The use of this kind of flow has as a consequence of risk of failing timing closure as well as the number of customization masks is not minimized. In this paper we propose an MPGA which preserves the gate-level netlist of an FPGA, preserves the placement and in addition has predefined routing resources. The study shows the considerations that we must make to guarantee the timing closure.

The paper is organised as follows. Section 2 gives an overview of a LUT-based FPGA architecture. Section 3 presents the proposed LUT-based MPGAs architecture and the physical layout. In Section 4 the MPGA interconnect model is presented. Section 5 explains the experimental methodology to perform timing analysis. Section 6 shows the experimental results. Section 7 concludes the paper.

2 Xilinx Virtex-II Pro FPGA Architecture

There are two major players in the FPGA market: Xilinx Inc. and Altera Inc. Both use architectures which are array based and have logic clusters at the lower hierarchy level. Both companies have two families one for high performance and high complexity, the other for low cost and higher volume. We use the Xilinx Virtex-II Pro [1] device for our studies.

The Xilinx Virtex-II Pro is a 130nm CMOS nine-layer (copper) FPGA device. Virtex-II Pro has an island style architecture. It consists of a two-dimensional array of Configurable Logic Blocks (CLBs) and programmable interconnect resources. Each CLB is a cluster of four identical sub-blocks called slice. The slice consists of two four-input LUTs, two FFs, gates (two AND2, one OR2, two XOR2), multiplexors, inverters and buffers. The LUTs are used to map four-input boolean logic. The gates are used to implement special functions as carry chains. Four multiplexors are used as mapped multiplexors (after synthesis) and the other multiplexors as configurable routing switches (i.e. to do programmable routing inside the slice). The inverters

0-7695-2533-4/06 $20.00 © 2006 IEEE   257

and the buffers are needed to implement the different clock types allowed in the FPGA. The four slices of one CLB share a common input/ output programmable crossbar for doing local signal communication between slices and signal communication with other CLBs in the array. The communication between CLBs is realised with a switch matrix. The switch matrix routes a signal to north, south, east or west direction. The topology of the switch matrix in the Virtex-II Pro is a special disjoint type. A CLB together with its associated switch matrix is called tile.

The programmable interconnect is responsible for connecting the inputs of the logic cells and the outputs together. The Virtex-II Pro has two hierarchical levels: local interconnect and global interconnect. The local interconnect is the programmable input/output crossbar commented before. The global interconnect consists of vertical and horizontal routing channels. A wire in a channel can expand two tiles (double lines), six tiles (hex lines) or all the row/column (long lines). The Virtex-II Pro has 40 double lines, 120 hex lines and 24 long lines in a channel. All this lines are routed within the switch matrix. There is a special type of wire that can connect point of two CLB without passing through the switch matrix: direct neighbour lines. The direct neighbour lines connect adjacent CLBs. This line goes directly from an output crossbar to an input crossbar.

The programmable resources are controlled by latches.

3 Zelix MPGA Architecture

The use of MPGAs is motivated to reduce the high area penalty of FPGAs, allow a less advanced process technology and simplify packaging due to power reduction. On the other hand flexibility is lowered by using production floor programmability. In an MPGA, all the mask layers that define the master device are pre-defined (partially can be pre-fabricated). The final metal layers are customised to implement the desired application circuit. In contrast to the FPGA, the programmable interconnect resources (i.e. SRAM cells) are replaced by metal layers and vias. This increases layout density and simplifies routing.

We named the LUT-based MPGA presented in this paper Zelix MPGA. The architecture has the same topology and the same gate-level logic elements as the Xilinx Virtex-II Pro. For the sake of simplification the embedded hardware blocks (as 18 Kb RAM memory modules) are currently not implemented in the MPGA.

Hence, the Zelix MPGA consists of a two-dimensional array of Mask-Configurable Logic Blocks (Mask-CLBs) and mask programmable interconnect resources (see Fig. 2(a)). The Mask-CLB has four slices same as the original.

The maskprogrammable interconnect in the Zelix MPGA is concepted in the same fashion as an FPGA, i.e. we have a predefined regular interconnect structure which consists

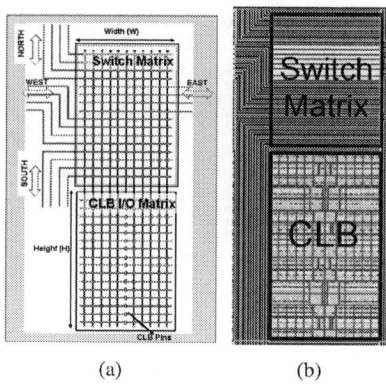

(a)   (b)

Figure 1. Zelix MPGA tile: (a) Routing Architecture Floorplan (b) Layout

of bundles of orthogonal metal segments. However, for the MPGA we did not choose a hierarchy of different segment lengths, but used a unified length of one, that is, one wire segment can only expand one tile. Longer lines easily can be created by simple metal bridges. A benefit of this regular interconnect structure is the possibility to use a simple FPGA routing tool. Another advantage is that physical phenomena can be limited to a set of cases. For instance, wire delays can be added. Also, effects of cross-talk can be modelled simpler. Finally, layout design is simpler as can be automated. The Zelix MPGA has two hierarchical levels for the interconnect: top-level interconnect and local interconnect inside a CLB. The local interconnect is a mask configurable input/output crossbar (Fig. 1(a)). As the configuration is done with vias, we have a fully-populated crossbar. The top-level interconnect is used for signal communication between CLBs which can mean next neighbours or long distance. This communication is done through the orthogonal segments which are concatentated at a switch matrix. This means that a segment can be prolongated into the same direction, different direction or forked. Another purpose of a switch matrix is to connect a CLB. The topology of the switch matrix is again a fully-populated crossbar. This has advantages for the routing density. Buffer insertion takes the advantage of the length-1 segment topology. For instance, in a wire of length two, a 2-segment line has only two points for inserting a buffer (input, output), but two 1-segment lines have three points (input, mid point, output).

An expression for the area of a tile can be extracted from the architectural parameters. Using Fig. 1(a), we see that the area is,

$$
\begin{aligned}
Area_{tile} &= W * H + N_{trcY} * k * H \\
&+ (2 * N_{trcX} + 2 * N_{trcY}) \\
&* (k * W + N_{trcY} * k^2)
\end{aligned}
\tag{1}
$$

Here W and H are the width and height of a CLB respectively. The numbers of tracks per horizontal and vertical channel are N_{trcX} and N_{trcY} respectively. k is the width of a wire and the minimum distance of one wire with other wire. W and H are fixed after physical layout of the CLB. k depends of the technology process used. N_{trcX} and N_{trcY} are architectural parameters and varies from one circuit mapped to other. The dependency of the number of tracks per channel mapping different circuits is discussed in Section 6.

3.1 Physical Layout

The technology used in the MPGA design is Infineon 130nm 4+2 copper metal layers CMOS process. We use a full-custom methodology with hand-crafted cells. Our VLSI layout takes advantage of the intrinsic hierarchical regularity of the MPGA architectures, i.e. first is designed the basic logic (LUTs, FFs, ANDs..), second the logic is assembled in slices, a set of slices is a Mask-CLB, a Mask-CLB with a switch matrix is a tile and finally an array of tiles is the MPGA.
We base all the layout studies on a tile level. The final MPGA array consists of a distributed arrangement of this tile.
Standard cells and hand-crafted cells are used. For reasons of qualification the standard cells are used as many as possible. Only in the LUT the use of hand-crafted cells was necessary. Circuit simulations with Synopsys Nanosim have been done to see that the Zelix logic has the same performance as Virtex-II Pro. Metal 5 is used for the power grid. Only Metal 3 and Metal 4 are needed for programming the MPGA. The clock network and the FFs control signals use Metal 1, Metal 2 and Metal 3. Fig. 1(b) is the layout of one tile. We can note that there are no active area for the input/output crossbar. The input/output crossbar is situated over the CLB and configured with the Metal 3, Metal 4 and Via34. The input/output CLB pins are arranged in the mid vertical line of the CLB. The two LUTs and two FFs occupy practically all the area of a slice. The implemented Zelix MPGA has 30 tracks per channels.
The total area for the tile layout is $48\mu m$ x $112\mu m$. About 57% of this area is dedicated to the routing resources.

3.2 Xilinx Virtex-II Pro Area Estimation

To calculate the area of the real Xilinx Virtex-II Pro, we use the presentation of [3]. This presentation has a die photograph of the Xilinx Virtex-II XC2VP7 device. No area for the photo die is given. Xilinx Virtex-II Pro has memory blocks of 18kb. We calculate the area of a tile using the area of a memory block. The area of four memory block is equal to 24 tiles. An internal Infineon tool give us the area

of a 18Kb dual-port embedded memory block. Therefore the area of a tile is $206\mu m$ x $145\mu m$. Zelix MPGA has a 82% of area reduction

4 Interconnect Model

The configurable nature of FPGA routing architectures means the use of programmable interconnect points (PIP). A PIP of a two-way interconnect consists of one pass-gate. It is common to have buffered PIP for reducing the associated signal degrading of pass transistors in series. The Virtex-II Pro routing architecture has all PIP buffered [1]. The direct neighbour, double and hex lines in the Virtex-II Pro are unidirectional. Therefore, they are buffered in only one direction. Long lines are bidirectional with buffers in two directions.
Fig. 2(a) shows an example of a net with our interconnect model. The interconnect parasitic model has five different elements: wire, input crossbar, output crossbar, cell input/output pins, input/output pads. All the cases are modeled as worst case. The length of the wire is the distance from one tile to the closest tile. For the input crossbar the worst case is assumed when one wire comes from the switch matrix and goes to the input cell, we assume the same parameters as the wire. The capacity effect of the cell input/output pins is depreciable in comparison with the wire or crossbar capacity. Values for the input/output pads capacity has been set. Infineon 130nm CMOS six-layer (copper) is used for setting the values of the interconnect elements. Infineon 130nm process technology is comparable with the IBM 130nm process technology used by Xilinx in the Virtex-II Pro.
The shrinking factor of the MPGA in respect to the FPGA represents a reduction in the interconnect delay. Moreover, in the MPGA there is no delay associated with the pass-gates and the buffers can be inserted when it is necessary. First we study the differences only assuming a wire without PIP and buffers. The following analysis is using 130nm technology parameters for Metal 2. We assume a 1-length of $173\mu m$ for the Virtex-II Pro. This number is calculated with the root square of area calculated in the previous section. A length wire of the MPGA is 64% smaller ($112\mu m$ is assumed in the MPGA). The delay for a 1-length line is about 5ps in the Virtex-II Pro and 2ps in the Zelix MPGA. The delay becomes more important when we considerer a long wire. For example, a wire that expands six tiles has an associated delay without buffer insertion of 72ps with FPGA values and 30ps in the Zelix MPGA. This delay factor is one of the reasons why Xilinx insert buffers in its pre-fabricated routing.
Circuit level simulations are performed with Synopsys Nanosim. The results are compared using STA analysis of our simple interconnect model. The STA tool used is the

0-7695-2533-4/06 $20.00 © 2006 IEEE

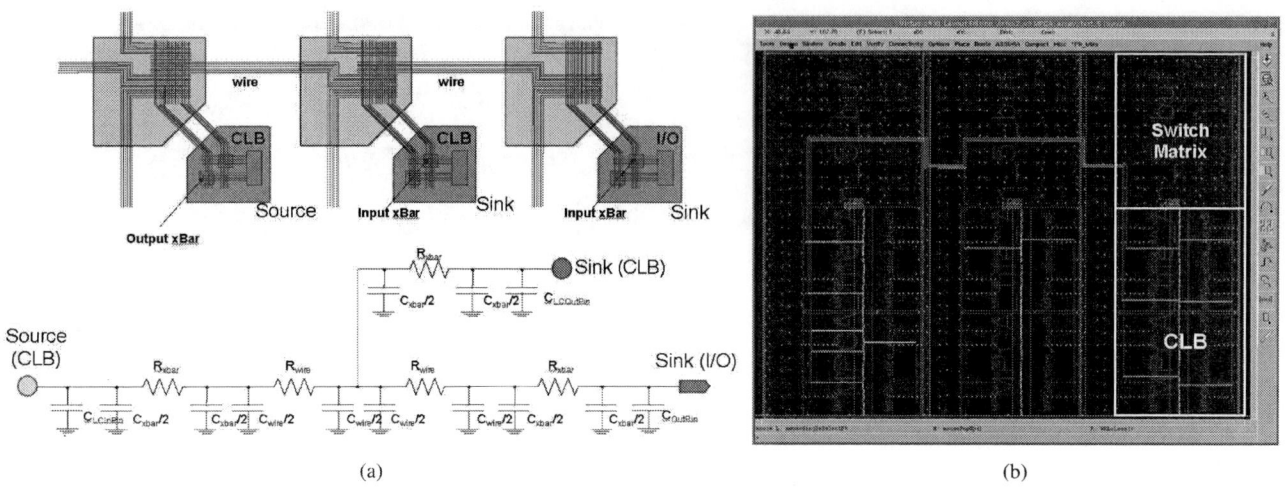

Figure 2. Example of the interconnect model: (a)Net1 Model (b)Net1, Net2 and Net3 layout

Synopsys PrimeTime and the parasitics are annotated with a dspf file. Three cases are studied: a fan-out one net that expands two tiles (Net1), a fan-out two net where the two in-pins expand one and two tiles (T2, T3) respectively (Net2), and a fan-out one net that is routed inside a tile (Net3). Figure 2(b) depicts the layout of these nets (only the routing metallisation is displayed). The simulations show that our simple interconnect model is similar to the circuit level simulations. No gate delays have been taken into account in the Zelix MPGA interconnect results. Circuit simulations show that the introduction of a buffer in the Net1 introduces a delay of 90ps, but the overall delay of the mapped circuit is improved by 10ps. We note that the delays in the MPGA must be smaller than in the FPGA. We conclude that the insertion of buffers are more necessary for "breaking" fanouts than for improving interconnect delays in the MPGA. This statement is corroborated in Section 6. Another issue is the buffer insertion strategy in the MPGA, i.e. if it is better to have all the 1-segment wires of the MPGA buffered or if is better to have a buffer insertion algorithm that insert buffers optimally. One of the features of the Zelix MPGA is that each 1-segment wire can have a physical buffer (in both directions). Therefore, the buffering of all the one-segments wires has no additional cost in terms of area. The option of a buffer insertion algorithm is more appropriate to have the smallest delay in the interconnect.

5 Experimental Methodology

The goal of the conversion is to have the same (or better) time performance with an MPGA as in an FPGA. In section 3 we have explained that in the Zelix MPGA logic is the same as the Xilinx Virtex-II Pro FPGA, so the logic delays are assumed equal. The delay difference is in the interconnect. The previous section has shown that the interconnect of the Zelix MPGA has potentially less delay than in the Virtex-II Pro. To verify that we reach the goal time, we use Static Timing Analysis (STA). In a STA, the timing analysis is carried out in an input-independent manner, and purports to find the worst-case delay of the circuit over all possible input combinations. Commonly STA is used as a part of the sign-off methodology in a digital silicon device. In our experiments we check the delay of all nets with STA. The same check with circuit level simulations is complex and time expensive, therefore prohibitive.

The STA analysis in the Virtex-II Pro is performed with Xilinx's design framework (ISE) and with Synopsys Prime-Time tool.

5.1 Zelix MPGA CAD Flow

The Zelix MPGA flow starts from the Xilinx STA gate-level netlist. This netlist models gates for signal routing by buffer. The buffers are used to annotate programmable interconnect delays. A Perl-script removes all the buffers (included the hardwired buffers). For our studies we import the same placement from the Xilinx ISE flow. The reason for using the Xilinx placement is that we want to compare our MPGA routing with the Virtex-II Pro routing. Further studies can look for a better MPGA placement. The Xilinx placement information is extracted from the Xilinx Description Language (XDL) file. For matching the names of the Xilinx gate-level netlist and the XDL file a deep analysis of the Virtex-II Pro architecture has been done. The obtained placement information is for tile level and not slice level. We use tile level placement because in the MPGA we expect no major differences for STA within placement inside a tile.

0-7695-2533-4/06 $20.00 © 2006 IEEE

The router tool has been programmed in C [5]. For routing the FPGA PathFinder algorithm has been programmed. Our variant of the Pathinfinder maze router algorithm implements a routability-driven router instead of a timing-driven routing. For using a timing-driven routing is needed a delay calculator. As the buffer insertion is done in a later stage in our flow, a delay model for the interconnect is not accurate. Moreover, because the regular interconnect architecture has a penalty in terms of area, we want to optimise the number of wires over the critical path. Notice that the conversion flow must guarantee the same time as in an FPGA, not a better performance. It is possible to set the number of tracks per channel in the MPGA router. The number of tracks per vertical or horizontal channels can be different (in the VPR FPGA tool vertical and horizontal channels are the same). This feature is important because the MPGA layout of a tile usually is rectangular, so in one direction there are more tracks per channels than in the other. The target MPGA architecture is described in a Perl-based language. Then the parser and abstraction steps transforms the architecture to an internal graph representation.

A Perl-script creates an interconnect RC tree with the routing information (a DSPF file).

Once that we have the interconnect information of the Zelix MPGA, we can proceed to do the STA analysis. The STA tool used is Synopsys PrimeTime. We guarantee that all the net delays are reported with a PrimeTime TCL-script. The STA analysis of the Xilinx Virtex-II Pro and the Zelix MPGA is compared. The delays in the Zelix MPGA are calculated from the output of one gate to other gate. In the Virtex-II Pro it is used the same path as the Zelix MPGA, i.e. if there are buffers in the path, we include it. When an MPGA-net has larger delay than the corresponding FPGA-net, there are two possible solutions: first look if the mapped circuit still satisfies the specifications (slack, critical path, ...) and second we can try to reduce the delay with buffer insertion. We noted that Xilinx reports for a special interconnect (as carry-chain) delay zero. We don't report this delay. This special interconnect is fan-out one and usually is intra tile or points to the direct neighbour tile. We consider that the MPGA has the same "zero" delay.

6 Experimental Results and Discussion

In this section we present the experimental results of four benchmarks circuits through the design flow described in Section 5. The benchmarks circuits used in these experiments are: an 8-bits FIR Filter (IFX_FIR), a 16-bits Biquad IIR Filter (IFX_IIR), a state machine of a Protocol Processor (IFX_PP) and a matching unit of an Associative Search Processor (IFX_ACE). The first two benchmarks are data processing oriented and the other two benchmarks are control flow oriented. Each circuit has been synthesised with

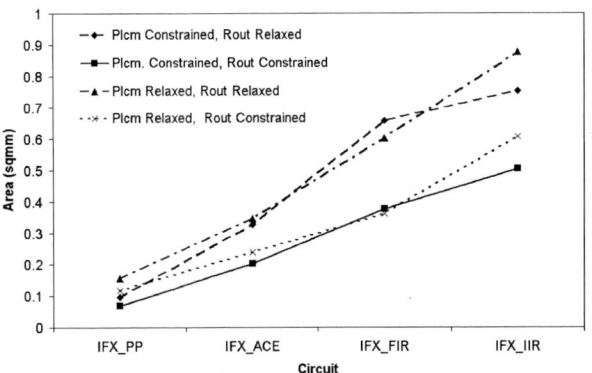

Figure 3. Total area results

the commercial Synplicity Synplify Pro tool. The Virtex-II Pro device used is 2vpfg256-7. The fourth column of Table 1 is the number of fpga-gates in the circuits. The fpga-gates cells are LUTs, FFs, Nand-gates, ... The last column shows the number of nets. The average fanout is around two or three. The biggest fanout for process-data nets (20-70) is concentrated in a small number of nets. The control (reset, set,..) and clock fanout depends on the number of FFs.

The investigation of the routing architecture is to find the number of tracks per channel (N_{trcX} and N_{trcY}) with a minimum area and less nets with delay larger than in the FPGA. To investigate this trade-off we experiment with different options in the CAD tools. First the placement is sparsed (or relaxed) over all FPGA array. The other option is to restrict the placement in a region of the FPGA. Note that the MPGA has the same FPGA placement. The Xilinx placement is slice level and not tile level. This can confuse people familiar with FPGA CAD research. Xilinx has not the usual packing-into-tile step common in FPGA CAD tools. Xilinx has packing for implementing special functions (as carry-chains), but no boolean logic. This results in a greater number of tiles in the relaxed placement. Lemieux shows in [4] that if we depopulate tiles of logic c ells the number of tracks per channel can be reduced. The other option concerns with the MPGA routing tool. We use the routing tool with no limit in the number of tracks per channel. The other option is reducing N_{trcX} and N_{trcY} until it is not possible to route the designed circuit, i.e. find the minimum N_{trcX} and N_{trcY}.

Table 1 shows the experimental results. We have nets with less delay than the FPGA when the routing is relaxed. The number of nets with larger delay is small compared with the total number of nets. This is because the smallest interconnect delay of the MPGA. We analyzed in detail the 9 nets of the IFX_FIR with placement constrained and routing relaxed. We see that the larger delay is because the nets have big fanout. The reset signal has fan-out 208 (the IFX_FIR

Table 1. Placement and Routing Results

Circuit	Placement Effort	Routing Effort	#cells	# Tiles	#tracks per channel[1]	Tile Area μm^2	#nets with larger delay	# Nets
IFX_PP	Low	Low	275	39	x=16 y=21	3968	2	381
IFX_PP	Low	High	275	39	x=8 y=7	2979	6	381
IFX_PP	High	Low	275	22	x=22 y=19	4336	3	381
IFX_PP	High	High	275	22	x=9 y=10	3136	5	381
IFX_ACE	Low	Low	654	82	x=19 y=22	4220	6	717
IFX_ACE	Low	High	654	82	x=6 y=9	2918	19	717
IFX_ACE	High	Low	654	56	x=36 y=29	5846	6	717
IFX_ACE	High	High	654	56	x=14 y=14	3594	21	717
IFX_FIR	Low	Low	1391	98	x=37 y=34	6136	6	1401
IFX_FIR	Low	High	1391	98	x=15 y=15	3696	20	1401
IFX_FIR	High	Low	1391	94	x=41 y=46	7000	9	1401
IFX_FIR	High	High	1391	94	x=18 y=18	4008	35	1401
IFX_IIR	Low	Low	1995	169	x=31 y=23	5188	16	2090
IFX_IIR	Low	High	1995	169	x=14 y=14	3594	22	2090
IFX_IIR	High	Low	1995	133	x=33 y=31	5668	17	2090
IFX_IIR	High	High	1995	133	x=16 y=16	3799	19	2090

has 208 FFs). The fan-out of the data signals is from 17 to 44. The solution to speed-up these nets is to break the net inserting buffers. We see that the number of tiles is bigger for the relaxed placement. When we have 150 tracks per channel, the area of the Zelix MPGA tile is approximately the same as the Virtex-II Pro FPGA. In Figure 3 the total area for each circuit is calculated with the area equation 1 for a tile and multiplying the result with the number of tiles. This calculation is not accurate in the sense that in the relaxed routing there are more empty tiles in the place area. But, with this total area we can have an idea of the area cost. We see that the best area efficiency is reached with the placement and routing constrained. The problem is that with the constrained routing, we also have more nets that cannot meet the required time. One solution is the insertion of buffers. Also, there is the possibility in the MPGAs of use more metal layers for the routing without increase the total area. The problem of more metallisation is the increase of the device cost and the buffers insertion becomes more difficult in the upper layers. Buffer insertion with a higher metallisation can be not possible because the lower metals blocks the possibility of arrive to the active area. One solution to this problem is use lower metals for buffering and the upper metals for the no buffers parts of the netlist. With the net average delay of all the mapped circuits, the worst average delay is when the routing is used with high effort. The best method to reduce delay is relaxing the routing effort.

7 Conclusion

Interconnect studies of a LUT-based MPGA have been presented. The goal of this MPGA architecture is the conversion from an FPGA with fast turnaround time. The Xilinx Virtex-II Pro FPGA is used as a test case. A simple model for the MPGA interconnect has been presented. This model is useful to perform static timing analysis. An MPGA is implemented in 130nm. Circuit level simulations show that our model is similar in terms of delay. The FPGA interconnect has more delay than the proposed MPGA. A special routing tool is described. The study has shown that high number of tracks in the MPGA is area prohibitive, but with better timing closure. Placement with hight effort is better for reduce timing delay. To improve timing delay it is possible to insert buffers in the critical nets.

References

[1] Virtex-II Pro^{TM} Platform FPGA Handbook. Xilinx Inc., San Jose, CA, 2002.

[2] HardCopy Series Handbook. Altera Inc., San Jose, CA, 2005.

[3] T. Taghavai, S. Ghiasi, A. Ranjan, S. Raje, and M. Sarrafzadeh. Innovate or perish: FPGA physical design. In Proc. International Symposium on Physiscal Design(ISPD'04), pages 148–155, Phoenix, USA, 2004.

[4] M. Tom and G. Lemieux. Logic block clustering of large designs for channel-width constrained FPGAs. In Proc. Design Automation Conference (DAC'05), Anaheim, USA, 2005.

[5] F. J. Veredas, M. Scheppler, and H. J. Pfleiderer. Automated conversion from a LUT-based FPGA to a LUT-based MPGA with fast turnaround time. In Proc. DATE'06, Munich, Germany, 2006.

---

[1]x: horizontal channel, y: vertical channel

A new Multilevel Hierarchical MFPGA and its suitable configuration tools

Zied Marrakchi, Hayder Mrabet and Habib Mehrez
Dept ASIM-LIP6
Université Paris 6, Pierre et Marie Curie
4, Place Jussieu, 75252 Paris, France
{zied.marrakchi, hayder.mrabet, habib.mehrez}@lip6.fr

ABSTRACT

In this paper we evaluate a new multilevel hierarchical MF-PGA. The specific architecture includes two unidirectional programmable networks: A downward network based on the Butterfly-Fat-Tree topology, and a special rising network. New tools are developed to place and route several benchmark circuits on this architecture. Comparison with the traditional symmetric, manhattan mesh architecture shows that MFPGA can implement circuits with fewer switches and a smaller area.

Keywords

FPGA, Hierarchical, Multilevel, Clustering, Routing

1. INTRODUCTION

The architecture of an FPGA is similar to a gate array and can be programmed to specify the function of the logic blocks and their interconnections. However these arrays suffer from lower density and speed compared with application specific integrated circuits ASICs. The most common FPGA architecture is the mesh which has a symmetrical grid of logic blocks and routing channels on all four sides of the logic blocks. Less work has been done for hierarchical FPGAs (HFPGAs), which contain a hierarchy of logic blocks and routing resources. HFPGA whose channels are fully populated with switches offers lower density than the other ones but have the advantage that routing has more predictable paths as well as lower delays. In previous papers [4] [9] different FPGAs with hierarchical depopulated interconnection structures were proposed to improve these shortcomings.

In our work we are also interested by this kind of architectures. We have explored the architecture of hierarchical HFPGA with regard to the particular issue of how the switch patterns of HFPGA can be partly depopulated. In fact the structure that we propose has a better ratio between logic and routing resources occupations than conventional mesh FPGAs. Hence the density of HFPGAs can be higher than conventional ones. Moreover, because fewer switches are needed for a connection path, the timing-skew problem is alleviated.

In this paper we give a description of our proposed architecture and the suitable techniques to program it.

2. ARCHITECTURE OVERVIEW

A standard hierarchical FPGA is denoted k-HFPGA in which a cluster has k subclusters. The structure can be represented by a tree. Figure 1 is an example of a 4-HFPGA where a cluster contains four subclusters. A vertex in this tree is used to represent a logic or a switch box. An edge between two vertices is used to represent a routing channel which consists of many tracks. The logic blocks are at the bottom of the tree while the switch boxes are those vertices above the logic blocks. We propose a modified multilevel hierarchical architecture denoted MFPGA which can be more interesant in terms of area and performances. Our architecture has the following particularities:

- The lowest level of the hierarchy contains the Logic blocks and the IO pads. Each logic element contains one 4 inputs Look-Up Table (4-LUT) followed by a bypass Flip-Flop.

- The routing architecture contains only unidirectional wires and the switch boxes are depopulated.

- In each level the ratio between parent tracks and child tracks is equal to k (k is the number of slaves in the cluster).

As we use unidirectional switches, we can distinguish two connecting networks as shown in figure 1.

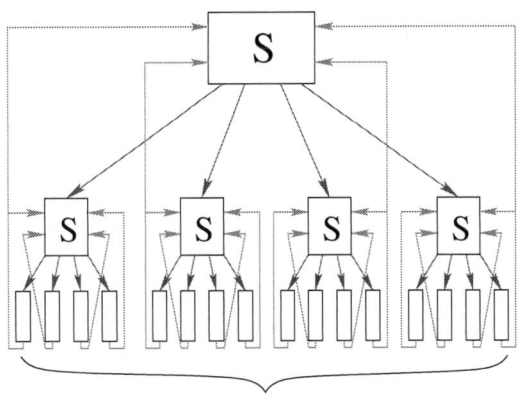

Logic Blocks & IO pads

Figure 1: Connection networks

0-7695-2533-4/06 $20.00 © 2006 IEEE   263

- A downward connecting network whose topology is equivalent to the butterfly fat tree. In this tree the edges come from the upper levels and reach the inputs of the logic blocks. The topology of this tree is equivalent to the one used in SPIN network [1] [2]

- An upward connecting network whose edges come from the leaves (outputs of logic blocks and input pads) to the switch boxes of each level.

2.1 The downward connecting network

Let us consider the case of a 2 levels tree with an arity equal to 4. In each level a cluster contains 4 slaves and a switch box. To depopulate the switch box, we divide it into four Mini Switch Boxes (MSB). In level 0 each MSB is in charge of connecting the upper level tracks and one input of each logic block as depicted in figure 2. Thus each MSB has 4 outputs which are equal to the number of logic blocks (slaves). The level 1 is constructed in the same manner, we

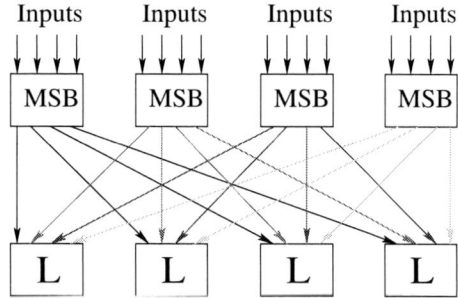

Figure 2: Top-down connecting tree in level 0

connect the switch box of each cluster of level 1 to 4 clusters belonging to level 0. As each cluster in level 0 has 16 inputs, we divide the switch boxes into 16 MSB and connect each one to one input of a cluster slave. Figure 3 shows the distribution of the interconnect in level 1. The previous

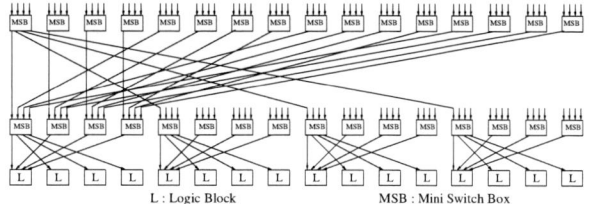

Figure 3: Top-down connecting tree in level 1

described butterfly fat tree has the following properties:

- From a track located in the top of a switch box we can reach any slave but in only one pin.

- From a track of a switch box we have only one path to reach a particular slave. Due to the regularity of the architecture, this path is easily predicted.

- In each level the interconnect resources are balanced between clusters.

2.2 The upward connecting network

The next question is how can the outputs of logic blocks and the input pads reach the inputs of other logic blocks. We propose to connect the output signals to specific switch boxes of upper levels. Thus for each logic block output (and input pad), we define a list of feedbacks, each one enables the output to reach a switch box in a particular level. An output of a source cell can reach its destination only if it has a feedback in the lowest common level or in a higher one. The way how we distribute the feedbacks on each level has an important impact on the number of different paths to reach a destination logic block from a source. In fact, if in each level the feedback is connected to an MSB with an index different from the one in the previous level, we can obtain two different paths to reach a cell. In our case we distribute

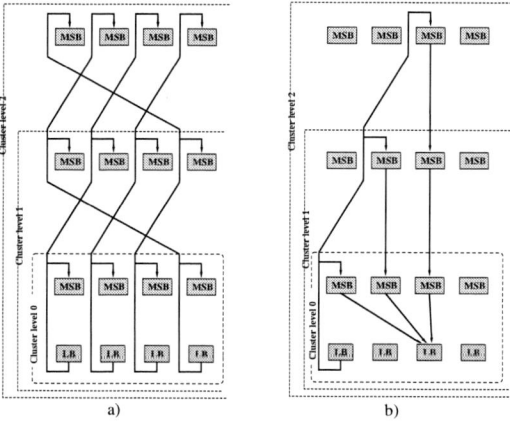

Figure 4: The rising connecting tree and different routing paths

the feedbacks as depicted in figure 4-a. We achieve a multi path structure which enhances the routability and gives us more different possibilities to reach a destination. As shown in figure 4-b, using two different feedbacks (located in two different levels), an output pin can reach a destination cell from two different paths but in two different pins. Since all the input pins in a LUT are logically equivalent, the router can complete a given connection using any one of the input pins of a LUT. Changing the ordering of the inputs in a LUT due to the connections performed by the router can be compensated by re-ordering the values of the LUT mask.

2.3 Connection with outside

As explained previously the input pads are located in the bottom of the architecture and are located inside the clusters of the level 0. These pads are connected like the outputs of the logic block in the rising network. The last point in this architecture concerns the location of the output pads. Those pads are also located inside clusters of level 0. To reduce the complexity of the routing architecture, those pads are not connected to the downward network. Thus they have their own network interconnect. This interconnect is local to the container cluster. As shown in figure 5 each output pad can be driven only by outputs of a logic block belonging to the same cluster. Such a layout simplifies the task of the router but adds more constraints to the placer.

0-7695-2533-4/06 $20.00 © 2006 IEEE

3. PLACEMENT

In this section, we present the placement technique used in the case of our hierarchical MFPGA. As explained previously our routing resources are limited and we have a few different ways to connect a source to a destination. Thus the placement of the cells has an important impact on the routability of the netlists. In fact most part of the effort will be devoted to the placement phase. Placement is done in two steps. First we apply a global placement. The aim of this phase is to balance the nets to route between clusters. It consists of a multilevel clustering followed by a multilevel refinement phase. After that, in each level we run a detailed placement to select slots that will be occupied inside clusters. As it will be explained in the following, this detailed placement is important to alleviate the routing congestion.

3.1 Multilevel clustering

DeHon [6] showed that for hierarchical FPGAs, 100% logic utilisation is not necessary benefical for overall device area minimisation. He presented some initial evidence to support this claim and presented a technique for depopulating gates in a hierarchical array. His results indicate that a careful partitioning of designs and depopulation of logic clusters can result in better FPGA resources utilisation. This remark was confirmed by results obtained by Singh [3] in the case of clustered mesh FPGA. In this work authors present the routability-driven bottom-up clustering technique. The aim of their technique is to alleviate routing congestion by absorbing as many nets into clusters as possible, and depopulating clusters according to Rent's rule in order to achieve spatial uniformity in the clustered netlist. In the following we explain how we have extended this technique to multilevel hierachical MFPGA clustering.

The clustering algorithm begins by choosing a logic block as a seed (the block with the highest separation) and assigning it to the first available slot in a cluster. We use the same objectif function proposed in [3]. First we identify low fanout nets and then we absorb them into a cluster.

In order to garantee spatial uniformity of the clustered netlist, we limit the number of available pins. Since in each level the interconnect is balanced between clusters, an attempt

is made to spread the logic evenly across all clusters in the level while limiting the number of pins available. An unclustered block can be absorbed into a cluster only if cluster size and a pins constraint (balancing constraint) are satisfied. The pins constraint can lead potentially to spatial uniformity and less than 100% cluster utilisation. As it is shown in [3], the described clustering technique can reduce the number of external nets and eliminate regions (clusters) with high congestion. Once clustering has been done, the original netlist is reduced to a new netlist with each node corresponding to a cluster. We propose to apply the same technique to construct super-clusters of clusters (see figure 6). Thus we use the same gain function to compute the attraction of each block to a cluster. We notice that pins constraint enforcing is inefficient when it is applied in high levels. This is due essentially to the bottom-up and the greedy aspect of our clustering technique. In fact in most cases when we impose pins constraints the tool needs more than the allowed clusters to achieve the clustering. To deal with such problem, we propose to create clusters in high levels without pins constraints enforcing. As presented in figure 6, once the multilevel clustering is achieved, we run a top-down refinement. The aim of this refinement is to move some blocks between clusters to reduce the number of pins per cluster in each level (pins balancing).

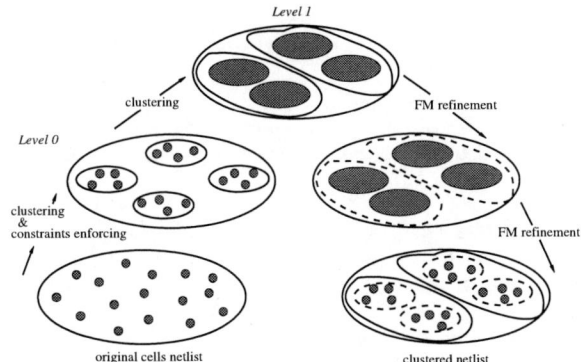

Figure 6: Multilevel clustering & refinement

3.2 Multilevel refinement

After the clustering phase, we obtain k clusters in each level. We can consider that the k clusters in a level present an initial solution to a k-way partitioning problem. During the refinement phase, cells will be moved between clusters (parts) to optimise an objectif function without violating the constraints imposed by the cluster size. In a level, cells are not allowed to move between all clusters, because this can decrease the quality of the solution obtained in the higher level. To prevent such bad effect, cells can only move between neighboring clusters. We call neighboring clusters, all clusters in a level belonging to the same supercluster. Thus in every level, neighboring clusters will be isolated and form a subgraph. In figure 6 those subgraphs are presented by the continued lines and partition by the dashed ones. A cell is allowed only to move across dashed lines. The objectif function is local to each subgraph and corresponds to the maximum number of pins of all clusters (parts) belonging to the same subgraph. An FM algorithm [7] will be applied

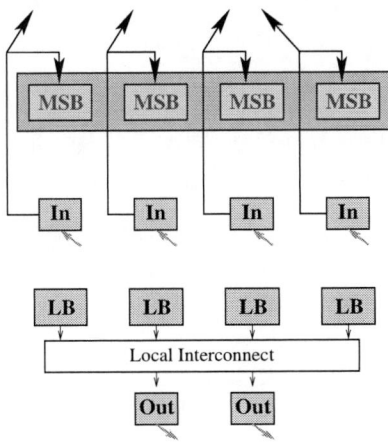

Figure 5: IO Pads connections

0-7695-2533-4/06 $20.00 © 2006 IEEE

on a subgraph to optimise the local objectif function. As described in [7] [10] this algorithm uses k(k-1) priority queues, one for each type of move. In each step of the algorithm, the moves with the highest gain are found from each of these k(k-1) queues, and the move with the highest gain that preserves or improves the balance, is performed. Note that the gain of a cell may be negative. After the move, all of the k(k-1) priority queues are updated. The balance criterion is used to select the cluster (part) from which a cell with the highest gain is to be moved. In our case, in each move we allow only one cluster to contain a number of cells exceeding the limit imposed by the architecture (the arity of the level). After all cells have been moved, the partition which has the best global gain and which respects the architecture arity is retained as the output result of this pass. The complexity of our k-way refinement is reduced since we do not apply it for all the graph but successively for each subgraph (in each subgraph there are small values of parts: Arity of the architecture). The hill-climbing capability of the FM algorithm serves a very important purpose. It allows movement of an entire cluster of vertices across a partition boundary. Note that it is quite possible as the cluster is moved across the partition boundary, the value of the objectif function increases, but after the entire cells moves across the partition, then the overall value of the objectif function comes down. As shown in figure 6, when we apply refinement, we begin from high levels down to lower ones.

3.3 Detailed placement

Now that we have obtained clusters with minimum number of pins and containing highly connected cells in each level, we proceed to the detailed placement. As in the proposed architecture we do not have full cross bar connection boxes, we can not place cells randomly inside the clusters. In fact the way cells are placed has an important impact on the routability. If during the detailed placement special properties of the netlist and the interconnect can be exploited, significant gains can be obtained in terms of routability and congestion reduction. The effect of the detailed placement on routability can be explained by the example shown in figure 7. In this example we have placed two logic blocks and

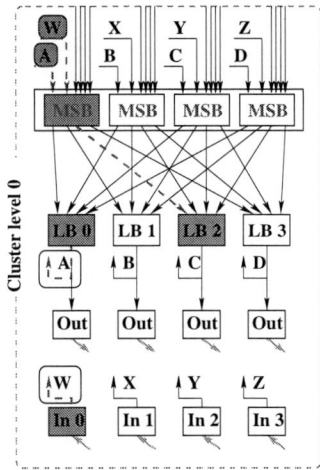

Figure 7: Impact of detailed placement on routability

an Input Pad in the same cluster. The logic block in position 0 LB0 and the Input pad in position 0 In0 drive the logic block placed in position 2 LB2. With the present placement we can not route both signals using only the switch box in level 0. In fact both signals reach the target block in the same pin. This problem can be resolved by simply changing the position of one of the driver blocks. This problem can also occur between two logic bloks located in two different clusters and trying to drive the same logic block. Our detailed placement is applied seperately in each level. For example we will present how this technique is applied in level 0. The same method can be extended to the other levels. In level 0 we notice that congestion is generated by cells that drive the same destination cell. Thus we introduce the notation of Cells Constraints Graph (CCG). Given a clustered netlist and a placement problem, a CCG denoted as $Gn = (V, E_n)$ consists of a set of vertices and edges, can be constructed from a netlist. In a CCG, a vertex is used to represent a cell of the netlist and an edge is constructed between two vertices which drive the same destination cell. Those cells are called adjacent cells.

A CCG is used to represent the routing constraints for a placement problem in a particular level. This means that the placement of vertices will depend on the positions of the adjacent vertices that have been already placed. Thus the placement of vertices will introduce constraints on the placement of the adjacent vertices. For each vertice we reserve a list of allowed positions (slots) inside the cluster. Initially each vertice has k possibilities (arity of the cluster). Our algorithm is as the following:

```
Loop over not placed vertices
   Eliminate occupied positions
   Loop over placed adjacent vertices
      Find positions to avoid
   End
END
```

To eliminate positions considering the adjacent placed vertices, we define a specific function that takes in account the following parameters:

- Clusters where the common destination, the vertice to place and the placed adjacent vertices are located.

- Type of the vertices to be placed and the adjacent placed ones (logic block/Inpud pad).

In our technique the order of placing vertices is very important and has a great impact on the efficiency of the method. We give priority to vertices that have the following properties:

- Vertices located far from their destinations. We know that these vertices have less paths to reach their destinations.

- Vertices with high number of adjacent vertices: more constraints.

Vertices are sorted depending on both previous properties and then placed by the algorithm following the method that we have described. The technique is expanded on all levels. When we apply the technique we begin from high levels down to lower ones.

4. ROUTING

The routing problem can be stated as assigning signals to routing resources in order to successfully route all signals. This goal is difficult to achieve in our architecture because of the lack of routing resources (depopulated switch boxes). In fact the number of paths to reach a destination from a source is significantly reduced and those paths depend on the location of cells and the number of levels in the architecture. Thus signals will compete for the same resources and the challenge is to find a way to allocate resources so that all signals can be routed. Despite this disadvantage we have a great advantage in our architecture since our unique path is predictable. Unlike the other architecture (mesh-connected arrays, Triptych ...) we do not need to define a directed graph to describe the routing architecture. This reduces the routing process complexity.

We have studied different routing techniques to find the most suitable one for our approach. For example the obstacle avoidance algorithm could be used: if in one path we find a used resource we can try another path by jumping to another level (using another feedback). This algorithm is easy to implement but it usually yields to many unroutable nets. Some rip-up and retry approaches have been proposed to ramedy the deficiencies of this approach [5].

To route our architecture we adopted a particular iterative rip-up algorithm based on the congestion negotiation called PathFinder [8]. PathFinder was applied to the mesh architectures and we have adapted it to our architecture. Since we have only one downward path to reach a destination, we have eliminated the breath-first search in the detailed routing part. Our detailed router corresponds to a function that determines directly the next wire to reach destination. In any way once we have chosen the corresponding feedback, only one path (only one next wire) can bring us to the destination.

Since the choice of the feedback imposes the path to follow, our negotiation must be done on the choice of the feedback that leads to a path with less congestion. According to this remark, we assign to each feedback an adjustable cost. The global router dynamically adjusts the congestion penalty for each feedback. During the first iteration of the global router each feedback has a cost equal to the index of the level where it is located. This encourages the use of lower level to reduce the path length and the number of switches to cross. During this iteration individual routing resources may be used by more than one signal. During subsequent iterations the penalty to use shared resources is gradually increased so that signals will negotiate effectively for resources. In fact costs of feedbacks of a source will change: a feedback belonging to a higher level can get a cost lower than a feedback located in a lower level. The implemented algorithm is described by the following:

```
While shared resources exist
  /*global router*/
  Loop over all signals i
    Loop until all sinks tij are found
      Rip up branch Bij
      Find feedback fij with lowest cost
      Bij <- fij
      /*detailed router*/
      Loop until new tij is found
        Find next_wire
          Add next_wire to Bij
      End
    End
  End
  /*backtrace*/
  Loop over nodes in Bij
  /*path from tij to si*/
    Update cost of fij
  END
END
```

The algorithm is based on two simple and basic functions that depend of our MFPGA routing architecture. The first one belongs to the global router and determines the feedback that the source will use to reach the destination. Knowing the source cell index, the sink cell index, this function return the best level to jump to, in other words the feedback with the lowest cost. The second function belongs to the detailed router and determines the next wire to use to reach destination knowing the actual wire index.

5. RESULTS

To validate and study the performances of our tools, we placed and routed some of the MCNC benchmark circuits. As shown in table 1 results are very promising since we were able to route circuits that occupies until 77% of the logic area. We have tested the effect of the refinement phase which was run after the multilevel clustering. So we have tried to rout resulting placed netlist in both cases:

- Multilevel clustering without FM refinement (column 10 of table 1).

- Multilevel clustering followed by FM refinement (column 11 of table 1).

We have noticed that in most cases the FM refinement alleviates congestion and leads to full routability. Nevertheless the router failed to route benchmarks with very high occupation like b9. In this case the router routes a large amount of the nets (until 98%). To improve the performances of the router we propose to:

- Better use routing resources by modifying the distribution of the rising interconnect or increasing the number of feedbacks in each level (An output can have more than one feedback in a level).

- Improve the placement and especially the detailed one (alleviate congestion).

To have an idea of the area efficienty of our architecture, we have compared switches and area requirements between our MFPGA architecture and the traditional mesh topology. The mesh is similar to the vpr422_challange_arch architecture with uniform routing with single-length segments and a subset switch box. Each Logic Block contains only one 4-LUT. One input appears in each side, and the output appears on the top and the right side. Both the inputs and the outputs are fully populated ($F_c = 1$). The IO pads are fully populated too.

We use the channel minimising VPR 4.3 router to route the mesh, and we vary the IO_ratio to achieve the optimal array size.

Benchmark		Mesh					MFPGA					
Name	LUTs	$\sqrt{N}$	W	IO ratio	Switches number	Area (λ^2) x10^3	Arch	Occup- ation%	R%	R% +ref %	Switches number	Area (λ^2) x10^3
b1	4	2	3	2	300	1284	4	100	100	100	32	288
cm138a	9	3	4	2	824	3344	4x4	56	100	100	512	2032
cm42a	10	4	3	1	948	4344	4x4	63	100	100	512	2032
pcle	29	6	5	2	3700	15316	4x2x2x4	46	100	100	3584	11968
decod	32	6	4	1	2768	11822	4x4x4	50	95	100	3584	11648
cc	33	6	5	2	3700	15316	4x4x4	52	92	100	3584	11648
count	37	7	5	2	4950	20577	4x4x2x2	58	98	100	4096	12608
my_adder	49	7	4	2	3960	16680	4x4x4	77	100	100	3584	11648
b9	61	8	5	4	7020	28656	4x4x4	96	90	98	3584	11648
i4	110	11	7	5	18298	71289	4x4x4x4	42	87	100	20480	46080
c2670	363	20	8	5	63968	249172	4x4x4x4x4	35	92	100	106496	299008
i9	471	22	8	2	72480	286356	4x4x4x4x4	46	85	100	106496	299008

Table 1: Benchmark statistics

VPR chooses the optimal size as well as the optimal channel width needed to place and route each benchmark. For the MFPGA we choose the structure that is large enough to support the benchmark circuit. MFPGA structures can be varied by changing the number of level, the arity of each level.

In both cases the number of switches consumed by each benchmark corresponds to the total number of switches used by the overall optimal target architecture.

We compare the area of both architectures using both a simple cost model based on routing switches count, and a more refined model that estimates effective circuit area. The mesh area is the sum of its basic cells area like SRAMs, Tri-states and multiplexers. The same thing with the MFPGA composed primarily of SRAMs and Multiplexers. We use the same cells library for both architectures.

Column 9 of table 1 shows the occupation average of each circuit in the target MFPGA. There is a low occupation average in the majority of the benchmarks. This is due to the depopulation of the interconnect. As mentioned previously we under-utilise the logic resources in this type of structure. In addition, the size of the smallest MFPGA that can contain the circuit under investigation is penalised due to the coarse granularity of this architecture. In spite of these constraints we achieve a gain in area efficiency compared to the mesh. Columns untitled "Switches number" and "Area" in table 1 show the difference in number of switches and the total area in the Mesh and the MFPGA structures.

It is clear from this comparison that the new architecture will be more efficient in terms of area if we can increase the Logic utilisation.

6. CONCLUSION

This paper described a new hierarchical multilevel MFPGA architecture and its suitable configuration tools. The preliminary results show that good balancing of the LUT and the interconnect utilisation reduces area compared with traditional Mesh architectures.

The new topology based on two hierarchical unidirectional networks seems to be more robust and can achieve better speed than symmetrical FPGA architectures.

The downward network is a predictable interconnect which has a very interesting impact on accelarating the routing phase.

The routing key of the proposed architecture is the upward network. Enhancing the routability needs to populate the upward network to increase paths between sources and destinations. This can leads to area increasing, but can be compensated by applying the Rent's rule to reduce the cluster inputs/outputs

7. REFERENCES

[1] A. G. A. Adrijean. Micro-network for soc: Implementation of a 32-port spin network. *Proc. DATE'03*, pages 1128–1129, march 2003.

[2] A. G. A. Adrijean. Spin: a scalable, packet switched, on-chip micro-network. *Proc. DATE'03*, pages 70–73, march 2003.

[3] M. M.-S. A. Singh. Efficient circuit clustering for area and power reduction in fpgas. *Proc. FPGA'02*, February 2002.

[4] A. A. Aggarwal and D. M. Lewis. Routing architecture for hierarchical field programmable gate arrays. *Pro. IEEE Custom Integrated Circuits Conference*, 1994.

[5] W. Dees and R. Smith. Performance of interconnection rip-up and reroute strategies. *Proc. DAC*, pages 382–390, June 1981.

[6] A. DeHon. balancing interconnect and computation in a reconfigurable array (or why you don't really want 100% lut utilisation). *Proc. FPGA'99*, 1999.

[7] C. M. Fiduccia and R. M. Mattheyeses. A linear-time heuristic for improving network partitions. *Proc. DAC*, pages 175–181, 1982.

[8] C. E. L. McMurchie. Pathfinder: A negotiation-based performance-driven router for fpgas. *Proc.FPGA'95*, 1995.

[9] Y. T. Lai and P. T. Wang. Hierarchical interconnection structures for field programmable gate array. *IEEE Trans. VLSI*, pages 186–196, 1997.

[10] L. A. Sanchis. Multiple-way network partitioning. *IEEE Trans. on computers*, 38(1), January 1989.

New non-volatile FPGA concept using Magnetic Tunneling Junction

Nicolas Bruchon, Lionel Torres, Gilles Sassatelli, Gaston Cambon

LIRMM
161 rue Ada
34392 Montpellier cedex 5
+33(0) 4 67 41 85 67

{name}@lirmm.fr

ABSTRACT

This paper describes a real time reconfigurable (RTR) micro-FPGA using new non volatile memory. Magnetic tunneling junctions (MTJ) used in Magnetic random access memories (MRAM) are compatible with classical CMOS processes. Moreover remanent property of such a memory could limit configuration time and power consumption required at each power up of the device. Each configuration memory point has to be readable independently from each other, which makes this approach radically different from the classical memory array one.

1. INTRODUCTION

Most FPGAs are currently SRAM based. Configuration memory is distributed throughout the device and organized as a shift register. Data are serially loaded, that is why configuration time, depending on chip size, may be long. Multiplying the number of shift registers can reduce this time and allows parallel load of the configuration bit stream and partial dynamic reconfiguration.

Some FPGAs and CPLDs use flash memory (Actel Proasic). This type of memory provides non volatility that allows devices to be ready to run at power up. Moreover, in embedded systems, the chip can be powered down when not in use in order to limit power consumption. Distribution of the memory all over the device raises new technological constraints and increases the number of masks.

Our solution consists in using magnetic memory cells. Indeed, MRAM announced features are interesting: high timing performances, high density integration, reliable data storage, low power consumption, endurance [1].

MTJ process requires dedicated steps (like Flash process), but these ones can occur after the standard CMOS process ("above IC"), MTJ are laid over the CMOS avoiding area overhead. Moreover only a few additional masks are required for this process, contrary to flash technology. Several architectures, based on unbalanced flip flop structure, are proposed and evaluated in the next sections.

The first section describes the MTJ read/write principle. The second section proposes a CMOS structure to read MTJ. Section three describes different kind of LUT SRAM/MRAM architecture for Run Time Reconfigurable (RTR) approaches. The last section describes the architecture of a prototype of such an FPGA which main advantages are density, non volatility and dynamic reconfiguration.

2. MTJ PRINCIPLE

2.1 What is it made of?

MTJ cell is made of 2 thin ferromagnetic layers separated by an ultra thin non-magnetic one (oxide layer). Relative magnetic orientation of these layers is used to store information. The magnetic orientation of one of these layers is pinned and the other one is free. In fact, both of these layers' magnetic orientation can be changed applying a magnetic field towards the junction. Magnetic remanence of the ferromagnetic elements provides non-volatility.

2.2 Read mechanism

Relative magnetic orientation of the layers exhibits two different values of equivalent resistance Rp (parallel – low resistance) and Rap (anti-parallel – high resistance) representing respectively the logic states '0' and '1'. The ratio between these two resistances is characterized by the Tunneling Magneto Resistance (TMR), which is defined by $TMR = \dfrac{\Delta R}{R} = \dfrac{R_{AP} - R_P}{R_P}$. This TMR can vary from 40% up to 220%[2] depending on the MTJ process used. The resistance of the MTJ is strongly related to the thickness of the oxide layer (ranging from 1 – 2 nm), varying from 3kΩ to 1MΩ [3]. This value also depends on the voltage applied to the junction as described in Figure 1, as a consequence, current is not a linear function of the voltage.

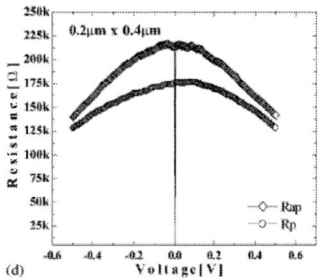

Figure 1. Resistance and current evolution with MTJ (CoFe) voltage variation [4]

In order to avoid the deviations introduced by these phenomena, the MTJ voltage has to remain within a -200mV/+200mV range. Resistance can be evaluated by sending current through the junction and measuring voltage at the nodes of the MTJ. Read lines (Figure 2) are used to provide current through the MTJ and to determine the value stored in the MTJ.

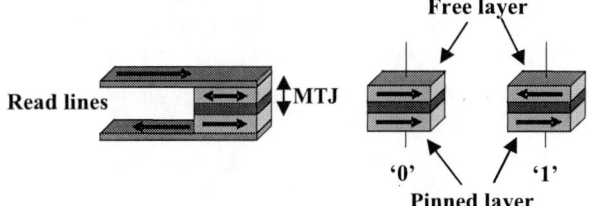

Figure 2. MTJ read mechanism

A structure is then required to determine whether the stored value is '1' or '0'. Single cell read structure are proposed and evaluated in section 3.

2.3 Write mechanism

To write information in the MTJ, a magnetic field should be applied through the junction. Two write lines are added to the structure to generate the magnetic field required to change the value stored in the MTJ. A current is applied on these lines [5], as shown on Figure 3.

0-7695-2533-4/06 $20.00 © 2006 IEEE

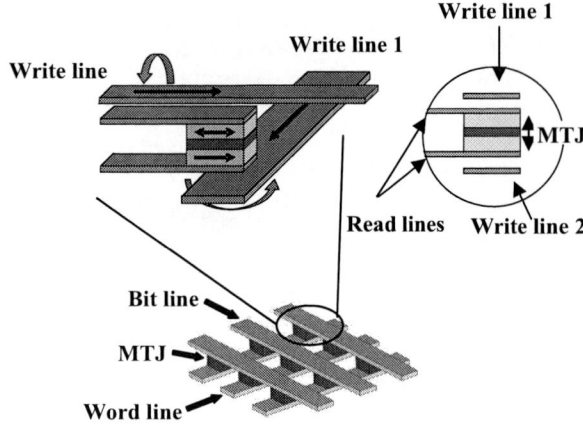

Figure 3. MTJ write mechanism & MRAM architecture [4]

Indeed, the current applied on the write lines generates a magnetic field around these lines. At the cross point of the write lines, the magnetic field (resulting from the composition of the two fields) is high enough to change the magnetic orientation of the layers. Contrary to the read lines, the write lines are not in contact with the MTJ. To minimize the current value sent on the write lines (and the power consumption), these ones should be as close as possible to the MTJ.

The pinned layer requires higher current than the free one to be oriented, which is why this layer is polarized only once (in order to limit power consumption) and then used as a reference. Indeed, during the operating mode, orientation of this layer can not be changed; currents that are used are not high enough to change this orientation.

2.4 About MRAM organisation

This section gives an overview of the classical MRAM organization, which can be used to define the MTJ based RTR FPGA.

In general, MRAM is made of 2 arrays of write lines. MTJ are positioned at the cross point as shown in Figure 3.

Instead of using a shift register as in classical SRAM-FPGA, configuration memory of our FPGA can be organized as a classical MRAM. Indeed, using this scheme, only a part of the memory could be changed, allowing dynamic and partial reconfiguration.

Magnetic RAM features seem to be interesting compared to other technologies for programmable applications.

Main drawback of this technology is the power consumption during the writing phase. Indeed, writing one MTJ requires a few mA. One solution proposed by MRAM designers considers that only the magnetic field generated by the current is used to write information on the MTJ. With the same current a lots of MTJ can be written that is why, it is important for our configuration memory (for the writing only) to be organised as a classical MRAM.

3. SINGLE CELL READ STRUCTURE

In order for each point to be readable independently from each other we evaluated a structure which is able to read a single memory point. Indeed, in programmable devices, each memory point can be used to drive logic or interconnect; it is strongly different from classical memory design architecture. The structure proposed (unbalanced flip-flop - UFF) is depicted in Figure 4 [6]. Transistors in this structure have to be sized as small as possible in order to be efficient (in term of silicon area, this structure will be duplicated to each configuration memory point).

In this structure, a reference cell and only one MTJ could be used instead of two MTJ, but TMR is too small and the structure too much sensitive to process variations to determine the correct logic value. Then, two MTJ cells storing complementary information are required. Indeed process dispersions of the magnetic junctions are important implying strong variation of the equivalent resistance value.

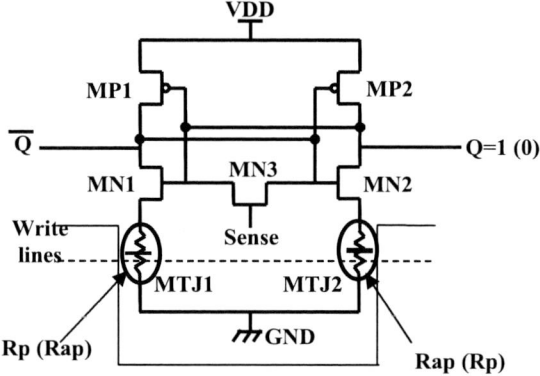

Figure 4. MTJ unbalanced flip flop (UFF)

3.1 Principle

Writing mode is described in the previous section; current is applied on the write lines creating a magnetic field that will write information on the free layer; of course this cycle does not have to occur during a read phase (Sense = '1'). Information to be read has to be established in the MTJs. During a read phase, the MN3 transistor acts as a short circuit, as a consequence, the two cross coupled inverters are pulled to a metastable operating state. Figure 5.a. shows the transfer curve (1) of two cross coupled inverters with one metastable state and two stable states. Figure 5.b. exhibits that the metastable state can be moved (curves (2) & (3)) by the two complementary MTJs moving it away from one of the stable states and bringing it closer to the other one. Then, when the Sense signal is released, the structure will move to the closest stable state.

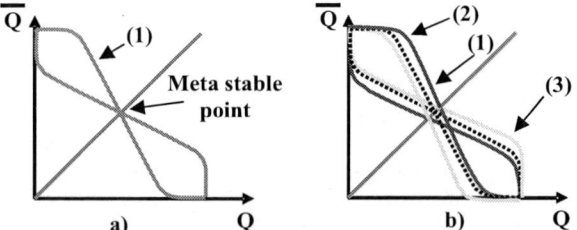

Figure 5. a) Transfer curve of cross coupled inverters, b) Transfer curve of cross coupled inverters with MTJ

0-7695-2533-4/06 $20.00 © 2006 IEEE 270

After this read cycle, the value considered is stored in the flip flop and can be used as many times as required. During this phase, information can be written in the MTJ during circuit runtime. Access times are equivalent to SRAM ones.

3.2 Read/Write process

Firstly, information is written into the MTJ (Figure 6.a), then a pulse is sent on the sense input (Figure 6.b). When the sense signal is high, the structure is pulled into the metastable state (Figure 5.a). When this signal is released, the structure switches to the closest stable state. New information can then be stored into the MTJ without altering the functioning of the device (values stored in the flip-flops remain unmodified, Figure 6.c).

3.3 Robustness

The structure must be as small as possible and robust enough to prevent process parameters dispersion influence. MOS transistor dispersion parameters that strongly impact the structure are the threshold voltage (Vth), the width (W) and the length (L) dispersions. The structure was evaluated using a CMOS 130nm technology. Dispersion parameters values are estimated from different CMOS process models [7-8-9].

Minimum resistance value of the MTJ required for the structure to work as expected can be evaluated replacing MTJ by resistors with dispersions of about 3% [10]. In our case, Monte Carlo simulations have been run with uniform distribution of 1σ of variation.

Structure with L=130nm, W=130nm, W=260nm and W=390nm were tested. Results of these experiments are summarized on the Figure 7 that shows the operating limit of the unbalanced flip flop structure.

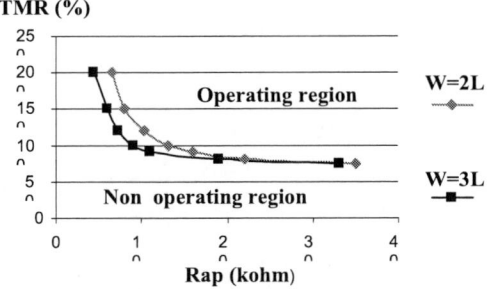

Figure 7. Minimum TMR and Rap required for W=2L & W=3L

3.4 Experiments

This structure has been evaluated; layouts have been designed (Figure 8) to determine parasitic parameters, structure limitations and their influence on structure functioning.

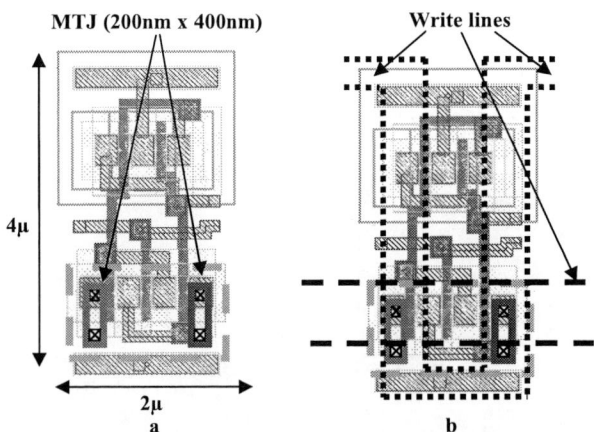

Figure 8. Unbalanced flip flop layout a) with MTJ b) with write lines (CMOS 130nm W=3L)

Monte Carlo simulations show that this structure is robust enough and that the output acts as expected as shown on Figure 9.

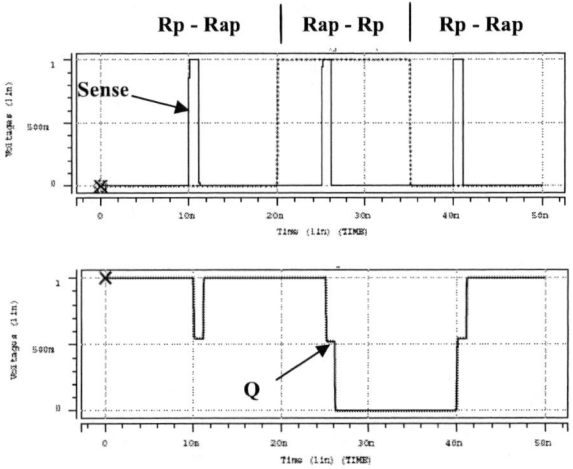

Figure 9. Output of the structure

One particularity of the MTJ process is that it occurs after the classical CMOS one. Indeed, the structure of MTJ is said "above IC". MTJ are laid over the CMOS as shown on Figure 10.

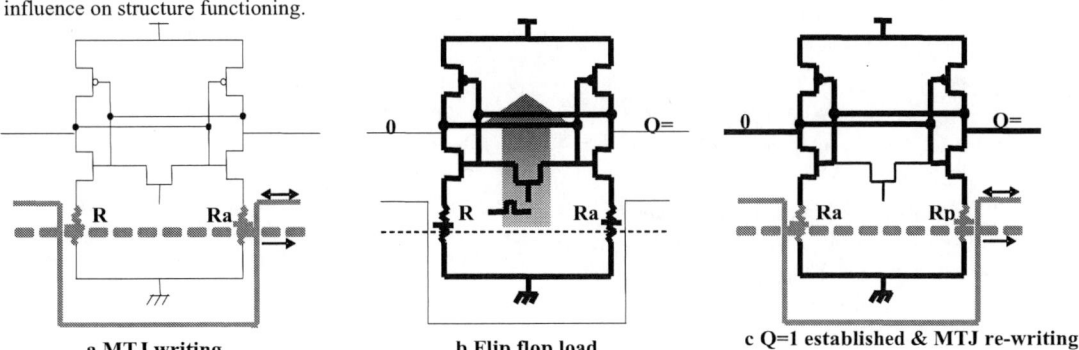

Figure 6. Unbalanced flip flop functioning

0-7695-2533-4/06 $20.00 © 2006 IEEE   271

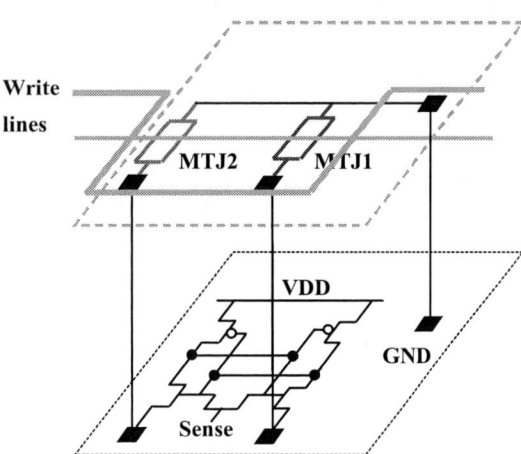

Figure 10. "Above IC" structure

The 4 first metal levels are used for the classical CMOS process and the last ones are used for the writing lines. Moreover, this structure can be used for any other technology that requires determining a resistance variation. Indeed, the unbalanced flip flop can be used with phase changing memories [11].

4. MRAM BASED CLB & SWITCH MATRIX

On programmable devices, configuration bitstream is often stored in SRAM cells in order to configure digital blocks: CLB (configurable logic block) and interconnect between these blocks. For instance, SRAM cells are used in switch matrix to drive pass gate transistor (as depicted in Figure 11).

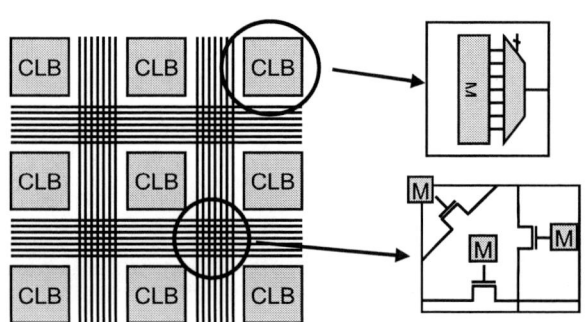

Figure 11. FPGA classical SRAM based architecture

Main elements used in FPGA are CLB (LUT) and switch. We propose new LUT/switch architectures based on MTJ.

4.1 LUT architecture

Lookup table's purpose is to implement Boolean functions. Indeed, truth table is stored into a RAM, and a multiplexing tree driven by the inputs sets the corresponding value to the output.

In classic FPGA, LUT are SRAM based, the number of transistors used to realize a LUT can be determined by $6 \times 2^N + 6 \times (2^N - 1)$ where N is the number of inputs of the LUT. For instance, a 4 inputs LUT requires 186 transistors). Non volatility can be provided using MTJ. Different structures of LUT have been proposed and evaluated.

4.1.1 RTR-LUT

The first one simply consists in replacing the standard SRAM point by a single unbalanced flip flop structure (Figure 12). Thanks to the fact that MTJ are laid over the CMOS, this design is smaller (in term of number of transistors). Indeed, the structure requires only 5 transistors for a single memory point whereas SRAM memory point is made of 6 transistors. The number of transistors required to build an N inputs LUT is given by $5 \times 2^N + 6 \times (2^N - 1)$. A 4 input LUT requires 170 transistors. Moreover it provides non volatility and allows masked reconfiguration (i.e. shadowed reconfiguration, where a configuration is loaded while the previous one still dictates circuit behaviour). Indeed, during a read cycle, information contained in the MTJ cells is loaded into the latch region of the unbalanced flip flop structure.

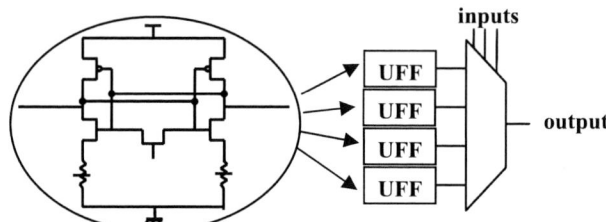

Figure 12. Unbalanced flip flop & RTR-LUT

4.1.2 Merged LUT

The second solution proposed considers that the structure shown on Figure 13 is a multiplexer. In this case the first multiplexer stage is merged with the LUT.

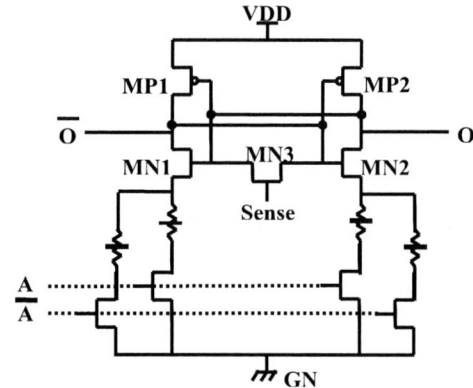

Figure 13. Flip flop based multiplexer

Indeed, A and Ā inputs determine the couple of MTJs that will be read during the next read cycle. This structure includes the RAM and the first multiplexing tree stage. The number of transistors is strongly reduced, indeed the number of transistors needed for an N inputs LUT is $(5+4) \times 2^{N-1} + 6 \times (2^{N-1} - 1) + 2$ that means 116 transistors for a 4 input LUT. But the structure is not as flexible as the previous one: each time A will change, a read cycle has to occur for the value to be available on the output; therefore this structure does not support shadowed reconfiguration as the previous one.

0-7695-2533-4/06 $20.00 © 2006 IEEE

Table 1. Number of transistors required for different LUT structures

Inputs number	LUT structure		
	SRAM	RTR	Merged
	nb of T	nb of T	nb of T
3	90	82	56
4	186	170	116
5	378	346	236

Choice of the structure to be used depends on the features required for the programmable device (dynamic and shadowed reconfiguration, area…).

4.2 Switch

Switch matrix are used to drive signals to the correct destination. In SRAM based FPGA, interconnection are made by transistors which gates are driven by SRAM.

One more time, the simplest way to provide non volatility on the interconnections is to replace SRAM points by unbalanced flip flop based structure with MTJs.). Area of a switch is about 60μm² using 130nm CMOS process.

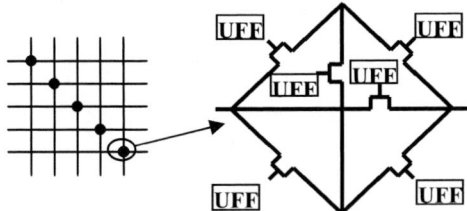

Figure 14. Switch matrix and switch point

5. FROM BASIC STRUCTURE TO μFPGA

5.1 Basic structures

In order to write information on the MTJs, current generators are required. We propose a bidirectional current generator that can provide a few mA Figure 15.

This is a three states current generator. Indeed the *Ini* signal allows sending current or not. *Vsen* value must be VDD2 or Vss.

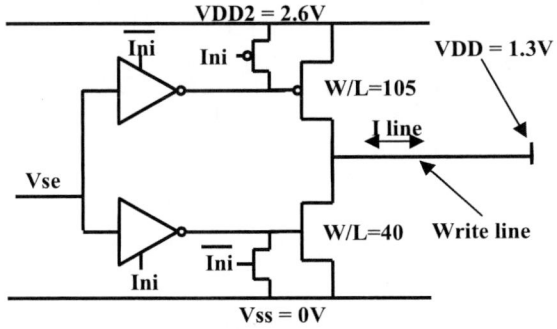

Figure 15. Current generator

As shown in Figure 16, such a generator can provide current up to 10mA. Moreover, current can be sent in both ways, depending on the Ini and Vsen signals.

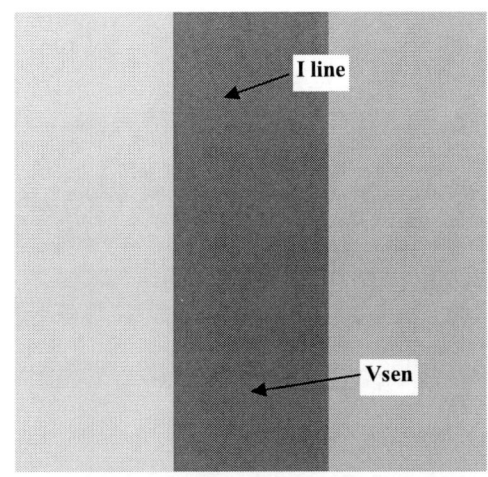

Figure 16. Current generator simulation

Switching matrix and LUT have been described in the previous paragraph. These structures layouts have been realized and tested (Figure 17) to determine dispersion influence on the functioning of a global structure.

Figure 17. Unbalanced flip flop & RTR-LUT layout

Then, these structures have been gathered on a same layout to realize CLB and switch matrix. A first μFPGA based on the RTR LUT is proposed.

5.2 μFPGA

Writing lines are organized as 2 perpendicular arrays (like in classical MRAM) allowing to write only the MTJ that is located at the cross point of the writing lines that are connected to the current generators. The configuration memory is addressable like a classical MRAM allowing reconfiguration of the selected MTJs. Current generators are located at the periphery of the die boundaries. (Figure 18)

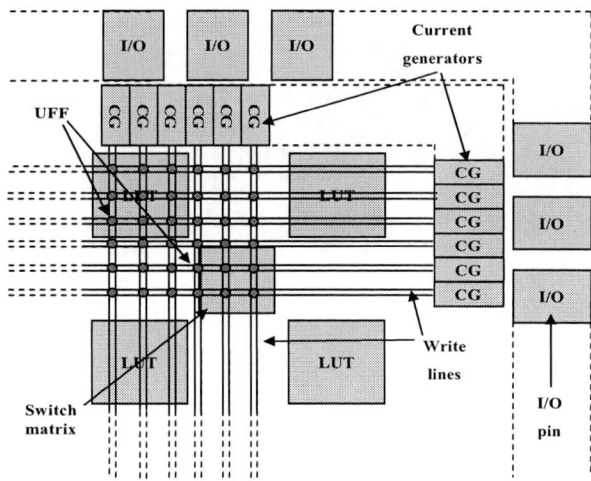

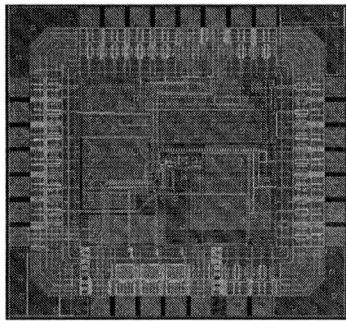

Figure 18. µFPGA floorplan

This µFPGA has been designed in a 130nm CMOS technology. A first silicon die has been proposed using RTR LUT, first silicon prototypes are currently fabricated and expected for characterization early 2006 (Figure 19).

Figure 19. First silicon die layout

6. CONCLUSION

This unbalanced flip flop structure seems to be interesting in FPGA design, because:

- After a sense cycle, the value is stored in the flip-flop and can be used as many times as required.

- Stored data are non-volatile due to residual property of the MTJ. This feature is particularly interesting for preventing additional power consumption upon device power up (configuration download).

Moreover MTJ state is not sensitive to SEU.

When the device is operating, Rap and Rp can be asynchronously written, because previous value is stored in the flip-flop and remain unmodified until a 'sense' cycle is initiated.

We propose three LUT structures which can be used depending on the context. Indeed RTR LUT is dedicated to structure requiring shadowed dynamical reconfiguration, whereas the merged LUT offers higher gain in term of number of transistors and silicon area. Contrary to RTR LUT that is totally combinational, merged and selection logic LUT require a sense cycle each time an input value changes – this implies either to implement additional logic for triggering a sense cycle upon input change, or to clock the unbalanced flip-flop array at a higher frequency than the one in use in the circuit.

Gathering of the basic structures allowed us to realize a first µFPGA prototype based on Magnetic tunneling junctions. Our first µFPGA main features are non volatility, real time partial reconfiguration.

7. References

[1] Min She, "Semiconductor flash memory scaling", PhD, University of California Berkeley, (2003)

[2] Stuart S.P.Parkin et al. Giant tunneling magnetoresistance at room temperature with MgO (100) tunnel barriers,nature materials, online publication www.nature .com/naturematerials, (2004)

[3] V. Javerliac, SPINTEC, private communication, (2005)

[4] H. Jinhee et al. "Characterization of the electrical and magnetic properties of sub-micron MTJ cells using scanning probe microscope interfaced with an external magnetic field generator", in Applied Surface Science 237, pp593-599, (2004)

[5] Thomas W. Andre et al., "A 4-Mb 0.18-microns 1T1MTJ Toggle MRAM With Balanced Three Input Sensing Scheme and Locally Mirrored Unidirectional Write Drivers" in IEEE journal of solid-state circuits, vol. 40, no. 1, pp. 301-309, (2005)

[6] W. Black & B. Das, "Programmable logic using giant-magneto-resistance and spin-dependent tunneling devices" J. Appl. Phys., vol. 87, pp6674-6679, (2000)

[7] Marcel J.M. Pelgrom et al. "Matching properties of MOS transistors" in IEEE journal of solid-state circuits, vol. 24, no. 5, pp1433-1440, (1989)

[8] T. Xinghai, "Intrinsic MOSFET parameter fluctuations due to random dopant placement" in IEEE transaction on VLSI systems, vol. 5, no. 4, pp369-376, (1997)

[9] J. Bhavnagarwala, "The impact of intrindic device fluctuations on CMOS SRAM cell stability", in IEEE journal of solid-state circuits, vol. 36, no. 4, pp 658-665, (2001)

[10] B.N. Engel et al. "A 4-Mb Toggle MRAM Based on a Novel Bit and Switching Method", in IEEE transaction on magnetics, vol. 41, no. 1, pp 132-136, (2005)

[11] A.L. Lacaita & al. "Electrothermal and phase-change dynamics in chalcogenide-based memories", in IEEE International Electron Devices Meeting, (2004)

Complexity and
System Organization

Profile Directed Instruction Cache Tuning for Embedded Systems

Kugan Vivekanandarajah, Thambipillai Srikanthan
Centre for High Performance Embedded Systems (CHiPES)
Nanyang Technological University
Singapore 639798
e-mail: {kugan,astsrikan}@ntu.edu.sg

Christopher T. Clarke
Dept of Electronic and Electrical Engineering
University of Bath
e-mail: {C.T.Clarke}@bath.ac.uk

Abstract— **Cache memories improve the performance due to the locality found within the loops of application. Because these loop characteristics are application dependent, the optimal cache hierarchy for performance and energy saving is also application dependent. Traditionally, cache simulations are employed to tune the cache hierarchy. In this paper we propose a simple yet effective loop profiler directed methodology for instruction cache hierarchy optimization. The proposed methodology utilizes the loop characteristics of the application which are readily available from the compiler making it easy to adopt the methodology in an existing design flow.**

I. INTRODUCTION

Low energy is one of the key design goals in current embedded systems. Customization of an architecture to the needs of the application is one of the most effective ways to tackle dynamic as well as static energy reduction without impacting the performance. This is especially advantageous in embedded systems which are often application specific. Architecture tuning aims to satisfy constraints such as size, performance, power and energy for a particular application. If the cache is larger than what is needed for the application, energy required to fetch from the larger cache may be unnecessarily high. Larger memory also dissipates higher leakage power. Conversely, if the cache is too small, excess energy may be wasted due to thrashing. The diversity among applications leads to very different cache requirements for different applications [22].

The major problem with the architecture tuning approach is that it is not feasible to perform exhaustive simulations for all possible combinations of cache parameters. Because of this multi-dimensional design space, the total number of possible designs can be very large. The design space will be even larger when design space exploration needs to be done for a cache hierarchy involving multiple cache levels, requiring higher design time. Typical cache parameters explored while evaluating the design space of the cache subsystems are the size, associativity, line size and cache write policy. In instruction caches, the parameter which dominates the power-performance metrics of the final system is the cache size [21]. Associativity and line size also allow the energy reduction to be tuned further but typically have a smaller impact compared to cache size. Techniques such as Balanced-Instruction cache reduces

the conflict misses with direct mapped cache configuration [20]. The goal of our work is therefore to identify the optimal instruction cache size for maximum energy reduction without compromising the overall energy-delay product.

In this paper, we propose a loop profiling based methodology for selecting a near-optimal instruction cache size, for a given application, without exhaustive simulations. The proposed methodology utilizes the loop profile data of the application to realize the instruction cache needed for a given application. The next section presents related research, before presenting our loop profiler directed methodology for tuning the instruction cache.

II. RELATED RESEARCH: DESIGN SPACE EXPLORATION FOR NEAR-OPTIMAL CACHE HIERARCHY DESIGN

Traditionally simulation based exhaustive search methods are used to explore the optimal cache designs, where a design is simulated with all the possible parameters and analyzed to find the optimal cache hierarchy. Unfortunately the time required for an exhaustive search is often prohibitive. When the design space becomes too large, iterative heuristics [21] are used to reduce the design space needed to be explored, so that the near-optimal cache configuration is reached without actually simulating the entire configuration. A crucial aspect to this design methodology is the simulation speed. Thus, some techniques focused on reducing the time required by using trace compression and /or simulating multiple configurations [5], [13] in a single pass. In trace compression, a reduced trace is obtained which approximates the behaviors of the original trace. Techniques are formulated to find approximate [7] or lossless trace reduction [11], [17], [19] to improve the simulation speed.

Another class of approach uses analytical design space exploration, where certain characteristics are extracted from the application and used to either find the miss rate for given cache parameters or, to find the optimal cache parameters for given set of requirements. For data caches, a source code analysis method has been proposed to determine the optimal data cache size by analyzing array access patterns of the application [10]. This was later extended to determine the best distribution of on-chip data memory into scratchpad memory and data caches

0-7695-2533-4/06 $20.00 © 2006 IEEE

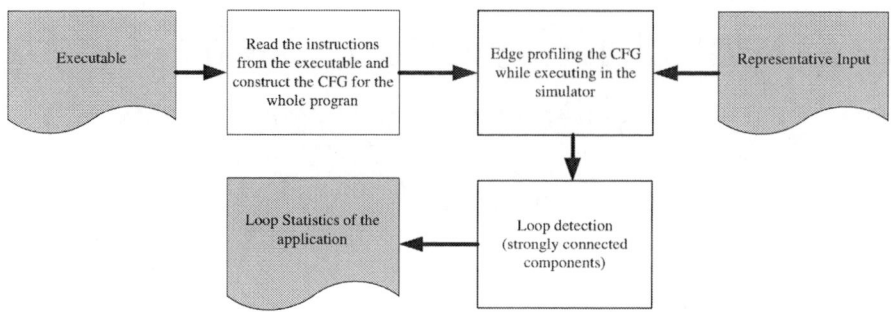

Fig. 1. Basic Operations of Loop Profiler

[9]. In addition, an analytical algorithm also was proposed to analyze the data memory references which is guaranteed to satisfy designer provided performance constraints [4]. Reuse distance [24] was also used to estimate the capacity miss rates of a fully associative cache [24]. These techniques are targeted towards data caches.

In contrast, our proposed methodology uses loop characteristics of the application such that capacity misses in the L1 cache is minimized. The next section presents our loop profiling methodology before presenting the cache size optimization methodology.

III. LOOP PROFILING

The central challenge facing computer architects, compiler writers and application programmers is to understand the program's dynamic behavior so that the architecture and the program can be optimized for the efficient execution of the program. Events that occur when a program executes are often indefinable, but they provide a basis for understanding the behavior and improving its performance. This is particularly useful in embedded systems where the application which is going to be executed in the system is known in advance. Thus, the architecture can be tuned for the efficient execution of the application.

Loop profiling is often done in embedded systems for different reasons such as hardware software partitioning [14] and optimization. Compilers frequently use profile-directed optimizations to find the hot-paths within the applications to perform optimizations. Instruction set customization also depends on identifying critical code segments for optimization. Thus, loop profiles for the embedded applications are readily available in most cases. When the loop statistics needed for the application is not available, the linked binary of the applications can be profiled to extract the loop characteristics of the application.

Several profiling tools have been developed to study the dynamic behavior of applications. Some tools like gprof [1] provide function level profiling and do not provide sufficient information. Tools such as ATOM [12] and SpixTools [3] are specific to a particular microprocessor family and these tools do not support the ARM ISA. For this reason, we implemented our own loop profiler to extract the loop statistics from the executable.

A loop profiler can be implemented either by instrumenting the binary to extracting the appropriate information or by extracting the information while executing the application in an instruction set simulator. In the second method, the binary is executed unmodified, leading to more accurate results although execution is slower,. Our loop profiler has been implemented using the second method. Our implementation, when coupled with SimpleScalar [2] instruction set simulator generates the loop profile of a given ARM program.

Figure 1 shows the basic steps in our loop profiling framework. We extract the Control Flow Graph (CFG) for the whole program, as a CFG for each procedure as well as a call graph for the procedures. We use the GNU *libbfd* library to read the various code segments from the executable. Constructing CFG's from schedule executable posses considerable challenges and has been studied in [16], [6]. The most notable difficulties are the indirect branches, delay slots and data inside code segments. For the indirect branches, if the target for the branches cannot be resolved, we resolve it at run time while executing the executable in the simulator. Then, the edge profile for the application is constructed when the program is executed in the simulator.

Once we have the CFG with the edge profile, we can calculate the loops (strongly connected components) in the flow graph using Tarjan's [15] algorithm. From the basic block list corresponding to the loop, we can extract the other information such as static instruction count of the loop, dynamic instruction count of the loop, total execution time of the loop, cache size required to contain this loop etc. from the list.

The next section presents the proposed loop profiler directed methodology for tuning the instruction cache.

IV. INSTRUCTION CACHE SIZE OPTIMIZATION METHODOLOGY

Cache memories work based on the principles of locality. If we look at the instruction accesses, the instructions are accessed consecutively till a control flow instruction (a taken branch instruction) is reached. Thus, the instruction access exhibits high spatial locality. The temporal locality in instruction

0-7695-2533-4/06 $20.00 © 2006 IEEE

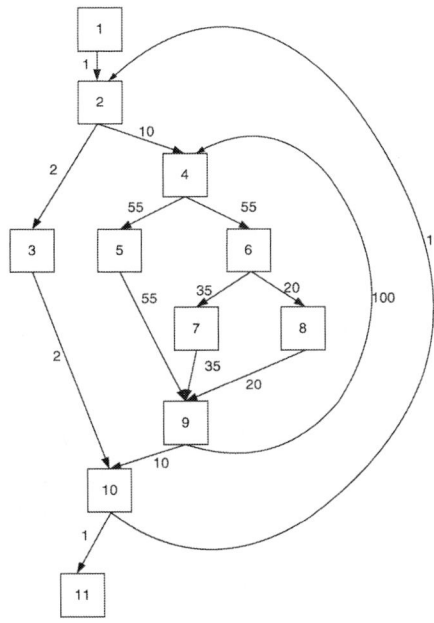

Fig. 2. CFG of a Typical Application

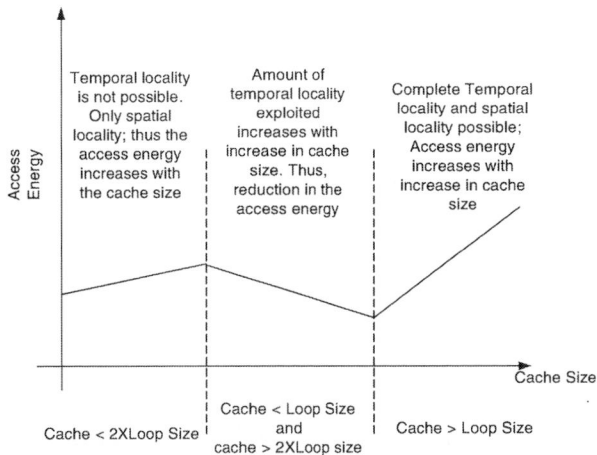

Fig. 3. Access Energy vs. Cache Size for a Single Loop Depending on the Relationship Between the Loop Size and the Cache Size

access comes either from the loops in the instruction stream or from the function (subroutine) calls to the same routine from different places. Thus, any instruction which is not part of any loop or function call need not be cached once all the instructions in the line are executed, as only spatial locality is possible. Any instruction that is part of loop or function call should be in the cache till they are needed for exploiting temporal locality.

For example, in the Figure 2, while executing the basic block 2, we no longer need basic block 1. This is because basic block 1 is not part of any loop. However, basic block 2 must be cached till the basic block 11 is reached. That is, if a basic block is part of a loop, it must be cached till the loop is exited. If a basic block is part of multiple loops, it must be cached till all the loops are exited. In the same way, if a basic block is part of a sub-routine it must be cached till the sub-routine is called for the last time. Our analysis with the embedded applications shows that subroutines that are not part of loops are often executed less frequently. Such sub-routines are called to handle unexpected conditions. Conversely, subroutines that are part of loops are executed more frequently.

However, in reality, we never design a system that has zero misses, excluding the misses resulting from the first time access, as this implies a huge cache. Thus, we cannot cache all the basic blocks in the loop. Instead, we need to cache the basic blocks in the hot-paths (the basic blocks in the path which are executed frequently) within the loop. For example if we look at the Figure 2, in the outer loop, the basic block 3 is executed only twice. Caching the non-frequent basic blocks will increase the cache size and access energy compared to the savings obtained with decreasing the main

memory access energy. Thus, we need to use a threshold value to filter out the basic blocks in a loop that are not executed frequently. In our experiments we use a threshold value of $0.5 \times \frac{Dynamic\_instruction\_count}{Static\_instruction\_count}$. This is based on the observation that on an average, each instruction executes for $\frac{Dynamic\_instruction\_count}{Static\_instruction\_count}$ times, thus, if an instruction executes less than the average number of times, we need not cache it. In our experiments, we use a threshold that is 50% of the average execution of an instruction. Some instructions in the hot path execute more frequently than the average, and some instructions in the non-frequent path execute much less frequently than the average value. Thus, in general, changing the threshold in the range of the average value does not affect the filtering of frequent basic blocks. So we use 50% of the average.

$$Threshold = 0.5 \times \frac{Dynamic\_instruction\_count}{Static\_instruction\_count} \qquad (1)$$

Our goal is to find the near-optimal cache which will minimize the energy consumed in the L1 cache and the external bus access. Therefore, to calculate the cache size needed for a given application, we have to calculate the relative access energy for each possible L1 cache size and select the L1 cache with the minimal access energy. We use an energy cost model for calculating the energy for a given $L1\_size$ with a particular loop. That is, for a given $loop\_size$, the L1 cache will have different energy cost depending on whether the loop is fully containable, partially containable or not containable at all. Here, we assume that L1 cache miss takes 50 times more energy for cache refill as compared to the L1 cache hit as shown in the literature [22].

Assuming that there are no conflict misses, as our cache is highly associative, an L1 cache of $L1\_size$ and loop of $loop\_size$, the $Cost(loop\_size)$ can be calculated as follows:
When the complete loop is cache resident : all the

0-7695-2533-4/06 $20.00 © 2006 IEEE   279

instructions will be hit in the L1 cache as temporal locality and spatial locality are possible:

$$C1 = 1.0 \times E_{L1} \times T_{ex} \qquad (2)$$

Here, T_{ex} stands for the execution spent in this particular loop as a fraction of the total execution time. This includes the effect of nested loops as discussed earlier. Since the complete loop is cached, it is multiplied by 1.0. And E_{L1} is access energy for L1 cache for a particular cache size. When the loop size is completely cache resident, increasing the cache size beyond will just increase the cost as E_{L1} will continues to increase (As shown in Figure 3)

When the partial loop is cache resident : When loop_size is greater than the L1_size and less than $2 \times L1\_size$, then the portion of the loop which can not be contained (NCR or Non Cache Resident part) is:

$$NCR = \frac{2 \times L1\_size - loop\_size}{loop\_size} \qquad (3)$$

Temporal locality is possible for the part which can be contained in the cache (1 - NCR times). For the rest (NCR), only spatial locality is possible. And also, off-chip accesses have to be made with 50 times more energy. Since the L1 cache has a 32byte line size and the instruction width is 4bytes, even when temporal locality is not possible, a maximum of seven out of eight instructions will hit in the L1 cache which is equal to 87.5%. That is for the part which can not be contained we still can have a maximum of 87.5% of the instruction accesses being a hit in the L1 cache.

When the loop is partially cache resident, increasing the cache size will exploit more and more temporal locality and cost will decrease till the full loop is cache resident (As shown in Figure 3)

$$C2 = (1 - NCR) \times E_{L1} \times T_{ex} + (0.875 + 0.125 \times 50) \times NCR \times E_{L1} \times T_{ex}$$
$$(4)$$

When no temporal locality is possible : If the loop_size is greater than $2 \times L1\_size$, there will not be any temporal locality possible. Since the complete loop is not cached, only spatial locality is possible so a maximum of 87.5% of the instructions will be hit in the L1 cache and 12.5% of the time the main memory access has to be made with 50 times more energy. When the loop size is such that the temporal locality is not possible, increasing the cache size will just increase the cost as E_{L1} will continues to increase (As shown in Figure 3)

$$C3 = (0.875 \times 1.0 + 0.125 \times 50 \times 1.0) \times E_{L1} \times T_{ex} \qquad (5)$$

Therefore, the cost for a cache size with a given loop is:

$$Cost(loop\_size) = \begin{cases} C1 & \text{if}(loop\_size < L1\_size) \\ C2 & \text{if}(loop\_size > L1\_size) \\ & \text{and } loop\_size < 2 \times L1\_size) \\ C3 & \text{if}(loop\_size > 2 \times L1\_size) \end{cases} \qquad (6)$$

With the loop profile and cost model, we can calculate the cost for each L1 size and then select the L1 cache with the lowest cost as the near-optimal L1 size. When we calculate the L1 cache size based on the loop statistics, we are essentially

Benchmark	Calculated cache size (KB)	Optimal cache size (KB)
adpcm.enc	1	1
adpcm.dec	1	1
gsm.untoast	4	1
gsm.toast	4	16
mpeg2.enc	2	4
mpeg2.dec	4	8
jpeg.dec	4	8
jpeg.enc	8	4
mipmap	2	1
osdemo	16	16
texgen	16	16
pegwit	8	8
pegwit.dec	8	8
pegwit.enc	16	16
epic	1	1
unepic	2	1
g721.enc	8	8
g721.dec	8	8

TABLE I

COMPARISON OF OPTIMAL VS. CALCULATED L1 CACHE SIZES (WITH THE ANALYTICAL MODEL) FOR MEDIABENCH APPLICATIONS

calculating the L1 cache size needed such that the capacity misses are reduced. In the next section we will provide the results for the L1 cache size optimization.

V. PERFORMANCE COMPARISON

For exhaustive simulations SimpleScalar [2] tool set with ARM Instruction Set Architecture (ISA) was used in these experiments. To measure the per-access energy, CACTI version 3.0 [18] was used with parameters set to 0.18μm technology.The processor model simulated was a single issue processor typically found in general purpose embedded processors where instructions are fetched from the cache hierarchy every clock cycle, if an instruction cache access results in a miss then it will result in a pipeline bubble. Also, the first level instruction access is in the critical instruction access path. 16 KB L1 cache with 32 byte line size and 8 way set-associative is used in all the experiments as the base case. Set of benchmarks from MediaBench [8] were selected as they characterize embedded applications working sets for high performance embedded processors. All the applications were compiled with gcc 2.93 and were run to completion.

Figure 4 compares the relative access energy obtained with the analytical model to that obtained with the simulation. Both the calculated and simulated values are normalized such that the minimum access energy is one. As can be seen from Figure 4, in general, the results obtained matches the analytical model quite closely. However, for smaller cache sizes the deviation is noticeable compared to larger cache sizes. This is because, in developing these models, we assumed that there are no conflicts between the instruction lines in the loop. And also, we select the basic blocks from the hot-paths are to keep in the cache with an approximate threshold based filtering technique. As a result, for some cases, the model does not follow the actual values closely for smaller L1 cache sizes, as it is evident

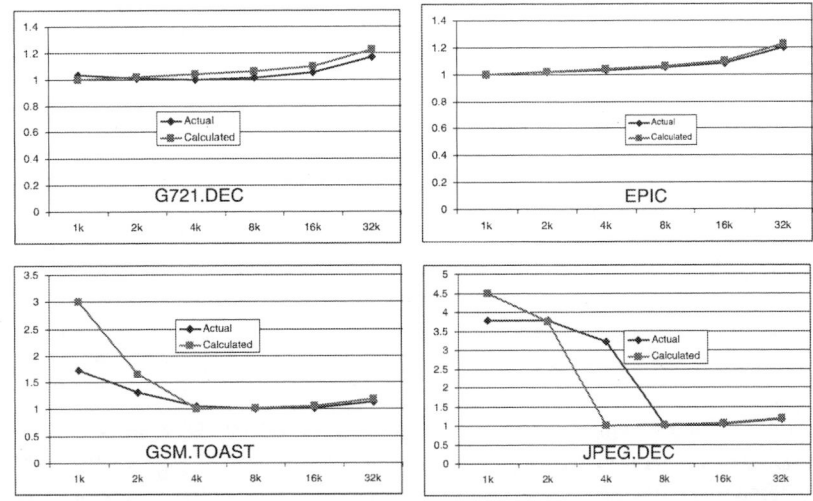

Fig. 4. Comparing Relative Access Energy with Calculated and Simulated Values for L1 Cache

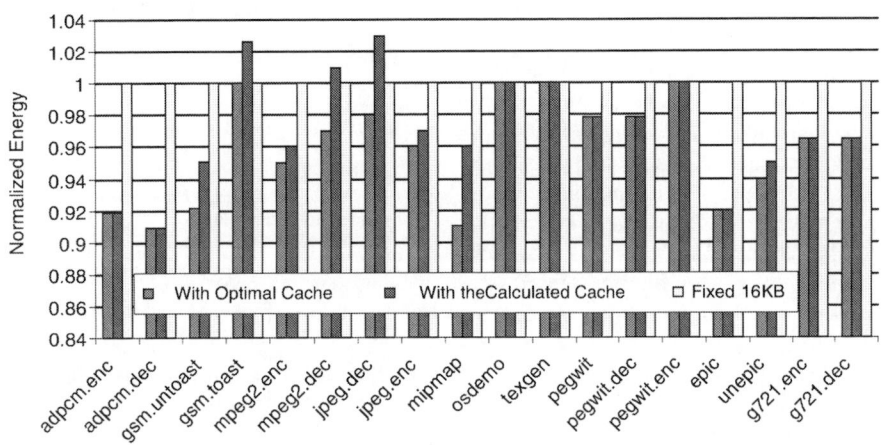

Fig. 5. Energy Reduction with Optimal vs. Calculated L1 Cache Sizes for MediaBench Applications

from JPEG.DEC. As a result, in few cases, when the difference between the optimal and next one is minimum, the model might find the next best configuration.

Table I compares the optimal cache size which is obtained by simulation with that obtained with the loop statistics. Here, as explained earlier, the optimal L1 cache sizes are calculated with the 32byte line size and 8-way associativity with various cache sizes. For calculating the energy consumption in the L1 cache and off-chip bus access, we used the method used in[22]. And also, since our focus in this section is L1 cache, we have excluded the filter cache altogether in the calculation.

Figure 5 compares the energy savings obtained with the optimal L1 cache size with that from the analytical model. As can be seen from the Figure 5, the analytical model settles for a cache size which results in energy reduction close to the optimal configuration. Overall, tuning the cache size with the proposed methodology results in up to 9% energy reduction

as compared to using a fixed cache size of 16KB. The average energy gain for the applications simulated is 3% as compared to the optimal value of 4%.

Although, the values calculated with the loop statistics are not always optimal, the energy savings are very close to those for the optimal case. Therefore, this along with the configurable cache architecture (discussed earlier) can be easily applied to get energy reduction in the instruction cache hierarchy. Hence, a methodology directed by simple loop statistics can be effectively used to deduce the cache size in an application specific manner.

It should be noted here that we do not include the leakage energy in the cost model we presented in this section. This is because, it is understood that using a custom L1 cache as against using a universal big cache will result in reduction in the leakage energy. As can be seen from the Figure 6 which is obtained with HotLeakage [23], the leakage power

0-7695-2533-4/06 $20.00 © 2006 IEEE

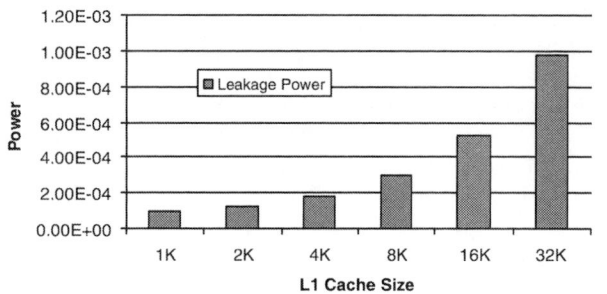

Fig. 6.   Leakage Power (W) for Various L1 Cache Sizes

is proportional to the L1 cache. Therefore, having a smaller L1 cache will reduce the leakage power. However, the leakage power can be very easily incorporated in to the model. Thus, the leakage energy also could be saved as compared to using a fixed L1 cache, which is large enough to contain the working set of all the applications.

VI. CONCLUSIONS

In this paper we have shown that by using readily available loop statistics, near optimal instruction cache size for maximum energy reduction can be realized. The proposed methodology employs simple analytical models, based on applications loop characteristics, to facilitate this. The methodology finds this cache configuration in matter of seconds, without actually simulating the cache configurations. As a result, the custom instruction cache design methodology can be used seamlessly for the design space exploration without increasing the design time.

REFERENCES

[1] *The GNU Profiler.* http://www.gnu.org, 2005.
[2] D. Burger, T. M. Austin, and S. Bennett. Evaluating future microprocessors: The simplescalar tool set. Technical Report CS-TR-1996-1308, 1996.
[3] R. F. Cmelik. Spixtools, introduction and user's manual. Technical Report TR-93-06, Sun Microsystems Inc, 1993.
[4] A. Ghosh and T. Givargis. Analytical design space exploration of caches for embedded systems. In *DATE '03: Proceedings of the conference on Design, Automation and Test in Europe*, pages 10650–10656. IEEE Computer Society, 2003.
[5] M. D. Hill and A. J. Smith. Evaluating associativity in cpu caches. *IEEE Trans. Comput.*, 38(12):1612–1630, 1989.
[6] D. Kastner and S. Wilhelm. Generic control flow reconstruction from assembly code. In *LCTES/SCOPES '02: Proceedings of the joint conference on Languages, compilers and tools for embedded systems*, pages 46–55, New York, NY, USA, 2002. ACM Press.
[7] S. Laha, J. H. Patel, and R. K. Iyer. Accurate low-cost methods for performance evaluation of cache memory systems. *IEEE Trans. Comput.*, 37(11):1325–1336, 1988.
[8] C. Lee, M. Potkonjak, and W. H. Mangione-Smith. Mediabench: A tool for evaluating and synthesizing multimedia and communicaton systems. In *International Symposium on Microarchitecture*, pages 330–335, 1997.
[9] P. R. Panda, N. D. Dutt, and A. Nicolau. Architectural exploration and optimization of local memory in embedded systems. In *ISSS '97: Proceedings of the 10th international symposium on System synthesis*, pages 90–97. IEEE Computer Society, 1997.
[10] P. R. Panda, N. D. Dutt, and A. Nicolau. Data cache sizing for embedded processor applications. In *DATE '98: Proceedings of the conference on Design, automation and test in Europe*, pages 925–926. IEEE Computer Society, 1998.

[11] T. R. Puzak. *Analysis of cache replacement-algorithms.* PhD thesis, 1985.
[12] A. Srivastava and A. Eustace. Atom: a system for building customized program analysis tools. *SIGPLAN Not.*, 29(6):196–205, 1994.
[13] R. A. Sugumar and S. G. Abraham. efficient simulation of multiple cache configurations using binomial trees. *Technical Report CSE-TR-111-91*, 1991.
[14] D. C. Suresh, W. A. Najjar, F. Vahid, J. R. Villarreal, and G. Stitt. Profiling tools for hardware/software partitioning of embedded applications. In *LCTES '03: Proceedings of the 2003 ACM SIGPLAN conference on Language, compiler, and tool for embedded systems*, pages 189–198. ACM Press, 2003.
[15] R. Tarjan. Testing flow graph reducibility. In *STOC '73: Proceedings of the fifth annual ACM symposium on Theory of computing*, pages 96–107, New York, NY, USA, 1973. ACM Press.
[16] H. Theiling. Extracting safe and precise control flow from binaries. In *RTCSA '00: Proceedings of the Seventh International Conference on Real-Time Systems and Applications (RTCSA'00)*, page 23, Washington, DC, USA, 2000. IEEE Computer Society.
[17] W.-H. Wang and J.-L. Baer. Efficient trace-driven simulation method for cache performance analysis. In *SIGMETRICS '90: Proceedings of the 1990 ACM SIGMETRICS conference on Measurement and modeling of computer systems*, pages 27–36. ACM Press, 1990.
[18] S. Wilton and N. Jouppi. Cacti: An enhanced cache access and cycle time model. Technical report, 1996.
[19] Z. Wu and W. Wolf. Iterative cache simulation of embedded cpus with trace stripping. In *CODES '99: Proceedings of the seventh international workshop on Hardware/software codesign*, pages 95–99. ACM Press, 1999.
[20] C. Zhang. Balanced instruction cache: Reducing conflict misses of direct-mapped caches through balanced subarray accesses. In *IEEE TCCA Computer Architecture Letters, 2005.* IEEE Computer Society.
[21] C. Zhang and F. Vahid. Cache configuration exploration on prototyping platforms. In *RSP '03: Proceedings of the 14th IEEE International Workshop on Rapid System Prototyping (RSP'03)*, page 164. IEEE Computer Society, 2003.
[22] C. Zhang, F. Vahid, and W. Najjar. A highly configurable cache architecture for embedded systems. *SIGARCH Comput. Archit. News*, 31(2):136–146, 2003.
[23] Y. Zhang, D. Parikh, K. Sankaranarayanan, K. Skadron, , and M. Stan. Hotleakage: A temperature-aware model of subthreshold and gate leakage for architects. *Technical Report CS-2003-05*, 2003.
[24] Y. Zhong, S. G. Dropsho, and C. Ding. Miss rate prediction across all program inputs. In *PACT '03: Proceedings of the 12th International Conference on Parallel Architectures and Compilation Techniques*, pages 79–90. IEEE Computer Society, 2003.

Complexity and Low Power Issues for On-chip Interconnections in MPSoC System Level Design.

Yuriy Sheynin, Elena Suvorova, Felix Shutenko
St. Petersburg State University of Aerospace Instrumentation.
St. Petersburg, Russia
sheynin@online.ru

Abstract

System level design for many-core chips includes general system architecture design, MPSoC design as a set of nodes and there interconnection. It requires adequate models and methodology to estimate MPSoC interconnection characteristics in complexity and power consumption to make decisions at the system level design stage of an MPSoC project. To determine performance and power consumption of MPSoC we suggest the simplified model of tentative wires for evaluating interconnection topology at early stages of design process, to correspond them with technology characteristics, with power dissipation and power consumption. With this approach we reason about many-core SoC interconnections, place and route, methodology for system level design of MPSoC interconnections.

1. Interconnections in many-core SoC

Multiprocessor systems-on-chip (MPSoC) are moving from simple multi-core structures with shared memory to many-core structures with distributed architecture.

Communication system of MPSoC can be based on bus interconnections, direct internodes interconnections and indirect internodes interconnections (using switches). Due to performance and power limitations buses may be used only in in future MPSoC with number of nodes less than a dozen, [1, 2].

In many-core MPSoC a chip includes a set of computing *nodes*, processing elements (PE) with some memory, and direct interconnections between the nodes.

Nodes and links between them form some mesh, a graph of nodes and links (it is common to call it

"interconnection topology"). Nodes exchange data and control information by packets, messages, transactions interchange. Characteristics of a mesh determine many important characteristics of an MPSoC as a whole. As the transistor size is scaling down, interconnect becomes more important in chip design. For instance, in 0.13 um technology already a delay caused by 1 mm interconnect is larger than the gate delay. Design moved from chip-wide synchronous circuitry to GALS (globally asynchronous, locally synchronous) many-core chips design.

Defining and optimization of interconnections for distributed architecture MPSoC is an important problem in MPSoC system level design. It is crucial to balance nodes and interconnection characteristics, along with MPSoC properties, tasks requirements and VLSI technology features. Problem of power dissipation and power density is also very important for systems-on-chip in deep submicron technologies.

The problem is that many characteristics of on-chip wiring, its routing and placement cannot be defined at the system level stage of an MPSoC design. Nevertheless, system level design should include development and balancing a structure of a many-core MPSoC, its nodes and interconnection topology between them. General models and based on them methodology are required for system-level interconnection design in prospective MPSoC.

Along with interconnection topology development in system level design one should select a standard for its links (and switching nodes, if any) implementation. Serial links as well as standards with parallel, multi-lane links are used in MPSoC now. Different standards require different chip resources for links and nodes implementation, have different maximal link length; different rates, different power consumption and power density.

0-7695-2533-4/06 $20.00 © 2006 IEEE

2. Models in MPSoC system level design

To estimate performance characteristics of MPSoC we use models, based on the transactor approach with MPSoC modeling at the transaction level. An elemental model of an MPSoC includes a model of a PE with transactors as interfaces and a transactor based interconnection model. Model of the interconnection is also based on transactors. One and the same PE model can be included in different MPSoC models: in meshes with different topologies, in networks based on different interconnections standards. To reuse a PE model in different network topology it should be designed with variable valence.

An MPSoC interconnection can be a mesh with switching nodes. In this case the switching node model also includes interface transactors that can be replaced to adapt the switching node model to different chip interconnection standards.

A system level model of an MPSoC can be used to estimate performance characteristics of the many-core SoC in design, to analyse and compare complexity of different MPSoC design variants. It can be used also for estimation of power consumption and power dissipation characteristics.

At the system level we define power as a function of number of messages that are sent through interconnection between the MPSoC nodes and energy that is expended in a message transfer from node to node. Thus a dependence of MPSoC power characteristics on interconnection graph parameters (mean graph diameter, graph thickness, etc.) is established.

Several on-chip schematic designs may correspond to one and the same interconnection topology. Geometric parameters of interconnection links have an effect on power consumption and power density.

An MPSoC system level model should take into account interconnection links geometric parameters.

Selection of an interconnection standard for an MPSoC project is also a part of its system level design. An MPSoC system level model should reflect specifics of particular interconnection standard, which is selected for the MPSoC, also.

3. Tentative wires model

As at the system level design stage designers do not have place and route characteristics of a MPSoC design one should use some system level simplified models for on-chip interconnections.

In the Thompson model, [2], a network is represented by a graph. Its nodes correspond to processing elements (PE); its edges correspond to wires. The graph is embedded in a 2-D grid. This grid consists of basic squares with a unit size. (the unit depends on the impelmebtation technology). All wires are considered to have a unit width also.

Every PE occupies a square of h units side (h^2 units area). The wires can run horizontally or vertically along grid lines. Two wires can cross each other at a grid point. Lines should not overlap or bend at the same grid point, which would form a knock-knee. With two layers of wires, one layer can be used to lay out all the vertical segments; another layer can be used to lay out all the horizontal segments. When a wire makes turn, its horizontal and vertical parts in different layers are connected by via holes. This model is used to estimate area of a network design. The network design area is defined as the area of the smallest rectangle that may contain all its nodes and wires.

In the multilayer Thompson grid model, [3], a network is represented just as in the classical Thompson model. The nodes of the graph are embedded in the first layer of the 2-D grid. The second and other layers are used for wires only. This model more accurately corresponds to situation when the technology allows or requires placement of nodes and wires in different layers.

However, these models are too crude for modern MPSoC system level design. For deep-submicron technology not only performance but power dissipation and other technology parameters are very important and should be taken into account at the system level design stage.

At the system level design stage a designer quite often doesn't know yet nodes' area, ports placement in nodes' designs, etc. It is possible that designer doesn't know for sure what technology will be used for the SoC implementation in a chip.

We suggest a simplified model of *tentative wires* for evaluating interconnection topology characteristics in SoC system design in correspondence to basic technological parameters.

In our model, like in the Thompson model, a MSoC graph is embedded in a 2-D grid of a chip, Figure 1. Nowadays width of wires is essentially less than linear size of PEs. Therefore we neglect by the width of wires. In the tentative wires model the PEs' squares are placed in the grid without intervals. Tentative wires join mechanical centers of the PEs' squares. This placement of tentative wires allows to put aside in the meanwhile real placements of interconnection ports in PEs designs. Wires can run like in the Thompson model.

For a MSoC interconnection graph the tentative wires model allows to estimate the average length of wires, the maximal length of wires, the summary length of wires (as functions of the node linear size) and the number of via holes.

This information allows to evaluate convenience of the MPSoC topology realization with the chosen technology, to estimate wiring delays and power dissipation. Designers can use the tentative wires model to estimate the length of wires and geometry characteristics of chip schematics design.

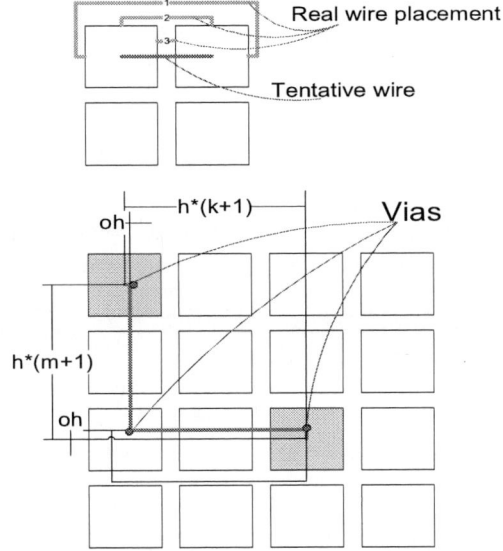

Figure 1. Tentative wires and vias

With tentative wires model a corresponding MPSoC system level model, which takes into account interconnection links geometric parameters and power parameters, can be developed and estimated.

4. MPSoC models with tentative wires

For three basic variants of placement of joined PEs the tentative wires length can be estimated:

1. Length of a tentative wire, L_u, between two PEs, that are vertical or horizontal neighbours is equal to the linear size of PE.

2. If the joined PEs are not neighbours but lie in one and the same row or in one and the same column then $L_u=(k+1)*h$, where k is the number of PEs between the joined PEs.

3. If the joined PEs lie in different rows and/or columns then $L_u = (k+1 +m +1)*h$, where k is number of PEs that lie between the joined PEs in row, m is number of PEs that lie between the joined PEs in column.

The simplified model of tentative wires can be used to estimate the number of via holes in the SoC interconnection design also. In our model we suppose that all parts of tentative wires that run vertically are placed in one metallization layer; all parts of tentative wires that run horizontally are placed in another metallization layer. It allows us to simplify our model. Instead of real via holes we use *tentative via holes*.

Number of tentative via holes can be estimated on the basis of the graph of system level MSoC representation. This number of tentative via holes can be grater than number of via holes in the real design.

If a tentative wire runs only horizontally or only vertically it is supposed to be placed in one metallization layer. If a tentative line consists of two parts (one of them runs horizontally, another runs vertically), it is supposed to have one tentative via hole. Number of tentative via holes for one tentative wire, N_u, is equal to number of tentative wire parts minus one. For a placement of the joined PEs:

1. If two joined PEs are vertical or horizontal neighbours the tentative wire between them runs vertically or horizontally. without via holes.

2. If the joined PEs are not neighbors but lie in one and the same column or in one and the same row, the tentative line can also run vertically or horizontally respectively.

But if in the row (or in the column) several pairs of joined PEs are placed, then some tentative lines try to be placed in one place and thus cause a collision. To avoid it we propose to consider these tentative lines as lines with three parts: begin part, main part and end part. If a tentative line runs vertically then the main part runs also vertically, the begin and the end parts run horizontally. If the tentative line runs horizontally then the main part runs also horizontally, the start part and the end part run vertically. The length of the main part is equal to the length of the tentative line. The length of the start part and of the end part are infinitesimal. These parts are needed to depose tentative line from the mechanical centers of the joined PEs. For such a tentative line $N_u=2$. Sometimes this value may be an overvaluation.

3. If the joined PEs lie in different rows and columns then $N_u=3$, because maximal number of parts for this tentative line is four. Similar to the variant 2, this line includes infinitesimal start and end parts. But it includes two main parts – a horizontal one and a vertical one.

With system level graph analysis done in the scope of the tentative wires model some information about the selected technology is required for further reasoning about interconnection characteristics. We need to know how wire width depends on wire length, how wiring delay and power dissipation depends on wiring geometry and via holes. Real number of via holes can depend on number of metallization layers used for wires in the selected technology.

With the simplified model of tentative wires we can evaluate number of metallization layers that are needed to place design onto chip and number of via holes

0-7695-2533-4/06 $20.00 © 2006 IEEE   285

(vias). We can evaluate also of network topology parameters for different technologies.

It will help designers to select appropriate interconnection graph for a multi-core/many-core SoC and to select appropriate technology for its implementation at the system level design stage.

5. Wires length and power characteristics in MPSoC interconnections

Geometry size of a chip to place the design (design layout) depends not only on interconnection pattern but on VLSI technology also. In many technologies the chip size is in some relation with the summary area used for interconnections on the chip. The latter parameter we can estimate with the tentative wires model. The summary length of tentative wires on a chip, L_s,:

$$L_s = \sum_{i=1}^{Q} L_{u_i} \qquad (1)$$

where L_{u_i} – the length of the tentative wire i.

Summary length of tentative wires is an important parameter for all VLSI technologies, it has strong affect on power consumption.

If for a selected VLSI technology maximum wire length is strongly constrained, additional parameter – max tentative wire length, L_m, is used.

$$L_m = \max_{i=1}^{Q} L_{u_i} \qquad (2)$$

Number of vias, N_s, for design is estimated as:

$$N_s = \sum_{i=1}^{Q} N_{u_i} \qquad (3)$$

where N_{u_i} – a number of vias for the wire i.

We can estimate power dissipation in interconnection system using our model of tentative wires. The ideal packet transmission energy is that dissipated by a dedicated wire linking source and destination, corresponding to the minimum physical activity required to complete the transmission [4]:

$$E_{ideal} = E_{wire} D \qquad (4).$$

Dissipated power in a wire depends on the line length, width, thickness and number of vias, and on VLSI technology to be used in chip production:

$$P_{u_i} = T_w L_{u_i} + T_v N_{u_i} \qquad (5).$$

where T_v – technology parameter that depends on vias implementation technology;

T_w – technology parameter; depends on technology library characteristics (materials of wire and dielectric, wire width and thickness).

$$T_w = \frac{\rho}{W * t} \qquad (6)$$

where ρ – constant, that characterizes materials; W – wire width; t – metal thickness, [5].

Summary power dissipation of the interconnection can be estimated as:

$$P_s = \sum_{i=1}^{Q} P_{u_i} \qquad (7)$$

All wires have resistance that represents the ability of the wire to carry a charge flow. Wires also have capacitance that represents charge, which must be added or removed to change the electric potential on the wire; wires also have inductance. A long delay of wires indicates a problem caused by their large resistance. Delay of an uninterrupted wire grows quadratically with the wire length. Designers can add repeating elements periodically along the wire. It makes total wire delay equal to the number of repeated segments multiplied by the individual segment delay; total wire delay is hence linear with total wire length [8]. Thus, in our model, L_{u_i} consists of several segments and a number of via holes.

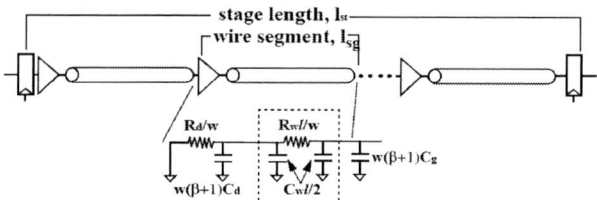

Figure 2. First-order model of a wire.

Figure 2 shows a first-order model of a wire. Here, L is the length of the wire, L_{st} is the length of one stage, L_{sg} is the length of a wire segment; w and βw are the widths of the repeater NMOS and PMOS transistors, (normalized by the minimum width). C_w, C_d, and C_g are the unit-length wire cap and drain and gate caps of the minimum-sized NMOS transistor respectively. R_w and R_d are the unit-length wire resistance and the resistance of the minimum-sized repeater respectively [9].

The wire can be approximated by a lumped capacitance of value C_w. In this approximation, the wire is modeled as a capacitive load. With this wire load model, it was shown that the propagation delay is linearly proportional to the wire length [10]. Thus, power, which is consumed for transmitting message from transmitter node to receiver node, is proportional

0-7695-2533-4/06 $20.00 © 2006 IEEE

to the distance between the nodes. Power consumption of the wire can be related by the propagation delay and tentative wire length between two nodes.

Let average diameter of the interconnection graph be equal to D_s. This means that, on the average, message passes D_s edges of the graph and D_s-1 nodes before reaching the target node. Average time of message transmitting is

$$T_s = (D_s-1) \cdot T_o + D_s \cdot T_r, ; \qquad (8)$$

T_o - the waiting period before processing message inside transit node, T_r - the average time of message transmition to the next transit node. Thus, the average message transmition time has direct proportional dependence on meshaverage diameter.

The energy consumed when transmitting a data packet depends on average energy dissipated when writing a packet into the input buffer, average energy dissipated when reading a packet from the input buffer, average arbitration energy, average link traversal energy and the number of nodes traversed by this packet [4].

The thermal resistance of via holes, which rises per watt of power dissipation, is a function of via diameter and the thickness of the plating on the walls of the via holes:

$$T_w = \frac{l}{K \cdot S} ; \qquad (10)$$

where: K - thermal conductivity (W/m-ºK); l - length (or thickness) of object (m); S - Cross sectional area (m^2).

Using (5), (6), (7), (10), length of a tentative wire, and number of vias we can estimate power dissipation of the MPSoC interconnection.

6. MPSoC nodes placement

One MPSoC topology can have different designs. Designs of nodes could be placed in different places on the chip. They could be placed all in one group of layers or in two different groups of layers, [6, 7].

Design of a PE takes 2-3 metal layers. With metal 6-7 layers available there are two main placements of PE on a chip:

1.   All the nodes are placed in one group of metal layers as a rectangular matrix.

2.   Nodes are divided into subsets; each subset is placed in its own group of metal layers.

Processing nodes can be placed in a rectangular matrix in different ways also, with different number of metal layers, line length, number of vias, etc.

There are some MPSoC nodes placement methods to reduce wire length and number of via holes. For interconnections, which graphs can be disassembled into subgraphs whith structure analogous to the original graph, clusterization method can be used, [11, 12]. An original graph is disassembled into subgraphs, they are disassembled again to subgraphs, etc., until next subgraphs are planar. Designs for these flat subgraphs are created. Next design for the original graph is constructed with designs of subgraphs. This method is usually used for hypercubes, butterfly and star topologies.

To reduce the number of long wires and the summary length of wires the *node interleaving* approach can be used. Initially this approach was developed for bidelta networks, [13, 14]. At the first phase, nodes are placed in a 2D grid. Every column of the grid includes one level of bidelta network. Long interconnection lines can exists between nodes in neighbouring columns if one of top nodes in one column connects with one of bottom nodes in the next or the previous column. In this case length of wire can be equal to chip size. To avoid too long wires nodes in columns are interleaved. This method allows to reduce maximum wire length to ¼ of chip size, [14].

We suggest to use nodes interleaving method for 3D-mesh and tore designs. Let us disassemble 3D-mesh M×N×L to 2D-meshes size M×N.

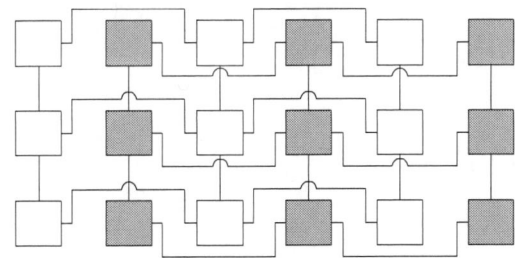

Figure 3.

We can interleave nodes of neighbor 2D-meshes, Figure 3. It allows to reduce wire length between the neighbour 2D-meshes.

White rectangles represent nodes of one of the odd 2D-mesh, gray ones represent nodes of the even 2D-mesh, that is neighbour of the former 2D-mesh.

We use interleaving for tore nodes placement in design. Customary tore placement without interleaving is shown at Figure 4. Length of wires between nodes in the first and the last column is equal to the chip size, and length of wires between nodes in the first and the last row is equal to chip size also.

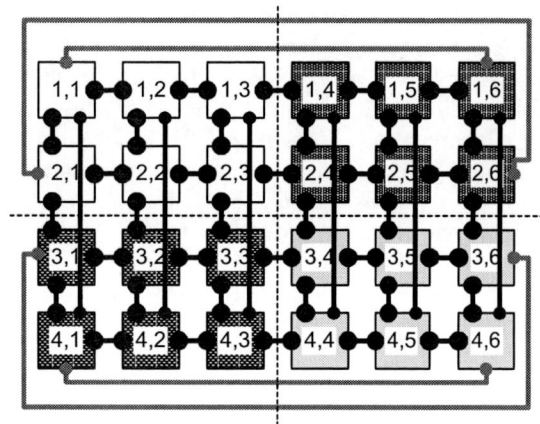

Figure 4.

To reduce the length of wires we use Placing nodes with interleaving, Figure 5.

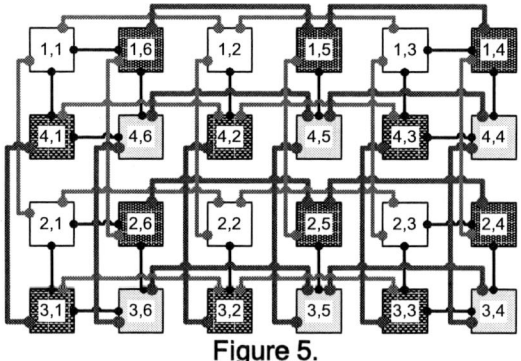

Figure 5.

7. Conclusion

In compliance with defined set of criteria we determine a methodology to select interconnection topology and design structure in respect to MPSoC requirements at the system level design stage:

1. Determination of graph thickness for interconnection topologies to assess possibility of using them for particular VLSI technology.

2. Construction of designs for the selected topologies. Computation of complexity and power consumption parameters for these topologies. Evaluation of scaling possibility.

3. Estimation of message transfer delays for the selected topologies. Selection of an interconnection with smallest mean diameter.

4. Placement of I/O nodes in the mesh (with minimal average message passing latency between internal nodes and I/O).

5. MPSoC nodes placement in the chip design, with placement optimization.

8. References

[1] Winegarden S., "Bus Architecture of a System on a Chip with User Configurable System Logic", *IEEE Journal of Solid State Circuits*. 2000. Vol. 35, No. 3. P. 425-433.

[2] Thompson, C. D., "Area-time complexity for VLSI" *Proc. ACM Symp. Theory of Computing*, 1979, pp. 81-88.

[3] Thompson, C. D., "*A complexity theory for VLSI,*" Ph.D. dissertation, Dep. Of Computer Science, Carnegie-Mellon Univ., Pittsburg, PA, 1980. 178 p.

[4] H. Wang, L.-S. Peh, S. Malik "Power-driven Design of Router Microarchitectures in On-chip Networks", MICRO-36, 2003

[5] N.Srivastava, Xiaoning Qi and K. Banerjee "Impact of On-chip Inductance on Power Distribution Network Design for Nanometer Scale Integrated Circuits", 0-7695-2301-3/05. IEEE 2005

[6] Brady, M. L. and Sarrafzadeh M., "Stretching a knock-knee layout for multilayer wiring", *IEEE Trans. Computers*, vol. 39, no. 1, Jan. 1990, pp. 148-151.

[7] Chi-Hsiang Yen, E. A. Varvarigos, B. Parhami, "The recursive grid layout scheme for VLSI layout of hierarchical networks", Proc. Merged Int'l Parallel Processig Symp. & Symp. Parallel and Distributed Processing. 1999. Apr. P.48-55.

[8] R. Ho et al. "The future of wires". Proceedings of the IEEE, 89(4):490–504, Apr. 2001.

[9] Seongmoo Heo & Krste Asanovic, "Replacing Global Wires with an OnChip. Network: A Power Analysis", CSAIL'2005

[10] I.-B. Dhaou, V.Sundararajan, H. Tenhunen and K. K. Parhi, "Energy efficient signaling in deep submicron CMOS technology", in Proc. ISCAS, 2001, Vol.5, pp.319 -324.

[11] C. H. Yeh, E.A. Varvarigos, B. Parhami, "Efficient VLSI layouts of hypercubic networks", Proc. Symp. Fron-tiers of Massively Parallel Computation. 1999. Feb. P. 98-105.

[12] C.Yeh, B.Parhami, E. Varvarigos, H..Lee, "VLSI layout and packaging of butterfly networks", Proc. ACM Symp.Parallel Algorithms and Architectures. 2000. P.196-205.

[13] Chen G., Lau F. "Layout of the Cube-connected Cycles without Long Wires", *The Computer Journal*. 2001. Vol. 44. P. 374-383.

[14] Kruskal C. P., Snir M. "A unified theory of interconnection network structur", *Theoretical Computer Science*. 1986. Vol. 48. P. 75-94.

Fast Configuration of an Energy-Efficient Branch Predictor

P. Hallschmid and R. Saleh
Department of Electrical & Computer Engineering
The University of British Columbia
Vancouver, BC, Canada
{peterh, res}@ece.ubc.ca

Abstract

Recent research in the area of Application Specific Instruction-set Processors (ASIPs) has focused on automatic configuration. In this paper, we propose a novel approach for selecting the size of the branch predictor pattern history table (PHT) to reduce the overall power dissipation for a specific application. This approach uses a fast configuration approach that dynamically measures aliasing for all PHT sizes in parallel and then uses a cost function that relates aliasing to power dissipation. Results show that by configuring the PHT using our approach, the overall power reduction closely matches that achievable with a "perfect" configuration.

1. Introduction

One level of configurability in an embedded System-on-Chip (SoC) is the Application Specific Instruction-Set Processor (ASIP) where the architecture of the processor is tailored to an application or application domain. Configurability of this type provides SoC designers with the ability to create processors with increased performance and reduced power consumption over general purpose processors.

So far, industrially available ASIPs such as Tensilica Xtensa [1] and ARC [2] have a base architecture with a specific set of configurable parameters. However, many other aspects of the architecture have a significant impact on performance and energy-efficiency but are normally fixed as prescribed by the base architecture. In this paper, we explore another aspect that is worth considering for automatic configuration, namely branch prediction.

Branch prediction has long been an important area of study for high-performance processors because it has a significant impact on the attainable instruction level parallelism (ILP). As processors become wider and pipelines become deeper, the penalty of a misprediction grows. To offset this, branch predictors have become more complex with each generation of processor design. As a consequence, branch predictors now accounts for a significant portion of the total power dissipation (more than 10% of the total processor [3]).

There are many ways in which the branch prediction unit (BPU) can be configured to reduce power dissipation for a specific application. One approach is to automatically choose a low-power branch prediction algorithm. While larger, more complex, branch predictors dissipate more power, they have better prediction rates so the processor takes fewer cycles to execute the application thus resulting in less overall power dissipation. This approach is not taken in this paper. Instead,

the branch predictor is configured by sizing its *pattern history table* (PHT) based on the application. Sizing the PHT has a significant impact on performance and is a less drastic departure from the base architecture.

An increase in PHT size improves the branch prediction rate thus reducing the number of cycles wasted due to branch prediction penalties. With fewer cycles needed to execute the application, the overall processor will consume less energy. On the other hand, an increase in the size of the PHT will increase the energy consumption of the BPU. This is because the PHT is typically implemented as a standard SRAM cell array so an increase in the number of entries will result in more cells, longer bit lines, and larger decoders. Larger bit line capacitances result in an increase in switching power, and an increase in the number of cells results in more leakage power. It is the goal of this paper to configure the PHT size with the correct trade-off between BPU complexity and execution time.

To solve the complete problem, there are many other variables that must be configured in an ASIP to minimize power dissipation. Together, these variables define the solution space. The most accurate way of solving for the optimal point of this solution space is a "brute force" search where every point is evaluated through simulation. However, if there are many variables all with many possible configurations, a "brute force" approach quickly becomes intractable.

To help reduce the potential solution space "explosion", this paper proposes a method of collapsing the dimension defined by the PHT size. This "fast" approach involves evaluating the total power dissipation of the processor with all PHT sizes in parallel during a single simulation. We use a *cost function* that uses run-time information to estimate how changes in PHT size increase the overall power dissipation of the processor (due to an increase in run-time) and how it increases the power dissipation of the BPU (due to an increase in its size).

2. Related Work

An initial investigation of the role of branch predictor organization on power was described in [3]. This work concluded that it is more worthwhile to spend extra power on a more complex branch predictor if it results in more accurate predictions and improves run-time. In spite of this conclusion, both methods proposed in [3] reduce power dissipation solely by reducing the capabilities of the branch prediction unit (BPU). The authors first suggest that power can be saved by banking the branch predictor in a similar way to the way it has been done in the past for caches. Second, they propose a prediction probe detector (PPD) that is used to switch off the BPU for non-branch instructions.

In [4], profiling is used to determine whether each branch instance is "biased" towards global or local predictability. Branch instructions are then encoded with a bit that specifies whether their direction should be predicted using a global type predictor such as GSelect [5] or a local type predictor such as Bimodal [6]. By doing so, they are able to eliminate the need for a meta-predictor which is normally used in hybrid branch predictors [7] to predict which of its predictors are more accurate for the current branch. Power is saved only by reducing the power dissipation of the BPU possibly at the expense of run-time.

The approach taken by [8] is similar to [4] in that the application is profiled to determine which of the gated parts of the BPU should be switched off at run-time. In [8], the branch target buffer (BTB) is resized and parts of the hybrid predictor are disabled. This work is the most similar to this paper in that they configure a table size ([8] sizes the BTB whereas this paper sizes the PHT). In both cases, power dissipation is reduced in part by improving run-time. The two approaches differ in that resizing of the BTB in [8] requires extra hardware for run-time support. Further, [8] sizes the BTB a priori using a "brute-force" trial-and-error approach where they simulate all possible configurations in the search space. Even with just a few dimensions in the search space, this approach quickly becomes impractical. In contrast, this paper uses a more structured approach which scales well as the number of dimensions of the search space increases.

In [9], compiler "hints" are embedded into instructions by a profiler to warn the processor when it should enable the BPU because a subsequent instruction is expected to be a branch instruction. In [10], a branch predictor prediction (BPP) is proposed that selectively switches off two of the three tables in a hybrid predictor using a small buffer that stores the preferred sub-predictor for the most recently encountered branches. In [11], power is saved by dynamically determining which branches are sufficiently well-behaved such that the BPU no longer needs to be accessed.

Previous work has focused on run-time solutions by gating part of the BPU. The processor architecture is fixed but has extra hardware in the pipeline and BPU specifically to reduce power. Conversely, extra bits are encoded into the instructions to provide the processor with "hints" on how to save power. In either case, extra hardware must be added. This work differs in that the architecture of the processor BPU is undecided until the synthesis tool/compiler profiles the application, compares all possible configurations using a cost function (Section 5) and finds the correct trade-off between BPU complexity and execution time. Extra hardware is not needed to provide run-time support.

3. Experimental Platform

Experiments were conducted using the StrongARM architecture [12]. By default, this paper assumes a Bimodal branch predictor with a PHT size of 2048 entries. Each entry is a 2-bit saturated counter that is incremented for branches that are taken and decremented for branches that are not taken. A branch is predicted as taken if this counter is "10" or "11 and as not-taken for "00" or "01". Each entry is indexed using the least significant bits of the branch address.

While the Bimodal predictor is used for this paper, we also study a 1024-entry GSelect predictor and a hybrid predictor. In GSelect, PHT entries are indexed by the branch address concatenated with a branch history register (BHR) which tracks the directions of all recent branches. By doing so, GSelect bases its prediction partially on the global history of all recent branches instructions and not just the local history of the current branch instruction. The hybrid predictor has both a Bimodal predictor for local predictions and a GSelect for global predictions. A second Bimodal predictor, called the meta-predictor, is used to predict which of these two sub-predictors will provide a more accurate result for the current branch instance. In this paper, the default hybrid predictor has sub-predictor sizes of 1024, 2048, and 2048 for the GSelect, Bimodal, and meta-predictor, respectively (based on default values from Panalyzer [13]).

Power and performance experiments were conducted using Panalyzer which is a detailed cycle-accurate power estimation tool that models internal switching power, I/O switching power, and leakage power. Panalyzer adds power estimation functionality onto Simplescalar [14], an instruction-level simulator. For this paper, Panalyzer was modified to support the hybrid predictor.

Benchmark applications are from the SPECcpu2000 Integer Suite [15] and the MediaBench Suite [16]. For SPECcpu2000, only integer applications were used because floating point benchmarks are generally easy to predict and have few dynamic branches. All benchmarks were compiled using the GNU cross-compiler toolset [17]. Each benchmark was simulated to a maximum of 50 million instructions which is long enough to saturate the PHT but short enough to make experimenting feasible.

4. Measuring Aliasing via Simulation

Experiments conducted in [18] found that two-level predictors such as GSelect are close to optimal when implemented with unlimited resources. Due to power, timing, and area constraints, PHTs must allow more than one branch and/or branch history to be mapped to each table entry. This is referred to as *aliasing* or *interference*. When *constructive aliasing* occurs, the BPU correctly predicts the branch direction by "coincidence" whereas *destructive aliasing* predicts the wrong direction [19]. Aliasing has been shown to be the dominating factor affecting branch prediction accuracy [20]. The larger the PHT, the smaller the chance that branches will interfere with one another.

An important claim of this paper is that aliasing can be used to predict how the overall power of the processor will be affected by the size of the PHT. First, we justify this claim by showing that there is a linear relationship between the degree of aliasing measured for a given PHT size and the hit rate of the BPU. Next, we show that there is a linear relationship between the BPU hit rate and the number of cycles of penalty needed to execute the application. Last, we show that the number of cycles of penalty has a linear relationship with the overall power dissipation of the processor. By combining these relationships, we can use aliasing measurements to estimate power and we can therefore choose the best PHT size to minimize power.

For a given simulation, aliasing statistics can be collected for all possible PHT sizes in parallel. A convenient implementation for this is the binary tree data structure shown in Figure 1 which was added to the update phase of the BPU in Panalyzer. The branch address is concatenated with the global branch history and then the r most significant bits are used to index the PHT (enclosed in a dashed box). Each row, r, of the figure represents a different size of the PHT, $S=2^r$, where $r=0...16$. The nodes of the tree at a given level r represent all of the possible table entries for a table of size, S.

When a branch is encountered during simulation, the branch address is concatenated with the global branch history to form the *full index*. The binary tree is then traversed from the root to a leaf based on the bits of the *full index* starting from the least significant bit. In this figure, the shaded nodes define the path taken for the index "01001101". As the tree is traversed, statistics are updated for exactly one node per level.

To measure aliasing, each node of the binary tree structure stores the last branch {address, history} pair from which it was mapped. When a new branch is mapped to a node, the current {address, history} pair is compared to that previously stored. If they differ then aliasing has occurred [19]. In addition to storing a {address, history} pair, each node keeps track of the number of times aliasing occurred. This approach for collecting statistics allows us to perform one simulation to simultaneously measure the aliasing rate for all PHT sizes in parallel by summing the total number of aliasing occurrences for each level (i.e., for each PHT size).

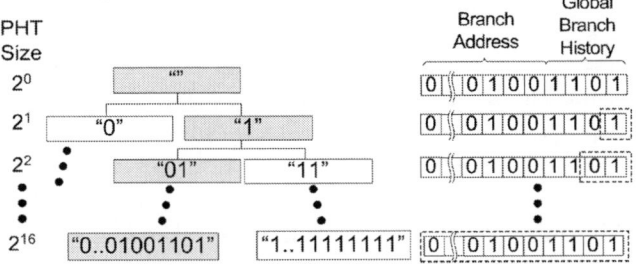

Figure 1: Data structure used to collect aliasing statistics for all PHT configurations in parallel.

5. Power Cost Function

Aliasing is measured per PHT size and can be used to predict the total power dissipation of the processor. To demonstrate this, the aliasing rate is shown to have a linear relationship with the hit rate of the BPU. The hit rate is then shown to have a linear relationship with the number of extra cycles needed to execute the application due to mispredictions. The number of extra cycles is then shown to have a linear relation-

ship with the total power dissipation of the processor. At the same time, switching and leakage power of the BPU is shown to have a linear relationship with its PHT size. All linear relationships are used to draw a relationship between power and aliasing as measured by simulations. Together, the slopes of these linear relationships form a set of parameters. These parameters can be combined to produce a cost function that can be used to predict the power dissipation of the processor for each PHT size.

5.1 Predicting Hit Rate

Figure 2 is a plot of the change in hit rate versus the change in aliasing rate as measured during simulations. For this plot, GSelect is used for branch prediction and the benchmark is *gcc*. A "least squares" regression has been added to the plot to show that the relationship is approximately linear (with slope, β).

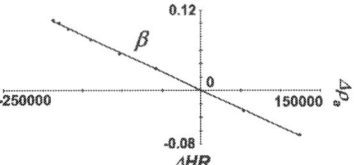

Figure 2: ΔHR as a function of $\Delta\rho_a$. Each data point represents a different GSelect PHT size for gcc. Data points are relative to the default size of 1024.

To predict the success rate of the branch predictor, an expression relating probabilities is derived that matches the results in Figure 2. The probability of aliasing is $\rho_a=P(aliasing)$. Given that a branch prediction lookup is aliased, constructive aliasing occurs with a probability $\rho_{ca}=P(correct\ pred.|\ aliasing)$ and destructive aliasing occurs with probability $1-\rho_{ca}=P(incorrect\ pred.\ |\ aliasing)$. When a PHT lookup is not aliased, the probability that the 2-bit saturated counter makes a correct prediction is $\rho_{cna}=P(correct\ pred.\ |\ non-aliasing)$. Accounting for the cases with and without aliasing, we use the following expression as the probability of a hit (i.e., hit rate):

$$HR = (1-\rho_a)\cdot\rho_{cna} + \rho_a\cdot\rho_{ca}. \qquad (1)$$

For an ideal branch predictor with infinite PHT resources, the hit rate is $HR=\rho_{cna}$. An expression for the change in prediction rate for all other configurations is:

$$\Delta HR = HR_2 - HR_1 = (\rho_{ca} - \rho_{cna})\cdot(\rho_{a,2} - \rho_{a,1}) \qquad (2)$$
$$= \beta\cdot(\rho_{a,2} - \rho_{a,1}) = \beta\cdot\Delta\rho_a$$

As stated earlier, only one simulation is needed to collect aliasing statistics for all configurations; however, a minimum of two simulations are needed to find the slope, β, of the hit-ratio versus aliasing curve. In Section 5.4, we provide results that show how "well-behaved" this approximation is over all benchmarks.

5.2 Predicting the Number of Cycles

Figure 3 is a plot of the change in number of cycles (due to misprediction penalty) versus the change in the hit rate when the PHT size is changed. As before, the benchmark application is *gcc* and all data points are relative to the values measured from a GSelect PHT size of 1024. The "least squares" approximation in this figure shows that the relationship can be approximated as linear. These results were consistent for all benchmarks.

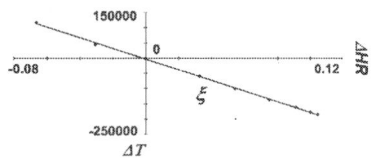

Figure 3: ΔT of execution as a function of ΔHR for different GSelect PHT sizes for *gcc*.

To predict the number of cycles, we assume the following relationship between the hit rate and the number of cycles of execution:

$$T = T_{ideal} + N_{CB} \cdot t_p \cdot (1 - HR) \qquad (3)$$

where T_{ideal} is the number of cycles needed to execute the application when branch prediction is 100% accurate, N_{CB} is the number of conditional branches encountered, t_p is the penalty for a branch misprediction, and HR is the hit rate. Using equation (3), the change in the number of cycles when comparing two different PHT sizes is:

$$\Delta T = T_2 - T_1 = t_p \cdot N_{CB} \cdot (HR_1 - HR_2) = \xi \cdot \Delta HR \qquad (4)$$

Equation (4) relates the change in the number of cycles needed to execute the program (due to branch penalties) to the change in the hit rate. ξ is the slope of this relationship.

5.3 Predicting Overall Power

The change in overall power dissipation of the processor can be divided into the change in power dissipated by the BPU (leakage and switching), ΔP_{BP}, and that dissipated by the rest of the processor, $\Delta P_{T\text{-}BP}$. As the size of the PHT changes, these two components are affected differently. An increase in the PHT size improves the branch prediction rate thus reducing the number of cycles devoted to branch penalties. With fewer cycles needed to execute the program, the entire processor will consume less energy. On the other hand, an increase in the size of the PHT will increase the energy consumption of the BPU even though there is a fixed number of lookups.

Figures 4 and 5 are plots of $\Delta P_{T\text{-}BR}$ versus the change in the number of cycles, ΔT, and ΔP_{BR} versus the change in the PHT size, ΔN, respectively. In both plots, a "least-squares" regression shows that the relationships are approximately linear. In this paper, these slopes are called α and γ.

The change in the overall power dissipation of the processor can be approximated by:

$$\Delta P = \Delta P_{BP} + \Delta P_{T-BP} = \gamma \cdot \Delta N + \alpha \cdot \Delta T \qquad (5)$$

where the first term is the change in power dissipation of the entire processor except for the BPU. It can be approximated as a linear function of the change in the number of clock cycles. The second term is the change in power dissipation of the BPU and can be approximated as a linear function of the PHT size.

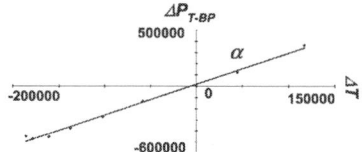

Figure 4: $\Delta P_{T\text{-}BP}$ versus ΔT. Each data point represents a different GSelect PHT size for *gcc*.

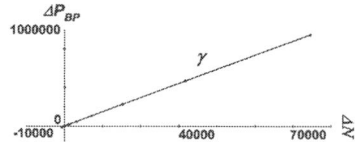

Figure 5: ΔP_{BP} versus ΔN for the gcc benchmark.

Equations (2), (4), and (5) can be combined to form a complete expression for the change in the total processor power dissipation as a function of aliasing and the PHT size:

$$\Delta P = \Delta P_{BP} + \Delta P_{T-BP} = \gamma \cdot \Delta N + \alpha \cdot \xi \cdot \beta \cdot \Delta \rho_a \qquad (6)$$

where α, ξ, β, and γ are slope parameters that have to be measured by simulating the processor with two different configurations. On the other hand, only one simulation is needed to determine $\Delta \rho_a$ for all PHT sizes in parallel. Together, these statistics are combined using equation (6) to estimate the power dissipation of a processor for a particular PHT size. Two curves are shown for the application *gcc* in Figure 6. One curve is the estimated total power dissipation (using equation (6)) as a function of the PHT size and the other curve is that measured using Panalyzer. This figure shows that equation (6) is fairly accurate at estimating the total power dissipation of the processor. In this example, our configuration tool would choose a PHT size of 8192 which corresponds to the minimum point in the curve.

5.4 Results: Quality of the Linear Relationships

Using Panalyzer, simulations were run on all benchmarks over all PHT sizes ranging from 16 to 65535 for the Bimodal and GSelect branch predictors. For each benchmark, "least-squares" linear regression slopes β, ξ, α, and γ were determined based on the statistics gathered for ΔHR versus $\Delta \rho_a$, ΔT versus ΔHR, $\Delta P_{T\text{-}BR}$ versus ΔT, and ΔP_{BR} versus ΔN, respectively. As an indication of the overall quality of fit for each regression, correlation coefficients (R^2) are given in Table 1. The closer a value to 1, the higher the quality of fit.

Most regressions in Table 1 have a high quality of fit except for *bzip2* and *parser*. *bzip2* and *parser* also coincide with having the lowest percentage of branch instructions and are

0-7695-2533-4/06 $20.00 © 2006 IEEE

therefore less likely to saturate the PHT. For these benchmarks, aliasing will likely be low regardless of the size of the PHT. Therefore, the "quality" of the linear relationship between ΔHR and $\Delta\rho_a$ will be dampened and overshadowed by "noise" generated by ρ_{ca} and ρ_{cna}. This inaccuracy has a negative impact on our ability to predict power dissipation. These results suggest that the accuracy of our power prediction methodology is less effective for applications with a low utilization of branch predictor resources.

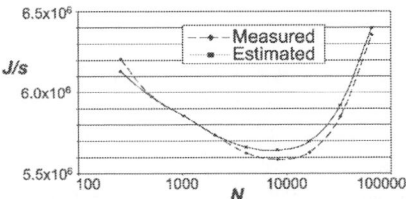

Figure 6: Estimated and measured total power dissipation for the *gcc*. The x-axis is the GSelect PHT size and the y-axis is power dissipation.

5.5 Results: Configured PHT (Bimodal and GSelect)

The prediction method proposed in this paper was used to configure a processor with a Bimodal predictor and a processor with a GSelect predictor. For each predictor, two simulations using Panalyzer were carried out to estimate the slope parameters β, ξ, α, and γ. At the same time, aliasing was measured for all PHT sizes in parallel. Using the cost function defined by (6), the total power dissipation was predicted for all configurations. Each benchmark was then simulated using the configuration with the lowest estimated power as determined by the cost function. Power dissipation results were then compared against that of the default branch predictors (i.e., as defined in Simplescalar and Panalyzer). These results are shown in Table 2.

Results indicate an average reduction in total power dissipation of 5.4% (maximum of 12.7%) for a processor with a GSelect predictor and 0.1% (maximum of 1.3%) for a processor with a Bimodal predictor. A processor with the "perfect" PHT size results in a 5.8% improvement and 0.7% improvement for the GSelect and Bimodal predictors, respectively. These results suggest that, by configuring the branch predictor using the cost function proposed in Section 5, the power reduction is close to that achievable by the "perfect" PHT configurations. It is important to note that a designer would typically use power as a guide rather than power-delay product so we report our results as such. Of course, the improvements would be even better if reported as power-delay product.

Table 2 shows power reductions for all benchmarks except for *bzip2* and *parser* which result in a small increase in the power dissipation. This result is similar to that of the "perfect" configuration which amounted to little or no power reduction for these benchmarks. This result is not because the "perfect" configuration matches (or nearly) matches the default configuration, but because the difference between all configurations is relatively small. The effects of this were also evident in the previous section which showed that the

miss rate of *bzip2* and *parser* did not "behave well" as a linear function of aliasing. The minimal power reduction of these two benchmarks does not show a deficiency in our approach because there was very little power reduction possible.

6. Cost Function for A Hybrid Predictor

To predict the power dissipation for more complex, hybrid predictors, a new estimate of the hit rate in equation (2) must be derived to account for all sub-predictors (ΔHR_{hybrid}). Once found, ΔHR_{hybrid} is then used with a modified version of equation (6) which accounts for the leakage and switching power of all sub-predictors:

$$\Delta P = \sum_{sp\in\chi} \Delta P_{BP,sp} + \Delta P_{T-BP}$$
$$= \sum_{sp\in\chi} \gamma_{sp}\cdot\Delta N_{sp} + \alpha\cdot\xi\cdot\Delta HR_{hybrid} \qquad (7)$$

where χ is the set of all sub-predictors.

For this paper, the hybrid predictor used consists of a GSelect sub-predictor for global predictions, a Bimodal sub-predictor for local predictions, and a Bimodal meta-predictor for choosing between the local and global predictors.

6.1 Results: Configured PHT (Hybrid)

For each sub-predictor, two simulations were carried out using Panalyzer to estimate the slope parameters, β, ξ, α, and γ. Aliasing was measured for all PHT sizes in parallel using the approach described in Section 4. Using the cost function defined by equation (7), the slope parameters and the aliasing measurements were used to predict the total power dissipation for all configurations in parallel. These predictions were then used to choose the best configuration for each benchmark. The final power dissipation results were then compared against a hybrid predictor with default sub-predictor sizes as defined by Simplescalar. These results are shown in Figure 7.

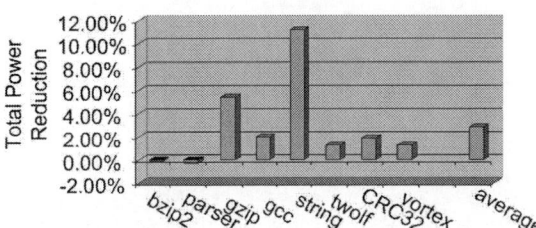

Figure 7: Total power reduction for the hybrid predictor. Average improvement is shown in the rightmost column.

Figure 7 shows that, by using the proposed cost function to configure the hybrid branch predictor, overall power dissipation is reduced by an average of 2.8% (maximum of 11.2%). As before, the only benchmarks with an increase in power dissipation were *bzip2* and *parser*. For these two benchmarks, the PHT size was most likely chosen incorrectly because of the non-linearity of the relationship between ΔHR and $\Delta\rho_a$ (see Section 5.4).

Table 1: The correlation coefficient (R^2) for all least square regressions. The closer a value to 1, the higher the overall quality of fit. Results are provided for both the GSelect predictor and the Bimodal predictor. The first column is the percentage of instructions executed that are branch instructions.

	branch %	Beta R^2		Zeta R^2		Alpha R^2		Gamma R^2		Avg R^2
		GSelect	Bimodal	GSelect	Bimodal	GSelect	Bimodal	GSelect	Bimodal	
bzip2	7.9	0.781	0.025	0.634	0.438	0.026	0.284	1.000	1.000	0.524
parser	9.6	0.996	0.624	0.985	0.971	0.008	0.728	1.000	1.000	0.789
gzip	9.9	0.984	0.747	0.996	1.000	0.999	0.993	1.000	1.000	0.965
gcc	15.7	0.999	0.998	0.999	0.999	0.994	0.998	1.000	1.000	0.998
string	16.0	0.992	0.983	0.999	0.978	0.986	0.977	1.000	1.000	0.989
twolf	16.3	0.993	0.978	0.996	1.000	0.995	0.993	1.000	1.000	0.994
CRC32	16.6	0.679	0.754	0.976	0.993	0.929	0.101	1.000	1.000	0.804
vortex	18.7	0.982	0.985	0.984	0.998	0.996	0.984	1.000	1.000	0.991
average	13.8	0.926	0.762	0.946	0.922	0.742	0.757	1.000	1.000	0.882

Table 2: Power reduction after using the proposed configuration approach for GSelect for the branch predictor (B.P), the total processor (Total), and the total processor minus the branch predictor (Total-B.P.). Also shown is the best improvement possible with the "perfect" configuration.

	Semi-Dynamic Profiling GSelect			Perfect GSelect	Semi-Dynamic Profiling Bimodal			Perfect Bimodal
	B.P.	Total - B.P.	Total	Total	B.P.	Total - B.P.	Total	Total
bzip2	66.3%	-2.0%	-1.8%	0.0%	95.1%	-0.8%	-0.3%	0.8%
parser	64.9%	-0.6%	-0.3%	0.1%	94.6%	0.7%	1.3%	1.6%
gzip	-249.0%	4.9%	4.3%	5.2%	95.0%	-0.3%	0.1%	0.3%
gcc	-581.8%	6.4%	4.6%	4.6%	48.7%	-0.4%	-0.1%	0.0%
string	-575.6%	7.8%	6.3%	6.4%	0.0%	0.0%	0.0%	0.4%
twolf	-527.7%	14.6%	12.7%	12.7%	-290.1%	2.3%	0.4%	2.1%
CRC32	-70.9%	9.3%	9.1%	9.1%	94.7%	-1.3%	-0.9%	0.0%
vortex	-559.5%	10.2%	8.6%	8.6%	-97.7%	0.7%	0.2%	0.2%
average	-304.2%	6.3%	5.4%	5.8%	5.0%	0.1%	0.1%	0.7%

7. Conclusions

In this paper, we have proposed a novel approach for sizing the branch predictor PHT to reduce overall power dissipation based on the application. Results show that our approach achieves an overall power reduction that closely matches that achievable with a "perfect" configuration. This approach uses a "fast" configuration methodology that measures aliasing for all PHT sizes in parallel and then uses a cost function to relate aliasing to power dissipation. This methodology significantly reduces the runtime needed for architecture optimization in ASIP tools by reducing the solution space by one dimension.

8. Acknowledgements

This work is sponsored by PMC-Sierra and NSERC (including Grant No. STPGP 257684).

9. References

[1] Tensilica [Online]. Available: http://www.tensilica.com
[2] ARC International [Online]. Available: http://www.arc.com
[3] D. Parkikh, K. Skadron, Y. Zhang, M. Barcella, and M. R. Stan, "Power Issues Related to Branch Prediction," *Proc. of HPCA*, 2002.
[4] M. Ekpanyapong, P. Korkmaz, and H. S. Lee, "Choice Predictor for Free," *Ninth Asia-Pacific Computer Systems Architecture Conference*, 2004.
[5] S.-T. Pan, K. So, J. T. Rahmeh, "Improving the Accuracy of Dynamic Branch Prediction Using Branch Correlation," *Proc. of ASPLOS V*, 1992.
[6] J. E. Smith, "A Study of Branch Prediction Strategies," *Proc. of ISCA*, 1981.
[7] S. McFarling, "Combining Branch Predictors," Digital Equipment Corporation, WRL Tech. Note TN-36, 1993.

[8] D. Chaver, L. Pinuel, M. Prieto, F. Tirado, and M.C. Huang, "Branch Prediction on Demand: an Energy-Efficient Solution," *Proc. of ISLPED*, 2003.
[9] M. Monchiero, G. Palermo, M. Sami, C. Silvano, V. Zaccaria, R. Zafalon, "Power-Aware Branch Prediction Techniques: A Compiler-Hints Based Approach for VLIW Processors," *Proc. of GLSVLSI*, 2004.
[10] A. Baniasadi, A. Moshovos, "Branch Predictor Prediction: A Power-Aware Branch Predictor for High-Performance Processors," *Proc. of ICCD*, 2002.
[11] A. Baniasadi, A. Moshovos, "SEPAS: A Highly Accurate Energy-Efficient Branch Predictor," *Proc. of ISPLED*, 2004.
[12] Intel StrongARM SA-1110 Data Sheet. 2000.
[13] N. S. Kim, T. Austin, T. Mudge, and D. Grunwald, "Challenges of Architectural Level Power Modeling," Book Chapter from Power Aware Computing, 2001, ed. R. Melhem and R. Graybill.
[14] D. Burger and T.M. Austin, The SimpleScalar Tool Set, Version 2.0. *Computer Architecture News*, June 1997.
[15] Standard Performance Evaluation Corporation. SPEC CPU2000 Benchmarks. http://www.specbench.org.
[16] C. Lee, M. Potkonjak, and W.H. Mangione-Smith, "MediaBench: a tool for evaluating and synthesizing multimedia and communications systems," *Proc. of Micro*, 1997.
[17] GNU GCC [Online]. Available: http://gcc.gnu.org
[18] I.-C.K. Chen, J.T. Coffey, and T.N. Mudge. "Analysis of Branch Prediction via Data Compression," *Proc. of ASPLOS*, 1996.
[19] P. Michaud, A. Seznec, and R. Uhlig, "Trading Conflict and Capacity Aliasing in Conditional Branch Predictors," *Proc. of ISCA*, 1997.
[20] S.Sechrest, C. Lee, and T. Mudge, "Correlation and Aliasing in Dynamic Branch Predictors," *Proc. of ISCA*, 1996.

Exploiting Software Pipelining for Network-on-Chip architectures*

Feihui Li   Mahmut Kandemir
Dept. CSE, The Pennsylvania State University
University Park, PA 16802, USA
{feli,kandemir}@cse.psu.edu

Ibrahim Kolcu
Computation Dept., UMIST
Manchester M60 1QD, UK
ikolcu@umist.ac.uk

Abstract

Recent developments in process technology have made it possible to produce chips consisting of a large number of processing elements. For factors such as scalability, performance, power-efficiency, the interconnection structure supporting such a chip needs to be an on-chip network architecture rather than a conventional bus-based system. Recent research has studied such network-on-chip (NoC) based systems from the performance and throughput, power/energy, reliability, predictability, synchronization, and concurrency perspectives. However, most of these studies are hardware based and it is not clear what type of compiler support would be best suited for these NoC based systems. Focusing on a mesh based NoC architecture that connects multiple processor cores, this paper explores the effectiveness of voltage/frequency scaling for processors and communication links with and without software pipelining, a compiler optimization for increasing parallelism. To our knowledge, this is the first paper that explores the influence of software pipelining in the context of the embedded NoC architectures.

1 Introduction

Increasing complexity of multi-core designs makes it difficult to provide point-to-point communication links for each pair of communicating blocks in a chip. As a result, a segmented network fabric, namely, the network-on-chip (NoC), is emerging as a promising solution to chip-level communication in embedded systems. Previous research [4, 19] has identified power consumption as one of the critical issues to be addressed in the context of the NoC based embedded designs. Among the studies published so far are on-chip network power modeling [5, 14], energy reduction techniques through communication link shut-down and voltage/frequency scaling of links [19, 18, 1], and energy-efficient task mapping and message routing [7, 2]. While there have been numerous system architecture designs and communication protocols proposed for the embedded NoC based systems, it is still not clear what type of software support should be provided for such systems. In particular,

in our opinion, the necessary compiler support for NoCs should be discussed and the suitability of existing compiler optimizations to the NoC based systems should be thoroughly evaluated.

This paper is a step towards understanding the needs of embedded NoCs as far as the required compiler support is concerned. Focusing on a mesh based NoC architecture that connects multiple processor cores together, this paper explores the effectiveness of *voltage/frequency scaling* for processors and communication links with and without *software pipelining,* a popular compiler optimization used for increasing parallelism [9, 16]. Specifically, we first show the amount of leakage/dynamic energy savings that could be achieved through voltage/frequency scaling when only instruction level parallelism (ILP) is exploited. After that, we focus on software pipelining, propose an approach, and identify its impact on leakage energy savings. Finally, we apply CPU/link voltage/frequency scaling to the software-pipelined codes and present the potential leakage and dynamic power savings. To our knowledge, this is the first paper that explores the influence of software pipelining in the context of the embedded NoC architectures.

We structure the rest of this paper as follows. Section 2 discusses the related research efforts. Section 3 describes the NoC architecture used in this paper. We go into the details of our approach in Section 4. The paper is concluded in Section 5.

2 Related Work

With the increasing number of integrated IP blocks in embedded designs, Network-on-Chip (NoC) architectures [3] have become a more promising solution compared to traditional bus based systems. Many NoCs architectures and related tools, e.g., [20, 17, 8], have already been proposed and developed.

There have been a large amount of efforts towards the optimization of NoC based systems, targeting the performance and/or power consumption. A methodology for designing and modeling power-aware, high performance NoCs have been proposed in [4]. Several task mapping/scheduling related works have studied the application parallelization on NoCs. Hu and Marculescu [7] propose an algorithm for mapping IP blocks onto an NoC in an energy- and performance-aware fashion. Asica et al [2] present a heuristic for task mapping on a mesh-based NoC built upon a

---

*This work is supported in part by NSF Career Award 0093082 and a grant from GSRC.

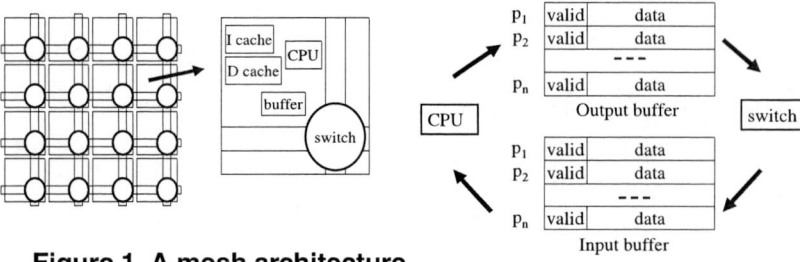

Figure 1. A mesh architecture.

Figure 2. The input/output buffer.

Figure 3. High level view of our approach.

multi-objective genetic algorithm (GA). Shin and Kim [18] design a GA-based algorithm that completes task assignment, tile mapping, routing path allocation, task scheduling and link speed assignment with voltage scalable links.

In NoC architectures, the energy consumed by the interconnection networks turns out to be a significant portion of the overall energy consumption. Prior proposals for reducing the power of interconnection networks include [19, 15]. However, the previous compiler related efforts [11, 13] on chip multi-processors focus mainly on improving performance. The energy-aware instruction scheduling algorithms [21, 10] are targeting VLIW or superscalar architectures, instead of NoCs.

Our approach is different from the previous efforts because we focus on the instruction level scheduling/mapping on an NoC. Based on RAW machine scheduling [11], our proposal is to map the schedule generated by software pipelining algorithm on a mesh-based NoC architecture and to explore the impact of software pipelining.

3 Architecture

In this paper, we employ a RAW-like mesh architecture shown in Figure 1. Each node in this architecture consists of a CPU, an instruction/data cache, a switch and a buffer. The CPU, together with its instruction/data caches, performs the local computation. The switch is used to connect the adjacent nodes, while the buffer stores the data exchanged between the nodes.

To reduce the communication latency among the different nodes, the buffer functions as an extension to the register file. It includes two parts, input buffer and output buffer, as shown in Figure 2. Each buffer contains $(m - 1)$ entries, where each entry corresponds to one of the other processors in the mesh (assuming there are m nodes in the mesh). A "valid" flag is associated with each buffer entry to indicate if the current data in the entry is ready for using (input buffer) or sending (output buffer). If the destination operand of an instruction is to be shared with another processor, the processor writes the result of this instruction into the corresponding entry in its output buffer. In the mean time, this entry's valid flag is set to 1. The switch immediately sends the data in the output buffer to the target node. After sending, the valid flag is reset. Following this, the processor can continue to write the next output data. At the receiver side, once the switch receives the data, it writes it into the input buffer directly so that the target processor can use it

right away. Therefore, it takes only two cycles between the data writing by a processor and the data reading by one of its neighbors. For example, if processor 0 writes a data at cycle 1, its neighbor, processor 1, can read this data at cycle 3. When an instruction writes into a valid entry, it is blocked until the previous data in the output buffer is sent out. Similarly, when an instruction reads an invalid entry in the input buffer, it is blocked until there is a valid data available. Note that the input/output buffer can be viewed as a special segment of the regular register file. In this organization, the data can be fed into the ALU directly, without an explicit copy instruction that copies the data in the buffer into a register (or the data copy can be performed simultaneously with computation).

4 Overview of Our Approach

Figure 3 shows the main flow of our compiler-directed power optimization framework for NoCs. Given a sequential code, our optimizing compiler extracts the inherent parallelism in the code and partitions/maps the code onto multiple processors of an NoC. From the resulting parallel code, a CPU/link speed (usage) analyzer statically estimates the *slacks* for CPUs and links (i.e., the idle periods that can be used for saving energy through frequency/voltage scaling). Based on the extracted information, it then inserts proper CPU/link voltage control instructions into the code and obtains the final parallel code that is optimized for power. Note that the CPU/link voltage control instructions are inserted only at a loop granularity due to the voltage scaling overhead involved. In our discussion of this section, for clarity, we assume that the frequency f (assuming that CPUs and communication links have the same frequency originally) can only be scaled to f/k, where k is an integer. We denote k by *scaling parameter*.

5 RAW-like Scheduling

5.1 Scheduling Support

A space/time scheduling algorithm for the RAW machine is proposed by the prior research [11]. The main steps of this algorithm include instruction space partitioning, global data partitioning, data and instruction placement, and communication code generation. This algorithm is based on list scheduling. The basic data structure it makes

0-7695-2533-4/06 $20.00 © 2006 IEEE

use of is a *data dependency graph (DDG)*, which is built from a given loop body. Although [11] did not specifically discuss inter-iteration dependences, we include it in our work so that it is possible to compare our proposed approach with the RAW-like scheduling. Figure 4(a) shows an example DDG, in which the vertices represent the instructions, and the edges capture the dependences among these instructions. The pair of values of the form $(diff, delay)$, associated with an edge characterizes the data dependency between the vertices incident on that edge, where $diff$ is the iteration difference between the two corresponding instructions and $delay$ is the minimal delay between them.

Figure 4(b) gives the result of the scheduling based on [11] for the DDG shown in Figure 4(a). Instead of explicitly giving the communication instructions like send/receive/route, we use arrows to represent the inter-processor communications for clarity. We see that, in this schedule, the communication from $P0$ to $P1$ is overlapped with the computation. Therefore, the execution time for a single loop iteration is as low as 5 cycles. If there are N loop iterations, the execution time is $5N$ cycles (neglecting the branching time).

We now make several simplifying assumptions so that we can calculate the energy consumption of our example DDG in Figure 4(a). First, we assume a 4-node mesh-based NoC architecture as shown in Figure 4(g). Second, we assume that the per cycle CPU energy includes both dynamic and leakage parts: ΔE_{CPU_dyn} and ΔE_{CPU_leak} (for the sake of illustration, we do not include the fact that the different types of instructions can typically consume different per cycle CPU energies). The unit communication link energy, ΔE_{link}, is the energy consumed by passing a unit data (one word) over a single communication link.

Based on these assumptions, we can calculate the overall energy consumption of N loop iterations as: $E = N * (\Delta E_{link} + 7\Delta E_{CPU\_dyn} + 20\Delta E_{CPU\_leak})$.

5.2 Voltage Scaling

In order to reduce power consumption, we can apply compiler-directed voltage/frequency scaling. Specifically, in our running example, we discover that only two processors and one communication link are being used in this loop. Therefore, the voltage of all the unused processors and links can be scaled down to zero, i.e., they can be shut down. Once a processor or a communication link is shutdown, its energy consumption is eliminated. We also observe that the sequence consisting of instructions 2 and 4 has a 3-cycle slack. Since the RAW-like mesh architecture utilizes an asynchronous global branch, this slack can be exploited to scale down the voltage/frequency of processor $P1$ and the communication link from $P0$ to $P1$. A voltage/frequency scale down instruction is inserted by the compiler at the beginning of this loop. When the frequency is scaled down from f to f/k, for illustrative purposes, we assume that the per cycle (original cycle time) CPU dynamic and leakage energies go down to $\Delta E_{CPU\_dyn}/k^3$ and $\Delta E_{CPU\_leak}/k^2$, respectively. The unit communication link energy becomes $\Delta E_{link}/k$ from ΔE_{link}. These assumptions are made based on CMOS energy formulas and link energy mod-

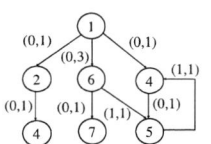
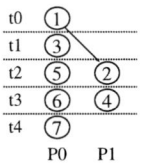

(a) An example DDG.

(b) Result of a RAW-like scheduling.

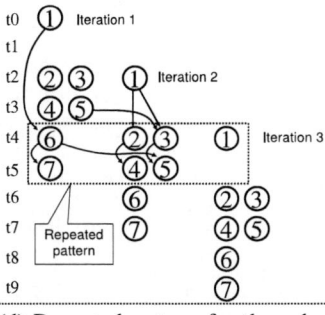

(d) Repeated pattern for the schedule in (c).

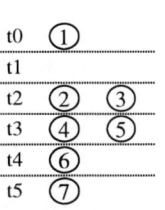

(c) Resulting schedule from the Lam's algorithm.

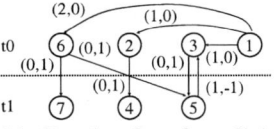

(e) Result after formalizing the repeated pattern in (d).

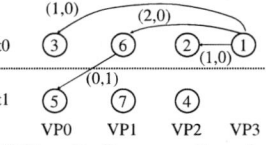

(f) Result after operation clustering.

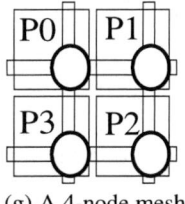

(g) A 4-node mesh.

(h) Result after cluster mapping.

Figure 4. An example.

els. Since k must be an integer from our assumption, we see that the best k for $P1$ and link $(P0 \rightarrow P1)$ is 2 here. Thus, after scaling down voltage, the overall energy consumption of the loop in our running example becomes: $E = N * (0.5\Delta E_{link} + 5.25\Delta E_{CPU\_dyn} + 6.25\Delta E_{CPU\_leak})$. In this example, we assume the simultaneous CPU/link voltage scaling algorithm proposed in [12]. This algorithm allocates the available slacks to CPUs and communication links so as to maximize the overall power savings.

6 Applying Software Pipelining to NoCs

6.1 Software Pipelining Basics

Software pipelining is a scheduling technique first proposed in the context of VLIW processors. The idea is to initiate a new loop iteration before the previous iteration completes. Thus, multiple loop iterations can run concurrently, thereby exploiting parallelism better. The objective of any given software pipelining algorithm is to achieve the *minimal initiation interval*, i.e., the time gap between the initiations of two successive loop iterations. There exist many algorithms proposed in the literature, e.g., [9, 16]. Generally speaking, any software pipelining algorithm should conform to both resource constraints in the architecture and precedence constraints in the program code. In this paper, since we need to partition and map the resulting schedule across multiple processors in the NoC, we use the Lam's algorithm [9], which gives a fixed schedule and a fixed initiation interval (though, in principle, any software pipelining algorithm in the literature could be used).

Lam's algorithm works on a specific type of DDG. This DDG is different from the traditional one in that each of its nodes represents an indivisible sequence of instructions. For clarity, the partial schedule for each node is not shown in this paper. Lam's algorithm generates a fixed schedule, which can be initiated at every initiation interval, as illustrated in Figure 4(c) for our running example. Figure 4(d) demonstrates how the different loop iterations are executed. In this example, the resulting initiation interval is 2. Software pipelining transforms the code into three parts: prelude, steady state, and postlude. In this paper, we are interested in the scheduling and mapping of the codes in the steady state, since they form the most time/power consuming part of a typical loop. In Figure 4(d), the computations enclosed in the dotted rectangle are the ones in the steady state.

6.2 Algorithms

6.2.1 Extracting Repeated Patterns

From the Lam's software pipelining algorithm, one can obtain the minimal initiation interval and the resulting schedule for steady state. The steady state in our work needs to keep track of the data dependencies among the operations[1], which are represented by the edges in Figure 4(d), since the dependencies across different nodes in an NoC indicate communications. We refer to these operations and their dependencies in the steady state as a *repeated pattern* in the rest of our discussion.

We see from Figure 4(d) that the edges can be across the repeated patterns or within a repeated pattern. In order to simplify our compiler analysis, we borrow the representation of DDGs and construct a formalized repeated pattern. Figure 4(e) shows the formalized repeated pattern corresponding to the repeated pattern in Figure 4(d). This formalized repeated pattern is different from the DDG shown in Figure 4(a) in two aspects. First, the repeated pattern contains timing information, which indicates the partial ordering of operations. For example, in Figure 4(e), operations 6, 2, 3, and 1 are executed in the same cycle, while operations 7, 4, and 5 are in the next cycle. Second, the pair associated with an edge in the repeated pattern has a different meaning than that of a DDG, though it is still in

---

[1]We use "operation" and "instruction" interchangeably when there is no confusion.

the format $(diff, delay)$. In the formalized repeated pattern, $diff$ indicates the repeated pattern iteration difference instead of the loop iteration difference used for the DDG representation. Also, $delay$ is the minimal delay within the repeated pattern. In the rest of this paper, when we mention a repeated pattern, we mean a formalized repeated pattern.

After constructing the formalized repeated pattern, our goal is to map it (both computations and communications) onto our mesh-based NoC architecture. Given a repeated pattern, such as the one in Figure 4(e), mapping it means that one needs to assign all of its operations into a schedule table of the same size (row: time steps; column: processors). The desired mapping is the one with the minimum overall execution cycles (including both computation and communication cycles), and with acceptable power consumption. This mapping is a type of instruction reordering. However, one cannot move the instructions vertically, i.e., changing its scheduling time, since this means changing the resulting schedule from software pipelining. Instead, the flexibility we have for operation reordering is to move operations horizontally, i.e., we can assign each operation to a new target processor (the communication cycles can be inserted after the operation mapping). We propose a two-step heuristic including *operation clustering* and *clustering mapping* to reorder the instructions. In operation clustering, we do not consider the actual network topology and cluster the operations into m *virtual processors*. We then map these virtual processors onto the physical processors in the actual NoC architecture under consideration. The details of these two steps are discussed below.

6.2.2 Operation Clustering

The goal of operation clustering is to minimize the inter-processor communications, which is desirable from both the performance and power perspectives. We observe that, for a repeated pattern, once we assign two operations, which are not scheduled in the same cycle, onto the same processor, we can eliminate the communication (edge) between them. However, it is impossible to remove the communication between the operations scheduled in the same cycle, since they must be scheduled onto different resources (processors) according to software pipelining. Based on this observation, we try to cluster instructions into m (the number of physical processors) virtual processors such that the number of inter-processor communications is minimized.

Algorithm 1 gives our operation clustering algorithm. First, we have an input $n \times m$ repeated pattern $\mathcal{R}$ that is generated by the software pipelining algorithm and our repeated pattern formalization step, as explained earlier. To cluster the operations in $\mathcal{R}$, we remove all the operations from $\mathcal{R}$ one by one and assign each operation into a location (cell) in an initially empty $n \times m$ table, $\mathcal{R}'$, in which the rows still correspond to time steps, while the columns represent the different virtual processors. This assignment is done column by column, that is, we first assign operations to virtual processor VP_0, and then to VP_1, and so on. At the beginning, the *vertical_degree* for each operation in $\mathcal{R}$ is calculated, which is the number of input/output edges that are not connecting the operations in the same cy-

Algorithm 1 Instruction clustering

INPUT: A $n \times m$ repeated pattern $\mathcal{R}$ (n: time steps; m: number of processors)
OUTPUT: An optimized $n \times m$ repetition pattern $\mathcal{R}'$ where the inter-processor communication load is minimized

```
1:  calculate vertical_degree(OP);
2:  for each virtual processor VP_i do
3:     remove OP with maximum vertical_degree from R;
4:     k = time(OP);
5:     assign OP to R'(k, i);
6:     for each time step j (j ≠ k) do
7:        update priority(OP) for each OP of R(j, 1 : m);
8:        remove OP with maximum priority from R(j, 1 : m);
9:        assign OP to R'(j, i);
10:    end for
11: end for
12: remove the edges within the same virtual processor;
```

cle. To fill each column of $\mathcal{R}'$, we always start by assigning the operation with the maximum *vertical_degree* in current $\mathcal{R}$ (see lines 3-5 in Algorithm 1; function *time* returns the schedule time step of the operation OP). This means that we give priority to the operations that have more potential communications. After that, we fill the other slots in this column in the time increasing order. When we are trying to fill slot $\mathcal{R}'(j, i)$ (j: time step; i: virtual processor ID), we can select operations only from $\mathcal{R}(j, 1 : m)$. This constraint guarantees that we do not change the vertical order of operations. As mentioned earlier, this change would lead to an undesired modification of the resulting schedule coming from software pipelining. We choose the operation with the maximum *priority* from $\mathcal{R}(j, 1 : m)$.

The job of the function *priority* in line 7 of Algorithm 1 is to assess all the operations in the queue $\mathcal{R}(j, 1 : m)$ for slot $\mathcal{R}'(j, i)$. The priority of an operation op is calculated as: $priority(op) = \sum communication(op, op_p)$, where op_p is any operation already scheduled onto the virtual processor VP_i. Function *communication* returns 1 if two operations have data dependencies between them; otherwise, it returns 0. Selecting the operation with the maximum priority from the queue $\mathcal{R}(j, 1 : m)$ to fill slot $\mathcal{R}'(j, i)$ means that we can eliminate most of the inter-processor communications. The asymptotic complexity of Algorithm 1 is $O(n * m^2)$, which is polynomial[2].

Figure 4(f) shows the result obtained using Algorithm 1. The edges between the operations within the same cluster (i.e., the operations assigned to the same virtual processor) are removed. Since we do not consider the actual underlying network topology in this step, we defer the consideration of communication cycles to the next step, cluster mapping.

6.2.3 Cluster Mapping

While operation clustering minimize the number of communications, cluster mapping is to reduce the number of hops taken during communication. For example, if two virtual processors have a lot of communications between them, it is better to map them onto two neighboring physical processors rather than to two distant physical processors. This

step needs to take the network topology into account.

Before going into our algorithm, we first discuss how to measure *communication load*, since it is one of the targets of our optimization in this step. We define a unit communication load as a unit data (one word) transfer over a single communication link. For example, if a one-word data is transferred from one processor to its neighbor, we say that the communication load is 1, whereas if a three-word data traverses over two links to reach its destination, the corresponding communication load becomes 6. In the cluster mapping step, we want to minimize the overall communication load.

Another optimization target is the overall execution time. According to the construction of the repeated pattern, we know that the time gap between an operation and its dependent can be calculated as: $time\_gap = d * diff + delay$, where d is the schedule length of the repeated pattern, and $(diff, delay)$ is the pair associated with the edge between these two operations. Considering the physical network and the cluster mapping, if the actual communication latency between the two corresponding operations is less than $time\_gap$, this means that there exist slacks for the sending operation or for the communication event. If the actual communication latency is equal to $time\_gap$, the communication is fully overlapped with the computations and there is no slack. However, if the actual communication latency is larger than $time\_gap$, i.e., the communication cannot be fully overlapped with the computation, extra cycles for the communication should be inserted into the repeated pattern just before the corresponding receive operation.

Based on the above discussion, we can calculate the overall communication load and the overall execution cycles given a specific cluster mapping. We search for an optimized cluster mapping by employing a swap-based greedy algorithm, which is similar to the one used by the RAW scheduler [11]. It first assigns operation clusters to arbitrary physical processors, and then searches for pairs of mappings that can be swapped to reduce the overall communication loads and the overall execution cycles (we find that these two optimizations are consistent in practice; i.e., the mapping with the minimum communication load has the minimum overall execution cycles). For our example, the result after this step is shown in Figure 4(h). We see that the communication load per repeated pattern is 5. The execution time (including both computation and communication) is 3 cycles. Compared to Figure 4(b), we see that the execution time is improved by 40% per repeated pattern. With a large number of loop iterations, the overall performance improvement should be very close to this improvement, i.e., 40%. Similar to the previous energy calculation in Section 5.1, the overall energy consumption for N repeated patterns is: $E = N * (5\Delta E_{link} + 7\Delta E_{CPU\_dyn} + 12\Delta E_{CPU\_leak})$. Compared to the energy consumption in Section 5.1, we see that exploiting software pipelining and our two-step mapping scheme can save a significant percentage of processor leakage energy, at the cost of increasing communication link energy.

---

[2]We assume that the degree of DDG, i.e., the maximum number of edges incident upon a vertex, is a constant.

6.3 Voltage Scaling

In our running example, from the schedule given in Figure 4(h), we see that only four communication links ($P1 \rightarrow P0$, $P1 \rightarrow P2$, $P0 \rightarrow P3$, and $P3 \rightarrow P2$) are in use (an XY-routing [6] is assumed). The voltages of the other four links are scaled down to zero. However, with this schedule, all of the four processors are used and thus none of them can be turned off. Further, we find that there exist multiple slacks: one cycle between operations 1 and 2, one cycle between operations 1 and 3, three cycles between operations 1 and 6 (across repeated patterns). These slacks are calculated by subtracting the actual network latency between two operations from the $time\_gap$ between them. Note that, in calculating slacks, we do not consider the network contention. However, when we scale down the voltages, we need to include the impact of network contention to avoid the potential performance degradation. Our approach is a conservative one in which we do not scale down the voltage of a communication link if there may be a contention on this link. In our example, there is a contention on communication link $P1 \rightarrow P2$, thus we do not scale down its voltage. The slacks over this link are updated accordingly by including the impact of contention. For example, the slack between operations 1 and 6 becomes two cycles from the original three cycles.

The slack between two operations can be classified as either the CPU slack or the communication link slack or both. As mentioned earlier, we use the algorithm given in [12] and select the most appropriate voltage/frequency for processors and links. That is, we choose the best slack distribution so that the power consumption is minimized.

For our example, we find that, to achieve the most energy saving in a performance aware fashion, the best solution is to insert one clock cycle into the result in Figure 4(h) (after time step $t2$) so that processor $P0$, $P1$ and $P2$ can be scaled by a parameter of 2, and the communication links $P1 \rightarrow P0$ and $P2 \rightarrow P3$ are also scaled by a parameter of 2 and 4, respectively. Thus, the overall energy consumption is: $E = N * (3.75\Delta E_{link} + 2.625\Delta E_{CPU\_dyn} + 7\Delta E_{CPU\_leak})$.

We see from the running example that exploiting software pipelining can still improve the performance significantly (20% in this case). The reason is that software pipelining extracts significant instruction level parallelism and utilizes the available on-chip resources very effectively. Applying voltage/frequency scaling based on it still gives better performance than the base case (no software pipelining, no voltage/frequency scaling) in this particular example.

7 Concluding Remarks

Emerging network-on-chip (NoC) based embedded systems bring several important issues such as performance and throughput, power and energy, reliability, predictability, synchronization, and management of concurrency. To address these challenges, it is critical to take a global view of the problem, which considers both hardware and software aspects. While the hardware aspects of NoC based designs have received a lot of attention lately, the software efforts for NoCs are still in their infancy. This paper is an attempt at exploring the role of a compiler in NoC based embedded designs. Specifically, it analyzes the impact of software pipelining when it is used in conjunction with voltage/frequency scaling.

References

[1] A. Andrei and et al. Simultaneous communication and processor voltage scaling for dynamic and leakage energy reduction in time-constrained systems. In *Proc. ICCAD*, 2004.

[2] G. Ascia and et al. Multi-objective mapping for mesh-based NoC architectures. In *Proc. CODES*, Sept. 2004.

[3] L. Benini and G. Micheli. Networks on chips: a new SOC paradigm. *IEEE computer*, 35(1), Jan. 2002.

[4] L. Benini and G. D. Micheli. Powering networks on chips: energy-efficient and reliable interconnect design for SoCs. In *Proc. ISSS*, 2001.

[5] X. Chen and L.-S. Peh. Leakage power modeling and optimization in interconnection networks. In *Proc. ISLPED*, Aug. 2003.

[6] J. B. Duato and et al. *Interconnection Networks*. Morgan Kaufmann Publishers, 2002.

[7] J. Hu and R. Marculescu. Energy- and performance-aware mapping for regular NoC architectures. *IEEE Trans. on Computer-Aided Design of Integrated Circuits and Systems*, 24(4), Apr. 2005.

[8] A. Jalabert and et al. xpipescompiler: A tool for instantiating application specific Netowrks-on-Chip. In *Proc. DATE*, 2004.

[9] M. Lam. Software pipelining: an effective scheduling technique for VLIW machines. In *Proc. PLDI*, 1988.

[10] C. Lee and et al. Compiler optimization on instruction scheduling for low power. In *Proc. ISSS*, 2000.

[11] W. Lee and et al. Space-time scheduling of instruction-level parallelism on a RAW machine. In *Proc. ASPLOS*, 1998.

[12] J. Luo and et al. Simultaneous dynamic voltage scaling of processors and communication links in real-time distributed embedded systems. In *Proc. DATE*, 2003.

[13] R. Nagarajan and et al. Static placement, dynamic issue (SPDI) scheduling for EDGE architectures. In *Proc. PACT*, Oct. 2004.

[14] C. S. Patel. Power constrained design of multiprocessor interconnection networks. In *Proc. ICCD*, Washington, DC, USA, 1997.

[15] V. Raghunathan and et al. A survey of techniques for energy efficient on-chip communication. In *Proc. DAC*, 2003.

[16] B. R. Rau. Iterative modulo scheduling: an algorithm for software pipelining loops. In *Proc. MICRO*, 1994.

[17] K. Sankaralingam and et al. Exploiting ILP, TLP, and DLP with the polymorphous TRIPS architecture. In *Proc. ISCA*, 2003.

[18] D. Shin and J. Kim. Power-aware communication optimization for networks-on-chips with voltage scalable links. In *Proc. CODES*, Sept. 2004.

[19] V. Soteriou and L.-S. Peh. Design space exploration of power-aware on/off interconnection networks. In *Proc. ICCD*, Oct. 2004.

[20] M. B. Taylor and et al. The RAW microprocessor: A computational fabric for software circuits and general purpose programs. *IEEE Micro*, 22(2), 2002.

[21] K. Yun and J. Kim. Power-aware modulo scheduling for higher performance vliw processors. In *Proc. ISLPED*, 2000.

System Level and
Circuit Analysis

An Efficient Algorithm for the Analysis of Cyclic Circuits

Osama Neiroukh[*]
Intel Corporation
osaman@ichips.intel.com

Stephen A. Edwards[†]
Columbia University
sedwards@cs.columbia.edu

Xiaoyu Song
Portland State University
song@ece.pdx.edu

Abstract

Compiling high-level hardware languages can produce circuits containing combinational cycles that can never be sensitized. Such circuits do have well-defined functional behavior, but wreak havoc with most logic synthesis and timing tools, which assume acyclic combinational logic. As such, some sort of cycle-removal step is usually necessary for handling these circuits.

We present an algorithm able to quickly and exactly characterize all combinational behavior of a cyclic circuit. It iteratively examines the boundary between gates whose outputs are and are not defined and works backward to find additional input patterns that make the circuit behave combinationally. It produces a minimal set of sets of assignments to inputs that together cover all combinational behavior. This can be used to restructure the circuit into an acyclic equivalent, report errors, or as an optimization aid.

Experiments show our algorithm runs several orders of magnitude faster than existing ones on real-life cyclic circuits, making it useful in practice.

1 Introduction

Cyclic circuits can be produced inadvertently during high-level synthesis and are also the most compact representation for certain circuits such as arbiters [11]. For certain input patterns, such circuits are well-behaved (functional), i.e., do not exhibit oscillations or state-holding behavior. Despite this, most circuit analysis tools forbid the presence of cycles. The central challenge of cyclic circuits is their data-dependent evaluation order, meaning their gates have no topological order. This causes difficulties for many tools such as static timing analyzers that rely on such a static order. Furthermore, applying regular logic simulation to these circuits is cumbersome.

[*]Neiroukh is sponsored by Intel Corporation
[†]Edwards is supported by an NSF CAREER award, a grant from Intel corporation, an award from the SRC, and from New York State's NYSTAR program.

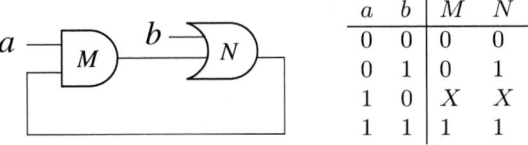

a	b	M	N
0	0	0	0
0	1	0	1
1	0	X	X
1	1	1	1

Figure 1: A trivial cyclic circuit and its truth table

Consider the small cyclic circuit in Figure 1. From its truth table, we see the circuit is well-behaved unless $a = 1$ and $b = 0$. For all other input patterns, the circuit behaves combinationally because the feedback loop is broken by a controlling input on one of the gates. A *partial assignment* is an assignment to one or more inputs to the loop; $\{a = 0\}$ is one such partial assignment. Our algorithm produces a set of partial assignments that provide a concise representation of the conditions under which a cyclic circuit is well-behaved. For example, the set of partial assignments $\{\{a = 0\}, \{b = 1\}\}$ constitutes necessary and sufficient conditions for combinational operation of the circuit in Figure 1: at least one of these must hold in order for the circuit to operate functionally.

In this paper, we present a novel algorithm that can rapidly identify all possible combinational behavior of a cyclic circuit. The algorithm takes a circuit containing one or more loops and produces a set of partial assignments that represent every condition under which the circuit behaves combinationally. Our algorithm relies on the fact that gates such as ANDs and ORs have controlling inputs (0 and 1 respectively) that break feedback loops to aggressively prune the search space. The set of partial assignments our algorithm produces can be used to rule out non-constructive operation of circuits produced by high level compilers such as Esterel [2], or they can be used to create an equivalent acyclic circuit [5].

2 An Example

Consider the cyclic circuit in Figure 2. In general, our algorithm analyzes a circuit one strongly-connected com-

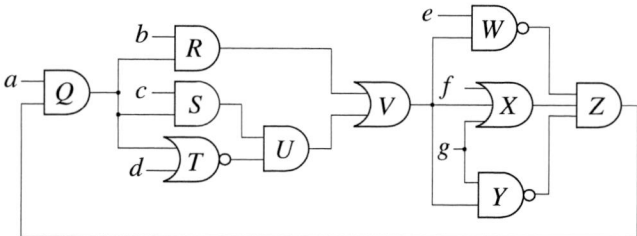

(a) A cyclic circuit

Assignment	Frontier	At Frontier	Acyclic
$\{a = 0\}$	$\{\}$		$\checkmark$
$\{b = 0\}$	$\{V\}$	$R = 0$	
$\{c = 0\}$	$\{V\}$	$U = 0$	
$\{d = 1\}$	$\{V\}$	$U = 0$	
$\{e = 0\}$	$\{Z\}$	$W = 1$	
$\{f = 1\}$	$\{Z\}$	$X = 1$	
$\{g = 0\}$	$\{Z\}$	$Y = 1$	
$\{g = 1\}$	$\{Z\}$	$X = 1$	

(b) First step: applying controlling values to each input in isolation

Gate	Assignment	Frontier	Acyclic
V	$\{b = 0, c = 0\}$	$\{\}$	$\checkmark$
V	$\{b = 0, d = 1\}$	$\{\}$	$\checkmark$
Z	$\{e = 0, f = 1, g = 0\}$	$\{\}$	$\checkmark$

(c) Second step: Merged partial assignments from first step

$\{a = 0\}$
$\{b = 0, c = 0\}$
$\{b = 0, d = 1\}$
$\{e = 0, f = 1, g = 0\}$

(d) Final result: A minimal set of partial assignments that produce all combinational behavior

Figure 2: Illustration of our algorithm.

ponent (SCC) at a time, but this example consists only of a single SCC.

Our goal is to find a small set of partial assignments of values to inputs that, together, "cover" all the combinational behavior of the circuit. That is, we want an input vector to be combinational if and only if it is a subset of one of our partial assignments.

Our algorithm begins by considering applying a controlling value to each input in isolation. Such a controlling value—a 0 input on an AND gate, a 1 applied to an OR gate—by definition forces the output of the gate to a given value regardless of the other inputs. Such inputs are required to "cut" the SCC and make it behave combinationally. We formalize this later in Theorem 1.

Figure 2b summarizes the results of these initial assignments. First, note that when the a input is 0, the circuit is always combinational because 0 is a controlling value on gate Q, effectively breaking the $Z \rightarrow Q$ feedback loop. We in-

clude the assignment $\{a = 0\}$ as part of our minimal cover and will not consider any further assignments that contain $\{a = 0\}$ (Theorem 2).

Consider what happens when we set $b = 0$. Although this is a controlling value for gate R (its output becomes 0 regardless of Q), by itself this is not enough to force the whole circuit to behave combinationally because a 0 on R is a non-controlling value on the OR gate V. We refer to all such gates as the *frontier* induced by a partial assignment (see Definition 4) because they define the boundary between combinational and possibly non-combinational behavior. Think of such gates as being the cause of a logjam; the next step in our algorithm is to break logjams.

The key step in our algorithm, and its main improvement over Edwards [5], attempts to break these logjams by looking for promising combinations of partial assignments that affect the same frontier gates. Only two gates, V and Z, appear in any frontier; we will attempt to set the outputs of these gates by judiciously combining sets of partial assignments that might completely define values at inputs of these gates.

To break the logjam at V, we consider subsets of the three partial assignments that affected its inputs, i.e., $\{b = 0\}$, $\{c = 0\}$, and $\{d = 1\}$. By definition, each of these set at least one of the inputs to V to a non-controlling value (0, because V is an OR gate). We can break the logjam by setting *all* of V's inputs to non-controlling values, i.e., by setting $R = 0$ and $U = 0$. To set $R = 0$, we need $b = 0$, but there are two ways to set $U = 0$: $c = 0$ and $d = 1$. Thus we decide to consider the partial assignments $\{b = 0, c = 0\}$ and $\{b = 0, d = 1\}$ in the next step.

Similar reasoning about frontier gate Z leads us to want to set $W = 1$, $X = 1$, and $Y = 1$. There appear to be two ways to do this by combining existing assignments, i.e., through $\{e = 0, f = 1, g = 0\}$ and $\{e = 0, g = 1, g = 0\}$. However, the latter one is nonsensical because we cannot set g to be both 0 and 1 simultaneously. We consider the partial assignments $\{g = 0\}$ and $\{g = 1\}$ to be *in conflict* and refuse to merge them.

Figure 2c lists the three new partial assignments we consider along with the frontier gate that induced them. Each partial assignment leads to an empty frontier and (therefore) an acyclic circuit. Our algorithm terminates and returns the partial assignments listed in Figure 2d.

3 Prior Work

In 1970, Kautz [7] showed that the minimal form of certain circuits contained combinational loops. Rivest [11] came to a similar conclusion, suggesting that combinational loops are more than just a nuisance. Stok [13] observed how they can arise from resource-sharing in high-level synthesis, motivating Malik's work [8] on analyzing combinational circuits, a forerunner of our work. Malik showed an equivalence between combinational cyclic circuits and least-fixed-points in three-valued simulation, an idea that Shiple, Berry, and Touati [12] applied to the Esterel language [2,3], whose hardware translation [1] often produces combinational cycles. Their approach uses a symbolic state-space traversal followed by an $O(n^2)$ replication procedure to remove cycles. Our algorithm pays more attention to both the structure and function of the circuit and, when coupled with the resynthesis technique of Edwards [5], produces smaller circuits. The BDD-based algorithm of Halbwachs and Maraninchi [6] takes a brute-force approach, ignoring the structure of the circuit. Namjoshi and Kurshan [9] take a very different approach, showing that any fixed-point is interesting, not just the least. Their analysis merely answers whether a circuit is combinational.

Recently, Riedel and Bruck [10] applied Rivest's observations to synthesize very compact combinational circuits that contain cycles. As part of their synthesis step, they check whether the circuit they generated is combinational using a fairly expensive BDD construction; our algorithm could potentially be used in that setting. More practically, the cyclic combinational circuits they generate have topologies complex enough to stymie the de-cyclification algorithm of Edwards [5], which our work builds on.

Our algorithm is a drop-in replacement for the first half of Edwards [5], which enumerates all the conditions under which a circuit is combinational then merges the resulting circuit fragments. Edwards's algorithm gets mired in considering using every input to an SCC to break a cycle; our algorithm is much more shrewd. When it finds a gate that might participate in a non-combinational cycle, it uses the behavior of simulations it performed earlier to work backward to identify primary inputs that will break the cycle. This reduces the number of input patterns the algorithm considers and hence greatly reduces its running time.

4 Notation and Definitions

This section defines the basic terminology necessary for explaining material in this paper.

We represent circuits with a *directed graph* (*digraph*). A digraph G is a pair (V, E) where V is a set of vertices and E is a set of edges. An *edge* is an element of $V \times V$ with distinct vertices. We represent a circuit as a digraph whose vertices correspond to gates and whose edges correspond to nets. A *controlling value* for a gate G is the value that applied to any input of G uniquely determines G's output independent of other inputs. To simplify our exposition, we only consider simple logic gates: NOT, AND/NAND, and OR/NOR. This is not a limitation as more complex gates can be represented as combinations of these gates.

Definition 1. *A strongly connected component* (SCC) *of a*

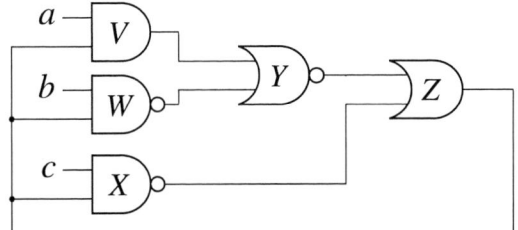

Figure 3: Cyclic circuit for illustrating definitions

digraph $G = (V, E)$ is a maximal subset of vertices $C \subseteq V$ such that any vertex in C is reachable from any other vertex in C. Inputs of an SCC are inputs of gates that are part of the SCC that are not driven by gates inside the SCC.

Figure 3 shows a circuit with a single SCC. Nets a, b, and c are inputs to the SCC. When analyzing a circuit, we first decompose it into SCCs using a standard algorithm [4]. If the input circuit contains more than one SCC, we consider each SCC separately in a topological order.

Our analysis methodology and logic simulation use a ternary domain consisting of $\{0, 1, X\}$ where X denotes an unknown digital value.

Definition 2 (Malik [8]). *A circuit is* combinational *for an input assignment if three-valued simulation starting with all internal nodes set to X resolves the output of every gate in the circuit to either 0 or 1 under the assignment.*

Literature on cyclic circuits also refers to this behavior as "well-behaved" and "constructive" [12]. Combinational behavior is equivalent to stating that the circuit behaves as if it were acyclic with no X's and no oscillations.

Definition 3. *A partial assignment (PA) is a set of assignments to one or more nets of a circuit.*

In this work, we shall be only concerned with partial assignments to inputs of SCCs. A valid PA for the circuit in Figure 3 is an assignment to one or more of the inputs $\{a, b, c\}$, such as $\{a = 0\}, \{b = 0, c = 1\}$, or $\{b = 1, c = 1\}$.

Definition 4. *The* controllability frontier *of a PA, or* frontier *for short, is the set of gates that have at least one input assigned but whose output is X.*

The frontier captures the notion of a boundary between gates whose output is defined and those whose output is not. When calculating the frontier for a PA, we use ternary simulation to propagate the SCC inputs as far as possible then check for cyclic behavior. For example, for Figure 3,

Partial Assignment	Frontier
$\{a = 1\}$	$\{V\}$
$\{a = 0, b = 1\}$	$\{Y, W\}$
$\{c = 0\}$	$\{\}$

5 Our Algorithm

Here, we describe our algorithm for rapidly extracting a cover for all combinational behavior of a cyclic circuit.

5.1 Theoretical Background

We start with a set of theorems that are key to the correctness and efficiency of our algorithm. The first two are due to Edwards [5].

Theorem 1 (Edwards [5]). *For a circuit with a strongly-connected component (SCC) to behave combinationally, at least one input to a gate in the SCC must be driven to a controlling value.*

Controlling assignments to SCC inputs for the circuit in Figure 3 are $a = 0$, $b = 0$, and $c = 0$. Theorem 1 tells us that at least one of these is required for combinational behavior. We use this property to seed our search space with a pool of PAs, each corresponding to a controlling assignment to an SCC input. Any combinational behavior is guaranteed to be present in combinations of one or more of these PAs.

Theorem 2 (Edwards [5]). *If a partial assignment* p *is combinational, then any further assignments that do not contradict any in* p *can also be computed combinationally by the circuit fragment implied by* p.

Consider the PA $\{c = 0\}$ applied to Figure 3. This breaks the connectivity of the SCC, making the circuit behave combinationally. This theorem indicates that additional assignments beyond $\{c = 0\}$ cannot reverse the combinational behavior already implied by this PA. This theorem allows us to avoid further consideration of acyclic PAs once we have identified them. This supports one of our objectives for the algorithm: generation of *minimal* PAs that capture all combinational behavior. We explain the notion of minimal PAs in Section 5.3.

This relates frontiers and combinational behavior:

Theorem 3. *A PA makes a circuit combinational if and only if its frontier is empty.*

Proof. If part: If the frontier is empty, then either no gates have any inputs assigned or none have an output of X. From Theorem 1, we know that at least one gate must be driven by a controlling value for combinational behavior. If none have an output of X, then the circuit under that PA is combinational by definition.

Only if part: This follows directly from definition of combinational behavior. ☐

Our algorithm records the frontier associated with each PA and uses them to look for opportunities to merge PAs to extend their frontiers.

Algorithm 1 Given a circuit, return a minimal set of PAs that together cover all combinational behavior.

```
 1:  A = ∅                    ▷ Set of acyclic PAs, the eventual result
 2:  K = ∅                    ▷ All known cyclic PAs, used for merging
 3:  Clear F                  ▷ A map from frontier gate → set of PAs
 4:  while circuit has SCCs
 5:      Find next SCC
 6:      P = controlling values for SCC inputs      ▷ Initial PAs
 7:      while P ≠ ∅
 8:          G = ∅                ▷ Frontier gates for this iteration
 9:          foreach p ∈ P          ▷ Consider each candidate PA
10:              simulate p
11:              if circuit is combinational under p then
12:                  add p to A
13:              else
14:                  add p to K   ▷ Remember the PA for merging
15:                  foreach gate g in the frontier induced by p
16:                      add g to G     ▷ Record the frontier gate
17:                      add p to F(g)  ▷ Remember p induced g
18:          P = ∅                ▷ Compute new candidate PAs
19:          foreach frontier gate g ∈ G
20:              if |F(g)| > 1 then     ▷ Need ≥ 2 PAs to merge
21:                  add each PA from mergeAtGate(K, g) to P
22:  return A
```

5.2 Searching for combinational behavior

Algorithm 1 is our technique for identifying all combinational behavior. The algorithm takes a circuit with any number of SCCs and produces a set of PAs under which the circuit is combinational. These PAs control SCC inputs.

The algorithm attacks one SCC at a time (line 4), finding a minimal set of covering partial assignments for each. For each SCC, it begins by considering partial assignments that place a single controlling value on each SCC input (line 6), then enters into a loop (lines 7–21) in which it alternates between testing whether any of the currently-considered partial assignments (the set P) induce combinational behavior (lines 10–17) and attempting to merge already-observed partial assignments (the set K) to generate a new set of PAs (lines 18–21). Its goal in this second phase is to break logjams by combining PAs to set the outputs of the latest set of frontier gates it has discovered. The map F records partial assignments that affect frontier gates: if g is a gate, then $F(g)$ is the set of all partial assignments that put at least one non-controlling value at an input of g.

Algorithm 1 is guaranteed to find all combinational behavior within in the subject circuit. Starting from individual controlling inputs into SCCs, our frontiers allow us to identify all opportunities where PAs can merge to extend controllability over more gates in an SCC. As we merge these PAs and continue the searching, other acyclic PAs are explored. We continue this cycle of search and merge terminating when we fail to generate new PAs.

Name	Assignment
p_0	$\{a = 1\}$
p_1	$\{b = 0, c = 1\}$
p_2	$\{c = 1, d = 1\}$
p_3	$\{e = 0\}$
p_4	$\{b = 1, f = 1\}$

(a) PAs

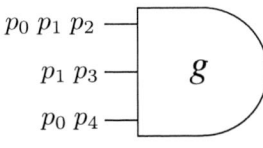

(b) How these PAs control the inputs of this gate

$$p_0 \cup p_1 = \{a = 1, b = 0, c = 1\}$$
$$p_0 \cup p_3 = \{a = 1, e = 0\}$$
$$p_2 \cup p_3 \cup p_4 = \{b = 1, c = 1, d = 1, e = 0, f = 1\}$$

(c) New PAs generated by merging

Figure 4: Merging PAs at a gate. If the five PAs in (a) control the three inputs on the gate (b), the merging algorithm (Algorithm 2) will generate three new partial assignments (c) by merging the five existing ones. By construction, each controls all three of the gate inputs.

5.3 Merging partial assignments

Here, we describe a key algorithm used by Algorithm 1: the generation of new partial assignments to break the logjam at a frontier gate. Given a set of PAs and a gate, Algorithm 2 generates a set of PAs that apply non-controlling values to *every* input of the gate, thus setting its output.

We store PAs in a simulated state that captures all assigned nodes and their values. Algorithm 1 only tries to merge PAs for a gate when at least two PAs set an input on the gate. Merging attempts to produce new PAs by propagating known values across these frontier gates to extend the set of gates whose output is not X.

Consider the example in Figure 4. This shows a 3-input gate g that is a frontier gate for partial assignments $p_0, p_1, \ldots, p_4$. Note that a gate can only be a frontier for a PA if that PA puts a non-controlling value on one or more of the gate's inputs. We wish to consider merging these PAs in order to extend the frontier beyond g. A desirable merge of PAs at a gate g must satisfy the following:

i) *Gate Cover*: The PAs to be merged must define every input of g.
ii) *Consistency*: The PAs to be merged must not contain conflicting assignments to inputs. In the example in Figure 4, partial assignments p_1 and p_4 cannot be combined due to a conflicting assignment for b.
iii) *Completeness*: PAs must be merged such that all permissible combinations are considered. The example in Figure 4 provides some degrees of freedom to cover every input that must all be considered. This ensures that our final PAs encapsulate both necessary and sufficient conditions for combinational behavior.

Algorithm 2 Return a set of PAs that apply non-controlling values to every input of a gate

```
1: function MERGEATGATE(K, g)
2:     R = ∅                              ▷ Generated set of PAs
3:     foreach input i of gate g
4:         p_i = PAs in K that set i and induce g as a frontier
5:         if p_i = ∅ then return ∅  ▷ Cannot control some input
6:         foreach P ∈ p_1 × p_2 × · · · × p_k    ▷ All combinations
7:             if the partial assignments in P do not conflict then
8:                 add minimize(merge(P)) to R
9:     return R
```

iv) *Minimality*: The merged PAs must not contain any PA that can be removed while satisfying the previous conditions. For example, the merge candidate $p_0 \cup p_3 \cup p_4$ is rejected since p_0 dominates p_4 (i.e., p_0 controls both the first and third gate input; p_4 only control the third). This condition is important for two reasons: it keeps the final output PAs as concise as possible by not including redundant conditions. Such redundancy burdens subsequent stages of the algorithm as it increases memory usage and makes testing of merge conditions against other candidate PAs more tedious.

We note that merging PAs is a sort of Binate Covering Problem (BCP) (it is covering because we must cover all gate inputs; it is binate because conflicts between PAs prevent certain combinations). However, the need for a complete enumeration is not a usual requirement in traditional BCP applications. In the context of merging PAs, the domain of the problem is rather small and makes enumeration tractable. We use an explicit enumeration algorithm with provisions for removing conflicts and minimizing merged PAs to eliminate any dominated PAs (Algorithm 2).

6 Experimental Results

We implemented our algorithm in C++ using the Standard Template Library and tested it on a number of cyclic circuits. We report execution times and the number of partial assignments considered compared to Edwards [5] in Table 1. The first four circuits come from Esterel programs [2] and contain simple loops. The rest are outputs of Riedel's *cyclify* [10] and are more complex. Our algorithm consistently runs more than two orders of magnitude faster than Edwards [5]. Also, our program is able to process many more candidate PAs in less time, which we attribute to removing the more expensive operations in Edwards, including the superset check against known-combinational PAs.

0-7695-2533-4/06 $20.00 © 2006 IEEE

Circuit	Netlist Gates	SCC Gates	Edwards [5]		Our Approach		Acyclic PAs
			PAs considered	runtime	PAs considered	runtime	
arbiter5	213	25	257	1.3	25	0.1	14
arbiter6	248	30	745	8	29	0.1	16
arbiter7	283	35	2205	69	33	0.2	18
arbiter8	318	40	6581	656	37	0.3	20
exp	124	69	54517	2868	23260	2	338
ex1	150	47	43777	2341	232	1	10
gary	177	32	-	-	290	0.6	11
planet	253	51	-	-	1489	0.3	22
s1488	272	61	-	-	588	0.2	89
table3	311	49	-	-	3604	1	38

Table 1: Experimental Results: Runtimes are in seconds; a dash indicates the algorithm did not terminate after one hour.

7 Conclusions

We presented a new algorithm for identifying all the combinational behavior of a cyclic circuit. The algorithm is useful for evaluating cyclic specifications that often arise from high-level synthesis [2, 3]. One application of our algorithm is transforming cyclic combinational circuits to an acyclic equivalent; it replaces the first half of the procedure described by Edwards [5].

The chief contribution of our work is a speed improvement of several orders of magnitude over Edwards [5] due to much more clever pruning of the search space. It is therefore able to deal with practical-sized cyclic circuits.

Our algorithm analyzes all possible inputs into SCCs without considering whether such patterns can in fact occur in the original circuit (i.e., whether they are controllability don't-cares). This saves us from performing an image computation on the surrounding circuit, making the analysis much faster. However, it is possible that considering the don't-care set would reduce the number of PAs we consider and further speed the search. We have yet to explore the trade-off between computing don't-cares and reducing the number of PAs.

Although our algorithm performs quite well, it can be improved further. The current performance bottleneck arises when merging PAs at a frontier gate to produce more PAs to consider. Most of our PAs are generated here and most are later discarded. A more clever approach, perhaps Espresso-based, might reduce both the number of new PAs generated and the time it takes to derive them.

Independent of these further refinements, we have presented a practical algorithm that is able to quickly characterize all the combinational behavior of a realistic-sized cyclic circuit. Our intended application is the construction of an acyclic equivalent of a cyclic circuit to make it palatable to existing synthesis tools, but we believe our algorithm has other important applications in analysis and formal equivalence verification of cyclic circuits.

References

[1] G. Berry. Esterel on hardware. *Philosophical Transactions of the Royal Society of London. Series A*, 339:87–103, Apr. 1992. Issue 1652, Mechanized Reasoning and Hardware Design.

[2] G. Berry. The constructive semantics of pure Esterel. Draft book, 1999.

[3] G. Berry. *The foundations of Esterel*. MIT Press, 2000.

[4] T. H. Cormen, C. E. Leiserson, R. L. Rivest, and C. Stein. *Introduction to Algorithms*. MIT Press, second edition, 2001.

[5] S. Edwards. Making cyclic circuits acyclic. In *Proc. Design Automation Conference*, pages 159–162, 2003.

[6] N. Halbwachs and F. Maraninchi. On the symbolic analysis of combinational loops in circuits and synchronous programs. In *Proc. Euromicro*, pages 345–348, 1995.

[7] W. Kautz. The necessity of closed circuit loops in minimal combinational circuits. *IEEE Trans. Comput.*, C-19:162–164, Feb. 1970.

[8] S. Malik. Analysis of cyclic combinational circuits. *IEEE Trans. Computer-Aided Design*, 13(7):950–956, July 1994.

[9] K. S. Namjoshi and R. P. Kurshan. Efficient analysis of cyclic definitions. In *Computer Aided Verification*, volume 1633 of *LNCS*, pages 394–405, Trento, Italy, July 1999.

[10] M. Riedel and J. Bruck. The synthesis of cyclic combinational circuits. In *Proc. Design Automation Conference*, pages 163–168, 2003.

[11] R. L. Rivest. The necessity of feedback in minimal monotone combinational circuits. *IEEE Trans. Comp.*, 26(6):606–607, 1977.

[12] T. Shiple, G. Berry, and H. Touati. Constructive analysis of cyclic circuits. In *Proc. European Design and Test Conf.*, pages 328–333, 1996.

[13] L. Stok. False loops through resource sharing. In *Proc. International Conference on Computer-Aided Design*, pages 345–348, 1992.

0-7695-2533-4/06 $20.00 © 2006 IEEE

Improving System Level Design Space Exploration by Incorporating SAT-Solvers into Multi-Objective Evolutionary Algorithms

Thomas Schlichter,[*] Martin Lukasiewycz, Christian Haubelt, and Jürgen Teich
Department of Computer Science 12
University of Erlangen-Nuremberg, Germany
{schlichter, haubelt, teich}@cs.fau.de

Abstract

Automatic design space exploration at the system level is the task of finding optimal or close to optimal mappings for a set of applications onto an optimized architecture. Especially, finding a feasible binding of processes onto resources that permit the communications imposed by data dependencies is known to be a $\mathcal{NP}$-complete task which demands the use of heuristic optimization approaches. Nearly all optimization approaches known from literature will fail in design spaces containing only a few feasible solutions. In this paper, we propose a novel approach based on the combination of Multi-Objective Evolutionary Algorithms and SAT-solvers to overcome these drawbacks. We will provide experimental results showing the efficiency of our novel methodology for synthetic and real life test cases.

1. Introduction

Modern embedded systems often consist of many communicating processor cores. The challenge in designing such heterogeneous multi-processor systems is to find optimal implementations with regard to multiple objectives while meeting several constraints. In order to allow an unbiased search, the task of *design space exploration* is performed before selecting (*decision making*) the actual implementation. Design space exploration is a very challenging constrained multi-objective optimization task [3]. The basic problem is the selection of appropriate hardware resources and the assignment of processes to the selected resources.

Due to data dependencies among the processes, nearly all possible implementations may be *infeasible*. Finding a feasible solution is known to be an $\mathcal{NP}$-complete task [3]. Thus, many researchers propose the use of Multi-Objective Evolutionary Algorithms (MOEAs) to solve these problems [8, 7]. In former work, we have proposed a method that integrates symbolic techniques using BDDs into MOEAs, that

guide the search towards the feasible region [11]. Nevertheless, BDDs cannot be used to solve the complete problem of always finding a feasible solution if there is one at all, as the size of a BDD grows exponentially with the number of variables.

In this paper we will present an even more sophisticated decoding technique based on another symbolic method known from the formal verification area: SAT-solver [1]. Typical for SAT-solvers, this approach is not memory limited, and for our test cases we could also see that the runtime is comparable to other methods.

The remaining of this paper is structured as follows: Section 2 contains the definition of the search space and the task of design space exploration. In Section 3, we describe how this optimization problem can be solved using Multi-Objective Evolutionary Algorithms. Section 4 compares a straight forward decoding technique with the new one based on a SAT-solver. We will provide experimental results showing the advantages of the SAT decoding technique in Section 5. Finally, Section 6 concludes the paper.

2. Problem Statement

In this paper, we consider the problem of design space exploration for embedded systems. Basically, the design space exploration problem is a constrained multi-objective selection and assignment problem.

To allow for a mathematical model of the search space, the concept of a so-called *specification graph* is needed. A specification graph specifies a multi-processor system by means of its applications, the possible architecture, and the relation between these two views. Here, we use a graph-based approach already proposed by Blickle et al. [3].

Definition 1 (Specification Graph [3]) *A   specification graph is a directed graph $g_s(V_s, E_s)$ that consists of a process graph $g_p(V_p, E_p)$, an architecture graph $g_a(V_a, E_a)$, and a set of mapping edges E_m. In particular, $V_s = V_p \cup V_a$, $E_s = E_p \cup E_a \cup E_m$, where $E_m \subseteq V_p \times V_a$.*

---

[*]supported by the Fraunhofer IIS, Germany

0-7695-2533-4/06 $20.00 © 2006 IEEE

Consequently, mapping edges relate the vertices of the process graph to vertices of the architecture graph. The edges represent user-defined mapping constraints in the form of a relation: "can be implemented by".

The goal of design space exploration is to find optimal solutions which satisfy the specification given by the specification graph. Such a solution is called a *feasible implementation* of the embedded systems.

Our optimization algorithm uses evolutionary algorithms to find optimal implementations. Usually there are several conflicting optimization goals (e.g. speed vs. energy consumption), thus there are many optimal implementations.

An implementation, consists of three parts: (1) the *allocation* that indicates which elements of the architecture graph are used in the implementation, (2) the *binding*, i.e., the set of mapping edges which define the binding of processes to resources of the architecture graph, and (3) the *schedule* assigning a start time to each operation in the process graph.

Before defining the term *implementation* formally, Blickle et al. [3] introduce the so-called *activation* of vertices and edges:

Definition 2 (Activation [3]) *The* activation *of a specification graph* $g_s(V_s, E_s)$ *is a function* $a : V_s \times E_s \mapsto \{0, 1\}$ *that assigns to each edge* $e \subset E_s$ *and to each vertex* $v \in V_s$ *the value* 1 *(activated) or* 0 *(not activated).*

For the sake of simplicity, it is assumed that all vertices $v \in V_p$ and all edges $e \in E_p$ of the process graph g_p are activated subsequently. So only the vertices $v \in V_a$ of the architecture graph and the edges $e \in E_a \cup E_m$ can be either activated or deactivated.

An *allocation* α of a given specification graph g_s is the subset of all activated vertices and edges of the architecture graph g_a, i.e., $\alpha = \{v \in V_a \mid a(v) = 1\} \cup \{e \in E_a \mid a(e) = 1\}$. A *binding* β of a given specification graph g_s is the subset of activated mapping edges E_m, i.e., $\beta = \{e \in E_m \mid a(e) = 1\}$.

In order to restrict the search space, it is useful to determine the set of *feasible allocations* and *feasible bindings*. A feasible binding guarantees that communications demanded by the process graph can be established in the allocated architecture. This property makes the resulting optimization problem $\mathcal{NP}$-complete [3].

Definition 3 (Feasible Binding) *Given a specification graph* g_s *and an allocation* α, *a* feasible binding *is a binding* β *that satisfies the following requirements:*

1. *Each activated mapping edge* $e \in \beta$ *ends at an activated vertex, i.e.,* $\forall e = (v_p, v_a) \in \beta : v_a \in \alpha$.

2. *For each process graph vertex* $v_p \in V_p$, *exactly one outgoing mapping edge* $e \in E_m$ *is activated, i.e.,*

$$\left| \{e \in \beta \mid e = (v_p, v_a), v_a \in V_a\} \right| = 1.$$

3. *For each process graph edge* $e \in (v_i, v_j) \in E_p$:

- *either both operations are mapped onto the same vertex, i.e.,* $\tilde{v}_i = \tilde{v}_j$ *with* $(v_i, \tilde{v}_i), (v_j, \tilde{v}_j) \in \beta$,

- *or there exists an activated edge* $\tilde{e} = (\tilde{v}_i, \tilde{v}_j) \in E_a \cap \alpha$ *in the architecture graph to handle the communication associated with edge* e, *i.e.,*

$$(\tilde{v}_i, \tilde{v}_j) \in E_a \cap \alpha \text{ with } (v_i, \tilde{v}_i), (v_j, \tilde{v}_j) \in \beta.$$

The term *feasible allocation* is used to indicate an allocation α that allows at least one feasible binding β.

Definition 4 (Implementation) *Given a specification graph* g_s, *a (feasible) implementation* ψ *is a triple* (α, β, τ) *where* α *is a feasible allocation,* β *is a corresponding feasible binding, and* τ *is a schedule.*

Now, the task of system synthesis can be formulated as a combinatorial *Multi-objective Optimization Problem*.

Definition 5 (System Synthesis) *The task of* system synthesis *is the following multi-objective optimization problem (MOP) where without loss of generality, only minimization problems are assumed here:*

minimize $f_1(x), f_2(x), \ldots, f_n(x)$,
subject to:
    x represents a feasible implementation ψ

where $f(x) = (f_1(x), f_2(x), \ldots, f_n(x)) \in Y$ *is the objective function,* Y *is the objective space,* $x = (x_1, x_2, \ldots, x_m) \in X$ *is the decision vector and* X *is the decision space.*

3. Design Space Exploration

In this section, we will summarize how to solve the system synthesis problem by using MOEAs [5]. As proposed in [3], the MOEA determines the optimal allocation and bindings with respect to different objective functions. Afterwards, the schedule of an implementation is computed by a list scheduler.

The specification graph consisting of the process graph, the architecture graph, and a set of mapping constraints is used by an MOEA to determine the optimal implementations. Due to several even conflicting objective functions, multi-objective optimization problems generally do not contain only one global optimum, but a set of so-called *Pareto points*. A Pareto-optimal implementation is a design which is not worse than any other feasible solution in the design space in all objectives.

The MOEA requires a meaningful *encoding* for an implementation that consists of an allocation and a binding. Obviously, if allocations and bindings may be randomly

a) Specification graph:

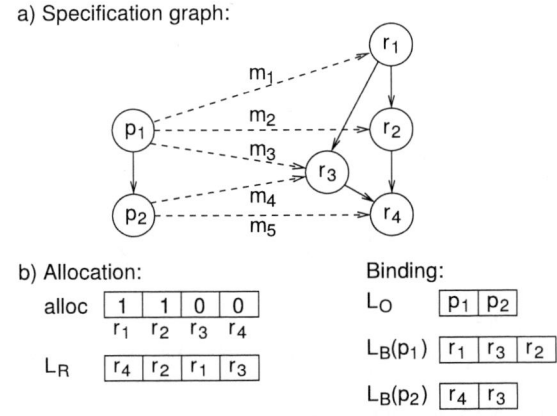

b) Allocation:

alloc

1	1	0	0
r_1	r_2	r_3	r_4

L_R

r_4	r_2	r_1	r_3

Binding:

L_O

p_1	p_2

$L_B(p_1)$

r_1	r_3	r_2

$L_B(p_2)$

r_4	r_3

Figure 1. a) An example specification graph g_S with b) an individual consisting of the allocation bitmap *alloc* **and different priority lists.**

chosen, a lot of them can be infeasible. Here, we use a repairing strategy to partially repair infeasible solutions.

The allocation of resources is directly encoded in the so-called *individual*, which contains the structures shown in Figure 1 b). The bit vector *alloc* encodes for each resource $v_a \in V_a$ if it is activated or not. As this simple encoding may result in many infeasible allocations, there is a ordered *repair allocation list* L_R, that contains all resources $v_a \in V_a$ and allows to repair the allocation. A first simple heuristic only adds new resources $v_a \in V_a$ to the allocation and reflects the simplest case of infeasibility that may arise from non-executable processes $v_p \in V_p$: Consider the set $V_B \subseteq V_p$ that contains all processes that can not be executed, because not a single corresponding resource vertex is allocated, i.e., $V_B = \{v \in V_p \mid \forall (v, \widetilde{v}) \in E_m : a(\widetilde{v}) = 0\}$. To make the allocation feasible (in this sense), we add for each $v \in V_B$ one $\widetilde{v} \in V_a$, until feasibility in the sense above is achieved.

Besides the allocation, also the binding is encoded in the individual. Therefore the *binding order list* L_O is used, which indicates in which order to bind the processes $v_p \in V_p$. For each of these processes the individual also contains a *binding priority list* L_B, which contains the resources connected to this process by the mapping edges $e \in E_m$. These priority lists allow to decode every individual to a feasible implementation, but this decoding is a hard task.

In this paper, this decoding of an individual is of special interest and will be discussed in more detail in the next section.

4. Integrating SAT-Solvers into the MOEA

Former methods used to determine a feasible implementation often fail in design spaces with just a few feasible bindings, because of their local view on the specification.

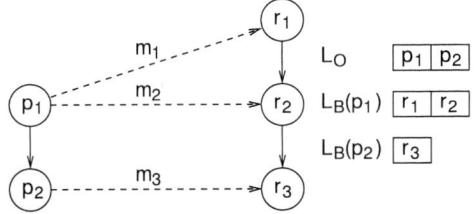

Figure 2. The sequential feasibility checker does not find a solution with the given binding order list L_O and binding priority lists L_B.

In the following, we will show that a) using a SAT-solver allows to have a global view on the specification and therefore guarantees to always find a feasible binding for a given allocation. b) Furthermore, if there is no feasible binding for a given allocation, the SAT-solver allocates additional resources to reach a feasible implementation.

The following subsections show the functionality of a simple heuristic called *sequential feasibility checker* that is used by existing design space exploration tools and will be used for comparison in Section 5. Here, we also present a symbolic representation of the complete specification and how a SAT-solver finds a feasible implementation using the information encoded in the individual.

4.1. Sequential Feasibility Checker

In this fairly simple constructive feasibility checker, each process is bound in the order it appears in the binding order list L_O of the individual. When binding a process, the different mapping edges are tested in the order of the binding priority list L_B. If the tested mapping edge is feasible with the binding performed up to this test, it is used.

So, which mapping edges are activated strongly depends on both the binding order list L_O and the binding priority list L_B. As only the yet bound processes are considered, the feasibility checker might be trapped, even if the allocation is feasible. An example is shown in Figure 2. Using the given binding order list L_O and the given binding priority lists L_B, the sequential feasibility checker does not find the valid binding $\beta = \{(p_1, r_2), (p_2, r_3)\}$. This happens because p_1 is first bound to r_1 what does not conflict with any already activated mapping edge (as there is none of that kind). However, that binding prohibits a feasible mapping of process p_2. As the sequential feasibility checker does not modify a chosen binding, it fails to find the valid binding.

4.2. Symbolic Representation of the Specification

The problem of finding a feasible implementation can be transformed into a satisfiability problem [6, 11, 10] by assigning Boolean variables to mapping edges and resources.

0-7695-2533-4/06 $20.00 © 2006 IEEE

These Boolean variables indicate if the associated element is activated. In order to apply a common SAT-solver, the function has to be given in conjunctive normal form (CNF). This can be done by the following interpretation of the three requirements from Definition 3:

The first requirement states that each activated mapping edge $m = (p, r)$ has to end at an activated resource r:

$$\bigwedge_{m=(p,r)\in E_\mathrm{m}} \overline{\mathrm{a}(m)} \vee \mathrm{a}(r) \qquad (1)$$

The second requirement states that exactly one mapping edge is activated for each process p, what can be split up into two statements. At least one mapping edge m has to be activated for each process:

$$\bigwedge_{p\in V_\mathrm{p}} \bigvee_{m=(p,r)\in E_\mathrm{m}} \mathrm{a}(m) \qquad (2)$$

And at most one mapping edge has to be activated for each process.

$$\bigwedge_{m_i=(p,r_i),m_j=(p,r_j)\in E_\mathrm{m}: r_i\neq r_j} \overline{\mathrm{a}(m_i)} \vee \overline{\mathrm{a}(m_j)} \qquad (3)$$

The third and last requirement states that communicating processes have to be mapped to the same or to an adjacent resource. This can be expressed by the following equation:

$$\bigwedge_{\substack{m_i=(p_i,r_i)\in E_\mathrm{m}:\\(p_i,p_j)\in E_\mathrm{p}}} \left[\overline{\mathrm{a}(m_i)} \vee \bigvee_{\substack{m_j=(p_j,r_j)\in E_\mathrm{m}:\\r_i=r_j\vee(r_i,r_j)\in E_\mathrm{r}}} \mathrm{a}(m_j) \right] \qquad (4)$$

There exists a feasible implementation iff the conjunction of Equation (1) - (4) is satisfiable. A SAT-solver can be used to determine a variable assignment representing a feasible solution. In the following section we will present a decoding algorithm using a SAT-solver. The advantage of this SAT-solver decoding technique is that it guarantees to find a feasible solution if such exists.

4.3. Davis-Putnam Decision Strategy Adjustment

To be able to decode an individual, the SAT-solver must regard the encoded information. Here we describe how to manage this by using a complete SAT-solver based on the Davis-Putnam (DP) backtrack search algorithm [4].

First, a decision strategy selects an unassigned variable which will be set to a fixed Boolean value. If the resulting CNF is recognized as unsatisfiable, the SAT-solver tries to resolve this conflict by a backtracking procedure. This continues until all variables are assigned, and this assignment represents a feasible solution. For our system-synthesis problem we developed a problem-specific decision strategy that processes the encoded information of the individual in the following three steps:

1. The Boolean value 1 is assigned to all allocated resources. The order does not matter as an allocation of a resource does not prohibit a feasible binding.

2. The Boolean value 0 is assigned to every not activated resource in the reverse order of the allocation priority list L_R. This ensures the sequential addition of resources respecting the priority in L_R if the CNF is unsatisfiable with the given resource allocation.

3. The SAT-solver tries to find a feasible binding based on the allocation determined in step 1 and step 2. The Boolean variables representing mapping edges are set to 1 regarding the order of the binding order list L_O and the corresponding binding priority list L_B. If this assignment prohibits a feasible binding, the next variable in the binding priority list is tested. If the end of the binding priority list is reached without finding a feasible assignment, the backtracking algorithm will automatically return to step 2 and add additional resources.

5. Experimental Results

In this section, we present experimental results from using a) the sequential feasibility checker respectively b) the SAT decoding technique. The PISA (Platform independent Interface for Search Algorithms [2]) framework was chosen for optimization purposes. In the present work, the SPEA2 selection procedure [12] was applied. For the SAT decoding technique, we used the zChaff [9] SAT-solver.

5.1. Problem Instances and Parameters

We created nine different classes of MOP instances (specification graphs). First we created three classes that differ in size. These are generated from following parameters: (i) The number of available resources in the architecture graph is 25, 50, respectively 75. (ii) The number of processes in the process graph g_p is either 50, 100, or 150. (iii) Each process has $3\ldots6$, $4\ldots8$, respectively $5\ldots9$ random mapping edges. (iv) The number of edges in the process graph is given by a probability value. This value determines the probability that two processes are connected by an edge and is 25% for every problem class. All these values are typical in system level design.

Then for each of these three problem classes we created three different subclasses with feasibility probabilities 15%, 25%, and 45%. This probability is used when the edges E_a of the architecture graph are created. For each possible mapping edge $m_i = (p_i, r_i), m_j = (p_j, r_j) \in E_\mathrm{m}$ of two adjacent processes p_i and p_j, the resources r_i and r_j are connected with this probability to satisfy the data dependency.

0-7695-2533-4/06 $20.00 © 2006 IEEE

Smaller probability values result in less created edges, and less feasible solutions exist.

For each of these nine problem classes we created 10 different problem instances. The genetic algorithm was run 10 times for each problem instance with both techniques.

We chose the parameters for the MOEA as follows: The population size was set to 100. For recombination, 25 children were created from 25 parents by single-point crossover. The mutation rate was set to $|decision\ variables|^{-1}$. The mutation operation is either single bit flip or order-based mutation.

5.2. Quantitative Results

To allow a comparison between the different optimization runs X_a, we combined all non-dominated points of a problem instance into a reference set X_R. This comparison was done using the ε-*dominance* performance indicator which scales the normalized points of the reference set X_R with an ε-value until this set does not dominate the particular set X_a anymore. So, smaller ε-values are better, and an ε-value of 1 means that the optimization run exactly reached the reference set. A detailed discussion on performance indicators can be found in [13].

Each optimization run is 1000 generations, and for every generation the average time to reach this generation is calculated and used to draw Figure 3. For each of the nine different problem classes, the average ε-values are calculated from the 100 different optimization runs. An optimization run only contributes to the average values after it found at least one valid solution.

Figure 3 a) shows, that the sequential feasibility checker is competitive to the SAT decoding technique for problems with a small number of processes and many feasible solutions. The less feasible solutions exist, the worse performs the sequential feasibility checker compared to the SAT decoding technique.

In Figure 3 b) one can see that the sequential feasibility checker becomes worse for more processes. The spikes in the curve for the sequential feasibility checker with a feasibility of 15% are due to optimization runs that find the first feasible solutions quite late.

Figure 3 c) indicates that the SAT decoding technique even performs well for a large number of processes compared to the sequential feasibility checker. The curve of the sequential feasibility checker with a feasibility of 15% shows that only a single run out of 100 was able to find any solutions at all.

If we compare the times for the different optimization runs, we can see that the runs of both decoding techniques take a longer time for bigger problems. This is due to the increased number of mapping edges, processes and resources which increase the runtime of the different quality functions. We can also see that the SAT-solver requires more time to compute 1000 generations than the sequential feasibility checker. This has two reasons:

1. The SAT decoding technique requires slightly more time to compute a binding for each individual (but one has to keep in mind that it always finds a feasible binding by adjusting the resource allocation properly).

2. As the SAT decoding technique finds more feasible solutions in each generation, the quality functions have to be executed for this increased number of individuals.

But even with these increased execution times the SAT decoding technique generates better solutions earlier for big problems that contain few feasible solutions. 150 processes is not the upper limit for the SAT decoding technique, internal test have shown that even 500 processes can be handled.

5.3. Real Life Example

Besides these synthetic test cases we also used the SAT decoding technique to optimize a real life example. This example describes an adaptive light control system of modern cars which has shown to be a very hard optimization task. The system is modeled by 234 communicating processes which can be mapped onto 1103 resources via 1851 overall mapping edges.

We tested the sequential feasibility checker and the SAT decoding technique to optimize this system. Due to space limitations, we omit exact figures and discuss the most important results only. If the sequential feasibility checker is used with a random initial population, the MOEA requires about 100 generations to find the first feasible solutions. Even if all resources are activated for all individuals of the initial population, the MOEA requires about 14 generations to find feasible solutions. The SAT decoding technique provides feasible solutions always beginning with the first generation. And even for later generations, our tests have shown that the SAT decoding technique always generated significant better solutions than the sequential feasibility checker after the same number of generations and even after the same optimization time.

6. Conclusions

In this paper, we have shown how to integrate SAT-solvers into Multi-Objective Evolutionary Algorithms in order to improve the convergence of the multi-objective optimization. We have presented experimental results from the area of automatic design space exploration of embedded systems which show the efficency of SAT-solvers in the necessary implementation decoding. From these experiments,

0-7695-2533-4/06 $20.00 © 2006 IEEE

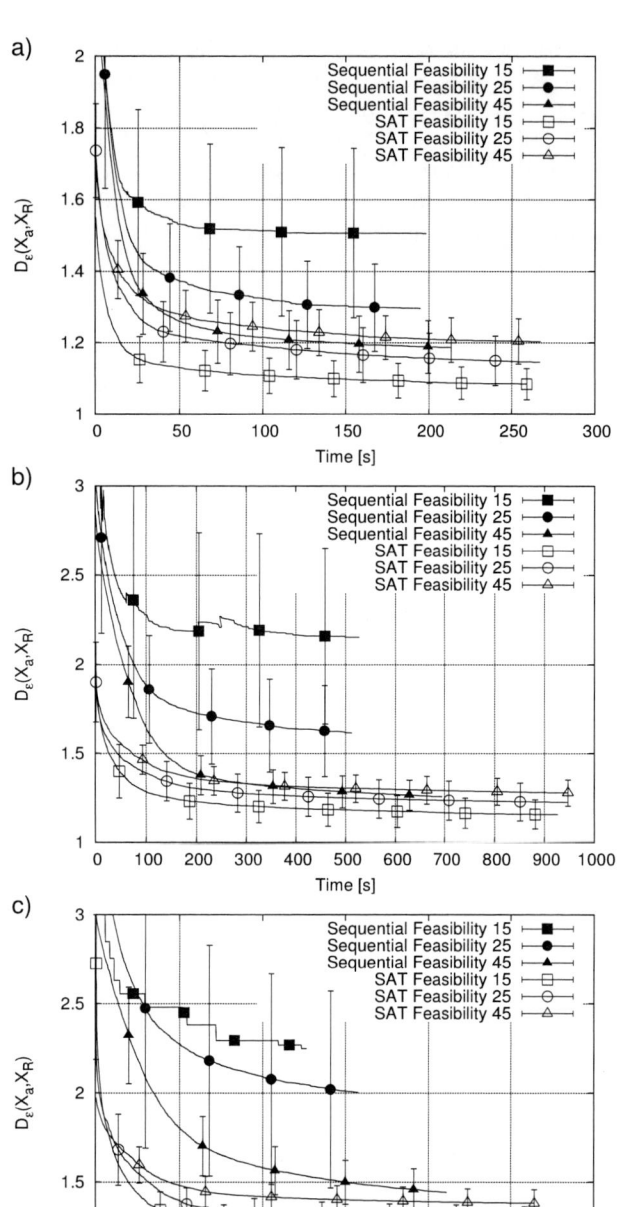

Figure 3. The mean ε-dominance over the average time for a)50, b)100, c)150 processes. The vertical bars indicate the standard deviation.

we conclude that our proposed method is especially well suited for problems with a search space containing many infeasible solutions.

Although the focus in this paper is on automatic design space exploration, there is the potential to generalize our results to other constrained combinatorial optimization problems. This issue will be investigated in future work.

References

[1] A. Biere, A. Cimatti, E. Clarke, and Y. Zhu. Symbolic Model Checking without BDDs. *Lecture Notes in Computer Science*, 1579:193–207, 1999.

[2] S. Bleuler, M. Laumanns, L. Thiele, and E. Zitzler. PISA - A Platform and Programming Language Independent Interface for Search Algorithms. In *Lecture Notes in Computer Science (LNCS)*, volume 2632, pages 494–508, Faro, Portugal, Apr. 2003.

[3] T. Blickle, J. Teich, and L. Thiele. System-Level Synthesis Using Evolutionary Algorithms. In R. Gupta, editor, *Design Automation for Embedded Systems*, 3, pages 23–62. Kluwer Academic Publishers, Boston, Jan. 1998.

[4] M. Davis and H. Putnam. A computing procedure for quantification theory. *J. Assoc. Comput. Mach.*, 7:201–215, 1960.

[5] K. Deb. *Multi-Objective Optimization using Evolutionary Algorithms*. John Wiley & Sons, Ltd., Chichester, New York, Weinheim, Brisbane, Singapore, Toronto, 2001.

[6] R. Feldmann, C. Haubelt, B. Monien, and J. Teich. Fault Tolerance Analysis of Distributed Reconfigurable Systems Using SAT-Based Techniques. In P. Y. K. Cheung, G. A. Constantinides, and J. T. de Sousa, editors, *Field-Programmable Logic and Applications, Lecture Notes in Computer Science (LNCS)*, volume 2778, pages 478–487, Berlin, Heidelberg, Sept. 2003. Springer.

[7] C. Haubelt, S. Mostaghim, F. Slomka, J. Teich, and A. Tyagi. Hierarchical Synthesis of Embedded Systems Using Evolutionary Algorithms. In R. Drechsler and N. Drechsler, editors, *Evolutionary Algorithms for Embedded System Design*, Genetic Algorithms and Evolutionary Computation (GENA), pages 63–104. Kluwer Academic Publishers, Boston, Dordrecht, London, 2003.

[8] V. Kianzad and S. S. Bhattacharyya. CHARMED: A Multi-Objective Co-Synthesis Framework for Multi-Mode Embedded Systems. In *Proceedings of the 15th IEEE International Conference on Application-Specific Systems, Architectures and Processors (ASAP'04)*, pages 28–40, Galveston, U.S.A., Sept. 2004.

[9] M. W. Moskewicz, C. F. Madigan, Y. Zhao, L. Zhang, and S. Malik. Chaff: Engineering an Efficient SAT Solver. In *Proceedings of the 38th Design Automation Conference (DAC'01)*, 2001.

[10] S. Neema. *System Level Synthesis of Adaptive Computing Systems*. PhD thesis, Vanderbilt University, Nashville, Tennessee, May 2001.

[11] T. Schlichter, C. Haubelt, F. Hannig, and J. Teich. Using Symbolic Feasibility Tests during Design Space Exploration of Heterogeneous Multi-Processor Systems. In *Proceedings of Application-specific Systems, Architectures and Processors (ASAP)*, Samos, Greece, July 2005.

[12] E. Zitzler, M. Laumanns, and L. Thiele. SPEA2: Improving the Strength Pareto Evolutionary Algorithm for Multiobjective Optimization. In *Evolutionary Methods for Design, Optimisation, and Control*, pages 19–26, Barcelona, Spain, 2002.

[13] E. Zitzler, L. Thiele, M. Laumanns, C. M. Fonseca, and V. Grunert da Fonseca. Performance Assessment of Multiobjective Optimizers: An Analysis and Review. *IEEE Transactions on Evolutionary Computation*, 7(2):117—132, Apr. 2003.

0-7695-2533-4/06 $20.00 © 2006 IEEE

System Level Design

Optimisation of the SHA-2 Family of Hash Functions on FPGAs

Robert P. McEvoy, Francis M. Crowe, Colin C. Murphy and William P. Marnane
Department of Electrical & Electronic Engineering,
University College Cork, Ireland
{robertmce, francisc, cmurphy, liam}@rennes.ucc.ie

Abstract

Hash functions play an important role in modern cryptography. This paper investigates optimisation techniques that have recently been proposed in the literature. A new VLSI architecture for the SHA-256 and SHA-512 hash functions is presented, which combines two popular hardware optimisation techniques, namely pipelining and unrolling. The SHA processors are developed for implementation on FPGAs, thereby allowing rapid prototyping of several designs. Speed/area results from these processors are analysed and are shown to compare favourably with other FPGA-based implementations, achieving the fastest data throughputs in the literature to date.

1. Introduction

In today's modern world of e-mail, internet banking, on-line shopping, and other sensitive digital communications, cryptography has become a vital tool for ensuring the privacy of data transfers. Hash functions operate at the root of many popular cryptographic methods in current use, such as the Digital Signature Standard (DSS), Transport Layer Security (TLS) and Internet Protocol Security (IPSec) protocols, numerous random number generation algorithms, encryption algorithms, all-or-nothing transforms, and password storage mechanisms.

Hash functions map messages of arbitrary length to a string of fixed length, called the 'message hash' or 'message digest'. This compression process is known as 'hashing'. In 2002, the National Institute of Standards and Technology (NIST) published the Secure Hash Standard [9], which detailed three new Secure Hash Algorithms SHA-256, SHA-384, and SHA-512. Since then, SHA-224 has been added to the standard, forming the 'SHA-2' family of hash functions. The SHA-2 family supersede the SHA-1 algorithm, whose security has been damaged by recent attacks [5].

This paper investigates architectures for hardware acceleration of the SHA-2 algorithms on FPGAs. FPGAs are suitable for use as cryptographic accelerators due to their low cost (relative to ASICs) and their flexibility when adopting security protocol upgrades. The reconfigurability of FPGAs also allows rapid prototyping of various VLSI designs.

This paper is organised as follows. The following section gives an overview of the algorithms in the SHA-2 standard. Section 3 surveys known techniques in the literature for hardware optimisation of SHA-2 operations. Novel SHA-2 architectures are presented in Section 4, whose speed/area performance results are discussed in Section 5. Conclusions are given in Section 6.

2. The Secure Hash Algorithms

Full descriptions of the SHA-224, -256, -384 and -512 algorithms can be found in the official NIST standard [9]. SHA-256 produces a 256-bit message hash; SHA-224 a 224-bit message hash etc. An overview of SHA-256 is given here, and then the differences between SHA-256 and the other members of the SHA-2 family are outlined. The SHA-256 algorithm essentially consists of 3 stages: (i) message padding and parsing; (ii) expansion; and (iii) compression.

2.1. Message Padding and Parsing

The binary message to be processed is appended with a '1' and padded with zeros until its length $\equiv 448 \bmod 512$. The original message length is then appended as a 64-bit binary number. The resultant padded message is parsed into N 512-bit blocks, denoted $M^{(1)}, M^{(2)}, \ldots, M^{(N)}$. These $M^{(i)}$ message blocks are passed individually to the message expander.

2.2. Message Expansion

The functions in the SHA-256 algorithm operate on 32-bit words, so each 512-bit $M^{(i)}$ block from the padding stage is viewed as 16 32-bit blocks denoted

---

0-7695-2533-4/06 $20.00 © 2006 IEEE   317

$M_t^{(i)}$, $0 \leq t \leq 15$. The message expander (also called the message scheduler) takes each $M^{(i)}$ and expands it into 64 32-bit W_t blocks, according to the equations:

$$\sigma_0(x) = ROT_7(x) \oplus ROT_{18}(x) \oplus SHF_3(x) \qquad (1)$$

$$\sigma_1(x) = ROT_{17}(x) \oplus ROT_{19}(x) \oplus SHF_{10}(x) \qquad (2)$$

$$W_t = \begin{cases} M_t^i & 0 \leq t \leq 15 \\ \sigma_1(W_{t-2}) + W_{t-7} + \sigma_0(W_{t-15}) + W_{t-16} & 16 \leq t \leq 63 \end{cases}$$
$$(3)$$

where the function $ROT_n(x)$ denotes a circular rotation of x by n positions to the right, whilst the function $SHF_n(x)$ denotes the right shifting of x by n positions. All additions in the SHA-256 algorithm are modulo 2^{32}.

2.3. Message Compression

The W_t words from the message expansion stage are then passed to the SHA compression function, or the 'SHA core'. The core utilises 8 32-bit working variables labelled A, B, ..., H, which are initialised to predefined values $H_0^{(0)}$–$H_7^{(0)}$ (given in [9]) at the start of each call to the hash function. Sixty-four iterations of the compression function are then performed, given by:

$$\begin{aligned} T_1 &= H + \sum_1(E) + Ch(E,F,G) + K_t + W_t \\ T_2 &= \sum_0(A) + Maj(A,B,C) \\ H &= G & G &= F \\ F &= E & E &= D + T_1 \\ D &= C & C &= B \\ B &= A & A &= T_1 + T_2 \end{aligned} \qquad (4)$$

where

$$Ch(x,y,z) = (x \; AND \; y) \oplus (\bar{x} \; AND \; z) \qquad (5)$$

$$Maj(x,y,z) = (x \; AND \; y) \oplus (x \; AND \; z) \oplus (y \; AND \; z) \qquad (6)$$

$$\sum_0(x) = ROT_2(x) \oplus ROT_{13}(x) \oplus ROT_{22}(x) \qquad (7)$$

$$\sum_1(x) = ROT_6(x) \oplus ROT_{11}(x) \oplus ROT_{25}(x) \qquad (8)$$

and the inputs denoted K_t are 64 32-bit constants, specified in [9]. After 64 iterations of the compression function, an intermediate hash value $H^{(i)}$ is calculated:

$$H_0^{(i)} = A + H_0^{(i-1)}, H_1^{(i)} = B + H_1^{(i-1)}, \ldots, H_7^{(i)} = H + H_7^{(i-1)}$$

The SHA-256 compression algorithm then repeats and begins processing another 512-bit block from the message padder. After all N data blocks have been processed, the final 256-bit output, $H^{(N)}$, is formed by concatenating the final hash values:

$$H^{(N)} = H_0^{(N)} \; \& \; H_1^{(N)} \; \& \; H_2^{(N)} \; \& \; \ldots \; \& \; H_7^{(N)}$$

2.4. SHA-2 Algorithm Differences

Apart from having different output hash lengths, the only other difference between SHA-224 and SHA-256 is the set of K_t constants used. Similarly, the SHA-384 and SHA-512 algorithms differ only in their hash lengths and initial conditions. The SHA-512 algorithm has a similar structure to the SHA-256 algorithm, where: (i) it processes messages in blocks of 1024 bits rather than 512 bits; (ii) it uses 64-bit operations instead of 32-bit operations; (iii) the K_t constants are different; (iv) it iterates its compression function 80 times rather than 64 times; and (v) Equations (1), (2), (7) and (8) have different degrees of rotation and shifting, but are otherwise similar in structure.

3. Known SHA Optimisation Techniques

Several hardware-based SHA-2 designs have appeared in the literature [1-4, 6-8, 10, 12-13]. The longest data path (critical path) in the SHA core is the calculation of working variable A, which involves addition (modulo 2^{32} for SHA-224/256, modulo 2^{64} for SHA-384/512) of 7 operands (see Equation (4)). The following techniques have been proposed to speed up calculations in the SHA core:

- Using Carry-Save Addition (CSA) [2, 3, 4, 6, 7]. CSA separates the sum and carry paths, thus minimising the delay caused by traditional carry propagation. Since CSAs accept 3 input operands, A can be calculated in the SHA core using just 5 CSAs (as in [3]).

  A further 2-operand addition is required following the carry-save adders, to recombine the sum and carry paths. This extra addition stage is generally a fast carry look-ahead addition (CLA). In this work, the platform being targeted is a Xilinx™ Virtex II™ FPGA. By investigating different types of adders, it was found that the FPGA's built-in Carry-Propagate Adder (CPA) actually has a lower delay than a CLA on this platform. Hence, the designs in this paper use CPAs to terminate chains of CSAs and recombine the sum and carry paths, rather than CLAs.

- Unrolling [1, 6]. An unrolled architecture implements multiple rounds of the core compression function in combinational logic, thereby reducing the number of clock cycles required to compute the hash. For example, if the core was unrolled once, then the hash should be calculated in half the number of clock cycles. This comes at the cost of an increase in area and a decrease in clock frequency.

- (Quasi-) Pipelining. Dadda et al. [2, 3, 7] have proposed fast pipelined SHA-2 architectures, which use

0-7695-2533-4/06 $20.00 © 2006 IEEE

registers to break the long critical path within the SHA core. Pipelining is not trivial, however, since registers cannot be added at will (due to the inherent feedback in the SHA algorithm). External control circuitry is required to enable the registers correctly. Nevertheless, these 'quasi-pipelined' designs achieve very short critical paths, allowing message hashes to be calculated at high frequencies and high data throughputs.

- Delay Balancing. Dadda et al. also investigated using delay balancing in conjunction with CSA [3]. A CLA adder is used to combine the sum and carry paths as discussed above, but the sum and carry signals are first registered. This moves the CLA adder out of the critical path, but requires an extra register and accompanying control circuitry.

- Moving addition to the message expansion stage [13]. In this architecture, the first addition in the critical path (i.e. $K_t + W_t$) is moved to the message expansion stage, since both K_t and W_t are available before the other operands in Eq. (4). However, quasi-pipelining [2, 3, 7] achieves similar segregation of the operands, where the resultant paths are even shorter.

- Use of Block RAM (BRAM) for storage of constants [8]. Reconfigurable hardware devices such as FPGAs often have on-board memories which can be pre-loaded. Storing the K_t constants in these memories frees up space in the device which can then be used to implement extra logic. The free space also leads to improved routing and, thus, a general speed-up in circuit operation.

- Use of a Parallel Counter. 5-to-3 Parallel Counters (PCs) are used as adders in [4] instead of 3-to-2 CSA adders. The PCs reduce the number of bits at each position in the sum from 5 to 3. Equation (4) is realised in [4] with a cascade of two 5-to-3 PCs, followed by a CSA and a carry-propagate adder (CPA), giving a reduction in the critical path.

This paper investigates combining some of these optimisation techniques in order to increase the data throughput. In particular, the designs presented incorporate CSA, use of BRAM, quasi-pipelining, and unrolling. To the best of the authors' knowledge, this is the first time a pipelined-unrolled architecture for the SHA-2 family has been presented.

4. Hardware Architectures

4.1. Dadda et al.'s Constructions

In [2, 3], Dadda et al. investigate several quasi-pipelined architectures for the SHA-2 core. These architectures, orig-

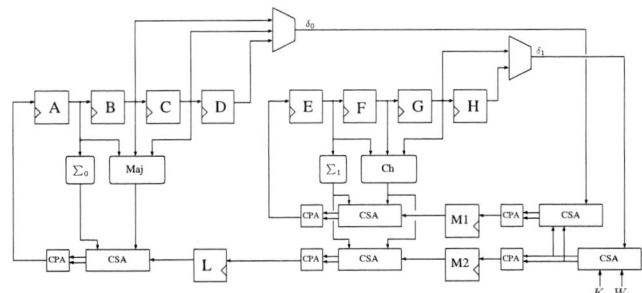

Figure 1. Dadda et al.'s [2] SHA-2 Core

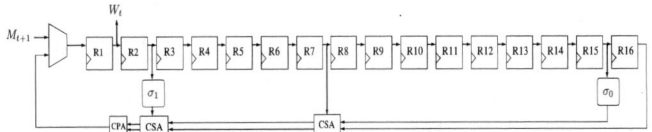

Figure 2. Basic SHA-2 Message Expander [2]

inally proposed for ASIC implementation, were captured using VHDL and synthesised for an FPGA implementation (see Section 5). Of these designs, it was found that the scheme illustrated in Figure 1 gave the shortest critical path and, therefore, could be operated at the highest frequency.

As well as registers to store the working variables A–H, the quasi-pipelined core (hereafter referred to as the "basic" core) includes 3 more registers (M1, M2, L) and 2 multiplexers. The select lines for the multiplexer outputs (δ_0, δ_1) and the clear signals for all the registers are controlled by external circuitry, as described in [2]. Since registers E–H receive their updates one clock cycle before registers A–D, the hash results $H_4^{(i)}$–$H_7^{(i)}$ are ready one clock cycle before $H_0^{(i)}$–$H_3^{(i)}$. It was found that the critical path in Figure 1 was from the multiplexer select line for δ_1 to the output of register M1.

In [2], an architecture for the message expander was presented which uses CSAs to reduce the number of required adders. Since the critical path in this design (see Figure 2) is shorter than that of the core, it is the core that determines how fast the overall hash algorithm can be executed. The authors also present a design employing delay balancing to further shorten the critical path in the message expander. However, this scheme is unsuitable for unrolling.

4.2. Combined Unrolled-Pipelined Design

This work investigates combining Dadda et al.'s fast quasi-pipelined architecture with the unrolling technique, with the aim of increasing the achieveable data throughput. Since the SHA processors were being designed for implementation on a reconfigurable platform, this allowed six different combinations to be investigated with relative ease.

0-7695-2533-4/06 $20.00 © 2006 IEEE

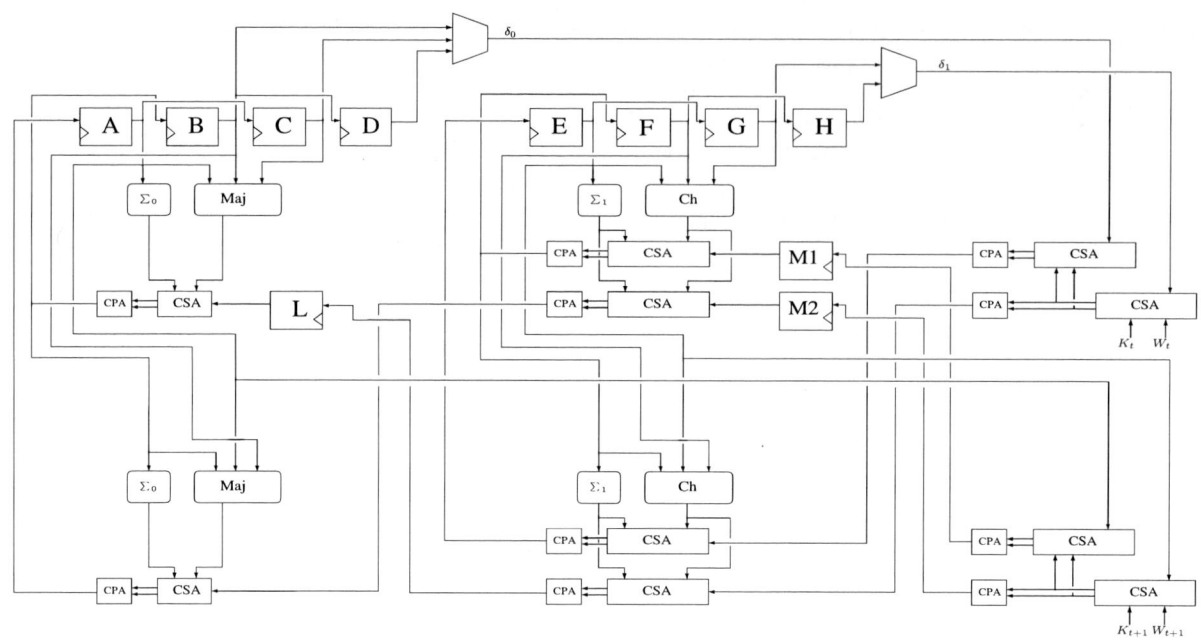

Figure 3. 2x-Unrolled-Pipelined SHA-2 Core

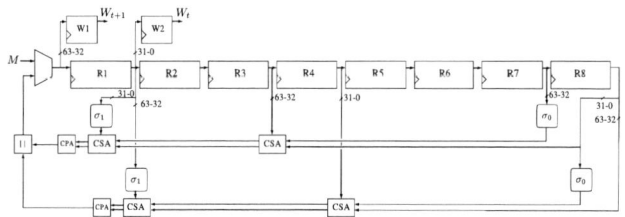

Figure 4. 2x-Unrolled SHA-2 Message Expander

For both SHA-256 and SHA-512, basic (i.e. not unrolled), once unrolled ('2x'), and three times unrolled ('4x') cores were implemented.

Core. The quasi-pipelined SHA core (Figure 1) was unrolled by adding extra combinational logic to the datapaths, such that 2 (or 4) calculations of Equation (4) could be performed in one clock cycle. Figure 3 shows the corresponding 2x-unrolled architecture that was developed. Although the logical functions from the basic core (Figure 1) are duplicated, the registers and multiplexers are not. The registers are also loaded differently in the unrolled core than in the basic core. This unrolled circuit can only be used once the pipeline has been initially filled, i.e. once registers A–D have received their first values via the protocol in [2]. While the pipeline is being filled (and emptied), the basic core of Figure 1 must be used. The K_t constants are stored in BRAM to conserve space on the

FPGA, and are addressed via a counter.

Expander. Since the 2x-unrolled core requires two W_t words to be available simultaneously, alterations to the basic message expander (Figure 2) were required, as shown in Figure 4. Firstly, the chain of 16 registers was modified to a chain of 8 registers of twice the size, e.g. for the SHA-256 message expander the registers changed from 32-bit to 64-bit. The multiplexer was also altered (in the SHA-256 case) to select 64-bit words rather than 32-bit words. This was necessary so that two new 32-bit W_t words would be produced on each clock cycle. Secondly, the combinational logic was duplicated so that two calculations of Equation (3) could be performed in one clock cycle. The results of these calculations are concatenated and fed back to the multiplexer. Finally, to prevent the message expander from contributing to the critical path in the core, the W_t words are registered before being passed to the unrolled SHA core. Note again that this unrolled circuit can only be used once the pipeline in the core has been filled.

Padding block. Most SHA designs in the literature assume that the message padding stage of the algorithm can be performed in software. However, here we aim to produce fast stand-alone implementations of the hashing algorithm, which could be used in constrained, embedded, or System-on-a-Chip (SoC) environments. Therefore, a message padding block was also included to implement the padding and parsing described in Section 2.1. This was realised as a synchronous finite state machine (FSM).

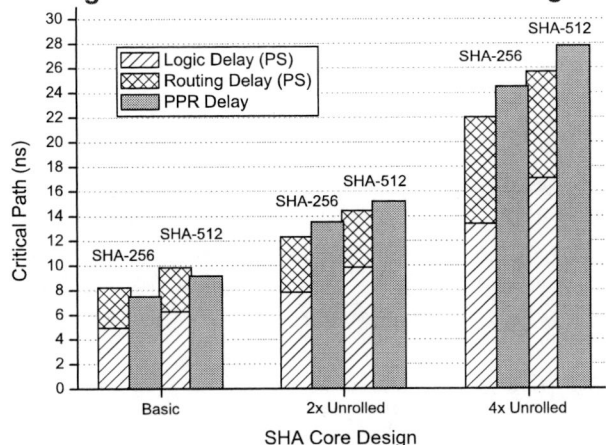

Figure 5. Critical Paths for SHA-2 Designs

Control Circuitry. External circuitry was designed to control the register load/clear signals in the core and the multiplexer select lines, as well as controlling the movement of data between the message padder, message expander and core. A second FSM was utilised for this purpose. The state machine works together with a counter to iterate the compression function (Equation (4)) appropriately, and to add the current hash value to the previous hash value. All control signals from the FSM were registered to ensure that they could not contribute to the critical path.

5. Results and Discussion

Six different SHA processors were captured at the RTL level using VHDL. For SHA-256, one processor had a basic quasi-pipelined core (as in [2]), one had a 2x-unrolled core, and another had a 4x-unrolled core. Three similar architectures were also developed for SHA-512 (and by extension SHA-224 and SHA-384). The design was synthesised using Xilinx ISE™ tools v5.1 for implementation on a Xilinx Virtex II™ *xc2v2000-bf957* FPGA. Post Synthesis (PS) and Post Place and Route (PPR) results for the critical paths in each SHA processor are displayed in Figure 5. The PS results are given in order to convey which proportions of the critical path are due to logic delay, and which to routing.

It is clear from Figure 5 that the critical path inside the core of each SHA processor increases with the degree of unrolling. In the basic designs (i.e. SHA processors based on cores from [2]), the PPR results are actually better than the PS results. This is because the synthesis tool only provides an estimate of the delay of the critical path. Since the percentage area occupied on the FPGA by the basic SHA processors is low, the PPR routing delay is not as high as was estimated by the synthesis tool.

In the 2x- and 4x-unrolled designs the PPR critical paths are longer than the PS estimates, as expected. The un-

rolled designs are larger than the basic designs and, as the FPGA device fills, it becomes more difficult for the Place and Route tool to route the signals along short paths. In the 4x-unrolled designs, approximately one third of the delay in the critical path is due to routing. It is reasonable to expect that these timing results could be further reduced by employing manual routing of the signals in the critical path.

The maximum clock frequency at which the processors can be operated is related to data throughput by:

$$Throughput = \frac{Block\ Size \times Max.\ Frequency}{\#\ Clock\ Cycles} \quad (9)$$

The block size for SHA-256 is 512 bits while the block size for SHA-512 is 1024 bits. Although the unrolled designs process messages in fewer clock cycles than the basic designs, the longer critical path in the unrolled designs means that the maximum clock frequency decreases. If the factor by which clock frequency decreases is less than that by which the number of clock cycles decreases, then an overall increase in data throughput will be achieved.

As noted in Section 2, the SHA-256 algorithm executes in 64 clock cycles, whilst SHA-512 executes in 80 clock cycles. Since the quasi-pipelined architectures include start-up and wind-down clock cycles, the basic quasi-pipelined processors execute the algorithms in 68 and 84 clock cycles respectively. In the new pipelined-unrolled designs presented in this paper, one must wait for the pipeline in the core to fill before activating the unrolled architecture. Similarly, at the end of a compression cycle, the unrolled architecture cannot be used, because two different halves of the intermediate hash $H^{(i)}$ appear in two consecutive clock cycles. Therefore, during start-up and wind-down, the control circuitry 'switches off' the unrolled section of the core and reverts to the basic configuration of Figure 1.

The number of clock cycles required by the compression algorithm in each architecture is shown in Table 1, along with the data throughput results (TP) attained from the six SHA-2 processors. The results show that unrolling the quasi-pipelined SHA-256 design provides no data throughput advantage. As the core is unrolled (from basic to 2x), the critical path length increases by a factor of 1.8. This is larger than the factor by which the number of clock cycles

Table 1. SHA-2 Processor Results

Design	Freq. (MHz)	Clk Cycles	TP (Mbps)	Area (slices)	TP/ Area
SHA-256					
Basic	133.06	68	**1009**	1373 (12%)	0.735
2x-unrolled	73.975	38	996.7	2032 (18%)	0.491
4x-unrolled	40.833	23	908.9	2898 (26%)	0.314
SHA-512					
Basic	109.03	84	1329	2726 (25%)	0.488
2x-unrolled	65.893	46	**1466**	4107 (38%)	0.357
4x-unrolled	35.971	27	1364	5807 (54%)	0.235

0-7695-2533-4/06 $20.00 © 2006 IEEE

decreases (1.79). Therefore, it is the basic quasi-pipelined SHA-256 design (of [2]) that performs best. However, in the SHA-512 processor, the best throughput results were obtained from the 2x-unrolled design. Going from a basic quasi-pipelined SHA-512 design to the 2x-unrolled design reduces the number of clock cycles by a factor of 1.83, but the critical path increases by a factor of only 1.66. Therefore, an overall increase in throughput is obtained, at the cost of a 51% area increase. Note that here the area in FPGA CLB slices refers to the area of the entire processor, i.e. padding block, expander, compressor and control circuitry. Although the efficiency (TP/Area) of the 2x-unrolled SHA-512 processor is not as high as that for the basic processor, it may still be of use in high speed applications where area is not constrained.

Throughput comparisons amongst FPGA-based SHA-2 designs have recently been drawn in the literature [10, 12]. The highest reported data throughput for SHA-256 was 693 Mbps [13], and 1034 Mbps for SHA-512 [6], both targeting Virtex-E FPGAs, and using 1261 and 3517 slices respectively. For completeness, synthesis of our designs on the Virtex-E platform resulted in a (basic) SHA-256 processor with a throughput of 812 Mbps (1340 slices), and a (2x-unrolled) SHA-512 processor with a throughput of 1243 Mbps (3597 slices). In particular, these improvements arise due to the fast new quasi-pipelined core of [2], implemented for the first time on FPGAs.

Therefore, to the best of the authors' knowledge, the FPGA-based SHA-2 architectures presented in this paper achieve the best data throughput results to date. By comparison, software implementations using functions from the MIRACL cryptographic library [11] achieved average throughputs of only 59.8 Mbps for SHA-256 and 27.7 Mbps for SHA-512. Software implementations were run on a 3 GHz Pentium 4 processor, with 1 GB RAM. This highlights the considerable advantage of using VLSI hardware implementations to accelerate cryptographic algorithms and protocols.

6. Conclusion

In this paper, several hardware optimisation techniques for the SHA-2 hash family were explored. A new architecture was proposed for the SHA-2 core, which combines a fast quasi-pipelined design with unrolling. This is the first time a pipelined-unrolled SHA design has been investigated in the literature. This new design was then applied to both the SHA-256 and SHA-512 algorithms by constructing VHDL modules for SHA processors. These full processors include hardware for the SHA core, message scheduler, and message padder, thereby facilitating easy reuse in a VLSI SoC environment. Six processor designs were presented in the paper, each investigating different degrees of unrolling. Finally, the performances of these designs were considered from data throughput and area perspectives, and compared with other leading VLSI designs in the literature that target reconfigurable hardware. It was shown that the data throughput rates achieveable by the SHA processors presented in this paper out-perform these other designs.

Acknowledgement

This research was supported by the Embark Initiative, operated by the Irish Research Council for Science, Engineering and Technology (IRCSET).

References

[1] F. Crowe, A. Daly, and W. Marnane. Single-chip FPGA implementation of a cryptographic co-processor. In *Proceedings of the International Conference on Field Programmable Technology, FPT 2004*, pages 279–285, December 2004.

[2] L. Dadda, M. Macchetti, and J. Owen. An ASIC design for a high speed implementation of the hash function SHA-256 (384, 512). In *ACM Great Lakes Symposium on VLSI*, pages 421–425. ACM, 2004.

[3] L. Dadda, M. Macchetti, and J. Owen. The design of a high speed ASIC unit for the hash function SHA-256 (384, 512). In *DATE 2004*, pages 70–75. IEEE Computer Society, 2004.

[4] T. Grembowski, R. Lien, K. Gaj, N. Nguyen, P. Bellows, J. Flidr, T. Lehman, and B. Schott. Comparative analysis of the hardware implementations of hash functions SHA-1 and SHA-512. In *ISC*, volume 2433 of *Lecture Notes in Computer Science (LNCS)*, pages 75–89. Springer, 2002.

[5] A. K. Lenstra. Further Progress in Hashing Cryptanalysis (white paper). http://cm.bell-labs.com/who/akl/hash.pdf, February 2005.

[6] R. Lien, T. Grembowski, and K. Gaj. A 1 Gbit/s partially unrolled architecture of hash functions SHA-1 and SHA-512. In *CT-RSA 2004*, volume 2964 of *LNCS*, pages 324–338. Springer, 2004.

[7] M. Macchetti and L. Dadda. Quasi-pipelined hash circuits. In *Proceedings of the 17th IEEE Symposium on Computer Arithmetic, ARITH-17*, pages 222–229. IEEE, 2005.

[8] M. McLoone and J. McCanny. Efficient single-chip implementation of SHA-384 and SHA-512. In *Proceedings of the International Conference on Field Programmable Technology, FPT 2002*, pages 311–314, December 2002.

[9] NIST. Secure Hash Standard, FIPS PUB 180-2, 2002.

[10] A. Satoh and T. Inoue. ASIC-Hardware-Focused Comparison for Hash Functions MD5, RIPEMD-160, and SHS. In *ITCC (1)*, pages 532–537. IEEE Computer Society, 2005.

[11] Shamus Software Ltd. Multiprecision Integer and Rational Arithmetic C/C++ Library (MIRACL). http://indigo.ie/~mscott/.

[12] N. Sklavos and O. Koufopavlou. Implementation of the SHA-2 Hash Family Standard Using FPGAs. *The Journal of Supercomputing*, 31:227–248, 2005.

[13] K. K. Ting, S. C. L. Yuen, K.-H. Lee, and P. H. W. Leong. An FPGA based SHA-256 processor. In *FPL*, volume 2438 of *LNCS*, pages 577–585. Springer, 2002.

A Novel Approach to Performance-Oriented Datapath Allocation and Floorplanning

Vijay Sundaresan and Ranga Vemuri
Department of ECECS, University of Cincinnati,
Cincinnati, OH 45221-0030, USA.
vsundare, ranga@ececs.uc.edu

Abstract

In recent years, integrating high-level synthesis and physical design has become essential for successful timing closure. In this paper, we present a novel probabilistic approach to perform integrated datapath allocation and floorplanning. In contrast to existing methodologies that tradeoff high-level design space searched with accuracy of physical estimates, the proposed methodology performs exhaustive high-level synthesis (datapath allocation) with heuristic floorplanning. Exhaustive search guarantees optimal allocation results. Heuristic floorplanning imparts flexibility to update the floorplan while constructing the allocation solution. This improves accuracy of the floorplan estimates used. To tackle the increased design space, we propose novel probabilistic-gain measures that predict the impact of an allocation decision on the constraints, prior to the allocation decision being made. The probabilistic-gain measures are used to select allocation decisions that will most likely lead to optimal solutions early, thereby minimizing time-consuming floorplan evaluations of decisions that will most likely lead to sub-optimal solutions. Experimental results confirm the effectiveness of the proposed approach.

1. Introduction

With increasing design complexity and decreasing time-to-market requirements, methodologies at higher levels of abstraction have become a necessity. High-level synthesis (HLS) allows quick realization of design implementations and efficient tradeoff analysis. However, for HLS solutions to be reliable, good quality metrics that consider physical information of logic and interconnects must be available [2] [10]. This is critical for deep sub-micron processes, where interconnect timing, area and power requirements tend to overshadow logic requirements.

There are numerous methodologies that perform HLS with physical information. Most methodologies tradeoff high-level design space explored with accuracy of physical estimates. Some methodologies perform HLS while exploring (updating) floorplan solution space. Updating floorplan improves accuracy of floorplan estimates. However, these

methodologies perform exploration in a constructive heuristic [9] or probabilistic heuristic [1] framework that does not guarantee optimal high-level results. There are a few methodologies that perform exhaustive high-level exploration with floorplan estimates [5]. However, these methodologies do not explore floorplan design space efficiently. All high-level design decisions are made with pessimistic estimates from a *preset* floorplan that could lead to rejection of many potentially optimal high-level synthesis solutions. Exhaustive search of high-level design space guarantees optimal high-level results. High-level solution construction leads to changes in the netlist used by the floorplanner. Therefore, updating floorplan during high-level solution construction leads to more accurate floorplan estimates. However, the important issue of increased design space must be addressed.

In most methodologies, physical information is used only to evaluate the impact of high-level decisions, *after* the decision is made [5, 9, 1, 11]. This is inefficient as the quality of a decision is not known until after the evaluation is complete. There are very few floorplan-aware high-level synthesis methodologies that predict the impact of a high-level decision, *prior* to the decision being made. [4] tries to predict the amount of routing (area) between resources, to generate a *tolerant* floorplan that can withstand changes in the netlist. Tolerant floorplan probabilistically accounts for routing area in the floorplan, to reduce the number of floorplan updations necessary during netlist optimization.

In this paper, we present a novel probabilistic approach to integrated datapath allocation and floorplanning (Figure 1). We propose a branch-and-bound exhaustive-constructive datapath allocation algorithm. Accurate timing and area estimates are obtained from a floorplan. Heuristic simulated-annealing floorplanner allows exploration in the floorplan solution space as well. A framework that supports efficient interaction between exhaustive-constructive datapath allocation and heuristic floorplanning, while maintaining consistency of partial allocation decisions during floorplanning is proposed. We propose novel probabilistic-gain measures that predict the impact of an allocation decision on optimization constraints thereby aiding prioritized search,

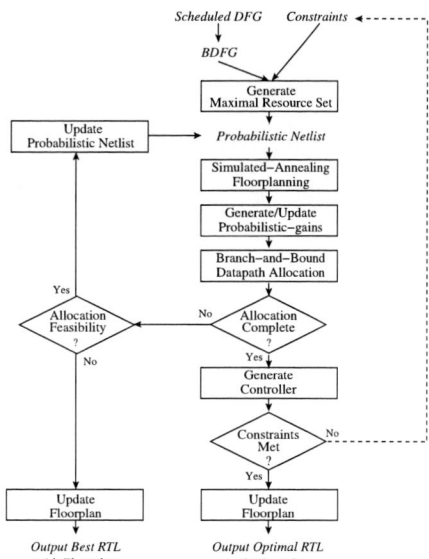

Figure 1. Integrated datapath allocation and floorplanning

efficient pruning, and hence, quick convergence.

This paper is organized as follows. Section 2 presents the framework for integrated HLS and floorplanning. Section 3 presents the probabilistic-gain measures. Section 4 details the exhaustive-constructive datapath allocation algorithm. Section 5 details the heuristic simulated annealing floorplanning based estimation. Section 6 describes the experimental results. Section 7 concludes the paper.

2. Integrated HLS and Floorplanning

The behavioral specification in VHDL is converted to a data flow graph (DFG). The DFG is then scheduled using time-constrained Force-Directed List Scheduling (FDLS) algorithm [7]. Figure 2(a) shows a sample behavioral code and Figure 2(b) shows its corresponding scheduled-DFG, scheduled in two control steps.

A Data Flow Graph is defined as $DFG = (V, E)$; where $V = v_i \mid i = 1, 2, \cdots n$ operations, and $E = e_k \mid k = 1, 2, \cdots m$ edges or data transfers between operations.

The DFG captures the *behavior* of the circuit in terms of operations and data transfers, and is hence suitable for datapath allocation. However, it is unsuitable for floorplanning as it *lacks structure* information. Therefore, we transform the DFG into a Bergamaschi's Behavioral Network Graph-style Data Flow Graph (BDFG) [8] [9]. BDFG is generated from the scheduled-DFG by introducing special State-Cut Nodes (SCNs) at control step boundaries. Control step boundaries are data transfers that connect operations scheduled at different control steps. State-cut nodes correspond to registers that store data at control step boundaries. Intro-

ducing SCNs reduces all edges between the various BDFG nodes (includes functional and storage operations) to interconnects (wires) as opposed to data transfers in the DFG. BDFG captures *behavior as well as structure* of the design. Figure 2 shows a sample BDFG (e), its corresponding RTL datapath (f) and a floorplan implementation (g).

A Bergamaschi's Behavioral Network Graph-style DFG is defined as $BDFG = (VS, DT)$; where $VS = BDFG$ nodes (operations) that include data flow operations and SCNs. $DT = $ edges between BDFG operations.

From the BDFG, a maximal resource set (R) is generated by selecting a unique resource for each operation in the BDFG. The resources are obtained from a pre-characterized component library with their shape functions (area-aspect-ratio-delay characteristics) defined. The shape function is assumed to be discrete delay-aspect ratio function. Figure 2(c) shows a sample resource set and Figure 2(d) shows an example of a shape function.

2.1. Cost model

In this framework, operation chaining, multi-cycling and pipelining are supported. We assume a one-hot encoded moore-type controller model, where registers define the control boundaries (control step). The minimum clock delay or clock period would then be the worst-case register-to-register (register-transfer) delay in the datapath. Clock period (δ) is modeled as follows:

$$\delta = t_{FF} + t_{RR} + t_{WL};$$

$t_{FF} = t_{mux} + t_{FU}$; where t_{mux} is the input multiplexer delay and t_{FU} is the functional unit propagation delay.

$t_{RR} = t_{mux} + t_{rs} + t_{rp}$; where, t_{mux} is the input multiplexer delay, t_{rs} is the setup time of the register and t_{rp} is the propagation delay for the register.

t_{WL} is wire delay between resources. $t_{WL} = \frac{rcL^2}{2}$; where r and c are unit resistance and capacitance respectively, and L is the wirelength modeled using half-perimeter wirelength model. Although pessimistic, the wirelength estimation task is simple and efficient.

2.2. Problem Formulation

Given a BDFG, maximal resource set, area constraint, aspect ratio constraint and a maximum clock period constraint, obtain an optimal RTL consisting of an assignment of BDFG operations (DFG operations and SCNs) to resources, and its corresponding floorplan such that the constraints are satisfied.

2.3. Target Architecure

Datapath Allocation Solution Representation

0-7695-2533-4/06 $20.00 © 2006 IEEE

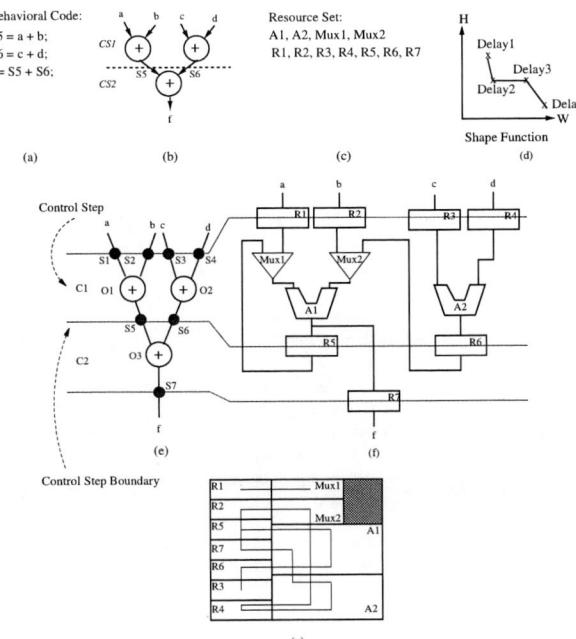

(a) (b) (c) (d)

(e)

(f)

(g)

Figure 2. (a) Behavioral code (b) DFG (c) Resource set (d) Shape function (e) BDFG (f) RTL datapath (g) Floorplan

The datapath allocation problem is represented as a Binding Graph (BG) [2].

A Binding Graph is defined as a bipartite graph, $BG = (V, R, E)$; V represents all BDFG operations (functional and storage), R represents all resources in the maximal resource set, E represents edges between operation and resources that can support the operation.

Floorplan Solution Representation

The floorplan problem is represented as a Probabilistic Netlist (PN). This is same as resource connection graph described in [4].

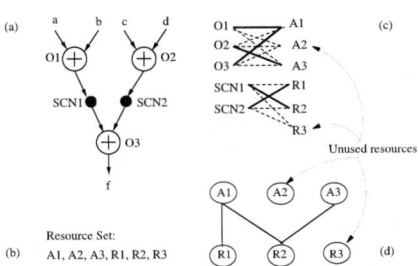

Figure 3. (a) Binding Graph (b) Resource Set (c) Binding Solution (d) Probabilistic Netlist (probabilities are not shown)

A Probabilistic Netlist is defined as an undirected graph, PN = (R, I); where R represents a set of weighted nodes

corresponding to resources in the binding graph. I represents the set of weighted interconnections between resources. Weight of the resource node is the resource probability and weight of interconnection in I is the resource-to-resource connection probability.

Figure 3(c) shows an example of a binding graph generated from a BDFG (Figure 3(a)) and a resource set (Figure 3(b)). Figure 3(d) shows an example of a Probabilistic Netlist. Resource probability and resource-to-resource connection probabilities are explained in the following section.

3. Probabilistic-Gains

3.1. Resource Probability

Let T represent the number of operation (resource) types the resource set can support and r represent the number of resources in each resource type. For the sake of simplicity, we will assume that each resource can support only one operation. However, we can easily extend this methodology to include resources that can support multiple operations.

Resource probability $P(R_j)$ is defined as the probability that some BDFG operation will be bound to resource R_j.

Let, P_{ij} be the probability that an operation O_i *will be* bound to a resource R_j. Then, assuming that for any operation there is no binding preference among the resources,

$P_{ij} = \frac{1}{r}$; where r is the number of resources that can support the operation. $1 - P_{ij}$ is the probability that an operation O_i *will not* be bound to resource R_j. Subsequently, $\prod(1 - P_{ij})$; for all operations O_i, is the probability that *none* of the BDFG operations *will* be bound to resource R_j.

Hence, $P(R_j) = 1 - \prod(1 - P_{ij})$; for all operations O_i, is the probability that *some* BDFG operation *will* be bound to resource R_j.

3.2. Resource-to-Resource Connection Probability

Resource-to-Resource Connection Probability $P(R_p \to R_q)$: is the probability that some BDFG data transfer would be bound as an interconnection between two resources R_p and R_q.

$P_{px} * P_{qy}$ denotes the resource probability that operations O_x *will* be bound to R_p and O_y will be bound to R_q.

$1 - P_{px} * P_{qy}$ is the probability that the data transfer $O_x \to O_y$ *will not* be bound to $R_p \to R_q$. Subsequently, $\prod(1 - P_{px} * P_{qy})$; for all data transfers $(O_x \to O_y)$, is the probability that *none* of the BDFG data transfers *will* be bound to the resource-to-resource interconnection.

Hence, $P(R_p \to R_q) = 1 - \prod(1 - P_{px} * P_{qy})$; for all data transfers $(O_x \to O_y)$, is the probability that *some* BDFG data transfer *will be* bound as an interconnection between the two resources R_p and R_q.

0-7695-2533-4/06 $20.00 © 2006 IEEE   325

The intuition behind calculating resource probability ($P(R_j)$) and resource-to-resource probability ($P(R_{p1} \rightarrow R_{q2})$) is that $P(R_j)$ represents the probability of the resource being present in the final RTL netlist (floorplan) and $P(R_{p1} \rightarrow R_{q2})$ represents the probability of an interconnection (wire) being present between two resources in the final RTL netlist (floorplan). These probabilities quantitatively define the impact of a resource or an interconnection on the optimization constraints i.e. the timing and area of the final RTL netlist (floorplan). Thus, by synergistically optimizing the shape and location of high probability resources, wirelengths of interconnects with high resource-to-resource connection probability, and by selecting high probability resources that have high positive impact on the optimization constraints for allocation, the methodology has better chance of reaching a constraint-satisfying optimal solution.

3.3. Probabilistic-Gains

Probabilistic-gain measure models the gain of binding an operation (functional or storage) to a resource with respect to optimization constraints. We define probabilistic-gain of binding an operation O_i to resource R_j as:

$$P(O_i \rightarrow R_j) = \frac{A_{Reqd}}{A_{R_j}/P(R_j)} * \frac{\delta_{Reqd}}{\Sigma(T(Res \rightarrow R_j)/P(Res \rightarrow R_j))}$$

A_{Reqd} is the area constraint. A_{R_j} is the *actual* (not probabilistic) area of the resource R_j obtained from the floorplan. $A_{R_j}/P(R_j)$ measures the probabilistic impact of binding $O_i \rightarrow R_j$ on the area constraint, if the resource is present in the final floorplan.

δ_{Reqd} is the required register-transfer delay (clock period). $P(Res \rightarrow R_j)$ is the probability of a connection existing between resources R_j and Res. $T(Res \rightarrow R_j))$ is the worst-case register-transfer delay between resources R_j and Res. $\Sigma(T(Res \rightarrow R_j)/P(Res \rightarrow R_j))$ measures the probabilistic impact of binding $O_i \rightarrow R_j$ on worst-case register-transfer delay (clock period) of data transfers connected to operation O_i in BDFG.

4. Datapath Allocation Algorithm

We propose an exhaustive-constructive branch-and-bound algorithm to solve the datapath allocation problem. Figure 4 shows the pseudo-code of the algorithm. From the BDFG and maximal resource set, an initial binding graph (BG) is generated, where all the BG operations are connected to all the BG resources that can support the operation. This captures the entire datapath allocation solution space. Using the BDFG and BG, a probabilistic netlist (PN) is then generated. The heuristic floorplanner takes the PN as input to generate a floorplan. The *actual* resource shapes and wirelength information are extracted from the floorplan. Probabilistic-gains for all operation-resource pairs are calculated and stored in the probabilistic-gain ma-

```
Branch–and–Bound (BDFG, R, Constraints) {
    Best_Cost = INFINITY;
    generate Binding Graph;
    generate Probabilistic Netlist;
    generate Initial Floorplan;
    while (all operations in BDFG are not bound) do
        select best unmarked operation–resource pair;
        bind operation to resource;
        bind input variables of the operation to registers ;
        bind output variables of the operation to registers;
        check Feasibility;
        if (Feasible) then
            update Binding Graph;
            generate Interconnects;
            update Floorplan;
            if (lower_bound <= Cutoff) then
                make binding permanent;
                update probabilistic–gains;
            else // ** branch terminated ** //
                mark operation–resource pair;
                backtrack;
            end if;
        else // ** infeasible – branch terminated ** //
            mark operation–resource pair;
            backtrack;
        end if;
    end while;
    // ** obtaining cost of allocation solution ** //
    update floorplan;
    if (Current_Cost < Best_Cost) then
        Best_Solution = Current_Solution;
        Best_Cost = Current_Cost;
        update Cutoff;
    end if;
    return Best_Solution;
end Branch–and–Bound
```

Figure 4. Branch-and-Bound Datapath Allocation Algorithm

trix. Operation-resource pair with maximum probabilistic-gain is selected. In case of ties, the operation-resource pair with higher resource probability and resource that is located closer to the center of the floorplan is selected. Probabilistic-gains prioritize the allocation solution space with respect to optimization constraints. These lead to identification of constraint-satisfying solutions early. These solutions are then used to prune the remaining solution space, leading to quick convergence.

In the BG, binding an operation O_i to resource R_j is performed by removing all edges from the node O_i to resource nodes other than R_j. This is illustrated in Figure 3(c), where dotted lines indicate removed edges. After binding operation O_i, the input variables connected to O_i and subsequently the output variables connected to O_i are bound to registers with highest operation-resource probabilistic-gain. The register-transfer containing the operation is now considered to be bound. The actual register-transfer delay can now be estimated and checked against the maximum clock period constraint. If the constraint is satisfied, BG is updated to remove any edges from other operations with overlapping lifetimes, to already bound resources.

After each binding decision, *feasibility* of the allocation solution is checked to see if the partial allocation solution will lead to a complete allocation solution. This step is simplified to identifying operations in the BG that has no edges connected to it. If the solution is infeasible, the partial allocation solution will never lead to a complete allocation solution as the operation with no edges connected to it can never be bound to a resource. The branch is terminated (pruned) and the algorithm backtracks to the previous binding decision and proceeds along a different search path. If solution is feasible, BG is parsed to get a list of unconnected (unused) resources i.e. resources in BG with no edges connecting them. This is illustrated in Figure 3(c). Resource $A2$ and register $R3$ are unused and can be removed from the BG and the probabilistic netlist as no operation can be bound to these resources. As binding continues, more and more resources become unused and can be removed from the floorplan. This reduces the allocation solution space and floorplan size. Interconnect (Multiplexer) generation/pruning is performed after each binding decision. Each bound BG resource node is analyzed after binding. Resource nodes with more than one edge to *bound* operations are identified and multiplexers are generated and added to these nodes. In Figure 3 (c), resource $A1$ is reused. A multiplexer is generated for resource $A1$. The multiplexer information is added to resource for further calculations.

Resource probabilities and resource-to-resource interconnection probabilities are then updated. The floorplan is updated. The algorithm proceeds by binding another operation and associated register-transfer. This continues until all register-transfers in BDFG are bound.

Cutoff point for bounding allocation search space: Once the register-transfer is *bound*, the floorplan is updated. Accurate register-transfer delays are estimated from the floorplan. In this framework, bounding (pruning) of allocation solution space is reduced to checking if all the *bound* register-transfers satisfy the maximum clock period constraint. If the constraint is violated then the search path is terminated and the algorithm backtracks to another search path. If the constraint is not violated, cluster constraints are added to the register-transfer resources in the probabilistic netlist (PN), so that the floorplanner can maintain the proximity of the resources and hence, maintain the legality of the datapath allocation decisions.

5 SA Floorplanning based Estimation

The floorplanner implements a heuristic simulated annealing (SA) floorplanning algorithm with clustering and range constraints [3]. The floorplanner takes the probabilistic netlist (PN) as input and generates/updates the floorplan.

The requirements for a heuristic evaluator in the proposed framework are twofold. First, the floorplanner must be able to modify the floorplan to explore the floorplan solution space during allocation solution construction. Range constraints with respect to the center of the floorplan, are generated for operations (including SCNs) in bound register-transfers. Range constraints make sure that high probability resources remain close to the center of the floorplan. This pushes low probability resources to the periphery. Hence, removal of unused (zero probability) resources will not lead to a drastic change in floorplan. Second, to maintain consistency during exhaustive-constructive allocation search, the floorplanner estimates must remain consistent during the entire allocation search process. That is, once a register-transfer is bound, the register-transfer delay estimates for this register-transfer must not violate cutoff (maximum clock period constraint) for any successive binding decisions or floorplan updates. Clustering constraints are generated for all the bound register-transfers. Clustering constraints make sure that the resources bound to a register-transfer remain in close proximity, thereby making sure that they do not invalidate the allocation decision made.

5.1. Cost Function

The cost function used to evaluate allocation solutions with estimates from the floorplanner is shown below. The same cost function is used by the floorplanner to optimize the floorplan after a binding decision.

$Cost = \alpha * (A + AS) + \beta * (WL_{prob})$; where A = actual floorplan area, AS = is the aspect ratio violation, WL_{prob} is the probabilistic wirelength; α = Number of decisions made / Number of decisions to be made; β = Area and wire length relative weightage. The number of decisions made is the number of operations in BG with only one edge. Number of decisions to be made is the total number of operations in the BG.

Initially, $\alpha = 0$, as no allocation decisions are made and all resources are probabilistically present in the floorplan. As α moves close to 1, the area constraint is imposed. For a complete allocation solution, $\alpha = 1$, meaning all decisions have been made. All resource probabilities and resource-to-resource interconnection probabilities in the floorplan equate to 1. Hence, estimations from the floorplanner for a complete allocation solution (final floorplan) would be the actual area and wirelength estimate.

6 Experimental results

Experiments were conducted on 8 data-dominant designs, with the largest being 4x4 matrix multiplication (used in 3D applications). All designs were synthesized using a commercial $0.18u$ technology library. The proposed methodology was implemented in C++/STL. Experiments were conducted on a Sun Blade 100 workstation with 512MB RAM. We compared the proposed methodology with a public-domain HLS tool ASSERTA [6]. ASSERTA

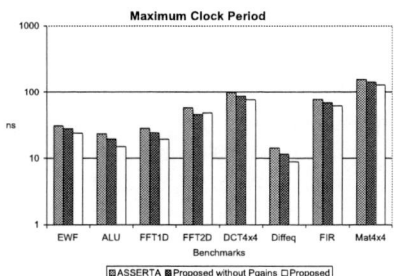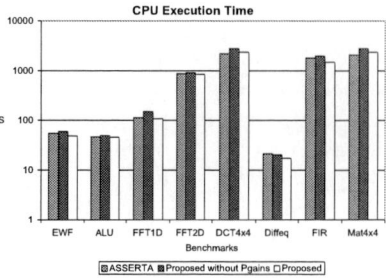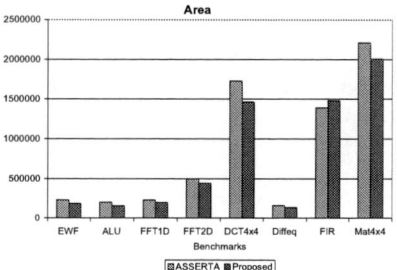

Figure 5. Experimental Results I

performs physical-unaware heuristic resource set generation, operation binding, register optimization, followed by interconnect and controller generation, sequentially. Each design was synthesized using a library of pre-characterized components. Logic synthesis and timing-driven placement with the same constraints was performed on the synthesized netlists from both the approaches using commercial synthesis and placement tools.

Resource Set vs Execution Time and Delay (Diffeq)		
Resource Set	Execution Time	Delay
10 Reg, 2 Add, 3 Mult, 2 Minus	7.8	10.45
12 Reg, 2 Add, 4 Mult, 2 Minus	8.3	11.12
14 Reg, 4 Add, 5 Mult, 2 Minus	10.1	9.63
16 Reg, 4 Add, 6 Mult, 2 Minus	17.4	8.84

Figure 6. Experimental Results II

Figure 5(a) reports the maximum clock period estimated from ASSERTA, the proposed methodology without using probabilistic-gains, and the proposed methodology while using probabilistic-gains. The corresponding execution times are reported in Figure 5(b). The proposed methodology without using probabilistic-gains shows on average, a 14% improvement on the maximum clock period, and when using probabilistic-gains shows a 25% improvement. Maximum clock period estimated after synthesis for each tool was compared with clock period estimated after placement. ASSERTA results deviated by as much as 171%, where as results from the proposed methodology match closely, with a maximum deviation of only 26.5%, thus illustrating the need for physical information during HLS. Without probabilistic-gains, the proposed methodology actually shows a increase in execution time by 15%. When the methodology uses probabilistic-gains, there is an average reduction in CPU time by 4%. Figure 5(c) shows the area obtained for each benchmark. There is an average area reduction of 14%. Figure 6 shows the effect of size of maximal resource set on the execution time of the proposed methodology, for benchmark *Diffeq* with same constraints. A non-linear variation in execution time can be seen with change in resource set. This is due to the increased search space that the methodology has to explore. However, a reduction in maximum clock period can also be seen, indicating the need to solve all sub-tasks in datapath allocation task simultaneously.

7 Conclusions

In this paper, we presented a novel method to perform integrated datapath allocation and floorplanning. A framework to perform exhaustive-constructive datapath allocation combined with heuristic floorplan estimation was presented. Novel probabilistic-gain measures to guide the allocation algorithm towards optimal constraint-satisfying solutions early were proposed. Compared with a physical-unaware HLS methodology, we have shown that our methodology produces significantly better results, with reductions in maximum clock period, area and runtime.

References

[1] A. Stammermann et. al. Binding Allocation and Floorplanning in Low Power High-Level Synthesis. In *IEEE Intl. Conf. on Computer-Aided Design (ICCAD)*, page 544, 2003.

[2] D. Gajski et. al. *High-Level Synthesis: Introduction to Chip and System Synthesis*. Kluwer Academic, 1992.

[3] E. F. Y. Young et. al. Placement Constraints in Floorplan Design. *IEEE Trans. on Very Large Scale Integration Systems (TVLSI)*, 12:735 – 745, July 2004.

[4] K. Bazargan et. al. Fast and Accurate Estimation of Floorplans in Logic/High-Level Synthesis. In *10th Great Lakes Symp. on VLSI (GLSVLSI)*, pages 95–100, 2000.

[5] M. Xu and F. Kurdahi. Layout-Driven RTL Binding Techniques for High-Level Synthesis Using Accurate Estimators. *ACM Trans. on Design Automation of Electronic Systems (TODAES)*, 2(4):312–343, 1997.

[6] N. Narasimhan. *Formal Synthesis: Formal Assertions Based Verification in a High-Level Synthesis System*. PhD thesis, University of Cincinnati, October 1998.

[7] P. G. Paulin et. al. Force Directed Scheduling for the Behavior Synthesis of ASICs. *IEEE Trans. on Computer-Aided Des. of Integ. Ckts. and Systems*, 8(6):661–679, June 1989.

[8] R. Bergamaschi. Behavioral Network Graph Unifying the Domains of High-Level and Logic Synthesis. In *36th Design Automation Conf. (DAC)*, pages 213–218, 1999.

[9] W. E. Dougherty and D. E. Thomas. Unifying Behavioral Synthesis and Physical Design. In *37th Design Automation Conf. (DAC)*, page 756, 2000.

[10] Y. L. Lin. Recent Developments in High-Level Synthesis. *ACM Trans. on Design Automation of Electronic Systems (TODAES)*, 2(1):2–21, 1997.

[11] Z. Gu et. al. Incremental Exploration of the Combined Physical and BehavioralDesign Space. In *42nd Design Automation Conf. (DAC)*, pages 208–213, 2005.

0-7695-2533-4/06 $20.00 © 2006 IEEE

CHESS: A Comprehensive Tool for CDFG Extraction and Synthesis of Low Power Designs from VHDL

Nagarajan Ranganathan, Ravi Namballa, and Narender Hanchate
Department of Computer Science and Engineering
University of South Florida, Tampa, Florida, 33620.

Abstract— In this paper, a new tool CHESS, is designed and developed for control and data-flow graph (CDFG) extraction and the high level synthesis of low power designs from behavioral level VHDL descriptions. The tool optimizes latency, area and power during the different phases of synthesis and provides several solutions to evaluate the trade-offs during design. Unlike the case of DFGs, not much work has been reported in the literature for low power synthesis of CDFGs. The tool consists of three individual modules: (i) CDFG extraction, (ii) scheduling and allocation of the CDFG, and (iii) binding, which are integrated to form a comprehensive high-level synthesis system. The first module for CDFG extraction includes a new algorithm in which compiler-level transformations are applied, followed by a series of behavioral-preserving transformations on the given VHDL description. The CDFG is fed to the scheduling module for resource optimization under the given set of time constraints. The scheduling algorithm is an improvement over the Tabu Search based algorithm described in [1] in terms of execution time. The improvement is achieved by pre-identification of mutually exclusive operations in the CDFG extraction phase, which, otherwise, is normally done during scheduling. The third and the final module of the proposed tool implements a new binding algorithm based on a game-theoretic formulation. The problem of binding is formulated as a non-cooperative finite normal game, and Nash equilibrium function is applied to achieve a power-optimized binding solution. Experimental results for several high-level synthesis benchmarks are presented which establish the efficacy of the proposed synthesis tool.

I. INTRODUCTION

The VLSI design flow involves various levels of design abstraction with its own trade-offs in the design-effort required and the accuracy of implementation. However, due to the increased complexity of circuits at lower levels of abstraction, most of the design tools are used at the higher levels of abstraction [2]. System level specification, which is the most abstract form of representation of the design, mostly gives its description in plain English. Behavioral description, the next level, gives a functional description of the design while avoiding the structural details of the design. Behavioral synthesis requires transformation of the Hardware Description Language (or VHDL code) into a internal representation which extracts registers, combinational logic equations and macros like '+', '*', etc. Most systems use something like the control flow graph and/or the data flow graph or the combination of the two, called CDFG as their intermediate format. A CDFG captures all the control and data flow information of the original VHDL description while preserving the various dependencies. The necessity and significance of the CDFGs in high-level synthesis has been highlighted in [2], [3]. [4]–

[6] provide a formal approach to the scheduling problem with a CDFG.

Scheduling in high-level synthesis groups the operations into control-steps based on their type and the dependencies in such a way that the operators in the same control step could be executed simultaneously. A wide variety of scheduling approaches which are directed at either reducing the execution time or resource minimization can be found in [7]–[10]. Time-constrained scheduling approach has the objective of realizing designs with minimum possible hardware while meeting time constraints, and is often adopted for designs targeted towards applications in real-time systems. Resource constrained scheduling algorithms are used in applications where the design is restricted by the silicon area. The goal of these algorithms is to minimize the number of control steps while satisfying the resource constraints. Allocation is the process of determining the functional units of each type for performing the operations while binding includes the process of assigning each such operation to a particular functional unit. Allocation and binding for functional units consists of grouping operations in such a way that each group consists of mutually exclusive operations while the total number of groups is minimized. Identifying these mutually exclusive operations during scheduling or binding phase is difficult and time consuming [1]. There are many pioneering works in the area of high-level synthesis for resource allocation and binding [7], [11]–[13].

II. MOTIVATION

Most of the previous works on high-level synthesis target data-dominated designs, but, are not adequate enough to handle control-dominated designs. Control-flow intensive behaviors with inherent loops and conditionals are quite possible in network-centered systems. Hence, there is need for a comprehensive high-level synthesis tool which is capable of handling both data intensive and control intensive flow graphs efficiently. The high-level synthesis process requires the compilation of the behavioral description of the design into a graphical representation, capturing the control and data dependencies. The derivation of such a control and data flow graph has been done independently and separately for control and data flows, leading to two different flow graphs. In addition, they are mostly derived manually, which makes this process time-consuming and error-prone at least in the earlier stages of synthesis. Hence, there is a need to integrate the data and control flows together and develop an automatic extraction

0-7695-2533-4/06 $20.00 © 2006 IEEE

tool which can create a single flow graph to represent the control and data dependencies. The traditional design automation tools were developed with the objective of reducing area or improving the speed of designs or both. Power dissipation in VLSI circuits is also an important design concern due to the portable wireless devices and mobile computing. Thus, we have addressed the problem of power optimization in the binding phase of our integrated synthesis system.

The identification of mutually exclusive operations is typically performed during the scheduling or allocation phase of the high-level synthesis. This identification process adds tremendous overhead to the running time of the scheduler [1]. We have noticed that the mutually exclusive operations can be easily identified while extracting the control and data dependencies from the behavioral description. This is because the mutually exclusive property and the control and data dependencies are inter-related. Hence, we have moved the identification process to be handled by the CDFG extraction tool, so as to gain a significant improvement in running time of the scheduler. This kind of improvement can only be achieved in a tool which can perform the CDFG extraction, scheduling and binding all together, inside the tool. Hence, this has motivated us to develop a comprehensive high-level synthesis system that could be used for both data-flow and control-flow intensive designs.

The integrated high-level synthesis system developed in this work is targeted at both control-flow intensive and data-flow intensive behaviors, and incorporates several additional features like optimization for resources and power consumption. Our system uses a single graph representation of control and data flow dependencies as its intermediate form, and also incorporates an automated tool for transforming the VHDL description of the design into the corresponding CDFG. Such a CDFG is generated in several formats to accommodate different implementation approaches. The CDFG is fed to a tabu search based scheduler to achieve a resource optimized Scheduled CDFG (S-CDFG). The scheduled CDFG is fed to the binding phase, where the functional units are assigned to the tasks of the CDFG with power reduction as the objective for binding the units.

III. CDFG Extraction From VHDL: Module 1

The flow-graph representation of the behavioral description is obtained by parsing the input VHDL code. Figure 1 depicts the various steps involved in this methodology. The *lexical analysis* phase translates the source program into a stream of tokens, where each token is a sequence of characters. This stream of tokens is further subjected to *syntactic analysis* which imposes a hierarchical structure on them to verify the syntax of the program. These two phases were generated using Lex and YACC on standard VHDL syntax. The parse tree is a conceptual visualization of the syntactic structure of the program. Explicit codes are required to extract such a parse tree from the YACC code, which was done using C++ in our tool. The parse tree is further compressed to obtain a *syntax tree* in which the operators appear as the interior nodes, and

the operands of an operator are the children of the node for that operator. The syntax tree is transformed, through another C++ code, into the final control and data flow graph that depicts the total flow of the control and data in the original description. The various data dependencies that are inherent in the flow graph can be revealed through one full scan of the graph. In [14], we have explained the intermediate steps, their algorithmic implementations, transformation of various VHDL constructs, and various output formats generated by CDFG extractor in much greater detail.

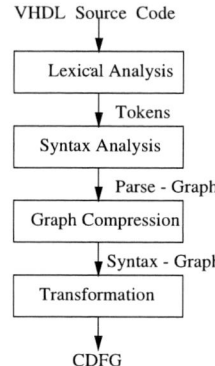

Fig. 1. Extraction of the CDFG from VHDL code

IV. Tabu-Search based Scheduler: Module 2

In this Section, we describe (i) scheduling approach based on the Tabu search method, aimed at reducing the number of resources under the given time constraints, (ii) resource allocation. The scheduling algorithm is an adaptation of the Tabu-search based scheduling algorithm described in [1], that uses a mathematical formulation based on penalty weights. The major difference is in the calculation of mutually exclusive operations, which is performed during the CDFG extraction phase in our high-level synthesis system. The tabu-search based scheduling algorithm is applied on the CDFG representation obtained from the Module 1, to find the optimum number of resources that would be needed for the design. The mathematical formulation of our scheduling approach is based on penalty weights rather than on cost evaluation as in [1]. The penalty weights include the real costs of the hardware units, which makes it possible to take into consideration different area parameters of the design. The objective function for our scheduling approach is given as $f = \sum_i W_i$, where each weight W_i is a cost-based penalization of a worst case assignment of operation nodes requiring the greatest amount of a given hardware resource. The scheduling approach, adopted from [1], is formulated as follows: Given the number of control steps K,

$$\text{Minimize } f(y1, y2, ..., yn) = \sum_i W_i, \quad (1)$$

satisfying the constraints,

$$S_i \leq y_i \leq L_i, for 1 \leq i \leq n, \quad (2)$$

$$y_i + d_i \leq y_j, for (i, j) \| i-> j, 1 \leq i and j \leq n, \quad (3)$$

0-7695-2533-4/06 $20.00 © 2006 IEEE   330

where, n is the total number of nodes of the control and data flow graph, y_i is the control step for node i, S_i is the ASAP step and L_i is the ALAP step for node i, d_i is the propagation delay for node i, and, $i-> j$ denotes node j is data dependent on node i. The penalty weights W_i are selected such that minimizing these weights would optimize the assignment of operation nodes. The weights will reducing the number of resources of a given type, while utilizing the mutual exclusion among nodes in the control step due to control and data dependencies.

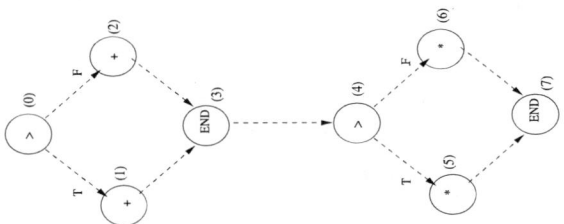

Fig. 2. Mutually exclusive nodes

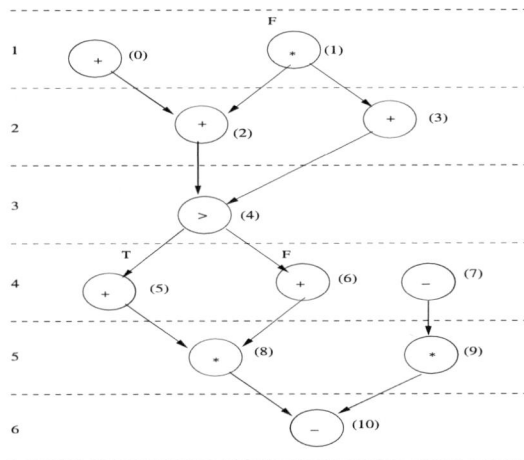

Fig. 3. Penalty weights

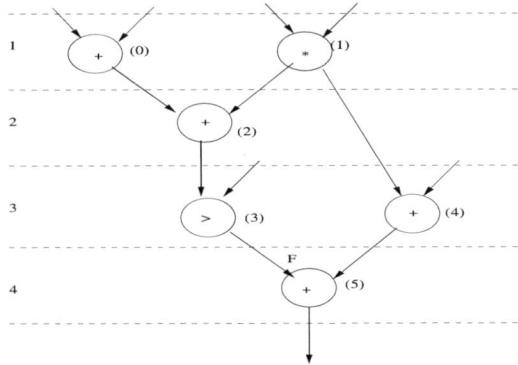

Fig. 4. Life-time and number of buses from CDFG

Definition: Two nodes, N_i and N_j, in a CDFG, are said to be mutually exclusive, if and only if, they cannot exist in any execution path of the CDFG at the same time.

Consider the CDFG with a two conditional statements shown in figure 2. Four possible execution paths exist for the given CDFG, which are, (0) (1) (3) (4) (5) (7), (0) (2) (3) (4) (5) (7), (0) (1) (3) (4) (6) (7) and, (0) (2) (3) (4) (6) (7). The nodes (1) and (2) cannot exist together in any of these paths, and hence, they are mutually exclusive. Similarly, nodes (5) and (6) are mutually exclusive. However, node (2) is not mutually exclusive with node (7), and neither is node (3) with node (6). We can observe that nodes (1) and (2) can be scheduled in the same control step with a single adder allocated to them, since we know that only one of them would be executed. Hence, it is advantageous to schedule two mutually exclusive operations of same type in the same control step. The mutual exclusion property could be exploited for potential resource sharing among such operations. The CDFG representation helps in determining all possible mutual exclusions among the operations in the CDFG, and hence, the mutual exclusive operations are identified during CDFG extraction phase. This saves the additional overhead of scanning the flow graph again and hence, reduces the overhead of determining the mutual exclusive operations during the scheduling phase.

The factors which are considered to penalize any assignment that requires more hardware resources are: (i) For each operation type m, the control step that has maximum number of operation nodes of type m is penalized, since that dictates the maximum number of resources executing operations of type m. Such a penalty weight is calculated as $W_1 = \sum_m (C[m] * maxK_{ms} \| 1 \le S \le k)$ where, K_{ms} is the maximum number of operations of type m that can be executed in control step s, $C[m]$ is the cost of function unit executing operation of type m, and po is the maximum number of operation types. The value of K_{ms} is obtained by considering the number of non mutually exclusive operations of type m in each control step. For the CDFG depicted in figure 3, there are a maximum of 2 multiplications in the control step 5, and a maximum of 2 additions in control step 2. Hence, the penalty weight associated would be W = 2 * cost[*] + 2 * cost[+]. (ii) The next weight penalizes the control steps which have a maximum number of non mutually exclusive operations of same equivalence class. Two operations of type i and j are said to belong to the equivalence class m, if and only if, both of them can be executed by a functional unit of type m. (iii) Two operations of type i and j belonging to equivalence class m have to be executed by 2 functional units of type i and j when assigned to the same control step, but, they could be executed by a single functional unit of type m, when assigned to different control steps. The penalty for assigning them to the same control step is calculated as, $C[i] + C[j] - C[m]$. For the CDFG in figure 3, the cost of assigning both addition and subtraction in same control step 4 is given as 1 * (C[+] + C[-] - C[+/-]). (iv) Another penalty weight is associated with the life times of weights. Whenever a node j uses the output of node i, the number of control steps separating i and j is the lifetime of the variable storing that value, The penalty factor for storing such values is calculated as: $W = \sum_i \sum_j [Su_{ji} * (y_j - y_i - d_i) + Su_{ij} * (y_i - y_j - d_j)] * cost[storage]$, where

0-7695-2533-4/06 $20.00 © 2006 IEEE   331

$Su_{ij} = 1$, if $i- > j$, 0, otherwise. In figure 4, the output of node (1) is used by nodes (2) and (4). Node (4) is 2 control steps away from node (1), and hence, it requires the output to be stored, leading to a weight of 1 * cost[storage]. (v) The final penalty weight is associated with the number of buses needed, which depends on the maximum number of distinct inputs used in a control step. In figure 4, the penalty weight is 4 * cost[bus].

Algorithm 1 Scheduling algorithm

Algorithm:

determine ASAP and ALAP times for each node in CDFG

$S \leftarrow$ generate initial solution

while fewer than $MAX_ITERATIONS$ have passed without any improvement on the best solution **do**

    $S_n \leftarrow$ choose best solution from neighborhood of S

    if S_n not visited in previous iterations **then**

        if S_n better than best solution found **then**

            save S_n as the best solution

            $S \leftarrow S_n$

        else

            discard S_n

        end if

    end if

end while

A pseudo-code for the Tabu-search based scheduling algorithm is shown in algorithm 1. The ASAP and ALAP time steps are determined initially and either of these solutions can be chosen as the initial solution S. The neighborhood of the solution is evaluated based on the penalty weights. The neighborhood of a solution is obtained by moving nodes from their current control step to another control step within the ASAP and ALAP values of that node. The nodes considered for movement are chosen from the control steps that contribute most to the penalty weights in the current solution. Such nodes are moved to control steps where there are lesser number of nodes of same operation type and of same equivalence class. The iterations are repeated until a fixed number of iterations pass without any improvement in the solution. This whole procedure is repeated by changing the time constraints until a good schedule is obtained. With large solution space explored by the Tabu search method, the scheduling algorithm would be able to perform global optimization of the number of resources used.

V. BINDING FOR POWER OPTIMIZATION: MODULE 3

The binding algorithm is a modification of the approach proposed by Murugavel et al. in [15] for minimizing the average power of a circuit during scheduling and binding. We have modified and extended this approach to include control constructs so as to apply it on the scheduled CDFG (S-CDFG) obtained from Module 2. The *binding matrix(B)* is a matrix ordered by the control steps and functional units with each entry being either a '0' or a '1'. An entry $B(i, j)$ takes the value '1' if functional unit j is bound to an operation

in control step i. A scheduled CDFG and its corresponding binding matrix are shown in figure 5. Power optimization is achieved through the concept of functional unit sharing, in which, neighborhood operations with at least one common input are assigned to the same functional unit so that the number of changing inputs are reduced.

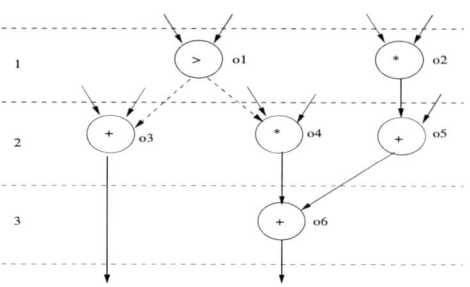

(a) Scheduled CDFG

Control	Functional Units				
step	comp1	mult1	add1	add2	add3
1	1	1	0	0	0
1	0	1	0	1	1
1	0	0	0	0	1

(b) Corresponding binding matrix

Fig. 5. Scheduled CDFG and its binding matrix

Applying Game and Auction theories, the low-power binding problem is transformed into a problem of finding the Nash equilibrium for the bidding strategies of the auction problem [15]. The available functional units are considered as the sellers and the operations as buyers in the auction problem. For an operation O_i, each functional units submits its power consumption for that operation as its sealed bid value and the lowest bidder is bound to that operation. The game-theoretic algorithm described in [15] was aimed at data-flow descriptions, and hence, their approach is applicable for scheduled data flow graphs (S-DFGs). However, since the CHESS tool is intended to deal with both control and data flow descriptions, we extend the algorithm described in [15] to make it applicable to a scheduled CDFG (S-CDFG). The extension stems from the inclusion of control nodes in the graph and by considering the corresponding functional units as sellers in the bidding problem. Also, the approach for S-CDFGs includes the flexibility of binding the same functional unit to two mutually exclusive operations of same type in a control step. The cost functions for each functional unit are evaluated in the same way as described in [15].

The cost function or power consumption associated with each module are calculated initially and are assumed to be available to all the operations in the S-CDFG. It is calculated as the difference between the switching cost (P_{sw}) and the cost due to changing inputs (P_{inp}), which are determined through simulations. The binding strategy is applied individually to each control step independent of the other control steps, since a module can be assigned to only one operation in each control step. Given a set of modules $M_i = (m_1, m_2, ..., m_{x_i})$ and the set of operations, $O_i = (o_1, o_2, ..., o_{y_i})$, where x_i is the number of modules compatible to operations of type i, and y_i is the number of operations of that type, we consider the feasible sets of module-operation pairs, and arrive at the most optimal binding using Nash equilibrium. The problem of binding is now reduced to finding a feasible set A^*, such that, $C(A^*) = \min C(A)$, and, the set A^* is the Nash equilibrium. The binding problem is solved for the Nash equilibrium using Gambit [16], a computational game solver. The pseudo-code for the Game theoretic binding algorithm is given in algorithm 2. The algorithm takes the S-CDFG and gives a binding matrix for that graph. Since the bidding strategy is applied separately for each control step, the Nash equilibrium has to be calculated for each set of applicable modules in each control step. Thus, the number of times the Nash equilibrium is applied depends on the number of control steps in the S-CDFG as well as the number of modules applicable for an operation in that control step.

Algorithm 2 Binding Algorithm

Input: Scheduled Control Data Flow Graph(S-CDFG), Set of modules, Power and Delay Values

Output: Binding Matrix B

  Algorithm:

  for all control step i **do**

    consider all the modules that can execute that operation

    select the best one

    for all set of compatible modules M in control step i **do**

      calculate cost matrix

      $C(O, M) \leftarrow$ calculate the power and delay based cost matrix

      find the Nash equilibrium solution

      $NE \leftarrow$ determine the Nash equilibrium using M, O and CM

      $B \leftarrow$ represent the NE as a binding matrix B

    end for

  end for

VI. EXPERIMENTAL RESULTS

The experimental results for the CDFG conversion tool are provided in [14]. The scheduling algorithm was also implemented in C++ on a 400MHz Sun Sparc station running SunOS with 256MB RAM. The algorithm takes, as its inputs, the CDFG, the types of functional units available, and the time constraints. For illustrating the efficiency of this algorithm, we concentrate on two of the most-widely used benchmark circuits, the diffeq and the elliptic filter, and compare the results to some of the existing systems. The following types of functional units were assumed to be available for scheduling the operations - adder, subtractor, adder/subtractor, multiplier, divider, and comparator. The scheduling results for differential equation benchmark are shown in Table I. HAL [7] and SPLICER [17] could schedule the diffeq benchmark with two multipliers, an adder, a subtractor and a comparator, but with CHESS, we were able to combine the adders and the subtractors by using a single ALU from the library that performs both the operations. This could be achieved because of the notion of the equivalence classes of operations used during the search for an optimized design. With six control steps, only one multiplier, one comparator and a single ALU performing addition and subtraction are required. Such a schedule was generated in 15 iterations in a CPU time of around 1 second.

TABLE I

COMPARISON OF SCHEDULES FOR THE DIFFERENTIAL EQUATION

BENCHMARK

Scheduling Algorithm	No. of control steps	No. of buses	No of FUs				
			(+)	(-)	(+/-)	(*)	(<)
ASAP	4	10	2	1	-	3	1
ALAP	4	8	1	1	-	2	1
HAL [7]	4	3	1	1	-	2	1
SPLICER [17]	4	6	1	1	-	2	1
CHESS (Solution 1)	4	6	-	-	1	2	1
CHESS (Solution 2)	6	4	-	-	1	1	1

TABLE II

COMPARISON OF SCHEDULES FOR THE ELLIPTIC FILTER BENCHMARK

Scheduling Algorithm	No. of control steps	No. of buses	No of FUs	
			(+)	(*)
ASAP	17	8	4	2
ALAP	17	8	3	2
HAL [7]	19	6	2	2
FDLS [18]	17	-	3	2
CHESS (Solution 1)	17	8	3	3
CHESS (Solution 2)	18	6	2	2
CHESS (Solution 3)	19	5	2	1
CHESS (Solution 4)	28	4	1	1

The scheduling results for elliptic filter are shown in Table II. Several design alternatives have been generated for 17, 18, 19 and 28 control steps. Only two adders, two multipliers and six buses were required for the schedule in 18 control steps. The schedule with 17 control steps does not provide significant improvement over other approaches because of the limited solution space explored in this case, 17 being the critical path length. With increasing control steps, the solution space is extended, and after a number of trials, the best solution is obtained with 28 control steps, where only a single adder and a single multiplier are required. These solutions were produced in a CPU time of around 6 to 14 seconds. To overcome this problem of excessive memory requirement, instead of saving a whole solution, we save only the modification characterizing the moves at each iteration. The graph shown in figure 6 shows the improvement in memory utilization obtained with this optimization approach. It can seen that the memory requirement has nearly been reduced by 30% with our approach.

The algorithm described in Section 4 was coded in C++ and tested upon the scheduled CDFGs obtained previously. The Nash equilibrium was calculated at every iteration of the algorithm using the *Gambit* tool for game theory [16]. To provide for alternative binding strategies for the operations, a library of FUs was developed using the Cadence design tool in $0.35\ \mu$ MOSIS SCN3M SCMOS technology. These cells

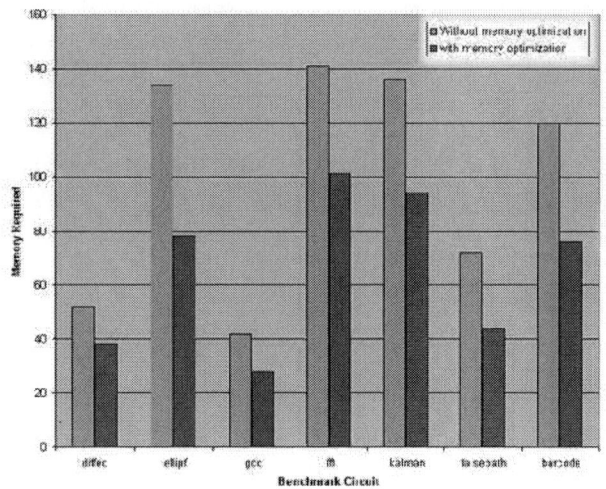

Fig. 6. Improvement in Memory Requirement

were characterized for power and delay through simulations. Each simulation was performed using the *hspice* simulation tool upon 100,000 random input vectors that were generated using MATLAB. Some other values were borrowed from [19]. The scheduled CDFGs of the benchmark circuits were subject to the game theoretic binding strategy. The total power for the S-CDFGs were calculated using a random binding strategy and a greedy approach. Later, the power values obtained through our binding strategy are compared against these, as illustrated in Table III. We can see that, on an average, our tool could achieve a 18% reduction in the average power consumed over the greedy approach, for sufficiently large benchmark circuits. It is observed that the reduction in power obtained through our approach is less for smaller circuits as compared to larger ones, because, the smaller circuits, when formulated with the game-theoretic approach, provide lesser number of strategies to be explored while optimizing the cost function, as in the case of the gcd benchmark circuit.

TABLE III

COMPARISON OF BINDING RESULTS

Benchmark	Random	Greedy	CHESS	%Red. over	%Red. over
Circuit	(mW)	(mW)	(mW)	Random	Greedy
FIR	24.005	18.300	15.005	37.49	18.005
IIR	16.290	15.304	10.290	36.83	32.76
diffeq	24.191	9.641	9.304	61.54	3.04
ellipf	34.595	28.890	25.595	26.02	11.41
fft	43.804	25.105	22.005	49.76	12.34
kalman	33.585	27.780	24.890	25.87	13.85
gcd	2.624	2.216	2.216	15.55	0

VII. CONCLUSIONS

We have presented a comprehensive tool for high-level synthesis for synthesizing both control and data intensive applications efficiently. Our tool includes an automatic conversion module, which converts the behavioral VHDL description into the corresponding CDFG representation so as to capture the entire flow information into a single flow graph. We have

demonstrated, through several examples, that the tool is quite efficient in converting various VHDL constructs into their corresponding flow-graph representations. Our intention in this work was to develop a high-level synthesis system which can handle both control and data intensive applications effectively, in one single integrated system. We have modeled our synthesis system to perform timing optimization during scheduling phase and power optimization during binding phase. However, the scheduling and binding modules of our tool can be replaced with any other efficient scheduling and binding algorithms which are capable of performing delay optimization or power optimization or both together, depending on the need of the design. In this work, we have achieved delay optimization by sharing the resources across the mutually exclusive operations and power optimization by exploiting the neighborhood operations with at least one common input.

REFERENCES

[1] S. Amellal and B. Kaminska, ""Functional Synthesis of Digital Systems with TASS"," in *IEEE Transactions on Computer Aided Design of Integ. Circuits and Systems*, vol. 13, no. 5, May 1994, pp. 537–552.

[2] N. S.Govindarajan and R.Vemuri, ""Dependecny analysis and operation graph generation for high-level synthesis from behavioral VHDL"."

[3] D. R.P.Dick and W.Wolf, ""TGFF: Task graphs for free"," in *Proc. of Int. Workshop on Hardware/Software Codesign*, Mar. 1998, pp. 97–101.

[4] S.Devadas and R.Newton, ""Algorithms for hardware allocation in data path synthesis"," *IEEE Trans. on CAD*, vol. CAD-7, no. 8, pp. 768–781, Jul. 1989.

[5] R.Composano, ""Path based scheduling for synthesis"," *IEEE Trans. on CAD*, vol. 10, no. 1, pp. 85–93, Jan. 1991.

[6] J. C.T.Hwang and Y.C.Hsu, ""A formal approach to scheduling problem in data path synthesis"," in *IEEE Int. Conf. on CAD*, vol. 10, Apr. 1991, pp. 464–475.

[7] P. G. Paulin and J. P. Knight, ""Algorithms for High-Level Synthesis"," in *IEEE Design and Test of Compters*, vol. 6, no. 6, December 1999, pp. 18–31.

[8] S. Gupta and S. Katkoori, ""Force-Directed Scheduling for Dynamic Power Optimization"," in *IEEE Computer Society Annual Symposium on VLSI*, 2002, pp. 68 –73.

[9] A. Sllame and V. Drabek, ""An Efficient List-Based Scheduling Algorithm for High-Level Synthesis"," in *Euromicro Symp. on Digital System Design*, September 2002, pp. 316 –323.

[10] T. Kim and C. Liu, ""A New Approach to the Multiport Memory Allocation Problem in Datapath Synthesis"," *VLSI Integration*, vol. 19, no. 3, pp. 133 –160, 1995.

[11] S. Hong and T. Kim, ""Bus Optimization for Low Power Data Path Synthesis Based on Network Flow Method"," in *ICCAD*, 2000.

[12] J. Chang and M. Pedram, ""Register Allocation and Binding for Low Power"," in *DAC*, 1995.

[13] C. Tseng and D. Siewiorek, "Automatic syntheis of data path on digital systems," in *IEEE Trans. Computer Aided Design of Integrated Circuits and Systems*, vol. 5, no. 3, July 1986, pp. 379–395.

[14] "Removed to facilitate blind review."

[15] A. K. Murugavel and N. Ranganathan, ""A Game-Theoretic Approach for Power Optimization during Behavioral Synthesis"," *16th Int. Conf. on VLSI Design*, pp. 452 –458, January 2003.

[16] R. McKelvey, A. McLennan, and T. Turocy, *"Gambit: Software Tools for Game Theory"*, California Inst. of Tech. and Univ. of Minnesota and Texas A&M Univ., September 2002.

[17] B.M.Pangrle and D.D.Gajski, ""Design tools for intelligent silicon compilation"," in *IEEE Trans. on Computer Aided Design*, vol. CDA-6, November 1987, pp. 1098 –1112.

[18] P. G. Paulin and J. P. Knight, ""Force Directed Scheduling for the Behavioral Synthesis of ASIC's"," in *IEEE Trans. Computer Aided Design*, vol. 8, June 1989, pp. 661–679.

[19] V. Raghunathan, S. Ravi, A. Raghunathan, and G. Lakshminarayana, ""Transient Power Management through High-Level Synthesis"," in *IEEE Int. Conf. on Computer Aided Design*, 2001, pp. 545 –552.

System Exploration of SystemC Designs

Christian Genz   Rolf Drechsler

Institute of Computer Science, University of Bremen, 28359 Bremen, Germany
{genz,drechsle}@informatik.uni-bremen.de

Abstract

Due to increasing design complexity new methodologies for system modeling have been established in VLSI CAD. The SystemC methodology gains a significant reduction of design cycles by introducing an executable specification and a top down refinement strategy. But still the size and the complexity of SystemC models grow, making it harder to understand the basic ideas architects and their designs intend. This extends the familiarization phase for coworkers and project partners. In modern design flows, this can become a significant problem.

In this work we present an approach for interactive system exploration of SystemC designs and its implementation. The aim of our approach is to facilitate the orientation towards complex SystemC models without the need for simulation based techniques. Our tool accomplishes system exploration by allowing to navigate hierarchically through SystemC designs. It uses schematic visualization at different levels of abstraction to display the structure and the behavior of the design. Further support is given for different schematic views, a source code view, crossprobing, path fragment navigation and module exploration.

1 Introduction

The system description language SystemC has become a standard in system level design. SystemC's ability to reach higher abstraction levels than *hardware description languages* (HDLs), while still being able to represent HW-structures and the pragmatic approach of implementing executable C++ specifications make SystemC attractive for industry and academia. The methodology introduced by SystemC aims to close the design gap, which implies a reduction of time to market.

Since SystemC has been introduced, the *electronic design automation* (EDA) community heads for extensive tool and library support. In conjunction with attached libraries SystemC is capable of functional simulation, simulation based verification [10] (SCV library), transaction level modeling [2] (TLM library) and modeling of analog and mixed-signals [18] (AMS library). Among other tasks current tools support waveform tracing [15], synthesis [7], co-simulation [3], bounded model checking [8] and analysis [11, 1, 16]. In contrast to the libraries and

tools supporting SystemC verification and analysis, the number of tools supporting system exploration of SystemC specifications is disproportionately smaller. Caused by this lack of tool support SystemC potentials are not fully exploited. Related work is discussed after the presentation of our approach in Section 6.

In this paper we introduce the tool *ViSyC*, that provides functionality for system exploration of SystemC models. In the following the term *system exploration* will be used as a concept that helps architects elaborating system designs. Simulation based attempts, as they are frequently used in design space exploration, assume an extended knowledge of the design for being able to interpret the results of elaboration. Unlike to these techniques system exploration only requires an essential understanding of VLSI CAD to explore the design. The advantages of our approach are:

- hierarchical visualization
- crossprobing
- path fragment navigation
- module exploration

Our tool supports the understanding of large *system on chip* (SoC) designs. They can be represented at different levels of abstraction via detail hiding, that encapsulates implementation details into three different schematic views. Thus, the internal structure, the behavior and the interface of a module are separated to simplify the representation of each single view. All three views preserve the hierarchy information that is given by the structure of the SystemC model. All elements of the schematic view are linked with a corresponding position in the source code view via bidirectional crossprobing, enabling intuitive path fragment navigation. The path fragment navigation is a technique that enables tracing of signals, ports, and operational elements by following their connected inputs or outputs. The functionality of the tool covers SystemC analysis and the generation of a database for visualization. Hence, the analysis is very complex, it is separated into two phases. The first phase is a transformation from an *abstract syntax tree* (AST) to an *intermediate representation* (IR). The second analysis phase is an interpretation of the function `sc_main`. The interpretation allocates internal memory images of the instantiated modules and builds up connectivity between these instances. We use the tool GateVision from Concept Engineering[1] as a backend for graphical system exploration. It

---

[1] www.concept.de

0-7695-2533-4/06 $20.00 © 2006 IEEE

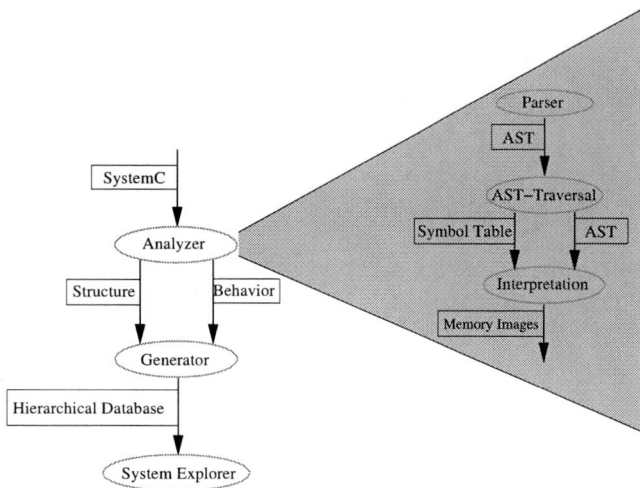

Figure 1. Overall flow within the tool

enables the designer to navigate through the database interactively. The overall flow is shown in Figure 1.

The paper is structured as follows: Section 2 introduces the reader to the standards of SystemC and its methodology. The way we analyze SystemC models is described in Section 3. Section 4 describes the background of visualization and the bidirectional connection between the source code and its schematic counterpart. In Section 5 our tool is applied to evaluate a system level description by the exploration of selected modules. Section 6 clarifies the alternatives to our approach, and discusses related work. It gives direct comparison to existing approaches and also discusses limitations. In Section 7 a brief summary of the paper is given and directions of future work are discussed.

2 SystemC

SystemC has been published by the *Open SystemC Initiative* (OSCI). OSCI is a non-profit association that has been formed by several industrial, academical and individual partners. The aim of OSCI is the standardization of SystemC as an open source standard for system level design. Since the SystemC library is open source, various kinds of modifications and extension libraries are publicly available, too [2, 18, 10, 14].

The system description language SystemC provides hardware constructs, implemented in a C++ class library. The hardware models specified using SystemC can be compiled on a large number of supported architectures using a standard C++ compiler. The compiled executables can be cycle accurate simulations as well as untimed algorithmic descriptions of the given design. The executable specifications can be used for evaluation, debugging and refinement purposes without the usage of a commercial simulator. Depending on the abstraction level the simulation speed can be a multiple of a functional equivalent HDL model. Because of its unrestricted C++ conformance each SystemC model can be combined with other software libraries. This allows system engineers to take advantage

of HW/SW Co-Design and to refine their SoC designs with a high level of flexibility. Another benefit of SystemC, coming with its C++ conformance, is a wide range of abstraction levels that can be used to simplify huge system designs. Complex communication protocols and control logic can easily be separated from functional parts of the specification. For this reason SystemC offers techniques that can raise or lower the level of abstraction. The TLM library implements such a technique to support SystemC's efficient refinement methodology. For more details see [12].

SystemC combines HDL typical features, like concurrency as it appears in hardware, with software paradigms, like object orientation. Those features distinguish SystemC from VHDL, Verilog and SystemVerilog and enable system description capabilities. SystemC allows real polymorphism which includes the application of arbitrary memory access using pointers and dynamic memory allocation. Even the concept of virtual functions that binds overloaded class members to function pointers, is applicable in system descriptions. Special benefits, like channels, make SystemC ideal for describing complex communication protocols and their easy reuse.

3 SystemC Analysis

SystemC analysis aims to transform a specification into an abstract representation. Recent efforts in SystemC analysis follow two approaches to evaluation: simulation and parsing. The simulation approach has the severe drawback that it does not simultaneously cover all paths of the *control flow diagram* (CFD). For this reason we implemented an analyzer that is able to analyze SystemC programs without simulation. The analyzer maps relevant information to an IR that contains the structure and the behavior of the model. The analyzer can be split into syntactical analysis, which is done by the parser, and semantical analysis. Hence, the semantical analysis is quite complex, it is split again into AST traversal and a following interpretation phase. The AST traversal collects the structure of the program while the interpretation phase collects the behavior and dynamic objects.

3.1 Extraction of Syntactical Information

The parser is realized by use of PCCTS [13]. The input is a standard C++ source file, implementing a SystemC model. The output of the syntactical analysis is an AST, where an example can be seen in Figure 2. The AST is an acyclic directed graph. Hierarchy information is stored in the nodes of the AST. Except for root and leaf nodes, each node in the AST has exactly one parent and one or two children. Each down edge references a leaf, or an adjacent node, that has further details concerning the parent node. Each right edge also references a leaf, a following statement or a following declaration.

Besides referencing its children, each node of the AST refers to an appropriate token structure. The token structure includes a token type, a token value and further details of the read word. The advantage of the AST is to have a datatype, that enables the analyzer to split the

0-7695-2533-4/06 $20.00 © 2006 IEEE   336

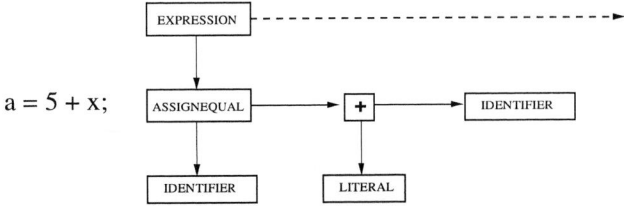

a = 5 + x;

Figure 2. AST example

analysis into several passes. Each AST node holds location information including the source file, the line of code and the character position of the underlying token. These are very important details in the process of visualization, that builds connections between symbolic objects and their corresponding source code fragments.

Most SystemC models use preprocessor directives to preserve a clear design style and to reuse existing structures and algorithms. Unfortunately common preprocessor frontends [17] manipulate the input stream of the parser. Frequently used directives like `#define` are eliminated at its point of definition and inserted at its points of usage in the source code. The parser and its backends have no opportunity to identify such substitutions. To avoid these problems we implemented a preprocessor frontend for recognition of preprocessor directives during syntactical analysis. The usage of widespread keywords like `SC_MODULE` is implemented by the grammar, used in the parser, not by preprocessor substitution. The SystemC grammar is an advantage of our approach that avoids many effects, that are caused by C++ workarounds in the SystemC library. One example of these workarounds in SystemC is the usage of internal module names, which are not necessarily equal to the declaration names. Another example is the limitation of parameters, that can be used in named port bindings.

3.2 Extracting Structural Information

With the completion of syntactical analysis the corresponding AST is available. To come to knowledge about the structure of the model we traverse this AST to collect details of the types and the modules, that are used in the design. During the AST traversal all occurrences of named declarations like types, functions and variables are stored in a symbol table. The symbol table is a hierarchical container that offers efficient functionality for searching named declarations and error checking. All elements hold a reference to their corresponding sub-AST to enable crossprobing later on. Initially the symbol table consists of an empty scope, which gets filled up during AST traversal, and common SystemC data types like `sc_int<n>`. To avoid multiple declarations of the same type preprocessor directives including the SystemC header files can be ignored.

Hierarchy in the symbol table is given by scopes, that emerge when traversing the AST of a function, a type declaration (`struct`/`class`/`module`) or a block statement. Each level of the hierarchy is associated to a private name-space with a disjoint set of variables, types and functions. A SystemC evaluation approach based on execution is aware of everything but of declared names, since objects

are referenced by their address through compilation. To provide object names for analysis such an approach is forced to attach additional name information, organized in a single namespace. Because no element of this namespace has knowledge of the parent scope, hierarchy information is lost. The deficit of hierarchy causes all equally named objects of different scopes to be misleadingly identified as the same object during evaluation. Another problem caused by introducing a single namespace is the disability to ensure the equivalence between declaration name and additional name of an object, which may lead to confusing effects. With scopes the symbol table prohibits the reuse of equal names at the same hierarchy level and reuse is allowed at an other level only.

The structure of the design is basically formed by the declarations and memory allocations caused by global variables, stack variables, or heap variables. Since SystemC implements C++ programs, the structure of a model can be build using dynamic memory allocation to instantiate modules and signals. Hence, dynamic modules and other SystemC objects that are allocated using the `new` operator, are not passed to the symbol table. Because our approach does not rely on simulation, we face this problem with an additional phase that interprets the specification partially as explained in Section 3.3.

3.3 Extracting Behavioral Information

The behavior of a SystemC model is given by statements that are placed in function definitions. Each function definition in the symbol table references an AST that holds such a statement block. The behavioral analysis starts at `sc_main`. Each statement of the function is analyzed and interpreted. As the top level routine which defines the starting position of the model, the `sc_main` function allocates top modules, signals and builds up connectivity.

Instances of static and dynamic variables are stored to a memory image. Each variable in the memory image has a value that can be read or written by the operational AST that is interpreted. Like other variables, that are typed as a struct or a class, modules have a constructor. The module initialization includes the interpretation of the constructor AST. This is where the process type is set and where the function to be executed is called. The block of this function is traversed recursively and stored into the memory image as a sequence of operations. At the end of this procedure all modules and their ports are known and interconnected by a continuous sequence of expressions.

4 SystemC Visualization

From the computed memory images the tool generates a binary file that holds a database. The generation of this database runs a mapping mechanism that converts SystemC objects to viewable symbols, dependent on the class the objects are derived from. Because SystemC follows an object oriented paradigm it is able to define an arbitrary number of derived types. Since the number of viewable symbol classes is restricted by the exploration backend,

0-7695-2533-4/06 $20.00 © 2006 IEEE

different types have to be grouped and mapped to a single symbol. The database can be displayed with an interactive GUI for design exploration from Concept Engineering.

Besides a schematic view, that uses the symbols of the database, the GUI offers a source code view. Each symbol in the schematic view corresponds to a passage of the source code view. *ViSyC* implements a bijective function that does a mapping between symbols and their corresponding source code passages. The GUI uses a crossprobing technique to implement arbitrary navigation between symbols, their declaration and their instantiation. Another technique that benefits from the bijective map is the path fragment navigation. The path fragment navigation uses the link between corresponding symbols and source code passages and enables system designers to follow data paths or control flow paths intuitively.

In order to get a short and compact representation of the SystemC specification, we need to extract the whole hierarchy. Once having this hierarchy, the circuit can be described at various levels of detail. It is important to note that all hierarchy information is given by the design and not generated by our tool. By using the extracted hierarchy without modifications *ViSyC* conserves the semantic equivalence between the source model and its output, the schematic view. The semantic equivalence again is a premise for crossprobing, that ensures each object of the model to have a counterpart in the symbolic view and vice versa.

The visualization of the hierarchy follows the different scopes of the design. Only the top level view shows all modules connected to their signals. The black box of a module or a channel is its most abstract symbolic representation and can be explored. By entering the module all members including ports and submodules are shown including their connections. The connections between ports are established by sequences of operational symbols, that have been extracted from the processes behavior. Loop statements and conditional statements are also represented by black boxes including conditional information. Statements that consist of one or more expressions are mapped to a sequence of symbols. This way arithmetical expressions and logical expressions can be displayed in a continuous flow. An expression that calls a function is mapped to a black box that is explorable.

Mapping signals to viewable symbols is done by a mechanism that creates single wires or buses according to the signal's width. Each wire or bus can be connected to an arbitrary number of ports. When a channel has been chosen instead of a bit signal, the symbol is a black box again. The exploration of the black box shows the implementation of the channel. Standard C++ or SystemC data types are not explored, because these basic types are known.

5 Evaluation of a RISC-CPU

To demonstrate the different abstraction levels of our approach, we perform system exploration on the SystemC source code of a RISC-CPU. The source code is part of the OSCI SystemC package and it is freely available [12]. The RISC-CPU consists of twenty eight source files, that define the ten modules. The SystemC model is implemented using *register transfer level* (RTL). For evaluation of the model we navigate through three different abstraction levels that clarify the structure and the behavior of the model.

5.1 Top View

The top view of a SystemC description assembles the modules that are instantiated globally or in `sc_main` and their connecting signals. The top level of the RISC-CPU can be seen in Figure 3. All modules, as well as other SystemC objects, are annotated with their declaration name. The internal SystemC name, that is attached to modules, ports and signals is not displayed. The application window in Figure 3 is separated into two fields. The left field of the window displays a hierarchical list that enumerates all instantiated modules. All entries of the list give access to a source code passage that declares the type of the corresponding module. The right field of the window displays the schematic view. The objects of this view are linked to their instantiation in the source code.

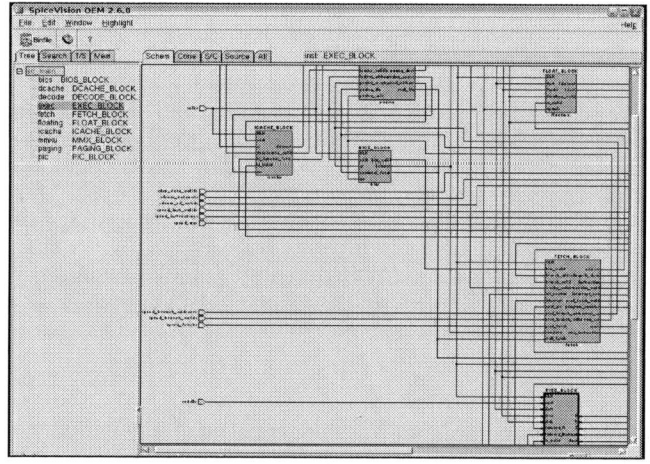

Figure 3. Top view of the RISC-CPU

5.2 Module View

To get a more detailed impression of the structure it is possible to explore single modules. The `exec` module of the CPU is an operational unit that keeps the instruction set. The input ports of the module are aligned to the left whereas the output ports are aligned to the right in alphabetic order. Between input and output ports the behavior of the module is displayed by schematic symbols. The process of the `exec` module is expanded by the entry function that is handled as an `SC_CTHREAD`, shown in Figure 4. Basically the entry function implements a selection statement that is sensitive to the opcode signal. The selection statement is represented by its functional blocks and a set of multiplexers. By use of the opcode the multiplexers are able to assign a calculated value of the respective functional block to its corresponding output port.

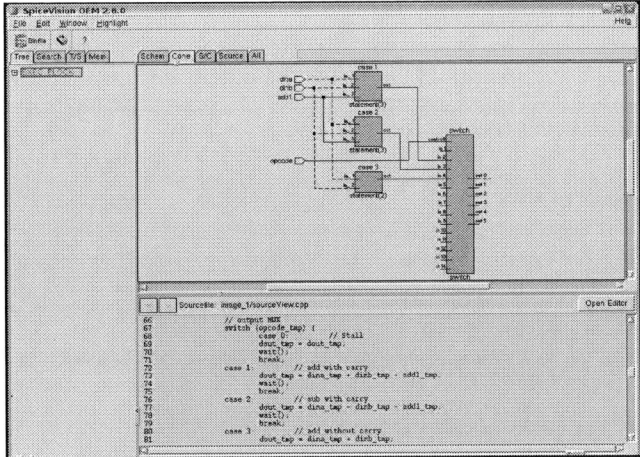

Figure 4. Module view of the RISC-CPU

5.3 Word Level View

The word level consists of a sequence of logic and algorithmic operations. It represents the behavior level of the CPU. This view can use logic and algorithmic operators only. Because the symbolic mapping depends on the hardware implementation of these operators and is not given in the system description, they are not explorable like modules. Figure 5 shows the case 0 of the selection statement that implements an arithmetic addition with carry-bit.

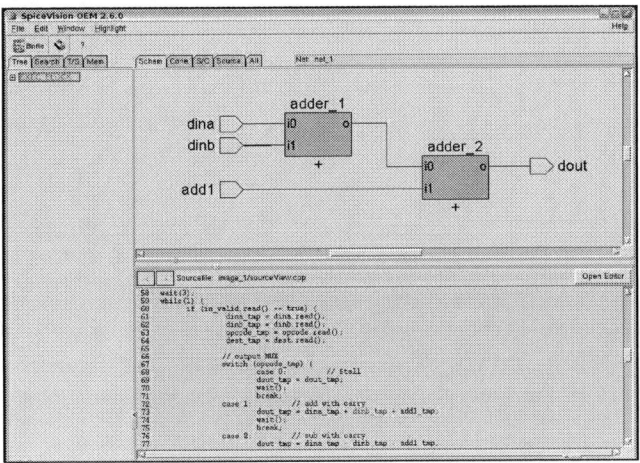

Figure 5. Behavior of the RISC-CPU

6 Discussion

Few approaches have been presented in the field of SystemC analysis and visualization. These approaches differ from ours in the degree of detail of the visual representation and in the type of analysis. None of them accomplishes system exploration, as it is described in Section 1. This section discusses the related work to show the advantages and restrictions of our approach.

6.1 Related Work

One of the first approaches that accomplishes SystemC design visualization has been introduced in [9]. The implementation of Große et al uses the SystemC kernel to analyze SystemC models during execution. An interactive graphical backend facilitates the design visualization. Even though models can be specified using the C++ features, analysis and visualization are limited to SystemC objects. Only the data flow can be viewed, no behavioral information is available. Since this approach has to execute the model without further information of declarations, it is not aware of detailed positional information regarding the objects. Hence, crossprobing facilities are very restricted.

The tool gSysC [5], an extension library for SystemC, is a recent work at the University at Lübeck. When using gSysC instead of SystemC, the underlying application is able to capture run time information of the executed specification analogously to [9]. Additional functionality of gSysC offers the graphical evaluation of state variables and the visualization of a flat module list. Similar to [9] this approach uses run time evaluation, but the hierarchy information is completely lost.

ParSyC [7] is a synthesis tool that does RTL synthesis on a SystemC subset. The subset is a combination of restricted C++, essential SystemC data types and SystemC control mechanisms. The output of the synthesis is unoptimized BLIF (as used in [6]). Instead of run time evaluation this approach uses a separate analyzer to extract structure and behavior of the source model. To reduce the complexity of input models ParSyC prohibits features like pointers, polymorphism and the declaration of templates. Hence, ParSyC does not achieve our requirements regarding SystemC analysis.

Another approach that includes an extra SystemC parser has been published by Snyder and is called SystemPerl [16]. SystemPerl is a Perl library that summarizes four major packages with SystemC support. The parser extracts the netlist interconnectivity and other information from an unpreprocessed file. But SystemPerl is not sufficient for our requirement of an analyzer frontend because it does not extract code from the body of a process. Like SystemPerl also SystemCXML [1] is a free tool with SystemC analysis capabilities. An interpretation of the SystemC source is done by Doxygen [19], that generates a respective XML representation. The XML files are used by SystemCXML to capture structural information of the model into an IR. Unfortunately the analysis is not detailed enough to capture the behavioral information of the model.

Pinapa [11] implements a free SystemC analyzer built upon a modification of the GNU C++ compiler and a modified SystemC library. Pinapa enables a fine grained analysis by executing the model in combination with an AST traversal procedure. Being based on execution, Pinapas analysis covers only parts of the design that are accessed in execution. Thus the quality of analysis depends on the reliability of a stimuli generator.

0-7695-2533-4/06 $20.00 © 2006 IEEE

6.2 Discussion of our Approach

Besides reliably interpreting SystemC descriptions, instead of simulation, our approach accepts SystemC as a language, which enables us to ignore some of its workarounds (see Section 3.1). The advantage of our interpretation over simulation approaches is that we are not dependent on a stimuli generator for an analysis that covers the complete model. Our analysis is even able to handle system descriptions without a stimuli generator. The backend of our analysis decides whether a part of the description is part of the model. With this information our visualization backend is able to hide unnecessary details that would only confuse the designer.

The lack of all tools listed in Section 6.1 and ours as well is that they are bound to one single SystemC version, or to a modification. Without the inclusion of the SystemC library ViSyC is not aware of any changes in the future of SystemC. To assure models to run that are build on a future SystemC version, this version has to be compatible to the current release (SystemC-2.1.0). But considering the development of SystemC as a difficult process that consumes much time and effort, this lack may be negligible.

ViSyC enables the exploration of SystemC designs. Visualization is a technique for representing a complex context in a symbolic way. The schematic view, used in our tool, is an aid that lets the user decide which kind of abstraction he/she wants to use for exploring the design. While pure visualization is able to represent one abstraction level only, our system exploration is interactive and allows GUI supported tracing of each symbol in the design. Besides the interaction, a bidirectional correlation between source code and symbols in schematic view is supported. This allows the tracing of code to its connected symbols and vice versa.

7 Conclusion

In this paper we presented an approach for interactive system exploration of SystemC models. Our approach is implemented as the tool *ViSyC*[4], that generates three schematic views from the source code of a model. The schematic views extend the hierarchical structure as well as the behavior of the model and enable system designers to navigate through different abstraction levels. Each part of a schematic view corresponds to a source code passage in the source code view. Via bidirectional crossprobing these connections can be traced intuitively.

Our approach does not rely on stimuli generators or a modified SystemC library and covers the complete model. It supports extensive information extraction from SystemC designs. This makes *ViSyC* a clever extension to SystemC and offers a platform for other development steps. Our future research on *ViSyC* will concentrate on debugging facilities. In detail our next steps will include a schematic run time evaluation that enables *ViSyC* to visualize selected configurations of a running system model.

Acknowledgements

The authors would like to thank Lothar Linhard and Gerhard Angst from Concept Engeneering for the friendly support and the cooperation.

References

[1] D. Berner, H. Patel, D. Mathaikutty, J.-P. Talpin, and S. Shukla. SystemCXML: An extensible SystemC front end using XML. Technical Report 06, FERMAT@Virginia Tech, Apr. 2005.

[2] L. Cai and D. Gajski. Transaction level modeling: an overview. In *IEEE/ACM/IFIP international conference on Hardware/software codesign and system synthesis (CODES+ISSS)*, pages 19–24, 2003.

[3] CoWare Inc. ConvergenSC. http://www.coware.com.

[4] R. Drechsler, G. Fey, C. Genz, and D. Große. SyCE: An integrated environment for system design in SystemC. In *IEEE International Workshop on Rapid System Prototyping*, pages 258–260, 2005.

[5] C. Eibl, C. Albrecht, and R. Hagenau. gSysC: A graphical front end for SystemC. In *European Conference on Modelling and Simulation*, pages 257–262, 2005. Source available at http://www.iti.uni-luebeck.de/ albrecht/gSysC/.

[6] Electronics Research Laboratory, University of California at Berkeley. *OCTTOOLS-5.2 Part II Reference Manual*, Mar. 1993.

[7] G. Fey, D. Große, T. Cassens, C. Genz, T. Warode, and R. Drechsler. ParSyC: An efficient SystemC parser. In *Workshop on Synthesis And System Integration of Mixed Information technologies (SASIMI)*, pages 148–154, 2004.

[8] D. Große and R. Drechsler. CheckSyC: An efficient property checker for RTL SystemC designs. In *IEEE International Symposium on Circuits and Systems*, pages 4167–4170, 2005.

[9] D. Große, R. Drechsler, L. Linhard, and G. Angst. Efficient automatic visualization of SystemC designs. In *Forum on Specification and Design Languages*, pages 646–657, 2003.

[10] C. Ip and S. Swan. A tutorial introduction on the new SystemC verification standard, 2003. Available at http://www.systemc.org.

[11] M. Moy, F. Maraninchi, and L. Maillet-Contoz. PINAPA: An extraction tool for SystemC descriptions of systems-on-a-chip. In *ACM international conference on Embedded software (EMSOFT '05)*, pages 317–324, 2005.

[12] OSCI. SystemC. http://www.systemc.org.

[13] T. Parr. *Language Translation using PCCTS and C++: A Reference Guide*. Automata Publishing Company, 1997.

[14] H. Patel and S. Shukla. Towards a heterogeneous simulation kernel for system level models: A SystemC kernel for synchronous data flow models. *IEEE Transactions in Computer-Aided Design*, 24(8):248–253, Aug. 2005.

[15] W. Snyder. Dynotrace. Available at http://www.veripool.com/dinotrace/.

[16] W. Snyder. SystemPerl home page. http://www.veripool.com/systemperl.html.

[17] The GNU Project. The gnu compiler collection. http://www.gcc.gnu.org.

[18] A. Vachoux, C. Grimm, and K. Einwich. SystemC-AMS requirements, design objectives and rationale. In *Design, Automation and Test in Europe*, pages 388–393, 2003.

[19] D. van Heesch. Doxygen. Available at http://www.doxygen.org.

Power Aware
LVSI Design

Reliability-Aware SOC Voltage Islands Partition and Floorplan

†Shengqi Yang, †Wayne Wolf, ‡N. Vijaykrishnan and ‡Yuan Xie

†Department of Electrical Engineering, Princeton University, Princeton, NJ, 08544

‡Microsystems Design Lab, The Penn State University, University Park, PA, 16802

{shengqiy|wolf}@princeton.edu {vijay|yuanxie}@cse.psu.edu

Abstract—**Based on the proposed reliability characterization model, reliability-bounded low-power design as a methodology to balance reliability enhancement and power reduction in chip design, for the first time, is illustrated. Voltage island partitioning and floorplanning for System-On-a-Chip (SOC) design is used as a case study for this reliability aware methodology. The proposed methodology partitions all SOC components into different voltage islands with power reduction and guaranteed system reliability. Experiments show that for a typical SOC the algorithm execution time is within several minutes while achieving 12% to 23% power reduction. Extended SOC algorithm partitions and floorplanns the voltage islands within 2.5 to 29.7 minutes and achieved 9.74% to 18.50% power reduction.**

I. INTRODUCTION

Continuous decrease in transistor feature size enables more and more devices to be fabricated in a single chip and various functional modules to be realized as a Systems-On-a-Chip (SOC). This scaling trend poses two critical issues for chip design, i.e., power consumption and system reliability.

Due to the big increase in power density and the wide use of portable systems, power consumption, including dynamic power and leakage power, is becoming a critical design metric. Voltage islands architecting [1]–[3] has emerged as a new technique for core-based low-power SOC design and can effectively cut down all kinds of power sources. A voltage island is a group of on-chip cores powered by the same voltage source, independently from the chip-level voltage supply. The use of voltage islands permits operating different portions of the SOC at different voltage levels in order to optimize the overall chip power consumption at core-level.

Accompanying the feature size scaling down, chip reliability, here it means immunity to soft errors, is becoming another important issue. Soft errors or transient errors are circuit errors caused due to excess charge carriers induced primarily by external radiations [4], [5]. Radiation directly or indirectly induces a localized ionization which upsets internal data states [6]–[8]. Soft errors are particularly troublesome for pipeline latches, and memory-based elements, such as caches, register files, branch target buffer, etc., as the stored values of the bits may be changed. Recent experiments [9] showed that soft error rate of combinational circuits is comparable to or more than those of SRAMs with similar sizes. Soft errors can have much serious impact and lead not only to corrupt data, but also to a loss of functionality and system critical failures. As the technology scales down, both the supply voltage and the node capacitance scale down accordingly. The above two factors make a single device node more sensitive to soft error.

Supply voltage is lowered, as a most effective way to reduce system power consumption, at some specific location (voltage island) of a SOC. However, lowered supply voltage makes the device nodes more sensitive to soft errors because less charge is required to flip a bit at the node. Systems become less reliable. Design methodology, which can not only reduce the power consumption but insure the Mean Time To Failure (MTTF) constrain without violation of the application deadline time, are desired to help tradeoff the requirements from both reliability and power reduction in chip designs. In the following section, for the first time we will illustrate this reliability-bounded low-power design methodology by using SOC voltage island partitioning as a case study and implementing the corresponding algorithm.

The remainder of this paper is organized as follows. Section II describes the problem. Section III details the algorithm implementation. Section IV discusses the experimental results and Section V concludes this paper.

II. PROBLEM DEFINITION

In this section, reliability enhancement and power reduction tradeoff problem is defined for the SOC voltage island partitioning.

Figure 1 shows an example of core-based SOC design. There are six cores for this SOC. Each core is associated with a list of voltages, that can be used to operate the core, and a list of reliability levels corresponding to each voltage. We will discuss how to get the reliability levels in the following section. For instance, the core C_3 can operate at v_2, v_3, and v_4. If it operates at v_3, the reliability level under this voltage is r_{32}. The chip-level voltage is assumed to be v_1 which is the highest voltage compared with v_2, v_3 and v_4. For core C_6, a distinct voltage island is not created for it because it only can operate at the chip-level supply voltage. In order to minimize the power consumption, one obvious way is to operate each core at its lowest voltage. This means that at least two voltage islands should be created: one for C_1 and C_2 which operate at v_2; one for C_3, C_4 and C_5 which operate at v_4. For the above answer, the power reduction is maximized. However, system reliability is totally ignored in the design procedure. A bad scenario is: C_3, for example an SRAM-based component in civilian aviation avionics system, is very sensitive to soft errors, that occasional bit flips will cause a failure of the application execution, i.e., a flight tragedy. In this case, reliability is so important for C_3 that we cannot neglect it just in order to reduce power. Instead a little power

---

Acknowledgments: This work was supported by National Science Foundation under Grant numbers CCR-0324869 and CCR-0329810.

0-7695-2533-4/06 $20.00 © 2006 IEEE

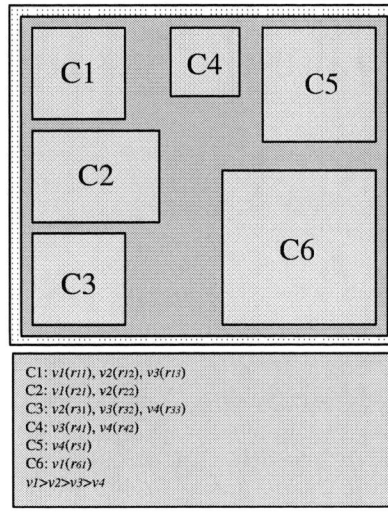

C1: v1(r11), v2(r12), v3(r13)
C2: v1(r21), v2(r22)
C3: v2(r31), v3(r32), v4(r33)
C4: v3(r41), v4(r42)
C5: v4(r51)
C6: v1(r61)
v1>v2>v3>v4

Fig. 1. A Core-based SOC.

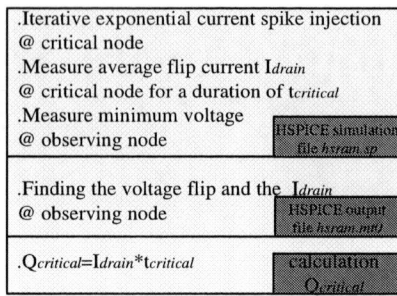

Fig. 2. $Q_{critical}$ calculation procedure.

reduction will be sacrificed to guarantee the core's reliability.

Now, Let us generalize the reliability and power consumption tradeoff problem for the SOC voltage islands partitioning. We assume given an SOC consisting of a set of cores C_i. Each core is associated with a table, named as PR-table, which specifies the legal voltage levels Vi, the corresponding average power consumption values Pi, and the corresponding average reliability levels Ri. The legal voltage levels of a core can be characterized by the core designers. For example, the designers may keep changing the core supply voltage, as long as the core-level timing assertions are satisfied. And this gives out a list or a range of voltages that can be used to operate the core. Once the legal voltages are selected, the average power consumption and the average reliability level corresponding to each voltage can also be characterized either through experimentation, simulation or analytical estimation. For a voltage island Λ_i, it consists of a few cores which are operated at the unique voltage, denoted as $\Lambda_i(v)$. $\Lambda_i(v)$ is selected from a set, denoted as ΛV_i, containing a list of supply voltage levels. ΛV_i is equal to the intersection of the legal voltage levels of all cores in this island. As an example, if we create an island for C_1 and C_2 in Figure 1, $\Lambda V_i = \{v_1, v_2\}$, while $\Lambda i(v)$ is equal to one of them, either v_1 or v_2, depending on the final tradeoff of power consumption and system reliability. For the whole SOC system, it can be represented as $SOC = \sum_i \Lambda_i + \sum_i C_i$, where $\sum_i C_i$ only denotes those cores not assigned to any voltage island and therefore operated at chip-level power supply voltage.

We use the PR-tables as the algorithm inputs and assume some cores in the SOC, which have a same voltage level in their V_is, can possibly be grouped together to form a voltage island. The constraints or rules for this grouping are power consumption, system reliability and application deadline time. For the convenience of discussion, all other constraints, like chip area, will be considered in the extended algorithm. The problem of power and reliability tradeoff island planning for core-based SOC design can be summarized as partitioning SOC cores into a set of voltage islands and under this partition reducing the total system power consumption as much as possible without violation of system reliability level and application deadline time.

III. MODELING AND ALGORITHM

In this section, we first illustrate how to characterize component reliability level, and then elaborate the algorithm for reliability-aware SOC voltage island partitioning and floorplanning. Note that the focus of this paper is on how to incorporate the reliability issue into low power design, specifically into the voltage island partitioning and floorplanning.

A. Characterization of Component Reliability Level

For a soft error to occur at a specific node in a SRAM-based element, the collected charge Q at that particular node should be more than a critical charge $Q_{critical}$. If this happens, a pulse is generated and latched on by the feedback mechanism of the inverter, which results in a bit flip at that node. Similar phenomenon happens in combinational logic. The rate at which soft errors occur is given as Soft Error Rate (SER). In order to estimate the Component Reliability Level (CRL), we first select the most sensitive node to soft error in either a SRAM-based component or a logic component, and calculate the node $Q_{critical}$. In general, the sensitivity of a circuit to soft error depends not only on the most sensitive node, but also on the average sensitivity of all circuit nodes. For simplicity, we only consider the most sensitive node. In order to measure $Q_{critical}$ for a particular node, we define it as:

$$Q_{critical} = \int_0^{t_{critical}} I_{drain}(t)dt \quad (1)$$

where $I_{drain}(t)$ is the drain current induced by the charged particle, and $t_{critical}$ is the flipping time. In HSPICE experiment, we model the particle striking current $I_{drain}(t)$ as an exponential current waveform to account for funneling and diffusion charge collection. The current was injected at the most sensitive node and measured up to a point where the circuit commits a bit flip. Finally the current pulse was integrated to get the critical charge of that node. The $Q_{critical}$ calculation procedure for an SRAM is shown in Figure 2. By using this procedure, the $Q_{critical}$ value for a 6-T standard SRAM (65nm Berkeley Predictive Technology (BPT)) is $15.1fC$ and $51.2fC$ for a ripple-carry adder, for example.

Then, $Q_{critical}$ is used to estimate the SER which is expressed in Equation (2) where N_{flux} is the intensity of the Neutron Flux, CS is the area of the cross section of the node and Q_s is the charge collection efficiency. We reasonably assume that N_{flux} and CS are same for different components

0-7695-2533-4/06 $20.00 © 2006 IEEE

TABLE I
CRL FOR ADDER, MULTIPLIER, AND SRAM.

Component	CRL	Area (Unit)
Adder	1.0000	1
SRAM	0.9877	2
Mult	0.9625	2

under the same technology. With this assumption, Q_s can also be set same for different components. We select a component, say C_N, and define its SER to be one, i.e., $SER_N = 1$. For other components, the SERs can be expressed as Equation (3), where $Q_{criticalN}$ and $Q_{criticali}$ mean the critical charges for C_N and C_i.

$$SER \propto N_{flux} \times CS \times exp(\frac{-Q_{critical}}{Q_s}) \quad (2)$$

$$SER_i = exp(\frac{Q_{criticalN} - Q_{criticali}}{Q_s}) \quad (3)$$

The next step is to relate the SER of each component to its reliability level (CRL). Mathematically, component reliability is defined as the probability with which this component can perform the intended function successfully for a period of time, namely [t1, t2], given that it worked properly at time t1. To calculate the reliability of a component, we should first determine its failure rate which is the probability with which it will fail in the next time unit. Although SER does not directly mean failure rate because sometimes soft error will not result an application failure, it can be modified and used to calculate the CRL as Equation (4), where α is a parameter within [0, 1] representing the probability a soft error finally results a failure. As discussed in [10], some masking, such as logical masking, temporal masking and electrical masking, can prevent soft error from resulting an actual failure. α reflects this fact and calculation of α is complicated. For simplicity, here we assume α is equal to one.

$$CRL_i = exp(-\alpha \cdot SER_i \cdot t) \quad (4)$$

In order to build the components library with reliability characterizations, Cadence Virtuoso was used to construct the components layout and HSPICE was for getting the $Q_{critical}$s. Table I shows the normalized CRL for adder, SRAM and carry-save multipler as examples under 65nm BPT. The CRL and area of the adder is set to one. For larger components which contain a lot of basic units, each basic unit like an adder in the netlist can be characterized individually. After this, by analyzing the interconnection of gates in the netlist, the overall soft error susceptibility of the whole component can be determined. Specifically, we use the following formula (5) to describe CRL:

$$CRL_i = \prod_{j=1}^{n} R_j \quad (5)$$

where n means there are n basic units inside the components. This equation reflects, to have a successful execution of an entire component, all the basic units must succeed.

Finally, for a SOC, the system Reliability Level ($RL(SOC)$) can be defined as Equation (6), which is the product of the reliability values for different voltage islands. While the reliability value for a voltage island $RL(\Lambda_i)$ is defined in Equation (7), where component C_{ij} belongs to Λ_i. For CAR, it reflects the application program properties. If a component or some part of the component is never accessed by the application program, soft errors inside this component or inside the specific part of that component cannot result in a program failure; while if a component is accessed very frequently, not only do soft errors which already exist in the component have much possibility to result in a program failure, but also more soft errors can happen in this component during the access time. We use Equation (8) to calculate the Component Access Rate (CAR). Of course, there are some other factors, for example whether a component is protected or not, which affect the final reliability level. Here the system reliability level calculation is a simplified case.

$$RL(SOC) = \prod_{i=1}^{m} RL(\Lambda_i) \quad (6)$$

$$RL(\Lambda_i) = \prod_{j=1}^{n} CAR_{ij} * CRL_{ij} \quad (7)$$

$$CAR_i = \frac{i_{th} \; component \; access \; time}{application \; execution \; time} \quad (8)$$

B. Reliability-aware SOC Voltage Island Partitioning Algorithm

For core-based SOC voltage island partitioning, let us first consider a simplified reliability-bounded low-power design algorithm which generates solutions meeting the requirements from system reliability and application deadline time with reasonable power reduction. There are several other important factors, for example area overhead, by adding which in the following extended algorithm the final optimal SOC voltage island partition and floorplan can be completed. Here, those additional factors are not considered to make the simplified algorithm clearly illustrate the idea of reliability-bounded low-power design. The algorithm is show in Figure 3.

In the algorithm for SOC island partitioning, the inputs are the lower bound of system reliability level $RL(SOC)^*$, the application execution time DT in the SOC system with each core operated at its highest voltage, the upper limit of application deadline time DT^*, the net information, and the characteristic parameters for each SOC component including the applicable voltage levels V_i, the power consumption value P_i, the component reliability level R_i, and CAR_i. Among the component parameters, V_i is a vector; while P_i, R_i, and CAR_i are scalars achieved with respect to the highest component supply voltage. The list of voltages, V_i, for a specific component can be characterized by the designer. Once a prototype of the SOC system is constructed, the targeted application programs can be tested in this prototype at some supply voltage. Then P_i and CAR_i for each component can be estimated and calculated on average. For example, if the component C_i is tested under the supply voltage v_{test} and the average power consumption is $P_i(v_{test})$, the $P_i(v)$ can be calculated according to Equation (9), where v and v_{test} belong to the set V_i. If C_1 works at v_1 as illustrated in Figure 1, C_2 at v_1, C_3 at v_2, C_4 at v_3, C_5 at v_4 and C_6 at v_1, the average access time of each component, AT_i, can be achieved by

0-7695-2533-4/06 $20.00 © 2006 IEEE

```
Input    :  Ci(Vi, Ri, Pi, CARi), RL(SOC)*, DT*
Output   :  All possible solutions which meet the criterion

Algorithm :

1:  For each SOC component Ci Do
2:      calculate CRLi and CPi
3:  End For
4:  construct a link list (VILL) for possible voltage island partitions
5:  For each partition in VILL Do
6:      update the CARi for each component
7:      calculate the deadline time DT for this partition
8:      If DT > DT* Then
9:          delete this partition from VILL
10:     Else
11:         calculate the reliability level RL(SOC)
12:         If RL(SOC) < RL(SOC)* Then
13:             delete this partion from VILL
14:         Else
15:             calculate the power reduction
16:         End If
17:     End If
18:     output the head of VILL
19: End For
20: sort VILL by the value of power reduction
21: output all the possible solutions
```

Fig. 3. Reliability and power consumption tradeoff algorithm for core-based SOC voltage island partitioning

running the application program in the SOC under the above specified voltage level for each component. The equation for calculation of CAR_i can be extended from Equation (8) to Equation (10). Later in the final implementation of the SOC system, if core C_1 operates at v_2 instead and the supply voltages for other cores do not change, the CAR_1 can be calculated as Equation (11) which scales the access time according to the frequency change caused by the voltage change. Here v_{th} means the threshold voltage. As a result, the voltage level at which CAR_i is achieved must be specified and input to the algorithm. For R_i, it is the reliability level for the core operated at the highest voltage.

$$P_i(v) = \frac{v^2}{v_{test}^2} \times P_i(v_{test}) \qquad (9)$$

$$CAR_i = \frac{AT_i}{\sum_{j=1}^{6} AT_j} \qquad (10)$$

$$CAR_1 = \frac{AT_1 \left(\frac{v_1 - v_{th}}{v_2 - v_{th}}\right)^{1.3} \times \frac{v_2}{v_1}}{AT_1 \left(\frac{v_1 - v_{th}}{v_2 - v_{th}}\right)^{1.3} \times \frac{v_2}{v_1} + \sum_{j=2}^{6} AT_j} \qquad (11)$$

$$CRL_i(v) = \frac{v}{v_{highest}} \times R_i \qquad (12)$$

First, the algorithm scales the R_i with respect to each voltage in V_i and keeps this number in a vector, CRL_i. The calculation is shown in Equation (12), where v and $v_{highest}$ are in the set V_i with the latter to be the highest supply voltage for core C_i. Here a linear dependence between reliability and operation voltage is assumed according to the idea of $Q_{critical}$. More complex analytical models can be applied for CRL_i calculation without significant change in the algorithm complexity. Similarly, the algorithm calculates the component power consumption CP_i at each voltage level according to Equation (9), by substituting the CP_i for P_i. Both CRL_i and CP_i are vectors. Second, the algorithm constructs a link

list VILL for all possible voltage island partitions. For each possible partitioning, the CAR_i should be refreshed for each component according to Equation (11). In the following steps, the application deadline time is first checked to see whether it exceeds the upper limit DT^* under this partitioning. If it does, then delete this partition from the link list. Otherwise the second checking is made for the system reliability level according to Equations (6). If the system reliability level under this configuration is less than $RL(SOC)^*$, then delete this partition from the list. The partitions remaining in the link list after both checks must meet the requirements from application deadline time and system reliability. Finally, the algorithm calculates the power reduction and the number of voltage islands for each possible voltage island partition.

The above algorithm is only for partitioning the SOC voltage islands by using three metrics, system reliability level, application deadline time, and power reduction. After this procedure which clearly illustrated the idea of reliability-bounded low-power design, all the remaining partitions guarantee the SOC system to be fast enough, reliable and power-saving. However, there are a few other significant factors which should be considered in order to find the final optimal solution for the SOC system. Here we consider application deadline time overhead, further reliability degradation, power consumption overhead and area overhead. The first three factors are caused by off-island interconnections between components in different voltage islands. While the fourth factor is caused by dead space inside a voltage island. In order to consider these factors, a floorplan algorithm must be applied subsequently in order to get the desired interconnection information.

Once the SOC system and its application are determined, the critical path and the corresponding components within this path can be identified. If floorplanning algorithm is applied after the partitioning, the locations of all components are set. In the case that the components in the critical path, for example C_1 and C_2, are partitioned into two different voltage islands, a lot of off-island communications between C_1 and C_2 are created. For this case, the application execution time increases compared with the case with two components in the same island. In order to reflect this performance overhead, a penalty is added to the deadline time. This penalty is proportional to the wire length between voltage islands in the critical path. As a result, the application deadline time under a specific partition is expressed as Equation (13), where δ is the penalty factor. If all the components in the critical path are located in the same island, δ is equal to zero. Otherwise, a positive fractional number which is less than one is added with respect to the number of islands which contain the critical components.

Furthermore, to make the communications between two voltage islands feasible, voltage converters are used between them. Soft errors and power consumption can also happen inside these converters. Correspondingly after the floorplanner applied (without floorplanner, the desired information is not available), the system reliability and system power consumption can be updated and expressed in Equations (14) and (15). Samely, ϵ and η are penalty factors for system reliability and power consumption, respectively. Both of them are positive fractional numbers and proportional to the islands number. More voltage islands mean more power consumption in the converters and less reliable system. For the area overhead,

```
Input     : component information, net information
            all kinds of constraints
Output    : optimal voltage island partition

Algorithm :

 1:  construct a link list for all possible voltage island partitions VILL
 2:  For each node in VILL Do
 3:      floorplan all voltage islands
 4:      calculate deadline time
 5:      If DT <= DT* Then
 6:          calculate system reliability level
 7:          If RL(SOC) >= RL(SOC)* Then
 8:              calculate area overhead
 9:              If (area overhead meets the requirement) Then
10:                  calculate the number of VIs
11:                  If (number of VIs meets the requirement) Then
12:                      calculate the power reduction
13:                  Else
14:                      delete this node
15:              Else
16:                  delete this node
17:          Else
18:              delete this node
19:      Else
20:          delete this node
21:  End For
22:  find the node in VILL with maximal power reduction
23:  output this node
```

Fig. 4. Extension of power consumption and reliability tradeoff algorithm for core-based SOC voltage island partitioning

a voltage island bringing C_1 and C_6 together would have a lot of dead space inside this island. This kind of area overhead should be minimized if it cannot be avoided. The information for calculating the above factors can be obtained after the floorplanning of the partitioned voltage islands. A simple floorplanner based on sequence pair representation and evaluation [2] is realized in the extended algorithm. By adding these factors, the algorithm is augmented and shown in Figure 4.

$$DT = (1 + \delta) \times DT \qquad (13)$$

$$RL(SOC) = (1 - \epsilon) \times RL(SOC) \qquad (14)$$

$$P(SOC) = (1 + \eta) \times P(SOC) \qquad (15)$$

In the extended algorithm, all possible voltage island partitions are constructed and saved in a link list. For each possible partition, the algorithm first floorplans all partitioned voltage islands bounded by the specified constraints. For example, a voltage island may not be allowed for a few components with one located on the periphery of the SOC chip because this may violate the constraint of proximity to the power pins. Furthermore, a clock generator/distributor is better to be put in the center of the chip in order to evenly distribute the clock signal to all components. After the floorplanning, the algorithm checks the deadline time and system reliability level according to Equations (13) and (14). Further, area overhead (dead space) and number of voltage island for this partition are also calculated. If the limit of either of the above four factors is not met, this partition node is deleted. Finally, the power consumption for the nodes meeting all the four requirements is calculated according to Equation (15). And the node or partition with the maximal power reduction is output.

IV. EXPERIMENTAL RESULTS

The reliability-bounded low-power design algorithms are realized in C programs. The platform for the experiments is

an Intel Celeron 2.0Ghz CPU with Red Hat Linux 9.0 as the operating system. To the best of our knowledge, this is the first work on reliability-bounded low-power design. We used two MCNC benchmarks [11] and constructed a few synthetic benchmarks $s10$, $s15$, $s30$ and $s50$ as summarized in Table II. For voltage setting, there are totally seven voltage levels available with range from $0.99V$ to $1.49V$. The second column shows the number of cores in each benchmark. Each core in the SOC system has a list of applicable voltage levels. The third column gives the average number of voltage levels for each core in the benchmark. For these benchmarks, we generated the power consumption values assuming that they are proportional to the corresponding core areas, and then scaled them appropriately according to Equation (9). The initial reliability level for each core is generated according to Equation (5). Assignments of the lower bound of system reliability level $RL(SOC)^*$ and the upper limit of system deadline time DT^* affect the number of final optimal solutions, but have little effect on the algorithm execution time.

In Table II, the fourth and the fifth columns show the results for a particular optimal partition in a benchmark with the maximal power reduction and the corresponding number of voltage islands under this partition. The sixth and the seventh columns show the results for another specific optimal partition with the minimal power reduction and the corresponding number of voltage islands. This is the extreme case with guaranteeing system reliability level with highest level and sacrificing power reduction. All the power reduction numbers are achieved by comparing the power under this specific partition and the power consumed by the SOC with all cores operated at their individual highest voltage levels. The eighth column shows the algorithm execution time. And the final column illustrates the reliability bound used to generate the optimal solutions. 95% means that the system reliability under one optimal partition should not be lower than 95 percent of the system reliability under the specific partition with all chip components operated at their highest voltage.

The results in Table II show that our algorithm is fast. It typically finishes in minutes with the actual running time depending on the size of a design and the constraints imposed on it, including number of cores and number of voltage levels for each core. Since in this work, the focus is on how to introduce the reliability issue in the partition and floorplan, the algorithm is not optimized by using advanced techniques, instead here we used heuristic algorithm. Further improvement in algorithm execution time is expected in the follow-up work on algorithm optimization.

Bounded by the system reliability and performance (application deadline time), the extended SOC algorithm searches for a final optimal solution whose area overhead and number of voltage island are within the tolerance range with maximal power reduction. Table III shows the results. Obviously, the extended algorithm is still very fast. Columns fourth and fifth compare the power consumption with reliability consideration and without consideration. Notice that the time unit is minute.

V. CONCLUSIONS

For the first time, we have developed a reliability characterization methodology and implemented a algorithm to illustrate the idea of reliability-bounded low-power design by using voltage island partitioning in SOC as a case study.

TABLE II

RESULTS FOR THE RELIABILITY-BOUNDED LOW-POWER SOC VOLTAGE ISLAND PARTITION ALGORITHM.

Benchmark	Number of cores	Average number of voltage levels	Max power reduction	Number of VIs	Min power reduction	Number of VIs	Execution time(sec)	Reliability bound
s10	10	2.0	13.87%	3	5.84%	2	1	95%
s15	15	3.7	12.01%	3	5.21%	3	58	95%
s30	30	2.3	23.10%	4	6.40%	3	431	95%
s50	50	1.9	15.68%	6	2.84%	4	480	95%
ami33	33	2.2	12.91%	4	7.8%	4	456	95%
ami49	49	1.7	15.20%	5	5.70%	4	479	95%

TABLE III

RESULTS FOR THE SOC EXTENDED ALGORITHM.

Benchmark	Number of cores	Execution time (min)	Power reduction w reliability	Power reduction wo Reliability	Number of VIs	Reliability bound
s10	10	2.5	10.77%	16.69%	3	95%
s30	30	19.8	18.50%	20.85%	4	95%
s50	50	29.7	9.74%	13.10%	5	95%
ami33	33	22.7	12.93%	16.37%	4	95%
ami49	49	25.8	8.98%	9.85%	4	95%

Three metrics, i.e. system reliability level, power reduction and application deadline time, were used to determine the quality of candidate voltage island partitions. Experiments show that for a SOC the algorithm execution time is within a few minutes while achieving 12% to 23% power reduction. For the extended SOC algorithm, it partitioned and floorplanned the voltage islands within 2.5 to 29.7 minutes and achieved 9.74% to 18.50% power reduction.

REFERENCES

[1] D. E. Lackey, *et al.*, "Managing Power and Performance for System-on-chip Designs Using Voltage Islands," in *Proc. Int. Conf. Computer-Aided Design*, Nov. 2002, pp. 195 – 202.

[2] J. Hu, Y. Shin, N. Dhanwada, and R. Marculescu, "Architecting Voltage Islands in Core-based System-on-a-chip Designs," in *Proc. Int. Symp. Low Power Electronics and Design*, Aug. 2004, pp. 180–185.

[3] J. N. Kozhaya and L. A. Bakir, "An Electrically Robust Method for Placing Power Gating Switches in Voltage Islands," in *Proc. Conf. Custom Integrated Circuits*, Oct. 2004, pp. 321 – 324.

[4] N. Seifert, D. Moyer, N. Leland, and R. Hokinson, "Historical Trend in Alpha-particle Induced Soft Error Rates of the AlphaTM Micropro-cessor," in *Proc. IEEE Int. Symp. Reliability Physics*, May 2001, pp. 259–265.

[5] N. Seifert, *et al.*, "Frequency Dependence of Soft Error Rates for Sub-micron CMOS Technologies," in *International Electron Devices Meeting, Technical Digest*, Dec. 2001, pp. 14.4.1 –14.4.4.

[6] V. Degalahal, N. Vijaykrishnan, and M. J. Irwin, "Analyzing Soft Errors in Leakage Optimized SRAM Design," in *Proc. Int. Conf. VLSI Design*, Jan. 2003, pp. 227–233.

[7] V. Degalahal, R. Rajaram, N. Vijaykrishnan, Y. Xie, and M. J. Irwin, "The Effect of Threshold Voltages on Soft Error Rate," in *Proc. Int. Conf. Quality Electronic Design*, Mar. 2003.

[8] R. Ramanarayaman, V. Degalahal, N. Vijaykrishnan, M. J.Irwin, and D. Duarte, "Analysis of Soft Error Rate in Flip-flops and Scannable Latches," in *Proc. IEEE Int. Conf. SOC*, Sept. 2003, pp. 231–234.

[9] M. Zhang and N. R. Shanbhag, "A Soft Error Rate Analysis (SERA) Methodology," in *Proc. Int. Conf. Computer-Aided Design*, Nov. 2004, pp. 111–118.

[10] C. Weaver, J. Emer, S. S. Mukherjee, and S. K. Reinhardt, "Techniques to Reduce the Soft Error Rate of a High-Performance Microprocessor," in *Proc. Int. Symp. Computer Architecture*, June 2004, pp. 264–275.

[11] "Mcnc floorplan benchmarks." [Online]. Available: http://www.cse.ucsc.edu/research/surf/GSRC/MCNCbench.html

Ultra-Low Energy Computing with Noise: Energy-Performance-Probability Trade-offs[*]

Pinar Korkmaz   Bilge E. S. Akgul   Krishna V. Palem

Center for Research on Embedded Systems and Technology

School of Electrical and Computer Engineering

Georgia Institute of Technology

Atlanta, Georgia 30332

{korkmazp, bilge, palem}@ece.gatech.edu

Abstract

Noise susceptibility and power density have become two limiting factors to CMOS technology scaling. As a solution to these challenges, probabilistic CMOS (PCMOS) based computing has been proposed. PCMOS devices are inherently probabilistic devices that compute correctly with a probability p. This paper investigates the trade-offs between the energy, performance and probability of correctness (p) of a PCMOS inverter. Using simple analytical models of energy, delay and p of a PCMOS inverter, the optimum energy delay product (EDP) value for given probability and performance constraints is found. The analytical models are validated using circuit simulations for a PCMOS inverter designed in a $0.13\mu m$ process. The results show that operating the PCMOS inverter at lower supply voltages is more preferable in terms of minimizing EDP. Our analysis is useful in optimal (in terms of EDP) circuit design for satisfying application requirements in terms of performance and probability of correctness. An analysis of the impacts of the variations in the temperature and the threshold voltage on the optimal EDP values is also included in the paper.

1 Introduction

As CMOS technology scales down into the nano-meter region, significant challenges to sustaining Moore's law have emerged. Two of these challenges are achieving noise immunity (see Shepard [20], Natori and Sano [14]) and low-energy consumption (see [9, 12]). The conventional approaches to overcome these challenges encountered in the semiconductor roadmap view noise as an impediment to scaling (see Kish [10], Sano [19], Meindl [12]). As a paradigm shift from the conventional approaches, we have innovated PCMOS based computing in [4, 5] and [16], wherein noise is viewed as a resource rather than as an impediment for realizing ultra low-energy computing in the context of probabilistic applications. In [5] and [11], we characterized the energy consumed per switching step and the associated probability of correctness for a PCMOS inverter. It was also shown that PCMOS characteristics can

be exploited at the application level for energy and performance benefits [4], wherein energy and performance benefits offered by PCMOS are quantified for a range of probabilistic applications.

In this work, we extend our characterization of a PCMOS inverter to include a succinct analysis of design trade-offs associated with its speed (or performance), energy and p. The characterization is achieved by using simple analytical models for energy, propagation delay, and p. In addition, we performed circuit simulations using BSIM3 models to verify our analytical model. In this paper, differing from our previous work [5, 11], we also consider leakage energy (in addition to the switching energy) of a PCMOS inverter, since the leakage energy is significant [17] especially for smaller feature sizes and for designs with low threshold voltages.

Lowering the supply voltage decreases the energy consumption, but also decreases p, which might be undesirable (depending on the value of p required by the application). Decreasing the supply voltage also decreases the switching speed of the circuit. Therefore, to meet the performance requirement demanded by the application, the threshold voltage should also be lowered. However, in this case, the static energy dissipation increases due to the increased leakage currents. Therefore, to study the trade-offs between energy consumption, performance and p, the parameters that we vary are the supply voltage (V_{dd}) and the threshold voltage (V_{th}). We also vary the RMS value of the noise to study the trade-offs between p, energy consumption and performance. In [9], Hegde and Shanbhag presented information-theoretic lowerbounds on the energy consumption of noisy gates. Their work is similar to our work since they also investigated the optimum values of V_{dd} and V_{th} that minimize the energy consumed by noisy gates. However, in their work, the primary focus is on computing reliably in the presence of noise, while our focus is on investigating the trade-offs between p (which is an independent design parameter), performance and energy. We also find optimal values of V_{dd} and V_{th} that satisfy p and performance requirements of an application, and minimize the EDP of PCMOS gates.

Supply voltage and threshold voltage scaling have been extensively studied (see [1, 2, 7]) in both the strong inversion and the subthreshold regions. The impact of V_{dd} and

---

[*]This work is supported in part by DARPA under Seedling # F30602-02-2-0124.

V_{th} on the energy and performance can be captured through the energy-delay product (EDP) metric that is commonly used to show the trade-offs between the two. In our work, we use the EDP metric to show the trade-offs between the energy, performance and p, with the goal of finding the optimal V_{dd}-V_{th} operation region for a PCMOS inverter. In particular, given a noise RMS value, a range of values for p and a performance constraint, we find the V_{dd} and V_{th} values that minimizes EDP. In addition, we consider the sensitivity of our analysis with respect to the variations in both the temperature and the threshold voltage. We show that our optimal V_{dd} and V_{th} operating points can change due to these variations.

Section 2 describes the PCMOS inverter. In Section 3, we describe our optimization procedure for finding the optimal values of V_{dd} and V_{th} that minimize EDP under given constraints on performance and a range of p values. In Section 4, we describe the impacts of V_{th} and temperature variations on the optimal values of V_{dd} and V_{th}. Finally, in Section 5, we conclude the paper.

2 Characterization of the Probabilistic Behavior of a PCMOS Inverter

A CMOS inverter is a digital switch that executes the *inversion* function with one input and output. For a deterministic inverter, $Y(t_2) = \overline{X(t_1)}$ where Y and X denote the binary values of the output and the input of the inverter, respectively, t_2 denotes the point in time when the switching ends, and t_1 denotes the point in time when the switching starts. For a probabilistic inverter, on the other hand,

$$Y(t_2) = \begin{cases} \overline{X(t_1)} & \text{with probability} \quad p \ (1/2 < p < 1) \\ X(t_1) & \text{with probability} \quad 1-p \end{cases}$$
(1)

where p denotes the probability of correctness for such an inverter. The probability p results due to the noise destabilizing the inverter. In this paper, we consider the case when thermal noise coupled to the output of the inverter is destabilizing the inverter. A comprehensive characterization of the PCMOS inverters in case of different couplings of noise can be found in [11].

We established in our prior work [5, 11] that a PCMOS inverter exhibits an exponential relationship between its p and the energy it consumes per switching, E. In addition, we showed that the relationship between the noise RMS value and the switching energy E is quadratic. The characterization of p and E derived from analytical modeling of noise susceptible CMOS inverters, has been extensively studied and verified using HSPICE simulations and physical measurements [11].

3 Trade-offs Between Energy, Performance and Probability of Correctness of a PCMOS Inverter

In this section, we explore the resulting values of energy, performance, and p for a range of values of V_{dd}: $0.30 \le V_{dd} \le 1.4$ and a range of values of V_{th}: $0.12 < V_{th} < 0.33$ for a PCMOS inverter realized in a 0.13μm process.

We consider an interval of p values, such as $0.90 < p < 0.95$ as seen from Figure 1, for which the design is being optimized. This interval of p values correspond to the bit error rate of the PCMOS device—in our case, the inverter—being optimized. Such a range of p values could reflect (1) the hardware-level degree of reliability of the device and (2) the application-level error tolerance range—and hence the quality—expected to be satisfied.

The hardware-level reliability, captured by the range of p values, is of interest for error redundancy mechanisms, such as NAND multiplexing studied by Norman et. al [15]. In their multiplexing scheme, a device is replicated N times, and the output values are compared according to a threshold, $\delta = p \in (0.5, 1)$, such that if the number of 0s (or 1s) is greater than $N \cdot \delta$, the output is decided to be 0 (or 1), whereas if it is in the interval $(N \cdot \delta, N \cdot (1 - \delta))$, the output is undecided. Therefore, the individual p of the devices can show variation. Such a scheme would imply that given a range of p values, such as $(0.7, 1)$ for example, corresponding to the variation of the p of the device, our aim would be to optimize the individual performance of the devices in terms of energy and speed while preserving that the optimum EDP point still corresponds to the p interval.

As for the application-level error tolerance, a wide range of applications from the digital signal processing or image processing as well as the networking domains require a reliability threshold, which in turn reflects the application-level quality. The digital signal or image processing domain of applications have a certain range of error tolerance, typically characterized by signal-to-noise ratio (SNR) or distortion [8], whereas for networking, the reliability measure of communication channels are characterized through bit error rate and packet loss rate [6].

3.1 Modeling Energy, Performance and Probability of Correctness of a PCMOS Inverter

This section presents the models we used for propagation delay, leakage energy, switching energy and probability of correctness of a PCMOS inverter.

The propagation delay (t_g) of an inverter in the subthreshold region is described by

$$t_g = \frac{K_s C_L V_{dd}}{I_o} e^{\frac{V_{dd} - V_{th}}{n\phi_t}}$$
(2)

where K_s and I_o are fitted parameters (obtained using circuit simulations performed in HSPICE). C_L is the capacitive load for the inverter.

We find the propagation delay of an inverter in the strong inversion region, using a simple α-power law model [18]

$$t_g = K \frac{V_{dd}}{(V_{dd} - V_{th})^\alpha}$$
(3)

where K is a parameter fitted using circuit simulations. α is the velocity saturation constant which is also fitted using circuit simulations.

In modeling the leakage energy, gate leakage, and other leakage components, such as pn-junction leakage and gate-induced drain leakage, are neglected. We consider only the subthreshold leakage component for simplicity. Based on the BSIM3 v3.2 [23] equation for leakage energy consumed per switching cycle (during t_g) is described by

$$E_L = V_{dd} I_{th} \left(1 - e^{-\frac{V_{DS}}{n\phi_t}} \right) \cdot e^{\frac{-V_{th} - V_{off} + dibl \cdot V_{DS}}{n\phi_t}} \cdot t_g L_{DP}$$
(4)

where I_{th} denotes the channel current when $V_{GS} = V_{th}$, n is the body effect coefficient, V_{off} is an empirically determined BSIM3 parameter, $dibl$ is the DIBL (Drain Induced Barrier Lowering) factor, ϕ_t is the thermal voltage $\frac{kT}{q}$ and L_{DP} denotes the logic depth. We use a value of 25 for L_{DP}. This value is estimated based on the logic depth for the implementations [4] of the hyperencryption and probabilistic cellular automata algorithms. The values of V_{off} and $dibl$ are derived from curve fitting based on circuit simulations.

The switching energy, total energy per switching cycle, and EDP are described by (5) to (7)

$$E_{SW} = aC_L V_{dd}^2 \tag{5}$$

$$E_T = E_{SW} + E_L \tag{6}$$

$$\text{EDP} = E_T t_g \tag{7}$$

where a denotes the activity factor. In this paper, we assume that a is 10%. This value of a is chosen based on the activity factor of the PCMOS inverters used in the implementation of probabilistic applications [4] such as probabilistic Bayesian inference, random neural networks, and probabilistic cellular automata. The probabilistic content (as a percentage of total number of operations) of these applications varies from 0.25% in the case of the Bayesian inference to 19.7% in the case of the randomized neural network.

The probability of correctness for a PCMOS inverter is found using

$$p = 0.5 + 0.5 erf\left(\frac{V_{dd}}{2\sqrt{2}\text{RMS}}\right) \tag{8}$$

We note that RMS denotes the standard deviation of the thermal noise that is coupled to the output of the inverter. In modeling the thermal noise, we follow the approach of [21], where the noise source is assumed to be a random process characterized by a Gaussian distribution. The details of derivation of (8) can be found in [11].

3.2 Optimal V_{dd} and V_{th} Operating Points

In this section, we employ the performance and p constraints imposed by an application on a PCMOS inverter to derive the optimal V_{th} and V_{dd} operating points that minimize the EDP of the inverter, for a given RMS value of noise. Such an optimization can be useful for architectural blocks (which implement probabilistic applications) whose minimum operating frequency needs to be greater than f_{min}, and whose reliability needs to be in a range, say, p_{min} to p_{max}. We now present the specific minimization problem under consideration, and the algorithm we have developed to solve the problem.

3.2.1 EDP Minimization Problem

We use the EDP metric to show the trade-offs between energy, performance and p of a PCMOS inverter. The performance of the PCMOS inverter is measured in terms of its maximum switching frequency, denoted as f_g, and is equal to the reciprocal of t_g.

In this section, we show normalized EDP (NEDP) contours, each denoting the ratio of the minimum EDP to the EDP corresponding to specific values of V_{dd} and V_{th}. To find the minimum EDP, we first find the values of V_{dd} and

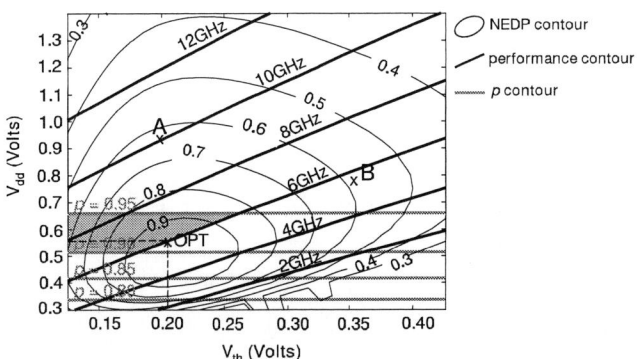

Figure 1. Constant NEDP, performance and p contours for a PCMOS inverter coupled with noise having an RMS value of 0.2V

V_{th} at which EDP is minimized. To find these values of V_{dd} and V_{th}, we differentiate (7) with respect to V_{dd} and V_{th} and equate the resulting equations to 0.

Figure 1 shows the NEDP, performance and p contours for a PCMOS inverter coupled with noise having an RMS value of 0.2V. In Figure 1 the rounded curves are contours of constant NEDP, the horizontal lines are contours of constant p, and the sloped lines are the contours of constant frequency (or performance). It is seen from the figure that, NEDP is high at lower values of V_{dd} and V_{th}. However, for very low values of V_{th}, NEDP becomes smaller due to the increased leakage energy. Figure 1 also shows that p increases as V_{dd} increases. We note that, the NEDP curves have small kinks near the border of subthreshold region, which are due to the discontinuity of the analytical model (equations (2) and (3)) at the boundary of subthreshold region. It is also seen from the figure that the higher the value of V_{dd} with respect to V_{th}, the higher the performance.

Given these trade-offs, our objective is to find the optimal V_{dd} and V_{th} operating points that minimize the EDP within the given constraints. The problem is stated as follows.

Minimize:

$$
\begin{aligned}
\text{EDP} = \ & V_{dd} I_{th}\left(1 - e^{-\frac{V_{DS}}{n\phi_t}}\right) \cdot e^{\frac{-V_{th} - V_{off} + dibl \cdot V_{dd}}{n\phi_t}} t_g^2 L_{DP} \\
& + aC_L V_{dd}^2 t_g
\end{aligned}
\tag{9}
$$

subject to:

$$f_g \geq f_{min} \tag{10}$$

$$p_{min} \leq p \leq p_{max} \tag{11}$$

where t_g is the propagation delay of the inverter given by (2) or (3) depending on the operation region of the inverter.

The solution to this problem is found using a two-dimensional search algorithm whose pseudo-code is shown in Figure 2. As seen from Figure 2, for the given values of p_{min}, p_{max}, f_{min}, and noise RMS value:

1. We assign a sufficiently large value to EDP_{min}.

2. We increase p from p_{min} to p_{max} in sufficiently small steps. For each p:

    (a) We find the corresponding value of V_{dd} using (8).

    (b) Using (2) and (3), and given f_{min}, we find the maximum possible value of V_{th} (V_{thmax}).

0-7695-2533-4/06 $20.00 © 2006 IEEE   351

1. Start: Input: p_{min}, p_{max}, f_{min}, noise RMS and $S \gg 1$
2. $V_{thmin} = 0.125$; $V_{thstep} = V_{thmin} / S$; $p_{step} = p_{min} / S$;
3.   i =0; j=0; EDP$_{min}$ = 1;
4. p Loop: repeat
5. $i = i+1$; p(i) = p_{min} + p_{step} ;
6.   compute $V_{dd}(i)$ using (8);
7.   compute V_{thmax} using (2) and (3);
8.   EDP$_{min}(i)$ = 1;
9. V_{th} Loop: repeat
10. $j = j+1$; $V_{th} = V_{thmin}$ + V_{thstep};
11.   compute EDP using (9);
12.   if EDP < EDP$_{min}(i)$
13.   EDP$_{min}(i)$ = EDP;
14. $V_{thopt} = V_{th}(j)$;
15. $V_{ddopt} = V_{dd}(i)$;
16.   until $V_{th}(j) > V_{thmax}$;
17.   if EDP$_{min}(i)$ < EDP$_{min}$
18.   EDP$_{min}$ = EDP$_{min}(i)$;
19.   until p(i) > p_{max};
20.   report EDP$_{min}$, V_{ddopt} and V_{thopt}

Figure 2. The pseudocode for the algorithm to obtain the optimal V_{dd} and V_{th} values that minimize EDP

    (c) We assign a sufficiently large value to the minimum value of the EDP of this step (EDP$_{min}(i)$).

    (d) We increase V_{th} from a given minimum value of V_{th} (V_{thmin}) to V_{thmax} in sufficiently small steps. For each V_{th}:

        i. We compute EDP using (9). If this value of EDP is lower than EDP$_{min}(i)$, then the values of EDP$_{min}(i)$, V_{thopt}, and V_{ddopt} are updated as shown in steps 13, 14, and 15 of the pseudocode.

    (e) If EDP$_{min}(i)$ is smaller than EDP$_{min}$, then we update the value of EDP$_{min}$ as shown in line 18 of the pseudocode.

Referring to Figure 1, for example, if the performance constraint is set at 6GHz, the search algorithm searches for the optimal V_{dd} and V_{th} operating points in the region to the left of the 6GHz line. Furthermore, if p_{min} and p_{max} are 0.90 and 0.95, respectively, then the search is performed within the shaded area shown in the figure. The algorithm finds that for the optimal EDP point, the values of V_{dd} and V_{th} are 0.552V and 0.201V, respectively, as shown by the point OPT in Figure 1.

3.2.2 Simulation Results

In this section, we compare our analytical results with the simulation results for EDP, performance and p.

We performed circuit simulations in HSPICE using BSIM3 models for a CMOS inverter in a $0.13\mu m$ process to measure the inverter's static and dynamic energy consumption, propagation delay and p. We measured the static energy and the switching energy separately. We have assumed an activity factor of $a = 10\%$ in calculating the switching energy consumption.

In modeling the thermal noise that is coupled to the inverter, the noise source is assumed to be a random process characterized by a Gaussian distribution. The details of modeling the noise, the coupling of the noise and calculation of p in the circuit simulations can be found in [11].

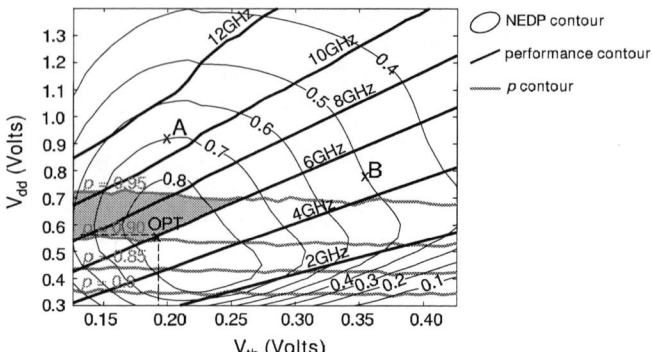

Figure 3. Constant NEDP, performance and p contours from circuit simulations

The results of the simulations are shown in Figure 3. When compared to Figure 1, the most striking difference is in the shape of the NEDP contours. The NEDP contours shown in Figure 3 are wider in the V_{dd} domain and narrower in the V_{th} domain compared to the NEDP contours in Figure 1. This difference results from the inaccuracy of the analytical model in estimating the propagation delay. The analytical delay model achieves an average error of 8.47%, but the standard deviation of the error is 9.47%. The analytical model overestimates the propagation delay for low values of V_{th}, and underestimates it for high values of V_{th}. Hence, as seen from Figures 1 and 3, at a fixed value of V_{th} and when V_{th} is small, the NEDP value from the analytical model is smaller than the NEDP value from the simulations. We can also see from Figures 1 and 3 that the analytical NEDP is higher than the simulated NEDP for higher values of V_{th}. This results from the underestimation of the propagation delay by the analytical model. For example, points denoted as A and B in both figures correspond to these two cases, where A represents the case when the analytical NEDP is lower, and B represents the case when the analytical NEDP is higher. As seen from Figures 1 and 3, analytical and simulation results for the performance contours are also deviating from each other due to the differences between the delay estimation in the analytical model and the simulations.

Comparing the p contours of Figures 1 and 3, we observe that the p contours found using simulations are traversing higher values of V_{dd}. This is caused by the fact that the transistors of the inverter used in simulations are not symmetrical, whereas the analytical model considers the case when the transistors are symmetrical. We have a more accurate model (see [11]) for the case when the transistors are not symmetrical. However, the more accurate model requires the midpoint voltage of the CMOS inverter, which we have not derived in the subthreshold region. Thus, we have chosen to use the model in (8) for simplicity in this paper. Furthermore, in Figure 3, the p contours are not exactly horizontal, but have a negative slope (which is very small in magnitude). This weak dependency of p on V_{th} is due to the dependency of p on the midpoint voltage of the inverter.

Comparing Figures 1 and 3, we see that the feasible region for the search example provided in Section 3.2.1 in case of simulations is slightly different from the feasible region in case of the analytical model. The optimal values of V_{dd} and V_{th} found from simulations are 0.55V and 0.196V (as opposed to 0.552V and 0.201V). We note that to find the optimal operating points in case of simulations, we use a variant of the algorithm in Figure 2. We replace the steps for calculations of $V_{dd}(i)$, V_{thmax}, and EDP by search steps. The search step traverses the simulation results, and finds the closest values for $V_{dd}(i)$, V_{thmax}, and EDP.

As seen from Figures 1 and 3, the supply and threshold voltages for optimal EDP are closer to the probability contour $p = p_{min}$, that is, operating the PCMOS inverter at lower supply voltages is more preferable in terms of the EDP. However, the supply voltages can not be reduced further beyond the point where the performance constraint is satisfied.

4 Variations in Temperature and V_{th}

So far we have assumed that we have full control on the threshold voltages and the operating temperature. However in reality, the threshold voltage might change due to process variations and changes in the operating temperature. In addition, the chip temperature changes due to heat dissipation. Neglecting the coupling between the chip temperature and the power dissipation [1], in this section, we demonstrate the impact of the variations in the temperature and in the threshold voltage on the energy and performance of the PCMOS circuits in terms of the EDP contours derived in previous sections.

The operating temperature of a circuit can be anywhere between 25°C and 125°C. The threshold voltage at temperature T can be calculated [1] using

$$V_{thT} = V_{th} - k(T - T_{amb}) \quad (12)$$

where V_{thT} is the threshold voltage at temperature T, T_{amb} is the ambient temperature (25°C), and k is the threshold voltage temperature coefficient whose typical value for a 0.13μm process is 0.7mV/K [22].

The temperature also affects the RMS value of the noise, since we consider a thermal noise source. For simplicity, we assume that the noise source is a resistive noise source and therefore, we calculate the RMS value of noise at temperature T using

$$\text{RMS}_T = \text{RMS} \cdot \sqrt{\frac{T}{T_{amb}}} \quad (13)$$

Figure 4 shows the effect of increasing temperature from 25°C to 35°C. In the figure, the dashed contours correspond to the case when T is 35°C and the solid contours correspond to the case when T is 25°C. As seen from the figure, the NEDP and the performance contours are shifted to right when the temperature is increased. This results from the decrease in the threshold voltage due to the increase in the temperature. Furthermore, p contours are shifted higher in the V_{dd} domain, that is, to obtain the same value of p, a higher V_{dd} value is required at a higher temperature. Hence, at a fixed value of V_{dd}, p decreases as T increases. This decrease in p is due to the increased RMS value of noise due to the increased temperature. Due to these variations in NEDP, performance and p, the optimal values of V_{dd} and V_{th} also change. For example, with p_{min} and p_{max} values of 0.90 and 0.95, and performance constraint of 6GHz, the optimal values of V_{dd} and V_{th} are now 0.568V and 0.211V as opposed to the values of 0.552V and 0.201V found previously

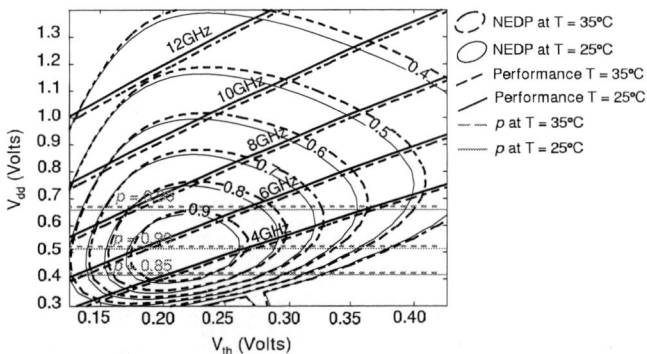

Figure 4. Constant NEDP, **performance and** p **contours for a** PCMOS **inverter at temperatures** $T = 25°$**C and** $T = 35°$**C**

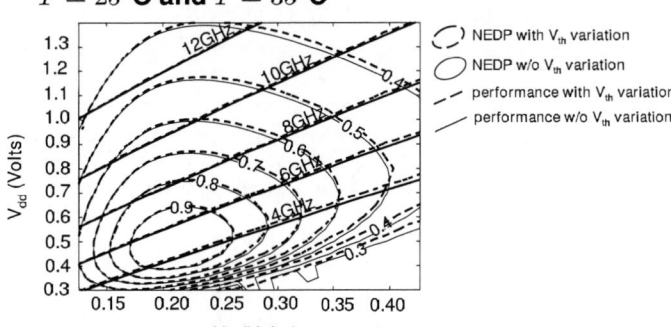

Figure 5. Constant NEDP **and performance contours for a** PCMOS **inverter with** V_{th} **variations in case when mean values of** EDP **and performance are considered**

in Section 3.2.1. As seen in this example, the change in optimal values of V_{dd} and V_{th} is small. However, if T is increased even further, we observe significant changes in the optimal values. For example, if T is 85°C, then the optimal V_{dd} and V_{th} values become 0.658V and 0.261V, corresponding to 19.2% and 22.9% difference when compared to the original values of 0.552V and 0.201V for $T = 25°$C.

Figure 5 depicts the effect of the variations in the threshold voltage. In the figure, the dashed contours correspond to the case when there is variation in the threshold voltage, and the solid contours correspond to the case when there is no variation in the threshold voltage. Empirical evidence suggests that the variation in V_{th} can be modeled by Gaussian distribution [13]. So, we model the threshold voltage variation as a Gaussian distribution. The mean value of the Gaussian distribution is 0 and its standard deviation σ is equal to 10% of the threshold voltage [3]. We only show the NEDP and performance contours since there is negligible effect of V_{th} on p. As seen from the figure, the NEDP contours are shifted upwards, that is, for a fixed value of V_{th}, to obtain the same value of NEDP, a higher V_{dd} value is required. This results from the fact that the change in EDP is larger when there is a positive change in V_{th} (eg. $V_{th} + \sigma$) compared to a negative change in V_{th} of the same magnitude (eg. $V_{th} - \sigma$). We note that, in Figure 5, we only show the mean values of NEDP and performance. As a result, the differences in NEDP and performance seem to be very small. Due to averaging,

0-7695-2533-4/06 $20.00 © 2006 IEEE   353

the kinks in the NEDP contours have also disappeared. Furthermore, the optimal values of V_{dd} and V_{th} change slightly. For example, with p_{min} and p_{max} values of 0.90 and 0.95, and performance constraint of 6GHz, the optimal values of V_{dd} and V_{th} are now 0.562V and 0.206V corresponding to 1.81% and 2.49% difference compared to the optimal values of V_{dd} and V_{th} obtained in case when there is no V_{th} variation. If we consider a worst case scenario such as the mean plus one standard deviation, the difference in NEDP and performance becomes more significant.

We can conclude from the results of this section that EDP and performance are dependent on the variations in the T and V_{th}. Similarly, our V_{dd} and V_{th} values for optimal EDP are also dependent on these variations. Thus, an analysis for optimizing the EDP of a PCMOS inverter should consider the variations in T and V_{th}. Furthermore, we can conclude that threshold voltage control is necessary when the variations in T and V_{th} can not be modeled accurately.

5 Conclusions

In this paper, we have shown the design trade-offs between energy, performance and p of a PCMOS inverter using simple analytical models of energy, delay and p. We have also found the values of V_{dd} and V_{th} for optimal EDP under given constraints on p and performance. We have observed that operating the PCMOS inverter at lower supply voltages is more preferable in terms of minimizing EDP. We have also performed circuit simulations to validate our analytical models. From these simulations we have observed that the shapes of EDP surfaces and the location of the optimal EDP point are dependent on the models used for energy and delay. As an example, for a $0.13\mu m$ technology, given a minimum p requirement of 0.90, a maximum p requirement of 0.95 and a minimum frequency requirement of 6GHz, our analytical analysis yielded a supply voltage of 0.552V and threshold voltage of 0.201V, while our simulation results yielded a supply voltage of 0.55V and threshold voltage of 0.196V, which are reasonably close to each other. Our analysis can be helpful in circuit design for applications with a minimum performance requirement and a specific range of p.

We have also included an analysis of the impact of the variations in threshold voltage and temperature on EDP and performance contours as well as on optimal value of EDP. We have found that accurately estimating the variations in temperature and threshold voltage is important for accurately optimizing the EDP of a PCMOS inverter. This analysis can further be extended to include the variations in V_{dd}.

References

[1] A. Basu, S. C. Lin, V. Wason, A. Mehrotra, and K. Banerjee. Simultaneous optimization of supply and threshold voltages for low-power and high-performance circuits in the leakage dominant era. In *Proc. Design Automation Conf.*, pages 884–887, 2004.

[2] B. H. Calhoun, A. Wang, and A. Chandrakasan. Modeling and sizing for minimum energy operation in subthreshold circuits. *IEEE J. of Solid-State Circuits*, 40:1778–1786, 2005.

[3] Y. Cao, P. Gupta, A. B. Kahng, D. Sylvester, and J. Yang. Design sensitivities to variability: Extrapolations and assessments in nanometer VLSI. In *IEEE Intl. ASIC/SOC Conf.*, pages 411–415, 2002.

[4] L. N. Chakrapani, B. E. S. Akgul, S. Cheemalavagu, P. Korkmaz, K. V. Palem, and B. Seshasayee. Ultra efficient embedded SoC architectures based on probabilistic CMOS (PCMOS) technology. *To appear in Proc. of DATE Conference*, March 2006.

[5] S. Cheemalavagu, P. Korkmaz, K. V. Palem, B. E. S. Akgul, and L. N. Chakrapani. A probabilistic CMOS switch and its realization by exploiting noise. In *Proc. of the IFIP VLSI-SOC 2005*, Australia, October 2005.

[6] F. Gong and G. M. Parulkar. An application-oriented error control scheme for high-speed networks. *IEEE/ACM Trans. Netw.*, 4(5):669–683, 1996.

[7] R. Gonzalez, B. M. Gordon, and M. A. Horowitz. Supply and threshold voltage scaling for low power CMOS. *IEEE J. of Solid-State Circuits*, 32:1210–1216, August 1997.

[8] R. Hegde and N. R. Shanbhag. A low-power architecture for phase-splitting passband equalizer. In *IEEE Workshop on Signal Processing Systems, SIPS 97 - Design and Implementation*, pages 385–394, November 1997.

[9] R. Hegde and N. R. Shanbhag. Toward achieving energy efficiency in presence of deep submicron noise. *IEEE Trans. VLSI Syst.*, 8:379–391, August 2000.

[10] L. B. Kish. End of Moore's law: thermal (noise) death of integration in micro and nano electronics. *Physics Letters A*, 305:144–149, December 2002.

[11] P. Korkmaz, B. E. S. Akgul, and K. V. Palem. Characterizing the behaviour of a probabilistic CMOS switch through analytical models and its verification through simulations. Technical Report CREST-TR-05-08-01, Available at http://www.crest.gatech.edu/palempbitscurrent/.

[12] J. D. Meindl. Low power microelectronics: retrospect and prospect. *Proc. IEEE*, 83:619–635, April 1995.

[13] S. Narendra, D. Blaauw, A. Devgan, and F. Najm. Leakage issues in IC design: trends, estimation, and avoidance. In *Proc. of Intl. Conf. on Computer Aided Design*, November 2003.

[14] K. Natori and N. Sano. Scaling limit of digital circuits due to thermal noise. *J. of Applied Physics*, 83:5019–5024, May 1998.

[15] G. Norman, D. Parker, M. Kwiatkowska, and S. Shukla. Evaluating the reliability of nand multiplexing with prism. *IEEE Trans. on Computer-Aided Design of Integrated Circuits and Systems*, 24(10):1629–1637, 2005.

[16] K. V. Palem. Energy aware computing through probabilistic switching: A study of limits. *IEEE Trans. Comput.*, pages 1123–1137, September 2005.

[17] K. Roy, S. Mukhopadhyay, and H. Mahmoodi-Meimand. Leakage current mechanisms and leakage reduction techniques in deep-submicrometer CMOS circuits. *Proc. of the IEEE*, 91:305–327, 2003.

[18] T. Sakurai and A. R. Newton. Delay analysis of series-connected MOSFET circuits. *IEEE J. of Solid-State Circuits*, 26, 1991.

[19] N. Sano. Increasing importance of electronic thermal noise in sub-0.1mm Si-MOSFETs. *IEICE Trans. on Electronics*, E83-C:1203–1211, August 2000.

[20] K. L. Shepard. Design methodologies for noise in digital integrated circuits. In *Proc. Design Automation Conf.*, pages 94–99, June 1998.

[21] K.-U. Stein. Noise-induced error rate as a limiting factor for energy per operation in digital ICs. *IEEE J. Solid-State Circuits*, 12:527–530, Oct. 1977.

[22] Y. Taur and T. Ning. *Fundamentals of modern VLSI devices*. Cambridge Univ. Press, 1998.

[23] U.C. Berkeley BSIM Homepage. http://www-device.eecs.berkeley.edu/ bsim3/.

0-7695-2533-4/06 $20.00 © 2006 IEEE

Delay and Energy Efficient Data Transmission for On-Chip Buses

Madhu Mutyam[†*]

[†]International Institute of Information Technology
Gachibowli, Hyderabad
India
mutyam@cse.psu.edu

Melvin Eze[*] N. Vijaykrishnan[*] Yuan Xie[*]

[*]Dept. of Computer Science and Engineering
Pennsylvania State University
University Park, PA 16802
{eze,vijay,yuanxie}@cse.psu.edu

Abstract

On-chip buses in deep sub-micron designs consume significant amounts of power and have large propagation delays. Thus, minimizing power consumption and propagation delay are the most important design objectives. In this paper, we propose a technique for delay and energy efficient data transmission for on-chip buses and evaluate the effectiveness of our technique by focusing on the L1 cache address/data buses of a microprocessor using the SPEC2000 CINT benchmark suit. We show that our technique achieves 31% (30%) of delay improvement along with energy savings of 13% (9%) over the base case for data transmission on address (data) bus.

1 Introduction

As VLSI fabrication technologies scaled down to the deep sub-micron region, the inter-wire capacitance (C_I) becomes significant compared to the wire-to-substrate capacitance (C_L). As C_I is the dominant capacitance in deep sub-micron era, it has two significant effects: large propagation delay due to opposite transitions on adjacent wires [5, 17, 19] and power dissipation associated with driving on-chip buses [17]. There are techniques to minimize the effects of opposite transitions on adjacent wires and/or power consumption due to capacitive coupling [6, 7, 8, 10, 9, 11, 12, 13, 14, 16, 18, 20]. Many of the techniques use *spatial redundancy* to minimize power or delay and hence have large area overhead. For example, *crosstalk avoidance codes* [18] eliminate opposite transitions on adjacent wires completely but require 46 wires to encode 32-bit data and a coding technique proposed in [16] is both area and energy-efficient and eliminates opposite transitions on adjacent wires completely but it requires 48 wires for encoding 32-bit data.

In this paper we exploit the *variable cycle transmission*

technique [9] to apply *temporal redundancy* in the process of minimizing both delay and energy for on-chip data transmission. Our encoding scheme requires *two* extra wires to encode *any* bit-width data and achieves 31% (30%) of delay improvement along with 13% (9%) of energy savings over the base case for data transmission on address (data) bus.

2 Analytical Models for Delay and Energy Consumption

In order to measure the actual effects of inter-wire capacitance in deep sub-micron technologies, analytical models for delay and energy consumption in deep sub-micron buses have been proposed in [12, 14, 15]. Throughout the paper we consider these analytical models. Let d_t be a n-bit data present on the bus. By considering $\lambda = \frac{C_I}{C_L}$, the propagation delay of the RC circuit, for transmitting a n-bit data d_{t+1}, can be calculated by [14]

$$T(d_t, d_{t+1}) = max\{T_k(d_t, d_{t+1}) \mid 1 \le k \le n\}, \text{ where}$$

$$\frac{T_k(d_t, d_{t+1})}{C_L R_T} = \begin{cases} ((1+\lambda)\Delta_1 - \lambda\Delta_2)\Delta_1, k=1 \\ ((1+2\lambda)\Delta_k - \lambda(\Delta_{k-1}+\Delta_{k+1}))\Delta_k, 1<k<n \\ ((1+\lambda)\Delta_n - \lambda\Delta_{n-1})\Delta_n, k=n \end{cases}$$

where R_T is the total resistance and $\Delta_k = d_{t+1}^k - d_t^k$. Similarly, the total energy (due to self and coupling transitions) consumed during the transition from d_t to d_{t+1} is given by [12, 15]

$$E(d_t, d_{t+1}) = \Sigma_{k=1}^n E_k(d_t, d_{t+1}), \text{ where}$$

$$E_k(d_t, d_{t+1}) = \begin{cases} C_L((1+\lambda)\Delta_1 - \lambda\Delta_2)d_{t+1}^1, k=1 \\ C_L((1+2\lambda)\Delta_k - \lambda(\Delta_{k-1}+\Delta_{k+1}))d_{t+1}^k, 1<k<n \\ C_L((1+\lambda)\Delta_n - \lambda\Delta_{n-1})d_{t+1}^n, k=n \end{cases}$$

For example, if $d_t = 010$ and $d_{t+1} = 101$, then $T(d_t, d_{t+1}) = C_L R_T(1+4\lambda)$ and $E(d_t, d_{t+1}) = C_L(2+4\lambda)$. On the other hand, if $d_t = 000$ and $d_{t+1} = 111$, then $T(d_t, d_{t+1}) = C_L R_T$ and $E(d_t, d_{t+1}) = 3C_L$. From these two examples (and hence from the above two equations), it is clear that the transition patterns, such as $0 \to 1$ (or $\uparrow$),

---

0-7695-2533-4/06 $20.00 © 2006 IEEE   355

Crosstalk Class	Relative Delay on the middle wire	Transition Patterns
1	0	$- - -, - -$ ↑,↑ $- -, - -$ ↓,↓ $- -,$ ↑ $-$ ↑,↑ $-$ ↓↓ $-$ ↑,↓ $-$ ↓
2	$C_L R_T$	↑↑↑,↓↓↓
3	$C_L R_T (1 + \lambda)$	$-$ ↑↑,↑↑ $-, -$ ↓↓,↓↓ $-$
4	$C_L R_T (1 + 2\lambda)$	$-$ ↑ $-, -$ ↓ $-,$ ↓↑↑,↑↓↓,↑↑↓,↓↑↑
5	$C_L R_T (1 + 3\lambda)$	$-$ ↑↓,$-$ ↓↑,↓↑ $-,$ ↑↓ $-$
6	$C_L R_T (1 + 4\lambda)$	↓↑↓,↑↓↑

Table 1. Crosstalk classes (here $\lambda = \frac{C_I}{C_L}$)

$1 \to 0$ (or ↓), $0 \to 0$ (or $-$), and $1 \to 1$ (or $-$), of data not only dictate the propagation delay but also the energy consumption.

Since the propagation delay is determined by the transition pattern of the data, transition patterns are classified into six different classes [9, 14] based on the relative delay a wire w.r.t. its adjacent wires, which we call as *crosstalk classes*. Table 1 shows different crosstalk classes. From this table, it is clear that the worst-case crosstalk delay for transmitting data items, without any coding, is $C_L R_T (1 + 4\lambda)$.

3   Related Work

Bus encoding techniques to eliminate crosstalk classes 5 and 6 have been proposed in [6, 18]. Note that the technique proposed in [18] can also eliminate part of the crosstalk class 4. From Table 1, it is clear that as a result of eliminating crosstalk classes 5 and 6, the worst-case crosstalk delay becomes $C_L R_T (1 + 2\lambda)$. Though the worst-case delay can be reduced by 50%, these techniques are difficult to implement and have large area overhead (for a 32-bit data, the technique in [18] requires 46 wires without memory and 40 wires with memory and the technique in [6] requires 52 wires). Similar to the ideas of [6, 18], a new coding technique is proposed in [16] to obtain 50% reduction in delay along with 10% reduction in energy, but it also requires 48 wires to encode 32-bit data.

A bus encoding technique to minimize power consumption and eliminate crosstalk classes 5 and 6 is proposed in [10], which requires 55 wires to encode 32-bit data. Coupling-driven bus encoding technique [8] reduces power consumption by 30% on an average, but it may not eliminate crosstalk class 4 and above and hence it is not much advantageous from the delay perspective. Another bus encoding technique to minimize both energy and delay is proposed in [7], which can eliminate only crosstalk classes 4 and 6 (but not crosstalk class 5) so that the worst-case delay is still $C_L R_T (1 + 3\lambda)$ and it requires 55 wires to encode 32-bit data.

Odd/even bus invert technique is proposed in [20] to minimize coupling energy. With half cycle delay of one of the adjacent wires, it can eliminate opposite transitions on adjacent wires and achieve 30% power savings. Though this technique is energy efficient, from the delay perspective it is not advantageous because the half cycle delay can result in crosstalk class 4 patterns such as $-$ ↑ $-$ and $-$ ↓ $-$. As a result, the net delay for receiving the data becomes $C_L R_T (2 + 4\lambda)$, which is more than that of crosstalk class 6. Another technique to minimize power consumption due to coupling transitions is proposed in [13], but it may not be easy to implement due to its complex codec circuitry.

A technique close to the topic of this paper is proposed in [9], where instead of considering the worst-case delay of $C_L R_T (1 + 4\lambda)$, the authors considered variable delay which is determined based on the data present on the bus and next data to be transmitted. Based on the crosstalk class of the next data w.r.t. the present data on the bus, the number of cycles required to transmit the next data is controlled dynamically and hence obtain significant performance benefits. This technique requires one extra wire, which acts as a shield wire between the actual data and the *ready*[1] signal. Since the data is transmitted as it is without any coding, this technique does not provide any energy benefits.

Contrast to the above mentioned works, here we propose a technique which eliminates crosstalk classes 5 and 6 with the help of temporal redundancy and uses variable delay for data transmission so that both delay and energy minimization can be achieved. Unlike the technique in [9], where for crosstalk class 6 it takes the delay of $4C_L R_T (1 + \lambda)$, our technique takes the delay of $3C_L R_T (1 + \lambda)$ so that we can even get better performance benefits over the technique proposed in [9].

4   Our Approach

Basic idea behind our approach is to analyze the crosstalk class of next data w.r.t. the data present on the bus and based on the crosstalk class we either transmit the original data or encode the data into two new data items and transmit the encoded data. In either case, we use the necessary delay for data transmission.

4.1   Area and Energy Efficient Crosstalk Analyzer

From Section 2, we know that the crosstalk class of next n-bit data w.r.t. the present data on the bus is determined by the maximum delay among all the wires $k, 1 \leq k \leq n$. So, in order to know the crosstalk class, first we have to find the delay of each of the n wires. In [9], a crosstalk analyzer is designed in such a way that whenever a next data is received, the analyzer determines the crosstalk class of the next data by comparing it with the present data on the bus. This process incurs extra hardware and consumes more energy as none of the computed data, in the process of finding

---

[1] the authors considered 33-bit data with 32-bit actual data and one *ready* signal

0-7695-2533-4/06 $20.00 © 2006 IEEE

```
Input: present data $d_t$ and next data $d_{t+1}$
Output: encoded data items $d'_{t+1}$ and $d''_{t+1}$
if(crosstalk_class $\geq$ 5)
  begin
    $d'_{t+1} = d_{t+1}$; $d''_{t+1} = d_{t+1}$
    for($i = n; i > 0; i--$)
      if(($XORP[i+1]\&XORN[i+1]\&(d_t[i] \oplus d_{t+1}[i]))\|$
        ($XORP[i]\&XORN[i]\&(d_t[i-1] \oplus d_{t+1}[i-1]))$))
        $d'_{t+1}[i] = 1$; $d''_{t+1}[i] = 0$
  end
```

Table 2. Encoder

```
Input: present data $z_{t+1}$ and previous data $R_1$ and $R_2$
Output: decoded data $d_{t+1}$
if(ready)
  if(temporal_redundancy)
    for($i = n; i > 0; i--$)
      if($R_2[i] == z_{t+1}[i]$)
        $d_{t+1}[i] = z_{t+1}[i]$
      else
        $d_{t+1}[i] = \sim R_1[i]$
  else
    $d_{t+1} = z_{t+1}$
```

Table 3. Decoder

a crosstalk class, is *reused* in the next stage. Generally, in the process of finding a crosstalk class, the analyzer checks whether or not there is any opposite transition on the adjacent wires (this is needed for crosstalk classes 4, 5, and 6). If d_1 is the present data and d_2 is the next data, then to know whether or not a wire k, $1 \leq k < n$, has an opposite transition w.r.t. wire $k + 1$, it is enough to take the logical AND of $d_1[k] \oplus d_1[k + 1]$, $d_2[k] \oplus d_2[k + 1]$, and $d_1[k] \oplus d_2[k]$. In this process, we can compute the XOR values of adjacent bits of d_2 only once but use them twice (i.e., for finding crosstalk classes of d_2 w.r.t. d_1 and d_3 w.r.t. d_2). As a result, the hardware and the energy consumption for the crosstalk analyzer is significantly reduced compared to the similar circuit given in [9]. We obtain the hardware reduction at the cost of two $(n - 1)$-bit registers, i.e., $XORP$ and $XORN$, to store the XOR values of adjacent bits of present n-bit data and next n-bit data, respectively. Whenever, a next data d_2 is received, we only compute $XORN$ and move the previous $XORN$ (that of d_1) to $XORP$ with the possible exception when the next data d_2 is in crosstalk class 5 or 6 w.r.t. the present data d_1. Since temporal redundancy is used for crosstalk classes 5 and 6, in such situations, $XORP$ gets the XOR values of adjacent bits of d''_2 (defined in the next subsection).

Our design requires 35 wires for transmitting a 33-bit data, which consists of 32-bit actual data and one *ready* signal. The additional wires are: one *temporal_redundancy* signal for crosstalk class 5 or 6 and one *shield* wire between the *ready* signal and the actual data to prevent opposite transitions. There is no need of a shield wire between the *temporal_redundancy* signal and the *ready* signal as the *temporal_redundancy* signal makes $0 \rightarrow 1$ transition only when the *ready* signal makes $0 \rightarrow 1$ transition and it makes $1 \rightarrow 0$ transition when the *ready* signal makes $1 \rightarrow 1$ transition.

4.2 Variable Cycle Transmission with Temporal Redundancy

We now present our bus encoding technique, i.e., variable cycle transmission with temporal redundancy (VCTR technique). In order to facilitate variable cycle transmission, we use two different clocks with periods $C_L R_T (1 + \lambda)$

and $2C_L R_T (1 + \lambda)$. Whenever a next data d_{t+1} is received, our crosstalk analyzer determines the crosstalk class of d_{t+1} w.r.t. the present data d_t on the bus. If the crosstalk class of d_{t+1} w.r.t. d_t is from the set $\{1, 2, 3, 4\}$, d_{t+1} is transmitted with a delay of $C_L R_T (1 + \lambda)$ (or $2C_L R_T (1 + \lambda)$) if the crosstalk class is ≤ 3 (or 4) and the *ready* signal is *set* and the *temporal_redundancy* signal is *reset*. If the crosstalk class is either 5 or 6, unlike delaying the transmission by the required number cycles (as proposed in [9]), we encode d_{t+1} into two new data items d'_{t+1} and d''_{t+1} (using the coding technique as given in Table 2) in such a way that the crosstalk class of d'_{t+1} w.r.t. d_t is from the set $\{1, 2, 3, 4\}$ and that of d''_{t+1} w.r.t. d''_{t+1} is 3 (formal proof is given in the next section). Note that the encoder (given in Table 2) takes different conditions for the boundary values, i.e., when $i = 1$, it checks only $(XORP[2]\&XORN[2]\&(d_t[1] \oplus d_{t+1}[1]))$ and when $i = n$, it checks $(XORP[n]\&XORN[n]\&(d_t[n - 1] \oplus d_{t+1}[n - 1]))$. Thus, we transmit d'_{t+1} with a delay of $2C_L R_T (1 + \lambda)$ and d''_{t+1} with a delay of $C_L R_T (1 + \lambda)$. As a result, the net delay for transmitting two new data items becomes $3C_L R_T (1 + \lambda)$. During the transmission of d'_{t+1}, we *reset* both *temporal_redundancy* signal and *ready* signal, whereas for d''_{t+1} transmission, we *set* both *temporal_redundancy* signal and *ready* signal. Clearly, for both crosstalk classes 5 and 6, our approach takes $3C_L R_T (1 + \lambda)$ cycles whereas the technique given in [9] takes $3C_L R_T (1 + \lambda)$ and $4C_L R_T (1 + 1\lambda)$ cycles, respectively. So, our approach not only provides energy benefits (due to elimination of coupling transitions of crosstalk classes 5 and 6) but also performance benefits.

Decoding the data at the receiver end is a simple process (as shown in Table 3). At the receiver end, we store the value on the bus in a n-bit register R_1 when the *ready* signal is *set* and the *temporal_redundancy* signal is *reset* and the value on the bus is stored in another n-bit register R_2 when both *ready* signal and *temporal_redundancy* signal are *reset*. We use the values of these two registers to decode the actual data. When the *ready* signal is *set* and the *temporal_redundancy* signal is *reset*, there is no need of decoding (because, this situation arises only when the transmitted data is in crosstalk class from the set $\{1,2,3,4\}$), so we consider whatever the data present on the bus as the next data. On the other hand, if both *ready* signal and *temporal_redundancy* signal are *set*, we decode the data using the decoding algorithm given in Table 3. Our decoder is a simple bit-wise comparison circuit so the delay due to the decoder can be ignored.

4.3 Correctness of our approach

We now prove that our encoding technique works correctly for any data and minimizes the delay and the energy consumption. Let $CC(d_i, d_j)$ denote the crosstalk class of d_i w.r.t d_j, $T(d_i, d_j)$ denote the delay for transmitting d_j after d_i, and $E(d_i, d_j)$ denote the energy consumption during the transition from d_i to d_j.

Theorem 1 *For any pair of n-bit data d_t and d_{t+1}, if $CC(d_{t+1}, d_t) \geq 5$, then d_{t+1} can be encoded as d'_{t+1} and d''_{t+1} such that*

- $CC(d'_{t+1}, d_t) < 5$ *and* $CC(d''_{t+1}, d'_{t+1}) = 3$
- $T(d_t, d'_{t+1}) + T(d'_{t+1}, d''_{t+1}) \leq T(d_t, d_{t+1})$
- $E(d_t, d'_{t+1}) + E(d'_{t+1}, d''_{t+1}) \leq E(d_t, d_{t+1})$

Proof. To prove $CC(d'_{t+1}, d_t) < 5$:

Let $\overline{tp(a,b)}$ be the transition pattern over $\{\uparrow, \downarrow, -\}$ of length n when a n-bit data a is transmitted after another n-bit data b. Assume that $tp(d_{t+1}, d_t) = p_1 \cdots p_n$, $p_i \in \{\uparrow, \downarrow, -\}$, $1 \leq i \leq n$. Since $CC(d_{t+1}, d_t) \geq 5$, there must be at least 2 wires which have opposite transitions w.r.t. one or both of their adjacent wires. For simplicity, assume that there are exactly 2 wires, i.e., wire i and wire $i+1$, which have opposite transitions w.r.t. each other. Note that the proof can be easily extended to any number of wires having opposite transitions w.r.t. one or both of their adjacent wires. So, there is no pattern of type $\downarrow\uparrow$ or $\uparrow\downarrow$ in both $p_1 \cdots p_i$ and $p_{i+1} \cdots p_n$. Let $tp(d'_{t+1}, d_t) = p'_1 \cdots p'_n$ and $tp(d''_{t+1}, d'_{t+1}) = p''_1 \cdots p''_n$. Since there is no pattern of type $\downarrow\uparrow$ or $\uparrow\downarrow$ in both $p_1 \cdots p_i$ and $p_{i+1} \cdots p_n$, according to our encoding technique, for $1 \leq j \notin \{i, i+1\} \leq n$, $p'_j = p_j$ and $p''_j = -$, and for $j \in \{i, i+1\}$, $p''_j = \downarrow$, and if $p_j = \uparrow$, $p'_j = p_j$, and if $p_j = \downarrow$, $p'_j = -$. Hence,

$$tp(d'_{t+1}, d_t) = p'_1 \cdots p'_{i-1} p'_i p'_{i+1} p'_{i+2} \cdots p'_n$$
$$= p_1 \cdots p_{i-1} p'_i p'_{i+1} p_{i+2} \cdots p_n$$

We know that either $p'_i = p_i$ or $p'_{i+1} = p_{i+1}$. In either case, $p_1 \cdots p_{i-1} p'_i p'_{i+1} p_{i+2} \cdots p_n$ does not have a pattern of type $\downarrow\uparrow$ or $\uparrow\downarrow$. Hence,

$$CC(d'_{t+1}, d_t) < 5.$$

To prove $CC(d''_{t+1}, d'_{t+1}) = 3$:

We know that

$$tp(d''_{t+1}, d'_{t+1}) = p''_1 \cdots p''_{i-1} p''_i p''_{i+1} p''_{i+2} \cdots p''_n$$
$$= - \cdots - \downarrow\downarrow - \cdots -$$

From Table 1, it is clear that

$$CC(d''_{t+1}, d'_{t+1}) = 3.$$

To prove $T(d_t, d'_{t+1}) + T(d'_{t+1}, d''_{t+1}) \leq T(d_t, d_{t+1})$:

Since $CC(d'_{t+1}, d_t) < 5$ and $CC(d''_{t+1}, d'_{t+1}) = 3$,

$$T(d_t, d'_{t+1}) = 2C_L R_T (1 + \lambda)$$
$$T(d'_{t+1}, d''_{t+1}) = C_L R_T (1 + \lambda)$$

But $CC(d_{t+1}, d_t) \geq 5$. So,

$$T(d_t, d_{t+1}) \geq 3C_L R_T (1 + \lambda)$$

Hence,

$$T(d_t, d'_{t+1}) + T(d'_{t+1}, d''_{t+1}) \leq T(d_t, d_{t+1}).$$

To prove $E(d_t, d'_{t+1}) + E(d'_{t+1}, d''_{t+1}) \leq E(d_t, d_{t+1})$:

For simplicity, as in the case of the first proof, assume that there are exactly 2 wires, i.e., wire i and wire $i+1$, which have opposite transitions w.r.t. each other. Let d^i_t be the value of ith-wire at time instant t. Assume that $d^i_t d^{i+1}_t = 01$ and $d^i_{t+1} d^{i+1}_{t+1} = 10$ so that $d'^i_{t+1} d'^{i+1}_{t+1} = 11$ and $d''^i_{t+1} d''^{i+1}_{t+1} = 00$. Note that the proof can be easily extended to the case where $d^i_t d^{i+1}_t = 10$ and $d^i_{t+1} d^{i+1}_{t+1} = 01$. Then

$$E(d_t, d_{t+1}) = C_L(x_1 - \lambda d^{i-1}_{t+1}) + C_L(\lambda d^{i+2}_{t+1} + x_2)$$
$$+ C_L(1 + 3\lambda - \lambda(d^{i-1}_{t+1} - d^{i-1}_t)),$$

where $C_L(x_1 - \lambda d^{i-1}_{t+1})$ is the energy consumption of first $(i-1)$ wires (negative component is the effect of wire i on wire $(i-1)$), $C_L(1 + 3\lambda - \lambda(d^{i-1}_{t+1} - d^{i-1}_t))$ is the energy consumption of wire i and wire $(i+1)$ including the effects of their other adjacent wires, and $C_L(\lambda d^{i+2}_{t+1} + x_2)$ is the energy consumption of last $(n - i - 1)$ wires including the effect of wire $(i+1)$ on wire $(i+2)$.

$$E(d_t, d'_{t+1}) = C_L(x_1 - \lambda d^{i-1}_{t+1}) + C_L(x_2) + C_L(1 + \lambda$$
$$- \lambda(d^{i-1}_{t+1} - d^{i-1}_t) - \lambda(d^{i+2}_{t+1} - d^{i+2}_t))$$
$$E(d'_{t+1}, d''_{t+1}) = C_L(\lambda d^{i-1}_{t+1}) + C_L(0) + C_L(\lambda d^{i+2}_{t+1})$$

If we assume that $E(d_t, d'_{t+1}) + E(d'_{t+1}, d''_{t+1}) > E(d_t, d_{t+1})$, after simplification, we obtain the following inequality

$$\lambda(d^{i-1}_{t+1} - d^{i+2}_{t+1} + d^{i+2}_t) > 2\lambda.$$

Since $\lambda > 0$,

$$d^{i-1}_{t+1} - d^{i+2}_{t+1} + d^{i+2}_t > 2,$$

which is always false. Hence,

$$E(d_t, d'_{t+1}) + E(d'_{t+1}, d''_{t+1}) \leq E(d_t, d_{t+1}).$$

Layer	H (μm)	W (μm)	S (μm)	T (μm)	C_L (fF/mm)	C_I (fF/mm)
MET 5	1.04	0.52	0.52	0.7	34.317	80.478

Table 4. Wire parameters

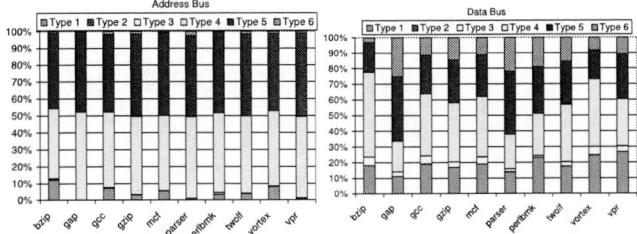

Figure 1. Distribution of transition patterns

Ori. code	000	001	010	011	100	101	110	111
CPC code	0000	0001	0100	0101	0111	1100	1101	1111

Table 5. CPC codes

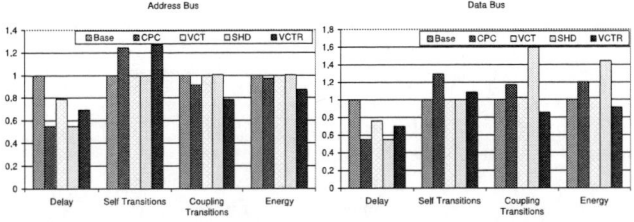

Figure 2. Normalized delay and energy

5 Experimental validation of our approach

We now experimentally validate our approach. We design the crosstalk analyzer, the encoder, and the decoder in Verilog and synthesize it using the Synopsys Design Compiler with $0.16\mu m$ *oki* technology library. We model three different wire lengths, i.e., 2mm, 5mm, and 10mm, to capture short, medium, and long wires, respectively, in the target technology. The Berkeley interconnect model [1] is used to calculate the ground and coupling capacitance of the interconnects. In our experimental results, we use the parameters of metal 5 (shown in Table 4).

We use Simplescalar 3.0 [4] and the SPEC2000 CINT [2] benchmark suite to simulate the performance of different on-chip buses between the processor datapath and L1 I-cache/D-cache. Transition pattern distributions of both address and data buses are given in Figure 1.

We compare the variable cycle transmission with temporal redundancy (VCTR) technique with the base case (i.e., without any coding), the variable cycle transmission (VCT) method [9], the crosstalk prevention coding (CPC) technique [18], and the shielding (SHD) method [3]. We consider two delays, $C_L R_T(1 + \lambda)$ (single cycle) and $2C_L R_T(1 + \lambda)$ (two cycles; note that while absolute delay is only $C_L R_T(1 + 2\lambda)$, but a variable cycle transmission scheme requires two-cycle delay to be a multiple of the single cycle period) for the VCTR technique, and based on the output of the crosstalk analyzer, we consider the appropriate delay. Note that the delay for the base case is $C_L R_T(1+4\lambda)$, for the CPC technique and shielding method the delay is $C_L R_T(1 + 2\lambda)$, and for the VCT technique the delays are multiples of $C_L R_T(1 + \lambda)$ based on the crosstalk class type.

In the VCT method, the delay between the next data item and the present data item is determined by the crosstalk class of the next data w.r.t. the present data item. Note that in this method, no encoding is applied on the transmitted data. Though by using the CPC method, one can get 46-bit crosstalk-free codes for any 32-bit data, it is very difficult to implement the method. As a result, here we consider 3-to-4 CPC coding, where every 3-bit data is encoded with 4-bit CPC code. Table 5 gives a set of CPC codes for 3-bit data. In order to prevent adjacent bits of different 4-bit encoded words to form crosstalk classes 5 and 6, we use shield wires between every pair of 4-bit encoded words. Hence, for a 33-bit bus we use 55 wires. In the SHD method, one shield, i.e., V_{dd} or $Ground$, wire is inserted between every two wires so that the crosstalk classes 5 and 6 are completely eliminated and hence the worst-case delay becomes $C_L \cdot R_T(1 + 2\lambda)$.

Main characteristics of different methods are given in Figure 2. The values present in this table are averaged across all the benchmarks and normalized w.r.t. the base case. Compared to the VCT technique, our technique achieves better performance in both address and data buses because of two reasons: 1) in the case of crosstalk class 6, the VCT technique considers a delay of $4C_L R_T(1 + \lambda)$, whereas our technique encodes the data into two data items such that the net delay becomes $3C_L R_T(1 + \lambda)$; 2) after encoding, for each 01 (or 10), d''_{t+1} gets 00 so that this can minimize the chances of getting crosstalk classes 5 or 6 for d_{t+2} w.r.t. d''_{t+1} (though we have not experimentally evaluated). The second reason is especially true, if same adjacent wires are repeatedly getting opposite transitions. This can be seen in address bus, where few LSB bits will change most of the time. That is the reason why, though there are less number of crosstalk class 6 transitions (from Figure 1), our technique outperforms the VCT technique. From the delay perspective, our technique achieves 31% (30%) savings for address (data) bus with just two extra wires. Though the CPC technique and SHD technique give more delay savings, they requires 22 and 32 extra wires, respectively. From the energy perspective, our technique achieves 13% (9%) savings for address (data) bus compared to the base case. Since coupling transitions play a major role in the total energy, though self transitions for our technique are more than that of the CPC technique in the address bus case,

0-7695-2533-4/06 $20.00 © 2006 IEEE

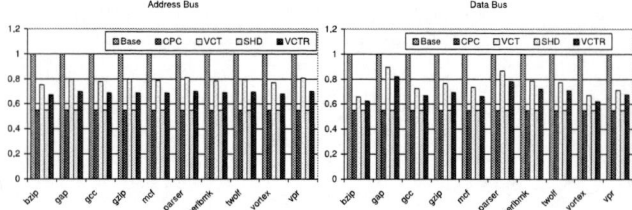

Figure 3. Benchmark-wise normalized delay

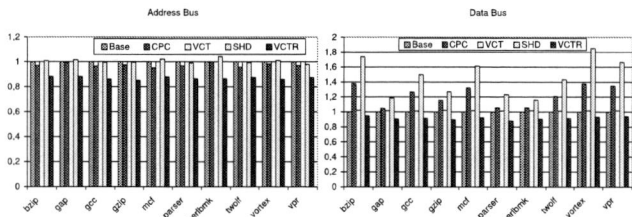

Figure 4. Benchmark-wise normalized energy

because of less number of coupling transitions, the energy consumption of our technique is less than that of the CPC technique. Since no encoding is applied in the VCT technique, its energy consumption is same as that of the base case. In the case of data bus, the CPC technique consumes more energy than that of the base case. The SHD technique consumes more energy than the base case because of more coupling transitions. Benchmark-wise normalized delay and energy consumption (w.r.t. the base case) of different techniques are given in Figure 3-4.

Table 6 gives the normalized area (w.r.t. the base case) and the power overhead. One can see from the table that our technique incurs an area overhead of 33%, 16%, and 11% for $2mm$, $5mm$, and $10mm$ bus, respectively. This is due to the codec (i.e., crosstalk analyzer, encoder, and decoder) and two extra wires. The absolute area overhead of our codec is $0.018mm^2$. One can observe that as the length of the interconnects increases the relative area overhead of our technique decreases. Our codec can run at a maximum clock speed of $660\,MHz$ and has power overhead of $2.16mW$.

6 Conclusion

By combining the ideas of variable cycle transmission and temporal redundancy, we proposed a technique for delay and energy efficient data transmission for on-chip buses and evaluated our technique by focusing on the L1 cache address/data buses of a microprocessor and by using the SPEC2000 CINT benchmark suit. From the results, it is clear that our method achieves 31% (30%) of delay improvement along with 13% (9%) energy savings over the

Method	# of wires	Normalized Area 2mm	5mm	10mm	Extra Power
Base	33	100	100	100	—
VCT	34	119	109	106	3.35mW
VCTR	35	133	116	111	2.16mW
CPC	55	177	171	169	1.06mW

Table 6. Codec overhead

base case for data transmission on address (data) bus.

7 Acknowledgements

Madhu Mutyam was supported by a BOYSCAST fellowship from the Department of Science and Technology (DST), New Delhi, India. This work was also supported in part by grants from NSF (Career No. 0093085), SRC (Grant No. 00541), and GSRC (MARCO/DARPA).

References

[1] Berkeley predictive technology model. http://www-device.eecs.berkeley.edu/~ptm/interconnect.html

[2] SPEC CPU2000 Benchmark. http://www.spec.org

[3] R. Arunachalam, E. Acar, and S.R. Nassif. Optimal shielding/spacing metrics for low power design. In *Proceedings of IEEE Computer Society Annual Symposium on VLSI*, pages 167-172, 2003.

[4] D.C. Burger and T.M. Austin. The SimpleScalar tool-set, version 2.0, Technical Report 1342, Department of Computer Science, UW, 1997.

[5] F. Caignet, S. Delmas-Bendhia, and E. Sicard. "The Challenge of Signal Integrity in Deep-submicrometer CMOS Technology". *IEEE*, 89(4), 2001, pp. 556-573.

[6] C. Duan, A. Tirumala, and S.P. Khatri. "Analysis and Avoidance of Crosstalk in On-chip Buses". In *Hot Interconnects*, 2001, pp. 133-138.

[7] Z. Khan, T. Arslan, and A.T. Erdogan. "A Novel Bus Encoding Scheme from Energy and Crosstalk Effeciency Perspective for AMBA based Generic SoC Systems". In *VLSI Design*, 2005, pp. 751-756.

[8] K. Kim, K. Baek, N. Shanbhag, C. Liu, and S. Kang. "Coupling-driven Signal Encoding Scheme for Low-power Interface Design". In *ICCAD*, 2000, pp. 318-321.

[9] L. Li, N. Vijaykrishnan, M. Kandemir, and M.J. Irwin. "A Crosstalk Aware Interconnect with Variable Cycle Transmission". In *DATE*, 2004, pp. 102-107.

[10] C.-G. Lyuh and T. Kim. "Low Power Bus Encoding with Crosstalk Delay Elimination". In *ASIC/SOC*, 2002, pp. 389-393.

[11] K.N. Patel and I.L. Markov. "Error-correction and Crosstalk Avoidance in DSM Busses". In *SLIP*, 2003, pp. 9-14.

[12] P. Sotiriadis and A. Chandrakasan. "Low Power Bus Coding Techniques Considering Inter-wire Capacitances". In *CICC*, 2000, pp. 507-510.

[13] P. Sotiriadis and A. Chandrakasan. "Bus Energy Minimization by Transition Pattern Coding (TPC) in Deep Sub-micron Technologies". In *ICCAD*, 2000, pp. 322-328.

[14] P. Sotiriadis and A. Chandrakasan. "Reducing Bus Delay in Sub-micron Technology using Coding". In *ASPDAC*, 2001, pp. 109-114.

[15] P. Sotiriadis and A. Chandrakasan. "A Bus Energy Model for Deep Submicron Technology". *TVLSI*, 10(3), 2002, pp. 341-350.

[16] S.R. Sridhara, A. Ahmed, and N.R. Shanbhag. "Area and Energy-efficient Crosstalk Avoidance Codes for On-chip Buses". In *ICCD*, 2004, pp. 12-17.

[17] D. Sylvester and C. Hu. "Analytical Modeling and Characterization of Deep-submicrometer Interconnect". *IEEE*, 89(5), 2001, pp. 634-664.

[18] B. Victor and K. Keutzer. "Bus Encoding to Prevent Crosstalk Delay". In *IC-CAD*, 2001, pp. 57-63.

[19] J. Yim and C. Kung. "Reducing Cross-coupling among Interconnect Wires in Deep-submicron Datapath Design". In *DAC*, 1999, pp. 485-490.

[20] Y. Zhang, J. Lach, K. Skandron, and M.R. Stan. "Odd/even Bus Invert with Two-phase Transfer for Buses with Coupling". In *ISLPED*, 2002, pp. 80-83.

0-7695-2533-4/06 $20.00 © 2006 IEEE

Power-Oriented Delay Budgeting for Combinational Circuits

Jialin Mi   and   Chunhong Chen

Department of Electrical and Computer Engineering, University of Windsor, Ontario, Canada
cchen@uwindsor.ca

Abstract

In this paper we propose an approach of providing the best power-delay tradeoff for combinational circuits. This is done by so-called power-oriented delay budgeting which is to combine the delay-budgeting technique with aggressive power optimization. We discuss the impacts that both discrete cell library and circuit topology may have on the potential power reduction. Experimental results show that up to 65% (an average of 35%) power savings can be achieved without any delay penalty.

1. Introduction

In the design flow of today's digital systems, circuit area, delay, and power consumption are three major concerns. Generally speaking, in order to reduce the area or power of a circuit, an increased delay is the price one has to pay, or vice versa. Recent research shows that this delay penalty can be avoided or reduced to a minimum because the circuit delay is independent of the non-critical paths where there are some delay budgets available [1-5]. With this arises the delay-budgeting problem which is to determine the delay budget for each component in an appropriate way. At high-levels (such as RT-level), these delay budgets can be used for timing-closure driven design to ensure the timing constraints being met in subsequent synthesis. At low-levels (such as gate-level), the delay budgets enable a decrease in power consumption or area cost.

Traditionally, the delay budgeting problem was formulated as a slack distribution/assignment problem for which many algorithms have been proposed [3-7]. [6] presented the zero-slack algorithm (ZSA) to assign delay budgets to signal nets for performance-driven layout. In [4], a delay upper-bound budgeting method was proposed for the net-based timing-driven placement. In [3], Kuo et al. formulated the delay-budgeting problem as a Lagrange-Multiplier -based slack assignment problem for the timing-closure-driven design. Chen et al. [1, 7] proposed the notion of potential slack and described a maximal-independent-set-based slack assignment algorithm. More recently, [2] targeted the capability of optimizing the power-delay tradeoff with different cell libraries, but did not consider the delay budgets which may provide further power optimization and thus promise a better tradeoff. An optimal delay budgeting algorithm was also reported in [5].

In a typical CMOS standard cell library (e.g., TSMC 0.13μm), the cells with same logic function may have several layouts with specific information about power, area

and delay. The basic idea of power-oriented slack assignment is that the slack of cells on the non-timing-critical paths provides the potential of reducing the power without violating timing constraints. This can be done by relaxing the delay to certain cells, where the layouts with longer delay and less power can be applied. However, the slacks of cells are dependent in general, and not all slacks can be used as the delay budgets. For the power savings to be maximized, some gates need to be assigned more budgets than others, depending on both circuit topology and individual power-delay characteristics of the cells. Compounding the problem is the discreteness of the cell library which may make the budgets meaningless if they are not large enough.

Previous research either looks at a pure delay budgeting problem for layout synthesis without keeping power optimization in mind, or explores a pure power-delay tradeoff without investigating the potential ability of optimizing the power consumption through delay budgets. In this paper, we present a power-oriented delay budgeting algorithm which is intended to provide the best power-delay tradeoff for any given combinational circuits. Experiments on a variety of benchmark circuits show that the potential power savings from the current synthesis flow are significant with an average improvement rate of 35%.

2. Problem Description

A combinational circuit can be described as a DAG (directed-acyclic graph) $G = (V, E)$, where each node denotes a gate (implemented with a standard library cell), and an edge e from node u to v means that v is an immediate fanout of u. Each node v (except primary inputs and outputs) is associated with a delay $d(v)$ (including the cell delay and wire delay). Once the arrival times for all primary inputs and the required times for all primary outputs are given, a propagation procedure for slack calculation can be performed to find the arrival time, required time and slack for all nodes [6, 1]. The delay-budgeting is to assign an additional delay (called budget) for each node so that a specific objective can be optimized while keeping non-negative slacks for all nodes. This objective can be chosen by the designers, depending on the applications. In general, a maximum sum of budgets over all nodes is one of interesting objectives. If the power reduction acts as such an objective, the problem turns out to be power-oriented delay-budgeting.

In the following, we briefly discuss the power-delay curves of the nodes by presenting four different models (the impact of graph topology on the delay-budgeting process will be investigated in the next section):

- **Model (a):** All nodes have uniform and linear power-delay characteristics. This is an ideal case where the potential power reduction is proportional to total delay budget, and the power-oriented delay-budgeting problem is equivalent to the traditional delay-budgeting.

- **Model (b):** All nodes' power-delay curves are linear (or piece-wise linear) but not uniform (i.e., they may have different slopes). This requires the delay to be weighted in the budgeting process. The nodes with steeper power-delay curves should be assigned a higher weight with the expectation of more power savings to be obtained.

- **Model (c):** The power-delay curves for all nodes are uniform but nonlinear. In this case, different nodes may have different weights depending on their current delay, and the nodes may also need to update their weights during the budgeting process to reflect the nonlinearity.

- **Model (d):** All nodes have different power-delay characteristics which are generally nonlinear. This represents a more practical case for real design, where not only do the nodes have to be weighted, but their weights are also expected to be updated during the delay assignment process. Finding an optimal or near-optimal solution to this case is a non-trial task and requires computationally efficient algorithms.

In the sections that follow, we first review the basic idea of state-of-the-art algorithm for delay budgeting. Then, we propose an improved greedy algorithm for the power-oriented delay budgeting problem with the goal of optimizing the power-delay tradeoff curves. This is followed by extensive experiments with benchmark circuits.

3. Power-Oriented Delay Budgeting

With the observation that the budget assignment strongly depends on the topological structure of an underlying design, the authors of [1] and [7] presented a maximal-independent-set based algorithm (MISA) to solve the budgeting problem. It was based on the fact that all immediate fanins (or fanouts) of a node are always independent of each other in terms of slack sensitivity, unless there are reconvergent directed edges. Therefore, a local potential slack was defined to be the largest among the slack of a node, the slack sum of its immediate fanins and that of its immediate fanouts, and serves as a criterion for the budget assignment.

In a standard library, however, the number of different cells for a logic gate is quite limited, and not all obtainable delay budgets can be used for power minimization. To deal with the nonlinearity of power-delay curves and the discreteness of target library as discussed above, we define a local potential power savings with delay budget (LPPDB) for each node as a weighted delay budget. The weight is set to zero if the budget is less than the delay of next available

gate. Otherwise, it is set to the slope of the node's power-delay tradeoff curve which can be obtained using the following power and delay models:

$$\begin{cases} d(v) \propto fanout\_area(v) / area(v) \\ p(v) \propto switching(v) \times (fanout\_area(v) + area(v)) \end{cases} \quad (1)$$

where $area(v)$ and $fanout\_area(v)$ are the size of the gate corresponding to node v and the sum of sizes of all its immediate fanouts, respectively, and $switching(v)$ is the switching activity for node v, available from gate-level description. All possible pairs of (p, d) from (1) form the power-delay curve of node v.

We define the potential power savings with delay budget (PPDB) for node v to be the maximum among the LPPDB of node v, the sum of v's immediate fanins' LPPDBs and the sum of v's immediate fanouts' LPPDBs. PPDB can be used as a new criterion for the power-oriented delay budgeting. The detailed pseudo-code of the proposed greedy algorithm is given as follows:

Greedy Algorithm {
Input: Graph $G = (V, E)$ and a given library
Output: Power Savings (PS)
Begin {
$PS \leftarrow 0$;
Do {
  for each node v
    Find PPDB and a candidate set:
    in $\leftarrow$ sum of LPPDBs for all immediate fanins of v
    out $\leftarrow$ sum of LPPDBs for all immediate fanouts of v
        self $\leftarrow$ LPPDB for v
    $PPDB(v) \leftarrow \max\{in, out, self\}$;
    $candidate(v) \leftarrow$ a set of nodes with $PPDB$;
  end for
  Find v_m such that $PPDB(v_m) = \max\{PPDB(v) \mid v \in V\}$;
  $PS \leftarrow PS + PPDB(v_m)$;
  Update slacks of all (transitive) fanins/ fanouts of nodes in $candidate(v_m)$ from G;
  }
  While $(\max\{PPDB(v) \mid v \in V\} > 0)$
}

In what follows, we demonstrate how the greedy algorithm works on an example graph shown in Fig. 1, using the library information given in Table 1. Assuming a minimum delay of each node is used to start with, the algorithm first identifies node v_2 that has a maximum $PPDB = 8$, and its candidate set which is $candidate(v_2) = \{v_3, v_4, v_5\}$ (i.e., the immediate fanouts of v_2). Then, the algorithm brings PS from 0 to 8, and updates slacks of (transitive) fanins/ fanouts of $candidate(v_2)$ (i.e., v_1, v_2, v_6, v_7 and v_8). During the second and third iterations of the algorithm, the PS is updated to $(8 + 4) = 12$ and $(12 + 2) = 14$, respectively. After that, $max\{PPDB\}$ turns out to be 0, which breaks the algorithm out of the *While* loop. The details of each iterative step are shown in Table 2. In this example, the original power dissipation is: $(4 + 3 + 4 + 6 + 4$

0-7695-2533-4/06 $20.00 © 2006 IEEE

+ 6 + 10 + 6) = 43 units. After the power-oriented delay budgeting, the *PS* of 14 units represents $14/43 \approx 33\%$ power savings.

Note that in the greedy algorithm, a large slack does not represent high delay budget, nor does a high delay budget necessarily provide big potential power savings. In the above example, if we choose $\{v_3, v_4, v_8\}$ instead of $\{v_3, v_4, v_5\}$ to assign the delay budgets in the first iteration, the final *PS* turns out to be 16 units. In general, to guarantee a maximum PPDB, an optimal algorithm is needed which can recursively check PPDBs of all transitive fanins and fanouts for each node, leading to a prohibitively expensive computation cost. The above algorithm is greedy but much faster, making it practical for large circuits.

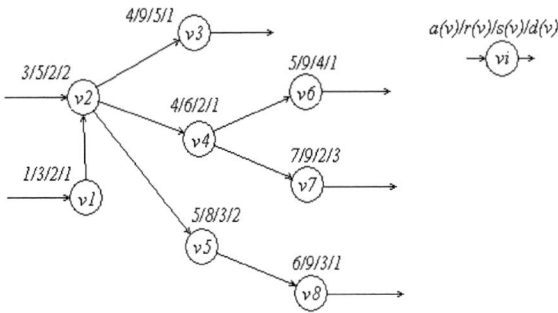

Figure 1. An eight-node subcircuit example

Table 1. The available library information for example graph of Fig. 1

node	available cells in a library (delay units | power units)		
v1	1|4	2|2	4|1
v2	2|3	3|2	6|1
v3	1|4	2|2	4|1
v4	1|6	2|3	6|1
v5	2|4	4|2	8|1
v6	1|6	2|3	3|2
v7	3|10	6|5	10|3
v8	1|6	3|2	6|1

Table 2. Iterations of the greedy algorithm on Fig. 1

iteration 1	v1	v2	v3	v4	v5	v6	v7	v8
arrival time	1	3	4	4	5	5	7	6
required time	3	5	9	6	8	9	9	9
slack	2	2	5	2	3	4	2	3
delay budget	1	1	3	1	2	2	0	2
PPDB	2	8	3	4	4	4	3	4
candidate(s)	v1	v3, v4, v5	v3	v6, v7	v8	v6	v4	v8

iteration 2	v1	v2	v3	v4	v5	v6	v7	v8
arrival time	1	3	7	5	7	6	8	8
required time	2	4	9	6	8	9	9	9
slack	1	1	2	1	1	3	1	1
delay budget	1	1	0	0	0	2	0	0
PPDB	2	2	1	4	1	4	0	0
candidate(s)	v1	v1	v2	v6, v7	v2	v6	v7	v8

iteration 3	v1	v2	v3	v4	v5	v6	v7	v8
arrival time	1	3	7	5	7	8	8	8
required time	2	4	9	6	8	9	9	9
slack	1	1	2	1	1	1	1	1
delay budget	1	1	0	0	0	0	0	0
PPDB	2	2	1	1	1	0	0	0
candidate(s)	v1	v1	v2	v2	v2	v6	v7	v8

iteration 4	v1	v2	v3	v4	v5	v6	v7	v8
arrival time	2	4	8	6	8	9	9	9
required time	2	4	9	6	8	9	9	9
slack	0	0	1	0	0	0	0	0
delay budget	0	0	0	0	0	0	0	0
PPDB	0	0	0	0	0	0	0	0
candidate(s)	v1	v2	v3	v4	v5	v6	v7	v8

4. Experiments

We implemented the proposed greedy algorithm in Java programming language and tested it on benchmark circuits on top of *SIS* package [8]. The synthesis results with minimum delay from *SIS* were used as an input to our delay-budgeting problem. We also compared the results with those from [1].

For each benchmark circuit, we first estimated its original circuit delay (T_0) and power consumption (P_0). Then, we ran our algorithm using T_0 as the timing constraints to show how much power savings can be achieved by assigning the delay budgets to the nodes on non-critical paths. The results represent the potential power savings of each benchmark circuit without any delay penalty. In order to get the optimized power-delay tradeoff, we relaxed the timing constraints from T_0 to $4T_0$ with a step of $0.1T_0$, and measured the power consumption P after the algorithm was performed. The results reflect the possible power savings with certain percentage of delay penalty.

Table 3 summarizes the power savings with our algorithm for different timing constraints. It can be seen that (a) an average of 35% power reduction (with a maximum of 65%) is achieved without any delay penalty, and (b) the amount of power savings for a given delay penalty varies significantly from circuit to circuit, depending upon their topologic structures. Table 4 shows the power savings using the algorithm from [1] for the same benchmarks. We see that the proposed algorithm provides an improvement of about 5~10%, which is measurable considering the fact that the percentage is taken with respect to initial power consumption (i.e., P_0).

Table 3. Power savings with the proposed algorithm on a set of benchmark circuits

circuit	size (#gates)	power consumption P/P_0 (%)						
		T_0	$1.5T_0$	$2T_0$	$2.5T_0$	$3T_0$	$3.5T_0$	$4T_0$
c17	5	100.0%	64.4%	46.6%	46.6%	32.2%	32.2%	25.0%
b1	7	86.2%	51.2%	41.1%	37.9%	28.2%	28.2%	25.0%
mux	28	97.1%	49.8%	44.2%	39.7%	35.7%	27.9%	25.0%
decod	30	100.0%	71.1%	64.6%	58.6%	40.0%	30.2%	28.5%
cm150a	32	79.6%	47.5%	38.6%	33.5%	30.9%	26.9%	25.0%
x4ml	33	79.9%	63.8%	48.7%	41.0%	32.4%	28.7%	26.7%
cc	42	48.7%	38.9%	32.4%	28.6%	32.3%	28.9%	26.0%
count	96	43.1%	29.5%	27.2%	25.9%	25.6%	25.6%	25.0%
b9	103	51.9%	37.0%	31.3%	25.5%	25.3%	25.3%	25.0%
apex7	151	47.7%	33.8%	29.7%	27.8%	25.4%	25.2%	25.2%
x1	218	40.2%	32.7%	31.1%	26.8%	25.9%	25.3%	25.0%
alu2	277	62.9%	47.5%	37.0%	31.8%	30.3%	26.7%	25.9%
apex6	487	34.8%	29.1%	26.2%	25.5%	25.3%	25.3%	25.2%
i8	748	34.1%	29.3%	27.2%	25.8%	25.1%	25.0%	25.0%
average		64.7%	44.7%	37.6%	33.9%	29.6%	27.2%	25.5%

Table 4. Power savings using the algorithm from [1]

circuit	size (#gates)	power consumption P/P_0 (%)						
		T_0	$1.5T_0$	$2T_0$	$2.5T_0$	$3T_0$	$3.5T_0$	$4T_0$
c17	5	100.0%	84.4%	46.6%	46.6%	32.2%	78.4%	25.0%
b1	7	87.1%	51.2%	41.1%	37.9%	28.2%	44.0%	25.0%
mux	28	97.1%	49.8%	44.2%	68.1%	40.9%	27.9%	25.0%
decod	30	100.0%	72.8%	64.6%	57.8%	43.1%	30.2%	28.5%
cm150a	32	79.6%	49.0%	63.2%	64.2%	54.9%	27.6%	25.0%
x4ml	33	80.5%	77.4%	55.1%	51.4%	43.4%	33.0%	26.7%
cc	42	48.8%	41.3%	32.7%	29.9%	32.3%	28.9%	25.8%
count	96	45.4%	32.2%	27.8%	29.6%	28.6%	36.4%	25.3%
b9	103	54.3%	44.1%	32.6%	27.4%	25.3%	32.0%	25.0%
apex7	151	50.2%	44.6%	32.6%	28.4%	28.6%	28.5%	25.0%
x1	218	47.9%	39.4%	31.9%	33.1%	26.5%	25.6%	25.0%
alu2	277	65.5%	57.6%	51.5%	44.7%	39.5%	31.0%	26.0%
apex6	487	37.5%	31.9%	26.7%	25.8%	25.7%	25.6%	25.5%
i8	748	36.1%	34.9%	30.5%	32.2%	28.7%	26.2%	25.0%
average		66.4%	49.3%	41.5%	41.2%	34.1%	34.0%	25.6%

5. Conclusions

We have proposed a greedy algorithm to solve the power-oriented delay budgeting problem. We have shown that the topologic information of the circuits as well as cell library can be used for aggressive power optimization in order to obtain the best power-delay tradeoff. Experimental comparison has also been made between a pure delay-budgeting algorithm and the proposed algorithm in terms of power savings. Further study is needed to apply the delay-budgeting method to high-level design, and develop one single unified metric (instead of separately-treated metrics of area, delay and power) for future design.

6. References

[1] C. Chen, X. Yang, and M. Sarrafzadeh, "Predicting Potential Performance for Digital Circuits", *IEEE Transactions on Computer-Aided Design of Integrated Circuits and Systems*, vol. 21, no. 3, pp. 253-262, March 2002.

[2] M. Vujkovic and C. Sechen, "Optimized Power-Delay Curve Generation for Standard Cell ICs," in *Proc. of International Conference on Computer-Aided Design*, pp. 387-394, November 2002.

[3] C. C Kuo and A. C. Wu, "Delay Budgeting for A Timing-Closure-Driven Design Method", in *Proc. of*

International Conference on Computer-Aided Design, pp.202-207, November 2000.

[4] M. Sarrafzadeh, D. Knol, and G. Tellez, " A Delay Budgeting Algorithm Ensuring Maximum Flexibility in Placement," *IEEE Transactions on Computer-Aided Design of Integrated Circuits and Systems*, vol. 16, no. 11, pp. 1332-1341, Nov 1997.

[5] Elaheh Bozorgzadeh, et al, "Optimal Integer Delay Budgeting on Directed Acyclic Graphs," in *Proc. of the 40th Design Automation Conference*, pp. 920-925, June 2003.

[6] R. Nair, C. L. Berman, P. S. Hauge, and E. J. Yoffa, "Generation of Performance Constraints for Layout," *IEEE Transactions on Computer-Aided Design of Integrated Circuits and Systems*, vol. 8, no. 8, pp. 860-874, August 1989.

[7] C. Chen, X. Yang, and M. Sarrafzadeh, "Potential slack: an effective metric of combinational circuit performance," in *Proc. of International Conference on Computer-Aided Design*, pp. 198-201, November 2000.

[8] E. M. Sentovich et al., "SIS: A System for Sequential Circuit Synthesis," Technical Report UCB/Erl M92/41, Univ. of California, Berkeley, May 1992.

0-7695-2533-4/06 $20.00 © 2006 IEEE

VLSI Circuits
and Optimization

Routing-Tree Construction with Concurrent Performance, Power and Congestion Optimization

Cengiz Alkan and Tom Chen
Electrical and Computer Engineering
Colorado State University, Fort Collins, Colorado

Abstract— We present a routing-tree construction algorithm that considers multi-objectives of performance, power and congestion concurrently. Congestion is measured with balanced usage of routing resources among layers. Simultaneous buffer insertion and layer assignment tends to produce routing-trees with shorter overall length. Applying the proposed simultaneous algorithm on a subset of routes on a commercial 64-bit microprocessor yielded 9% less repeater usage and 1.5% shorter overall routing-tree length with improved overall performance at the same time, compared to sequential routing-tree construction approach.

I. INTRODUCTION

The scaling to *Deep Sub-Micron* (DSM) feature sizes for CMOS VLSI has led to decreased device cost and increased performance. But, starting from $250nm$ process technology node, interconnect delay dominates gate delay [22] due to poor scalability of interconnects. Performance of future VLSI chips will no longer be determined by gate delays rather by interconnect delays. Consequently, such rapid scaling requires performance-driven layout techniques. *Routing-Tree* (RT) synthesis is an increasingly important part of the interconnect design.

Typical goal of a performance-driven RT construction is to eliminate *Delay Violation* (DV) to any sink node. Delay violation is defined as the difference between actual delay and required delay for a given interconnect. When RT construction is performance-driven, layer assignment and buffer insertion has to be considered concurrently within the RT construction process since they also have a significant impact on delay. Traditionally, layer assignment and buffer insertion are often performed on finished RT [3, 23, 18, 4, 9, 20], and RT is constructed without the knowledge of buffer insertion and layer assignment. This can yield suboptimal solutions. A performance-driven RT construction algorithm without either buffer insertion or layer assignment or both will try to eliminate delay violations by trading-off minimum length RT topology resulting in star-like and longer routing-trees. Attempts have been made recently to incorporate buffer insertion into RT construction [2, 5, 10]. But none attempts to consider all three methods, performance-driven routing-tree construction, layer assignment, and buffer insertion, concurrently.

Routing-tree construction is NP-Hard [21] and various heuristics are usually employed. Iterative 1-Steiner algorithm [21, 15] starts from a Minimum Spanning Tree (MST) and iteratively modifies the original tree each time introducing a new edge which minimize the total tree length. Minimum Rectalinear Steiner-Tree (MRST) minimizes the total tree length which is preferable in terms of routing resource demand. But, they will not necessarily satisfy timing constraints due to the fact that minimum length routing-tree does not necessarily mean the length from source to a critical sink node is minimized.

Jaewon proposed a Linear Programming (LP) based algorithm [13] which has constraints on the source-sink lengths. Even with minimum critical source-sink path length, routing-tree may not be suitable for VLSI routing since delay at any sink is not only a function of source-sink length but also the topology of rest of the tree.

Critical sink routing-tree construction algorithms [17, 23, 16] progressively builds the routing-tree which, at each iteration, a node is connected (or modified if the starting tree is a sub-optimal tree) so that maximum delay is minimized (or maximum delay violation or total delay is minimized). Since delay of a routing-tree has to be calculated in order to decide which connection is next in the suboptimal (or partially completed) tree, resistance and capacitance of the tree edges should be known which is a function of routing metal layer that tree is being routed. Therefore, layer assignment is generally done without the knowledge of routing-tree topology resulting in suboptimal layer assignment.

RTs generated without respect to buffer insertion also yield suboptimal star-like tree topologies. Q-Trees [10] attempts to further optimize the suboptimal buffered trees by *hannan grafting* with a possible buffer replacement. Other algorithms such as P-Tree [24] and C-Tree [25] incorporate multi-objectives of wire sizing, delay/area tradeoff simultaneously, but assume fixed layer assignment and ignore the impact of buffer insertion on RT topology.

Yildiz's preferred Steiner-trees [12, 1] assign edges of the routing-tree based on length. Upper layers become cost effective for long wires while shorter wires are routed on lower layers due to the additional cost of vias. Although this scheme is desirable, it is not really performance-driven. It is conceivable that a short net can be critical. Furthermore. it can result in overcrowded layers since it does not consider congestion. Cho's layer assignment algorithm [7] uses a Network Interface Graph (NIG) to model the potential cross-talk between nets. Hence, nets with high probability of interference are placed on

0-7695-2533-4/06 $20.00 © 2006 IEEE

separate layers. NIG does not consider timing, hence it can assign critical nets to disadvantaged layers. Saxena [11] suggests assigning nets that failed timing to upper routing layer based on layer area quotas during routing process. Most of the existing RT construction algorithms use Ginneken-type dynamic programming based algorithms [18, 9, 8] for repeater insertion during or after RT construction.

In summary, traditional routing flow first builds a routing-tree topology. Layer assignment is performed before or after the routing process. Buffers are usually inserted last for any nets with a delay violation. Therefore, routing-trees are constructed with incomplete information about layer assignment or buffer insertion. The resulting routing-trees are often suboptimal. Thus, performance-driven routing-tree construction tends to increase the length with star-like tree structures in order to meet timing requirements. This will greatly increase routing resource demand and potentially increase use of repeaters. Moreover, resulting routing-tree topology may also uneven layer resource usage due to failed nets being reassigned to other layers during the rip-up and reroute process.

In this paper, we present a new routing-tree construction flow that concurrently performs layer assignment and buffer insertion during the process of routing-tree construction. The proposed algorithm generates more compact routing-trees with less buffers and less congestion at different layers while producing better delays. Delay calculation during RT construction uses AWE-based methods (Asymptotic Waveform Evaluation) [14] for accuracy.

Although a straight-forward RT construction method is used in this paper, other RT construction algorithms can be adopted. It is worth repeating that concurrent implementation of RT construction, layer assignment, and buffer insertion will benefit in terms of delay vialotion, wire length and number of buffers comparing the same RT construction algorithm that performs these tasks separately.

Remaining part of this paper is organized as follows: Section II describes the proposed routing-tree construction algorithm with simultaneous layer assignment and buffer insertion. Experimental results using a 64-bit microprocessor core is given in Section III. Finally summary and concluding remarks are presented in Section IV.

II. ROUTING-TREE CONSTRUCTION WITH SIMULTANEOUS LAYER ASSIGNMENT AND BUFFER INSERTION

The proposed algorithm of RT construction with Layer Assignment and Buffer Insertion (LABI) is shown in Figure 1. For a given netlist, at each iteration, algorithm grows all the routing-trees one node at a time using *GrowPartialTree* subroutine. *GrowPartialTree* chooses the most timing critical sink node from unconnected sink nodes. If a net is a non critical net, i.e. there is no delay violation, *GrowPartialTree* chooses the sink node which minimizes the total tree length. Layer assignment is then performed on the netlist based on current delay violation values of partially constructed nets. Layer assignment also considers distribution of nets into layers such

```
1   LABI(S)
2   begin
3       S = All Nets
4       InitialLayerAssignment(S)
5       n_net = unconnectedSinks(S), net ∈ S
6       while (n_net! = φ, ∀ net ∈ S)
7           foreach net ∈ S
8               GrowPartialTree(net)
9           do
10              LayerAssignment(S)
11              foreach net ∈ S
12                  InsertBuffer(net)
13              while (Buffer inserted for any net in S)
14          foreach net ∈ S
15              foreach buf ∈ net
16                  RemoveBuffer(net, buf)
17                  if (DV_net > 0)
18                      ReinsertBuffer(net, buf)
19  end
```

Fig. 1. Simultaneous buffered routing-tree construction with layer assignment

that layer usage is similar at each layer. For all the nets with a delay violation, buffers are inserted in the next step. Layer assignment and buffer insertion are applied to the netlist until there is no improvement on any net. Once all the partially constructed nets are layer and buffer optimized, algorithm continues growing nets one more time.

The intuition of full layer and buffer optimization on partially constructed trees is that when buffer insertion and layer assignment are applied to the partial tree, it will update the timing of the partially constructed tree allowing next sink node to connect to a better location. In other words, by inserting buffers or improving the delay of the tree with layer assignment, sink nodes can potentially connect to the lower edges in the tree topology (or to a closer edge to the sink), effectively reducing the total tree length.

A simple example to compare the RT construction process is given in Figure 2 where *rat* is *required arrival time* and *dv* is *delay violation*. Traditional methods, where RT construction, layer assignment and buffer insertion are performed separately, produce star-like topology as shown in Figure 2 a). Proposed method, buffers inserted into the partial tree in Figure 2 b), enables the subsequent connections to better locations yielding more compact and higher performance tree topology.

It is clear from this example that simultaneous consideration of different objectives during RT construction can produce better routes. No single task, i.e building tree, layer assignment, and buffer insertion, takes in its entirely without considering its impact on other tasks and the overall routing quality. The following subsections give more details about different aspects of the RT construction process.

A. Routing-Tree Construction

Proposed algorithm for *GrowPartialTree* shown in Figure 3 uses a greedy approach to progressively build the routing-tree with minimum delay violation. Initially, the tree includes only the source node. At each iteration, all the unconnected sink

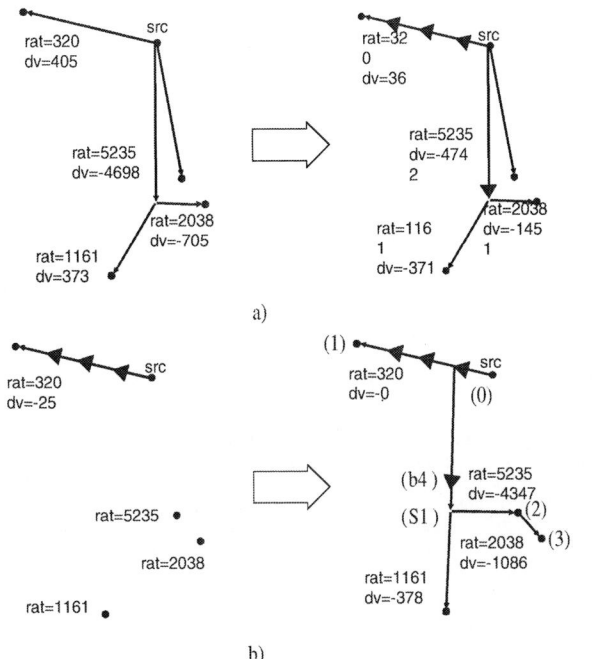

a)

b)

Fig. 2. a)RT construction, layer assignment and buffer insertion b)Concurrent RT construction, layer assignment and buffer insertion

```
1   GrowPartialTree(T)
2    #one iteration
3   bestDV = -INF
4   bestLen = INF
5   foreach unconnected node n ∈ T
6     bestDVPin = INF
7     bestLenPin = INF
8     foreach edge e ∈ T
9       JoinNodeAtEdge(T, n, e);
10      awe(T);
11      if (DV(T) < 0  &  Len(T) < bestLenPin)
12        bestLenPin = Len(T)
13        bestEdge = e
14      elsif (DV(T) < bestDV)
15        bestDVPin = DV(T)
16        bestEdge = e
17      DisjoinNode(T, n)
18    if (DV(T) < 0  &  bestLenPin < bestLen)
19      bestLen = bestLenPin
20      bestNode = n
21    elsif (besDVPin > bestDV)
22      bestDV = bestDVPin
23      bestNode = n
24   JoinNodeAtEdge(T, bestNode, bestEdge)
```

Fig. 3. Performance driven progressive construction

B. Buffer Insertion and Removal

Buffer insertion problem is NP-hard [9, 18]. Equal distance buffer insertion generally yields good *delay reduction-complexity* trade-off [23]. This is due to the fact that if the distance between buffers d_i is equal to one other, $\sum d_i^2$ is minimized provided that $\sum d_i$ is constant [6]. The proposed algorithm inserts a buffer which maximizes the delay violation reduction by evaluating delay reduction with respect to possible buffer locations. Algorithm evaluates buffer locations at Steiner points as well as on routing segments.

```
1   InsertBuffer(T)
2   bestDV = INF
3   foreach edge e ∈ T
4     b = InsertBufferAtEdge(T, e)
5     awe(T)
6     if (DVImp(T) > 10ps  &  DV(T) < bestDV)
7       bestEdge = e
8       bestDV = DV(T)
9     RemoveBuffer(T, b)
10  InsertBufferAtEdge(T, bestEdge)
```

Fig. 4. Buffer Insertion

nodes are tentatively connected to the existing tree and the most critical sink node is chosen and permanently connected to the existing tree. In the proposed algorithm, preference is given to the most timing critical sink node to be connected first. In the existing RT construction algorithms, preference has been given to the least timing critical sink [16, 23]. Although there is no clear preference based on theoretical advantages given in the literature, our own experiments indicate a slight advantage of giving preference to the most timing critical sink. This can be seen from the example in Figure 2 b): The buffer insertion (buffer 4) isolates the most critical path (path 0-1) in the early stages allowing less critical sinks (sink 2,3) to connect a location (s1) which is not on the critical path. Progressiveness of the proposed RT construction allows concurrent buffer insertion and layer assignment as an integral part of the RT construction algorithm. The proposed algorithm falls back to the minimum length tree construction if there is no delay violation.

For a partially constructed tree with i pins, maximum number of edges is $2i - 3$ and the remaining number of pins is $n - i$ where n is the total number of pins. Possible permutations of tree to connect the rest of the pins can be expressed as:

$$\sum_{i=1}^{n} (2i - 3)(n - i) \qquad (1)$$

Hence, complexity of the proposed RT construction algorithms is $O(n^3)$ due to the inner search loop for the most critical pin.

At each iteration, a buffer is tentatively inserted into each candidate buffer location. The buffer location with the best delay violation improvement is chosen. Iteration loop stops when the amount of delay improvement is below a preset threshold for all possible buffer locations, or the timing constraints for the net are met. The complexity of the overall buffer insertion algorithm is $O(n)$.

0-7695-2533-4/06 $20.00 © 2006 IEEE

C. Layer Assignment

Layer assignment can be used to improve routing quality in three different ways:

- Delay for the same length wires is better on the upper layers. Hence, critical nets can be assigned to the upper layers to improve timing.

- Coupling is usually less between layers. Therefore, crosstalk can be reduced by placing high-interference nets into separate layers.

- If a routing area on a particular layer is highly congested, some nets can be assigned to a different layer to reduce the congestion.

The proposed algorithm performs an initial layer assignment for all the nets before the routing-tree construction starts. Although an accurate timing, congestion and cross-talk information are not available prior to routing, an estimated metric can be used to assign a net to a layer to improve the routing. Our algorithm uses estimated layer usage to calculate the utilization of the layer. Layer usage is defined as:

$$U_l = \sum_{i \in B} b_i^l / pitch^l + \sum_{j \in N+S} d_j^l \qquad (2)$$

Layer usage incurred from blockages is calculated as the sum of the track lengths covered by the blockage. b_i^l is the area of the blockage i at layer l. $pitch^l$ is the distance between routing tracks at layer l. Layer usage incurred from existing routing N and current routing S is simply calculated by adding the horizontal (or vertical) distance d_j^l of the edges of each net.

Total wire length in addition to blockages on a layer seems to be the best criteria for congestion we can measure at this stage without actual routing information. However, localized congested areas will cause overflows during the actual routing process. Initial layer assignment expected to be re-adjusted during global routing to clear the local congested areas.

The proposed assignment algorithm relies on sort list created based on routing capacity of each metal layers as shown in 6. Each metal pair occupies a segment of the sort list. Lower metal layers sit towards the bottom of the sort list. Initially, all nets are randomly inserted into the list. The nets in the list are sorted based their delay violations (DV). If the layer assignment of both net is the same, the net with higher delay violation bubbles-up, if the nets are in different layers, layer usages are first compared:

- If the upper layer's usage is bigger than lower layer's usage, both net are assigned to the lower layer.

- If the layer usage is similar, nets are swapped tentatively to compare the maximum delay violation. The assignment that yields the best delay violation is chosen.

Process is repeated until there is no legal swap left in the list. Fig 5 gives the layer assignment algorithm. Resulting layer as-

```
1   LayerAssginment(S)
2   #S=net list
3   #initialize layer usages
4   U_l = 0,1 ∀ layers
5   U_l = U_l + addBlockages()
6   U_l = U_l + addExistingRouting()
7   do
8       x = S.end()
9       y = x − 1
10      do
11          x = x − 1
12          y = y − 1
13          if x and y are on the same layer
14              if DV_x < DV_y
15                  swap(x,y)
16          elsif x and y are on neighboring layers
17              if U_{l_x} > 1.1 * U_{l_y}
18                  l_x = l_y
19              elsif U_{l_x} < 0.9 * U_{l_y}
20                  l_x = l_y
21              elsif max(DV_{x_{l_y}}, DV_{y_{l_x}}) <
22                      max(DV_x, DV_y)
23                  swap(x,y)
24                  updateLayerUsage(U_{l_x})
25                  updateLayerUsage(U_{l_y})
26      while y ≠ S.begin()
27  while swapped
```

Fig. 5. Performance-Driven Layer Assignment

signment moves critical nets to the upper metal layers. However, this is counter balanced by the requirement that the layer usage distribution be as homogeneously as possible among different layers to obtain less coupling and less congestion when routing is actually performed. It is worth noting that nets in the list are swapped throughout the routing process in concurrent fashion with repeater insertion and RT construction.

Although the worst case complexity of sort algorithm is $O(n^2)$, sort is usually completed around $O(n)$ except during the first iteration where nets are randomly inserted into the list since delay (hence criticality) is unknown initially. In the subsequent iterations, list is mostly sorted and swap operation

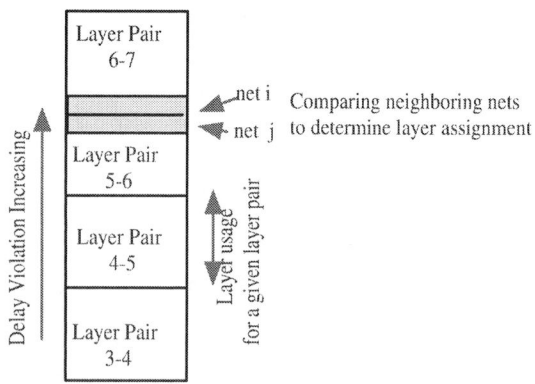

Fig. 6. DV sort list for layer assignment

does not occur frequently.

III. Experimental Results

The proposed algorithm was applied to construct s for a subset of the global nets in a 64-bit microprocessor core. The subset has a total of 13204 unconnected nets. In order to understand the behavior of the proposed RT construction algorithm, different types of multiple-pin nets, i.e. 2-pin nets and 3-pin nets, were extracted from the unrouted nets to form different categories for comparison. There are a total of 9938 2-pin nets and 1339 3-pin nets. Since there are not enough nets with more than 3 pins in the unrouted set. Therefore 1000 randomly generated 4 to 6-pin nets were used for comparison. For the nets in the microprocessor core, the timing constraints were extracted from a static timing database. For randomly generated nets, the timing constraints were generated based on timing of the corresponding minimum length Steiner-tree plus a random perturbation. The routing uses five upper metal layers.

RT construction were performed on each set of routes using both proposed simultaneous routing-tree construction and sequential routing-tree construction.Sequential flow of routing-tree construction is performed as follows: First, nets are assigned to layers randomly. Then, performance-driven Steiner-trees are constructed using the same routing-tree algorithm in this paper. Finally buffers are inserted for nets with a delay violation. Last two steps are repeated in a loop until there is no improvement on any net in the netlist in order to achieve the best possible solution using sequential flow. This is a typical flow in a highly hierarchical custom design.

The results were compared based on the following metrics:

- $\#B$: number of buffers inserted.

- $\#DV$: number of nets with delay violation.

- $maxDV$: maximum delay violation.

- L : total length of the routing-trees.

- CPU : CPU time.

Results in Table I show that the number of delay violations are consistently better in each test case for the proposed algorithm. The number of buffers required for achieving s better delay violations are also consistently less for the proposed algorithm, ranging from 9% less buffers for μP core to around 29% for 6-pin nets. The proposed algorithm also achieved around 1.5% shorter overall routing tree length. For 2-pin nets, the proposed algorithm is reduced to the traditional method since there is only one possible connection for 2-pins nets. The proposed algorithm performs better overall as the pin count increases. This is due to fact that critical pins are optimized early in the routing process and less critical pins are connected to better locations on the existing later in the routing process. The observation has been illustrated earlier in the paper in Figure 2.

Table II shows how the wires are distributed among layers for the μP core test case. The proposed algorithm yields

TABLE I
SEQUENTIAL AND SIMULTANEOUS CONSTRUCTION RESULTS

| | | $\#B$ | $\#DV$ | $maxDV$ | L | CPU |
				(ps)	(m)	(s)
9938/2-pins	Sequential	2339	277	205	22.6	27
	Concurrent	2339	277	205	22.6	40
1339/3-pins	Sequential	1129	184	208	6.5	1.9
	Concurrent	979	181	201	6.4	2.8
1000/4-pins	Sequential	685	63	348	16.5	2.2
	Concurrent	490	53	214	16.5	3.8
1000/5-pins	Sequential	783	53	267	19.7	3.2
	Concurrent	699	40	244	19.5	4.1
1000/6-pins	Sequential	1041	56	470	23.1	7.3
	Concurrent	735	41	309	22.5	8.2
13204 nets	Sequential	3690	535	221	32.1	69
μP core	Concurrent	3335	533	201	31.7	159

more homogeneous layer utilization. This can potentially yield better congestion optimization for the global routing. Since the actual routing has not been performed at this stage, wire lengths are used as a congestion metric instead of overflows on routing tiles.

TABLE II
WIRE DISTRIBUTION

	M3	M4	M5	M6	M7
Sequential	12%	28%	23%	27%	10%
Concurrent	17%	22%	24%	22%	15%

IV. Summary and Conclusions

A new multi-objective based routing-tree construction technique considering performance, power and congestion has been presented. The proposed algorithm incorporates buffer insertion and layer assignment simultaneously into the RT construction process. Experiments on 2-pin to 6-pin nets shows that the proposed algorithm consistently produces shorter tree length, less number of buffers and less number of delay violations. The higher fanout of a net, the better improvement the proposed algorithm can provide.

V. Acknowledgment

The research presented in this paper was funded partially by HP through an HP Graduate Fellowship. The authors would

0-7695-2533-4/06 $20.00 © 2006 IEEE

like to thank HP for their generous support throughout this research

REFERENCES

[1] Ameya R. Agnihotri, Patrick H. Madden, "Congestion reduction in traditional and new routing architectures," *Proceedings of the 13th ACM Great Lakes Symposium on VLSI*, pp. 211 - 214, April 2003.

[2] Hai Zhou; D.F. Wong; I-Min Liu; A. Aziz, "Simultaneous routing and buffer insertion with restrictions on buffer locations," *Computer-Aided Design of Integrated Circuits and Systems, IEEE Transactions on*, pp. 819 - 824, July 2000.

[3] S. Muddu; E. Sarto; M. Hofmann; A. Bashteen, "Repeater and interconnect strategies for high-performance physical designs," *Integrated Circuit Design, 1998. Proceedings. XI Brazilian Symposium on*, pp. 226 - 231 , Oct 1998.

[4] Jiang Hu; C.J. Alpert; S.T. Quay; G. Gandham, "Buffer insertion with adaptive blockage avoidance," *Computer-Aided Design of Integrated Circuits and Systems, IEEE Transactions on*, pp. 492 - 498, April. 2003.

[5] Li-Da Huang; Minghorng Lai; D.F. Wong; Youxin Gao, "Maze routing with buffer insertion under transition time constraints," *Computer-Aided Design of Integrated Circuits and Systems, IEEE Transactions on*, pp. 91 - 95, Jan 2003.

[6] P. Saxena; B. Halpin, "Modeling Repeaters Explicity Within Analytical Placement," *Design Automation Conference, 2003. Proceedings*, pp. 699 - 704, June 2004.

[7] C.D. Cho; S. Raje; M. Sarrafzadeh; M. Sriram; S.M. Kang, "Crosstalk-minimum layer assignment," *Custom Integrated Circuits Conference, 1993., Proceedings of the IEEE 1993*, pp. 29.7.1 - 29.7.4 , May 1993.

[8] M. Hrkic; J. Lillis, "Buffer tree synthesis with consideration of temporal locality, sink polarity requirements, solution cost, congestion, and blockages," *IEEE Transactions on CAD*, pp. 481 - 491, April 2003.

[9] J. Lillis; Chung-Kuan Cheng; T.-T.Y. Lin, "Optimal wire sizing and buffer insertion for low power and a generalized delay model," *Solid-State Circuits, IEEE Journal of*, pp. 437 - 447, March 1996.

[10] A.B. Kahng; Bao Liu, "Q-Tree: a new iterative improvement approach for buffered interconnect optimization," *VLSI, 2003. Proceedings. IEEE Computer Society Annual Symposium on*, pp. 183 - 188, Feb. 2003.

[11] P. Saxena; C.L. Liu, "Optimization of the maximum delay of global interconnects during layer assignment," *Computer-Aided Design of Integrated Circuits and Systems, IEEE Transactions on*, pp. 503 - 515, April 2001.

[12] M.C. Yildiz; P.H. Madden, "Preferred Direction Steiner trees," *Computer-Aided Design of Integrated Circuits and Systems, IEEE Transactions on*, pp. 1368 - 1372, Nov. 2002.

[13] Jaewon Oh; Iksoo Pyo; M. Pedram, "Constructing lower and upper bounded delay s using linear programming," *Design Automation Conference Proceedings 1996, 33rd*, pp. 401 - 404, June 1996.

[14] N. Gopal; D.P. Neikirk; L.T. Pillage, "Evaluating RC-interconnect using moment-matching approximations," *Computer-Aided Design, 1991. ICCAD-91. Digest of Technical Papers, 1991 IEEE International Conference on*, pp. 74 - 77, Nov. 1991.

[15] A.B. Kahng; G. Robins, "A new class of iterative Steiner tree heuristics with good performance," *Computer-Aided Design of Integrated Circuits and Systems, IEEE Transactions on*, pp. 893 - 902, July 1992.

[16] K.D. Boese; A.B. Kahng; B.A. McCoy; G. Robins, "Near-optimal critical sink constructions," *Computer-Aided Design of Integrated Circuits and Systems, IEEE Transactions on*, pp. 1417 - 1436, Dec. 1995.

[17] Jiang Hu; S.S Sapatnekar, "A timing-constrained simultaneous global routing algorithm," *Computer-Aided Design of Integrated Circuits and Systems, IEEE Transactions on*, pp. 1025 - 1036, Sept. 2002.

[18] Van Ginneken, L.P.P.P., "Buffer placement in distributed RC-tree networks for minimal Elmore delay," *Circuits and Systems, 1990., IEEE International Symposium on*, pp. 865 - 868, May 1990.

[19] T. Deguchi; T. Koide; S. Wakabayashi, "Timing-driven hierarchical global routing with wire-sizing and buffer-insertion for VLSI with multi-routing-layer," *Design Automation Conference, 2000. Proceedings of the ASP-DAC 2000*, pp. 99 - 104, Jan. 2000.

[20] C.C.N. Chu; D.F. Wong, "A quadratic programming approach to simultaneous buffer insertion/sizing and wire sizing," *Computer-Aided Design of Integrated Circuits and Systems, IEEE Transactions on*, pp. 787 - 798, June 1999.

[21] J. Griffith; G. Robins; J.S. Salowe; Tongtong Zhang, "Closing the gap: near-optimal Steiner trees in polynomial time," *Computer-Aided Design of Integrated Circuits and Systems, IEEE Transactions on*, pp. 1351 - 1365, Nov. 1994.

[22] www.src.com, "International Technology Roadmap for Semiconductors," *Semiconductor Research Corporation*, 2003.

[23] Jiang Hu; S.S. Sapatnekar, "Algorithms for non-Hanan-based optimization for VLSI interconnect under a higher-order AWE model," *Computer-Aided Design of Integrated Circuits and Systems, IEEE Transactions on*, pp. 446 -458, April 2000.

[24] J.Lillis; C.K.Cheng; T-T.Y.Lin; C-Y.Ho, "New Performance Driven Routing Techniques With Explict Area/Delay Tradeoff and Simultaneous Wire Sizing," *33th Design Automation Conference Proceedings*, pp. 395 -400, June 1996.

[25] D.Wang; E.S.Kuh, "A new General Connectivity Model and Its Applications to Timing-Driven Steiner Tree Routing," *Electronics, Circuits and Systems, 1998 IEEE International Conference on*, pp. 71 -74, Sept 1998.

0-7695-2533-4/06 $20.00 © 2006 IEEE

Clock Gated Static Pulsed Flip-Flop (CGSPFF) in Sub 100 nm Technology

A. S. Seyedi, S. H. Rasouli. A. Amirabadi, and A. Afzali-Kusha

Low-power High-Performance Nanosystems Lab., Elec. and Comp. Eng. Dept., University of Tehran, Tehran, Iran
a.seyedi@ece.ut.ac.ir, hrasouli@ece.ut.ac.ir, a.amirabadi@ece.ut.ac.ir ,afzali@ut.ac.ir

Abstract

In this paper, a new flip flop called Clock Gated Static Pulsed Flip-Flop (CGSPFF) is proposed. The dynamic power consumption in CGSPFF is reduced by avoiding unnecessary input pulse transitions with clock gating. Two transistors in the main block of the flip-flop are eliminated to achieve low leakage power as well. Using the new clock pulse generator leads to a higher operational speed and lower power consumption compared to the previously proposed flip-flops. The results of the simulation show that the PDP of the proposed flip flop is reduced by at least 58.3%.

1. Introduction

The power consumption of circuits and systems is critically important in modern VLSI especially for low power applications and, hence, the power optimization techniques are applied at different levels of the digital design. The design of low power logic is one of the most important tasks to minimize the power consumption of digital circuits. The power consumption includes both a dynamic and static parts. The dynamic part is mainly attributed to charging and discharging of the capacitances. The other part is associated with the leakage current of the transistors. In addition to the dynamic power consumption, the high leakage current in deep sub-micron regimes has become a significant contributor to the power dissipation of CMOS circuits as the CMOS technology scales down [1]. The subthreshold leakage power is expected to become a significant fraction of the total power in the sub-100 nm CMOS technology where reducing the subthreshold leakage power of the circuit is crucial. One of the sources of the power consumption in digital systems is the clock tree which may consume up to 45% of the system power [2]. Hence the reduction in the clock tree power consumption can leads to a considerable reduction in the system power consumption. In addition, among the logic elements, latches and flip-flops (FF) are critical to the performance of the digital system [3]. In addition to the power consumption of these elements, there are other important parameters determining the performance of FFs that should be optimized. Some of the parameters include T_{clk-q} (delay from *clk* to output of DFF) and C_{clk} (load capacitance of the clock). Several flip-flops have been proposed in the literature for improving the speed and/or reducing the power consumption (see, e.g. [4], [5], [6]). Hybrid Latch Flip-Flop (HLFF) has been proposed based on generating explicit transparency window where the transition is allowed [4]. HLFF is a static, single edge-triggered FF where the existence of redundant transition in internal node in HLFF leads to more power consumption. It is similar to latch because it can provide a soft clock edge which allows for a slack passing to minimize the effect of the clock skew on the cycle time [7]. Semi-Dynamic Flip-Flop which has been known as a fast flip-flop is a single edge-triggered FF and faster than HLFF [5]. For this FF, the existence of 1-1 glitch leads to wasting of the power. The number of transistors in this logic is also greater than that of HLFF. Two structures of low-power Dual-edge triggered Static Pulsed Flip-Flops (DESPFF, DSPFF) have been presented in [6]. They are composed of a dual edge pulse generator and a static flip-flop with equal toggling delays. In low switching activities, an additional power consumption due to unnecessary flip flop triggering by the clock tree is observed. Furthermore, it has a rather high leakage power consumption.

In this paper, a static pulsed flip flop with non redundant transition clock is proposed that has a lower dynamic and leakage power consumption and delay compared to those of the previous flip flops. This paper

0-7695-2533-4/06 $20.00 © 2006 IEEE

is organized as follows. In Section 2, the structure of some high performance flip-flops are explained and compared. The proposed static pulsed flip flop is described in Section 3. Subthreshold current is discussed in section 4 while Section 5 contains the results and discussion. The paper ends with the conclusions given in Section 6.

2. Review of some flip fop structures

The structure of Hybrid Latch Flip-flop (HLFF) is shown in Figure 1 [4]. HLFF has very simple structure but the unnecessary internal transition increase the total power consumption of flip-flop. Every time, the input is high a glitch is generated, regardless of the previous state of the output [4]. Furthermore, the transistors in the stack degrade the performance of the logic.

In Figure 2 [5] the circuit diagram of Semi-Dynamic Flip-Flop (SDFF) is illustrated that is faster than HLFF due to a lower number of transistors in the stack. However, the total number of transistors is greater than that of HLFF. Similar to HLFF, unnecessary internal node transitions exist in SDFF. To see the first drawback of this FF more clearly, suppose that the input has the high logic value in two successive clocks. Node Q has high logic value and node X is pre-charged to V_{dd} when clock is low. At the rising edge of clock, there is a short circuit path from Q to Gnd until the node X is discharged. This leads to a 1-1 glitch which causes more power consumption.

Edge-triggered latches can overcome the race problem existing in simple latch structures with creating a narrow sampling window. This idea has been applied to the static flip-flop structures to propose the static pulsed FFs shown in Figure 3 [6]. They are composed of a pulse generator that generates pulse signal at both rising and falling edge of the clock. The pulse signal applied to the static flip-flop creates a narrow transparency window in which data inputs can affect the state of static flip-flop. It is observed that when the input remains the same in two successive clocks, the flip-flop does not need to be triggered because the output should not be changed either.

In the case of DESPFF as well as DSPFF, the transitions in the clock pulse generator occur for these cases too. These transitions are redundant and cause unnecessary power consumption. If the clock to the clock pulse generator can be deactivated when the input remains unchanged in the consecutive clock cycles, then the power can be saved. This may be achieved through designing a non-redundant transition clock pulse generator as will be explained next.

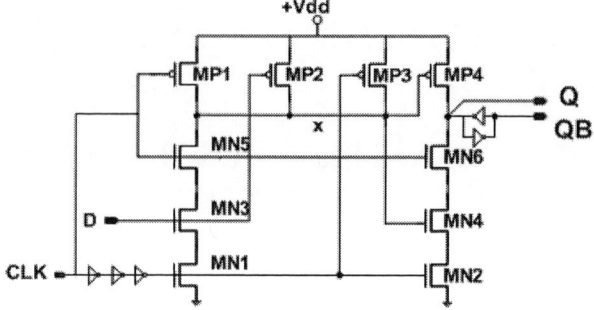

Figure 1. Structure of HLFF [4].

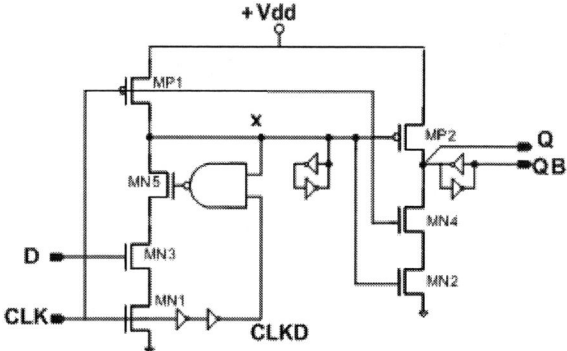

Figure 2. Circuit Diagram of SDFF [5].

3. Proposed static flip flop

To avoid unnecessary transitions in the Static pulsed flip- flops, we propose a Clock Gated Static Pulsed Flip-Flop (CGSPFF) whose circuit is shown in Figure 4. In this flip-flop, the node transitions occur only when the inputs are different in two successive clocks. The operational principle of this circuit is explained here.

Note that CLK2, CLK3, and CLK4 are generated from CLK1. The comparison between the input D and the output Q is performed by an XOR gate (transistors MN8 and MN9). When D = Q, XOR = 0 while XOR = 1 when D ≠ Q. The output of the XOR gate is used as an enable signal for the clock pulse generator of the static flip-flop similar to the works introduced in [8]. The gating logic is a simple NMOS transistor (MN10). When D is equal to Q, the clock signal is disabled via MN10 and by turning on MN11, CLK1 pulls down while CLK3 and CLK4 are low and high, respectively. In this state, MP12 and MN13 are OFF while MP14 and MN15 are ON making PULS low. On other hand, when D and Q are different, MN10 enables the input clock, CLK1 changes to high but CLK3 remains low

0-7695-2533-4/06 $20.00 © 2006 IEEE   374

for a period equal to the delay of two inverters while CLK4 in that time remains high. This enables MP14 and MN15 to pass CLK1 to PULS that indicate a low-to-high transition in PULS. After CLK3 changes its state and becomes high, CLK4 becomes low and, hence, MP14 and MN15 turn OFF while MN13 turn ON. Thus, PULS is pulled down by MN13 generating the narrow pulse required for the operation of flip-flop.

The main block of CGSPFF is similar to DSPFF except for the elimination of MN6 and MN7. Eliminating transistors leads to reduction in leakage power consumption. Based on the simulation results, using simpler structure for the pulse generator and the main block, improves the performance and reduces the leakage power.

4. Subtholod current

The subthreshold or weak inversion conduction current between the source and drain in an MOS transistor occurs when the gate voltage is below threshold voltage (V_{th}) [1]. The weak inversion typically dominates the modern device off-state leakage due to their lower threshold voltage [1]. The weak inversion current can be expressed as [9]

$$I_{ds} = \mu_0 C_{ox} \frac{W}{L}(m-1)(v_T)^2 \times \exp[\frac{(V_g - V_{th})}{mv_T}]$$

$$\times (1 - \exp[\frac{-V_{DS}}{v_T}]) \qquad (1)$$

where $v_T = KT/q$ is the thermal voltage, C_{ox} is the gate oxide capacitance, μ_0 is the zero bias mobility, and m is the subthreshold swing coefficient (also called body effect coefficient).This coefficient is given by

$$m = 1 + \frac{3t_{ox}}{W_{dm}} \qquad (2)$$

where W_{dm} is the maximum depletion layer width and t_{ox} is the gate oxide thickness [1]. As is obvious from (1), if $V_{DS} = 0$, then subthreshold current will be zero. Based on this feature, we present a brief discussion on the leakage behaviors of the previous flip-flop structures. In HLFF (Figure. 1) and SDFF (Figure 2), when the node X is high, a voltage equal to V_{dd} is applied across the first branch in the pull down network (consisting of MN1, MN3 and MN5). On the other hand, when the node X is low then Q (output) will be high and the output pull-down tree sustains a voltage equal to V_{dd}. This high V_{DS} leads to a high voltage drop across the transistors causing a large leakage currents and powers. The situation is even worse in the case of

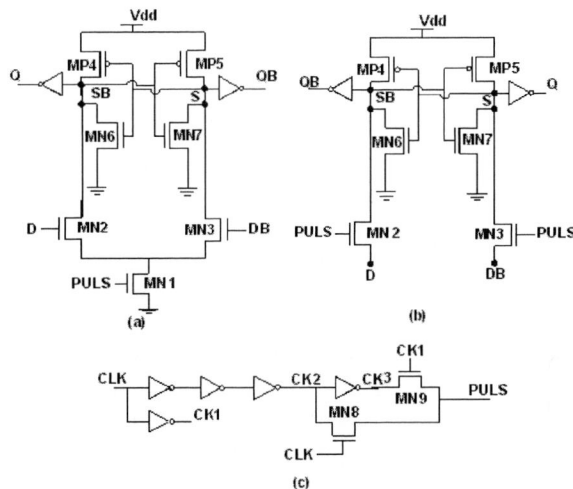

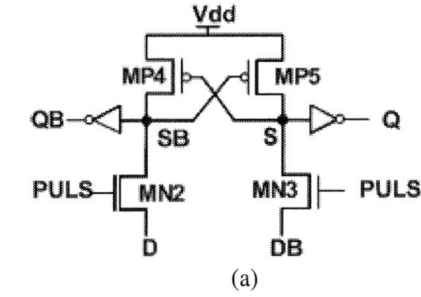

Figure 3. Structures of (a) DESPFF (b) DSPFF (c) Pulse clock generator [6].

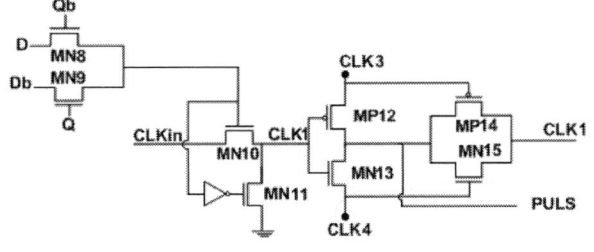

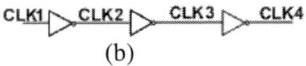

Figure 4. Structure of (a) CGSPFF, (b) Pulse clock generator.

SDFF where this voltage exists across two transistors compared to the case of HLFF where three transistors exist in the output pull down network. Let us explain the situation in DESPFF (Figure 3(a)). Suppose that D is low, and then the voltage of node SB and, hence, V_{DS} of MN6 is equal to V_{dd}. In the case that D is high, the V_{DS} of MN7 will be equal to V_{dd} and, hence, only one transistor has a high V_{DS} drop. As a result, the leakage

0-7695-2533-4/06 $20.00 © 2006 IEEE   375

current will be high. The situation in DSPFF is similar to DESPFF due to existence of MN6 and MN7.

The subthreshold current in CGSPFF is very low which is due to the fact that the V_{DS} of each transistor in the pull-down network will be zero. Assuming D is high (DB is low), node SB will be high (S will be low) and, thus, both the drain and the source of MN2 (MN3) have high (low) logic values leading to an approximately zero V_{DS} for these transistors. Therefore, the last part in (1) as well as subthreshold leakage current will be zero. When D and Q are low, the voltage across the output pull-down tree will be zero giving a zero subthreshold leakage current too. Compared to other flip flops, the subthreshold current in CGSPFF is expected to very low.

5. Results and Discussion

To evaluate the efficiency of the proposed flip-flop, we have simulated the flip-flops discussed in this work by HSPICE using a 70nm CMOS process [10]. The simulations were performed for a V_{dd} of 0.7V at the clock frequency of 100 MHz. The load capacitance of 0.005pF was assumed. Transient behavior of the proposed flip flop is shown in Figure 5.

The results of the simulations are for various flip-flops are given in Table I. As is evident from the Table, CGSPFF has a lower power–delay product (PDP) than those of others. The improvement of the power delay product shown in the last column of the table is computed from

$$\alpha = (1 - \frac{PDP\ (CGSPFF\ )}{PDP\ (DFF\ )}) \times 100 \qquad (3)$$

The results show that the power-delay product of CGSPFF is best among all the flip-flops discussed here.

The subthreshold leakage power of different flip-flops is presented in Table II. Again, CGSPFF has the least leakage power among all the flip-flops discussed in this paper. Figure 6 shows the power consumption as a function of the data switching activity for DESPFF and CGSPFF. The proposed flip flop has a lower power consumption for all the switching activities.

6. Summary and conclusion

A flip-flop called Clock Gated Static Pulsed Flip-Flop (CGSPFF) is proposed that has a better performance compared to previous flip-flops. The unnecessary internal node transition existing in the previous static pulsed flip-flops in this logic is avoided using clock gating which reduces the clock tree power. Furthermore the leakage power in CGSPFF is very low

compared to previous flip-flops. The simulation results revealed that the improvements in the PDP of CGSPFF compared to other flip-flops are between 58% and 69%.

Table 1. Comparing Various Parameters of DFFs.

Flip-Flop	No. of Tr.	No. of Clked Tr.	Clk-Q (ps)	Power (µW)	PDP (fj)	α
HLFF	20	4	132	1.87	0.247	68.4%
SDFF	23	5	124	1.9	0.236	66.7%
DSPFF	22	4	145.7	1.3	0.189	58.7%
DESPFF	23	4	151.8	1.23	0.187	58.3%
CGSPFF	24	1	80.1	0.982	0.078	-

Table 2. The Leakage Power of Different Flip-Flops.

Flip-Flop	SDFF	HLFF	DESPFF	DSPFF	CGSPFF
Leakage Power (nW)	86	49	41.1	43.8	19.0

7. References

[1] Roy, K.; Mukhopadhyay, S.; Mahmoodi-Meimand, H."Leakage current mechanisms and leakage reduction techniques in deep-submicrometer CMOS circuits," in Proc. IEEE Volume 91, Issue 2, 2003 PP. 305 – 327, Feb. 2003.

[2] G. Palumbo, F. Pappalardo, and S. Sannella, "Evaluation on power reduction applying gated clock approaches", ISCAS 2002, Page(s):IV-85 - IV-88 vol.4.

[3] M. Hamada, T. Terazawa, T. Higashi, S. Kitabayashi, S. Mita, Y. Watanabe, M. Ashino, H. Hara, T. Kuroda, "Flip-Flop Selection Technique for Power-Delay Trade-off," in IEEE Int. Solid-State Circuits Conf., pp. 270–271, Feb. 1999.

[4] Nedovic, N.; Oklobdzija, V.G, "Hybrid latch flip-flop with improved power efficiency," IEEE Symp. on Integrated Circuits and Systems Design, pp. 211-215, sep. 2000.

[5] F. Klass, "Semi-Dynamic and Dynamic Flip-Flops with Embedded Logic," in Symp. VLSI Circuits Dig. Tech. Papers, pp. 108–109, June 1998.

[6] Aliakbar Ghadiri and Hamid Mahmoodi, "Dual-Edge Triggered Static Pulsed Flip-Flops",

Proceedings of the 18th International Conference on VLSI Design, (VLSID'05).

[7] E. Partovi, R. Burd, U. Salim, F. Weber, L. DiGregorio, and D. Draper, "Flow-Through Latch and Edge-Triggered Flip-Flop Hybrid Elements," in IEEE Int. Solid-State Circuits Conf., Feb. 1996.

[8] Y. Xia and A.E.A Almaini, "Differential CMOS edge-triggered flip-flop with clock-gating", ELECTRONICS LETTERS 3rd January 2002, Vol. 38 No. 1.

[9] Y. Taur and T. H. Ning, "Fundamentals of Modern VLSI Devices. New York," Cambridge Univ. Press, 1998, ch. 3, pp. 120–128.

[10] Berkeley Predictive Technology Model, http://www-device.eec.

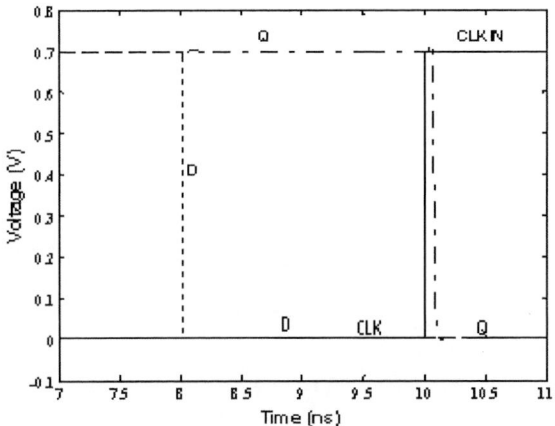

Figure 5. Transient behavior of the proposed FF.

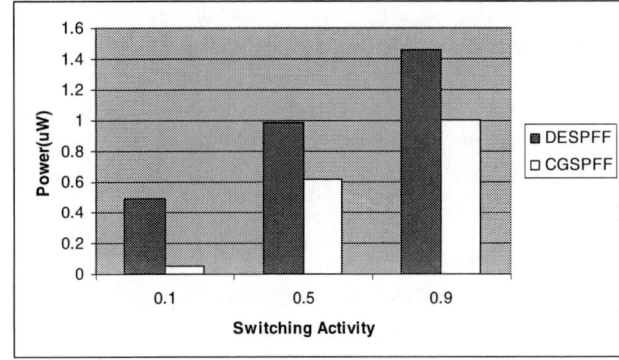

Figure 6. Power versus data switching activity at 100 MHz.

Performance and Power Analysis of Globally Asynchronous Locally Synchronous Multi-Processor Systems

Zhiyi Yu, Bevan M. Baas

ECE department, University of California, Davis

{zhyyu, bbaas}@ece.ucdavis.edu

Abstract

This paper investigates the performance and power dissipation of Globally Asynchronous Locally Synchronous (GALS) multi-processor systems. We show that communication loops are a source of significant throughput degradation in communications links and that there is no degradation whatsoever under certain conditions for one-way links, and that it is possible to design GALS multi-processors without this performance penalty. Independent clock domains and unbalanced computation in the GALS multi-processor allow scaling of the clock frequency and supply voltage to achieve high energy efficiency. The synchronization overhead between independent clock domains results in a less than 1% performance reduction compared to a globally synchronous system over a number of DSP and numerical applications. Clock and voltage scaling can achieve an approximately 40% power savings with no reduction of performance. These results compare favorably with the 25% power savings and more than 10% performance reduction reported for GALS uniprocessors.

1. Introduction

Clocking circuits have become increasingly difficult to design with larger chip sizes, higher clock rates, larger relative wire delays, and larger parameter variations [1]. Additionally, high speed global clocks consume a significant portion of power budgets. The Globally Asynchronous Locally Synchronous (GALS) clocking style separates processing blocks such that each part is clocked by an independent clock domain. The approach is a promising strategy to address these design challenges. Previous GALS work includes performance and power analysis in an ASIC system [2] and a uniprocessor system [3, 4, 5]; and the clock domain analysis for a clustered array processor [6].

Modern deep submicron fabrication technologies are not able to sustain historical increases in clock frequencies, but do enable very high levels of integration such as chips with

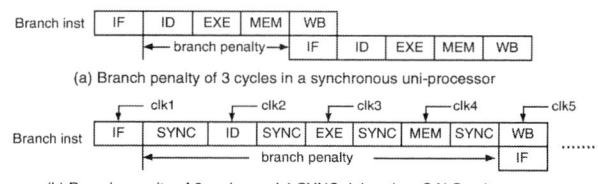

Figure 1. Pipeline control hazard penalties of a 5 stage synchronous uniprocessor and a 5 stage GALS uniprocessor

more than a billion transistors [7]. Multiple-processor chips now show a promising future [7, 8]. In this context, array processors—which combine multiple processors in an array—are increasingly attractive. An array processor can also provide high energy efficiency since parallel computing improves performance and may allow the clock frequency and voltage to be reduced.

1.1. Performance reduction and energy efficiency of the GALS uniprocessor

In addition to multiple clock generators, GALS systems require synchronization circuits between clock domains to reliably transfer data. Small clock domains normally simplify clock trees, but unmatched clocks between different domains and synchronization circuitry introduce communication delays.

We define a *GALS uniprocessor* as an architecture where the processor itself is partitioned into multiple clock domains. The GALS overhead increases the delay between pipeline stages and reduces processor performance. Figure 1 shows the control hazard of a simple DLX RISC processor [9]. The lower subplot shows a GALS uniprocessor where each pipeline stage is in its own clock domain. During the cycle with the taken branch, the synchronous processor has a 3-cycle control hazard, while the GALS system has a $3 + 4 \times SYNC$ cycle penalty, significantly reducing system performance. Reported performance reductions of the GALS uniprocessor include 10% [3], 7%–11% [4] and 4% [5].

0-7695-2533-4/06 $20.00 © 2006 IEEE

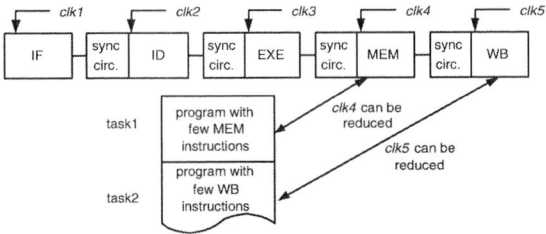

Figure 2. Clock scaling in a GALS uniprocessor

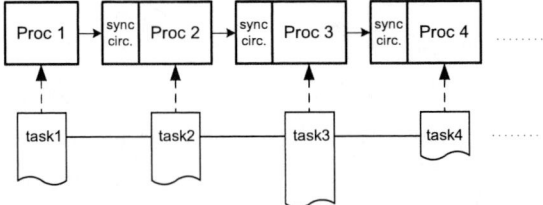

Figure 3. Mapping multiple tasks to a GALS array processor

GALS uniprocessors normally control their independent clock domains adaptively to achieve high energy efficiency, by reducing the clock frequency and voltage of modules that are less heavily used. Figure 2 illustrates this concept where the frequency of the MEM module clock *clk4* is reduced when executing task1 since it has few MEM instructions. Then in task2, the frequency of *clk5* is reduced. Unfortunately, reducing the clock of some modules reduces performance. The *static scaling* method sets the frequency before execution and reduces the energy by approximately 16% with an approximately 18% reduction in performance [3]. The *dynamic scaling* method changes the frequency at runtime and achieves 20%–25% energy savings along with a 10%–15% performance reduction [4, 5].

1.2. The GALS array processor

While the GALS uniprocessor puts synchronization logic between pipeline stages, the array processor puts synchronization circuits between different processors as shown in Fig. 3.

Clock frequency selection in GALS uniprocessors is based upon the utilization probability of each function module. Similarly, GALS array processors base their frequency selections on the activity of its computational tasks, as shown in Fig. 3.

2. A GALS Array Processor

Two array processor designs were implemented for comparison: one is fully synchronous and the other is a GALS array processor. Both contain multiple uniform simple processing units, as shown in Fig. 4a. Both processors contain

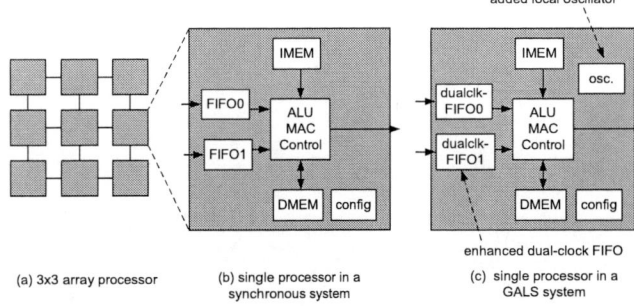

Figure 4. Two array processors: one using synchronous processors and the other using GALS array processors

small instruction and data memories, two 32-word FIFOs, a 16-bit datapath, and execute 32-bit instructions. Each processor communicates with its four neighboring processors.

Figure 4b shows a single synchronous array processor. It utilizes simple fully synchronous FIFOs. Figure 4c shows a single GALS array processor. In order to support the GALS methodology, a frequency configurable oscillator is added as the processor's local clock, and the synchronous FIFO is enhanced with features to allow it to perform as a dual-clock FIFO [10]. The dual-clock FIFO writes and reads data in independent clock domains and reliably transfers data across the domains. For increased characterization capability, a configurable number of synchronization registers are inserted at the clock domain interface to avoid metastability. The local oscillator occupies approximately 0.5% of the processor's area. The area overhead of the dual-clock FIFO is also around 0.5%. In addition, the GALS system has a simplified clock tree.

A 6×6 GALS array processor chip has been implemented [11]. The synchronous array processor is emulated by special configurations in the GALS array processing chip.

3. Performance Analysis of the GALS Array Processor

Several applications are mapped and simulated onto the RTL model of both synchronous and GALS array processors to investigate their performance. The synchronous system uses a global clock and has no synchronization registers. The GALS system uses a local oscillator and two synchronization registers.

The mapped applications we consider include: an 8-point DCT using 2 processors, an 8×8 DCT using 4 processors, a zig-zag transform using 2 processors, a merge sort using 8 processors, a bubble sort using 8 processors, a 5×5 matrix multiplier using 6 processors, a 64-point complex FFT using 8 processors, a JPEG encoder using 9 pro-

0-7695-2533-4/06 $20.00 © 2006 IEEE   379

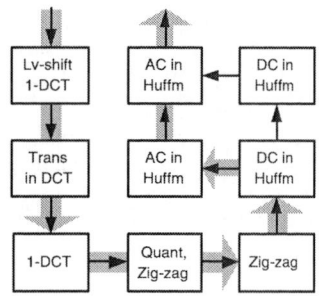

Figure 5. JPEG encoder core using 3x3 processors

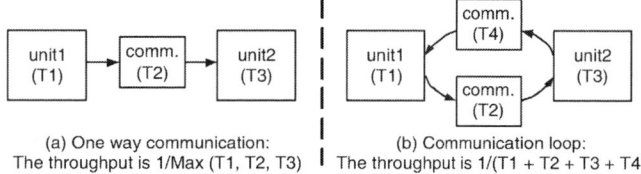

(a) One way communication:
The throughput is 1/Max (T1, T2, T3)

(b) Communication loop:
The throughput is 1/(T1 + T2 + T3 + T4)

Figure 6. System throughput in a) one way communication path, and b) communication loop path

cessors, and an IEEE 802.11g/802.11a wireless LAN transmitter using 22 processors [12].

The JPEG encoder core composition as shown in Fig. 5 is one application example. The main functional blocks include a level shifter, an 8×8 DCT, quantization, zig-zag reordering, and a Huffman encoder. The 8×8 DCT is processed using two 1-dimensional DCTs. The second DCT data transpose is avoided by changing the quantization table order and zig-zag order. Four processors are used for the Huffman encoding.

3.1. Comparison of application performance

The first two lines of Table 1 show the computation time in clock cycles when mapping these applications onto the synchronous and GALS array processors. The third line lists the relative performance penalty of the GALS array processor. The performance of the GALS system is nearly the same as the synchronous system with an average of less than 1% performance reduction, which is much smaller than the 10% performance reduction of a GALS uniprocessor [3, 4].

3.2. Performance effects of GALS clocking

3.2.1 Importance of the communication loop delay

The performance penalty of a GALS system comes from its increased communication delay. More specifically, simple *one way communication* does not affect system performance, but *communication loops*—in which two units wait for information from each other—can degrade performance.

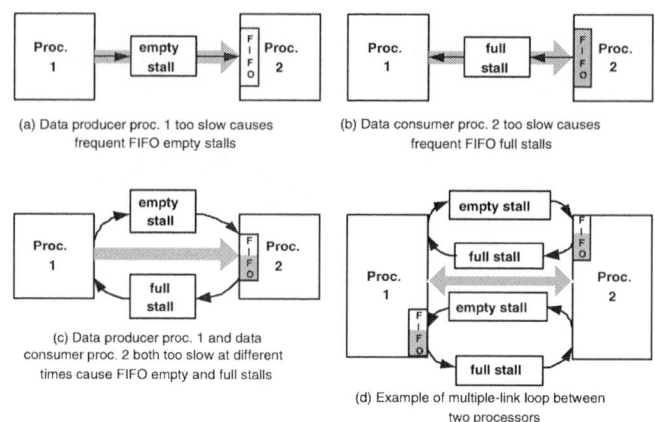

(a) Data producer proc. 1 too slow causes frequent FIFO empty stalls

(b) Data consumer proc. 2 too slow causes frequent FIFO full stalls

(c) Data producer proc. 1 and data consumer proc. 2 both too slow at different times cause FIFO empty and full stalls

(d) Example of multiple-link loop between two processors

Figure 7. Examples of stalls and stall loops in a GALS array processor

In a *one way communication path* as shown in Fig. 6a, the system throughput is dependent on the slowest unit and is not related to the communication—assuming communication is not the slowest unit, which is true in our case. However, throughput is significantly impacted when the communication has feedback and generates a loop, as shown in Fig. 6b. If unit 1 and unit 2 both need to wait for information from each other, the throughout will be dependent on the sum of unit execution time and communication time. Then the communication time affects the performance of both synchronous and GALS systems, but the GALS system has a larger performance penalty due to its larger communication time.

A similar conclusion can be drawn from the GALS uniprocessor. In instructions without pipeline hazards, the GALS uniprocessor has the same performance as the synchronous uniprocessor since it has only *one way communication*. However, during instructions such as taken branches (where the new *PC* needs the feedback from the execution result), a *communication loop* is formed. Thus the GALS style brings a performance penalty in some cases. Other pipeline hazards also generate similar communication loops.

3.2.2 FIFO stall loops in GALS array processors generate communication loops

In both synchronous array processors and GALS array processors, *FIFO stalls* effect performance since one processor must wait for the information from another processor when they enter a stall status. GALS systems have a larger performance penalty than synchronous systems since they have larger latencies for FIFO stall information.

A *FIFO-empty stall* occurs when a processor reads an empty FIFO and must wait (stall) until data is available, as illustrated in Fig. 7a. A *FIFO-full stall* occurs when a processor writes a full FIFO and must wait until there is

0-7695-2533-4/06 $20.00 © 2006 IEEE   380

Table 1. Clock cycles (1/*throughput*) of several applications mapped onto a synchronous array processor and a GALS array processor, with 32-word FIFOs

	8-pt DCT	8×8 DCT	zig-zag	mergesort	bubblesort	matrix	64 FFT	JPEG	802.11
Synchronous array	41	498	168	254	444	817.5	11439	1439	87857
GALS array	41	505	168	254	444	819	11710	1443	88989
GALS Perf. reduction	0%	1.4%	0%	0%	0%	0.1%	2.3%	0.3%	1.3%

writable space, as shown in Fig. 7b.

Pure FIFO-full stalls or FIFO-empty stalls alone as in Fig. 7a,b generate one way communication and have no effect on system throughput. The situation shown in Fig. 7c is one example of the FIFO stall communication loop. Processor 2 has FIFO-empty stalls and must wait for processor 1, and processor 1 has FIFO-full stalls and must wait for processor 2. When FIFO-full stalls and FIFO-empty stalls both exist (obviously at different times) in a link, they produce a communication loop and this reduces system performance—for both synchronous and GALS systems, albeit with less of a penalty for a synchronous system. Another situation which generates a FIFO stall communication loop is shown in Fig. 7d. In this case, processor 1 and processor 2 send data to each other, and each processor has both FIFO full stalls and FIFO empty stalls. An example of this case exists in our FFT application where some processors are used as data storing coprocessors and they send and receive data to computation processors.

Simulation results in Table 1 show that the GALS array processor has nearly the same performance as the synchronous array processor. This performance reduction is much less compared to the GALS uniprocessor's reduction. This implies that the chance of the FIFO stall loop in an array processor for our applications is much smaller than the probability of a pipeline hazard in a uniprocessor. These results match well with our model. In Table 1, the 8-pt DCT, zig-zag, mergesort and bubblesort have no GALS performance penalties since they have only one-way FIFO stalls. The 8×8 DCT and JPEG have situations like Fig. 7c and have an approximately 1% performance penalty. The 64-point FFT and 802.11g/a applications have situations like Fig. 7c and Fig. 7d and their performance penalty is slightly larger.

3.3. FIFO size affects synchronous and GALS system performance

FIFO stalls are highly dependent on the FIFO size. When the FIFO is large enough, there will be no FIFO-full stalls, and at the same time, the number of FIFO-empty stalls can be greatly reduced. With a sufficiently large FIFO, the communication loop in Fig. 7c will be broken due to the missing FIFO-full stalls. The likelihood of the communication loop in Fig. 7d will also be highly reduced, but is still pos-

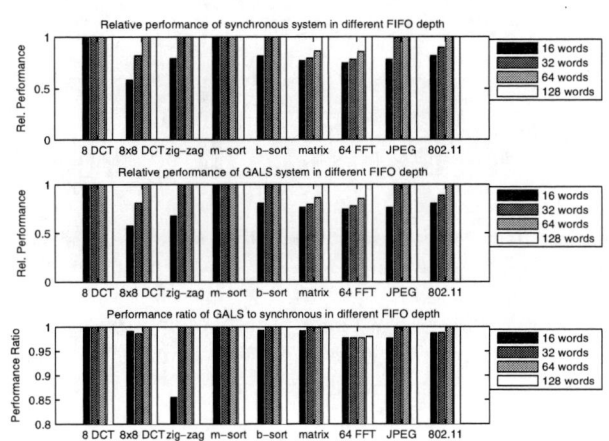

Figure 8. Performance of synchronous and GALS array processors with different FIFO sizes

sible since FIFO-empty stalls alone can still form a loop. Reduced FIFO stall loops increase the system performance and reduce the GALS performance penalty.

The top and middle subplots of Fig. 8 show the performance of the synchronous and GALS systems with different FIFO sizes, respectively. Whether using a synchronous or GALS style, increasing the FIFO size will increase system performance because of reduced FIFO stall loops. Also, a threshold FIFO size exists above which the performance won't change. The threshold is the point when the FIFO-full stall becomes non-existent due to having a large enough FIFO size, and increasing the FIFO size further gives no benefit. The threshold is dependent on the application as well as the mapping method. In our case, the thresholds for the 8×8 DCT and 802.11g/a are 64 words; JPEG and bubble sort are 32 words; the 8-pt DCT and merge sort are less than or equal to 16 words.

The bottom subplot of Fig. 8 shows the performance ratio of the GALS system to the synchronous system. The ratio normally stays at a high level larger than 95%. When increasing the FIFO size, the ratio tends to increase due to fewer communication loops. The ratio normally reaches 1.0 at the threshold, which means the FIFO communication loops are all broken and the GALS system has the same performance as the synchronous system. The exception in the examples is the FFT in which the GALS system always

0-7695-2533-4/06 $20.00 © 2006 IEEE

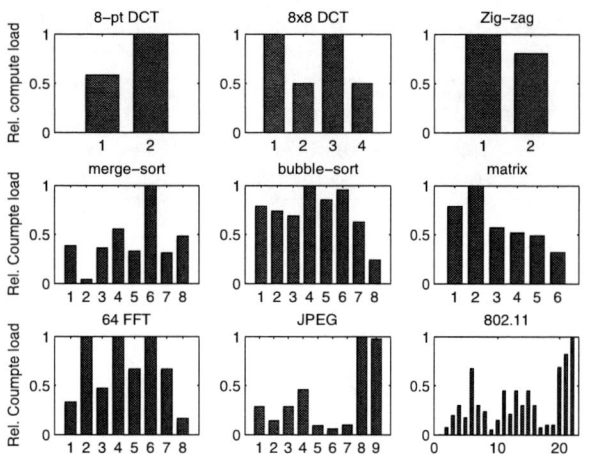

Figure 9. Relative computational load of different processors in nine applications illustrating unbalanced loads

has a noticeable performance penalty of approximately 2%. The reason can be seen from Fig. 7d where the FIFO-empty stall alone can generate the stall loop without a FIFO-full stall.

4. Energy Efficiency Analysis of Independent Clock Frequency Scaling

Several researchers have reported the high power efficiency of the GALS style due to its simplified clock tree [2, 6]. We focus on another power consumption benefit of the GALS system due to the flexibility of clock frequency and supply voltage scaling. The work presented here addresses static scaling methods.

The Synchroscalar [13] system utilizes processors with rationally-related clocks. While the approach avoids the extra hardware of asynchronous communication, its clocks are not as flexible as GALS clocks.

4.1. Unbalanced processor computation loads give power saving potential

Traditional parallel programming methods normally seek to balance computational loads in different processors. On the other hand, when using adaptive clock methods, unbalanced computational loads are no longer a problem, and in fact give an opportunity to reduce the clock frequency and supply voltage of some processors to achieve further power savings without degrading system performance [14]. Releasing the constraint of a balanced computational load enables the designer to explore wider variations in other parameters such as program size, local data memory size and communication methods. Figure 9 shows the unbalanced computational load among processors when mapping our applications onto an array processor.

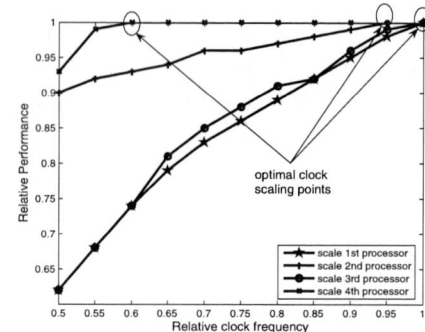

Figure 10. Throughput changes with statically configured processor clocks for an 8×8 DCT

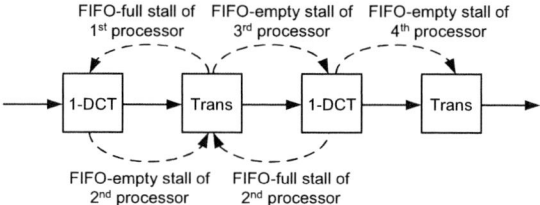

Figure 11. Relationship of processors in an 8×8 DCT

4.2. Computational load and position affect optimal clock frequency

The optimal processor clock frequency in a GALS array processor depends strongly on its computational load, and also depends on its position and relationship with respect to other processors.

Figure 10 shows the system throughput versus the clock frequencies of four processors in the 8×8 DCT. The computational load of the four processors is 408, 204, 408 and 204 clock cycles respectively. The throughput changes with the scaling of the 2nd and 4th processor much more slowly than the scaling of the 1st and 3rd processors. This illustrates the clear point that a processor with a light computational load is more likely to maintain its performance with a reduced clock frequency. Somewhat counterintuitively, however, the 2nd and 4th processors have the same light computational load, but the throughput changes with the 4th processor scaling much more slowly than the 2nd processor's scaling. Minimal power consumption is achieved with full throughput when the relative clock frequencies are 100%, 95%, 100%, and 57% of full speed respectively.

The reason for the different behavior of the 2nd and 4th processors comes from their different positions and FIFO stall styles as shown in Fig. 11. The 2nd processor has both FIFO-full stalls and FIFO-empty stalls, while the 4th processor has only FIFO-empty stalls.

0-7695-2533-4/06 $20.00 © 2006 IEEE

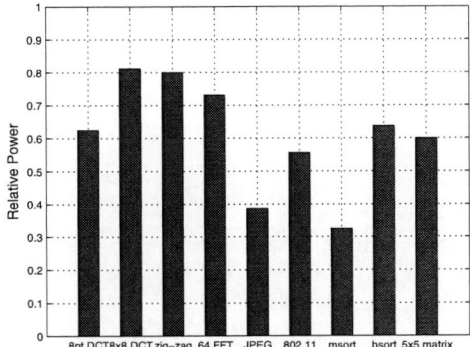

Figure 12. Estimated relative power of the GALS system with static clock/voltage scaling compared to a synchronous system

4.3. Estimating power reduction

Reducing the clock frequency allows for a reduction in voltage to obtain further power savings. The relationship between clock frequency and voltage, and power and voltage can be modeled by a simple linear relation and a square relation respectively. However, these relationships become much more complex in the deep submicron regime because of other parameters such as leakage power. For this analysis, we use a model derived from measured data from a 0.18 μm technology [15] to estimate power consumption.

Using the optimal clock frequency for each processor as described in Sec. 4.2, and the power-frequency-voltage model, we estimate the relative power consumption of the GALS array processor compared to the synchronous array processor after using static clock frequency and supply voltage scaling for several applications. The result is shown in Fig. 12. The GALS system achieves an average power savings of approximately 40% without affecting the performance. This power savings is much higher than the GALS uniprocessor which was reported to save approximately 25% power when operating with a performance reduction of more than 10% [3, 4, 5].

5. Summary and acknowledgments

It has been shown that communication loops are a source of significant throughput reduction in communication links and that there is no reduction under certain conditions for one-way links. A key advantage of the GALS array processor compared to the GALS uniprocessor is that communication loops occur far less frequently and therefore the performance penalty is significantly lower. The proposed GALS array processor has a throughput penalty of less than 1% with a power dissipation reduction of 40% over a variety of DSP and numerical workloads. These results compare well with a reported 25% power reduction and a 10%

performance reduction with GALS uniprocessors.

Data presented in this paper are based on the fabricated GALS processor and its synchronous mode of operation [11]. While results will certainly vary over different applications and specific architectures, we expect the general conclusion that multi-processor GALS systems have smaller performance reductions and larger power reductions, should still hold.

The authors thank E. Work, T. Mohsenin, other VCL processor co-designers, R. Krishnamurthy, M. Anders, S. Mathew; and support from Intel, UC MICRO, NSF Grant No. 0430090, and a UCD Faculty Research Grant.

References

[1] S. Borkar et al., "Parameter variations and impact on circuits and microarchitecture," in *DAC*, 2003, pp. 338–342.

[2] T. Meincke et al., "Globally asynchronous locally synchronous architecture for large high-performance asics," in *ISCAS*, May 1999, pp. 512–515.

[3] A. Iyer et al., "Power and performance evaluation of globally asynchronous locally synchronous processors," in *ISCA*.

[4] E. Talpes and D. Marculescu, "A critical analysis of application-adaptive multiple clock processor," in *ISLPED*.

[5] G. Semeraro et al., "Energy-efficient processor design using multiple clock domains with dynamic voltage and frequency scaling," in *HPCA*, 2002, pp. 29–40.

[6] A. Upadhyay et al., "Optimal partitioning of globally asynchronous locally synchronous processor arrays," in *GLSVLSI*, 2004, pp. 26–28.

[7] S. Naffziger et al., "The implementation of a 2-core multi-threaded Itanium family processor," in *ISSCC*, 2005.

[8] M. B. Taylor et al., "The raw microprocessor: A computational fabric for software circuits and general purpose programs," *IEEE Micro*, pp. 25–35, 2002.

[9] D. A. Patterson et al., *Computer Architecture – A Quantitative Approach*, Morgan Kaufmann, second edition, 1999.

[10] R. Apperson, "A dual-clock FIFO for the reliable transfer of high-throughput data between unrelated clock domains," M.S. thesis, UC Davis, 2004.

[11] Z. Yu et al., "An asynchronous array of simple processors for DSP applications," in *ISSCC*, Feb. 2006.

[12] M. Meeuwsen et al., "A full-rate software implementation of an IEEE 802.11a compliant digital baseband transmitter," in *SiPS*, 2004, pp. 297–301.

[13] J. Oliver et al., "Synchroscalar: A multiple clock domain, power-aware, tile-based embedded processor," in *ISCA*.

[14] T. Njolstad et al., "A socket interface for gals using locally dynamic voltage scaling for rate-adaptive energy saving," in *ASIC/SOC*, Sept. 2001, pp. 110–116.

[15] K. Nowka et al., "A 32-bit powerpc system-on-a-chip with support for dynamic voltage scaling and dynamic frequency scaling," *JSSC*, pp. 1441–1447, Nov. 2002.

Implementing Register Files for High-Performance Microprocessors in a Die-Stacked (3D) Technology

Kiran Puttaswamy[†] and Gabriel H. Loh[‡]
Georgia Institute of Technology
School of Electrical and Computer Engineering[†]
College of Computing[‡]
Email: kiranp@ece.gatech.edu, loh@cc.gatech.edu

Abstract— **3D integration is a new technology that will greatly increase transistor density while providing faster on-chip communication. 3D integration stacks multiple die connected with a very high-density and low-latency interface which provides increased device density and the ability to place and route in the third dimension. While past studies have explored 3D integrated on-chip caches, this research explores the implementation of register files, which have very different capacity and bandwidth requirements. Partitioning the register file across multiple die reduces the lengths of many critical wires, which provides both latency and energy benefits. In particular, a 3D implementation of 256-entry physical register file in a two-die stack achieves a 24.1% latency improvement with a simultaneous energy reduction of 58.5%, while a four-die version achieves a 36.0% latency improvement with a 58.2% energy reduction. Our results demonstrate that 3D integration is a promising approach for improving both the performance and power of wire-dominated circuits.**

I. INTRODUCTION

The semiconductor industry faces an increasing number of challenges and obstacles that must be overcome to keep pace with Moore's Law [1] and industry projections [2]. Some challenges include poor scaling of wire RC delays [3–6], increasing power consumption [7–10], limits in manufacturing techniques, and others. Three-dimensional stacked-die integration has the potential to address many of these problems.

The current rate of transistor density increase is a doubling in the number of devices approximately every eighteen months. Unfortunately, the performance of many high-end microprocessors are increasingly limited not by computation delay, but rather by communication delay [5, 6]. Although transistor size and speed continue to improve, the relative speed of wires has not improved at the same rate.

3D integration can greatly reduce the impact of wire delays. Two functional unit blocks connected by a long global route in a planar implementation can instead be vertically stacked to drastically reduce the communication distance by routing in the third dimension. Wire-dominated functional unit blocks can be folded on top of themselves to reduce the effects of intra-block wiring. Reducing the amount of wire can also have a significant impact on power consumption as interconnect power is already estimated to consume about one half of a chip's power. Since many of a high-performance processor's critical circuits are dominated by wire delay [11], 3D integra-

tion may have significant power and performance benefits for future microprocessors [12].

While previous research has studied the 3D implementation of on-chip cache structures [13–15], this study focuses on the physical register file. Although both the caches and the register file are SRAM structures, the register file presents different challenges and opportunities for optimization in a 3D organization. In particular, the high port-count of physical register files found in modern superscalar processors exacerbates the wire delay problem.

Products using 3D integration are already available in the embedded market, including SRAM stacked on a small microprocessor [16] and 3D-stacked DRAM [17]. However, the focus of this research is on using 3D for implementing high-performance microprocessors, which have different targets and constraints for performance, clock speed, area and power.

The rest of the paper is organized as follows. Section II provides a short background on 3D integration technology. Section III describes register files implemented in a conventional 2D (planar) process. Section IV explains our designs for register files implemented in 3D. Section V details our experimental methodology. Section VI presents the results and analysis of our 3D register files. Section VII summarizes our contribution.

II. 3D DIE-STACKING TECHNOLOGY

There are currently several proposed methods for vertically integrating multiple die. All of these must address how the die are bonded, how the die are aligned, how the die are electrically connected, and how the die are thinned. Cu-Cu bonding takes two die and deposits stub vias on the top metal layer of each die, and then aligns the two die [18, 19]. Figure 1(a) shows a "face-to-face" organization. Under proper heat and pressure (thermocompression), the via stubs fuse together providing both the die-to-die interconnects as well as the physical mechanism for holding the die together. After the bonding, the bottom die is thinned with chemical-mechanical polishing (CMP) down to only ~10μm, which allows low impedance backside vias to be etched through providing I/O and power/ground connections.

When stacking two separate die, there are three possible organizations. Figure 1(a) already showed a face-to-face (F2F)

0-7695-2533-4/06 $20.00 © 2006 IEEE

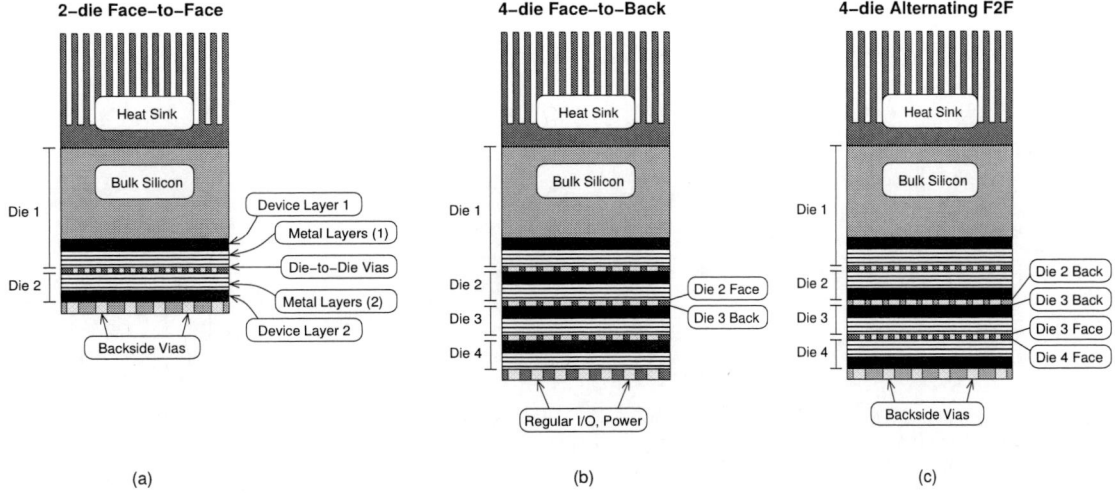

Fig. 1.   (a) A 2-die face-to-face (F2F) 3D structure, (b) a 4-die face-to-back 3D structure, and (c) a 4-die alternating F2F 3D structure.

organization. The other two topologies are face-to-back (F2B) and back-to-back (B2B). In a F2F organization, the vias are simply masked and deposited on top of the top metal layer using conventional metal deposition technologies. Therefore, the vias can be as dense as regular on-die interconnects, and the realizable pitch is only limited by the accuracy of aligning the two die. If vias must be etched through the back side of a die, then the pitch will be less dense due to the need to etch through $\sim$5-10μm of the backside silicon. The advantage of a F2B organization is that an arbitrary number of die can be stacked; Figure 1(b) shows a four-die F2B stack. With a F2F topology, after stacking two die, the only available surfaces are the backsides, and therefore backside vias cannot be avoided. Figure 1(c) shows a four-die stack that combines two F2F 2-die stacks with a B2B interface between the pairs of die.

The die-to-die (d2d) vias are perhaps the most critical 3D design parameter from the perspective of implementing circuits in 3D. The pitch and latency of the d2d vias will dictate the granularity at which a circuit can be partitioned across the different die. Current academic implementations of 3D processes support d2d via pitches from 3μm to 10μm depending on the technology [20]. The embedded industry is already using 2.4μm d2d vias in 3D structures [21].

The physical characteristics of the d2d vias determine the signal propagation delay between the two die. Fortunately, the thinning of the die reduces the distance that a d2d via must cross to connect the two die. As mentioned earlier, the individual die are thinned to only $\sim$10μm, and in a F2F organization the d2d vias only need to cross the distance separating the two top metal layers. Depending on the technology, the d2d via height may be $<$5μm to $\sim$20μm [20]. While the B2B vias are larger, the embedded community is already using B2B vias with a 6μm pitch, and has reported that the next technology generation will support $<$4μm B2B vias [21].

We believe that the F2F organization is more desirable. The F2F vias are denser than F2B vias, and if more than two die are

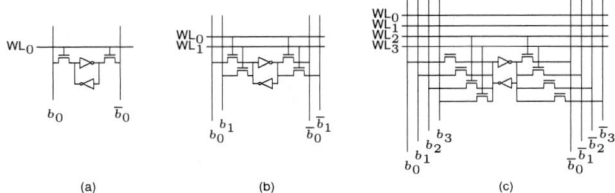

Fig. 2.   A schematic view demonstrating the quadratic area increase of a simple SRAM bitcell as the number of ports increases from (a) one to (b) two to (c) four ports.

stacked together, the density of the B2B vias are no worse than the F2B vias. Furthermore, the F2F vias do not pass through the device layer, and therefore do not impose any floorplanning and placement constraints on the underlying devices.

III. PLANAR (2D) REGISTER FILES

Register files are structurally similar to on-chip caches in that both consist of regular arrays of 6T SRAM cells. However, on-chip caches have much larger capacity and require both tag and data arrays. Register files typically have lower capacity requirements and do not have a tag array. However, modern superscalar processors capable of issuing many instructions per cycle require a large number of read and write ports from the register file, and the size of an SRAM cell increases dramatically with increasing port requirements. Figure 2 shows SRAM cells with (a) one port, (b) two ports, and (c) four ports. While Figure 2 illustrates a simplistic SRAM cell design (for example read-port isolation transistors have been omitted for clarity), it is still useful to demonstrate the quadratic increase in area with respect to the port count. The area explosion forces all of the wordlines and bitlines to also increase in length that in turn increases both access latency and power consumption.

The physical register file is a critical component of

0-7695-2533-4/06 $20.00 © 2006 IEEE

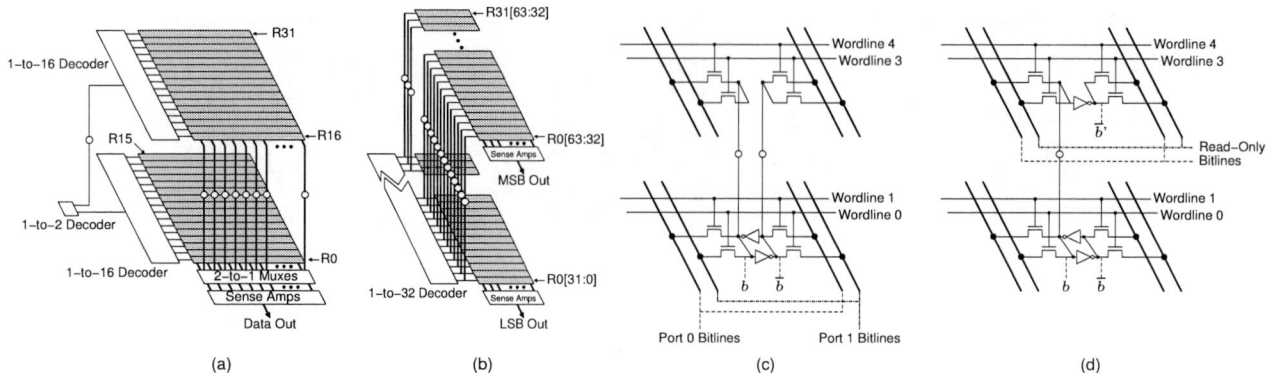

Fig. 3. 3D register file organizations achieved through (a) register-partitioning, (b) bit-partitioning, (c) port-splitting, and (d) a port-splitting alternative that only uses one die-to-die via per bitcell. A ○ represents a die-to-die via.

modern processors in terms of impact on clock frequency and instructions per second (IPC) rates. As a result, many microarchitecture-level proposals have been made to deal with the size, latency and power of the physical register file, including register caching [22] and register file banking [23]. While these techniques can reduce the average latency of register file access, they significantly complicate of the processor data and control paths. Increases in processor clock frequency and the relative decrease of the speed of wires only exacerbate the problem. Another example of the poor scaling of register file latency and area can be found in the Alpha 21264's 4-issue integer execution core which would normally require an 8-read port, 4-write port register file. Instead, the designers chose to duplicate the entire contents of the register file such that each copy only needs half as many read ports [24]. Two full copies of a moderately ported register file proved to be smaller and faster than a single highly-ported structure. Since the register file is dominated by wire, 3D may provide an effective means for controlling the latency, power and area of the large physical register files required by modern high-performance processors.

IV. 3D REGISTER FILES

There are many possible designs for register files in a 3D integration technology. We propose three different strategies for partitioning the register file across multiple die.

IV-A. Register-Partitioned (RP) 3D-Register Files

A two-die register-partitioned (RP) 3D register file takes half of the register entries and places them on the second die. Figure 3(a) illustrates a 32-entry register file where the bottom die contains registers R0 through R15, and the top die contains R16 through R31. A result of this topology is that the vertical distance (along the bitlines) has been halved, which can greatly reduce the latency and power associated with toggling the bitlines. The row decoder's height has also been halved, which reduces the length of the critical path associated with accessing the farthest entry in the register file. The overall footprint of

the register file has also been halved, which may enable more compact processor floorplans.

To implement a 4-die RP 3D register file, the register entries would simply be partitioned such that one quarter of the entries reside on each die. The row decoder can be further decomposed in a manner similar to the 2-die version.

Note that while Figure 3(a) shows a 32-entry register file, with a single read port, the physical register file of modern high-performance processors may have 80 [24] or 128 [25] entries. As the number of in-flight instructions increase, the physical register file size will also grow. The large number of read and write ports also exacerbates the area and wire lengths of the register file. Furthermore, the width of each register has increased over time from 32 bits to 64 bits, and even 128 bits for some instruction set architectures (e.g., Intel's SSE3 ISA extension provides 128-bit registers).

IV-B. Bit-Partitioned (BP) 3D-Register Files

The bit-partitioned (BP) 3D register file stacks different bits of the same register across the different die. The BP register file can be viewed as the dual of the RP organization: one folds the register file upon itself in the horizontal direction while the other folds in the vertical direction. Figure 3(b) shows a 2-die 64-bit 32-entry bit-partitioned register file where the bottom die stores the least significant bits of the register values and the top die stores the most significant bits. The bit-partitioned register file reduces the wire length and gate loading on the wordline, which provides both latency and energy benefits.

While Figure 3(b) shows the bits of each register partitioned by significance, one could instead store the bits in odd positions on one die and the bits in even positions on the other die. Choosing one over the other does not impact the area, latency or power of the 3D BP register file. However, the choice should be made to match the datapaths throughout the rest of the processor. For example, if one implements a 3D integer ALU partitioned by significance (X[0:31]+Y[0:31] on one die, X[32:63]+Y[32:63] on the second die) [26], then the register file bit-partitioning should also be arranged by significance to avoid unnecessary d2d routing between the register file outputs

0-7695-2533-4/06 $20.00 © 2006 IEEE

and the ALU inputs.

The BP 3D register file requires that the row decoder outputs be fanned out to the different die. This extra communication incurs a small overhead, but the latency reduction due to the halving of the wordline length still provides a significant net benefit. For a 4-die organization, the overhead of row decoder output fan out increases, but only slightly because the fan-out trees scale well with increasing die count.

IV-C. Port-Split (PS) 3D-Register Files

For on-chip caches, the individual SRAM cells are very small to maximize the capacity of the cache, while the area-per-bit for a register file cell is dominated by the wordlines and bitlines for implementing multiple read and write ports. Tsai et al. suggested that the relative size of a 6T SRAM cell and a d2d via make it difficult to take an individual 6T cell and split it across multiple die [14]. However, register file SRAM cells have a substantially larger footprint (due to the high port count) which may provide the opportunity to allocate one or two d2d vias for each cell. Figure 3(c) shows a 2-die port-split (PS) SRAM cell where each die contains the bitlines, wordlines and access transistors for half of the ports (either read or write). Two d2d vias are requires per bit-cell to route the outputs of the chained inverters to the second die.

The PS register file provides substantial benefits in terms of area footprint reduction. Stacking the wordlines on top of each other halves the height, while stacking the bitlines halves the width. A 50% reduction in both dimensions leads to an overall footprint reduction of 75% for the SRAM array. The total register file reduction is slightly less because structures like the row decoder and sense amps may not observe as large of a compaction benefit. This substantial area reduction also translates into latency and energy savings because both bitline and wordline lengths have been halved.

Depending on the register file design parameters and the relative size of d2d vias, it may not be possible to allocate two die-to-die vias per bit cell. Figure 3(d) shows an alternative implementation of a 2-die port-split (PS) 3D register file cell where only a single d2d via is used to route the data bit b to the second die. On the upper die, an extra inverter is required to recompute the complement bit $\bar{b}$. This shows how in some situations, logic duplication may be used to tradeoff against excessive inter-die communication.

A limitation of the single-via configuration is that the ports on the top die can only support read operations because there is no path to access the "true" $\bar{b}$ storage node.[1] This limitation is likely not critical as the number of write ports is typically much less than the number of read ports.

With two d2d vias per cell, a third alternative would be to split the back-to-back inverters across the two die. This would place all of the b bitlines on (say) the bottom die,

and all of the $\bar{b}$ bitlines on the top cell. We do not evaluate this configuration as it has several disadvantages. First, the wordlines must be replicated across both die (similar to the BP register file configuration) which eliminates the wirelength reduction in one dimension. Second, splitting the differential bit-lines across more than one die may require designing sense amplifiers that are themselves partitioned across more than one die. The technique also does not scale to beyond a two-die organization.

IV-D. Hybrid Configurations

Register files implemented across four (or more) die can use a combination of the partitioning strategies described above. This may be particularly useful in an alternating F2F/B2B die-stacking organization where the available d2d via density changes between pairs of die. In a 4-die stack with alternating F2F interfaces, one could first use register-partitioning to assign half of the registers to dies 0/1 and the other half to dies 2/3, which limits the usage of the coarser B2B vias to the periphery of the main SRAM array. Then among each pair of F2F die, port-partitioning could be employed to exploit the denser F2F interface within the SRAM array.

V. CIRCUIT LATENCY AND ENERGY SIMULATION

We use HSpice to obtain the critical path latency and overall energy consumption of SRAM register files. We generate a custom netlist for the different register file configurations based on parameters such as the number of entries, bit-width per entry, number and type of ports, and 2D vs. 3D organization. Our HSpice simulations use the Berkeley 70nm BSIM transistor models [27] and wire parameters extrapolated to 70nm from a TSMC 180nm technology. We sweep through a range of transistor sizings to minimize the latency of our register file configurations.

We use a distributed RC-ladder model for all of the wires in the circuits. For a two-die stack, the distance between the top metal layers on the two die is very small, and the pitch of the D2D vias are of the same order as the top level metal [28]. While the pitch of current manufacturable d2d vias is already only $2.4\mu m$ and $4\mu m$ for F2F and B2B interfaces [21], respectively, we assume that the d2d sizes will continue to scale at least for a few generations. For our circuits, we used a $1.0\mu m$ pitch for the F2F vias, and a $2.0\mu m$ pitch for the B2B vias. Furthermore, we assume a Cu-Cu 3D integration technology, which means that the d2d vias are made of the same copper as the traditional on-die interconnects, and therefore have similar unit resistance and capacitance. We assume that the d2d via lengths (die-to-die distance) is $10\mu m$ and $20\mu m$ for the F2F and B2B interfaces, respectively. We include an additional resistance equal to a top-level via to simulate the contact resistance where the two halves of the d2d via are bonded together.

We evaluated physical register file designs for a high-performance superscalar processor. In particular, we consider a four-wide superscalar machine which means the processor core can execute four instructions per cycle. Each of the

---

[1]One could conceivably build a single-ended write port that only changes the value of node b and simply relies on the SRAM cell itself to override the $\bar{b}$ node, but this would increase the write latency and substantially increase the duration of the short-circuit interval where both PMOS pull-up and NMOS pull-down circuits are active.

TABLE I

ACCESS LATENCIES OF PHYSICAL REGISTER FILES FOR A 4-WIDE SUPERSCALAR PROCESSOR, AND THE RELATIVE CHANGE IN LATENCY COMPARED TO THE BASELINE 2D/PLANAR IMPLEMENTATION. THE BEST 2-DIE AND 4-DIE CONFIGURATIONS ARE BOLDED.

RF Size	Base 2D	3D/2-die		
		RP	BP	PS
128	591 ps	557 ps -5.75%	**492 ps -16.8%**	508 ps -14.0%
256	784 ps	**595 ps -24.1%**	660 ps -15.8%	654 ps -16.6%

RF Size	3D/4-die				
	RP/RP	BP/BP	PS/PS	BP/RP	BP/PS
128	532 ps -9.98%	435 ps -26.4%	477 ps -19.3%	460 ps -22.2%	**422 ps -28.6%**
256	565 ps -27.9%	589 ps -24.9%	611 ps -22.1%	**502 ps -36.0%**	551 ps -29.7%

TABLE II

ENERGY PER ACCESS OF PHYSICAL REGISTER FILES FOR A 4-WIDE SUPERSCALAR PROCESSOR, AND THE RELATIVE CHANGE IN LATENCY COMPARED TO THE BASELINE 2D/PLANAR IMPLEMENTATION.

RF Size	Base 2D	3D/2-die		
		RP	BP	PS
128	4.46 nJ	**3.50 nJ -21.5%**	4.55 nJ +2.02%	3.82 nJ -14.4%
256	11.24 nJ	**4.67 nJ -58.5%**	10.43 nJ -7.21%	9.03 nJ -19.7%

RF Size	3D/4-die				
	RP/RP	BP/BP	PS/PS	BP/RP	BP/PS
128	**2.84 nJ -36.3%**	4.35 nJ -2.47%	3.78 nJ -15.2%	3.38 nJ -24.2%	3.82 nJ -14.3%
256	**3.69 nJ -67.2%**	9.71 nJ -13.6%	8.49 nJ -24.5%	4.70 nJ -58.2%	8.24 nJ -26.7%

four instructions requires two read ports and one write port. Furthermore, to retire four instructions per cycle, an additional four read ports are needed to read the physical register contents before updating the architected (committed) register file. The total port requirement is 12 read ports and 4 write ports. We assume a register file layout where one half of the bitlines flank either side of the decoders, thus reducing the critical path length of the wordline.

In some microarchitectures, the physical register file also contains the instruction status information required for maintaining in-order retirement of instructions; each of our registers contains 160 bits of data [29]. We simulate a 128-entry register file which is representative of modern processors as well as a 256-entry version to model future register file demands.

VI. RESULTS

The 3D implementations of the physical register file have the potential to substantially reduce both the latency and power associated with long wires. In the results described below, we compare a baseline 2D/planar implementation of a physical register file to a variety of 3D organizations. In particular, we evaluated both 2-die and 4-die versions employing the different stacking strategies explained earlier in the paper.

The overall access latency reduction due to 3D varies depending on the stacking approach. Table I lists the access latencies in picoseconds for each of the register file configurations considered. For the 128-entry, 2-die, 3D register file, the BP approach provides the greatest latency reduction. The wordline is heavily loaded by the two access transistors per column, and so splitting the wordline across two die reduces a major component of the wire latency. When considering a 256-entry register file, the height of the overall structure increases, thus making the row decoder and bitline/sense-amp delay more critical. In this situation, the RP approach provides greater benefit. Although the PS organization has a substantially smaller footprint, it does not provide the fastest

performance due to fact that the access transistor loading on the wordlines has not been reduced.

There are more design options when implementing structures across four stacked die. Table I also lists the latency benefits for implementing the register file in a 4-die stack. The notation X/Y indicates that the register file is split across the F2F boundary using partitioning scheme X, and split across the B2B boundary using partitioning scheme Y. For example, the BP/RP organization places one half of the registers on dies 0/1 and the other half on dies 2/3 (RP across the B2B interface), and then within each pair of die each register's bits are split half on one die and half on the other. The best performing register file organizations make use of hybrid partitioning strategies. This makes sense as the application of one partitioning technique may reduce the latency of a critical wire delay by a significant amount such that it is no longer the worst delay in the circuit. A second different partitioning can then address the new worst delay. Our results show that a 128-entry register file can be sped up by 28.6% and a 256-entry register file observes a 36.0% improvement. This latency reduction may potentially be converted into a clock frequency increase or a reduction in the number of cycles necessary to access the register file.

The performance improvements of the hybrid configurations highlight the generality of the 3D stacking techniques. For different processor configurations, the critical wire delay components within the register file will likely be different. The benefits of 3D are not limited to specific processor microarchitectural parameters; that is, for each situation, a circuit designer can choose the appropriate 3D stacking strategy to provide a latency reduction.

In addition to reducing the access latency of the physical register file, the 3D organization can also reduce the power consumption of the overall structure. Table II lists the energy required for one read operation from the physical register file. The 3D configuration that minimizes energy consumption is not necessarily the configuration that has the lowest latency. For example, the wordline switching delay in the 128-entry scheduler is critical for performance, but from a energy consumption perspective, the many parallel output bits require far more energy. Our physical register file width of 160 bits would require toggling 320 bit lines and switching 160 sense amplifiers. The RP approach effectively halves the bitline length for each doubling of the number of stacked die. The shorter bitlines greatly reduces the loading on the sense amp, which in turn may be sized smaller to consume even less energy. For all 3D configurations evaluated, the RP always provides the greatest energy reduction. For the 2-die BP configuration, the energy consumption actually increases slightly compared to the baseline configuration. This is simply due to the fact that the BP organization does not change the bitline length or loading, and we had resized some of the transistors in the sense amps to further improve the latency at the cost of a slight increase in power.

VII. SUMMARY

3D integration presents an opportunity to greatly reduce the impact of the poor scaling of wire delay. We have studied one particular microarchitectural module and demonstrated how a 3D circuit implementation can simultaneously provide both performance and power benefits by reducing the lengths of critical wires. There are many opportunities for further research in the application of 3D integration to the design of high performance microprocessors, and also to other areas such as embedded processors, DSPs, and other computing devices. Much more work is required at many levels, including circuit design, microarchitectures, and CAD tool support.

ACKNOWLEDGMENTS

Funding and equipment for this project have been provided by Intel Corporation and a grant from the Microelectronics Advanced Research Corporation (MARCO).

REFERENCES

[1] G. E. Moore, "Cramming More Components Onto Integrated Circuits," *Electronics*, April 1965.
[2] Semiconductor Industry Association, "The National Technology Roadmap for Semiconductors," 1999.
[3] V. Agarwal, M. S. Hrishikesh, S. W. Keckler, and D. Burger, "Clock Rate Versus IPC: The End of the Road for Conventional Microarchitectures," in *Proceedings of the 27th International Symposium on Computer Architecture*, Vancouver, Canada, June 2000, pp. 248–259.
[4] R. Ho, K. W. Mai, and M. A. Horowitz, "The Future of Wires," *Proceedings of the IEEE*, vol. 89, no. 4, pp. 490–504, April 2001.
[5] S. Borkar, "Design Challenges of Technology Scaling," *IEEE Micro Magazine*, vol. 19, no. 4, pp. 23–29, July 1999.
[6] R. Ronen, A. Mendelson, K. Lai, S.-L. Lu, F. Pollack, and J. P. Shen, "Coming Challenges in Microarchitecture and Architecture," *Proceedings of the IEEE*, vol. 89, no. 3, pp. 325–340, March 2001.
[7] R. Gonzalez and M. Horowitz, "Energy Dissipation in General Purpose Microprocessors," *IEEE Journal of Solid-State Circuits*, vol. 31, no. 9, pp. 1277–1284, September 1996.
[8] M. J. Flynn, P. Hung, and K. W. Rudd, "Deep-Submicron Microprocessor Design Issues," *IEEE Micro Magazine*, vol. 19, no. 4, pp. 11–22, July 1999.
[9] D. Brooks, P. W. Cook, P. Bose, S. E. Schuster, H. Jacobson, P. N. Kudva, A. Buyuktosunoglu, J.-D. Wellman, V. Zyuban, and M. Gupta, "Power-Aware Microarchitecture: Design and Modeling Challenges for Next-Generation Microprocessors," *IEEE Micro Magazine*, vol. 20, no. 6, pp. 26–44, November 2000.

[10] V. Srinivasan, D. Brooks, M. Gschwind, P. Bose, V. Zyuban, P. N. Strenski, and P. G. Emma, "Optimizing Pipelines for Power and Performance," in *Proceedings of the 35th International Symposium on Microarchitecture*, Istanbul, Turkey, November 2002, pp. 333–344.
[11] S. Palacharla, "Complexity-Effective Superscalar Processors," Ph.D. dissertation, University of Wisconsin, 1998.
[12] D. Nelson, C. Webb, D. McCauley, K. Raol, J. R. II, J. DeVale, and B. Black, "A 3D Interconnect Methodology Applied to iA32-class Architectures for Performance Improvements through RC Mitigation," in *Proceedings of the 21st International VLSI Multilevel Interconnection Conference*, Waikoloa Beach, HI, USA, September 2004.
[13] K. Puttaswamy and G. H. Loh, "Implementing Caches in a 3D Technology for High Performance Processors," in *Proceedings of the International Conference on Computer Design*, San Jose, CA, USA, October 2005.
[14] Y.-F. Tsai, Y. Xie, N. Vijaykrishnan, and M. J. Irwin, "Three-Dimensional Cache Design Using 3DCacti," in *Proceedings of the International Conference on Computer Design*, San Jose, CA, USA, October 2005.
[15] P. Reed, G. Yeung, and B. Black, "Design Aspects of a Microprocessor Data Cache using 3D Die Interconnect Technology," in *Proceedings of the International Conference on Integrated Circuit Design and Technology*, Austin, TX, USA, May 2005, pp. 15–18.
[16] Tezzaron Semiconductor, "WWW Site," http://www.tezzaron.com.
[17] Samsung Electronics Corporation, "Samsung Electronics Develops World's First Eight-die Multi Chip Package for Multimedia Cell Phones," January 10 2005, press Release from http://www.samsung.com.
[18] P. Morrow, M. J. Kobrinsky, S. Ramanathan, C.-M. Park, M. Harmes, V. Ramachandrarao, H. mog Park, G. Kloster, S. List, and S. Kim, "Wafer-Level 3D Interconnects Via Cu Bonding," in *Proceedings of the 21st Advanced Metallization Conference*, San Diego, CA, USA, October 2004.
[19] R. Reif, A. Fan, K.-N. Chen, and S. Das, "Fabrication Technologies for Three-Dimensional Integrated Circuits," in *Proceedings of the 3rd International Symposium on Quality Electronic Design*, San Jose, CA, USA, March 2002, pp. 33–37.
[20] S. Das, A. Fan, K.-N. Chen, and C. S. Tan, "Technology, Performance, and Computer-Aided Design of Three-Dimensional Integrated Circuits," in *Proceedings of the International Symposium on Physical Design*, Phoenix, AZ, USA, April 2004, pp. 108–115.
[21] S. Gupta, M. Hilbert, S. Hong, and R. Patti, "Techniques for Producing 3D ICs with High-Density Interconnect," in *Proceedings of the 21st International VLSI Multilevel Interconnection Conference*, Waikoloa Beach, HI, USA, September 2004.
[22] R. Balasubramanian, S. Dwarkadas, and D. Albonesi, "Reducing the Complexity of the Register File in Dynamic Superscalar Processors," in *Proceedings of the 34th International Symposium on Microarchitecture*, Austin, TX, USA, December 2001, pp. 237–248.
[23] J. H. Tseng and K. Asanović, "Banked Multiported Register Files for High-Frequency Superscalar Microprocessors," in *Proceedings of the 30th International Symposium on Computer Architecture*, San Diego, CA, USA, May 2003, pp. 62–71.
[24] R. E. Kessler, "The Alpha 21264 Microprocessor," *IEEE Micro Magazine*, vol. 19, no. 2, pp. 24–36, March–April 1999.
[25] G. Hinton, D. Sager, M. Upton, D. Boggs, D. Carmean, A. Kyler, and P. Roussel, "The Microarchitecture of the Pentium 4 Processor," *Intel Technology Journal*, Q1 2001.
[26] J. Mayega, O. Erdogan, P. M. Belemjian, K. Zhou, J. F. McDonald, and R. P. Kraft, "3D Direct Vertical Interconnect Microprocessors Test Vehicle," in *Proceedings of the ACM Great Lakes Symposium on VLSI*, Washington, DC, USA, April 2003, pp. 141–146.
[27] Y. Cao, T. Sato, D. Sylvester, M. Orshansky, and C. Hu, "New Paradigm of Predictive MOSFET and Interconnect Modeling for Early Circuit Design," in *Proceedings of the 2000 Custom Integrated Circuits Conference*, Orlando, FL, USA, May 2000, pp. 201–204.
[28] Y. Deng and W. Maly, "2.5D System Integration: A Design Driven System Implementation Schema," in *Proceedings of the Asia South Pacific Design Automation Conference*, Yokohama, Japan, January 2004, pp. 450–455.
[29] J. P. Shen and M. H. Lipasti, *Modern Processor Design: Fundamentals of Superscalar Processors*. McGraw Hill, 2005.

VLSI Circuits
and Technologies

Leakage-Aware SPM Management[*]

Guangyu Chen, Feihui Li, Ozcan Ozturk, Guilin Chen, Mahmut Kandemir
Computer Science and Engineering Department
Pennsylvania State University
University Park, PA 16802, USA
{gchen,feli,ozturk,guilchen,kandemir}@kandemir@cse.psu.edu

Ibrahim Kolcu
Computation Department
UMIST
Manchester M601QD, UK
i.kolcu@umist.ac.uk

Abstract

Increasing use of scratch-pad memories (SPMs) in embedded systems makes it imperative to consider optimizations tailored to their needs. Since these memories are managed by software, they present unique opportunities to the designer/compiler writer as far as energy optimizations are concerned. This paper proposes and quantifies the benefits of a compiler-directed energy optimization scheme for banked SPMs used to store the data manipulated by an application program. In contrast to most of the prior efforts on SPMs, which focus mainly on performance and dynamic energy consumption, the approach proposed in this paper is leakage oriented. Specifically, it tries to reduce the amount of SPM space (the number of banks) used to strike a balance between leakage and dynamic energy savings, with the goal of minimizing the total energy consumption due to data accesses. This paper presents an ILP (integer linear programming) based formulation of this problem and evaluates the proposed approach using a set of eight embedded application codes.

1 Introduction and Motivation

Many embedded systems employ fast, software-managed on-chip memory components either as an alternative or in addition to conventional on-chip cache memories. These software-managed memories are referred to as the Scratch-Pad Memories (SPMs) and have been a popular research subject in the last five years or so. There have been three research directions based on SPMs. The first direction includes the comparison of these architectures against conventional caches from a performance angle. The second direction studies code generation and data allocation issues to increase the potential benefits of SPMs. The third direction on the other hand concentrates on the potential energy savings these software-controlled memories can bring when compared to conventional caches.

However, most of the prior energy-related studies focus exclusively on dynamic energy savings, which are achieved as a side-product of the performance improvements brought by SPMs. For example, an approach that carefully decides what data should be placed into the SPM (based on data reuse characteristics of the application at hand) can reduce both execution cycles and dynamic energy consumption. Both these benefits are due to the reduction in the number of off-chip memory accesses. Reducing execution cycles has also a positive impact on leakage consumption, but even this comes as a side-benefit of performance optimization.

In contrast to these previous efforts, this paper looks at the data SPM management from a leakage perspective and shows that it is possible to improve energy savings beyond what could be achieved using the current SPM management techniques, by trading off dynamic power with leakage power. The main idea behind our approach is to shut off the selected portions of the available SPM space under compiler control. Doing so reduces the number of data elements that can be kept in the SPM (which can in turn causes an increase in dynamic energy consumption), but at the same time, this also reduces leakage consumption in the SPM. Clearly, the crucial issue for such an approach to be successful in practice is to select the right portion(s) of the SPM to shut off. In the proposed approach, this is achieved by employing an ILP (integer linear programing) based approach under compiler guidance, which decides the contents of a banked SPM at specific program points by performing the necessary dynamic energy-leakage energy tradeoffs.

The remainder of this paper is organized as follows. Section 2 explains the architectural model we assume. Section 3 gives the details of our ILP-based leakage-aware SPM management approach. A discussion of the prior related studies is given in Section 4. Our concluding remarks are provided in Section 5.

---

[*]This work is supported in part by NSF Career Award #0093082 and a fund from GSRC.

0-7695-2533-4/06 $20.00 © 2006 IEEE

2 Architectural Modeling

Figure 1 shows the architecture of an embedded system with a banked SPM used for storing data. The SPM is divided into n banks and each bank can be turned off by the application/compiler to conserve leakage energy. When an SPM bank is turned on (i.e., when it is active), it consumes leakage power P_0; when it is turned off, on the other hand, it does not consume leakage energy.[1] Further, we use E_S and E_M to denote per access dynamic energies for the SPM and main memory, respectively. Since the SPM is tightly coupled with the processor core and is smaller than the off-chip main memory, it consumes less energy per access than the off-chip main memory, i.e., $E_S < E_M$.

Between the SPM and the main memory is a DMA (direct memory access) channel so that the SPM and the main memory can exchange data in large chunks directly, without using the data path in the processor core. The dynamic energy consumed for transferring a data block (chunk) with size S between the SPM and the main memory using the DMA channel can be expressed as:

$$E_{DMA}(S) = E_0 + E_1 S,$$

where E_0 is the energy cost for DMA initialization, and E_1 is energy cost for transferring a byte of data.

Let us further assume that a data block of size S is loaded into the SPM at the entry of loop nest $\mathcal{L}$ and remains in the SPM throughout the execution time of this loop nest. The energy saving achieved by loading this data into the SPM and accessing it from there (as compared to accessing data directly from the main memory without using the SPM) can be computed as:

$$\begin{aligned} \Delta E &= KE_M - (E_{DMA}(S) + \lceil \frac{S}{B} \rceil TP_0 + KE_S) \\ &= K(E_M - E_S) - E_{DMA}(S) - \lceil \frac{S}{B} \rceil TP_0, \end{aligned}$$

where K is the number of accesses to this data block, T is the execution time of this loop nest, and B is the size of an SPM bank. According to this equation, loading data to the SPM can save energy because the SPM has lower per access energy than the main memory. However, the projected energy savings, $K(E_M - E_S)$, can be offset by the energy overheads due to transferring data between the main memory and the SPM (i.e., $E_{DMA}(S)$), and the leakage consumption in the SPM (i.e., $\lceil S/B \rceil TP_0$).

Normally, employing an SPM for storing/accessing data brings savings in energy consumption, as shown by the previous research. Our goal in this work is to explore whether we can increase these energy savings further through explicit leakage control.

---

[1]We assume that the leakage-saving circuitry employed kills the contents of the SPM bank when it is shut off. Our approach can also work with leakage-saving techniques that preserve the contents of the bank.

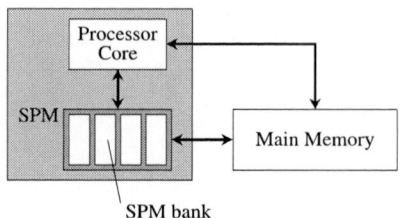

Figure 1. Banked SPM architecture.

3 Compiler Support

3.1 Mathematical Framework

Let us first describe the mathematical framework used by our compiler. Given vectors $\vec{U} = (u_1, u_2, ..., n_n)^{\mathrm{T}}$ and $\vec{V} = (v_1, v_2, ..., v_n)^{\mathrm{T}}$, we write $\vec{U} \le \vec{V}$ if and only if $u_i \le v_i$ holds for all $i = 1, 2, ..., n$. For ease of discussion, define vector set $[\vec{U}, \vec{V}]$ as:

$$[\vec{U}, \vec{V}] = \{\vec{I} \mid \vec{U} \le \vec{I} \le \vec{V}\}$$

We use

$$\mathcal{L} : \vec{I} \in [\vec{U}, \vec{V}]\mathcal{B},$$

where $\vec{I} = (i_1, i_2, ..., i_n)^{\mathrm{T}}$, $\vec{U} = (u_1, u_2, ..., u_n)^{\mathrm{T}}$, $\vec{V} = (v_1, v_2, ..., v_n)^{\mathrm{T}}$, and $\mathcal{B}$ is a sequence of instructions, to denote a loop nest of the following type:

$$\mathcal{L}\text{: for } i_1 = u_1 \text{ to } v_1$$
$$\text{for } i_2 = u_2 \text{ to } v_2$$
$$\ddots$$
$$\text{for } i_n = v_n \text{ to } v_n \; \{ \, \mathcal{B} \, \}$$

We refer to vectors $\vec{I}$, $\vec{U}$, and $\vec{V}$ as the iteration vector, the lower bound vector, and the upper bound vector, respectively.

The body (denoted as $\mathcal{B}$) of the loop nest may contain multiple array access instructions. In this paper, we are interested only in the affine array accesses. An affine access to array A can be written as $A[M\vec{I} + \vec{o}]$, where $\vec{I}$ is the iteration vector of the loop nest, M is a constant matrix (namely, the access matrix), and $\vec{o}$ is a constant vector (namely, the offset vector). For example, $A[5i_i + i_2][5i_1 + i_3 + 5]$, the second reference to array A in the loop nest shown in Figure 2, can be expressed as:

$$\begin{pmatrix} 5 & 1 & 0 \\ 5 & 0 & 1 \end{pmatrix} \begin{pmatrix} i_1 \\ i_2 \\ i_3 \end{pmatrix} + \begin{pmatrix} 0 \\ 5 \end{pmatrix}.$$

Let us assume that $\mathcal{A}_1, \mathcal{A}_2, ..., \mathcal{A}_k$ are the k accesses to array A in the body of loop nest "$\mathcal{L} : \vec{I} \in [\vec{U}, \vec{V}]\mathcal{B}$" and that access $\mathcal{A}_j$ is in the form of $A[M_j\vec{I} + \vec{o}]$. The set of array

```
//— P₁ —
L₁: for i₁ = 0 to 999
   //— P₂ —
   L₂: for i₂ = 0 to 9
      //— P₃ —
      L₃: for i₃ = 0 to 9
         B = B + f(A[5i₁ + i₂][5i₁ + i₃])
            +g(A[5i_i + i₂][5i₁ + i₃ + 5]);
```

Figure 2. An example loop nest.

elements touched by $\mathcal{A}_j$ during the execution of loop nest $\mathcal{L}$ can be expressed as:

$$H(\mathcal{L}, \mathcal{A}_j) = \{A[M_j \vec{I} + \vec{o}] \mid \vec{I} \in [\vec{U}, \vec{V}]\}.$$

Therefore, the set of elements of array A accessed by loop nest $\mathcal{L}$ can be written as:

$$G(\mathcal{L}, A) = \bigcup_{j=1}^{k} H(\mathcal{L}, \mathcal{A}_j).$$

We further define:

$$F(\mathcal{L}, A) = [\vec{J}_1(\mathcal{L}), \vec{J}_2(\mathcal{L})],$$

such that

$$[\vec{J}_1(\mathcal{L}), \vec{J}_2(\mathcal{L})] \supseteq G(\mathcal{L}, A) \quad \wedge$$
$$\forall [\vec{J}_1', \vec{J}_2'] \supseteq G(\mathcal{L}, A) : [\vec{J}_1(\mathcal{L}), \vec{J}_2(\mathcal{L})] \subseteq [\vec{J}_1', \vec{J}_2'].$$

Note that $F(\mathcal{L}, A)$ gives the minimum polyhedral region that contains all the elements of array A that *might* be accessed during the execution of loop nest $\mathcal{L}$. Its size can be computed using a polyhedral tool such as the Omega Library [1].

3.2 Flexibility in SPM Management

For a given loop nest accessing a certain array, the elements of this array can be loaded into the SPM at different points with different performance/energy costs. For example, in the loop nest shown in Figure 2, the elements of array A can be loaded into the SPM at points $\mathcal{P}_1$, $\mathcal{P}_2$, or $\mathcal{P}_3$. Figures 3(a), (b) and (c) show three versions of the same code with the SPM loading instruction inserted at different points. Figure 3(d) compares the energy costs of these three alternate SPM loading strategies. We assume, for the sake of explanation, that the SPM in question consists of four 1KB-banks and there is no other array competing for the SPM space with A. If we load the elements of array A into the SPM at point $\mathcal{P}_1$, we need to load all the elements that are accessed by this loop nest at once. In this case, we can compute the region of array A that needs to be loaded into the SPM as $F(\mathcal{L}_1) = \left[\binom{0}{0}, \binom{5094}{5009}\right]$, which contains

a total of 25,525,950 elements. Since $F(\mathcal{L}_1)$ cannot be fit into the SPM space, we cannot load the data to the SPM at point $\mathcal{P}_1$. If the insert the SPM loading instruction at point $\mathcal{P}_2$, this instruction will be executed 1000 times, and each time it loads a data block of 1500 elements. As a result, we need to keep two 1KB-banks in the active mode. Since the loop nest in Figure 2 accesses the array 2×10^6 times, as compared to directly accessing the data elements from the main memory, the energy saving achieved in this case due to using the SPM is:

$$\Delta E_2 = 2 \times 10^6 (E_M - E_S) - 10^3 (E_0 + 1500 E_1) - 2 P_0 T,$$

where E_M is the per access energy for the off-chip main memory, E_S is the per access energy for the SPM, P_0 is the leakage power consumption of an SPM bank, and T_2 is the execution time of this loop nest.[2] Similarly, if we insert the SPM loading instruction at point $\mathcal{P}_3$, we need to load 10,000 blocks, each of which contains only 15 elements; therefore, only a single SPM bank needs to be kept in the active mode, and the resulting energy saving can be computed as:

$$\Delta E_3 = 2 \times 10^6 (E_M - E_S) - 10^5 (E_0 + 15 E_1) - P_0 T.$$

Since $\Delta E_3 - \Delta E_2 = P_0 T - 99000 E_0$, if the execution time T is longer than $99000 E_0 / P_0$, we have $\Delta E_3 > \Delta E_2$, that is, loading the data elements into the SPM at point $\mathcal{P}_3$ saves more energy than doing so at $\mathcal{P}_2$. On the other hand, if the execution time T is shorter than $99000 E_0 / P_0$, $\mathcal{P}_2$ is a better place for loading data into the SPM.

3.3 ILP Based Approach

We now present our ILP (integer linear programing) based approach to determine the optimal point for each nest to load the SPM. Since a typical nest does not have too many loops, our ILP based approach is very fast in practice. Given an n-level loop nest $\mathcal{L} : \vec{I} \in [\vec{U}, \vec{V}]$, we use $\mathcal{L}_k$ to label the kth loop in this loop nest. Particularly, $\mathcal{L}_1$ represents the outer-most loop and $\mathcal{L}_n$ represents the inner-most loop. For ease of discussion, we use $\mathcal{P}_k$ $(k = 1, 2, ..., n)$ to denote the point in the code right before the loop $\mathcal{L}_i$. We assume that this loop nest accesses m arrays: $A_1, A_2, ..., A_m$. We use a binary variable $\theta_{j,k}$ $(1 \leq j \leq m, 1 \leq k \leq n)$ to indicate whether to load array A_j at point $\mathcal{P}_k$ or not as follows:

$$\theta_{j,k} = \begin{cases} 1, & \text{load } A_j \text{ into the SPM at point } \mathcal{P}_k; \\ 0, & \text{otherwise.} \end{cases}$$

---

[2] Since inserting the SPM loading instructions at different points may incur different performance overheads, the overall execution time of the loop nest is dependent on the point where the SPM loading code is inserted. However, as compared to the overall execution time, the performance overheads due to data loading is small; so, we can assume that the overall execution time of the loop nest is independent of the insertion point of the SPM loading instructions without causing significantly error in our analysis.

```
A'[0..5094][0..5009] ⇐ A[0..5094][0..5009];
    // load the elements of A into SPM resident array A'.
L₁: for i₁ = 0 to 999
  L₂: for i₂ = 0 to 9
    L₃: for i₃ = 0 to 9
      B = B + f(A'[5i₁ + i₂][5i₁ + i₃])
         +g(A'[5iᵢ + i₂][5i₁ + i₃ + 5]);
```

(a)

```
L₁: for i₁ = 0 to 999
  A'[0..99][0..14] ⇐ A[5i₁..5i₁ + 99][5i₁..5i₁ + 14];
    // load the elements of A into SPM resident array A'.
  L₂: for i₂ = 0 to 9
    L₃: for i₃ = 0 to 9
      B = B + f(A'[i₂][i₃])
         +g(A'[i₂][i₃ + 5]);
```

(b)

```
L₁: for i₁ = 0 to 999
  L₂: for i₂ = 0 to 9
    A'[0..14] ⇐ A[5i₁ + i₂][5i₁..5i₁ + 14];
      // load the elements of A into SPM resident array A'.
    L₃: for i₃ = 0 to 9
      B = B + f(A'[i₃]) + g(A'[i₃ + 5]);
```

(c)

Loading point	Number of loads	For each SPM load			Leakage energy
		Data block	Size	E_{DMA}	
$\mathcal{P}_1$	1	$\binom{0}{0}, \binom{5094}{5009}$	25,525,950	N/A	N/A
$\mathcal{P}_2$	1,000	$\binom{5i_1}{5i_1}, \binom{5i_1+99}{5i_1+14}$	1,500	$E_0 + 1500E_1$	$2P_0T$
$\mathcal{P}_3$	100,000	$\binom{5i_1+i_2}{5i_1}, \binom{5i_1+i_2}{5i_1+14}$	15	$E_0 + 15E_1$	P_0T

(d)

Figure 3. Loading the elements of array A into the SPM at different points in the code shown in Figure 2. (a) loading at $\mathcal{P}_1$; (b) loading at $\mathcal{P}_2$; (c) loading at $\mathcal{P}_3$. Table (d) compares the costs of the alternate loading schemes shown in (a), (b), and (c). The column labeled as "E_{DMA}" gives the dynamic energy spent for loading each block into the SPM. We assume that the SPM consists of four 1KB-banks and there is no other array competing for the SPM space with A.

Based on the energy cost and gains, we can choose whether to load the elements of array A_j into the SPM or not. If we choose to load the elements of array A_j into the SPM, we also need to choose the point to insert the SPM loading instruction for this array. *The job of our ILP solver is to determine this optimal point to maximize overall energy savings.* Since an array is loaded at only one point, we have the following constraint:

$$\sum_{k=1}^{n} \theta_{j,k} \leq 1, \quad \forall j.$$

If we load the elements of array A_j at point $\mathcal{P}_k$, the set of elements of array A_j that need to loaded can be computed using function $F(\mathcal{L}_k, A_j)$ as discussed in Section 3.1. Since the total size of the data loaded into the SPM cannot exceed S_{SPM}, the total size of the SPM, we have the following constraint:

$$\sum_{j=1}^{m} \sum_{k=1}^{n} \theta_{j,k} |F(\mathcal{L}_k, A_i)| \leq S_{SPM}.$$

Note that $|F(\mathcal{L}_k, A_i)|$ is a constant that can be computed during the compilation time.

The leakage energy consumption of the SPM can be estimated as:

$$E_{leakage} = \frac{P_0 T \sum_{j=1}^{m} \sum_{k=1}^{n} \theta_{j,k} |F(\mathcal{L}_k, A_i)|}{B},$$

where P_0 is the leakage power consumption of an SPM bank, B is size of a SPM bank, and T is the overall execution time of loop nest $\mathcal{L}$. Note that T can be estimated at compilation time using techniques such as the one proposed in [17].

By analyzing the application code, our compiler computes K_j, the number of accesses to A_j. Therefore, we can compute the total savings in dynamic energy due to array accesses (as compared to accessing arrays from the main memory without using the SPM) as:

$$\Delta E_{dynamic} = \sum_{j=1}^{m} \sum_{k=1}^{n} \theta_{j,k} K_j (E_M - E_S),$$

where E_M and E_S are the per access dynamic energy consumptions, as defined in Section 2.

By analyzing the structure of the loop nest, the compiler also computes X_k, the number of times that the code at point $\mathcal{P}_k$ is executed. Based on this, we compute the overheads due to the DMA initialization and data transfer between the SPM and the main memory as:

$$\Delta E_{overhead} = \sum_{j=1}^{m} \sum_{k=1}^{n} \theta_{j,k} X_k (E_0 + E_1 |F(\mathcal{L}_k, A_i)|),$$

where E_0 is the energy cost for a DMA initialization, and E_1 is the energy cost for transferring a byte of data.

Now, we can compute the overall energy savings achieved by our SPM management as:

$$E_{save} = \Delta E_{dynamic} - E_{overhead} - E_{leakage}.$$

We use a commercial ILP solver [2] to determine the values of variables $\theta_{j,k}$ ($1 \leq j \leq m$, $1 \leq k \leq n$) such that E_{save} is maximized while all the constraints listed above are satisfied.

0-7695-2533-4/06 $20.00 © 2006 IEEE

4 Related Work

Scratch-pad memories have been widely targeted in recent research efforts. The related studies mainly focus on the management strategies such as static versus dynamic and instruction SPM versus data SPM. So far, data SPM related efforts have mainly focused on single application scenarios, i.e., the SPM space available is assumed to be managed by a single application at any given time. Angiolini et al [4] propose an algorithm to optimally solve the SPM location mapping problem. They map the segments of external memory to physically partitioned banks of an on-chip SPM. In [5], a compiler strategy has been used to automatically partition the data among the memory units. A unified hardware/software solution to support scratch-pad memories is presented in [6]. Panda et al [10] propose a technique for efficiently exploiting on-chip scratch-pad memory in order to minimize the total execution time. Verma et al [16] use the scratch-pad memory for storing instructions and they propose a cache aware SPM allocation algorithm. A compiler controlled dynamic on-chip scratch-pad memory management framework has been proposed in [7]. Both loop and data transformations have been applied. Steinke et al [15] propose an algorithm integrated into a compiler to analyze the application code and to select the program and data parts to be placed into the scratch-pad memory.

Prior research efforts on cache resizing have mainly focused on two design aspects: cache organization and resizing strategy. While cache organization could be selective-ways (varies the cache's set-associativity), selective-sets (varies the number of cache sets), or hybrid, resizing strategy could be static (cache size is set prior to an application's execution) or dynamic (resizing happens both with in and across applications). Yang et al [18] evaluate different design choices for resizable caches, and test the effectiveness of cache resizing in reducing the energy-delay product. They also propose a hybrid selective-sets-and-ways cache organization. Ranganathan, et al [13] propose a statically resizing selective-ways data cache. In this organization, unused parts are used for maintaining instruction reuse statistics that can be used to increase performance. Albonesi [3] proposes a statically resizing selective-ways cache. On the other hand, in [19], Yang et al investigate the possibility of dynamically resizing selective-sets and try to reduce leakage energy. Zhou et al [21] propose an adaptive mode control technique similar to cache decay. Powell et al [12] resize an instruction cache by measuring the miss rate at run-time and keeping it under a certain threshold. This threshold is preset prior to execution and remains the same throughout the execution. Note that this approach increases hardware cost and area. Cache decay [9] has been proposed by Kaxiras et al as mechanism for reducing leakage consumption. Pokam and Bodin [11] propose a configurable cache to offer embedded compilers the opportunity to reconfigure it according to the dynamic behavior of a program (i.e., at a phase basis), rather than on a per-application basis. Prior efforts mentioned so far use conventional RAM-tag caches, in which the tag and data arrays are organized as RAM structures. Cache resizing has also been employed in CAM-tag caches. Zhang and Asanovic [20] use hierarchical bitlines to divide each cache subbank into small way partitions. In this approach, switching and leakage power are only dissipated in active ways.

5 Conclusions

Effective utilization of on-chip storage space is important from both performance (execution cycles) and energy perspectives. While conventional hardware managed on-chip cache memories have been widely used in the past, several factors, including lack of data access time predictability and limited effectiveness of compiler optimizations, indicate that they may not be the best option for portable/embedded devices. Prior research studied static and dynamic approaches that demonstrate performance and energy benefits of SPMs. However, most of the previous energy-oriented efforts have focused exclusively on dynamic energy consumption. In contrast, this paper shows that it is possible to tradeoff dynamic energy with leakage energy by shutting off the select portions of the SPM, leading to further reductions in overall energy consumption. Our ILP-based scheme has been implemented and tested using eight embedded applications against a state-of-the-art SPM management scheme.

References

[1] The Omega Project: Frameworks and Algorithms for the Analysis and Transformation of Scientific Programs. http://www.cs.umd.edu/projects/ omega/

[2] Xpress-MP, http://www.dashoptimization.com/pdf/ Mosel1.pdf, 2002.

[3] D. Albonesi. "Selective cache ways: On-demand cache resource allocation". In Proc. of the 32nd International Symposium on Microarchitecture, Nov. 1999.

[4] F. Angiolini, L. Benini, and A. Caprara. "Polynomial-time algorithm for on-chip scratchpad memory partitioning". In Proc. of the International Conference on Compilers, Architecture and Synthesis for Embedded Systems, San Jose, CA, 2003.

[5] O. Avissar, R. Barua, and D. Stewart. "An optimal memory allocation scheme for scratch-pad based embedded systems". ACM Transactions on Embedded Computing Systems, 626, November 2002.

[6] P. Francesco et al. "An Integrated Hardware/software Approach for Run-time Scratchpad Management", In Proc. of the 41st Conference on Design Automation, San Diego, CA, 2004.

[7] M. Kandemir et al. "Dynamic management of scratch-pad memory space". In Proc. of the 38th Conference on Design Automation, Las Vegas, NV, 2001.

[8] M. Kandemir and A. Choudhary. "Compiler-directed scratch pad memory hierarchy design and management". In Proc. of the 39th Conference on Design Automation, New York, NY, USA, 2002.

[9] S. Kaxiras, Z. Hu, and M. Martonosi. "Cache decay: Exploiting generational behavior to reduce cache leakage power". In Proc. the 28th International Symposium on Computer Architecture, June 2001.

[10] P. R. Panda, N. Dutt, and A. Nicolau. "Efficient Utilization of Scratch-Pad Memory in Embedded Processor Applications". In Proc. of the European Conference on Design and Test, 1997.

[11] G. Pokam and F. Bodin. "Energy-efficiency Potential of a Phase-based Cache Resizing Scheme for Embedded Systems". In Proc. of the 8th Workshop on Interaction between Compilers and Computer Architectures, 2004.

[12] M. Powell, S. Yang, B. Falsafi, K. Roy, and T. N. Vijaykumar. "Reducing leakage in a high-performance deep-submicron instruction cache". IEEE Transactions on Very Large Scale Integrated Systems, volume 9, number 1, pages 77–90, 2001.

[13] P. Ranganathan, S. Adve, and N. P. Jouppi. "Reconfigurable caches and their application to media processing". In Proc. of the 27th International Symposium on Computer Architecture, pages 214-224, June 2000.

[14] P. Shivakumar and N. Jouppi, CACTI 3.0. http://research.compaq.com/wrl/people/jouppi/CACTI.html

[15] S. Steinke, L. Wehmeyer, B. Lee, and P. Marwedel. "Assigning Program and Data Objects to Scratchpad for Energy Reduction". In Proc. of the Conference on Design, Automation and Test in Europe, Paris, France, 2002.

[16] M. Verma, L. Wehmeyer, and P. Marwedel. "Cache-Aware Scratchpad Allocation Algorithm". In Proc. of the Conference on Design, Automation and Test in Europe, Paris, France, 2004.

[17] M. Wolf, D. Maydan and D. Chen. "Combining loop transformations considering caches and scheduling". In Proc. of the 29th annual ACM/IEEE International Symposium on Microarchitecture, Paris, France, 1996.

[18] S-H. Yang, M. D. Powell, B. Falsafi, and T. Vijaykumar. "Exploiting choice in resizable cache design to optimize deep-submicron processor energy-delay". In Proc. of International Symposium. on High Performance Computer Architecture, Feb. 2002.

[19] S.-H. Yang, M. D. Powell, B. Falsafi, K. Roy, and T. N. Vijaykumar. "An integrated circuit/architecture approach to reducing leakage in deep-submicron high-performance i-caches". In Proc. of the 7th IEEE Symposium on High-Performance Computer Architecture, Jan. 2001.

[20] M. Zhang and K. Asanovic. "Fine-grain CAM-tag cache resizing using miss tags". In Proc. of the International Symposium on Low Power Electronics and Design, August, 2002, Monterey, California, USA

[21] H. Zhou, M. C. Toburen, E. Rotenberg, and T. M. Conte. "Adaptive mode control: A static-power-efficient cache design". IEEE Transactions on Embedded Computing Systems, volume 2, number 3, pages 347–372, 2003.

Dependability Analysis of Nano-scale FinFET circuits

Feng Wang*, Yuan Xie*, Kerry Bernstein[†],Yan Luo[‡]

*Computer Science Engineering Department,
The Pennsylvania State University,University Park, PA, 16802
[†] IBM T.J. Watson Research Center, Yorktown Heights, NY, 10598, USA
[‡] Silicon Engineering Group, Synopsys .Inc, Shanghai, China
Email: fenwang@cse.psu.edu

Abstract

FinFET technology has been proposed as a promising alternative for deep sub-micro bulk CMOS technology, because of its better scalability. Previous work have studied the performance or power advantages of FinFET circuits over bulk CMOS circuits. This paper provides the dependability analysis of FinFET circuits, studying the soft error vulnerability of FinFET circuits and the impact of process variation. Our experiments compare FinFET circuits against bulk CMOS circuits in both 32nm and 45nm technologies, showing that FinFET circuits have better dependability and scalability, which is indicated by better soft error immunity and less impact of process variation. It is concluded that FinFET-based circuit design is more robust than the bulk CMOS based circuit design.

1. Introduction

Conventional bulk CMOS scaling beyond 45nm is severely constrained by short channel effects and vertical gate insulator tunneling [9]. Double-gate FinFET technology [3][7][12[15] has been proposed as a very promising candidate to circumvent the conventional bulk CMOS scaling constraint, by changing the device structure in such a way that MOSFET gate length can be scaled further even with thicker oxide, so that we can continue scaling beyond the limit of conventional bulk CMOS. Figure 1 shows a multi-fin FinFET structure, which will be explained in Section 3.

One of the grand challenges for nano-scale VLSI designers is guaranteeing dependability. Shrinking geometries, lower supply voltage, and higher frequencies, all have a negative impact on circuit dependability: the occurrences of soft errors increases due to these factors, and higher levels of device parameter variations change the design problem from deterministic to probabilistic. Consequently, reducing soft error rate and mitigating the impact of process variation are becoming increasingly critical.

Both logic and SRAM FinFET technologies have been previously demonstrated [3][7]. Previous work have shown the performance and power advantage of FinFET circuits over bulk CMOS [2][3][7][12][15]. However, we haven't seen a comprehensive analysis for dependability of FinFET circuits. In this paper, we provide the reliability and scalability analysis for FinFET circuits, showing that FinFET circuits have better soft error immunity, as well as less impact of process variation on the performance, comparing against the bulk CMOS counterparts.

The paper is organized as follows: Section 2 provides a survey of related work. Section 3 describes the FinFET structure and the device model parameters. Section 4 describes the methodology to compare the soft error vulnerability of FinFET circuits against bulk CMOS circuits. Section 5 discusses the impact of process variation on both FinFET circuits and bulk CMOS circuits. Experimental result for 32-nm and 45-nm FinFET and bulk CMOS circuits are presented in both Section 4 and Section 5. Finally, Section 6 concludes the paper.

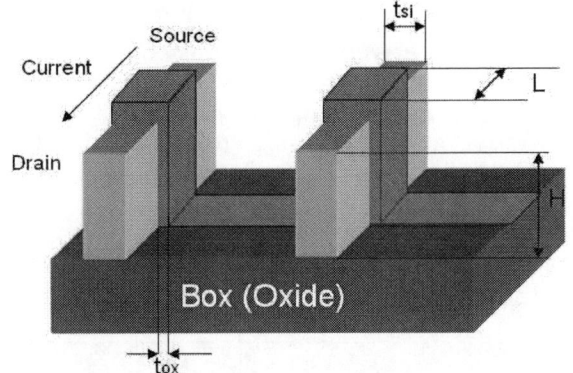

Fig. 1. Multiple-fin FinFET structure [15]

2. Related Work

FinFET device [5] was proposed as an elegant implementation of double-gate FET [9] , which enables continuous technology scaling. Rainey *et al.* [7] have demonstrated the feasibility of FinFET logic implementation in 180nm, by showing inverter-chain

0-7695-2533-4/06 $20.00 © 2006 IEEE

operation for the first time. FinFET SRAM behavior was investigated by Joshi *et al.* [3] and it has been shown to exhibit higher performance and lower power compared against conventional planar PD-SOI. IBM was the first to convert an existing microprocessor design in a 100 nm FinFET technology and ensure its functionality [6].

Even though FinFET device and circuits design consideration have been studied recently [1][2][3][6][7][12][15], most of the work focused on the performance or power advantage that FinFET circuits can provide. Soft error vulnerability and process variation analysis for FinFET circuits are necessary since reliability is also an important design goal for nano-scale VLSI design. In [15], the design space exploration for performance power and reliability is performed for a 65nm FinFET SRAM. Guo et.al [12] showed that FinFET-based SRAM design has a 30% improvement of noise margin over that of the bulk CMOS SRAM. In this paper, we study the soft error vulnerability as well as the impact of process variation for FinFET circuits.

3. FinFET Structure and Device Model Parameter

Fig. 1 shows the structure of multi-fin double-gate FinFET devices [15]. Double gate FinFET consists of two SOI gates connected together. The thickness (Tsi) of a single fin equals to silicon channel thickness. The current flows from the source to drain along the wafer plan. Each fin provides 2H of device width, where H is the height of the each fin. For the FinFET devices, widths are quantized into units of the fins. Large width of device is obtained by using multiple fins.

We use Synopsys circuit simulator HSPICE to study FinFET circuits behavior. The FinFET devices in 32nm and 45nm, and the corresponding bulk CMOS counterparts, are modeled using Predictive Technology Model [8]. Table 1 shows the nominal 45nm and 32 nm device parameters that is used in our simulation.

Table 1. Nominal 32nm and 42nm FinFET device parameters

Parameter	32 nm FinFET	45nm FinFET
Supply Voltage (Vdd)	0.9 V	1.0 V
Physical Gate Length(L)	32nm	45nm
Physical Oxide Thickness(Tox)	1.4nm	1.5nm
Body thickness(Tsi)	8.6nm	8.4nm
Fin-Height(H)	65nm	65nm

4. Soft Error Vulnerability Analysis for FinFET circuit

In this section, we describe the methodology to conduct soft error vulnerability analysis for FinFET circuits. Section 4.1 gives a brief introduction on soft error and the simulation method; Section 4.2 shows the soft error vulnerability analysis method for bulk CMOS and FinFET logic and SRAM. Section 4.3 presents the experimental results.

4.1 Soft Error Background and the Simulation Methodology

Alpha particles from the packaging materials, high energy neutrons from cosmic radiations and the interaction of cosmic ray thermal neutron are three major sources of the radiations that induce soft error, which is also called Single Event Upset (SEU) [10]. A soft error occurs, when a particle strike at a specific node in a circuit, which results in a transient voltage pulse can cause a bit flip in the SRAM cell or be captured by the latch. The critical charge is defined as the minimum charge collected for a specific voltage transient occurs at the gate output, which can cause a bit flip in the SRAM cell or propagate through logic chains to the primary output and be latched by storage elements or. This concept of critical charge is generally used to estimate the sensitivity of SEU for both the SRAM circuits and combinational circuits [10][11]. A smaller critical charge value indicates higher soft error rate (SER).

Similar to previous work [10][11], a double exponential current source model is used to model a SEU at a node:

$$I(t) = I_{peak} \times (e^{-t/\tau_a} - e^{-t/\tau_b}) \quad (1)$$

where $I_{peak} = \dfrac{Q}{\tau_a - \tau_b}$, in which Q is the charge collected as a result of particle strike, τ_a is the collection time-constant, and τ_b is the ion-track

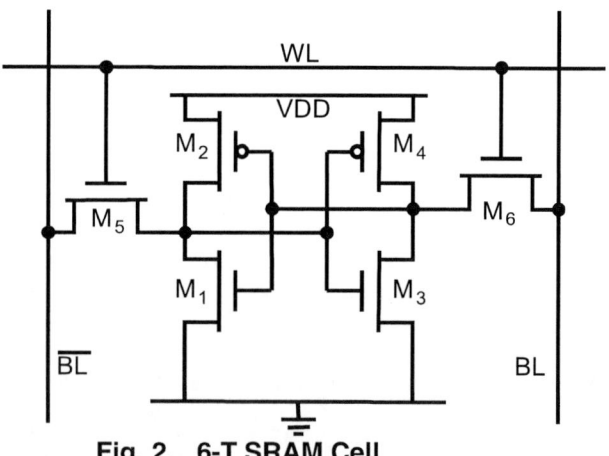

Fig. 2. 6-T SRAM Cell

0-7695-2533-4/06 $20.00 © 2006 IEEE

establishment time-constant [10]. τ_a and τ_b are the constants which depend only on process-related factors.

4.2 Soft Error Vulnerability Analysis

The analysis is performed on the SRAM cells design using FinFET and bulk CMOS. Conventional 6T-SRAM cell is shown in Fig.2. It consists of two NMOS pass gate transistors, two pull up PMOS transistors and two pull down NMOS transistors. To ensure robust read stability, the transfer ratio equals to 2 for both bulk CMOS based SRAM cell and FinFET-Based SRAM cell [16]. In other words, the pull down FinFET, M1 and M3, consists of two fins while the pass gate transistors, M5 and M6, consists of a single fin channel. For the bulk CMOS based SRAM cell, the W/L of pull down NMOS is twice of the W/L of pass gates [16]. To measure the critical charge, the current source shown in Equation (1) is injected at the storage node and the I_{peak} is swept until the bit flip occurs, the critical charge is then measured to be the soft error vulnerability metric. Note that here we assume that the charge collection for bulk CMOS device and FinFET device are the same.

FO4 (Fanout-of-4) logic gate is commonly used to characterize the specific technology. The analysis of the susceptibility of combination logic to soft errors is performed using FO4 inverter, which is an inverter with fan out of 4 inverters. The current pulse is injected at the output node of the inverter and the resultant transient voltage pulse is observed. We assume a soft error occurs when the transient voltage reaches a value of Vdd/2.

The analysis of soft error propagation in logic chain is performed using an inverter chain. 8 stage FO4 inverter chain is used in our experiments as shown in Fig.3. Current pulse is injected at the output of the first inverter and we measure the transient voltage propagated to the output of the logic chain. *Electrical masking* happens when the voltage transient resulting from a particle strike is attenuated by subsequent logic gates, because of the electrical property of the logic gate [4]. We assume soft error occurs, when the transient voltage pulse at the output of inverter 8 has a value of Vdd/2.

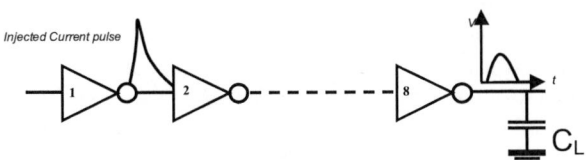

Fig. 3. 8 stage FO4 logic chain

4.3 Analysis Results

The soft error analysis for SRAM and combinational logic is performed for 32 nm and 45 nm technology using Predictive Technology Model [8].

Fig. 4 shows the critical charge reduces with the adoption of the FinFET devices to build the same designs, from the SRAM cells to logic chains. Comparing the critical charge of the FinFET based designs to that of the designs in bulk CMOS, we can see the average increases of 291% and 150% in the critical charge for 32nm and 45nm technology respectively.

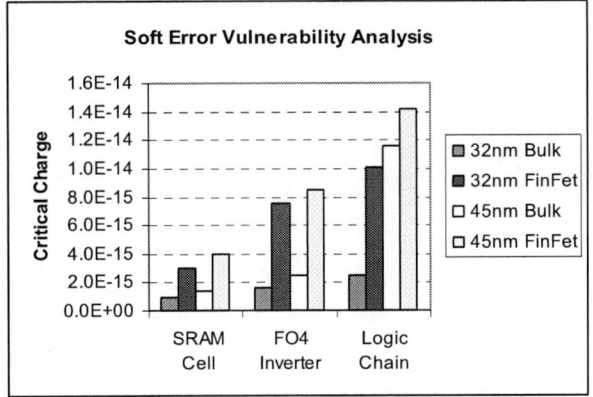

Fig. 4. Critical Charge of Different Designs in FinFET and Bulk CMOS in 32nm and 45nm technology

Fig. 5 shows the critical charge reduction when the design scales from 45nm to 32nm. Although the scaling of both FinFET and bulk CMOS causes a higher soft error susceptibility, the critical charge reduction of the FinFET device is much less than the bulk CMOS devices, when technology scales.

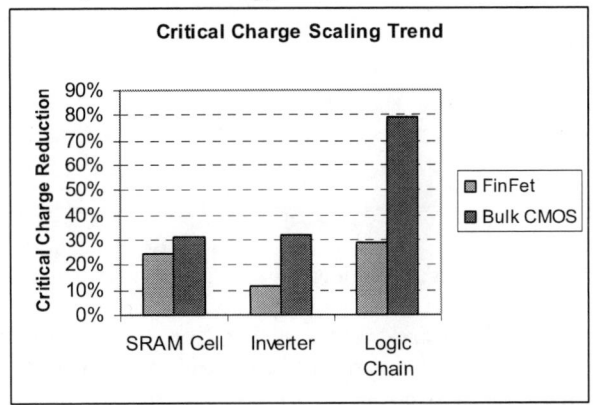

Fig. 5. Soft Error Vulnerability Scaling Trend for the FinFET Devices and Bulk CMOS devices.

Fig. 6 shows the effects the electrical masking effect of FinFET based and Bulk CMOS based inverter chains. In Fig. 6, the same height voltage pulse is

0-7695-2533-4/06 $20.00 © 2006 IEEE

generated at the output of stage 1 of logic chain, the FinFET based logic chain has less attenuation effects than that of the bulk CMOS, which increases the soft error susceptibility of FinFET logic chain in terms of the electrical masking effects. However, Figure 4 shows that it needs much higher collected charge (>4x) for a FinFET inverter to generate an identical transient voltage pulse, comparing with a bulk CMOS inverter. Therefore, even though the electrical masking attenuation in logic chain is getting worse when switch from bulk CMOS to FinFET, the FinFET circuits still have better soft error immunity than the bulk CMOS counterparts.

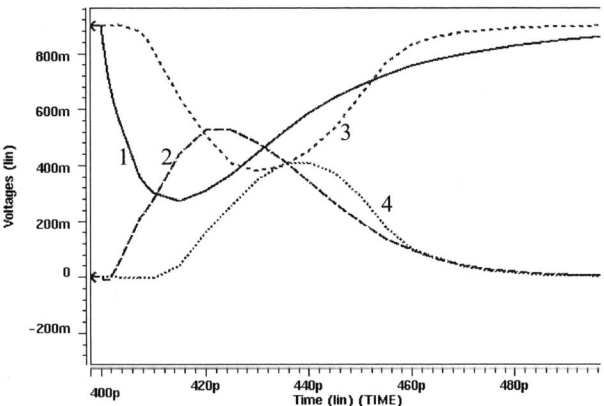

(a) Electrical masking effect in FinFET logic chain

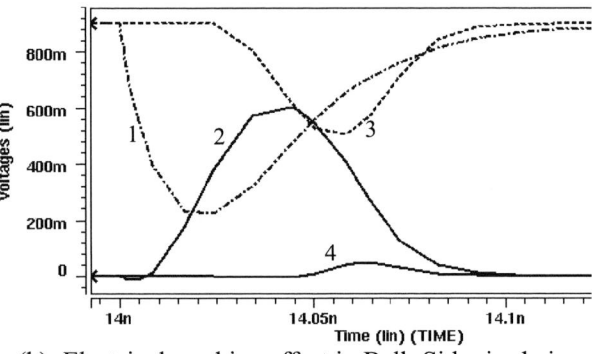

(b) Electrical masking effect in Bulk-Si logic chain

Fig. 6. Transient attenuation for an inverter chain based on (a) FinFET devices and (b) Bulk CMOS devices. The snap shot of output of stage 1, 2, 3 and 4 with the same height of the transient voltage pulse generated at the output of the stage 1

5. The Impact of Process Variation on FinFET Circuit

In this section, we study the impact of process variation on FinFET circuits and compare against that of bulk CMOS circuit. Section 5.1 gives a brief introduction on process variation, and Section 5.2 shows the impacts of process variation on SRAM and combination logic circuits.

5.1 The Impact of Process Variation

The challenges in fabricating small feature size transistors have resulted in significant variability in the transistor parameters such as channel length, gate oxide thickness and threshold voltage across identically designed neighboring transistors (intra-die variation) and across different identically designed chips (inter-die variation). The sources of variability results from different phenomena such as wafer misalignment, random dopant fluctuations, and imperfections in planarization steps. The divergence between designed and fabricated transistor parameters creates significant functional correctness concerns such as stability problems in memory cells, creation of new critical paths in the design and increased leakage currents. It has been shown that process variation can cause about 20x variation in chip leakage and 30% variation in chip frequency [13]. Process variations are of particular concern in memories since memories are typically designed using minimum feature sizes for density reasons, effects of process variation are most significant. Effects of variation on memory circuits can result in different reliability concerns of read stability, write failures, hold failures or access time increase.

Because the effect of other parameter can be translated into the effective variation in threshold voltage [14], threshold voltage fluctuation is considered as the major source of process variation when the performance impacts of the parameter fluctuations are investigated.

Traditional SRAM cell design using six transistors, as shown in Fig. 1, is used in this study. The access time distribution of the memory cell is used to evaluate the impacts of the process variation. The SRAM cell access time is defined as the time required for sufficient signal margin developed between bitline pair to trig the sense amplifier. A larger spread of the access time results in the large probability of the failure of the SRAM cells.

The distribution of the delay of the combination logic gate is used to analyze the impact of parameter fluctuation on logic circuits. Similar to the section 4, FO4 inverter gate is used to evaluate the delay of the combinational logic gate. The larger the spread in the delay distribution of the logic gates, the more vulnerable the circuits to the process variations.

5.2 Simulation Results

0-7695-2533-4/06 $20.00 © 2006 IEEE

Using HSPICE and Monte Carlo analysis, we consider both the intra- and inter-die variations, assuming both inter-die and intra-die 3σ Vt deviation of 15%. We obtained the normalized variance (normalize to the mean value) of the SRAM cell access time and logic gate delay distributions with 1000 Monte Carlo simulation runs. Note that the effect of other parameter variations can be translated into the effective variation in threshold voltage [14], and we assume that the bulk CMOS and FinFET device have the same range of parameter variation.

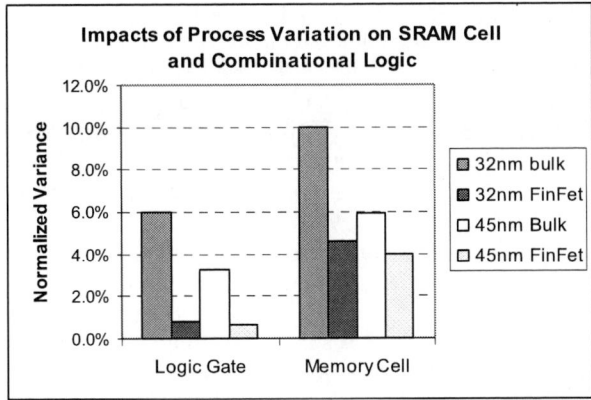

a) Normalized variance for logic gate delay distributions and for SRAM cell access time distributions

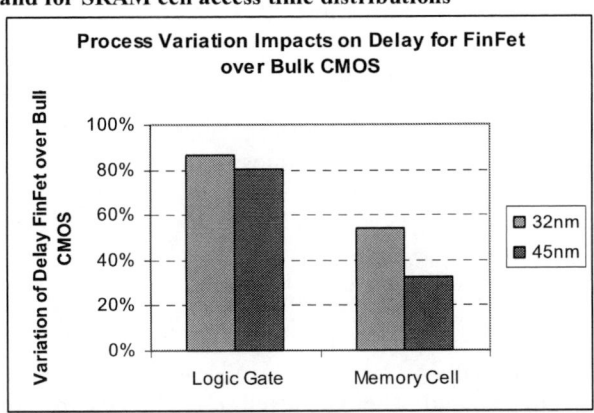

b) Variation reduction obtained by using the FinFET devices for SRAM cell and logic gates in both 32nm and 45nm technology

Fig. 7. The Impact of Process Variation on SRAM and Logic Gate

The delay variation performance of the FinFET Based design were compared to the designs in the bulk CMOS. Compared to the FinFET based design, the larger spread in the distribution of the bulk CMOS based design can be observed in Fig 7(a). Therefore, the designs in Bulk COMS is more vulnerable to the process variation than the FinFET based design. From Fig. 7(b), we can see FinFET based designs provide the average of 83% and 43% reduction of the delay

variation for logic gates and the memory cells over the designs in the Bulk CMOS.

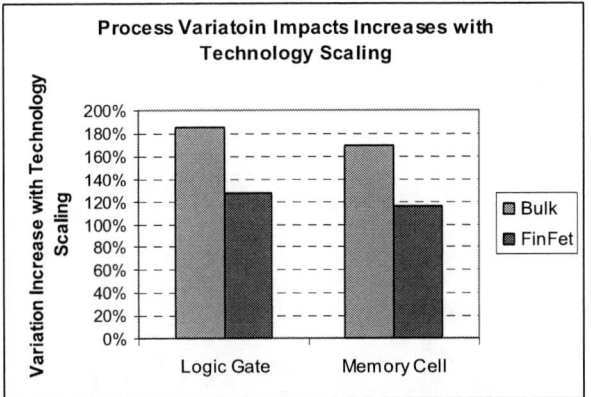

Fig. 8. Process Variation Increase with Technology Scaling from 45nm to 32nm.

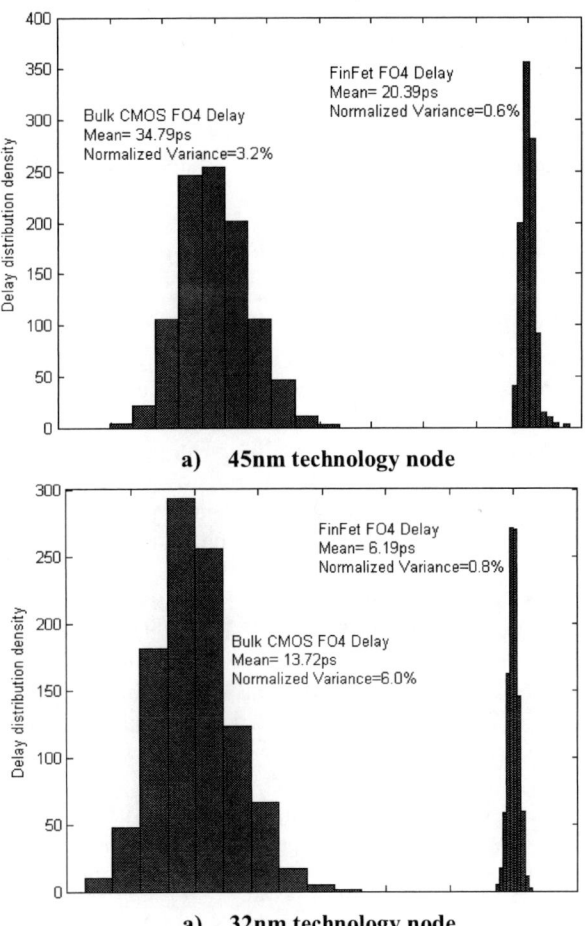

a)   45nm technology node

a)   32nm technology node

Fig. 9. Impact of process variations on the FO4 delay with 3σ Vt deviation of 15%

Similar to Section 4, we show how process variation effects will scale for both FinFET and Bulk CMOS. Fig. 8 shows that when technology scales from 45nm to 32 nm, both FinFET and bulk CMOS have a higher

0-7695-2533-4/06 $20.00 © 2006 IEEE   403

level of delay variation. However, the bulk CMOS is more vulnerable to the scaling. With the device scaling, we expect the FinFET based design become more robust than the designs in bulk CMOS.

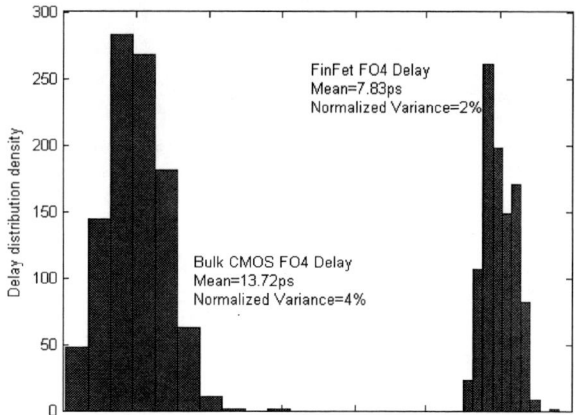

Fig. 10. Impact of process variations on the FO4 delay at 32nm technology with 3σ gate length (LG) deviation of 10%

Fig. 9 shows the impact of process variations on the FO4 delay with 3σ Vt deviation of 15%. It can be seen that the FinFET FO4 delay is much faster than the bulk CMOS, and FinFET circuits are less vulnerable to process variation. Fig. 10 shows a similar result for the simulation with a 3σ deviation of 15% for gate length.

6. Conclusion

Double-gate FinFET technology has been proposed as a promising alternative for bulk CMOS technology to continue the technology scaling. The performance or power advantages of FinFET circuits over bulk CMOS circuits have been studied by previous research. This paper provides the dependability analysis for FinFET circuits. The experiments compare the bulk CMOS circuits and FinFET circuits in 45nm and 32nm technologies, showing that FinFET circuits have better soft error immunity, and are less vulnerable to process variation. It is concluded that the FinFET-based circuit designs are much robust than the bulk CMOS counterparts.

Acknowledgement: The authors would like to thank Kevin Cao from Arizona State University for his help on FinFET PTM models. This work was supported in part by NSF Awards 0454123.

References:

[1] T. King, "FinFETs for Nanoscale CMOS Digital Integrated Circuits", *Proceedings of ICCAD*, 2005.
[2] K. Roy *et al.*, "Double-Gate SOI Devices for Low-Power and High Performance Applications", *Proceedings of ICCAD*, 2005.

[3] R.V. Joshi, R.Q. Williams, E. Nowak, K. Kim, J. Beintner, T. Ludwig, I. Aller, and C. Chuang, "FinFET SRAM for high-performance low-power applications", *Proceeding of the 34th European Solid-State Device Research conference*, pp. 69 – 72, 2004
[4] P. Shivakumar, M. Kistler, S.W. Keckler, D. Burger, and L. Alvisi., "Modeling the Effect of Technology Trends on Soft Error Rate of Combinational Logic", *International Conference on Dependable Systems and Networks* , pp. 389-398, June, 2002.
[5] Ron Ho, Kenneth W. Mai, and Mark A. Horowitz. The Future of Wires. In Proceedings of the IEEE, volume 89, pages 490–504, April 2001.
[6] R. Ho, K.W. Mai, and M. A. Horowitz, "The Future of Wires," *In Proceedings of the IEEE*, vol. 89, pp. 490–504, April 2001.
[7] B. A. Rainey, D. M. Fried, M. Ieong, J. Kedzierski, E. J. Nowak, "Demonstration of FinFET CMOS circuits", *Device Research Conference*, pp. 47-48,2002.
[8] Predictive Technology Model, http://www.eas.asu.edu/~ptm/
[9] Y. Taur, D. A. Buchanan, W. Chen, D. J. Frank, K. E. Ismail, S.-H. Lo, G. A. Sai-Halasz, R. G. Viswanathan, H.-J. C.Wann, S. J.Wind, and H.-S.P. Wong, "CMOS scaling into the nanometer regime," *Proc. IEEE*, vol. 85, no. 4, pp. 486–504, 1997.
[10] R. C. Baumann, "Soft errors in advanced semiconductor devices-part I: the three radiation sources", *IEEE Transactions on Device and Materials Reliability*, Vol. 1, pp. 17-22, March 2001.
[11] P. Hazucha, and C. Svensson, "Impact of CMOS Technology Scaling on the Atmospheric Neutron Soft Error Rate," *IEEE Transactions on Nuclear Science*, Vol. 47, No. 6, Dec. 2000.
[12] Z. Guo, S. Balasubramanian, R. Zlatanovici, T.-J. King, B. Nikolic, " FinFET-Based SRAM Design", *International Symposium on Low Power Electronics and Design*, pp. 2-7, 2005.
[13] S. Borkar *et al.*, "Parameter Variations and Impact on Circuits and Microarchitecture", *Proceedgins of DAC*, 2003.
[14] A. Agarwal, , B.C. Paul, S. Mukhopadhyay and K. Roy, "Process variation in embedded memories: failure analysis and variation aware architecture" *IEEE Journal of Solid-state Circuits*, Vol. 40, pp. 1804- 1814, September 2005.
[15] H. Ananthan, A. Bansal, and K. Roy, FinFET SRAM - device and circuit design considerations *Proceedings of the Sixth International Symposium on Quality Electronic Design (ISQED 2004)*, pp. 511-516, 2004.
[16] J. Rabaey, A.Chandrakasan, and B.Nikolic, "Digital Integrated Circuits", Prentice Hall, 2003.

0-7695-2533-4/06 $20.00 © 2006 IEEE

A Low-Power 2-Dimensional Bypassing Multiplier Using 0.35 um CMOS Technology

Chua-Chin Wang†, *Senior Member, IEEE*, Gang-Neng Sung
Department of Electrical Engineering
National Sun Yat-Sen University
Kaohsiung, Taiwan 80424
Email: ccwang@ee.nsysu.edu.tw

Abstract— **This paper presents a low power 8×8 digital multiplier design by taking advantage of a 2-dimensional bypassing method. The proposed bypassing cells constituting the multiplier skip redundant signal transitions when the horizontally partial product or the vertically operand is zero. Hence, it is a 2-dimensional bypassing architecture. Thorough post-layout simulations show that the power dissipation of the proposed design is reduced by more than 75% compared to the prior design with obscure cost of delay and area. A physical implementation of the proposed design using a standard 0.35 μm 2P4M CMOS process is also presented to justify the functionality as well as the low power performance of the 2-dimensional bypassing method.**

Keywords : low power multiplier, bypassing, CMOS, partial product, timing control

I. INTRODUCTION

Booming of battery-operated multimedia devices requires energy-efficient circuits, particularly digital multipliers which are building blocks of digital signal processors (DSP). Though many efforts have been focused on the improvement of adder and multiplier designs, [7], to challenge the GHz operations, the major trade-off of these GHz logic circuits is the high power consumption which is not a tolerable price to pay in recent mobile technologies [4]. Besides adders, digital multipliers are the most critical arithmetic functional unit in many DSP applications, e.g., Fourier Transform, DCT, filtering, etc. Array and parallel multipliers are very welcomed due to their high execution speed and throughput. However, the increasing capacitive wire load and operands' bit length result in very large power dissipation, [1], [2], [3], [5], [6]. Despite all of these difficulties, we still manage to reduce the power dissipation by an observation that the energy consumption of CMOS logic is proportional to the number of transitions, i.e., $P_{\text{diss}} \propto f \cdot C \cdot V^2$, where C is the load, V denotes the voltage swing, and f is the frequency of switches.

Many prior digital multipliers were aimed at transition or switch reductions to reduce power dissipation. A leapfrog multiplier was proposed in [6] by using a hardware bypassing approach to avoid the redundant computations by disabling the adder units whose partial product becomes zero. Another power saving approach is to skip the computation caused by th sign extension bits which are located at the left side of operands, e.g., [2]. What [5] proposed was close to a "bypassing" multiplier which skips the addition when the

partial product of a row is zero. [1] revealed another power-saving strategy by grouping the operands with the same sign and then computing them separately to avoid unnecessary transitions. All of these prior methods depend on certain decision logic given that a partial product is zero to either skip or shut down adding cells in a row-based manner. In other words, all of these prior works utilized a one-dimensional by-passing approach basically. We, thus, propose a 2-dimensional bypassing approach which detects the nullity of the partial products as well as the multiplicand at the same time to determine whether the adding cells on the corresponding row and those on the corresponding column are skipped or not, respectively. A 8×8 digital multiplier using the proposed 2-dimensional bypassing design is carried out by TSMC 0.18 μm 1P6M CMOS process. The post-layout simulations show that the power reduction compared to the prior multipliers is at least 75%.

II. 2-DIMENSIONAL BYPASSING MULTIPLIER

A basic guideline to reduce the power dissipation of a digital multiplier is to reduce its unnecessary switching activities. Hence, we proposed to detect the bitwise nullity of the multiplicand in the vertical direction and the partial product in the horizontal direction in an array multiplier to remove the unnecessary operations taken place in the corresponding adding cells.

A. Prior 1-dimensional bypassing algorithm

A typical array multiplication is based upon the following equation.

$$
\begin{aligned}
P &= P_{2n-1} \ldots P_1 P_0 \\
&= \sum_{i=0}^{n-1} \sum_{j=0}^{n-1} (X_i \cdot Y_j) 2^{i+j}
\end{aligned}
\tag{1}
$$

where P, $X = X_{n-1} \ldots X_1 X_0$, $Y = Y_{n-1} \ldots Y_1 Y_0$ are the product, the multiplier and the multiplicand, respectively. $P_k, k = 2n - 1, \ldots, 0$ denote the partial products, $X_i, i = n - 1, \ldots, 0$ and $Y_j, j = n - 1, \ldots, 0$ are respectively the bit representations of the multiplier and the multiplicand, and n is the bit length of the operands. A typical implementation of such a multiplier is the Braun's design which is given in

0-7695-2533-4/06 $20.00 © 2006 IEEE

Fig. 1. Every adding unit consists of an AND to carry out the multiplication and an FA (full adder) to accumulate the partial product. An $n \times n$ multiplication, thus, requires a total of $n(n-1)$ FAs and n^2 AND gates.

A simple thought to improve the power efficiency was proposed by [6]. As soon as X_i was found to be zero, the corresponding partial product (row direction) is automatically reset and bypassed to avoid triggering those adding units in the row. Hence, two MUXs (multiplexer) are required in the adding unit to realize the bypassing operation. Meanwhile, there is a possibility that bypassing operations will result in the truncation of the carry from the corresponding previous stages. A total of $2(n-1)^2$ MUXs must be included to resolve this problem. A 4×4 multiplier example using such an implementation is shown in Fig. 2.

B. 2-dimensional bypassing design

Besides the power saving by row-based bypassing, we propose a 2-dimensional bypassing which detects the bitwise nullity of the multiplicand bits, Y_j's, in addition to the state of the multiplier, X_i's. In other words, as soon as the Y_j is found to be zero, the results from the adding units residing in the previous column are automatically passed to the corresponding adding units in the next column. However, a conflict appears when one adding cell, AC_{ij}, encounters a scenario that $X_i = Y_j = 0$.

For instance, assume $i = 2, j = 1$ and $X_2 = Y_1 = 0$ in Fig. 3 which shows a 2-dimensional bypassing 4×4 multiplier design. Then, we expect the second row and the second column are bypassed if we directly apply the prior 1-dimensional bypassing method. If the carry out of the adding cell AC_{02} is "1", it should be propagated to the carry in of AC_{21} and then its carry out. However, the carry bit will be lost if AC_{21} is bypassed due to $Y_1 = 0$. Consequently, an error is occurred, since the carry out of AC_{21} will be zero. We, thus, propose to include a bypass logic in certain adding cells.

C. Adding cell with bypass logic

According to the illustrative example, a simple rule is introduced. If $X_i = Y_j = 0$ and the carry out of $AC_{(i-2)(j+1)}$ is 1, then adding cell, AC_{ij} can not be bypassed. Hence, an adding cell with the bypass logic is proposed in Fig. 4. Notably, all of the 3 tri-state buffers as well as the two MUXs are gated by the output signals of the embedded bypass logic. Notably, an adding cell with the bypass logic is represented with a gray box in Fig. 3 and the other figures in this work.

D. Domino effect in large multipliers

It is obvious that not every adding cell needs the bypass logic. For instance, those adding cells in charge of the calculation of LSBs of X and Y. It will be very area-efficient if we can identify which adding cells require the bypass logic to produce a correct multiplication result. Given $n = 4$, it can be easily concluded that AC_{21} is the only unit with the necessity

of a bypass logic. If $n = 5$ and the identical array structure is used, then $AC_{21}, AC_{22}, AC_{31}, AC_{32}$ need the bypass logic to attain correct results. By a similar induction, for any $n \times n$ multipliers, where $n \geq 4$, all of the adding cells, AC_{ij}, where $n - 2 \geq i \geq 2$ and $n - 3 \geq j \geq 1$, must contain the bypass logic to execute the correct multiplication. In other words, when $n = 4$, there is only one adding cell which must contain the bypass logic. If $n = 5$, then the 5×5 multiplier has a total of $(5-3) \times (5-3) = 4$ adding cells with bypass logic. If $n = 8$, a total of $(8-3) \times (8-3) = 25$ adding cells with bypass logic are required, as shown in Fig. 5. In short, the number of the required adding cells with bypass logic is as follows.

$$
\begin{aligned}
1 &= (4-3) \times (4-3), & n = 4 \\
4 &= (5-3) \times (5-3), & n = 5 \\
9 &= (6-3) \times (6-3), & n = 6 \\
\vdots &= \vdots
\end{aligned}
$$

Therefore, the following rule is concluded.

Theorem 1 : A total of $(n-3)^2$ adding cells with bypass logic are required to constitute a 2-dimensional bypassing multiplier, $\forall n > 3$.

Notably, if an earlier adding cell with the bypass logic is set to be activated, all of the following adding cells in the same column must be activated, too. Otherwise, a carry generated in the earlier adding cell will be lost in the bypassing chain. For instance, if the adding cell AC_{22} is activated, then the following adding cell, AC_{32}, must be activated to ensure a carry (=1) is propagated correctly from the carry in of AC_{22} to the carry out of AC_{32} and even further. Namely, it is a domino effect of activation of adding cells in the same column.

III. SIMULATION AND IMPLEMENTATION

TSMC (Taiwan Semiconductor Manufacturing Company) 0.35 2P4M CMOS process was adopted to carry out the proposed design. Fig. 6 and Fig. 7, respectively, shows the layouts of a normal adding cell and an adding cell with bypass logic. The area penalty is 33% increase for a single adding cell. The power dissipation of the proposed 8×8 multiplier using 2-dimensional bypassing method has been simulated at all PVT corners (process transistor models, power supply voltage variations, and temperatures). The critical path delay is 13.0 ns. The simulations are carried out by HSPICE Monte Carlo method with sweep = 30. The outcome is tabulated in Table I.

Fig. 8 is the diephoto of the proposed 8×8 multiplier using 2-dimensional bypassing. The core of the chip is merely 363×208 μm^2. The physical measurement on silicon by Agilent 16702B Logic Analysis System is given in Fig. 9. A comparison of the proposed design with several prior 8×8 multiplier designs is summarized in Table II. It is obvious that the proposed design possesses the edge of low power.

IV. CONCLUSION

We have proposed a low power digital multiplier design by taking advantage of a 2-dimensional dynamic bypassing method. The post-layout simulations by HSPICE Monte Carlo method justify the advantage of the proposed design in terms of power dissipation. By a small area penalty, we gain more than 75% power saving compared to the prior designs. Physical implementation and measurement of the proposed design using a standard 0.35 μm 2P4M CMOS process also justify the functionality as well as the low power performance of the 2-dimensional bypassing method.

ACKNOWLEDGMENT

The authors would like to thank National Science Council (NSC), since this research was partially supported by NSC under grant 92-2220-E-110-001 and 92-2220-E-110-004.

REFERENCES

[1] T. Ahn, and K. Choi, "Dynamic operand interchange for low power," *Electronics Letters,* vol. 33, no. 25, pp. 2118-2120, Dec. 1997.

[2] J. Choi, J. Jeon, and K. Choi, "Power minimization of functional units by partially guarded computation," *2000 International Symposium on Low Power Electronics and Design (ISLPED'00),* pp. 131-136, July 2000.

[3] J. Di, J. S. Yuan, and M. Hagedorn, "Energy-aware multiplier design in multi-rail encoding logic," *The 2002 45th Midwest Symposium on Circuits and Systems (MWSCAS-2002),* vol. 2, pp. 294-297, Aug. 2002.

[4] W. Hwang, G. D. Gristede, P. N. Sanda, S. Y. Wang, and D. F. Heidel, "Implementation of a self-resetting CMOS 64-bit parallel adder with enhanced testability," *IEEE J. Solid-State Circuits,* vol. 34, no. 8, pp. 1108-1117, Aug. 1999.

[5] S. Hong, S. Kim, M. C. Papaefthymiou, and W. E. Stark, "Low power parallel multiplier design for DSP applications through coefficient optimization," *1999 Twelfth Annual IEEE International ASIC/SOC Conference,* pp. 286-290, Sep. 1999.

[6] J. Ohban, V. G. Moshnyaga, and K. Inoue, "Multiplier energy reduction through bypassing of partial products," *2002 Asia-Pacific Conference on Circuits and Systems (APCCAS '02),* vol. 2, pp. 13-17, Oct. 2002.

[7] C.-C. Wang, C.-J. Huang, and K.-C. Tsai, "A 1.0 GHz 0.6-μm 8-bit carry lookahead adder using PLA-styled all-N-transistor logic," *IEEE Trans. of Circuits and Systems, Part II : Analog and Digital Signal Processing,* vol. 47, no. 2, pp. 133-135, Feb. 2000.

[8] C.-C. Wang, Y.-L. Tseng, P.-M. Lee, R.-C. Lee, and C.-J. Huang, "A 1.25 GHz 32-bit tree-structured carry lookahead adder using modified ANT logic," *IEEE Trans. on Circuits and Systems − I Fundamental Theory and Applications,* vol. 50, no. 9, pp. 1208-1216, Sep. 2003.

SS model	TT model	FF model
10.719 mW	12.376 mW	14.980 mW

TABLE I

POWER DISSIPATION OF THE 8 × 8 MULTIPLIER USING 2-DIMENSIONAL BYPASSING @ 77 MHZ DATA RATE

Braun's	100 mW
[6]	62 mW
[1]	67 mW
ours	47.91 mW (with pads power)

TABLE II

COMPARISON WITH THE PRIOR 8 × 8 MULTIPLIER DESIGNS

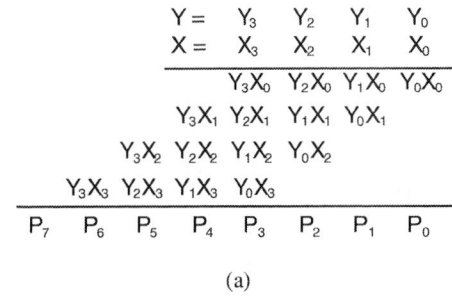

(a)

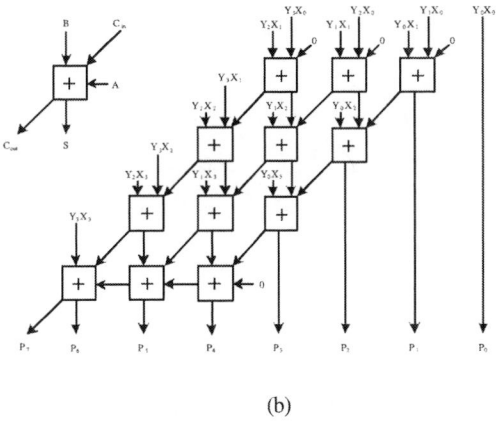

(b)

Fig. 1. Generic array multiplier

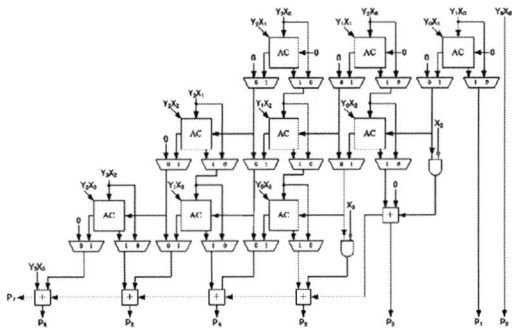

Fig. 2. Prior 1-dimensional bypassing multiplier design

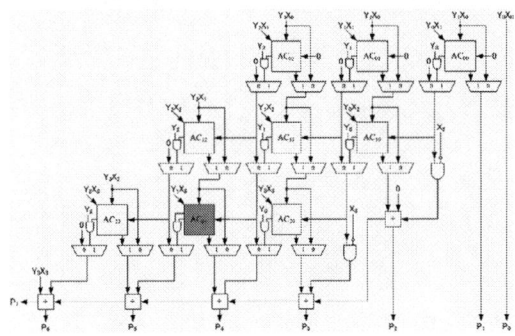

Fig. 3. Proposed 2-dimensional bypassing multiplier (4×4)

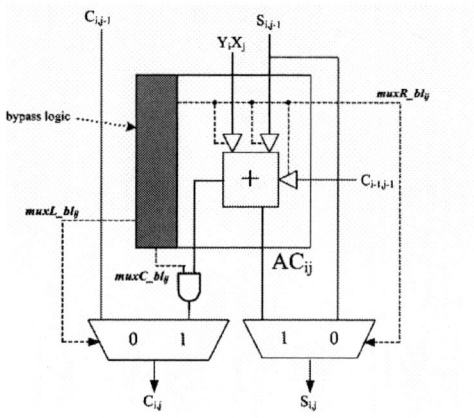

$$muxR\_bl_{ij} = Y_{i+1} \cdot X_j + \overline{X_i} \cdot \overline{Y_j} \cdot C_{i-2,j+1} \cdot X_{i+1} \big|_{i-2,\,n-2\xi,\,j\xi1}$$

$$muxR\_bl_{ij} = Y_{i+1} \cdot X_j + \overline{X_i} \cdot \overline{Y_j} \cdot C_{i-2,j+1} \cdot X_{i+1} + \overline{X_{i-1}} \cdot \overline{Y_{j-1}} \cdot C_{i-3,j} \cdot X_i \big|_{n\xi i\xi 3,\,n-2\xi,\,j\xi1}$$

$$muxL\_bl_{ij} = Y_{i+1} + \overline{X_i} \cdot \overline{Y_j} \cdot C_{i-2,j+1} \cdot X_{i+1} \big|_{i-2,\,n-2\xi,\,j\xi1}$$

$$muxL\_bl_{ij} = Y_{i+1} + \overline{X_i} \cdot \overline{Y_j} \cdot C_{i-2,j+1} \cdot X_{i+1} + \overline{X_{i-1}} \cdot \overline{Y_{j-1}} \cdot C_{i-3,j} \cdot X_i \big|_{n\xi i\xi 3,\,n-2\xi,\,j\xi1}$$

$$muxR\_Cl_{ij} = X_j + \overline{X_i} \cdot \overline{Y_j} \cdot C_{i-2,j+1} \cdot X_{i+1} \big|_{i-2,\,n-2\xi,\,j\xi1}$$

$$muxR\_Cl_{ij} = X_j + \overline{X_i} \cdot \overline{Y_j} \cdot C_{i-2,j+1} \cdot X_{i+1} + \overline{X_{i-1}} \cdot \overline{Y_{j-1}} \cdot C_{i-3,j} \cdot X_i \big|_{n\xi i\xi 3,\,n-2\xi,\,j\xi1}$$

Fig. 4. Scehmatic of the bypass logic

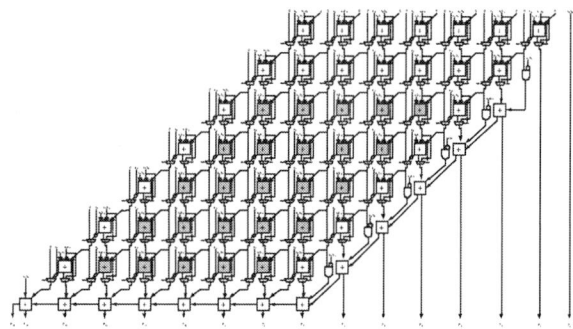

Fig. 5. A 8×8 multiplier using 2-dimensional bypassing

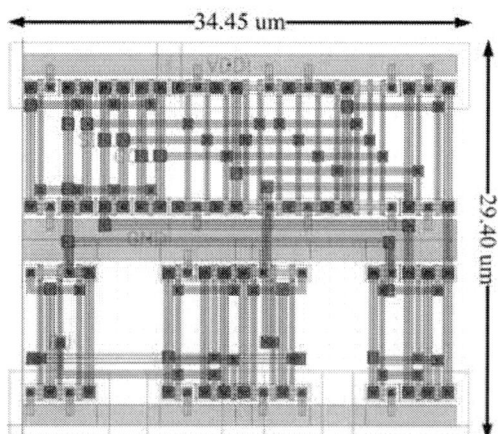

Fig. 6. Layout of a normal adding cell

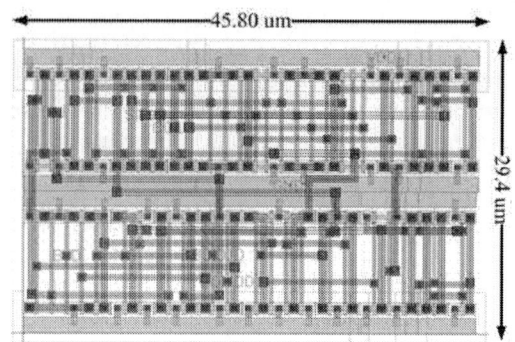

Fig. 7. Layout of the adding cell with bypass logic

Fig. 8. Die photo of the 8 × 8 multiplier using 2-dimensional bypassing

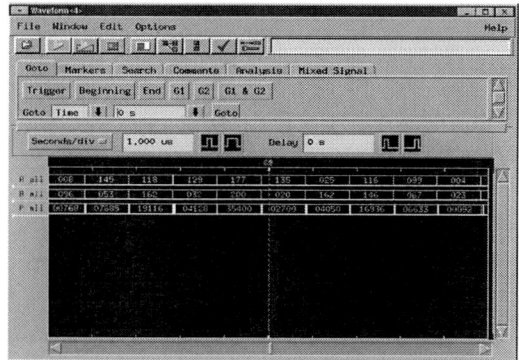

Fig. 9. Measurement of the low-power 8 × 8 multiplier on silicon

0-7695-2533-4/06 $20.00 © 2006 IEEE

Poster Papers

Multi-Level Buffer Block Planning and Buffer Insertion for Large Design Circuits

Ali Jahanian   Morteza Saheb Zamani

IT and Computer Engineering Department, Amirkabir University of Technology, Tehran, IRAN.
{ jahanian, szamani@ce.aut.ac.ir }

Abstract

Buffer insertion plays an increasingly critical role on circuit performance and signal integrity, especially in deep submicron region. Buffer insertion stage is very important for buffering efficiency. Early buffer insertion (e.g. at the floorplanning stage) may cause misestimation due to unknown cell locations, on the other hand buffer insertion after placement or during global routing may tend to be ineffective because the cell locations have been fixed and buffer resources may be distributed inappropriately.

In this paper, a new method for buffer insertion is presented which inserts buffers during placement based on the planning of buffers at the floorplanning stage and congestion considerations. Experiments show that by our method, performance and congestion control are improved in large circuits including large amount of buffers.

Keywords

Buffer planning, buffer insertion, incremental placement

1. Introduction

As technology continues to scale down, the sizes of transistors and modules are getting smaller, and a significant portion of circuit delay is coming from interconnects. In some advanced systems, as much as 80% of the clock cycle is consumed by interconnects [1].
Buffer insertion is widely recognized as a key technology for improving the performance of VLSI interconnects performance. As transistor count and chip dimension get larger and larger, more and more buffers are expected to be needed for high performance designs. It was projected that over 700K buffers will be inserted on a single chip in the 70nm technology [2].

Buffers can be inserted in floorplanning, placement or global routing stages in physical design flow. In large circuits with large number of buffers, buffer insertion in floorplanning stage may cause misestimation of critical paths and buffer requirements of various regions of the layout due to the unknown cell locations. In this stage, cells are placed on the center (or top-left) of clusters since there is no concise information about wire length of nets.

On the other hand, after placement the location of buffers and cells are fixed and there is no guarantee that buffer distribution is capable of meeting buffer requirements in clusters. Therefore, buffer insertion after placement may not efficiently use buffer resources. Besides, buffer insertion after placement may override or invalidate the floorplanning estimations because there is no information about floorplanning assumptions (such as global cut size, cluster congestion map, and critical path estimation) at the placement and global routing stages.

In this paper, a new approach for buffer planning and insertion is presented which estimates global region for each buffer in floorplanning stage and final locations of buffers will be fixed during placement in concerning with floorplanning assumptions about the location of buffers.

This approach uses the global view and the assumptions of the floorplanning stage (e.g. congestion map of global clusters, buffer block planning and critical paths) are incorporated into the placement optimization process concurrently.

In Section 2, a brief review on buffer insertion techniques is presented. In Section 3, the proposed method for buffer planning and buffer insertion is described and in Section 4, experimental results are reported. Finally, Section 5 concludes the paper.

2. Previous work

In [3], buffer insertion was performed by dynamic programming and improved in [4] by STree, PTree, SPTree, and CTree algorithms. In [5], a buffer insertion algorithm during the global routing was presented and in [6] a probabilistic approach for buffer insertion was proposed.

By increasing the number of required buffers in deep submicron technologies, conventional methods for buffer insertion are not efficient and buffer distributions must be planned in earlier phases of physical design such as the floorplanning stage. A good planning of module positions during the floorplanning stage so that buffers can be inserted wherever needed in the later routing stages can be useful. In [7], a new method for buffer planning was presented in which buffers were planned in the floorplanning stage with respect to interconnect delays.

In [8], add in the notion of independence to feasible regions so that these regions for different buffers of a net can be computed independently and in [9], a routability-driven floorplanner with congestion estimation and buffer block planning was presented.

These methods use the planning of buffers in the floorplanning stage without using these plans in later stages of physical design.

2. Algorithm Description

The proposed method has two major stages; "floorplanning with buffer planning" and "buffer insertion during placement". In the first stage, the design is floorplanned and a global region is estimated for each buffer using the algorithm is proposed in [7] and in the second stage, the final location of buffers are fixed during cell placement with respect to buffer planning performed in the first stage with congestion control. In this approach, the global information of the floorplanning stage are incorporated into the local optimizations of the cell placement process to improve buffer insertion quality.

Our experimental results show that this approach is well suited for large circuits with large number of buffers and the results are improved by increasing the size of input designs.

In the next subsections we propose these stages in more details.

3.1. Floorplanning with Buffer Block Planning

In the first stage of the algorithm, the design is floorplanned, critical paths and nets are estimated, buffers are planned for critical nets, and a feasible region are assigned to each buffer. This region is a region in which inserting a buffer meets the timing constraints. The tasks of this stage consist of seven phases:

Phase 1: The design is partitioned to clusters and these clusters are floorplanned by a slicing multi-level algorithm. We use hMetis algorithm for partitioning [10]. Balance factor of partitioning is very important factor. With larger balance factor, better partitions are obtained in terms of cut size, but this may create clusters with bad aspect ratios (i.e. very narrow or very tall clusters). We use 25 for balance factor which means that the size of partitioned clusters may different at most by 25%. In this phase of algorithm, the cells of each cluster are placed at the center of it.

Phase 2: The congestion level of each cluster is estimated using Rent's rule [11] that is a system level congestion estimation method based on the complexity of interconnects and terminals. Using Rent's rule algorithm, the interconnect complexity of design is describe based on the number of logic blocks in a module and the number of its external terminals.

Phase 3: After floorplanning, list of critical paths is estimated using a static timing analysis based on timing requirements and clusters locations. This list is a high level estimation of final critical paths at floorplanning stage.

Phase 4: Global nets are those nets which pass between clusters. These nets are the appropriate candidates for buffer insertion. We select those nets of the estimated critical path that travel between clusters as critical. If the design becomes larger, more accurate estimations will be achieved by this method, due to the smaller ratio of clusters dimensions to the layout dimensions.

Phase 5: Along all the global/critical nets, several buffers are inserted. The number and the distance between buffers along a net is estimated using the algorithm in [7].

Phase 6: For any planned buffer in Phase 5, a feasible region is designated in which timing constraints can be satisfied. Feasible region estimation is improved by increasing the size of design. When a buffer is placed inside its feasible region, it is guaranteed that the timing constraints are meet whereas placing buffers outside the feasible regions, maybe result in timing violation.

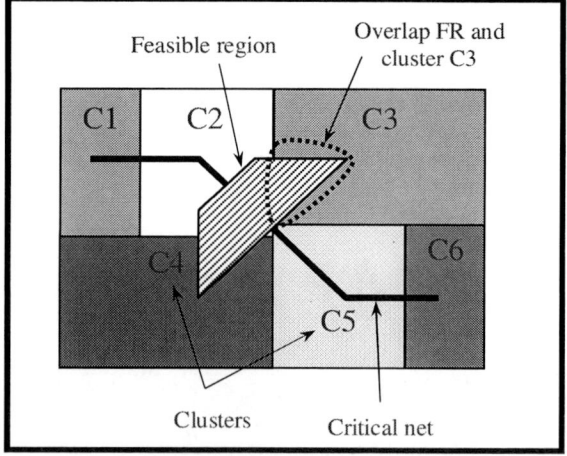

Figure 1. Floorplanned clusters, congestion level of clusters, a critical path and the overlapping area of a feasible region with cluster C3

Phase 7: Each planned buffer must be assigned to a cluster of design. We consider two factors to assign a buffer to a cluster: "The intersection area of the buffer's feasible region" and "the cluster and the congestion level of the cluster that overlap with the buffer's feasible region".

In Figure 1, clusters C1 to C6, an estimated critical path and a critical net is shown. The feasible region of critical net is the dashed trapezoid and the overlapping area of feasible region with the cluster C3 is highlighted by a dotted closed curve. The congestion level of each cluster is shown using gray scale colors where darker clusters are more congested.

For assigning buffers to clusters, *ClusterSuitability* factor is defined as follows:

$$ClusterSuitability = \alpha * \frac{1}{Cg} + \beta * Ov \qquad (1)$$

Where Cg is the congestion level of each cluster which is overlapped with feasible region and Ov is the amount of overlapping area between the feasible region and the cluster.

In (1), α and β are balance factors. If α is considerably grater than β, buffers are assigned to the cluster with lower congestion. Therefore, new buffers are inserted in clusters with lower congestion and result in better congestion in the final layout.

On the other hand, If β is considerably grater than α, buffers are assigned to the cluster with larger overlap area with the feasible region. As will be seen, placement algorithm will place any buffer in its feasible region. Therefore, with large β, buffer and cell placement optimization is improved without regarding congestion control.

3.2. Buffer Insertion During Placement

We modified Dragon placement algorithm [12] to place cells and buffers concurrently. Dragon has two phases, namely global placement (GP) and detailed placement (DP). In GP, the design is partitioned iteratively into four partitions and at the end of each partitioning level, a low temperature simulated annealing is performed to place the partitions and minimize the total wirelength. In DP, the cells in any partition are placed and the layout is legalized.

In our approach, buffers are marked as cells with positions restricted within their feasible region. In other words, placer is forced to place each buffer inside its feasible region. The Dragon's GP stage was modified such that after each 2-way partitioning, if a buffer is located outside its feasible region, it is assigned to another partition to fit into its feasible region. In addition, the cost function in the simulated annealing algorithm for GP was modified to fit the

buffers in their fixed regions. In our algorithm, perturbations that place buffers outside their feasible region are fined with a large penalty to force the placer to place buffers inside within their feasible region.

In this way, in the proposed method, the buffers and cells are placed with respect to the fixed region constraints planned in the floorplanning stage.

4. Experimental Results

In order to test our proposed algorithm, the conventional buffer insertion flow and our proposed flow were implemented using C++ language on a 2.4GHz Intel Pentium processor and the results of these two flows were compared. The conventional flow for buffer insertion (PPBI[1]) is shown in Figure 2.

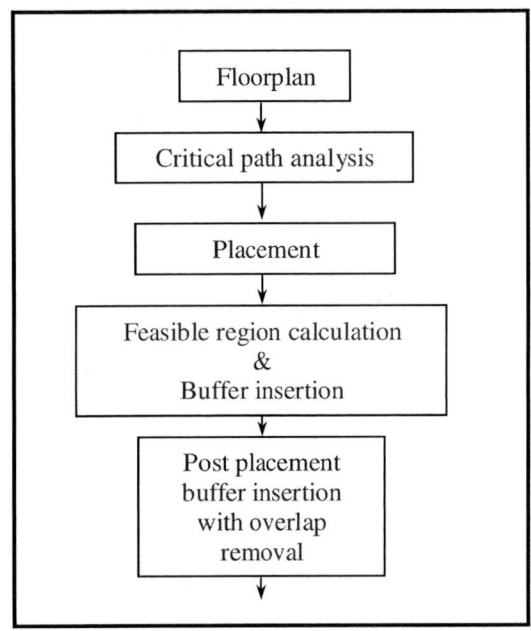

Figure 2. Post placement buffer insertion (conventional method for buffer insertion)

In the conventional flow, the design is partitioned, floorplanned and placed, and then the critical paths are extracted and several buffers are inserted to meet the timing constraints. In this flow, buffer insertion is performed by an incremental placer. The feasible region of each buffer is calculated and the buffer is inserted within its feasible region. This may result in some overlaps which are removed using simple bisection technique proposed in [13].

In Figure 3, our proposed flow for buffer insertion is shown. After floorplanning, critical paths were estimated and buffers are planned for the critical nets. For each buffer, its feasible region was calculated and

---

[1] Post placement buffer insertion

buffers where assigned to their feasible regions. this information was fed into placement environment. In placement, the design was placed considering these constraints (i.e. each buffer was restricted to its feasible region). We called this flow as BIDP[1].

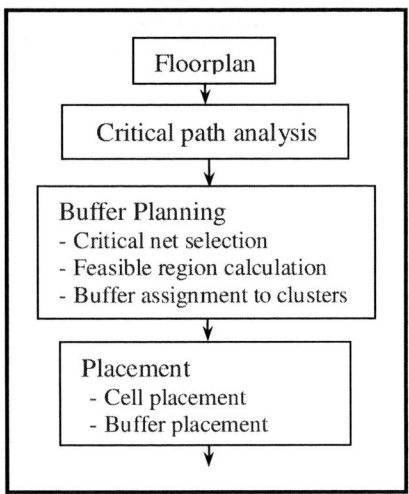

Figure 3. Buffer insertion during placement

These two flows were tested on three industrial circuits that are shown in Table 1. IBM benchmarks were not suitable because they do not have any information about internal memory cells and loops. MCNC benchmarks are very small to test our idea. In all experiments, we used $\alpha = 75\%$ and $\beta = 25\%$ in (1).

Table1. Input benchmarks

Benchmark	#Nodes	#Nets
Circuit No.1	6500	5000
Circuit No.2	15300	12000
Circuit No.3	32000	30000

The experimental results are shown in Table 2 and Table 3. As shown in tables 2. and 3., timing violation was not observed in BIDP, but it was seen that in PPBI 6% to 10% of buffers violate their timing constraints. Timing violations are occurred when a buffer cannot be placed inside its feasible region [7]. This is mainly because in PPBI, buffers are inserted after placement and maybe placed outside their feasible region during overlap removal phase but in BIDP, buffers are limited to their feasible region during placement and placer places them inside their feasible region. For large designs, the results of BIDP were better and when the number of buffers grew, the timing violation of BIDP remains zero, while timing violation of PPBI increase for large number of buffers.

---

[1] Buffer insertion during placement

Experiments show that for small circuits, area and wirelength of PPBI was better that BIDP.

Table 2. Normalized values for area, wirelength and congestion with PPBI flow

Circuit	Timing Failure	Layout Area	Total Wirelength	Congestion
No.1	12%	1	1	1
No.2	8.5%	1	1	1
No.3	6%	1	1	1

Table 3. Normalized values for area, wirelength and congestion with BIDP flow

Circuit	Timing Failure	Layout Area	Total Wirelength	Congestion
No.1	0%	1.2	1.09	0.98
No.2	0%	1.14	1.02	0.98
No.3	0%	1.08	0.97	0.92

As can be seen, the increase in area and wirelength in BIDP was small, specially for larger circuits. In fact, the total wirelength of for circuit No.3 in BIDP is been better than PPBI. This is because the placer in BIDP has wider optimization area to place cells and buffers than PPBI.

In order to evaluate the congestion level of the design after buffer insertion, we define **TotalCg** factor as (2). **TotalCg** describes the increase in congestion level for all clusters after buffer insertion.

$$TotalCg = \sum_{for\ all\ clusters} (Cg * Diff) \qquad (2)$$

where **Cg** is the congestion level of a cluster before buffer insertion and **Diff** is the increase in the congestion level. Experimental results show that **TotalCg** of BIDP is better than PPBI about 2% for circuits No.1 and No.2 and 8% for circuit No.3. In other words, congestion was more evenly distributed by BIDP flow. In BIDP buffers are kept outside the congested clusters during buffer planning. Therefore, it avoids increasing the congestion level of the congested clusters. However, PPBI inserts buffers regardless of the congestion levels of clusters.

5. Conclusion

In this paper, an effective buffer planning and insertion method was presented in which buffers are planned in the floorplanning stage and their detailed placement was performed later with respect to the planning of buffers obtained from the floorplanning stage. Buffer insertion is performed using both floorplanning global information (such as the congestion map of clusters and critical paths) and placement optimization, concurrently.

Experimental results showed that the performance and congestion control were improved for large circuits with large amount of buffers, while the total wirelength and area was slightly increased.

6. References

[1] D. Hill, and A. B. Kahng, "RTL to GDSII-From foilware to standard practice," *IEEE Design & Test of Computers*, 2004.

[2] Y. Ma, X. Hong, S. Dong, S. Chen, Y. Cai, C.K.Cheng, and J. Gu ,"An integrated floorplanning with an efficient buffer planning algorithm", *ISPD*, pp. 136-142, 2003.

[3] L. P. van Ginneken, "Buffer placement in distributed RC-tree networks for minimal Elmore delay", *ISCAS*, pp. 865–868, 1990.

[4] C. J. Alpert, et. al., "Buffered steiner trees for difficult instances", *ISPD*, pp. 4-9, 2001.

[5] M. Lai, and D. F. Wong, "Maze routing with buffer insertion and wiresizing", *DAC*, pp. 374–378, 2000.

[6] V. Khandelwal, A. Davoodi, A. Nanavati, and A. Srivastava, "A probabilistic approach to buffer insertion", *ICCAD*, pp. 560-567, 2003.

[7] J. Cong, T. Kong, and D. Z. Pan," Buffer block planning for interconnect-driven floorplanning", *ICCAD*, pp. 358-363, 1999.

[8] P. Sarkar and C. K. Koh, "Routability-driven repeater block planning for interconnect-centric floorplanning", *IEEE Transactions on CAD of Integrated Circuits and Systems*, Vol. 20, pp. 660-671, 2001.

[9] K. W.C. Wong, and E. F.Y. Young, "Fast buffer planning and congestion optimization in interconnect-driven floorplanning", *ASP-DAC*, 2003.

[10] G. Karypis and V. Kumar, "Parallel multilevel k-way partitioning scheme for irregular graphs", *SIAM Rev.*, Vol.41, No.1, pp. 278-300, 1999.

[11] D. Stroobandt, "A priori system-level interconnect prediction: Rent's rule and wire length distribution models", *SLIP*, pp. 3-21, 2001.

[12] M. Wang, X. Ynag, and M. Sarrafzadeh, "DRAGON2000: standard-cell placement tool for large industry circuits", *ICCAD*, pp. 260-263, 2000.

[13] W. Choi, and K. Bazargan, "Incremental placement for timing optimization", *ICCAD*, pp. 463-466, 2003.

0-7695-2533-4/06 $20.00 © 2006 IEEE

Effect of Glitches on the Efficiency of Components' Region-Constrained Placement as a Fast Approach to Reduce FPGA's Dynamic Power Consumption

Seyed E. Esmaeili[1], Nabil I. Khachab[1], Moustafa Y. Ghannam[2]

[1]EE Department
Kuwait University
P.O. Box 5969, Safat 13060, Kuwait
E-mail: khachab@eng.kuniv.edu.kw

[2]Department of Physics
The American University in Cairo
E-mail: mghannam@aucegypt.edu

Abstract

The effect of components' region-constrained placement on reducing internal nets total capacitance and the corresponding change in internal nets' total dynamic power consumption is investigated. Two logic circuits were specified as components covering around 80% of total FPGA busy gates. These components are multiplexers and adders along with multipliers. Each of these components was implemented on two of Xilinx FPGA's families, namely; Spartan II and Virtex. Gate-level power estimation for different region-constrained placements of each logic circuit was carried out using the Xilinx hierarchal power distribution analyzer, XPower.

1. Introduction

Current FPGAs play many important roles, ranging from small glue logic replacement to System-on-Chip designs. In rising to the challenge to reduce power, the semi conductor industry has adopted a multifaceted approach, attacking the problem on three fronts:

- Reducing chip capacitance through process scaling. This approach to reducing the power is very expensive.
- Reducing voltage. While this approach provides the maximum relative reduction in power it is complex, difficult to achieve and requires moving the systems industries to a new voltage standard.
- Employing better architectural and circuit design techniques. This approach promises to be the most successful because of its relatively small cost in comparison to the other two approaches.

1.1 Related Work

Several analysis of FPGAs power consumption have appeared in the literature. It was shown that power dissipation in FPGAs devices is predominantly in the programmable interconnection network. It was reported that as high as 50% to 70% of total power is dissipated in the interconnection network, with the remainder being dissipated in the clocking, logic, and I/O blocks. The dominant role of interconnect in total FPGA power consumption implies that characterization and management of net capacitance is a crucial part of a power–aware FPGA design. Leakage power reduction in FPGAs was proven to be achievable through region-constrained placement where the FPGA fabric was divided into small regions and the power supply to each region was switched on/off using a sleep transistor [1]. In this paper, the effectiveness of region-constrained placement of different logic components as a fast approach to reduce dynamic power consumption is investigated.

2. Theory and Implementation

Placement is essentially assigning a unique position inside the FPGA to each of the configurable logic blocks of the logical circuit to be implemented. Placement has essentially been used as a mean to reduce area and path delay. In the following, dynamic power reduction through region-constrained placement for different components is investigated.

A VHDL code was written to implement a 32-to-1 line multiplexer. In addition, a combination of a 16 by 16 bits add/shift multipliers and three carry look ahead adders were initialized as components to implement a simple function $F=2\times[(a\times b)+(a\times b)]$. The logical circuits had to be repeated in order to increase the percentage of the FPGA's busy gates to around 80%. The multiplexer had to be repeated 90 times and the adders/multipliers 3 times. For each of the logic circuits mentioned earlier, the components were randomly placed across the FPGA. A total of seven random placements were specified for each of these circuits' components.

3. Results Obtained

XPower divides the reported dynamic power consumption into the following five categories: Clock Power, Inputs Power, Logic Power, Outputs Power, and Signals Power. It should be noted that glitches are translated in XPower as higher activity rates which means higher power.

Plotting signals dynamic power consumption versus total signals capacitance has revealed that the relationship between internal nets capacitance and the corresponding dynamic power consumption changes with respect to the logic circuit being implemented. Only signals were selected because the only profound change in reported capacitance was in this category while the remaining four categories had approximately constant power as well as constant total capacitance. To further investigate this relationship, the highest reported capacitance for signals in the seven region-constrained placements for each of the logic circuits that were implemented was taken and the corresponding reduction of signals capacitance and its respective effect on dynamic power was calculated for both device families used in our design. The correlation coefficient between the change in the capacitance and the corresponding change in power consumption was calculated. A correlation coefficient > 0.75 was considered to indicate strong positive relation, whereas a -0.75 < coefficient < 0.75 would suggest that they are unrelated, and a coefficient < -0.75 would indicate strong negative relationship. Tables 1 and 2 show the correlation between the change in signals' capacitance and its dynamic power consumption. In these tables the percentage reduction in signals' capacitance was calculated with respect to the placement with the highest reported capacitance and hence there are only six placements reported in each table.

4. Conclusion

Although substantial reduction in internal nets (signals) capacitance is achievable through careful region-constrained placements of components, the expected gain of reduced dynamic power consumption of these internal nets is dependent on the logic circuit being implemented. In the case of the multiplexer, there was an observable high correlation between signals' capacitance reduction and the corresponding reduction in their dynamic power consumption. However, the correlation between reducing signals' dynamic power through capacitance reduction was diminished in the case of the multipliers and adders; where it is a known fact that glitches are common in arithmetic circuits, especially

in large multipliers where they often represent the major part of transitions [2] and where glitching power dissipation can amount to 20% of the total power dissipation [3]. Reducing the capacitance of internal nets through careful region-constrained placements of components must be accompanied with balancing data paths delays in order to reduce glitches as much as possible and thus increase the correlation between capacitance and dynamic power reduction.

Table 1. Signals' Capacitance-Dynamic Power Relationship for the Multiplexer

| | Spartan II | | | Virtex | |
| | Percentage Change in Signals' | | | Percentage Change in Signals' | |
Placement	Capacitance	Dynamic Power	Placement	Capacitance	Dynamic Power
P#1	-1.14%	-1.52%	P#1	-3.90%	-2.53%
P#3	-1.06%	-1.74%	P#2	-2.22%	-0.73%
P#4	-0.67%	-0.02%	P#3	-1.72%	-3.05%
P#5	-0.67%	-0.02%	P#5	-0.03%	0.15%
P#6	-0.35%	1.56%	P#6	-0.37%	2.77%
P#7	-0.35%	-0.25%	P#7	-3.24%	-3.52%
Correlation Coefficient	0.86		Correlation Coefficient	0.76	
Indication	Strong Positive Relation		Indication	Strong Positive Relation	

Table 2. Signals' Capacitance-Dynamic Power Relationship for the Adders/Multipliers

| | Spartan II | | | Virtex | |
| | Percentage Change in Signals' | | | Percentage Change in Signals' | |
Placement	Capacitance	Dynamic Power	Placement	Capacitance	Dynamic Power
P#1	-42.09%	113.79%	P#1	-7.11%	0.78%
P#2	-6.99%	-8.58%	P#3	-50.68%	89.62%
P#3	-42.09%	113.15%	P#4	-22.57%	24.32%
P#4	-19.26%	6.66%	P#5	-28.11%	16.36%
P#5	-25.03%	2.25%	P#6	-19.85%	20.77%
P#6	-42.09%	109.94%	P#7	-27.38%	10.03%
Correlation Coefficient	-0.94		Correlation Coefficient	-0.90	
Indication	Strong Negative Relation		Indication	Strong Negative Relation	

REFERENCES

[1] A. Gayasen, Y. Tsai, N. Vijaykrishnan, M. Kandemir, M. J. Irwin, T. Tuan, "Reducing Leakage Energy in FPGAs Using Region-Constrained Placement", *ACM*, 1-58113-829-6/04/2004.

[2] H. Eriksson, P. Larsson, "Glitch-Conscious Low-Power Design of Arithmetic Circuits", 0-7803-8251-X/04, *IEEE*, 2004.

[3] T. Seko, A. Nakamura, T. Kikuno, "Measurement of Glitches Based on Variable Gate Delay Model Using VHDL Simulator", 0-7803-5146, *IEEE*, 1998.

0-7695-2533-4/06 $20.00 © 2006 IEEE

Towards a Faster Simulation of SystemC Designs

Ali Habibi, Haja Moinudeen, Amer Samarah and Sofiène Tahar
Department of Electrical and Computer Engineering
Concordia University
1455 De Maisonneuve West, Montréal, Québec H3G 1M8
{habibi,haja_m,amer_sam,tahar}@ece.concordia.ca

Abstract

Accelerating simulation is one of the main reasons beyond the introduction of system level modeling. Here SystemC is one of the main players proven to speed-up simulation in comparison to classical HDL languages. However, the kernel architecture of the SystemC simulator treats the design as a black box. For instance, all active processes are executed without checking if they are relevant to the test plan. We illustrate the performance of our approach on a set of models built on top of the Master/Slave library part of the SystemC release and for two levels of abstraction: untimed functional (UTF) and bus-cycle-accurate (BCA).

1 Introduction

SystemC [?] was introduced as system-level modeling library -that is, modeling of systems above the RTL level of abstraction, including systems which might be implemented in software, hardware or some combination of the two. The challenge was to come up with an abstract design environment where design time is shorter and simulation is faster in comparison to the classical RTL design and verification flow. The classical way to use SystemC is to define test plans in order to guide the simulation towards a targeted functional coverage. Assertions are integrated with the design to serve as monitoring points for the evaluation of the design properties. This approach presents two main problems. First, the test plan is usually user-defined. Second, the simulation considers the whole design. A generic architecture of a System-on-Chip (SoC), at system level, consists of a set Intellectual Properties (IP). Several design properties are concerned with a subset from these IP blocks. Finding a way to identify a reduced model preserving the design's properties will result in simplifying the system's complexity; hence, getting a faster simulation.

In this paper, we illustrate an approach to accelerate SystemC simulation by reducing SystemC designs according to a set of predicates. We generate a so-called grouped FSM

of the design, where all the states are classified according to the grouping conditions. The reduced model is constructed directly from the grouped FSM by identifying the modules and processes involved in the system states required in the evaluation of a set of user-defined properties.

Several approaches have been proposed in order to attain faster SystemC simulation; we cite in particular [?], [?] and [?]. Both [?] and [?] propose to modify the discrete-event SystemC kernel by adding a static scheduling aiming to improve the simulation efficiency. This approach allows to find a better order of executing SystemC design processes. Its main problem is the fact that it does not consider the relevance of running a given process for a specific property checking or test scenario.

2 Used Approach Description

Figure 1 describes the used approach which consists of two steps: (1) generating the grouped system's FSM; and (2) generating the reduced model. Then, in order to evaluate the performance of our approach, both the original SystemC design and the reduced model are simulated.

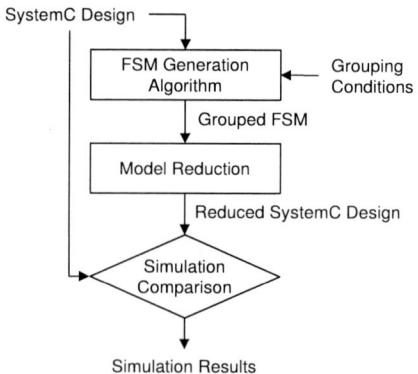

Figure 1. Used Approach.

The generation of the reduced model is performed in four

0-7695-2533-4/06 $20.00 © 2006 IEEE

steps: (1) Collecting, from the SystemC design, a set of state variables and processes to be considered in the state exploration; (2) Generating the grouped FSM; (3) Identifying the relevant states and transitions in the reduced model; and (4) Constructing the reduced model.

3 Application: Master/Slave Models

3.1 Models Description

We implemented a multi-point hierarchical master salve architecture using the SystemC master slave communication library [?]. Our architecture consists of a set of masters, slaves, and middle modules. Master module has master port connected to the master-slave channel, while middle module has input slave port and output master port. It also checks if it has to forward the data to the next connected module or not. Slave module has only input slave port.

Table 1. Benchmark Description.

Model	Abst. Lev.	Num. of Mod.	Num. of Proc.
Design 1	UTF	7	9
Design 2	BCA	16	27
Design 3	UTF	9	11
Design 4	BCA	21	34

We started building functional blocks of masters, middle and slaves modules as untimed functional (UTF) module. At the UTF abstraction level, execution order is modeled accurately while time is not modeled. All the processes at this level have a sensitive list based on triggering events between modules except the test bench. After validating the operation of that model at the functional level, we moved towards bus cycle accurate model (BCA) by invoking full handshaking among data exchange between those modules.

We consider a benchmark composed of four designs (Design 1, Design 2, Design 3 and Design 4). Both Design 1 and Design 2 are composed of 1 master, 2 middle modules, and 4 slaves. Design 3 and Design 4 are composed of 3 masters, 2 middle modules and 4 slaves. The designs are defined for two abstraction levels: (1) UTF for Designs 1 and 3; and (2) BCA for Designs 2 and 4. Table 1 shows the summary of the benchmark providing the number of modules and processes for each design.

3.2 Experimental Results[1]

In Table 2, the simulation results of our designs are shown for various grouping conditions. From the table, we infer that a significant improvement in simulation time has

---

[1]Experimental platform: Pentium IV 2.4 GHz / 1GB of memory.

been achieved (see column 5). The improvement varies(9% up to 47%) based on the grouping conditions [2].

Table 2. Simulation Comparison.

Model	δ (μs)	Group.	δ_{red} (μs)	Simu. Acc.(%)
Design 1	339.68	G1	186.80	45.00
		G2	293.60	13.57
		G3	276.40	18.63
Design 2	63.21	G1	48.84	22.73
		G2	33.39	47.18
		G3	48.39	23.45
Design 3	553.24	G4	369.06	33.29
		G5	499.68	9.67
		G6	377.50	32.67
Design 4	68.47	G4	51.97	24.10
		G5	60.92	11.03
		G6	50.19	26.70

4 Conclusion

We used an approach to reduce SystemC designs based on a set of grouping conditions in order to have faster simulation time. We performed simulations on a set of benchmark of SystemC Master Slave library. The experimental results show a significant improvement in simulation time of up to 47% for some configurations.

References

[1] S. Meftali, J. Venni, and J. Dekeyser. A Fast SystemC Simulation Methodology for Multi-level IP/SoC Design. In *Proc. of the International Workshop on IP Based System-on-Chip Design*, Grenoble, France, November 2003.

[2] Open SystemC Initiative. www.systemc.org, 2005.

[3] OSI. *Master-Slave Communication Library – A SystemC Standard Library – Version 2.0.1*. 2005.

[4] H. D. Patel and S. K. Shukla. Towards a Heterogeneous Simulation Kernel for System Level Models: A SystemC Kernel for Synchronous Data Flow Models. In *Proc. of the IEEE Symposium on VLSI*, pages 241–242, Lafayette, LA, USA, 2004.

[5] D. G. Pérez, G. Mouchard, and O. Temam. A New Optimized Implemention of the SystemC Engine Using Acyclic Scheduling. In *Proc. of Design, Automation and Test in Europe*, pages 552–557, Munich, Germany, 2004.

---

[2]We refer the reader to http://hvg.ece.concordia.ca/Research/SoC/ for a more detailed the benchmark and the reduced models.

An Optimized BIST Architecture for FPGA Look-Up Table Testing

Mahnaz Sadoughi Yarandi, Armin Alaghi, Zainalabedin Navabi

Department of Electrical and Computer Engineering, University of Tehran, Tehran, Iran
m.sadoughi@ece.ut.ac.ir, a.alaghi@ece.ut.ac.ir, navabi@ece.neu.edu

Abstract

This paper presents a complete BIST architecture for FPGA Look-Up Tables. This architecture can detect multiple faults, and can be configured to detect a fault in as few as five LUTs. Unlike an earlier work, our method places the Output Response Analyzer (ORA) within the FPGA.

1. Introduction

Field Programmable Gate Arrays (FPGAs) have been widely used for rapid prototyping and manufacturing of complex digital systems, such as microprocessors and high speed telecommunication chips [1]. Frequent reconfigurations of the same FPGA make it fault-prone [2].

Many components of an FPGA must be tested for ensuring reliable usage of this device. Algorithms for interconnect testing are discussed in Reference [3]. In Reference [4], general test algorithms for Arrays of RAMs/LUTs has been proposed. In this paper, we only consider test of LEs and focus on LUTs within LEs.

An earlier work on a *Built-In-Self-Test* (BIST) architecture for LUT testing proposes the use of an off-chip *Output Response Analyzer* (ORA). Our objective is to design a BIST with a good balance between test time, test area, and granularity with an internal ORA.

In this scheme, LUTs of the FPGA are partitioned into *Test Chains* in which some LUTs are configured as *Test Pattern Generators* (TPGs) and some as *Circuit Under Tests* (CUTs). This arrangement is discussed in Section 2. In our method, while a TPG is generating test vectors for CUT, LUTs configured as TPG are also tested. In Section 3, the effect of fault propagation on the *Response* signal is discussed. Section 4 presents the ORA hardware design for fault detection and discusses various costs associated with our architecture.

2. BIST Architecture and Methodology

Our proposed BIST chains a group of LUTs of FPGA logic elements to form TPGs, CUTs and their corresponding ORA. Each group is called as a *Partial*

BIST (Figure 1). LUTs must be configured by various test patterns in order to detect different kinds of faults [4]. With this architecture and test patterns loaded into LUTs, signals between *Partial Chain*s (PCs) toggle periodically. So each PC's output can be used as the clock input of the next PC. The clock frequency is divided by two from one PC to the next.

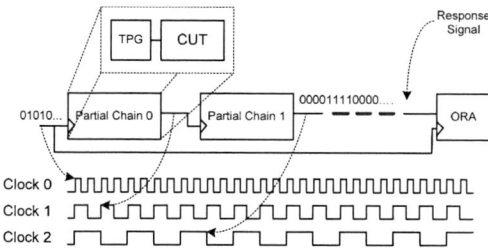

Figure 1. *Partial BIST* Architecture

3. Fault Propagation

If a single fault occurs in a PC, its output becomes faulty. This fault appears as a missing positive or negative pulse. A missing pulse results in width expansion of PC's output signal. In what follows, "width" is defined as the time between an edge to the next similar edge of a signal. With this, a clock pulse width is multiplied by two when it goes from one PC to the next. We refer to w_0 as the original clock and $(w_r)_i$ as the reference at i^{th} PC's output. Due to a fault (in the ith PC), the extra amount of width (w_c) will be carried through the *Test Chain* to the last PC's output (*Response* signal). w_c can be calculated by:

$$(w_c)_i = 2(w_r)_{i-1} = 2^i w_0 \qquad (1)$$

The *Response* signal of the *Partial BIST* has the accumulation of all PCs' faults that appear as a corrupt width. Let $(w_f)_n$ be this accumulated width in a *Test Chain* of n PCs:

$$(w_f)_n = \sum_{m=1}^{n} c_m . 2^m . w_0 \qquad (2)$$

where $c_m = 1$ for faulty PC and 0 otherwise. By examining this pulse width, multiple faults in different PCs are detected. Note that we can only detect a single fault in each PC, and can detect as many such faults as the number of PCs.

4. ORA Design

Our proposed ORA structure is based on fault propagation presented by Equation 2. In this expression, c_m value of 1 identifies the location of a fault. c_m values form a binary number that is embedded in the pulse width of the *Test Chain* output, and is extracted by our ORA to identify faulty PCs and their locations.

Counting pulse widths (total positive and negative pulses) with the original system clock yields the binary number that identifies various faults. Our ORA requires two counters to measure this accumulated width (Figure 2). One counter measures the duration of positive *Response* pulses and the other measures the duration of the negative pulses. The result is registered in an accumulator register.

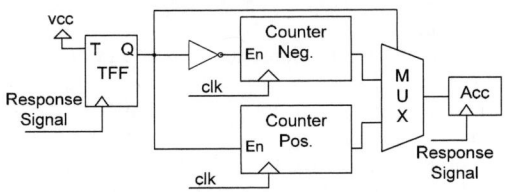

Figure 2. ORA Hardware Design

As the number of the PCs within a *Test Chain* increases, more counters will be needed. In this proposed ORA architecture, we reached the optimum number of 3 PCs per *Test Chain*. This was achieved experimentally by implementing the ORA on different Altera devices such as FLEX10k, Cyclone, Stratix, and ACEX1K.

Figure 3 shows a measure of fault detection efficiency for different number of PCs within a *Test Chain*. We define test efficiency as the ratio of *test area* to the area occupied by the ORA. *Test area* is the FPGA part that is being tested (total TPGs and CUTs).

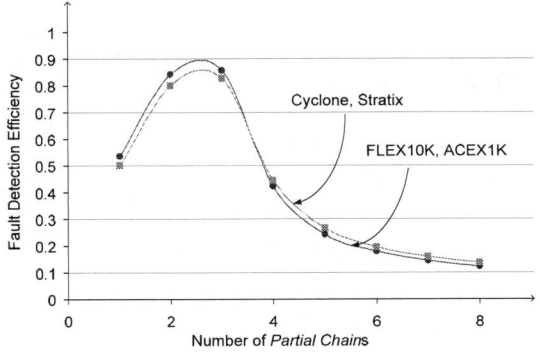

Figure 3. ORA Efficiency

By increasing the number of *Test Chain*s, more ORAs will be needed and the test area decreases. Using an optimum ORA, an increase in the number of *Test Chain*s makes them smaller, which results in less LUTs per PC and higher granularity.

The highest test area can be achieved by having only one *Test Chain* with three PCs, which results in a low granularity (0.81 faults detection in 1000 LUTs).

On the other hand, the best granularity can be achieved by having only 5 LUTs per a PC (4 LUTs for the TPG [5] and 1 for the CUT), which results in having 87 chains and 34.9% test area in an Altera device.

By switching the location of the *Test Chain*s and ORAs, the maximum granularity (200 faults detection per 1000 LUTs) and the maximum test area (100%) can be achieved simultaneously. This requires more number of configurations of the FPGA.

5. Conclusion

This paper presented a BIST architecture for FPGAs Look-Up Table testing. The method is general and can be applied to most LUT based FPGAs. Our method is based on altering periodic signal pulse widths in the presence of Look-Up Table faults. An ORA of a test chain detects one fault of all PCs. Although the focus has been on LUTs, many logic faults between LUTs and LE Flip-flops are also detected. More work for FPGA RAM, and PLL testings is required to complete this methodology.

Reference:

[1] S. Brown, R. J. Francis, J. Rose, and Z. G. Vranesic, *Field Programmable Gate Arrays*, Kluwer Academic Publishers, Boston, MA, 1992.

[2] C. Metra et al., "Novel Technique for Testing FPGA", Design, Automation and Test in Europe, pp 89-94, 1998.

[3] M. Renovell, J. Figueras, Y. Zorian, "Test of RAM-based FPGA: Methodology and Application to the Interconnect", in Proc. 15th VLSI Test Symp., pp. 230-237, 1997.

[4] W.K. Huang, F.J. Meyer, N. Park, F. Lombardi. "Testing memory modules in SRAM-based configurable FPGAs," *MTDT*, p. 79, 1997 IEEE International Workshop on Memory Technology, Design and Testing (MTDT '97), 1997.

[5] E. Atoofian, Z. Navabi. "A BIST Architecture for FPGA Look-Up Table Testing Reduces Reconfigurations," *ATS*, p. 84, 12th Asian Test Symposium (ATS'03), 2003.

[6] Altera Data Book, San Jose, USA, 2002.

Variation Aware Placement for FPGAs

Suresh Srinivasan and Vijaykrishnan Narayanan
Department of Computer Science and Engineering
Pennsylvania State University, University Park, PA - 16802

1 Introduction

Impact of variations in different process parameters like, gate length, threshold voltage, oxide thickness etc. have been discussed in different components of digital circuits extensively in recent times. The various manufacturing effects on different process parameters have been demonstrated in [1] and [2]. Degradation in the operating frequencies by nearly 2X and increase in leakage power consumption by a factor of 3, have been demonstrated in FPGAs in [3]. Although static tuning of device parameters may counter such impacts to some extent, there are still variations that continue to impact the performance and power consumed by the chip. Consequently, such variations call for changes in the existing design tools and incorporation of variation awareness in those tools. In this paper, we propose a variation aware placement scheme in FPGAs and demonstrate the effectiveness of the scheme on Xilinx FPGAs and regular island style FPGAs. Our approach provides leakage benefits close to 14% on an average over different benchmark designs.

2 Process Variation Impacts On FPGAs

Variations in gate length, effective channel widths and the resultant effects on the threshold can affect both delay and power consumption of the individual slices on FPGAs. To provide a comprehensive analysis we breakdown our analysis into the impact on memories and LUTs separately. Each of the components were laid out using micromagic and simulated using HSPICE, using the model files from BSIM4 for 65nm technology. Since SRAM cells are not present in the critical path while estimating the frequency of operation of any design on FPGAs, their performance and latencies do not affect the frequency of operation of the FPGAs. Moreover, most FPGA technologies typically tend to use high threshold gates in designing such SRAM cells to minimize the leakage power consumed the impact of the leakage power consumed by the FPGAs is minimal as well. Such an observation directs all our focus towards analyzing and optimizing the impact of variations of process parameters on the LUT multiplexers as discussed in the next section. Since the LUT multiplexers are in the critical path they may affect the frequency of operation of FPGAs. The delay varies significantly, in fact nearly by a factor of 1.3X, even with 20% variation in the threshold voltage from the nominal voltage of 0.2V. We observed a variation of nearly 2X in the delays of LUTs for 20% varia-

tion in threshold voltages of the transistors around the nominal threshold. It is however important to note that while the leakage power consumption increases by nearly 2X with 20% decrease in the threshold voltage, it does not change much with the increase in the threshold voltage.

Another important step in such variation aware analysis is to determine the actual delay and the power consumed by the blocks given a FPGA board. This may be statically performed by configuring appropriate delay or leakage sensors on teh FPGA. Programming leakage sensors on FPGAs may not be feasible to employ due to their analog nature. However, a digital ring oscillator based delay sensor, which may be configured at different parts of the FPGA separately to obtain the delays of different blocks, may be used to get an estimate of the delay category in which the block falls into. To avoid any overhead such delay estimator could be used before loading the configuration of the design onto the FPGA, and removed thereafter.

3 Placement Methodology

We propose two different schemes for modifying the placement algorithm to incorporate variation tolerance.

3.1 Block Discarding Policy (BDP)

We provide some hard constraints to the Xilinx placement algorithm, to skip certain blocks which may have adverse effects on the delay and the leakage power consumption of the design Since the utilization of the logic blocks is not typically 100% in most designs, this leverages an opportunity to impose some placement constraints. The constraint to skip any block is imposed by adding the *PROHIBIT* constraint in the User Constraint File (UCF) file. Due to process variations in the LUTs of different blocks, the delay and the leakage power are conflicting in their trends, which makes it hard to decide on the blocks to be skipped. Consequently, the decision to skip a block should be based on both the leakage power consumption and the delay of any slice in the FPGA. We set a threshold value (L,D) for the leakage power consumption and delay of a SLICE respectively, above which the SLICE is considered for discarding. We vary this threshold based on the flexibility in the design, keeping in mind the utilization of the device since it affects the number of blocks to be skipped. Such a constraint file is generated as a text output from a perl

0-7695-2533-4/06 $20.00 © 2006 IEEE

program, which determines the blocks to be skipped, based on the leakage and delay thresholds.

3.2 Variation Aware Placement

The block discard placement (BDP) scheme imposes a hard constraint and requires iterative determination of the number of blocks to be discarded based on the flexibility in the design. However, the variation aware placement scheme imposes a soft constraint that is incorporated by modifying the placement algorithm itself. We modified the placement algorithm in Versatile Place and Route tool (VPR) to incorporate the impact of new delays and leakage power consumed by the blocks selected for placiing the design. At first, we associate the exact delay numbers with each of the blocks in VPR (termed as sub-blocks in VPR), to get a precise estimate of the critical path, which automatically ensures the incorporation of the variation in delay of each of the blocks in our algorithm. VPR's placement algorithm works on the principle of simulated annealing and therefore at each step, while jumping from one solution to another, it has a cost function to evaluate the current solution. Therefore, in order to incorporate the leakage power of the blocks, we explicitly add a normlized leakage cost factor to the total cost of each of the blocks as estimated by the existing placement algorithm.

4 Experimental Setup and Results

The leakage and delay numbers for different threshold voltages of the transistors were generated using simulations explained in 2. To implement the BDP we generated the Gaussian distribution of threshold voltages using a library provided by GNU and obtained the leakage and delay of each of the SLICEs from SPICE simulations. The final threshold values (L,D) mentioned in section 3, are obtained when the design had the minimum possible leakage power consumption and placement and routing tools do not fail due to over constraining. The variation aware placement algorithm was implemented in VPR for island style FPGA devices as described in section 3. To demonstrate the effectiveness of the block discarding policy, we implemented various designs using the Xilinx ISE tool flow(ver 6.0), provided by Xilinx. There is an average increase in the total leakage power consumed by nearly 20% over different benchmarks due to process variations. Note that the leakage power consumption is always going to increase with any process variations, since the percentage increase in the leakage power of a single LUT is much higher than the decrease for the same deviation from its nominal threshold. We picked up the benchmarks with device utilizations lower than 90%, the reason being the fact that our whole algorithm is based on the flexibility in discarding the bad blocks while placing the cells. After applying the BDP based placement appropriately, nearly 14% leakage savings were obtained, averaged over all the benchmarks.

The variation awareness incorporated into the placement algorithm of VPR was tested over a set of blif benchmarks.

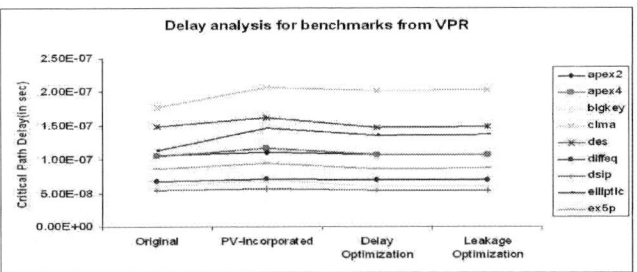

Figure 1. Impact on clock frequencies of different implementations placed using VPR

We compare the operating frequencies of four different simulations for each benchmark in figure 1. The first label in the X-axis of the graph in figure 1 shows the original cycle time of the design after placement. The next simulation marked as pv-incorporated, is the frequency of operation of the design implemented on the FPGA, once the new delay values due to process variations are incorporated. In this simulation the delays are associated to the blocks after placement and therefore it provides an estimate on the impact of process variations on operating frequencies. Once again it is evident from the figure that process variations no longer allow the designs to operate at their original frequencies but at a lower frequency on an average by 12%. The third simulation shows the cycle time obtained from timing driven placement algorithm of VPR without incorporating the leakage optimization while the final simulation is for the variation aware placement scheme with the leakage power optimization incorporated. With the new delays incorporated into the blocks, the placement algorithm finds a better placement which allows to operate the design at a frequency on an average 10% higher than the increased frequency due to process variations. Finally, leakage benefits are obtained without compromising on the operating frequencies by choosing low leakage paths in the non critical nets. We observe an average leakage power savings of 14%, which is similar to the savings obtained in Xilinx implementation of BDP.

This work was supported in parts by grants from GSRC and CAREER Awards 0093085.

References

[1] S. Borkar, T. Karnik and V. De "Design and reliability challenges in nanometer technologies" *Proceedings of Design Automation Conference (DAC)*, 2004.

[2] S. Nassif "Modeling and analysis of manufacturing variations" *Proceedings of IEEE Custom Integrated Circuits Conference*, 2001.

[3] H. Yan, P. Wong, L. Cheng, Y. Lin and L. He "FPGA device and architecture evaluation considering process variation " *Proceedings of International Conference on Computer Aided Design*, 2005.

0-7695-2533-4/06 $20.00 © 2006 IEEE

A Regular Layout Approach for ASICs

Cláudio Menezes, Cristina Meinhardt, Ricardo Reis
Instituto de Informática -UFRGS
{cmeinhardt, reis}@inf.ufrgs.br

Reginaldo Tavares
Universidade Estadual do Rio
Grande do Sul
regi@inf.ufrgs.br

Abstract

This paper presents a regular layout approach addressing the need of a more predictable circuit performance and DFM. Experiments were done using a regular matrix of cells composed by 2-input NAND gates. The regular layout approach considers some strategies to improve the predictability of connections and the routability of circuits. Initial experiments show that this approach should improve performance when compared with a standard cell approach.

1. Introduction

Layout regularity is one of the design-for-manufacturability techniques proposed to deal with DSM designs. Regularity improves the predictability of the physical implementation and lets a more accurate estimative of the circuit delay, especially the interconnection delay. This work presents a physical synthesis strategy exploring layout and connection regularity. The strategy is based on the use of a matrix of cells. We are working in two kinds of approaches, the design of circuits using a prediffused matrix of cells (a structured ASIC approach [1]) and the design of custom functional blocks or circuits. This paper addresses the approach for the design of custom circuits using matrices of cells. These cells can be just one type of cell, or a set of basic cells or programmable cell. The first experimentation that was done is the generation of a matrix of NAND cells. This makes possible the implementation of logic equations mapped to networks of NANDs and inverters, including the parameterization of the fanout. For this reason, it was developed a CAD tool, Martelo [2], which allows the automatic generation of a matrix of logic cells. The number of basic cells is not fixed, i.e., the matrix is built on the fly by the tool. Martelo generates matrices with logic cells in CMOS technology. The experiments showed that this is a promising technique and some results are presented.

2. Regular Layout Generation

- Martelo is developed to address the automatic regular layout generation based on the use of matrices of cells. The regularity of layout and connections are the keys on this synthesis approach. The regular layout is achieved by repeating a pattern of a single basic cell that has 4 transistors (two P and two N) implementing a NAND2. It forces a monotony layout scenario since the same cell layout is used several times. The main characteristics are:

- The layout is created by tiling the base cell in a 2 dimensional pattern arranged in columns and rows.

- The area occupied by a unique cell in the matrix is the same for all other cells.

- All NAND cells are aligned in vertical and horizontal directions and the rows of cells are alternately flipped (mirrored) in the X axis.

- The distance between cells is the minimum distance allowed by the layout rules and the number of tracks per cell is the same for all cells.

- It is just possible two cases of free spaces: dummy cells and extra tracks if they are really needed.

Only NANDs are used in this experiment, but inverters are also necessary and they are implemented using NAND gates.

The routing synthesis must guarantee that the final circuit will be totally routed. To reach this goal, router [3] and cell generator work together. There is a high density of contacts and vias because the cells are very close one to each other. Sometimes the router is not able to finish the routing. The matrix builder has 3 different strategies to overcome this problem: metal1adjacent contacts, track insertion or dummy space insertion. For technologies with 4 or more metal layers the example circuits were completely routed.

Metal1 Adjacent Contacts: the matrix builder identifies NAND cells placed side-by-side with a common connection and makes a straightforward connection in metal 1. This operation removes some nets and reduces the total number of nets to be routed. The input nets position can be changed. This strategy can be used in several ways, Chang[4] shows an approach to reconnect pins after the placement step. In our approach the pin assignment can be changed during the layout generation. The results from an initial experience indicate that on average this strategy can reduce in 21% the number of connections.

Track insertion: extra routing tracks are inserted in both horizontal and vertical directions when some area is blocked and the router cannot implement a connection. This operation increases area because a track crosses the matrix from one side to another side.

Dummy cell insertion: when a congested area is found, a dummy cell is placed in a position where new routing paths are created. Martelo adopts two approaches: first, dummy cells are inserted in the points indicated to insert vertical or horizontal tracks. Second, a strategic dummy insertion, where dummy cells are inserted in congestion areas. This approach should increase the number of possible metal1 connections between adjacent cells and obeys to one constraint: it is just possible to insert one dummy cell in a row of cells. This technique is associated with track insertion to achieve good results.

3. Layout Analysis

We compare area, delay and wire length results obtained by Martelo with the results obtained with Cadence tools. PKS_SHELL and SEULTRA Cadence tools were used for Standard Cell layout generation. In this context, we have two configurations of the logic synthesis for each circuit: one where the fan out is limited to one, named Fanout Free synthesis logic, and another where the maximum fan out is four. TABLE I shows the results. The synthesis logic fanout strategy is presented in the column fanout. Fanout Free logic synthesis is more expansive in number of gates, but it reaches smaller delays. We can observe that the Cadence results, identified with (1) in the table, have

better area and wire length than Martelo, identified with (2) in the table. But standard cell layouts present worse delays. In average, Martelo improves 12,4 % the delays, beyond the regularity advantages.

10. Conclusions

This paper presents a layout synthesis method that addresses regular and predictable layouts. Furthermore, this flow guaranties an automatic routability, by the insertion of extra tracks or dummy cells when needed due to congestions. The strategic dummy insertion technique should reduce the number of connections to be implemented by the router.

The matrix generated with Martelo ensures a layout regularity, but with some area penalty. On the other hand time analysis shows that the Martelo layout versions achieved better performance than standard cell ones. Fanout Free synthesis logic can improve the performance, despite the increase in area and wire length. A new version of the matrix generator is under construction to allow the use of other basic cells. This can reduce the number of transistors and area.

Finally, it is important to notice that our approach improves design for manufacturability (DFM) as we already know the printability of a layout after a first matrix implementation. Any need of a layout change for printability, will be solved in the step of matrix template development.

4. References

[1] B. Zahiri, "Structured ASICs: Opportunities and Challenges," ICCD 2003, 13-15 October, 2003. p. 404- 409.

[2] C. C. Menezes. Geração Automática de Leiaute através de uma Matriz de Células NAND – MARTELO. 2004. Máster Dissertation: UFRGS, Porto Alegre.

[3] G. Flach, R. Hentschke, R. Reis. Improving Maze Routers Routability by A New Rip-up and Reroute Approach. XX SIM, May, 2005. P: 87 – 91.

[4] K. Chang, I. L. Markov, V. Bertacco. "Post-Placement Rewiring and Rebuffering by Exhaustive Search for Functional Symmetries," ICCAD2005, November 2005.

TABLE I.   RESULTS OF THE SYNTHESIS WITH CADENCE (1) AND MARTELO (2)

Circuit	Fanout	#Cell	Area (1)	Area (2)	%	Delay (1)	Delay (2)	%	Wire length (1)	Wire length (2)	%
7seg	1	159	9800	12115,8	19,1	1,09	0,99	- 9,2	21,8	29,5	35,2
7seg	4	92	6427,2	7187,2	10,6	1,28	1,22	- 4,7	26,1	35,7	36,6
9symml	1	774	41979,2	55988,3	25,0	2,46	2,17	- 11,8	27,5	36,8	34,1
9symml	4	501	27654,8	34396,2	19,6	3,29	2,41	- 26,8	34,9	45,5	30,1
Lal	1	515	27182,1	38901,4	31,1	2,2	2,07	- 5,9	24,5	40,3	64,1
Lal	4	266	15824,6	22659,2	30,2	2,96	2,41	- 18,6	22,4	40,5	80,8
Frg	1	870	45234,8	60863,4	25,7	2,06	1,87	- 9,2	31,1	40,4	29,9
Frg	4	285	16263,6	19381,4	16,1	2,19	1,91	- 12,8	28,1	40,8	45,5

0-7695-2533-4/06 $20.00 © 2006 IEEE

Evaluating the Impact of Data Encoding Techniques on the Power Consumption in Networks-on-Chip

José C. S. Palma [1], Leandro Soares Indrusiak [2], Fernando G. Moraes [3],
Alberto Garcia Ortiz [2], Manfred Glesner [2], Ricardo A. L. Reis [1]

[jcspalma, reis]@inf.ufrgs.br, [lsi, agarcia, glesner]@mes.tu-darmstadt.de,
moraes@inf.pucrs.br

1. PPGC - II - UFRGS - Av. Bento Gonçalves, 9500, Porto Alegre, RS – Brazil	2. MES – TU Darmstadt – Karlstr. 15, 64283 Darmstadt - Germany	3. PPGCC - FACIN – PUCRS - Av. Ipiranga, 6681, Porto Alegre, RS – Brazil

Abstract

This work addresses the problem of power consumption in networks-on-chip (NoCs). It investigates the reduction of dynamic power consumption through the reduction of transition activity using data coding techniques. Power macromodels for various NoC modules were built, allowing the estimation of the power consumption as a function of the transition activity at each module's inputs. Such macromodels were embedded in a system model and a series of simulations were performed, aiming to analyze the trade-off between the power savings due to coding techniques versus the power consumption overhead due to the encoding and decoding modules.

A network-on-chip (NoC) is an infrastructure essentially composed of routers interconnected by communication channels. It is suitable to support the GALS paradigm, since it provides asynchronous communication, high scalability, reusability and reliability [1][2].

The growing market for portable battery-powered devices adds a third dimension (power) to the previously two-dimensional (speed, area) VLSI design space [3]. NoCs are not particularly efficient in power consumption. They solve the problem of high capacitances in wires and, consequently, the power consumption in long lines, but on the other hand the power consumed by the router modules is not negligible.

The dynamic power consumption in a NoC grows linearly with the amount of bit transitions in subsequent data packets sent through the interconnect architecture. Using the Hermes NoC architecture [4] as a case study we could show that bit transitions affect the dynamic power consumption by as much as 6400% for interconnect lines, 180% for router input buffers and 20% for router control logic. Based on

such results, this paper addresses the hypothesis that the power dissipation on a NoC-based system can be reduced by encoding the data sent through the network with coding schemes that reduce the average number of signal transitions [5]. Several such schemes were proposed in the late 90's, all of them addressing bus-based communication architectures. The contribution of this work is on the evaluation of such schemes in the context of NoC-based systems and on the trade-off analysis of the power savings obtained by the application of such coding schemes versus the power consumption overhead due to the additional encoding and decoding circuitry.

In Networks-on-chip, the data is transmitted in packets which are sent through routers, from one source to one target core. Encoding and decoding operations must be done in the source and target cores only, so to convert the original data to the encoded (and transmitted) one and vice-versa. In this work the encoder and decoder modules were inserted in the local ports of the routers, that is, between the integrated cores and the NoC interconnect structure.

For the latest CMOS technologies, static power accounts for the smallest part of the overall consumption. Accordingly, this work focuses only on NoC dynamic power consumption, using it as an objective function to evaluate the quality of data encoding schemes.

Dynamic power consumption is proportional to the switching activity arising from packets moving across the network. Interconnect wires and routers dissipate dynamic power. Several authors have proposed to estimate NoC power consumption by evaluating the effect of bits/packets traffic on each NoC component.

Average power per hop[1] (*APH*) is used here to denote the average dynamic power consumption in a single hop of a packet transmitted over the NoC.

Table 2 – Experimental results.

Stream	Bit transition reduction (reported in [5])	Bit transition reduction (simulated within the scope of this work)	Average Power consumption without encoding	Average Power consumption with encoding	Encoding power consumption	# of hops
HTML	9,3 %	10,5 %	18,58 mW	18 mW	23 mW	39
GZIP	16,3 %	8%	20,21 mW	19,7 mW	23,65 mW	46
GCC	15,6 %	4 %	19,8 mW	19,7 mW	23,5 mW	235
Bytecode	-	13 %	23,26 mW	22,24 mW	24,64 mW	24
Synthetic 1	-	75 %	31,39 mW	21,23 mW	25,9 mW	3
Synthetic 2	-	- 11 %	20,9 mW	21,92 mW	26,02 mW	-

APH can be split into three components: average power consumed by a router comprised of buffers, router wires and logic gates for switching (*APR*); average power consumed on a link between routers (*APL*); and average power consumed on a link between the router and the system core attached directly to it (*APC*). Equation (1) gives the average power consumption of a packet transmitted through a router, a local link and a link between routers.

(1) APH = APR + APL + APC

Moreover, our analysis showed that a better understanding of the average power consumption in the router (*APR*) can be achieved by dividing it into its buffer (*APB*) and control (*APS*) components. This is because the bit transition effect on power consumption at the router control is much smaller than its effect on the power consumption at the router buffer.

In regular tile-based architectures, tile dimension is close to the average core dimension, and the core inputs/outputs are placed near the router local channel. Therefore, *APC* is much smaller than *APL* and may be safely neglected without significant errors in total power dissipation. Therefore, Equation (2) computes the average router-to-router communication power dissipation, from tile τi to tile τj, where η corresponds to the number of routers through which the packet passes.

(2) RRPij = η ×(APB + APS) + (η – 1) ×APL

Considering now an approach with data encoding in the NoC local ports, two new parameters can be introduced to Equation (2): *APE* and *APD* (encoder and decoder average power consumption, respectively), producing the Equation (3).

(3) CRRPij = APE + η × (APB + APS) + (η – 1) × APL + APD

The parameters of the macromodel were acquired after the SPICE simulation of the communication infrastructure and encoding modules with different traffic patterns. Table 2 shows the experimental results obtained by system level simulation within Ptolemy II, using the macromodels described above. The first column describes the type of traffic. The second column reproduces the reduction of transition activity reported by Benini et al. in [7] (when available) and the third column presents the reduction of found on our experiments by reproducing the coding techniques proposed in [5]. Both cases report the results in terms of reduction in the number of transitions with respect to the original data streams.

The fourth column shows the power consumption (*APH*) without use of data coding techniques, while the fifth column shows the same measurement when coding techniques are used. Finally, the sixth and seventh columns presents the power consumption overhead due to the encoder and decoder modules (*APE + APD*) and the number of hops which are needed to amortize this overhead.

Experiments with the transmission of synthetically generated streams showed that the use of coding techniques can be advantageous if the average hop count for packages is larger than 3 (so the overhead of encoding and decoding is amortized), or larger than 24 hops for a java bytecode (power reduction of 1,02 mW/switch). By synthesizing a variety of traffic patterns, we could see that the power savings due to coding range from -1 to 18 mW/switch, thus pointing the direction for further research addressing the use of multiple encoding schemes to better match the transition activity patterns, and the use of NoC topology and routing information to help the decision whether a given stream should be encoded or not.

References

[1] A. Iyer and D. Marculescu. "Power and performance evaluation of globally asynchronous locally synchronous processors". 29th Annual International Symposium on Computer Architecture (ISCA), pp. 158-168, May 2002.

[2] W. Dally and B. Towles. "Route packets, not wires: on-chip interconnection networks". Design Automation Conference (DAC), pp. 684–689, June 2001.

[3] D. Singh, J. M. Rabaey, M. Pedram, F. Catthoor, S. Rajgopal, N. Sehgal and T. J. Mozdzen. "Power conscious cad tool and methodologies: A perspective". Proc. IEEE, vol.83, pp. 570-594, Apr. 1995.

[4] F. Moraes, N. Calazans, A. Mello, L. Möller and L. Ost. "HERMES: an infrastructure for low area overhead packet-switching networks on chip". The VLSI Journal Integration (VJI), vol. 38, issue 1, pp. 69-93, Oct 2004.

[5] L. Benini, A. Macii, E. Macii, M. Poncino, R. Scarsi. "Synthesis of low-overhead interfaces for power-efficient communication over wide buses". DAC, 1999. Proceedings. 36th, 21-25 June 1999 Page(s):128 – 133.

Dual-Mode High-Speed Low-Energy Binary Addition

Johannes Grad
Dept. of Electrical and Computer Engineering
Illinois Institute of Technology
Chicago, IL 60616 USA
jgrad@ece.iit.edu

James E. Stine
Dept. of Electrical and Computer Engineering
Oklahoma State University
Stillwater, OK 74078 USA
james.stine@okstate.edu

Abstract

Sparse tree adders are a common choice for the implementation of high performance binary addition. However, for constant supply voltage they have constant energy consumption regardless of the operating frequency. This paper presents a dual-mode sparse tree adder that offers a low-speed low-energy mode. This is achieved by disabling the prefix tree in the low-speed mode. Simulation results using extracted mask layouts show a reduction in energy consumption in the low-speed mode by a factor of 3.6 in static CMOS and a factor of 2.3 in domino logic.

1 Introduction

Addition is a key operation in many digital systems. In a microprocessor, for example, adders are in the critical path of both the ALU and the address computation unit. This means that adders typically operate at the highest frequency and become local thermal hot spots [2]. Adders, therefore, have to be optimized for both timing and energy consumption in order to meet the tight performance and energy constraints of deep submicron CMOS designs.

Most computing systems experience varying amounts of utilization. In times of low utilization the system clock can be reduced to achieve lower power consumption. In times of high utilization the system returns to the maximum clock frequency. The dynamic power consumption of a CMOS circuit can be expressed as follows:

$$P_{dyn} \propto C \cdot V_{dd}^2 \cdot f \qquad (1)$$

Here C is the switched capacitance and f is the operating frequency. While reducing f results in a reduction in power consumption, only the reduction of V_{dd} results in a reduction in energy consumption [3]. To further reduce the energy consumption the capacitance C has to be reduced.

This paper introduces an algorithm that lowers the switched capacitance of sparse tree adders at low-speed operation.

A conventional adder is a single-mode adder with constant energy consumption E_1 independent of f, at constant V_{dd}. The proposed dual-mode adder performs like a conventional sparse tree adder in the high-speed mode. In low-speed mode, however, the switched capacitance is reduced by turning off the prefix tree. As a result, all operations up to a certain frequency f_2 can be performed with lower energy consumption $E_2 < E_1$. The algorithm applies to both static and dynamic CMOS implementations.

2 A dual mode sparse tree adder

A block-diagram of the proposed dual-mode adder is shown in Fig. 1. In a conventional sparse tree adder the block carry-in signals would only come from the prefix tree, with no link between the sum blocks. While this architecture allows for high speed — the sum-blocks precompute the sum, rather than being idle — it also has significant power consumption, due to the large devices and long wires in the parallel prefix tree. When the overall system is in low-speed mode the fast carry generation by the parallel prefix tree is not necessary. Instead, the sum blocks can be connected in a serial fashion, such that each block carry-in is the carry-out of the previous block. This mode of addition is slower but also consumes less energy because no switching activity is taking place in the prefix tree.

The proposed implementation uses a carry-increment Ling architecture, although any sparse tree adder architecture can be used. The carry-out d_{j+8} has to be added as an output to the sum blocks and is available one complex gate delay after the block carry-in d_j arrives. Only the sum block in the least significant position has a larger delay, since the d_i have to be propagated until d_7 is available. However, this propagation has a delay of only one NAND gate per bit position [1]. As a result, even though the longest path in the low-speed mode consists of all sum blocks in series, the delay is only moderately worse.

0-7695-2533-4/06 $20.00 © 2006 IEEE

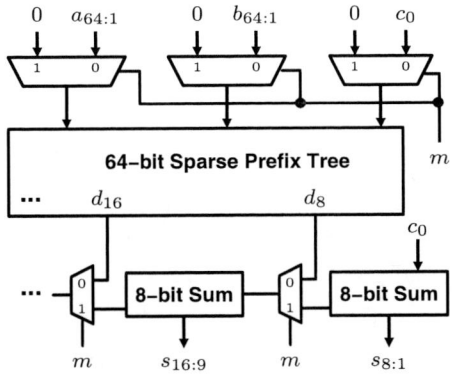

Figure 1. Implementation in static CMOS

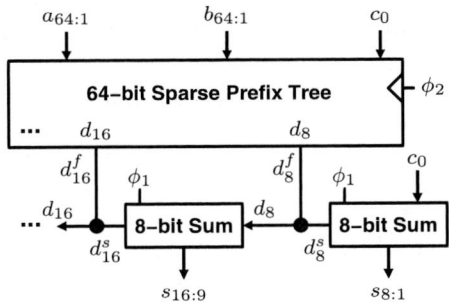

Figure 2. Implementation in domino logic

3 Implementation

The algorithm can readily be implemented in cell-based static CMOS logic as shown in Fig. 1. In high-speed mode, the prefix tree operates as usual; but in low-speed mode all prefix tree inputs are tied off with multiplexers. The circuit has to be optimized in high-speed mode by disabling all timing paths from the input m and by defining false paths through the multiplexers between the sum blocks during optimization. Once the circuit is implemented static timing analysis is performed to determine the achieved delay in both operating modes.

The implementation in domino logic is shown in Fig. 2. The clock ϕ_1 controls the sum blocks and is active every clock cycle. The prefix tree uses ϕ_2, which is only active for high-speed operation, and produces the fast carries d_j^f. In slow-speed mode ϕ_2 is inactive and the sum blocks produce the slow carries d_j^s. Since the two clocks are integer multiples of each other the signals from both clock domains can be joined in a wired-OR fashion, as long as only a single pull-down nfet device is used on the common wire. As a result, no multiplexer delay is inserted into the critical path.

Table 1. Static CMOS implementation

	Delay [ns]		Energy [pJ]	
High-speed mode	2.0		2.0	
Low-speed mode	8.0	(+4x)	0.56	(−3.6x)

Table 2. Domino logic implementation

	Delay [ps]		Energy [pJ]	
High-speed mode	220		70	
Low-speed mode	690	(+3.1x)	30	(−2.3x)

4 Results

Both adders, with sparse Kogge-Stone prefix trees, have been implemented in a 90nm bulk CMOS process and simulated using extracted layouts. The static CMOS implementation has been synthesized, placed and routed with Cadence Encounter at 0.9V using the worst-case library corner. The power consumption was measured by simulating $100,000$ random input vectors. The results are given in Table 1. The energy consumption drops by a factor of 3.6 in the low-speed mode. The domino logic implementation has been implemented and extracted using Cadence Virtuoso and Assura. The power consumption was measured with $10,000$ random vectors with Synopsys Nanosim at 1.2V using the "typical" device corner. The simulation results are summarized in Table 2. In low-speed mode the energy consumption drops by a factor of 2.3.

5 Conclusion

A dual-mode sparse tree adder has been presented. It allows for both high-speed/high-energy and low-speed/low-energy operation. Implementations in both static and dynamic CMOS have been presented. In a reference implementation on a 90nm process a reduction in energy consumption in the low-speed mode by a factor of 3.6 in static CMOS and a factor of 2.3 in domino logic is achieved.

References

[1] J. Grad and J. E. Stine. New algorithms for carry propagation. In *Proc. 15th ACM Great Lakes Symp. on VLSI*, pages 396–399, 2005.

[2] S. Mathew, M. Anders, R. K. Krishnamurthy, and S. Borkar. A 4-GHz 130-nm address generation unit with 32-bit sparse-tree adder core. *IEEE J. Solid-State Circuits*, 38(5):689–695, 2003.

[3] M. Vratonjic, B. R. Zeydel, and V. G. Oklobdzija. Low- and ultra-low power arithmetic units: Design and comparison. In *Proc. 2005 Int. Conf. on Computer. Design*, pages 249–252, 2005.

0-7695-2533-4/06 $20.00 © 2006 IEEE

A Flexible Architecture For Block Turbo Decoders Using BCH Or Reed-Solomon Components Codes

Erwan PIRIOU, Christophe JEGO, Patrick ADDE & Michel JEZEQUEL

GET/ENST Bretagne, CNRS TAMCIC UMR 2872, Brest, France
firstname.lastname@enst-bretagne.fr

Abstract

In this paper, the first flexible architecture dedicated to block turbo decoders is presented. The major innovation concerns the component code that is used by the block turbo code. In fact, our architecture is able to decode BCH and Reed-Solomon codes with single or double correction power. To the authors' knowledge, this is the first architecture implementing Reed-Solomon block turbo codes. This approach makes it possible to select the block turbo decoder architecture using optimum component codes in any circumstance. Our flexible elementary SISO decoder is dedicated to extended binary BCH codes (32,26) and (32,21) and to non-extended Reed-Solomon codes (31,29) and (31,27). Experimentation has been done on a Stratix-based NIOS development board.

1. Introduction

The use of error correction coding is one of the most powerful techniques available for the communication engineer to improve digital communication quality. Currently, the turbo code [1] family is considered to be an efficient coding scheme for channel coding. This family of codes consists of two key design innovations: parallel concatenated encoding and iterative decoding. The iterative decoding is applied to Soft Input Soft Output (SISO) decoders. The general concept of the iterative SISO decoding of concatenated convolutional codes has been extended to product codes [2] and Low Density Parity Check (LDPC) codes. Block Turbo Codes (BTC) is based on product codes that were constructed by the serial concatenation of two (or more) systematic linear block codes. In the last few years, many block turbo decoder architectures have been designed [3]. Product codes using binary BCH component codes have generally been chosen. The block turbo code concept was recently successfully extended to Reed-Solomon component codes [4]. This was motivated by the highest code rate property of Reed-Solomon codes and their efficiency for burst error correction. Moreover, the decoding algorithms of binary BCH codes and Reed-Solomon codes are similar because Reed-Solomon codes are a subclass of non-binary BCH codes. For these reasons, it is possible to design efficient BTC architectures using Reed-Solomon

component codes from the BCH-BTC architecture know-how. This paper presents a flexible block turbo decoder architecture using single or double error correcting component codes.

2. Block Turbo Code Performance

BCH and RS encoders take k data symbols and adds ($n-k$) redundant symbols to generate an n symbol codeword. The decoder can correct up to t symbols that contain errors in a received word. The flexible elementary SISO decoder that we have chosen is dedicated to extended binary BCH codes (32,26) and (32,21) and to non-extended Reed-Solomon codes (31,29) and (31,27). The main advantage of Reed-Solomon BTC using single or double error correcting component codes is in high code rate applications. Indeed, for high code rates (0.8-0.9), the RS BTC requires a much smaller code length. For example, table 1 shows that the RS BTC (31,29) and (31,27) enable the code length be reduced by 3.4. This feature is very interesting from the architecture point of view because it allows a reduction in RAM complexity. On the other hand, if the necessary code rate is about 0.5, BCH BTC using single or double error correcting component codes has the best trade off between performance and complexity [3].

block turbo code		n (bits)	k (bits)	code rate
BCH	$(32,26)^2$ t=1	1024	676	0.66
	$(32,21)^2$ t=2		441	0.43
Reed-Solomon	$(31,29)^2$ t=1	4805	4205	0.875
	$(31,27)^2$ t=2		3645	0.758
BCH	$(128,120)^2$ t=1	16384	14400	0.878
	$(128,113)^2$ t=2		12769	0.779

Table 1. Comparison of BCH and RS block turbo codes

The performance of block turbo codes with single error correction power is simulated for the AWGN channel using 16 test vectors and after 8 iterations. At a frame error rate of 10^{-4}, the distance between the code performance and the theoretical limit is lower than 1 dB for the three BTC: 0.75 dB for BCH$(128,120,4)^2$, 0.95 dB for BCH $(32,26,4)^2$ and 0.8 dB for RS $(31,29,2)^2$.

0-7695-2533-4/06 $20.00 © 2006 IEEE   430

3. Flexible architecture dedicated to block turbo decoding

Our flexible BCH/RS elementary decoder consists of four blocks: reception, processing, transmission and control. In the reception part, the n=32 binary or m-ary symbols of the received word are processed progressively. So, this block computes initial syndromes, positions of the m least reliable symbols and parity of the received word. The number of syndromes is contained between 1 and 4 and depends on the type of code (BCH or RS) and the error correction power (t=1 or t=2). Code features are given by the control unit to the block to select the appropriate syndrome cells. The main function of the processing part is to build and then to correct the test vectors obtained from the initial syndrome and to combine the least reliable bits. The number of tests chosen in our BCH/RS architecture is 16 because more vectors do not offer a significant performance gain for an increase in design complexity. Decoding the test vectors is done by an algebraic decoder. The flexible architecture that carries out the defined algebraic decoder is described in Figure 1. It can decode BCH or RS codes with single or double error correcting powers. The code features are given by the control unit to the algebraic decoder to configure the appropriate data paths as shown in Figure 1. If the code is a BCH code with a single error correction power then the correction position is given by the syndrome value. In other cases, the Peterson Gorenstein Zierler (PGZ) algorithm and a look up table estimate the error positions. Next, parity is computed for BCH codes or error values are computed for RS codes. Finally, if the code is not a BCH code with a single error correction power then the algebraic decoder verifies if the corrected test vector is a codeword. Moreover, the processing block has to produce a metric (Euclidean distance between test vector and received word) for each test vector. Finally, a selection function allows the maximum likelihood code word w_d and three concurrent words w_c to be selected. In the transmission part, the block performs different functions: computing the reliability for each binary symbol, computing the extrinsic information and correction of the received symbols. The n=32 binary or m-ary symbols of the codeword are corrected progressively. The three previous blocks are supervised by a control part. In the design, this task is achieved by a Nios embedded processor. This processor enables the data communication through an Avalon system bus between the matrix memory and the elementary decoder. Moreover, the processor sequences the iterative decoding process of all the matrix codewords. In parallel, the Nios embedded processor supervises the reconfiguration. The flexible BCH/RS architecture uses 3 pipeline blocks for decoding a word of 32 binary (m=1) or m-ary (m=5) symbols. Two different clocks are necessary: Ha for reception and emission blocks and Ht for processing block. The ratio between Ha and Ht depends on the code type (BCH or RS) and the error correction power (t=1 or t=2). Thus in all cases, the three blocks of the decoder have the same latency. 2mHa clock periods are necessary for the algebraic decoding of each one of the 16 test vectors. The latency of the elementary decoder is 64m Ha clock periods which corresponds to two words.

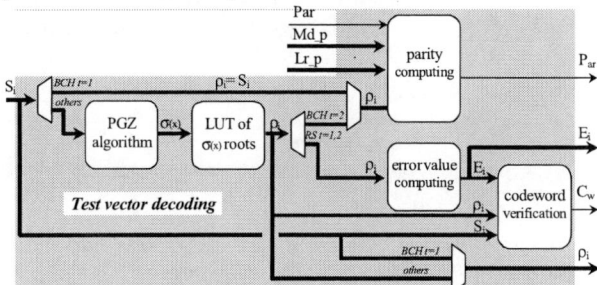

Figure 1. Algebraic decoder for the test vectors

The RS/BCH block turbo decoder was designed and mapped to an Altera's Nios® Development Kit, Stratix™ Professional Edition. Reception, processing and transmission blocks and the control unit were implemented into hardware modules and into a software module (Nios processor) respectively. 2498 and 4563 logic elements are necessary for the integration of the three functional blocks and the control unit respectively. The clock frequency of the Nios processor is 100 MHz but the data system bus transfer throughput between the processor and the hardware blocks is 25 Mbps.

This architecture meets the requirements of flexibility and adaptation for future digital communication applications. It considers a family of error correcting codes (block turbo codes) on which no previous work has been done in a flexibility context. Many potential applications such as mobile communications, optical transmission, data storage and xDSL can benefit from flexible architectures dedicated to block turbo codes.

References

[1] C. Berrou, A. Glavieux, P. Thitimajshima, *Near Shannon limit error correcting coding and decoding: Turbo Codes*, IEEE ICC93, vol. 2/3, May 1993.
[2] R. Pyndiah, A. Glavieux, A. Picart, S. Jacq, *Near optimum decoding of product codes*, GLOBECOM94, November 1994.
[3] R. Pyndiah, P. Adde, R. Zhou, *Block Turbo Codes: Ten years later*, IEE Seminar on Sparse Graph Codes, October 2004.
[4] R. Zhou, A. Picart, R. Pyndiah, A. Goalic, *Reliable transmission with low complexity Reed Solomon block turbo codes,* ISWCS Conference, pp 193-197, September 2004.

Transparent Management of Reconfigurable Hardware in Embedded Operating Systems

Krzysztof Kościuszkiewicz[a]   Fearghal Morgan[b]   Krzysztof Kępa[c]

[ab]*Electronic Engineering Department, National University of Ireland, Galway*
[c]*Computer Architecture Group, Wrocław University of Technology, Poland*
[a]*kokr@kn.ict.pwr.wroc.pl,* [b]*fearghal.morgan@nuigalway.ie,* [c]*krzysztof.kepa@pwr.wroc.pl*

Abstract

A reconfigurable System-on-Chip (rSoC) architecture incorporating embedded uClinux operating system and multiple co-processing nodes has been reported [4]. Our paper extends this platform to incorporate transparent management of reconfigurable hardware resources. This provides a convenient and flexible means for rapid application development of rSoC systems. The co-processor allocation and task management system is described. Details of the implementation are also included.

1. Introduction

Published work on rSoC parallel processing architectures describes a hybrid rSoC multiprocessing system consisting of multiple co-processing nodes interfaced to the main processor [4]. Each of the nodes can be individually programmed. Code and data memory in each node are mapped to virtual entries in a uClinux [2] operating system (OS) file system.

Ideally, as in the case of software-only development, a programmer, using a rSoC need not be concerned with the underlying hardware implementation. This paper proposes delegating the management of hardware reconfiguration to an embedded OS. This offers a convenient and flexible means for rapid development of rSoC applications. The hardware engineer can focus on development of IP cores, which meet a flexible but clearly defined interface specification. This cleanly separates development of hardware and software.

A number of solutions to enabling hardware/software interfacing mechanisms have been proposed [1, 3, 5]. Each approach requires significant knowledge of the underlying hardware architecture by software engineers.

This paper describes the modifications introduced to the uClinux kernel to provide transparent management of re-

configurable hardware. System implementation details are also given.

2. Review of the rSoC platform

The hybrid rSoC parallel processing architecture [4] consists of a central microprocessor (microBlaze) and an array of co-processing microcontroller nodes (picoBlaze). The main processor and each of the co-processors communicate over two unidirectional FIFO buffers. These are implemented using fast simplex link (FSL) peripherals on the microBlaze. Each node also contains a control module that examines incoming data and acts accordingly upon reception of a control word, making node reconfiguration possible.

Applications on the main processor run under control of uClinux OS; it provides hardware abstraction for applications. Processes executing on the main processor can access the co-processor nodes through character device files. This allows reprogramming and data processing delegation.

3. Model for software-hardware integration

Character device file level of abstraction mentioned in section 2 allows use of the same mechanism for accessing different devices which operate on byte streams. However the application using reconfigurable resources still has to deal with problems like multiple access to single device and details of co-processor implementation. Also introducing the co-processing requires significant changes to the application code.

Other approach was proposed by Wigley and Kearney [1]. Their solution was incorporated in MARC OS, which was designed to explicitly address reconfigurable computing issues. Application partitioning is done in a static way and during runtime, logical modules of applications are mapped to reconfigurable or software modules.

0-7695-2533-4/06 $20.00 © 2006 IEEE

Idea of hardware processes have been already shown in context of uClinux and reconfigurable logic [5]. Point-to-point communication is used in both IP core pipelines and linux software pipelines. Cryptography, DSP and multimedia processing fit well into this processing scheme. Complex tasks may be achieved by composition of basic blocks and each block of can be independently designed and reused.

From the programmer's point of view using reconfigurable hardware should not differ from communication with another process implementing the same functionality. Because of that we propose a method for integration of hardware processes into uClinux software pipelines.

4. uClinux kernel modifications

Two basic goals for successful and robust kernel-level implementation were identified. They include implementation of a module handling custom binary executable format for picoBlaze cores and a module containing routines that allocate code to co-processors and handle hardware IPC between software processes and picoware [4].

In the case of picoBlaze executables the code can not be executed on the main processor, because the picoBlaze instruction set is not compatible with microBlaze architecture. On a request to load this type of executable, identified by file header information, the handler will have to reprogram one of the available co-processors in the system in order to execute the application.

The co-processor management module provides a way to reprogram the co-processor nodes and handle data transfers between them and software processes running in the OS.

The first step performed by the module after loading of picoBlaze executable file is check for unallocated co-processors in the system. If none is found, execution of binary file fails. Otherwise one of the is allocated and re-programmed. Standard input and output descriptors of the current process are redirected to FSL links of newly allocated co-processor. If any of above descriptors is already connected to one of the FSL devices, then data transfers will be handled internally by the module. Afterwards the unnecessary data allocated to process that called the `exec()` function is freed and it is ensured, that the process will not be scheduled for execution (similarly to 'zombie' process state). After this step new tasklet is scheduled in the kernel. It tries to read data from the process input and write to the output. If any of these operations fail, then the co-processor is de-allocated.

Testing the status of file descriptors in the scheduled tasklets allows detection of picoware termination conditions. When a processes communicating with picoBlaze finishes execution, then operations on allocated file descriptors will fail. This way the kernel will know that this co-processor is no longer used. Such approach avoids complicated changes to OS scheduler algorithm.

5. Implementation and results

The platform described in section 2 has been implemented in Xilinx Spartan 3 FPGA. Changes described in section 4 were applied to uClinux version 2.4.27-uc2.

We have successfully tested the proposed model in the mentioned configuration. The OS kernel was able to allocate code loaded from picoBlaze binary executable files on different co-processing nodes. With four co-processors system behavior was correct. The proof-of-concept application contains configurable number of processes copying squared input values to output.

6. Conclusions and future research

This paper has outlined design and implementation of a flexible and powerful method of integrating reconfigurable hardware resources into a uClinux operating system. The concept of hardware-implemented processes is virtually transparent for the application developers. Using this method hardware IP blocks in the system have the properties of binary software applications. The implementation of hardware processes can be extended to mimic behavior of software applications in almost all aspects.

Further research will focus on testing and development of more efficient and flexible hardware platforms that will take advantage of runtime partial reconfiguration of the FPGA logic.

References

[1] G. Wigley and D. Kearney. The management of applications for reconfigurable computing using an operating system. volume 6 of *Conferences in Research and Practice in Information Technology*, pages 73–81, Melbourne, Australia, 2002. ACS. Seventh Asia-Pacific Computer Systems Architectures Conference (ACSAC2002).

[2] J. A. Williams. Microblaze uClinux project. 2004.

[3] J. A. Williams and N. W. Bergmann. Embedded linux as a platform for dynamically self-reconfiguring systems-on-chip. In *Engineering of Recofigurable Systems and Algorithms (ERSA 2004)*, Las Vegas, Nevada, USA, June 2004.

[4] J. A. Williams and N. W. Bergmann. Programmable parallel coprocessor architectures for reconfigurable system-on-chip. In O. Diessel and J. A. Williams, editors, *International Conference on Field-Programmable Technology*, pages 193–200, Brisbane, Australia, Dec. 2004.

[5] J. A. Williams, N. W. Bergmann, and X. Xie. FIFO communication models in operating systems for reconfigurable computing. In K. L. Pocek and J. M. Arnold, editors, *Field Programmable Custom Computing Machines (FCCM 2005), IEEE Symposium on*, Napa, California, USA, Apr. 2005.

0-7695-2533-4/06 $20.00 © 2006 IEEE

An open-source tool for simulation of partially reconfigurable systems using SystemC

Alisson V. de Brito, Elmar U. K. Melcher, Wilson Rosas
COPELE - Departamento de Engenharia Elétrica
Universidade Federal de Campina Grande - UFCG
P.O. Box 10053, Campina Grande, PB 58109970 Brazil
{alisson, elmar,wilson}@dee.ufcg.edu.br

Abstract

This paper presents a novel technique for simulation of dynamically and partially reconfigurable systems using SystemC. Its kernel was modified in order to enable the deactivation of any module at simulation time. An example of how to use this technique is presented. Simulations using our MPEG-4 decoder implementation is being developed.

1. Introduction

A subset of dynamic reconfigurable systems is named partially reconfigurable systems. It is characterized by reconfiguration of some elements of the system while the others keep working normally. There are FPGA [1,2,3,4], also named *Dynamically Programmable Gate Arrays* (DPGA [6]), and others specific architectures capable to perform partial reconfiguration [7,8].

There are other projects of system-level modeling for dynamically reconfigurable systems using SystemC (www.systemc.org), like the Adriatic project [9] and the OSSS+R project [10], but using others strategies and with different objectives.

In order to simulate partially reconfigurable systems, the simulator must perform some specialized operations. The switching of two modules from the same area is the basis of all other application of partial reconfiguration. This paper presents a simulation strategy using SystemC, which execute this operation. The base of this technique is disabling modules at simulation-time, while the system keeps running normally.

2. An open-source tool for simulation of partially reconfigurable systems using SystemC

During each simulation cycle, all available methods

and threads are executed. The routine *crunch* from *sc_simulation_context* class is responsible for executing all methods and threads. This routine verify if the method to be executed at current simulation cycle belongs to a module registered in a linked list named *configList*, if so, it cannot be executed at current simulation cycle.

Two routines *dr_sc_turn_on* and *dr_sc_turn_off* were implemented in order to use the blocking resources. These functions respectively add and remove modules names from *configList*. Once *dr_sc_turn_off* routine is called during simulation methods from the parameter module will not be executed until a call to routine *dr_sc_turn_on*.

3. A Simulation example

A simple example was implemented in order explain how to use the simulation routines *dr_sc_turn_on* and *dr_sc_turn_off*. It contains two modules (*moduleA* and *moduleB*), which generate two different waveforms (see Figure 1). The modules produce quadratic waveforms, while *moduleA* produces a waveform with 1ns of period, *moduleB* produces a signal with 3ns of period. These two modules are connected to the same signal, which is traced. This signal presents a mixed behavior, with the modules sending data at the same time. Using the routines *dr_sc_turn_on* and *dr_sc_turn_off*, the modules will work always at different moments, avoiding the overlapping of their signals and forming two scenarios, one with just *moduleA* configured and working and other with just *moduleB*. In simulation context, just one module is working at each instant.

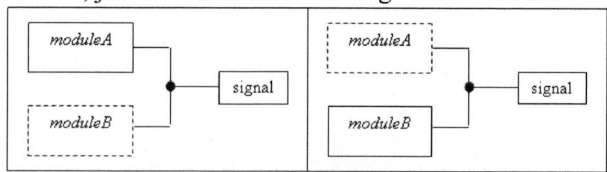

Figure 1. Two scenarios switched dynamically at simulation time. At each scenario just one module is configured.

The resulted waveforms from *moduleA* and *moduleB* showing how partial reconfiguration works with our technique is presented in Figure 5.

Figure 2 presents the waveforms generated from this simulation. It presents *moduleA* and *moduleB* signals being executed separately without partial reconfiguration in order to know how are their behaviors. The output signal produced during partial reconfiguration is named *reconf* and it aggregates the behavior of *moduleA* until 30ns, and the behavior of *moduleB* during remaining simulation time. This waveform shows that the system really had its behavior changed during simulation.

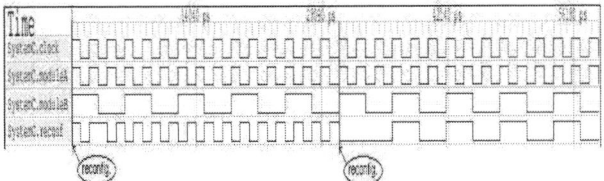

Figure 2. Waveforms from the example. The reconfiguration moments are pointed.

In order to analyze these simulation features, our MPEG-4 decoder was compiled and configured with Virtex II FPGA from Altera (www.altera.com) with all modules, and using fake modules. The static area reduction was about 50%.

The area reduction of a whole video processing depends on original video encoding process. If a video where four *intraframe* is received by decoder for each *interframe* the decoder need to be completely configured just 20% of the time, and the total chip area reduction is about 40%, but can increase if number of *intraframes* per *interframes* also increases.

Acknowledgments

This research work is supported by "Conselho Nacional de Desenvolvimento Científico e Tecnológico (CNPq)", Brazil.

References

[1] ATMEL. Application Note: Implementing Cache Logic with FPGAs. USA, Sep. 1999. Available at http://www.atmel.com/dyn/resources/prod_documents/DOC0 461.PDF

[2] Ferrandi, F., Santambrogio, M., Sciuto, D., A Design Methodology for Dynamic Reconfiguration: The Caronte Architecture. Proceedings of the 19th International Parallel and Distributed Processing Symposium (IPDPS'2005), 2005, Denver, CA, USA

[3] Hadley, J. D., Hutchings, B. L. Design Methodologies for Partially Reconfigured Systems, In Peter Athanas and Kenneth L. Pocek, editors, Proceedings of the IEEE Workshop on FPGAs for Custom Computing Machines, pages 78-84, Los Alamitos, California, April 1995. IEEE Computer Society, IEEE Computer Society Press.

[4] Markovsiy, Y., Caspi E., Huang R., Yeh J., Michael C., J. Wawrzynck. Analysis of Quasi-Static Scheduling Techiniques in a Virtualizad Reconfigurable Machine. Proceedings of FPGAs 2002, ACM, Monterey, California, USA. February 2002.

[6] Dehon, A., DPGA Utilization and Application, in FPGA'2006, pp 115-121, February, 1996.

[7] Becker, J. and Hartenstein, R., Configware and morphware going mainstream. Journal of Systems Architecture. 49, 4-6, 127-142, September, 2003.

[8] Becker, J., Vorbach, M., Architecture, Memory and Interface Technology Integration of an Industrial/Academic Configurable System-on-Chip (CSoC), IEEE COMPUTER SOCIETY. ANNUAL Symposium ON VLSI, Tampa, Florida, February 2002

[9] Qu, Y., Tiensyrja, K. Masselos, K., System-Level Modeling of Dynamically Reconfigurable Co-Processors, International Conference on Field Programmable Logic and Applications, Antwerp, Belgium, August-September, 2004.

[10] Schallenberg, A., Oppenheimer, F., Nebel, W. Designing for Dynamic and Partially Reconfigurable FPGAs with SystemC and OSSS, Forum on Specification and Design Languages (FDL '04), Lille, France, Sept. 2004.

Partial and Dynamic Reconfiguration of FPGAs: a top down design methodology for an automatic implementation

Florent Berthelot
CNRS UMR 6164
IETR-INSA
20 av des Buttes de Coesmes,35043 Rennes, France
florent.berthelot@ens.insa-rennes.fr

Fabienne Nouvel
CNRS UMR 6164
IETR-INSA, 20 av des Buttes de Coesmes
35043 Rennes, France
fabienne.nouvel@ens.insa-rennes.fr

Abstract

Dynamic and partial reconfiguration of FPGAs enables systems to adapt to changing demands. This paper concentrates on how to take into account specificities of partially reconfigurable components during the high level Adequation Algorithm Architecture process. We present a method which generates automatically the design for both partially and fixed parts of FPGAs.

1. Introduction

The recent and next generations wireless systems (802.11 up to 4G) are being designed to provide a wide variety of multimedia services and seamlessly switch between different wireless standards. Most Software Defined Radio solutions are based on Digital Signal Processors (DSPs) combined with Field Programmable Gate Array (FPGAs). The split between Hardware/Software components during the partitioning process leads to a compromise between system's performance of an hardwired solution and flexibility of a software solution. Reconfigurable devices, including FPGAs, can fill the gap between hardwired and software technology. Recently runtime reconfiguration of FPGA parts has led to the concept of virtual hardware. This technic allows to change only a specified part of the chip while other areas remain operational and unaffected by the reconfiguration [3]. In this paper we focus on reconfiguration methodology for FPGAs, based on AAA approach and associated tool SynDEx [1]. The way of modeling a partially runtime reconfigurable part of a FPGA with Syn-DEx is exposed. A case study is presented in last section followed by the conclusion.

2. Reconfiguration Levels

In the case of mobile communications, three main constraints have to be combined : high performance, low power consumption and flexibility. Three levels of reconfiguration can be considered :

- System Level Reconfiguration : In this case, the application is very often supported by an heterogeneous architecture. Either, these hardware solutions do not allow the reconfiguration of the datapath.

- Functional Level Reconfiguration : Some FPGAs support partial dynamic runtime configuration. A function can be replaced by another one while other parts stay operative. A runtime reconfiguration manager will control, monitor and execute the dynamic reconfiguration.

- Logic and RTL level Reconfiguration : The majority of the FPGAs are fine grain. Switching the configuration of a design is very quick, as the bitstream differences are smaller than the entire bitstream. Neither, this level is architecture's manufacturer dependant and do not allow the designer to adopt an open methodology.

Considering the SDR constraints, the functional reconfiguration level seems to be the best level for our partial and dynamically reconfiguration of systems.

3. Design flow for dynamic reconfiguration : AAA approach

Some partitioning methodologies based on various approaches are reported in the literature [2]. Among theses methods we have chosen the AAA approach. AAA methodology aims at finding the best matching between an algorithm and an architecture while satisfying time constraints. SynDEx automatically generates a distributed and

0-7695-2533-4/06 $20.00 © 2006 IEEE

optimized synchronized executive. The reader is referered to previous works [1] which describe all the steps of the methodology and extensions to FPGAs modelization. Figure 1 depicts the overall methodology flow. The mapping and scheduling of the operations and data transfers onto the operators and the communication media are carried out by a heuristic which takes into account durations of computations and inter-component communications.

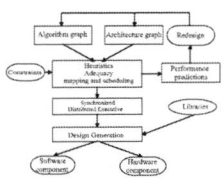

Figure 1. SynDEx methodology flow

To reduce the difficulty in managing such dynamic reconfigurable application and to provide reliable implementation, following issues must be adressed : automatic or manual partitioning of an application, specification of the dynamic constraints, automatic generation of the C or VHDL core controler. Considering these features in SynDEx, runtime reconfigurable parts of an component must be considered as vertices in the architecture graph. As shown in Figure 2, runtime reconfigurable parts of a FPGA (D1 and D2) and fixed parts (F1) can be represented as hardware operators of the architecture. (D1 and D2) will integrate both dynamic modules and the control to manage the reconfiguration.

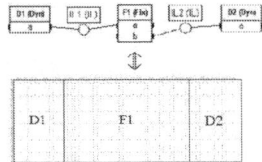

Figure 2. Model of runtime reconfigurable parts of a FPGA with SynDEx

Reconfiguration of the dynamic part is performed by two sub-parts: a Configuration manager and a Protocol configuration builder. The Configuration manager is in charge of the configuration bitstream which must be loaded on the reconfigurable part by sending configuration requests. Reconfigurations are performed as soon as the reconfigurable part is unused. Configuration requests are sent to the Protocol configuration builder which is in charge to construct a valid reconfiguration stream. Encapsulation of operators with a standard interface allows to reconfigure only the area containing the operator without altering the design around.

Functionalities involved in the general control of the dynamic area and control manager remain all the time in the static part. These two operators are automatically generated by SynDex and translate in VHDL with our libraries. In our design flow, the Xilinx Modular Design back-end flow is used to place and route each module and to generate the associated bitstream.

4. Case study : MC-CDMA reconfigurable transmitter

Our top-down design flow has been used to design a transmitter system based on MC-CDMA modulation scheme [4]. This transmitter is implemented on a prototyping board with a FPGA Xilinx Xc2v2000 which integrates an internal reconfiguration access port(ICAP) for partial reconfiguration .

The FPGA is divided in two parts. The first one is static and implements non reconfigurable logic, the second one is dedicated to the dynamic operator.The self reconfiguration operates at 20Mhz, one bistream byte is loaded each cycle by the ICAP. The reconfiguration time needed takes about 4ms. As the maximum net bit rate per user is about of 0.296Mbits/s, time to reconfigure the modulation is of the order of some data frames. Additional details and informations will be given during the poster presentation.

5 Conclusion

A methodology flow to manage automatically partially reconfigurable parts of a FPGA has been proposed. The AAA methodology and associated tool SynDEx have been used to perform mapping and automatic code generation for fixed and dynamic parts of FPGA. Either, SynDEx's heuristic needs additional developments to optimize time reconfiguration. Furthermore, complex design and architecture can support more than one dynamic part.

References

[1] F. Berthelot, F. Nouvel, and D. Houzet. Design methodology for dynamically reconfigurable systems. *JFAAA , Dijon France*, pages 47–52, January 2005.

[2] J. Harkin, T. McGinnity, and L. Maguire. Partitioning methodology for dynamically reconfigurable embedded systems. *IEE Proc-Comput. Digit. Tech*, 147, November 2000.

[3] E. L. Horta, J. W. Lockwood, D. E. Taylor, and D. Parlour. Dynamic hardware plugins in an fpga with partial runtime reconfiguration. *Design Automation Conference (DAC)*, 2002.

[4] S. Lenours, F. Nouvel, and J.-F. Helard. Design and implementation of mc-cdma systems for future wireless networks. *EURASIP JASP*, pages 1604–1615, August 2004.

0-7695-2533-4/06 $20.00 © 2006 IEEE

Self-Timed Thermally-Aware Circuits

David Fang, Filipp Akopyan, Rajit Manohar*
Computer Systems Laboratory
Electrical and Computer Engineering
Cornell University
Ithaca, NY 14853, U.S.A.

Abstract

Thermal management is becoming increasingly important in circuit designs with high power density. Circuits that overheat beyond specified operating conditions may suffer timing failures, or become damaged for various reasons, including thermal runaway. We present a novel application of a thermally sensitive circuit to automatically regulate the performance and power consumption of asynchronous circuits, with minimal implementation overhead, and free of interruption of operation.

1 Introduction

As the power density of modern integrated circuits continues to increase, power and temperature management become increasingly important and challenging [3]. In this paper, we demonstrate a mechanism that regulates the performance of asynchronous circuits using a simple thermally-sensitive circuit. Our approach does not require temperature measurement, rather, we leverage the temperature response of subthreshold devices to construct a temperature-sensitive delay element to directly regulate the speed of the system. Asynchronous circuits are capable of operating correctly in the presence of continuous and dynamic changes in delays [2], and can self-modulate their performance without interruption of operation.

2 Circuits

As a circuit heats up, the gate delays increase and the frequency naturally drops, reducing the circuit's dynamic power consumption and self-heat generation. However, the natural negative-feedback retardation of this self-heating rate is too weak to halt the increase in temperature [5].

In the subthreshold region, the source-drain current is exponentially dependent on temperature:

$$I_D = I_0 \cdot \exp\left(\frac{V_{gs} \cdot q}{\zeta \cdot k \cdot T}\right)$$ where I_D is the drain-source current,

---

*E-mail: {fang, filipp, rajit}@csl.cornell.edu

I_0 depends on channel width, channel length, diffusion constant of carriers, carrier density and electron charge [6], ζ is a nonideality factor, and T is temperature in K. We use the high temperature-sensitivity of the subthreshold transistors to construct a temperature-sensitive voltage source, shown boxed in Figure 1. Transistors M1 and M2 are biased differently to have contrasting thermal sensitivities—M1 operates in deep subthreshold (more temperature-sensitive), while M2 operates near-threshold—forming a temperature-sensitive, resistive voltage divider. The bias voltages and the sizes for M1 and M2 are tuned to achieve a desired temperature response for a given technology. Figure 1 shows a thermally-sensitive voltage source controlling the gate of a foot transistor in typical logic circuitry. As temperature increases, the foot transistor starves the current until the point where switching ceases, and the conditional staticizer retains the value of V_{int}.

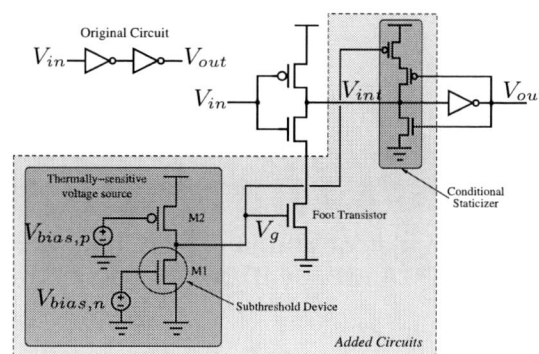

Figure 1. Temp.-sensitive delay element

We simulated the above temperature-dependent circuit using TSMC 0.18μm technology parameters, with 1.8 V nominal supply V_{dd}. Figure 2 shows the V_g switching sharply near 100°C. Above 113°C, V_{out} never switches—the current through the pull-down logic cannot overpower the pull-up staticizer, so the delay is infinite. Once the temperature drops sufficiently, operation will resume.

The peak current of the subthreshold transistor is on the order of tens of nA, comparable to leakage current for this

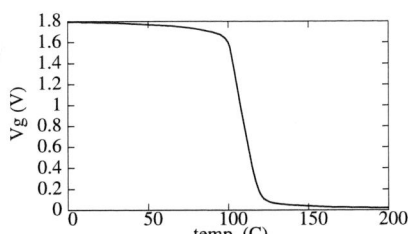

Figure 2. Foot Transistor V_g-T Dependence

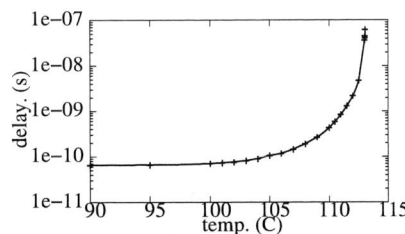

Figure 3. Delay through modified inverter

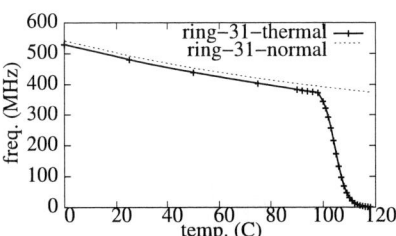

Figure 4. Freq. of ring oscillator vs. circuit temp.

technology, and is thus, negligible. For above-threshold devices, we use a conventional delay-derating model as a function of voltage and temperature [1].

3 Evaluation

In each scenario, we take the original digital circuit and then replace a few selected inverters with temperature-sensitive delay elements. The result is a circuit that regulates its speed based on the local temperature with minimal hardware overhead. Note that this applies to circuits of arbitrary complexity, because having a slow gate on a handshake cycle is sufficient to limit the local throughput of a complex asynchronous circuit. To demonstrate the effectiveness of our application of thermally-sensitive circuits we present examples of varying complexity.

Ring oscillator. A 31-stage ring oscillator with one stage modified to a thermally-sensitive inverter was simulated. The frequencies are plotted against temperature in Figure 4. As expected, the frequency quickly degrades beyond 100°C, where the delay grows exponentially. Before 95°C, the difference between the modified and unmodified oscillator shows that the scheme introduces minimal overhead.

FPGA. We simulated a 5x5 asynchronous FPGA (the design is described in [4]) running a function-block-intensive benchmark to demonstrate a circuit that reaches a self-heating equilibrium. We report the normalized throughput against the peak surface temperature in Table 1. The thermally-aware FPGA's average surface temperature stabilizes at around 100°C after 1 ms of simulated time at an operating frequency of 27% of the room-temperature throughput. In the same scenario, the FPGA without our thermal-aware modifications heats itself to destructive temperatures.

Automatic Dynamic Scheduling. Our technique can be used to dynamically schedule activity away from hot-spots. We designed a dynamic instruction scheduler that uses workload-driven non-deterministic scheduling to issue tasks to execution units A and B. Normally the scheduler issues an equal number of instructions to A and B. The two

execution units were modified to use the thermally sensitive delay. We examined the impact of unit A operating at a higher temperature than B. Table 2 shows that beyond 100°C, unit A practically stops operating, yielding almost all of the computation work to unit B. Once unit A cools down, it will begin to request data from the dispatcher more frequently. Note that in this example, the dispatch circuit is unmodified.

Table 2. Processed data items in a fixed time window for dynamic scheduling.

Temp.-A (°C)	40	60	100	101
Instructions processed by A	46	43	12	8
Temp.-B (°C)	40	42	45	45
Instructions processed by B	46	44	49	48

4 Conclusion

We presented a simple thermally-sensitive circuit that can be used to regulate gate delays in digital circuits. We demonstrated a few applications of the technique in self-regulating asynchronous circuits, as well as in automatically dispatching instructions away from hot-spots without modifying the dispatch circuit. A small number of strategically placed thermally-sensitive gates is sufficient to regulate the local throughput in an asynchronous system with minimal circuit overhead and redesign effort in a way that prevents the circuit from exceeding its thermal budget.

References

[1] J. M. Daga, E. Ottaviano, and D. Auvergne. Temperature effect on delay for low voltage applications. In *Proc. DATE*, pages 680–685, 1998.

[2] Alain J. Martin. The limitations to delay-insensitivity in asynchronous circuits. In William J. Dally, editor, *Proc. ARVLSI*, pages 263–278. Massachusetts Institute of Technology, 1990.

[3] Dennis Sylvester and Himanshu Kaul. Future performance challenges in nanometer design. In *Proc. DAC*, pages 3–8, New York, NY, USA, 2001. ACM Press.

[4] J. Teifel and R. Manohar. Highly pipelined asynchronous FP-GAs. In *Proc. FPGA*, February 2004.

[5] J. A. Tierno. *An Energy-Complexity model for VLSI computations*. PhD thesis, California Institute of Technology, 1995.

[6] Neil Weste and David Harris. *CMOS VLSI Design: A Circuits and Systems Perspective*. Addison-Wesley, 2005.

Table 1. Normalized throughput (f) of thermal-aware FPGA at different temperatures

°C	25	45	80	89	94	97	100
f	1.00	0.85	0.70	0.64	0.60	0.49	0.27

0-7695-2533-4/06 $20.00 © 2006 IEEE

A New Protocol Stack Model for Network on Chip

Masood Dehyadgari[*], Mohsen Nickray[*], Ali Afzali-kusha[*], Zainalabedin Navabi[**]

[*]Low-Power High-Performance Nano-systems Laboratory,
School of Electrical and Computer Engineering, University of Tehran
[**]Department of Electrical and Computer Engineering, Northeastern University
Email: {m.dehyadegari, m.nickray}@ece.ut.ac.ir, afzali@ut.ac.ir, navabi@ece.neu.edu

Abstract

In this paper, we present a communication protocol for Network on Chip architectures which have complex packet switched communication protocols. In order to manage this complexity and advance reusability, a layered approach is taken. It is a 4-layered protocol stack including application, transaction, data-link and physical layers. Our protocol stack supports the best effort traffic as well as guaranteed bandwidth using the virtual channels which logically share the physical links. In order to evaluate the design, a HDL implementation of this protocol stack is implemented and synthesized. The results show 0.5% of a Virtex II 2VP30 FPGA is employed by our proposed protocol stack for each resource network interface.

1. Introduction

Most of the NoC researchers have proposed layered stacks similar to OSI reference model. Our proposed protocol is a 4-layered protocol stack based on 7-layered OSI model as a reference model. The lower two layers of the OSI reference model, namely, physical and data link are used with minor modifications in their functionality for the on-chip packet switched communication. Because of the limited scope of on-chip communication, the top five layers are normally compressed into two layers [1].

Our protocol stack is not dependent on the topology and, hence, it can be applied to any of the topologies with slight changes. The proposed protocol has a modular structure whose parameters may be changed easily. The parameters which can be changed include error handling, flow control, and the number of virtual channels.

2. OSI Reference Model for NoC

Open Systems Interconnection (OSI) model is a reference model developed by ISO as a framework of standards for communication in the network of different equipments. The OSI model defines the communications process into 7 layers [1]. The OSI model is capable of being used for wide area networks and, hence, it is more complex than what is needed for a NoC.

The features for the implementations of computer network and NoC protocol stacks are given in Table 1. Inspired by these differences, we propose our protocol stack with four layers namely application, transaction, data link, and physical layer. In the proposed model, we compress the transport and the network layers into the transaction layer. The upper three layers of OSI model could be merged into one layer that is called application layer. The physical and data link layers are implemented with slight differences in their functionality when compared to the corresponding layers in the OSI model.

Table 1. compare Implementation type of computer network and NoC protocol stacks.

Protocol Stack Layers	Implementation Type	
	Computer Network protocol stack	NoC Protocol Stack
Application	Software	Hardware/Software
Presentation	Software	
Session	Software	
Transport	Software	Hardware
Network	Hardware/Software	
Data Link	Hardware/Software	Hardware
Physical	Hardware/Software	Hardware

3. Design of protocol stack layers

Each layer of proposed protocol stack divides into both transmit and receive parts.

3.1 Application Layer

This layer is the highest layer of the protocol stack. In a NoC environment, each IP core or Process Element (PE) can be as the application which communicates with the other applications using the lower layers.

3.2 Transaction Layer

The most important responsibilities of this layer are packetization, depacketization, providing end to end connection, assembling and deassembling of the packets, and the end to end error handling. In order to provide Quality of Service (QoS), we borrow the

---

0-7695-2533-4/06 $20.00 © 2006 IEEE

Virtual Channel (VC) and Traffic Class (TC) concepts from the PCI Express architecture [2][3]. Each virtual channel presents a complete set of resources that is dedicated to a channel. The VC allows the systems prioritize packet transfers in the network. Each packet, generated by an application, has a TC field which declares the number of the VC through which the packet should flow through. Figure 1 shows the transaction layer with two VC's.

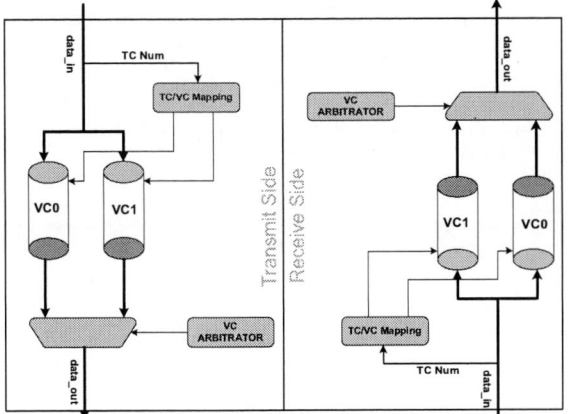

Figure 1. Transaction layer with two virtual channels.

3.3   Data Link Layer

This layer is responsible for the reliable transport of packets from one device to another device across the link and applying the flow control to prevent any congestion. For these purposes, it creates, sends, and receives ACK/NAK (acknowledge/negative acknowledge) packets and Flow Control (FC) packets between two switches or between a switch and a PE and prevent delivering bad packets to the transaction layer. Error handling methods are classified into two groups which are error detection and retransmission and error correction. The error detection and retransmission protocol is chosen for the data link layer of our proposed protocol stack. Most of the error correction method use SEC-DED (Single Error Correction-Double Error Detection) method [4]. Because of the signal integrity issue, the latter method is vulnerable in on chip communications. One should note that to implement the ACK/NAK protocol, some counters and extra logic which increase the area and the power consumptions of the data link layer are needed. The Data Link Layer (DLL) appends additional information, required for the error checking at the receiver device, to the packet. A 3 bit CRC with polynomial $x^3 + x + 1$ is used [5] in the proposed protocol stack.

3.4   Physical layer

The physical layer is responsible for providing the electrical media to connect different switches to each other or between a switch and a PE as well. An asynchronous handshaking protocol is used for transferring the packets between two switches or between a switch and a PE. The Bundled Data with four phase handshake are used. The Bundled Data protocol has two control wires associated with each 8-bit data channel. This leads to 10-bits for each 8-bit channel. The four-phase protocol requires 4 events per each handshake. In our work, the physical interface between switches includes two receive and transmit sections.

4.   Results and Discussion

We synthesized and performed the post place and route simulation for the transaction, data link layer, and physical layer. The Table 2 presented the amount of area occupied by these layers on the Virtex II Pro 2VP30 FPGA are used. The number of the slices used by these three layers is 500 about 0.5% of the total slices and so is the amount of flip-flops.

Table 2. synthesis result of the Transaction/Data Link/Physical layers.

Property	Value	Available
Slices	200/200/100	13696
Flip Flop	50/40/60	27392
LUTs	55/50/40	27392
BRAM	2/1/-	136

The transaction and data link layers operate correctly at 200MHz while the physical layer is an asynchronous circuit.

5.   References

[1]   OSI 7 layers reference Model for Network Communication:http://www.javvin.com/osimodel.html , retrieved on Nov. 2005.

[2]   PCI Special Interest Group, "PCI Express Base Specification Revision 1.0a," Apr 2003.

[3]   D. Mayhew and V. Krishnan, "PCI Express and Advanced Switching: Evolutionary Path to Building Next Generation Interconnects," *Proceedings of the 11 the Symposium on High Performance Interconnects (HOTI'03)*, pp.21 – 29, Jun. 2003.

[4]   H. Zimmer, A. Jantsch, Fault Modelling and Error-Control Coding in a Network-on-Chip, *Master Thesis*, Royal Institute of Technology (KTH), Sweden, 2002.

[5]   Philip Koopman, Tridib Chakravarty, "Cyclic Redundancy Code (CRC) Polynomial Selection For Embedded Networks," *The International Conference on Dependable Systems and Networks*, pp.145 – 154, Jul. 2004.

A Robust Synchronizer

Jun Zhou, David Kinniment, Gordon Russell, and Alex Yakovlev
Newcastle University, UK
{Jun.Zhou,David.Kinniment,G.Russell,Alex.Yakovlev}@ncl.ac.uk

Abstract

We describe a new latch circuit designed to give a high performance in low voltage synchronizer applications. By increasing the latch current only during metastability, we can more than maintain the value of the metastability time constant, τ, without significantly increasing the power. Our circuit also reduces the variation of τ with Vdd and temperature, so that it has a lower τ at 50% Vdd than the conventional jamb latch has at 60% Vdd.

1. Introduction

An important effect of scaling is the increase in both dynamic and static power dissipation. Currently proposed solutions to this problem include dynamic lowering of the voltage in selected sub-systems when high performance is not required. Unfortunately, reduced power supplies usually disproportionately affect the performance of synchronizers since the synchronizer τ depends on the small signal parameters in metastability rather than large signal switching times, and a 50% reduction in power supply voltage may result in over 100% increase in τ. In this paper we present a circuit that is both faster than a conventional jamb latch synchronizer, and less sensitive to Vdd.

2. Jamb Latch

The Jamb latch is a simple circuit commonly used as a synchronizer because of its relatively good performance [1], and its basic configuration is shown in Figure 1. Here, the latch is reset by pulling node B to ground, and then set if data is high and clock is low, by pulling node A to ground. Metastability occurs if the overlap of data and clock is at a critical value that causes node A to be pulled down, and node B up to about the same voltage. By extensive use of SPICE simulation using parameters for a TSMC 0.18μ process, we optimised the transistor sizes for the Jamb Latch to give a low value for τ. The

results are shown in Table 1, and show a value for τ at Vdd of 1.8v is 35.6ps.

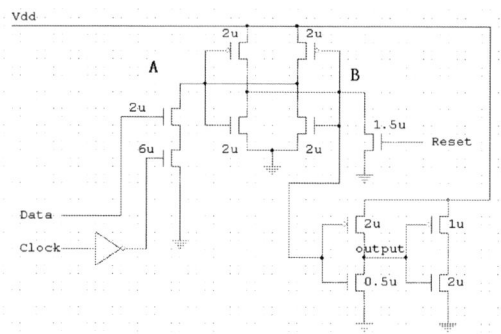

Figure 1 Jamb Latch

The minimum value of τ is limited in this circuit by the capacitance of reset/set transistors, which cannot be further reduced in size, otherwise the circuit will not reliably set or reset. The optimum value of τ is determined by both the capacitance and transconductance of the transistors when the circuit is in metastability. We found the best ratio between p-types and n-types is 1:1, a result also reported by others [1]. Table 1 also shows how τ varies with Vdd and temperature for the Jamb Latch. It can be observed from Table 1 that τ increases with Vdd decreasing and this reduction in speed becomes quite rapid where Vdd approaches the sum of thresholds of p and n-type transistors so that the value of τ is more than doubled at a Vdd of 0.9V, and more than an order of magnitude higher at 0.7V, -25 °C. For comparison we show the FO4 inverter delay in this technology that demonstrates τ is likely to track a processor clock period rather poorly, making design difficult. We also investigated the effect of increasing the width of all transistors by the same factor. In order to estimate the average energy used during metastability, we assume that the average metastability time is τ. As the width increases, the total switching energy increases in proportion but τ only decreases slowly as transistor sizes increase, and reaches a limit at around 31ps.

0-7695-2533-4/06 $20.00 © 2006 IEEE 442

Vdd(v)	τ(ps) at 27 °C	τ(ps) at -25 °C	FO4 at 27 °C	FO4 at -25 °C
1.8	35.6	29.5	91.6	78.6
1.7	36.8	30.5	94.2	80.9
1.6	38.3	31.8	97.3	83.6
1.5	40.1	33.3	100.9	86.8
1.4	42.4	35.4	105.3	90.6
1.3	45.4	38.0	110.5	95.3
1.2	49.4	41.8	117.0	101.1
1.1	55.2	47.3	125.2	108.6
1	63.8	56.2	136.0	118.5
0.9	78.5	72.8	150.7	132.1
0.8	106.8	114.6	171.9	151.2
0.7	252.8	522.2	203.3	182.7
0.6	754.7	1528.8	264.7	247.2

Table 1 Jamb latch τ vs Vdd

Vdd(v)	τ(ps) at 27 °C	τ(ps) at -25 °C
1.8	27.1	20.7
1.7	28.9	21.2
1.6	31.2	21.9
1.5	33.2	23.1
1.4	34.1	24.5
1.3	35.7	26.9
1.2	38.2	30.6
1.1	42.6	33.8
1	47.6	36.6
0.9	53.1	43.1
0.8	62.8	48.5
0.7	74.3	59.2
0.6	93.0	76.8

Table 2 Improved synchronizer τ vs Vdd

3. Improved Synchronizer

An improved synchronizer circuit that is much less sensitive to power supply variations is shown in Figure 2.

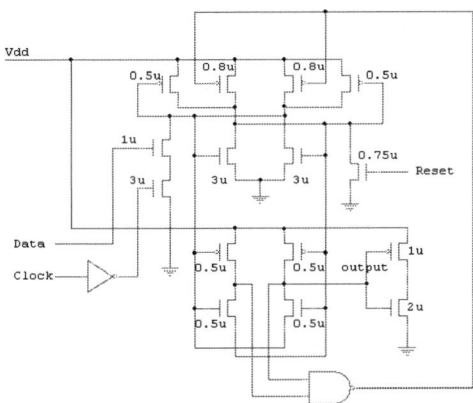

Figure 2 Improved synchronizer

Here the two 0.8µ p-type load transistors maintain sufficient current in the latch during metastability to keep the total transconductance high even at supply voltages less than the sum of thresholds of the p and n-type transistors. Two additional feedback p-types are added to the modified Jamb Latch in order to maintain the state of the latch when the main p-type loads are turned off. Because of these additional feedback p-types, the main p-types need only to be switched on during metastability, and the total power consumption is not excessive. A similar circuit has been described in [2], but few implementation details are given. In our implementation a metastability filter is used to produce the synchronizer output signals, which only go low if the two nodes have a significantly different voltage.

The outputs from the metastability filter are fed into a NAND gate to produce the control signal for the gates of two main p-types. In this circuit, the main p-types are off when the circuit is not switching, operating like a conventional jamb latch, but at lower power, when the circuit enters metastability the p-types are switched on to allow fast switching. The output is taken from the metastability filter, again to avoid any metastable levels being presented to following circuits. Now there is no need for the feedback p-types to be large, so set and reset can also be small. The optimum transistor sizes for the improved synchronizer are shown in Figure 2, and the resultant τ at Vdd of 1.8v is as low as 27.1ps because the main transconductance is provided by large n-type devices and because there are two additional p-types contributing to the gain. It also operates well at 0.6V Vdd and -25ºC, because it does not rely on any series p and n-type transistors being both switched on by the same gate voltage. The relationship between τ and Vdd for the improved synchronizer is shown in Table 2. The switching energy for this circuit is 0.1783pj, compared with 0.1438pj for the conventional Jamb Latch. At the same time as maintaining a low value for τ, the ratio between τ and FO4 is much more constant at around 1:3 over a wide range of Vdd and temperature.

4 References

[1] C.Dike and E.Burton. "Miller and Noise Effects in a Synchronizing Flip-Flop". IEEE Journal of Solid State Circuits Vol. 34 No. 6, pp.849-855, June 1999

[2] R. L. Cline. "Method and circuit for improving metastable resolving time in low-power multi-state devices" US patent 5,789,945, February 27, 1996

Low Power Layered Space-Time Channel Detector Architecture for MIMO Systems

T. Takahashi[1], A.T. Erdogan[1,2], T. Arslan[1,2], and J. H. Han[2]

[1]Institute for System Level Integration, Livingston EH54 7EG, UK
[2]School of Engineering and Electronics, University of Edinburgh, Edinburgh EH9 3JL, UK
E-mail: Tetsuya.Takahashi@sli-institute.ac.uk, {ate, ta, j.han}@ee.ed.ac.uk

Abstract

This paper presents the low power implementation of a Maximum Likelihood (ML) based detector used in the receiver part of a Multiple Input and Multiple Output (MIMO) systems. Low power is mainly achieved through complexity reduction of the ML detector. In particular, Manhattan metric approach is proposed for removing the need for the use of multipliers in the architecture, leading to significant complexity reduction in the ML detector implementation with only 0.7 dB loss in the Bit Error Rate (BER) performance. Results are presented showing that our ML detector achieves 29% saving in area and 34.4% saving in power consumption compared to conventional implementations.

1. Introduction

Multiple-input multiple-output (MIMO) is the leading technology for high throughput transmission in a multipath fading environment. MIMO systems provide a nearly linear increase of capacity with the number of transmit antenna elements, affording significant increase over conventional systems, and this leads to the increase of the transmit capacity and the Quality of Service (QoS) [1].

Concurrent transmission with multiple antennas is the basic technique for MIMO systems where the transmission is performed with the diversity in space and time domain. This technology is called Layered Space-Time (LST) architecture and was proposed by Foschini [2]. There are several LST architectures proposed, and in particular Bell Labs Layered Architecture Space-Time (BLAST) is well known and researched.

Transmit techniques have been developed to achieve high throughput. However, the receiver is also one of the key elements of the system. Since MIMO system uses the multipath environment as the source for higher transmission, the receiver has to deal with multi stream interferences. Moreover noise or fading problems also need to be solved in practice.

The MIMO system can be represented as follows:

$$\mathbf{r} = \mathbf{Hs} + \mathbf{n} \qquad (1)$$

where $\mathbf{r} = [r_1, \ldots , r_R]^T$ is the vector of received symbols, $\mathbf{H}$ is the RxT channel matrix, $\mathbf{s} = [s_1, \ldots , s_T]^T$ is the vector of transmitted symbols, and $\mathbf{n}$ is additive white Guassian noise (AWGN).

Since each received channel is convoluted with each other, the receiver has to differentiate the multiple data [2]. This requirement leads to higher complexity of the system and hence higher power consumption. Hence, there are some algorithms to reduce the complexity such as sphere decoding [3] and QR decomposition [4]. However, this paper proposes a new approach for reducing the complexity of MIMO channel detector. In particular, this paper proposes the Manhattan metric approach for eliminating the use of multipliers, leading to significant complexity reduction in the channel detector implementation.

2. Maximum likelihood detector (MLD)

The basic architecture of a MIMO detector consists of two parts: estimation and detection. In the estimation part, the candidates of received signal are estimated based on the modulated symbols and the channel matrix. With the estimated candidates, the actual received signals are detected in the detection part. There are several algorithms used to decide which candidate is matched best with the received symbol.

MLD is chosen in this paper because of its optimal BER performance. The ML detection algorithm can be expressed with the equation below:

$$\hat{s} = \arg\min_{s \in \wedge} \|\mathbf{Hs} - \mathbf{r}\|^2 \qquad (2)$$

0-7695-2533-4/06 $20.00 © 2006 IEEE

where the best candidate, $\hat{s}$, is selected which minimises the distance between the received symbol, **r,** and all the estimated candidates, **Hs**.

3. Complexity reduction

There are several reduction algorithms as mentioned before, and most of them are about the reduction of search steps. However, this paper presents a novel approach to reduce the number of multipliers. As mentioned above, ML algorithm searches the most probable one from all the candidates by calculating the distance between the received symbols and weighted candidates. Euclidean distance is traditionally used, which is the direct and exact distance, as shown in equation (3). However, this metric system requires the use of multipliers.

$$D = \sqrt{(Ax - Bx)^2 + (Ay - By)^2} \qquad (3)$$

$$D' = |Ax - Bx| + |Ay - By| \qquad (4)$$

In this paper, Manhattan metric approach is proposed to reduce the complexity of computing the distance between the received symbols and weighted candidates. This metric is based on the addition of the X component and the Y component, as expressed by equation (4). It is clear that Manhattan metric does not require any multiplications and this leads to significant reduction in computational complexity.

However, while Manhattan metric can reduce the complexity, it incurs some loss in BER performance. Fig. 1 illustrates the BER performance results obtained with MATLAB simulations. It is clearly shown that the BER performance of Manhattan metric system is about 0.7 dB lower than that of Euclidean metric system which represents 5% loss in BER. Nevertheless, this result is still much better than that of V-BLAST.

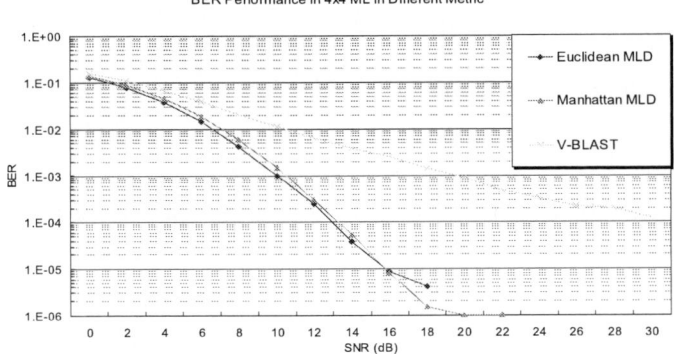

Figure 1. BER performance analysis

4. Simulations and results

In this paper, two architectures have been synthesised and compared: one based on Euclidean and one based on Manhattan metric calculations. Synthesis has been carried out with Synopsys Design Compiler using UMC 0.18um CMOS standard cell library and power estimation has been performed with Synopsys Power Compiler.

Table 1 shows the area and power comparisons of Euclidean and Manhattan based implementations for a clock frequency of 100 MHz. Clearly, Manhattan approach achieves 29% reduction in area and 34.4% in power consumption.

Table 1. Performance comparisons

	Euclidean	*Manhattan*	*Saving (%)*
Speed (MHz)	*100*	*100*	*-*
Area (um²)	*1028181*	*730184*	*29*
Power (mW)	*145.62*	*95.51*	*34.4*

5. Conclusion

In this paper, a low power architecture for the implementation of an ML detector for MIMO systems has been presented. Manhattan metric approach has been proposed for reducing the complexity of the detector with only 0.7dB loss in BER performance. It has been shown that the proposed approach can reduce area and power consumption by 29% and 34.4% respectively. In addition, the architecture achieves higher throughput compared to an Euclidean based implementation.

6. References

[1] *Bells Laboratory, "BLAST High-Level Overview", http://www1.bell-labs.com/project/blast/high-level-overview.html.*

[2] *G. J. Foschini, "Layered Space-Time Architecture for Wireless Communication in a Fading Environment When Using Multiple Antennas", Bell Labs Technical Journal, Vol. 1, No. 2, Autumn 1996, pp 41-59.*

[3] *D. Garrentt, L. Davis, S. T .Brinnk, B. Hochwald, and G. Knagge, "Silicon Complexity for Maximum Likelihood MIMO Detection Using Spherical Decoding", IEEE J. of Solid-State Circuits, vol.39, no.9, Sep. 2004, pp.1544-1552.*

[4] *H. Kawai, K. Higuchi, N. Maeda, M. Sawahashi, T. Ito, Y, Kakura, A Ushirokawa, and H, Seki, "Likelihood Function of QRM-MLD Suitable for Soft-Decision Turbo Decoding and Its Performance for OFCDM MIMO Multiplexing in Multipath Fading Channel", IEICE Trans. comms, no.1 Jan. 2005, pp 47-57.*

0-7695-2533-4/06 $20.00 © 2006 IEEE

Sensor-Driven Power Management:
Enhancing Performance and Reliability of Autonomously Powered Systems

Josef Haid and Dietmar Scheiblhofer

Infineon Technologies Austria AG

{josef.haid,dietmar.scheiblhofer}@infineon.com

Abstract

The number of system-on-chips supplied by renewable energy such as solar power, or electro-magnetic fields is steadly increasing. The amount of available energy changes spatio-temporally depending on the environmental conditions.

This work focuses on the requirements of power management for SOCs operating in instable, inhomogeneous energetic environments. Sensors monitoring the environment and on-chip conditions are proposed to be used extensively in order to enhance power utilization and system stability. A classification of sensor-driven power management policies is given. Policies are proposed that avoid system breakdown caused by internal or environmental changes.

Results show the impacts of response time and adaptivity of the sensor-driven power manager on power utilization and system stability.

1. Introduction

Enhancing power utilization and system stability of SOCs operating in inhomogeneous environments, such RF- or solar-powered devices [3], requires a deep understanding of the environment in which the device is operating. The environmental parameters are either known a priori, e.g. temperature conditions, or have to be gathered during runtime. Sensors, widely used in the area of ubiquitous and pervasive computing, can be utilized to accomplish these tasks.

A large number of today's VLSI systems already implement sensors for tracking critical environmental parameters, such as the voltage level. However, these sensors are mostly used to set the design in power-on or power-off state, ignoring the potential to enhance the system's performance by evaluating the data in a power manager. The paper addresses this topic and proposes a sensor-driven power management approach.

2. Sensor-Driven Power Management

Sensor-driven power management is employed to ensure the correct operation of a device in instable operating environments. The operating range is defined by a set of parameters $OP = \{op_0, ..op_i, .., op_n\}$, where n is the number of operating parameters. The set OP is defined by the chip's production technology as well as other aspects, such as actual power dissipation and on-chip voltage.

Naturally, the environment the chip is working in has influence on the operating parameters op_i. The set of environment paramters influencing OP is denoted as $E = \{e_0, .., e_j, .., e_m\}$. The elements in E are defined as $\forall_{j=0}^m e_j \in E : \Delta V(e_j) \to \Delta V(OP)$, where $\Delta V(e_j)$ is the change in value of the parameter e_j and $\Delta V(OP)$ is the change in at least one value $op_i \in OP$. In reality not all changes of a value e_j have a significant influence on the operating parameters OP of the device. This results in the definition of a subset of E, denoted as $E' \subseteq E$, so that $\forall_{j=0}^m e'_j \in E' : V(e'_j) + max_\epsilon(e'_j) \to \Delta V(OP) > min_\epsilon(OP)$, where $max_\epsilon(e'_j)$ defines the maximum allowed change in value of the parameter e'_j, and $min_\epsilon(OP)$ denotes the minimum change in the parameter set OP that is considered to be significant.

All the parameters e'_i relevant for power management, a subset of E', is defined and denoted as E'_{power} ($E'_{power} \subseteq E'$).

The power management policy (PMP) obtains inputs from sensors monitoring the parameters defined in E'_{power} and controls the system for optimization for a pre-defined set of criteria OC. This is done by exploiting the power management mechanisms M available on the chip. The second aspect of sensor-driven power management is the response time of the power management (t_{resp}) to a change in value of external parameters e_i and vice versa: $\frac{\Delta V(E)}{\Delta t} \longleftrightarrow \frac{\Delta V(OP)}{\Delta t}$ where $\Delta V(E)$ denotes the change in value of one or more parameters e_j, and $\Delta V(OP)$ denotes the change in value of one or more parameters op_i.

0-7695-2533-4/06 $20.00 © 2006 IEEE

3 Sensor-Driven PM Policies

As stated above, the constraints for PM policies in terms of response time are given by the environmental parameters and the robustness of the system. Most power management policies focus on meeting certain deadlines while reducing power consumption. Sensor-driven PM additionally stresses the aspect of system stability, i.e. keeping the system in its operating range OP while environmental parameters change in value. The following proposed classification reflects the different requirements of sensor-driven PM: *immediate action* policies, *adaptive* policies, and *hybrid* policies.

Immediate action PM policies are adequate if a change in value of one or more external parameters ($\Delta V(E'_{power})$) has an immidiate effect on internal operating parameter(s) op_i, and vice versa. A short response time, in the range of a few system clock cycles, is required to avoid system breakdowns caused by sudden changes in environmental paramters E'_{power} or changing operating conditions OP on the SOC.

Adaptive PM policies are designed to find the most appropriate working point in a given environment. The decisions are based on data that is continously collected from the sensors monitoring E'_{power}. The complexity of the policy depends on the number of parameters and the observation period. The policy may also take into consideration information not collected from sensors [1].

Immediate action policies avoid a system breakdown caused by fast and unexpected changes in external parameters. Their disadvantage is the restriction of finding the appropriate working point for the application in the given environment. Adaptive policies are able to perform these tasks based on collected past power information. *Hybrid* PM policies combine both approaches and generally lead to enhanced system robustness and performance.

4. Experimental Results

The SystemC simulator TPSim provides the framework for cylce-accurate microcontroller- based system power analysis [2]. The experimental system consists of a 8051-like CPU with cache, volatile and non-volatile memories as well as analog components, such as voltage regulators.

The power management policies are analyzed using standard benchmarks and real-world applications, consisting of a Dhrystone test (*dhry*), cryptographic algorithms implemented in native C (*aesSw*), and a simple operating system (*simpleOS*). The criteria for analysis are execution time (t_ex), power utilization (P_util), and the number of cycles the device operates in a state which violates the specified operating conditions OP (*Viol*).

The benchmarks are executed using two different environment settings, *const* and *var*. For the first setting (*const*) the power supply level is kept constant, For the second setting (*var*) the power supply level is varied and changes every $50\mu s$ between the default value and a value reduced by 33%.

Table 1 shows a comparison between different power management policies for different benchmarks at a supply level of 2.4mA. In general, immidiate power management policies provide good system stability and power utilization. The adaptive policy significantly increases the run-time and decreases stability due to long response time. The hybrid policy combines shows the best results for all benchmarks simulated.

benchmark,	const	var	const	var	const	var
setting,policy	t_ex [us]	t_ex [us]	Viol	Viol	P_util[%]	P_util[%]
aes,const,imm	3009	3867	0	0	91	92
aes,const,adapt	3218	4311	393	46521	90	87
aes,const,hybrid	3218	4563	0	0	86	86
dhry,const,imm	5618	7195	0	0	91	92
dhry,const,adapt	5889	8003	37	84264	90	86
dhry,const,hybrid	6159	8359	0	0	87	86
simpleOS,const,imm	3234	4155	0	0	91	92
simpleOS,const,adapt	3451	4676	241	46167	90	87
simpleOS,const,hybrid	3616	4827	0	0	86	86

Table 1. Analysis of PM policies for different benchmarks.

5. Conclusion

In this paper, we proposed different power management policies for sensor-driven power management. The proposed algorithms are shown to be very effective in power/energy-limited environments and can be combined with existing low power techniques [1] to further enhance performance, power utilization, and system stability.

References

[1] L. Benini and G. de Micheli. System-level power optimization: techniques and tools. *ACM Transactions on Design Automation of Electronic Systems (TODAES)*, 5(2):115–192, 2000.

[2] J. Haid and D. Scheiblhofer. Power-aware design space exploration of uc-architectures using systemc. In *Proceedings of the Austrochip 2005*, 2005.

[3] D. Li and P. H. Chou. Maximizing efficiency of solar-powered systems by load matching. In *ISLPED '04: Proceedings of the 2004 international symposium on Low power electronics and design*, pages 162–167, 2004.

Reducing Memory Requirements through Task Recomputation in Embedded Multi-CPU Systems

H. Koc and S. Tosun
Electrical Engineering and Computer Science Department
Syracuse University
e-mail: {hkoc,stosun}@ecs.syr.edu

O. Ozturk and M. Kandemir
Computer Science and Engineering Department
Pennsylvania State University
e-mail: {ozturk,kandemir}@cse.psu.edu

Abstract

As embedded applications are processing increasingly larger data sets, keeping their memory space consumptions under control is becoming a very pressing issue. Observing this, several prior efforts have considered memory space reduction techniques (in both hardware and software) based on data compression and lifetime-based memory recycling (garbage collection). In this work, we propose and evaluate an alternate approach to memory space saving in multi-CPU embedded systems such as chip multiprocessors. The unique characteristic of our approach is that it recomputes the results of select tasks in a given task graph (which represents the application), instead of storing these results in memory and accessing them from there as needed.

1 Introduction

Memory space consumption is an important metric to optimize for many embedded designs with tight memory constraints. While this is certainly true for both code and data memory, the rate at which the data sizes of embedded applications increase far exceeds the rate at which their code sizes increase. As a result, optimizing for data memory size is becoming increasingly more important than optimizing for code memory size. Several research papers aimed at reducing the data space requirements of embedded designs and proposed different techniques that can be adopted by optimizing compilers and design synthesis tools. These techniques range from compressing data [2, 8] to lifetime based memory reuse analysis [1, 6] to code restructuring for memory reuse [4, 5, 7].

This paper proposes a novel approach for reducing memory space consumption based on task recomputation. The basic idea is to reduce memory space demand by recomputing select tasks (in a task graph representation of the program) whenever their results are needed, instead of storing those results in memory (after their first computation) and

accessing them from memory. While this approach can reduce memory demand, performing frequent recomputations can also lead to an increase in overall execution latency. In other words, there is a clear tradeoff between performance and memory space consumption. Consequently, this approach should be applied with care to select tasks only. Working on a task graph representation of a given embedded application, we propose a fully-automated scheme that identifies the tasks to recompute in such a way that the potential negative impact on execution time is minimized.

Focusing on an embedded multi-CPU architecture, the proposed approach first identifies the critical paths in the task graph under consideration. It then marks all the tasks that sit in the critical paths as non-recomputable, meaning that these tasks are computed only once and their results are stored in memory for future use as long as they are needed. The remaining tasks, i.e., those that are not in the critical path, are marked as recomputable. The rest of our approach traverses the tasks marked as recomputable and selects a subset of them (to be recomputed) such that the overall increase in execution latency is bounded by a preset value. A particularly interesting optimization problem that can be instantiated from our general problem description is to minimize the memory space requirements (by maximizing task recomputation) without increasing the original execution latency (i.e., the latency that would be obtained when no task recomputation is used). This can be made possible by not allowing any path in the task graph to have a latency which is larger than that of the critical path.

We implemented our approach and tested it using several task graphs (both automatically generated and extracted from applications). Our experimental analysis shows that the proposed approach can be used as a practical tool for studying the performance/memory space consumption tradeoffs in embedded designs that accommodate a multi-CPU architecture. Specifically, for the example task graphs in our experimental suite, we found that our approach can reduce memory requirements by about 13.6% on average without any increase in original execution latency.

0-7695-2533-4/06 $20.00 © 2006 IEEE

Task Graph Label	Number of Nodes	Number of Edges	Data Size (No Opt.)	Data Size (Lifetime)	Latency
tg1	11	16	86	47	51
tg2	14	19	136	59	37
tg3	20	30	184	94	73
tg4	31	45	306	139	71

(a)

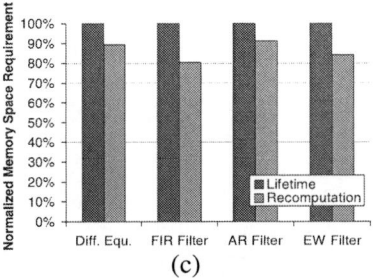

(b)   (c)

Figure 1. (a) Task graphs and their important properties. (b) Normalized memory requirements of our approach for the task graphs in (a). (c) Normalized memory requirements for the task graphs extracted from benchmarks.

In the next section, we give our experimental results obtained using several task graphs.

2 Experimental Evaluation

Our goal in this section is to present an experimental evaluation of the proposed recomputation based approach. For this purpose, we used both automatically-generated task graphs and task graphs extracted from benchmarks. For the first part, we used the TGFF tool [3] and generated several task graphs. Unless otherwise stated, we assumed 10 processors in our experiments. Also, in our experiments, each task has a latency between 7–20 units, and uses a memory size of 4–15 units. The important characteristics of our task graphs are given in Figure 1(a). The first column of this figure gives the name of the task graph and the next two columns give the number of nodes and edges in the graph. The fourth column of the figure shows the total size of the data manipulated by the nodes of the task graph when no memory space saving technique is employed. The next column gives the amount of data space requirements when a *lifetime based* memory space recycling is used. In this scheme, the memory space allocated for storing the results of a task is recycled when the data stored are no longer needed. When comparing these two columns of this figure, we see that a lifetime analysis based memory space recycling cuts the memory space requirements by 49.5% on the average. Our goal is to further increase memory space savings through task recomputation. Finally, the last column of the Figure 1(a) shows the execution latency of each task graph when *no* recomputation is used. In the rest of this section, all memory saving results are given as *normalized values* with respect to the corresponding values in the fifth column of Figure 1(a) (i.e., over the lifetime based approach). Similarly, the performance overheads (if any) incurred by our approach are given as *normalized values* with respect to the corresponding values listed in last column of this figure.

The bar-chart in Figure 1(b) shows the normalized memory space savings obtained by our recomputation-based approach for our task graphs. In these experiments, we use the version of our approach that does not increase the original execution latency. We see that our approach reduces the memory space requirements by 13.6% for the task graphs tested. These results clearly show the effectiveness of our approach in reducing memory space requirements and the important point we want to emphasize here is that these savings come at no performance cost.

In addition to these task graphs generated by the TGFF tool, we also performed experiments with the task graphs extracted from several embedded applications. The normalized memory requirements for this set of task graphs are given in Figure 1(c) for the case when no increase in the original execution latencies is tolerated. We see that our recomputation based approach is very effective in reducing memory space requirements of these task graphs as well, achieving an average memory saving of 13.9%.

References

[1] D. A. Barrett and B. G. Zorn. Using lifetime predictors to improve memory allocation performance. In *Proceedings of the ACM Conference on Programming Language Design and Implementation*, pages 187–196, 1993.

[2] L. Benini, D. Bruni, A. Macii, and E. Macii. Hardware-assisted data compression for energy minimization in systems with embedded processors. In *Proceedings of the Conference on Design, Automation and Test in Europe*, page 449, 2002.

[3] R. P. Dick, D. L. Rhodes, and W. Wolf. Tgff: task graphs for free. In *Proceedings of the International Workshop on Hardware/Software Codesign*, pages 97–101, 1998.

[4] M. D. Lam, E. E. Rothberg, and M. E. Wolf. The cache performance and optimizations of blocked algorithms. In *Proceedings of the International Conference on Architectural Support for Programming Languages and Operating Systems*, pages 63–74, 1991.

[5] W. Li. *Compiling for Numa Parallel Machines*. PhD thesis, Ithaca, NY, USA, 1993.

[6] H. Lieberman and C. Hewitt. A real-time garbage collector based on the lifetimes of objects. *Commun. ACM*, 26(6):419–429, 1983.

[7] L. Wang, W. Tembe, and S. Pande. Optimizing on-chip memory usage through loop restructuring for embedded processors. In *Proceedings of the International Conference on Compiler Construction*, pages 141–156, 2000.

[8] J. Yang, Y. Zhang, and R. Gupta. Frequent value compression in data caches. In *Proceedings of the 33rd Annual ACM/IEEE International Symposium on Microarchitecture*, pages 258–265, 2000.

0-7695-2533-4/06 $20.00 © 2006 IEEE

Compiler-Directed Management of Leakage Power in Software-Managed Memories*

G. Chen, F. Li, M. Kandemir, O. Ozturk
Computer Science and Engineering Department
Pennsylvania State University
University Park, PA 16802, USA
{ozturk, gchen, kandemir}@cse.psu.edu

I. Demirkiran
Electrical Engineering & Computer Science Department
Syracuse University
Syracuse, NY 13244
idemirki@eecs.syr.edu

1 Motivation

One of the problems associated with the ever-increasing level of on-chip integration in CMOS is excessive power consumption [3]. While dynamic power consumption currently is the dominating component of power, leakage energy consumption is becoming increasingly important and projected to be the main power roadblock in future CMOS designs [5]. Large on-chip memory components are particularly problematic from a leakage perspective since they accommodate a large number of transistors. Current proposals for reducing leakage consumption of memory components focus exclusively on cache architectures. While caches are being increasingly used in embedded computing, software-managed memories (SMMs) have also found their ways into commercial products. For example, both StrongArm [1] and IBM's Cell chip multiprocessor [2] contain software-managed memories, contents of which can be explicitly controlled by a compiler. Our goal in this paper is to demonstrate that an optimizing compiler can be very successful in reducing leakage energy consumption of on-chip SMMs for array-dominated embedded applications.

2 Experimental Results

There are two important components of our implementation: an optimizing compiler and a simulation environment. Our compiler-based approach is implemented as a separate phase within the SUIF infrastructure from Stanford University [7]. We found that the maximum increase in compilation time due to our approach was about 70%. Since compilation is essentially an offline activity, this is a small price to pay for large leakage savings in our opinion. Also, the code size increase due to modifications made by our approach was less than 2% for all our applications.

The second component of our framework, the simulation environment, is built upon the Wattch toolset [4]. In addition to implementing and simulating our approach, we also simulated two other schemes (both are proposed originally in the context of traditional cache memories) and compared their performances to ours. The first of these pure hardware-based schemes is the cache decaying mechanism proposed by Kaxiras et al [8]. In our implementation of this approach, each SMM segment is associated with a counter which is initialized to a certain value (in our experiments, we used the value that gives the largest energy savings). The value of a counter is decreased at certain intervals as long as none of the data elements in the corresponding segment is accessed. When the contents of a counter reaches zero, the entire segment is put in the low-leakage mode.[1] The second scheme against which we compare our approach is due to Flautner et al [6]. In this scheme, the entire target memory is periodically placed into the low-power mode. An SMM segment put in the low-leakage mode is reactivated when an access to it is made. These two hardware-directed schemes are referred to as "decay based" and "periodic" in the remainder of this section.

Unless otherwise stated, the configuration used in our experiments has a 64KB on-chip SMM and divided into segments of 128 bytes. The state-preserving circuit mechanism employed is from [6], and we assume a reactivation penalty of 2 cycles. When a segment is put in the low-power mode, it is assumed to consume 5% of its original leakage energy (per cycle), a reasonable value based on [6]. The table in Figure 1 gives the application codes used in our experiments. The common characteristic of these applications is that they are array dominated (i.e., although there are a few scalar references, most of data references in these applications are to array data). The last two columns of this

---

*This work is supported in part by NSF Career Award #0093082 and a fund from GSRC.

[1]Note that while the original cache decaying scheme turns off the cache line thereby destroying its contents, our approach just places the segment into low-power mode without killing its contents.

0-7695-2533-4/06 $20.00 © 2006 IEEE

Application Name	Algorithm Implemented	Dynamic Energy (mJ)	Leakage Energy (mJ)
Vcap	video capture and processing	122.5	144.1
Convolution	convolution filter	289.0	310.4
TM	image converion	177.6	204.8
IA	image understanding	222.8	253.8
H.263	H.263 decoder	118.6	127.0
ImgMult	image multiplication	147.3	169.7
Fact	face recognition	316.8	363.4
Pgp	encryption	267.5	280.1

Figure 1. Our application codes.

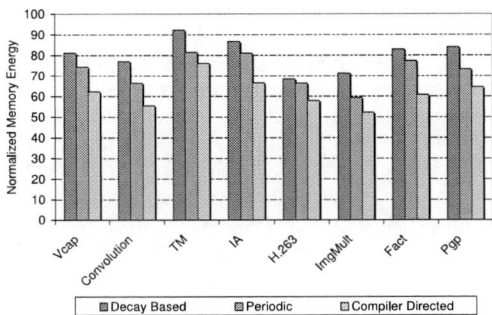

Figure 2. Normalized energy consumption due to data accesses.

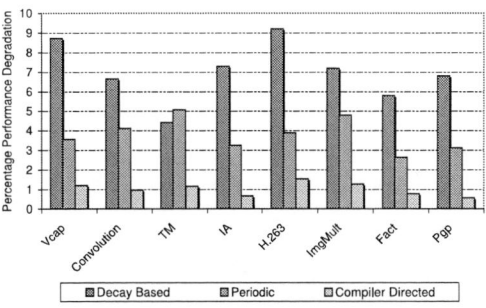

Figure 3. Percentage performance degradation.

table gives the dynamic and leakage energy consumptions – under 70nm – due to data accesses (the SMM energy) when the applications are executed (and used the software-managed memory) *without* any leakage optimization.

The bar-chart shown in Figure 2 gives the energy results. Each bar is given as a *normalized* value with respect to the last column of the Table in Figure 1 and includes not only the leakage consumption in the SMM but also the extra dynamic energy consumption caused in the off-chip memory and other parts of the system. For example, explicit mode transitioning instructions inserted by our scheme causes extra dynamic energy consumption which are captured in Figure 2. The energy consumption in other parts of the system are modeled using Wattch [4]. The average energy savings with our approach, decay based scheme and periodic scheme are 38%, 19.5% and 27.5%, respectively. The main reason that our approach saves more energy than both of the hardware-based schemes is its accuracy in determining the SMM segments to be accessed; the remaining segments are safely placed into the low-leakage mode. The decay-based scheme, on the other hand, waits for a decay period before putting a segment into the low-power mode and, during that period, the segment continues to leak. Similarly, the periodic scheme can keep an SMM segment in the active mode unnecessarily, as it waits for the next interval point to turn off the entire memory.

The performance degradations caused by the leakage-saving schemes are given in Figure 3. Each bar represents the percentage increase in execution time (with re-

spect to the original execution without any leakage optimization). The average performance overheads caused by our approach, decay based scheme and periodic scheme are 1%, 7% and 3.8%, respectively. In other words, our approach generates much better performance than the hardware-based schemes. The main reason for this result is the segment pre-activation optimization employed by our approach (i.e., we activate an SMM segment before it is actually accessed by the program). To sum up, when we look at the results shown in Figures 2 and 3 together, we can say that our compiler-based approach is preferable over both of the hardware-based schemes from both the energy and performance perspectives, and hence, it is a better option for on-chip SMMs.

References

[1] Intel Application Processors. http://developer.intel.com/ design/ pca/ applicationsprocessors/

[2] Introducing the IBM/Sony/Toshiba Cell Processor. http://arstechnica.com/ articles/ paedia/ cpu/ cell-1.ars

[3] L. Benini and G. De Micheli. System-Level Power Optimization: Techniques and Tools. ACM Transactions on Design Automation of Electronic Systems, 5(2), pp. 115–192, April 2000.

[4] D. Brooks, V. Tiwari, and M. Martonosi. Wattch: a framework for architectural-level power analysis and optimizations. Proc. of the International Symposium on Computer Architecture, 2000.

[5] A. Chandrakasan et al. Design of High-Performance Microprocessor Circuits, IEEE Press, 2001.

[6] K. Flautner, N. Kim, S. Martin, D. Blaauw, T. Mudge. Drowsy Caches: Simple techniques for reducing leakage power. Proc. of the 29th International Symposium on Computer Architecture, Anchorage, AK, May 2002.

[7] M. W. Hall, J. M. Anderson, S. P. Amarasinghe, B. R. Murphy, S.-W. Liao, E. Bugnion, and M. S. Lam. Maximizing multiprocessor performance with the SUIF compiler. IEEE Computer, December 1996.

[8] S. Kaxiras, Z. Hu, M. Martonosi. Cache decay: exploiting generational behavior to reduce cache leakage power. Proc. of the 28th International Symposium on Computer Architecture, Sweden, June 2001.

A PARALLEL ARCHITECTURE FOR HARDWARE FACE DETECTION

T. Theocharides
theochar@cse.psu.edu

University of Cyprus

N. Vijaykrishnan, M. J. Irwin
{vijay, mji}@cse.psu.edu

Penn State University

ABSTRACT

Face detection is a very important application in the field of machine vision. In this paper, we present a scalable parallel architecture which performs face detection using the AdaBoost algorithm. Experimental results show that the proposed architecture can detect faces with the same accuracy as the software implementation, on real-time video at a frame rate of 52 frames per second.

1. INTRODUCTION

Face detection is defined as the process of identifying all image regions that contain a face regardless of the position, the orientation and the environment conditions in the image. Face detection is a necessary operation in a wide range of applications. From control and security applications, to identification systems, face detection plays a primary role. It is the primary step towards face recognition [2] and serves as a fore step towards multiple applications such as identification, monitoring, tracking, etc. Face detection algorithms have been heavily researched, and have improved drastically both in terms of performance and speed. So far however, the operation has been extensively done in software [2, and 4]. State-of-the-art software face detection can achieve up to 15 image frames per second [5] under favorable circumstances, and as such is not quite suitable for real-time video deployment. A fast hardware implementation that can be integrated either on a generic processor or as part of a larger system, directly attached to the video source, such as a security camera or a robot's camera, is therefore desirable. There have been a few attempts at hardware implementations that implement face detection on multi-FPGA boards and multiprocessor platforms using programmable hardware [1, 3, 6-9], but do not achieve real time frame rate and high accuracy. A stand-alone accurate and real-time capable detection system which can interface with existing video interfaces is therefore more desirable. Such a system can either be mapped on an FPGA or as an Application-Specific Integrated Circuit (ASIC), as it can be placed on a camera, as a co-processor, as part of an embedded platform, or as an entirely stand alone system. In this paper, we present a scalable parallel architecture that implements one of the most widely accepted face detection algorithms, the AdaBoost face detection technique [5]. Accepted by the computer vision community as the state-of-the-art in terms of speed and accuracy [2], the AdaBoost algorithm uses the boosting classification approach [5]. The approach presents two significant advantages; it has the ability to quickly eliminate non-face regions, and the classification process itself is extremely fast. The classifiers used are called features, each feature consisting of a set of black

---

This work was supported in part by grants from NSF CAREER 0093085, and MARCO/DARPA GSRC-PAS.

---

and white rectangles. The result of a feature is the sum of the pixels under the white rectangle minus the sum of the pixels under the black rectangle. A group of features composes a stage, the outcome of which is the sum of the feature outcomes. The outcome of a stage determines whether the region of the image examined contains a face or not. The classification process benefits from using the integral and integral squared images, a transformation of the original image. A location in the integral image holds the sum of the intensity values of the pixels located above and to the left of the location in the original image. The integral image and the rectangle computation that benefits from it are shown in Figure 1.

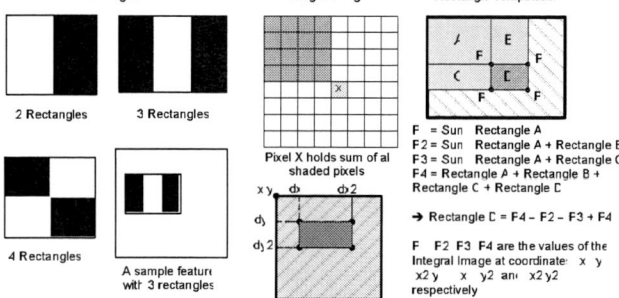

Figure 1: Rectangles, Features, integral image and rectangle computation

The size of the feature determines the size of the image region being searched. When the base size computes for all regions, the features are then enlarged in subsequent scales, and evaluated for each scale, thus able to detect faces of larger sizes. Additional algorithm details are omitted for space limitations, and readers are encouraged to refer to [5] for details. The algorithm outline is shown in Figure 2. Next we describe the proposed architecture.

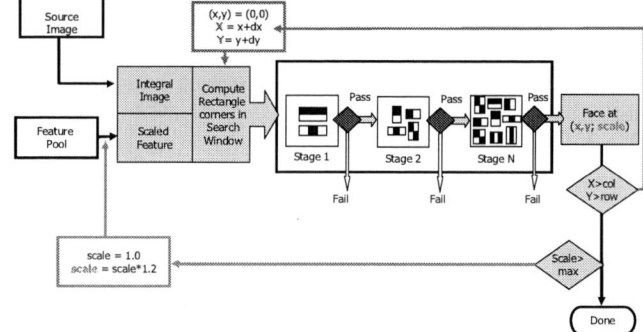

Figure 2: Algorithm Outline

2. PROPOSED ARCHITECTURE

It is evident that we need to provide parallel data access to the integral and integral squared images. As such, we propose the use of a grid array processor, as the structure of our architecture. The array is used as memory to store the computation data and as data

---

0-7695-2533-4/06 $20.00 © 2006 IEEE   452

transfer unit, to aid in accessing the integral image in parallel. Essentially the system consists of three major components: the collection and data transfer units (CDTUs), the multiplication and evaluation units (MEUs) and the control units (CUs). The CDTUs serve as data storage elements for the integral image. In our architecture, the size of the array depends on the size of the image. An image size of *mxn* pixels requires an *mxn* array of CDTUs. While this might seem that the architecture requires excessive resources, techniques such as image scaling can be used to reduce the input image frame and thus keep the size of the array within reasonable limits. Additionally, image partitioning can be used in conjunction with image scaling, to produce image sizes to fit the hardware budget. A floorplan of the system is shown in Figure 3, illustrating the location of each unit and the data movement across the system. Due to space limitations, we omit the detailed description of each unit; however we outline the computation next.

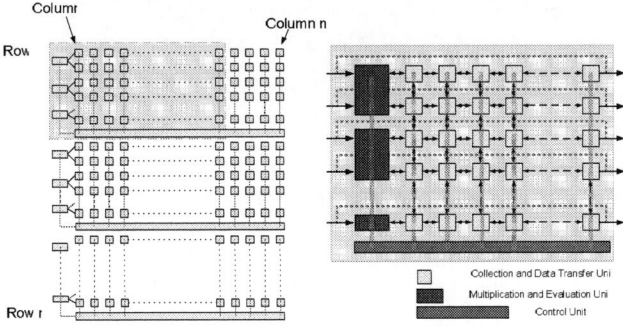

Figure 3: Architecture Floorplan

The operation essentially is partitioned into the following stages: configuration, computation of integral and integral squared images, and computation of image variance, computation of rectangles per feature, feature computation, stage evaluation and image evaluation. When a feature size increases (i.e. when a search window size increases) the computation is essentially treated as a new one. When the image has been searched at all search window sizes, the system is ready for the next image frame. In each case, all three units collaborate to perform the computation. Incoming pixels stream in the processor in parallel along all rows of the processor and are shifted in row-wise every cycle. First, the integral image is computed. The computation consists of horizontal and vertical shifts and additions. Incoming pixels are shifted inside the array on each row. Depending on the current pixel column, each of the computation units performs one of three operations; it either adds the incoming pixel value into the stored sum, or propagates the incoming value to the next-in-raw processing element while, either shifting and adding in the vertical dimension (downwards) the accumulated sum or simply doing nothing in the vertical dimension. The entire computation takes $2 * [(m + (m-1) + (n-1)]$ cycles, for an input image of n rows by m columns. Next, the rectangle computation happens. For each rectangle in each feature, each corner point is shifted towards the collection point. The points move one at a time, but in parallel for all rectangles in the array. At each collection point, the point is either added or subtracted to an accumulated sum, with the rectangle value computed when all points of each rectangle arrive at the collection point. Each point requires $dx+dy$ cycles to reach the collection point, where dx and dy are the offset coordinates of the point with respect to the upper left corner of the search window. When finished collecting the rectangle sums for a single feature, the collected sums are stored in the CDTU that represents the starting corner for each feature. Next, all the collected sums are then shifted leftwards towards the MEUs,

one sum at a time per MEU. From left to right, eventually all sums arrive in each MEU, where the rectangle sums are used with the training data of each feature, in order to evaluate the feature. The feature result is shifted in a toroidal fashion to the CDTU on the far right of the grid, to continue the computation. Eventually, when all feature results are computed, they are stored back into the CDTUs in the grid and the computation resumes with the next feature. When all stages complete for a single scale, the flagged locations contain a face. If the scale computed is the last one, the computation ends, and each search window with a face still has its flag bit set inside the representing CDTU. Each location that contains a face is shifted to the right and outside of the grid array, to the output of the processor for the host application to proceed.

3. PERFORMANCE EVALUATION

To evaluate the performance of the proposed implementation, we designed and verified the architecture using Verilog HDL and ModelSim®. We then synthesized the architecture using a commercial 90nm library and targeting a 500 MHz clock cycle. We use the Intel Open Computer Vision Library (CV) [8] for the training data. We evaluated our design using sample images which contain various numbers and sizes of faces. Our synthesized design indicates that the experimental architecture consumes an area of approximately $115mm^2$. We computed the average number of cycles per test frame and obtained a rough estimate of 52 frames per second.

4. CONCLUSION

This paper presented a parallel architecture that performs face detection using the AdaBoost algorithm. The architecture targets both parallel computation and parallel data movement, and is capable of processing on average 52 frames per second.

5. REFERENCES

1. R. McCready, "Real-Time Face Detection on a Configurable Hardware System", *International Symposium on Field Programmable Gate Arrays*, October 2000.

2. E. Hjelmås, B. K. Low, "Face Detection: A Survey", *Computer Vision and Image Understanding*, Vol. 83, No. 3, September 2001.

3. B. Srijanto, "Implementing a Neural Network based face detection onto a reconfigurable computing system using Champion", *MS Thesis*, University of Tennessee, August 2002.

4. Intel Open Source Computer Vision Library, Oct. 2005, http://www.intel.com/technology/computing/opencv/index.htm.

5. P. Viola and M. Jones, "Robust Real-time Object Detection", *International Journal of Computer Vision*, October 2002.

6. M. Reuvers, "Face Detection on the INCA+ System", Masters Thesis, *University of Amsterdam*, 2004.

7. T. Theocharides, et al, "Embedded Hardware Face Detection", *In the Proc. of VLSI Design*, Mumbai, India, January 2004.

8. V. Kianzad et al, "An Architectural Level Design Methodology for Embedded Face Detection", *Proceedings of the International Conference on Hardware/Software Codesign and System Synthesis*, New York City, September 2005.

9. H. Broers et al, "Face Detection and Recognition on a Smart Camera", *Proceedings of the Advanced Concepts for Intelligent Vision Systems Conference*, Belgium, August 2004.

A VLSI GFP Frame Delineation Circuit

Ciaran Toal, Sakir Sezer, Xin Yang

Institute of Communications and Information Technology, Queen's University Belfast, Northern Ireland Science Park, Queen's Road, Queen's Island, Belfast BT3 9DT
Ciaran.Toal@ee.qub.ac.uk

Abstract

This paper presents the design and study of a circuit architecture able to perform 16Gbps GFP frame delineation with single bit error correction using UMC 130nm standard cell technology. The design targets the development of a hard macro core for the design of next generation network processing platforms.

1. Introduction

GFP Frame delineation is specified by the ITU-T in recommendation G.7041. At the transmitter the Core Header Error Check (cHEC) field is composed of the third and forth bytes of the GFP frame core header. The cHEC field is calculated from the first 2 bytes of the core header i.e. the PLI field or Payload Length Indicator. When a GFP frame is received the cHEC is again calculated from the first 2 core header bytes and compared with the third and forth bytes. In the absence of errors, both values are identical and the frame boundary is assumed to be located.

The cHEC field is calculated as a remainder of the modulo-2 division of the PLI field with the CRC generator polynomial $G(x) = 1+x^5+x^{12}+x^{16}$. GFP frame synchronisation is a sequential process performed by a synchronisation state machine and a parallel cHEC calculate circuitry. The standard includes single-bit error correction capability.

In this paper, we present a high performance VLSI implementation. 16Gbps is a very impressive processing capability for a 130 nm implementation. The design is written in VHDL and can easily be ported to any other technology.

2. Hardware Implementation

The design requires 8 16-bit In/16-bit Out CRC units and 8 16-bit comparator units.

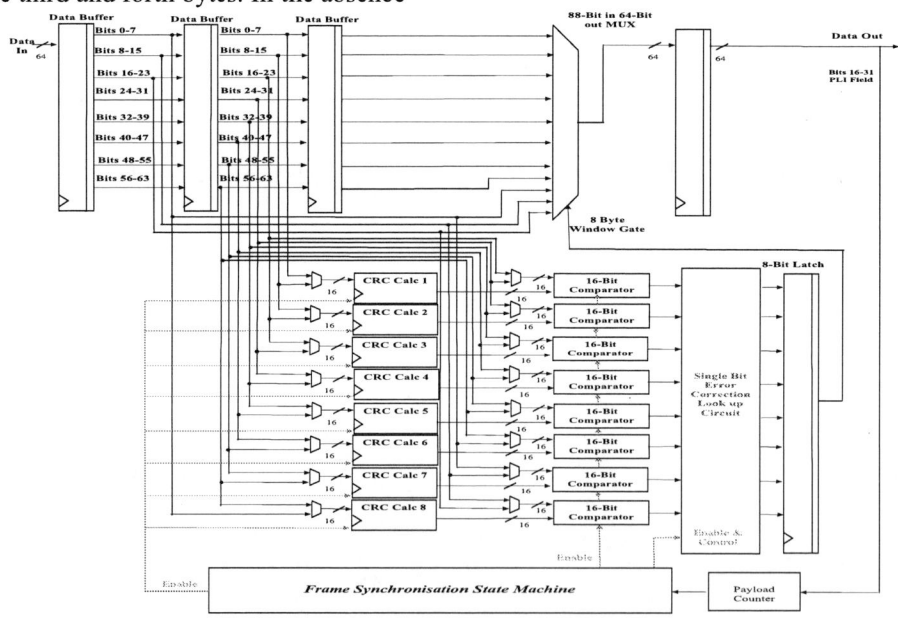

Figure 1. 64-Bit GFP Frame Delineation and Synchronisation Circuit

0-7695-2533-4/06 $20.00 © 2006 IEEE

Every clock cycle 8 new bytes of data are scanned in. The circuit is designed to locate a possible cHEC on all 8 input byte locations. The first positive match between the PLI field CRC remainder and the subsequent transmitted CRC field found by a comparator unit (i.e. a located cHEC) is latched. This latched signal controls what is essentially an 8-byte window gate enabling 8 consecutive bytes of a possible 11 to be routed through to the output. The 64-bit architecture is shown in figure 1. A generic parallel CRC core has been developed based on the work presented in [1]. This VHDL core synthesizes optimised parallel CRC matrices based on the divider polynomial and the input/output bus size, which are entered as generic values. This reusable IP core has been proven to be extremely useful for many designs in network processing.

The deployed error correction technique is based on a similar architecture presented by Shukla et al [2]. However, in our implementation the RS lookup table is based on a ROM implementation instead of a dual port RAM. Due to the fact that only a small number of entries (32) are required, ROM based logic synthesis on FPGA presents an efficient solution, overcoming memory addressing issues and resulting in a reasonably small circuit. As expected, due to the nature of the design, a high fan-out is observed, which is much greater than that introduced by the 8 parallel CRC circuits. The key advantage of synthesizing a ROM table is the portability to other technologies in the form of a technology independent IP core.

3. Synthesis Results and Circuit Analysis

The GFP frame delineation circuit was synthesised using Sysnopsys Physical Compiler and targeted to 130nm UMC standard cell technology using Cadence Encounter. The post layout synthesis results are presented in Table 1 and the chip layout presented in figure 2.

Table 1. Post layout synthesis results with UMC 130nm technology

Area (mm^2)	Clock Frequency (MHz)	Data Throughput (Gbps)	Total Power (Watts)
0.12	250	16	1.6×10^{-02}

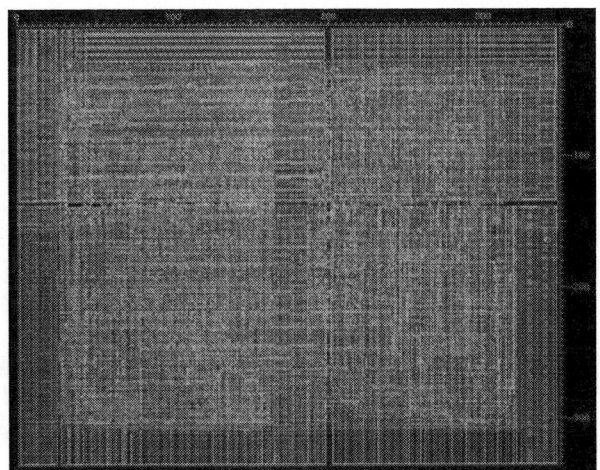

Figure 2. Circuit layout as hard macro

4. Conclusions

In this paper, we have presented the architecture and implementation of a fully operational parallel GFP frame delineation architecture. The system is designed in VHDL and offers potential as a high performance IP core that can be easily migrated across different technologies, including FPGA technology. As a 64-bit data-path implementation, based on UMC 130nm standard cell technology, the hard macro core is able to process line-rates of up to 16Gbps, meeting the next generation requirements of encapsulating 10Gbps protocols. GFP is expected to be the dominant encapsulation protocol for gigabit packet transmission over optical networks, including next generation 10G Ethernet over SONET/SDH. Frame delineation and synchronisation is one of the most computationally complex components of the GFP encapsulation protocol.

5. References

[1] T.-Bi-Pei and C. Zukowski, "High-speed parallel CRC circuits in VLSI", *IEEE Transaction on Communication*, vol. 40, pp. 653-657, April 1992.

[2] Sunil Shukla and Neil W. Bergmann, "Single Bit Error Correction Implementation in CRC-16 on FPGA", *Field Programmable Technology*, 2004.

0-7695-2533-4/06 $20.00 © 2006 IEEE

Effects of Parameter Variations and Crosstalk Noise on H-Tree Clock Distribution Networks

Itisha Chanodia and Dimitrios Velenis
Department of Electrical and Computer Engineering
Illinois Institute of Technology
Chicago, IL 60616
Email: chaniti@iit.edu, velenis@ece.iit.edu

Abstract— **The effects of parameter variations and crosstalk noise on the clock signal propagating along an H-tree clock distribution network are investigated in this paper. In particular, the effects of variations in power supply voltage(V_{DD}), and temperature on the delay and the transition time of the clock signal are evaluated. Furthermore, the effects of crosstalk between an H-tree structure and other interconnect wires are investigated. Different scenarios of capacitive coupling along different spatial locations of an H-tree are considered. The effects of coupling on the propagation delay, the transition time, and the waveform shape of the clock signal are demonstrated.**

I. INTRODUCTION

With the transition to deep submicrometer technologies, shrinking geometries have led to an aggravation of the effects that cause the delay of a signal to deviate from a target value. Deviations of the clock signal from a target delay can cause incorrect data to be latched within a register, resulting in a system malfunctioning. Delay variations of the clock signal can be caused by a number of factors that affect a clock distribution network, such as process and environmental parameter variations [1], [2], and interconnect noise [3].

In this paper the effects of parameter variations and crosstalk among interconnects on the delay and transition time of a clock signal propagating along an H-tree clock distribution network [4] are investigated. The design of a two level buffered H-tree clock distribution network is described in Section II. The effects of parameter variations on the clock signal are discussed in Section III. In addition, the spatial dependence of crosstalk effects on the delay and transition time of the clock signal is discussed in Section IV. Finally, some conclusions are presented in Section V.

II. DESIGN OF H-TREE CLOCK DISTRIBUTION NETWORK

In this paper the design of a two-level H-tree clock distribution network is considered. as illustrated in Figure 1. The clock signal is distributed from a primary clock source to 16 destination points within a die with dimensions 2cm × 2cm. The capacitive load of each of the clock buffers located at the destination points is 150fF. The clock lines on the first level of the H-tree are routed on metal five and on the second level of the H-tree on metal four. The tree is designed using $0.18 \mu m$ technology. The minimum propagation delay of the clock signal achieved by the H-tree design is 425ps. The

transition time of the clock signal at the destination nodes is 47ps.

III. PARAMETER VARIATION EFFECTS

The effects of parameter variations on the delay and transition time of a clock signal distributed along the H-tree structure described in Section II are investigated in this section.

A. Power supply voltage variations

Power supply variations affect the current drive of the clock buffers within a clock distribution network. A V_{DD} variation of ±5% from the nominal value of 1.8V is considered at the buffers along the second level of the H-tree. The delay of the clock signal varies by up to 4% of the nominal value due to the variations on the current drive of the clock buffers. Furthermore, the transition time of the clock signal at the output nodes of the H-tree is affected significantly. It is demonstrated that the transition time of the clock signal at the final nodes of the tree changes by up to 20%.

In addition, the effect of ±5% V_{DD} variation on the first level of the H-tree is investigated. The variations both in the clock delay and the transition time of the clock signal are minimal.

B. Temperature variations

The variation of temperature along an H-tree clock distribution network is considered in concentric temperature zones. The effect of temperature variations on the transition time and the delay of the clock signal arriving at the leaf nodes of the H-tree is listed in Table I. Three leaf nodes A, B, and C within three different temperature zones are considered. The

TABLE I

TEMPERATURE GRADIENT EFFECTS ON THE DELAY AND TRANSITION
TIME OF THE CLOCK SIGNAL

End Node	Temp. $^{\circ}C$	Delay (ps)	Variation (%)	Transition time (ps)	Variation (%)
A	25	454.7	7.1	48.21	4.8
B	75	458.9	8.2	52.03	13.1
C	100	466.9	9.8	55.02	19
A	100	473.5	11.5	53.69	17
B	50	469.2	10.4	49.22	7
C	25	461.7	8.6	47.12	2

0-7695-2533-4/06 $20.00 © 2006 IEEE

transition time and delay of the clock signal at nodes A, B, and C is listed in Table I for the two different temperature gradient scenarios. It is shown that the direction of the temperature gradient affects significantly both the delay and the transition time of the clock signal.

IV. CROSSTALK EFFECTS

A. Coupling Scenarios Along an H-tree

Crosstalk effects on interconnect lines can be evaluated using the *aggressor - victim* model for capacitively coupled lines [5]. Four different scenarios of capacitive coupling between opposite switching signals propagating along an *aggressor* line and an H-tree structure are investigated, as shown in Figure 1:

 i. *Scenario A*. Coupling at the first level of the H-tree with $1cm$ long aggressor line.
 ii. *Scenario B*. Coupling at the first level of the H-tree with $0.5cm$ long aggressor line.
 iii. *Scenario C*. Coupling at the horizontal segment of the second level of the H-tree with $0.5cm$ long aggressor line.
 iv. *Scenario D*. Coupling at the vertical segment of the second level of the H-tree with $0.5cm$ long aggressor line.

B. Effect of crosstalk on clock delay

The increase in the propagation delay of the clock signal due to crosstalk at different locations along an H-tree is considered. As demonstrated in simulating experiments, coupling scenario A considering a $1cm$ long line along the first level of the H-tree creates the highest increase in delay. In addition, coupling at the final segment of the H-tree, as described by scenario D, increases delay by up to 30 ps. The effect of coupling scenarios B and C on the signal delay is less because the coupling length is short and the clock signal is effectively restored by the clock buffers along the H-tree.

The effect of the driver size of the aggressor line on the delay of the clock signal is also investigated. The driving strength of the buffer determines the temporal relation between the signal transition on the aggressor line and the transition of the clock signal. It is shown that the effect of buffer size on delay is greater for coupling scenarios A and D.

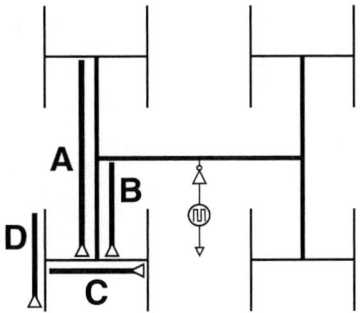

Fig. 1.   Spatial coupling scenarios along an H-tree structure

C. Effect of crosstalk on clock transition time

In addition to clock propagation delay, the effects of crosstalk on the transition time of the clock signal are investigated. The increase in the clock signal transition time at the destination nodes of an H-tree introduced by the different capacitive coupling scenarios is considered. It is demonstrated that the transition time of the clock signal is significantly affected by crosstalk for coupling scenario D, at the final stage of an H-tree. The effect of crosstalk on the earlier stages of an H-tree is alleviated by the inserted clock buffers that restore the waveform of the clock signal. However, in coupling scenario D there are no additional downstream buffers to restore the transition time of the clock signal at the output of the H-tree.

Furthermore, the effect of the buffer size of the aggressor line on the transition time of the clock signal is investigated. When coupling occurs at the internal segments of an H-tree the size of the driver of the aggressor line has no effect on the signal transition time at the destination nodes. The waveform of the clock signal is fully restored by the clock buffers along the internal segments of the H-tree.

V. CONCLUSIONS

The effects of parameter variations and crosstalk noise along a two-level buffered H-tree clock distribution network are investigated in this paper. It is shown that parameter variation effects on the first level of an H-tree can be compensated by the clock buffers close to the leaf nodes of the tree. However, the effects of parameter variations on the second level of the H-tree become more significant on the delay and transition time of the clock signal. Furthermore, different coupling scenarios at different segments of an H-tree are considered. It is shown that coupling on long segments of the H-tree has a significant effect on the total propagation delay of the clock signal. Furthermore, the transition time of the clock signal can be significantly increased when coupling occurs at the final segments of the tree that cannot be alleviated by the clock buffers. Crosstalk noise can be induced on the clock signal due to coupling along H-tree segments that are remotely driven by clock registers.

REFERENCES

[1] S. Sauter, D. Schmitt-Landsiedel, R. Thewes, and W. Weber, "Effect of Parameter Variations at Chip and Wafer Level on Clock Skews," *IEEE Transactions on Semiconductor Manufacturing*, Vol. 13, No. 4, pp. 395–400, November 2000.
[2] D. Velenis, M. C. Papaefthymiou, and E. G. Friedman, "Reduced Delay Uncertainty in High Performance Clock Distribution Networks," *Proceedings of the IEEE Design Automation and Test in Europe Conference*, pp. 68–73, March 2003.
[3] A. Vittal, L. H. Chen, M. Marek-Sadowska, K.-P. Wang, and S. Yang, "Crosstalk in VLSI Interconnections," *IEEE Transactions on Computer-Aided Design of Integrated Circuits and Systems*, Vol. 18, No. 12, pp. 1817–1824, December 1999.
[4] M. Nekili, C. Bois, and Y. Savaria, "Pipelined H-Trees for High-Speed Clocking of Large Integrated Systems in Presence of Process Variations," *IEEE Transactions on Very Large Scale Integration (VLSI) Systems*, Vol. 5, No. 2, pp. 161–174, June 1997.
[5] K. T. Tang and E. G. Friedman, "Delay and Noise Estimation of CMOS Logic Gates Driving Coupled RC Interconnections," *Integration, the VLSI Journal*, Vol. 29, No. 2, pp. 131–165, September 2000.

Author Index

Adde, Patrick 430
Adi, Wael 24
Afzali-kusha, Ali 373, 440
Aken'Ova, Victor 103
Akgul, Bilge E. S. 349
Akopyan, Filipp 438
Alaghi, Armin 420
Alkan, Cengiz 367
Alsharqawi, Abdelhalim 71
Amini, Esmail 199
Amirabadi, A. 373
Arslan, Tughrul 12, 85, 185, 444
Baas, Bevan M. 378
Becker, Jürgen 97, 109, 159, 251
Beerel, Peter A. 173
Benoit, P. 251
Bergmann, Neil W. 109
Bernstein, Gary 242
Bernstein, Kerry 399
Berthelot, Florent 436
Bridges, Seth 133
Bruchon, Nicolas 269
Cambon, Gaston 251, 269
Carvajal, Gonzalo 133
Chang, Chan-Hao 167
Chanodia, Itisha 456
Chen, Bo-Wei 128
Chen, Chunhong 237, 361
Chen, G. 50, 450
Chen, Guangyu 393
Chen, Guilin 393
Chen, I-Shun 128
Chen, Tom 367
Clarke, Christopher T. 277
Crocker, Michael 242
Crowe, Francis M. 317
de Brito, Alisson V. 434
Dehyadgari, Masood 440
Demirkiran, I. 450
Drechsler, Rolf 335
Edwards, Stephen A. 303
Ejnioui, Abdel 71
Erdogan, Ahmet T. 12, 185, 444
Ernst, Rolf 24
Esmaeili, Seyed E. 416
Eze, Melvin 355

Fang, David 438
Ferguson, Ian 85
Figueroa, Miguel 133
Fu, Ning 38
Fujimura, Toru 18
Fujita, Shinobu 231
Garcia Ortiz, Alberto 426
Genz, Christian 335
Ghannam, Moustafa Y. 416
Glesner, Manfred 77, 426
Golani, Pankaj 173
Grad, Johannes 428
Habibi, Ali 418
Haid, Josef 446
Hallschmid, P. 289
Han, J. H. 185, 444
Hanchate, Narender 329
Hanoun, Abdulrahman 24
Hariyama, Masanori 193
Haubelt, Christian 309
Heino, Pekka 117
Hsieh, Sheng-Ta 7
Hu, Xiaobo Sharon 242
Hübner, Michael 97, 159
Indrusiak, Leandro Soares 77, 426
Irwin, M. J. 452
Isoaho, Jouni 217
Jahanian, Ali 411
Jain, Rahul 91
Jansen, Pierre G. 211
Jantsch, Axel 205
Jego, Christophe 430
Jezequel, Michel 430
Jone, Wen-Ben 147
Joshi, Supreet 122
Jung, Markus 159
Kajitani, Yoji 18
Kameyama, Michitaka 193
Kandemir, Mahmut 50, 295, 393, 448, 450
Karakoy, M. 50
Karfa, C. 141
Karjalainen, Päivi H. 117
Katti, Rajendra 153
Kavaldjiev, Nikolay 211
Kaya, Ilhan 179
Kępa, Krzysztof 432

Keymeulen, Didier.............................. 85
Khachab, Nabil I. 416
Khan, Zahid 12
Kinniment, David 442
Kobayashi, Yasuhiro 193
Koc, H. 448
Kocak, Taskin 179
Kolcu, Ibrahim..........................295, 393
Kolze, Paige................................... 33
Korkmaz, Pinar.............................. 349
Kościuszkiewicz, Krzysztof 432
Kühnle, Matthias 97
Kursun, Volkan................................ 59
Li, Feihui 295, 393, 450
Li, Ming.. 147
Lin, Cheng-Wei 7
Lin, Sheng-Jang 128
Liu, Zhiyu...................................... 59
Lo, Feng-Hsiang 128
Loh, Gabriel H. 384
Lu, Zhonghai 205
Lukasiewycz, Martin 309
Luo, Yan....................................... 399
Mandal, C. 141
Manohar, Rajit............................... 438
Marculescu, Diana 167
Marnane, William P......................... 317
Marrakchi, Zied 263
Matamala, Esteban.......................... 133
Matsushita, Daisuke......................... 231
McEvoy, Robert P. 317
McKillen, C. 65
Mehrez, Habib 263
Meinhardt, Cristina 424
Melcher, Elmar U. K. 434
Menezes, Cláudio 424
Mi, Jialin..............................237, 361
Mineshima, Mitsutoshi 38
Moinudeen, Haja 418
Moraes, Fernando G. 426
Morgan, Fearghal............................ 432
Mrabet, Hayder 263
Mukherjee, Anindita 91
Muraoka, Koichi 231
Murphy, Colin C. 317
Mutyam, Madhu 355
Najibi, Mehrdad.............................. 199
Nakatake, Shigetoshi 18, 38, 44
Namballa, Ravi 329
Narayanan, Vijaykrishnan 422
Navabi, Zainalabedin..................420, 440
Neiroukh, Osama 303
Nickray, Mohsen 440
Niemier, Michael............................ 242

Nigussie, Ethiopia........................... 217
Nojima, Takashi 18, 44
Nouvel, Fabienne............................ 436
Ohba, Ryuji................................... 231
Okazaki, Koji 18
Ono, Nobuto 18
Oreifej, Rashad 71
Ozturk, O. 50, 448, 450
Ozturk, Ozcan................................ 393
Palem, Krishna V. 349
Palma, José C. S. 426
Paul, Kolin 91
Paulsson, Katarina 159
Pedram, Hossein 199
Pentakota, S. R. 141
Pfleiderer, Hans-Joerg 257
Piriou, Erwan 430
Plosila, Juha.................................. 217
Prudêncio, Romualdo Begale 77
Puttaswamy, Kiran........................... 384
Rachlin, Eric 225
Ranganathan, Nagarajan 329
Rasouli, S. H. 373
Reade, Chris 141
Reis, Ricardo A. L. 426
Reis, Ricardo 424
Robert, M. 251
Rosas, Wilson 434
Ruan, Xiaoyu 153
Russell, Gordon 442
Saleh, Resve 103, 289
Samarah, Amer 418
Sarkar, D. 141
Sassatelli, Gilles 251, 269
Savage, John E. 225
Scheiblhofer, Dietmar....................... 446
Scheppler, Michael 257
Schlichter, Thomas 309
Schuck, Christian............................. 97
Seyedi, A. S. 373
Sezer, Sakir 65, 454
Sharma, Dinesh.............................. 122
Sheynin, Yuriy 283
Shukla, Sunil 109
Shutenko, Felix 283
Smit, Gerard J. M. 211
Song, Xiaoyu 303
Soudan, Bassel............................... 24
Srikanthan, Thambipillai 277
Srinivasan, Suresh 422
Stefatos, Evangelos F. 85
Stine, James E. 428
Sun, Tsung-Ying................................ 7
Sundaresan, Vijay............................ 323

Sung, Gang-Neng 405
Suvorova, Elena 283
Tahar, Sofiène 418
Takahashi, T. 444
Tanamoto, Tetsufumi 231
Tavares, Reginaldo 424
Teich, Jürgen 309
Theocharides, T 452
Thompson, John S. 12
Toal, Ciaran 454
Torres, Lionel 251, 269
Tosun, S. 448
Uchida, Ken 231
Velenis, Dimitrios 456
Vemuri, Ranga 323
Veredas, Francisco-Javier 257
Vijaykrishnan, N. 343, 355, 452
Vivekanandarajah, Kugan 277
Wang, Chua-Chin 405
Wang, Feng 399

Wang, Hsiang-Min 7
Wehn, Norbert 3
Wolf, Wayne 4, 343
Wolkotte, Pascal T. 211
Xie, Yuan 343, 355, 399
Yakovlev, Alex 442
Yan, Minjun 242
Yan, Tan 44
Yang, Shengqi 343
Yang, Xin 65, 454
Yarandi, Mahnaz Sadoughi 420
Yasuda, Shinichi 231
Yin, Bei 205
Yu, Zhiyi 378
Zamani, Morteza Saheb 411
Zeng, Qing-An 147
Zhai, Bumei 257
Zhou, Jun 442
Zhu, Qing K. 33

Press Operating Committee

Chair
Roger U. Fujii
Vice President
Northrop Grumman Mission Systems

Editor-in-Chief
Donald F. Shafer
Chief Technology Officer
Athens Group, Inc.

Board Members
Thomas Baldwin, *Manager, Conference Publishing Services (CPS)*
Hal Berghel, *Associate Dean, University of Nevada at Las Vegas*
Mark J. Christensen, *Independent Consultant*
James Conrad, *Associate Professor, UNC-Charlotte*
Herb Krasner, *Senior Lecturer, University of Texas at Austin*
Phillip Laplante, *Associate Professor, Penn State University*
Ted G. Lewis, *Professor, Computer Science, Naval Postgraduate School*
Deborah Plummer, *Manager, Authored Books*
Linda Shafer, *Professor Emeritus, University of Texas at Austin*
Richard Thayer, *Professor Emeritus, California State University, Sacramento*

IEEE Computer Society Executive Staff
David Hennage, *Executive Director*
Angela Burgess, *Publisher*

IEEE Computer Society Publications
The world-renowned IEEE Computer Society publishes, promotes, and distributes a wide variety of authoritative computer science and engineering texts. These books are available from most retail outlets. Visit the CS Store at *http://computer.org/cspress* for a list of products.

IEEE Computer Society *Conference Publishing Services* (CPS)
The IEEE Computer Society produces and actively promotes conference publications for more than 200 acclaimed international conferences each year in a variety of formats, including soft-cover books, hard-cover books, CD-ROMs, video, and on-line publications. For information about the IEEE Computer Society's *Conference Publishing Services* (CPS), please e-mail: tbaldwin@computer.org or write to: *Conference Publishing Services* (CPS), IEEE Computer Society, P.O. Box 3014, 10662 Los Vaqueros Circle, Los Alamitos, CA 90720-1314. Telephone +1-714-821-8380. Fax +1-714-761-1784. Additional information about the IEEE Computer Society's *Conference Publishing Services* (CPS) can be accessed from our web site at: *http://www.computer.org/cps*.

IEEE Computer Society / Wiley Partnership
The IEEE Computer Society and Wiley partnership allows the CS Press authored book program to produce a number of exciting new titles in areas of computer science and engineering with a special focus on software engineering. IEEE Computer Society members continue to receive a 15% discount on these titles when purchased through Wiley or at: *http://wiley.com/ieeecs*. To submit questions about the program or send proposals, please e-mail dplummer@computer.org or write to: Books, IEEE Computer Society, 10662 Los Vaqueros Circle, Los Alamitos, CA 90720-1314. Telephone +1-714-821-8380. Additional information regarding the Computer Society's authored book program can also be accessed from our web site at: *http://www.computer.org/portal/pages/ieeecs/publications/books/about.html*.

Revised: 03 January 2006

CURRAN ASSOCIATES INC.
proceedings
.com

9780769525334